METAL IONS IN BIOLOGICAL SYSTEMS

METAL IONS IN
BIOLOGICAL SYSTEMS

Edited by

Astrid Sigel
and **Helmut Sigel**

Institute of Inorganic Chemistry
University of Basel
CH-4056 Basel, Switzerland

VOLUME 36

Interrelations Between Free Radicals
and Metal Ions in Life Processes

MARCEL DEKKER, INC.   NEW YORK · BASEL

ISBN: 0-8247-1956-5

This book is printed on acid-free paper.

Headquarters
Marcel Dekker, Inc.
270 Madison Avenue, New York, NY 10016
tel: 212-696-9000; fax: 212-685-4540

Eastern Hemisphere Distribution
Marcel Dekker AG
Hutgasse 4, Postfach 812, CH-4001 Basel, Switzerland
tel: 41-61-261-8482; fax: 41-61-261-8896

World Wide Web
http://www.dekker.com

The publisher offers discounts on this book when ordered in bulk
quantities. For more information, write to Special Sales/Professional
Marketing at the headquarters address above.

Current printing (last digit):
10 9 8 7 6 5 4 3 2

PRINTED IN THE UNITED STATES OF AMERICA

Preface to the Series

Recently, the importance of metal ions to the vital functions of living organisms, hence their health and well-being, has become increasingly apparent. As a result, the long-neglected field of "bioinorganic chemistry" is now developing at a rapid pace. The research centers on the synthesis, stability, formation, structure, and reactivity of biological metal ion-containing compounds of low and high molecular weight. The metabolism and transport of metal ions and their complexes is being studied, and new models for complicated natural structures and processes are being devised and tested. The focal point of our attention is the connection between the chemistry of metal ions and their role for life.

No doubt, we are only at the brink of this process. Thus, it is with the intention of linking coordination chemistry and biochemistry in their widest sense that the *Metal Ions in Biological Systems* series reflects the growing field of "bioinorganic chemistry". We hope, also, that this series will help to break down the barriers between the historically separate spheres of chemistry, biochemistry, biology, medicine, and physics, with the expectation that a good deal of future outstanding discoveries will be made in the interdisciplinary areas of science.

Should this series prove a stimulus for new activities in this fascinating "field", it would well serve its purpose and would be a satisfactory result for the efforts spent by the authors.

Fall 1973

Helmut Sigel
Institute of Inorganic Chemistry
University of Basel
CH-4056 Basel, Switzerland

Preface to Volume 36

It is commonly accepted that radical reactions play a major role in aging and that such reactions are involved in a large variety of diseases, such as for example, cancer, atherosclerosis, Alzheimer's disease, amyloid-osis, and osteoarthritis. It is believed that the initiators of the indicated radical-induced deleterious processes are mainly hydroxyl and super-oxide radicals. Hydroxyl radicals ($\cdot$OH) may be formed in biological systems by "Fenton-like" reactions, i.e., via iron(II) or other transition metal ions and hydrogen peroxide (H_2O_2), by the absorption of ionizing radiation or ultraviolet light, and probably also when ozone (O_3) is introduced in a biological system. Superoxide radicals ($O_2^{\cdot-}$) may form as side products of the electron transport chains in mitochondria and the endoplasmic reticulum and by some oxidation of heme during oxygen transport.

Many studies have been published about radicals derived from dioxygen metabolism, yet despite this wealth of information much fundamental knowledge remains obscure and despite considerable efforts it is still not possible to predict with certainty what the active oxidant species will be under a given set of conditions, especially under different types of physiological conditions. Hence, somewhat different and partially conflicting views are presented in some chapters. However, it is hoped that the chemical facts and analyses summarized provide a solid perspective for future research.

Aside from the deleterious effects of radicals indicated above, DNA damage as well as radical migration through the DNA helix are also considered. Lipid peroxidation, the formation of methemoglobin, methyl-mercury-induced generation of free radicals, and the role of thiyl radicals are also described, as are the potential use of organocobalt compounds for the pH-dependent formation of active radical species and their application in the defense of cancer.

Another biologically important radical species is nitrogen monoxide, also called nitric oxide (NO). Until about 10 years ago it was of interest to only a few fairly select groups of researchers, such as cardiovascular pharmacologists, environmental chemists, or inorganic chemists; only in the late 1980s and early 1990s did several lines of investigation converge to find that NO is a ubiquitous biological messenger molecule and this has radically changed our view of how cells communicate with one another. In this context the 'ingenious' use of nitrogen monoxide by blood-sucking insects, to assure that they receive a sufficient blood meal, warrants mentioning. In the two terminating chapters the potential clinical use of NO modulators as possible drugs for certain nitrogen monoxide-related diseases is outlined.

Astrid Sigel
Helmut Sigel

Contents

Contributors

Numbers in parentheses indicate the pages on which the authors' contributions begin.

Craig S. Atwood Genetics and Aging Unit, Neurosciences Center, Massachusetts General Hospital, Bld 149, 13th Street, Charlestown, MA 02129, USA (FAX: +1-617-724-9610; <atwoodc@helix.mgh.harvard.edu>) (309)

Jacqueline K. Barton Division of Chemistry and Chemical Engineering and the Beckman Laboratory of Chemical Synthesis, MIC 127-72, California Institute of Technology, Pasadena, CA 91125, USA (FAX: +1-626-577-4976; <jkbarton@cco.caltech.edu>) (211)

Brian P. Booth Department of Pharmaceutics, School of Pharmacy, State University of New York at Buffalo, 513 Hochstetter Hall, Box 601200, Buffalo, NY 14260-1200, USA (FAX: +1-716-645-3693; <bbooth@acsu.buffalo.edu>) (723)

Bruce P. Branchaud Department of Chemistry, University of Oregon, Eugene, OR 97403, USA (FAX: +1-541-346-4645; <bbranch@oregon.uoregon.edu>) (79)

Ashley I. Bush Genetics and Aging Unit, Neurosciences Center, Massachusetts General Hospital, Bld 149, 13th Street, Charlestown, MA 02129, USA (309)

George V. Buxton Centre for Joint Honours in Science, University of Leeds, Leeds, L26 9JT, UK (FAX: +44-113-2332-689; <cbs6gvb@lucs-03.novell.leeds.ac.uk>) (103)

Josiane Cillard Laboratoire de Biologie Cellulaire et Végétale, IN-SERM U456, Faculté de Pharmacie, 2, av. Pr. Léon Bernard, F-35043 Rennes Cedex, France (FAX: +33-2993-36992) (251)

Simon P. Fricker AnorMed Inc., #100, 20353 64th Avenue, Langley, BC, V2Y 1N5, Canada (FAX: +1-604-530-0976; <sfricker@anormed.com>) (665)

Jon M. Fukuto Department of Pharmacology, UCLA School of Medicine, Center for the Health Sciences, Los Angeles, CA 90095-1735, USA (FAX: +1-310-825-6267; <jon@pharm.medsch.ucla.edu>) (547)

Ho-Leung Fung Department of Pharmaceutics, School of Pharmacy, State University of New York at Buffalo, 513 Hochstetter Hall, Box 601200, Buffalo, NY 14260-1200, USA (FAX: +1-716-645-3693; <hleung @acsu.buffalo.edu>) (723)

Edith Butler Gralla Department of Chemistry and Biochemistry, University of California Los Angeles, 405 Hilgard Avenue, Los Angeles, CA 90095-1569, USA (FAX: +1-310-206-7197; <egralla@chem.ucla.edu>) (125)

Hans-Jürgen Hartmann Anorganische Biochemie, Physiologisch Chemisches Institut, Eberhard-Karls-Universität, Hoppe-Seyler-Str. 4, D-72076 Tübingen, Germany (FAX: +49-7071-296391) (389)

Xudong Huang Genetics and Aging Unit, Massachusetts General Hospital, Bld 149, 13th Street, Charlestown, MA 02129, USA (309)

Shana O. Kelley Division of Chemistry and Chemical Engineering and the Beckman Laboratory of Chemical Synthesis, 127-72, California Institute of Technology, Pasadena, CA 91125, USA (FAX: +1-626-577-4976) (211)

Willem H. Koppenol Laboratorium für Anorganische Chemie, ETH Zürich, Universitätsstrasse 6, CH-8092 Zürich, Switzerland (FAX: +41-1-632-1090; <koppenol@inorg.chem.ethz.ch>) (597)

Joseph R. Landolph Departments of Microbiology and Pathology, USC/Norris Comprehensive Cancer Center, Institute for Toxicology, 1441 Eastlake Avenue, Room 8314, MS 83, Los Angeles, CA 90033-0800, USA (FAX: +1-213-764-0105; <landolph@hsc.usc.edu>) (447)

Ilia Ya. Levitin Institute of Organoelement Compounds, Russian

Academy of Sciences, 28, Vavilov Street, Moscow 117813, Russia (FAX: +7-095-135-6432 or -5085; <gbkc@ineos.ac.ru>) (485)

Stefan I. Liochev Department of Biochemistry, Duke University Medical Center, Durham, NC 27710, USA (FAX: +1-919-684-8885; <liochev@acpub.duke.edu>) (1)

Thomas J. Lyons Department of Chemistry and Biochemistry, University of California Los Angeles, 405 Hilgard Avenue, Los Angeles, CA 90095-1569, USA (FAX: +1-310-206-7197) (125)

Colin L. Masters Department of Pathology, University of Melbourne, Parkville, Victoria 3052, Australia, and Neuropathology Laboratory, Mental Health Research Institute of Victoria, Parkville, Victoria 3052, Australia (365)

G. Mike Makrigiorgos Joint Center for Radiation Therapy and Dana Farber Cancer Institute, Department of Radiation Oncology, Harvard Medical School, 330 Brookline Avenue, Boston, MA 02215, USA (FAX: +1-617-667-9599; <makri@jcrt.harvard.edu>) (521)

Dan Meyerstein Department of Chemistry, Ben Gurion University of the Negev, P. O. Box 653, Beer-Sheva 84105, and College of Judea and Samaria, Ariel, Israel (FAX: +972-7-6469067; <danmeyer@bgumail.bgu.ac.il>) (41)

Robert D. Moir Genetics and Aging Unit, Massachusetts General Hospital, Bld 149, 13th Street, Charlestown, MA 02129, USA (309)

William R. Montfort Department of Biochemistry, University of Arizona, Tucson, AZ 85721, USA (621)

Isabelle Morel Laboratoire de Biologie Cellulaire et Végétale, IN-SERM U456, Faculté de Pharmacie, 2, av. Pr. Léon Bernard, F-35043 Rennes Cedex, France (FAX: +33-2993-36992) (251)

Quinto G. Mulazzani Istituto di Fotochimica e Radiazioni d'Alta Energia del C. N. R., Via P. Gobetti 101, I-40129 Bologna, Italy (103)

Gerd Multhaup ZMBH-Center for Molecular Biology, University of Heidelberg, Im Neuenheimer Feld 282, D-69120 Heidelberg, Germany (FAX: +49-6221-546849 or -545891; <g.multhaup@mail.zmbh.uni-heidelberg.de>) (365)

Hans Nohl   Institute of Pharmacology and Toxicology, Veterinary University of Vienna, Veterinärplatz 1, A-1210 Vienna, Austria (FAX: +43-1-25077-4490; <hans.nohl@vu-wien.ac.at>) (289)

Sergei P. Osinsky   Institute of Experimental Pathology, Oncology and Radiobiology, Kiev 252022, Ukraine (485)

José M. C. Ribeiro   Department of Entomology, University of Arizona, Tucson, AZ 85721, USA and Department of Medical Entomology, Laboratory of Parasitic Diseases, NIAID, National Institutes of Health, Bethesda, MD 20892, USA (621)

José-Luis Sagripanti   Molecular Biology Branch, Division of Life Sciences, Office of Science and Technology, Center for Devices and Radiological Health, Food and Drug Administration (HFZ-113), 12709 Twinbook Parkway, Rockville, MD 20852, USA (FAX: +1-301-594-6775; <jus@cdrh.fda.gov>) (179)

Theodore A. Sarafian   Department of Pathology, UCLA Center for Health Sciences, Room 13-385, Los Angeles, CA 90095-1732, USA (<tsarafia@pathology.medsch.ucla.edu>) (415)

Odile Sergent   Laboratoire de Biologie Cellulaire et Végétale, INSERM U456, Faculté de Pharmacie, 2, av. Pr. Léon Bernard, F-35043 Rennes Cedex, France (FAX: +33-2993-36992; <osergent@univ-rennes1.fr>) (251)

Christian Sievers   Anorganische Biochemie, Physiologisch Chemisches Institut, Eberhard-Karls-Universität, Hoppe-Seyler-Str. 4, D-72076 Tübingen, Germany <FAX: +49-7071-296391) (389)

Klaus Stolze   Institute of Pharmacology and Toxicology, Veterinary University of Vienna, Veterinärplatz 1, A-1210 Vienna, Austria (FAX: +43-1-25077-4490; <klaus.stolze@vu-wien.ac.at>) (289)

Mohammad A. Tabrizi-Fard   Department of Pharmaceutics, School of Pharmacy, State University of New York at Buffalo, 513 Hochstetter Hall, Box 601200, Buffalo, NY 14260-1200, USA (FAX: +1-716-645-3693) (723)

Rudolph E. Tanzi   Genetics and Aging Unit, Massachusetts General Hospital, Bld 149, 13th Street, Charlestown, MA 02129, USA (309)

Joan Selverstone Valentine Department of Chemistry and Biochemistry, University of California Los Angeles, 405 Hilgard Avenue, Los Angeles, CA 90095-1569, USA (FAX: +1-310-206-7197; <jsv@chem.ucla.edu>) (125)

Mark E. Vol'pin† Institute of Organoelement Compounds, Russian Academy of Sciences, 28, Vavilov Street, Moscow 117813, Russia (485)

F. Ann Walker Department of Chemistry, University of Arizona, Tucson, AZ 85721, USA (FAX: +1-520-621-8645; <awalker@u.arizona.edu>) (621)

Ulrich Weser Anorganische Biochemie, Physiologisch Chemisches Institut, Eberhard-Karls-Universität, Hoppe-Seyler-Str. 4, D-72076 Tübingen, Germany (FAX: +49-7071-296391; <ulrich.weser@uni-tuebingen.de>) (389)

David A. Wink Tumor Biology Section, Radiation Biology Branch, National Cancer Institute, Bethesda, MD 20892, USA (547)

Contents of Previous Volumes

---

*Out of print.

---

*Out of print.

---

*Out of print.

Volume 33. Probing of Nucleic Acids by Metal Ion Complexes of Small Molecules

Comments and suggestions with regard to contents, topics, and the like for
future volumes of the series are welcome.

The following Marcel Dekker, Inc. books are also of interest for any reader
dealing with metals or other inorganic compounds:

Handbook on Toxicity of Inorganic Compounds
edited by Hans G. Seiler and Helmut Sigel, with Astrid Sigel
In 74 chapters, written by 84 international authorities, this book covers the
physiology, toxicity, and levels of tolerance, including prescriptions for
detoxification, for all elements of the Periodic Table (up to atomic number

103). The book also contains short summary sections for each element, dealing with the distribution of the elements, their chemistry, technological uses, and ecotoxicity as well as their analytical chemistry.

Handbook on Metals in Clinical and Analytical Chemistry
edited by Hans G. Seiler, Astrid Sigel, and Helmut Sigel
This book is written by 80 international authorities and covers over 3500 references. The first part (15 chapters) focuses on sample treatment, quality control, etc., and on the detailed description of the analytical procedures relevant for clinical chemistry. The second part (43 chapters) is devoted to a total of 61 metals and metalloids; all these contributions are identically organized covering the clinical relevance and analytical determination of each element as well as, in short summary sections, its chemistry, distribution, and technical uses.

Chapter 1

The Mechanism of "Fenton-Like" Reactions and Their Importance for Biological Systems. A Biologist's View

Stefan I. Liochev

Department of Biochemistry, Duke University
Medical Center, Durham, NC 27710, USA

1

1. INTRODUCTION

It seems appropriate to begin this chapter by defining terms. Reaction
(1) is the Fenton reaction.

$$Fe(II) + H_2O_2 \longrightarrow Fe(III) + HO^\bullet + HO^- \tag{1}$$

The term "Fenton-like reactions" is often used when transition metal complexes replace Fe(II). Here I would like to give a more expanded definition of this term, i.e., all reactions describing the reduction of peroxides of any kind by transition metal complexes.

These reactions play a central role in oxidative stress and in the overlapping phenomenon of toxicity caused by transition metal ions (complexes). The chemistry of the Fenton-like reactions (FLRs) is described in Sect. 2 of this chapter and the role of FLRs in causing oxidative stress in vivo is discussed in Sect. 3. In both cases, more attention is paid to the processes that precede FLRs and allow them to occur than to the damage caused by the hydroxyl radical (HO$^{\bullet}$) and by other products of FLRs. In Sect. 3 the processes of adaptation of the cells to oxidative stress are also discussed, with emphasis on the fact that to a significant extent the biological significance of these adaptations is to prevent FLRs from occurring and to repair the damage when they do occur.

It should be stated that this field has grown so vast that discussing, or even citing, most of the important contributions is virtually impossible; therefore, the reader is often referred to reviews. Even so, important aspects might, and will, remain uncovered, whereas problems of special interest to this author will be discussed in more detail. It is hoped that such problems might also be of interest to the reader.

2. THE FENTON REACTION AND FENTON-LIKE REACTIONS

2.1. The Classical Fenton Reaction

Although Fenton [1] was the first to describe a reaction dependent on both Fe(II) and H_2O_2, the mechanism of the Fenton reaction was initially studied and essentially discovered by Willstätter, Haber, and Weiss[2,3] and subsequently by Barb et al. [4]. Because this chemistry is so important for understanding the nature of oxidative stress, it deserves detailed exposition.

When H_2O_2 and Fe(II) are the only reactants, reaction (1) is just the first reaction of a complex process. In reaction (2) superoxide radical (O_2^-) is generated through the oxidation of H_2O_2 by HO$^{\bullet}$.

$$HO^{\bullet} + H_2O_2 \longrightarrow O_2^- + H^+ + H_2O \qquad (2)$$

The finding [2,3] that led to the proposal of the second most famous reaction in the free radical field was that at high H_2O_2/Fe(II) ratio more H_2O_2 was decomposed than Fe(III) formed, indicating a chain mechanism.

Reaction (3), which is now called the Haber-Weiss reaction, causes iron-independent decomposition of H_2O_2 and moreover regenerates HO$^\bullet$. Thus the chain process consists of reactions (2) and (3) [2,3].

$$O_2^- + H_2O_2 \longrightarrow O_2 + HO^\bullet + HO^- \tag{3}$$

It should be noted that $HO_2^\bullet$ and not O_2^- is the reactant in the original papers. However, since at neutral pH O_2^- predominates due to the dissociation of $HO_2^\bullet$ to H^+ and O_2^-, the above reactions are written to accommodate this fact.

However, by 1951 Barb et al. [4] concluded, and later Rush and Bielski [5] confirmed, that reaction (3) does not occur and that O_2^- reduced Fe(III) rather than H_2O_2.

$$O_2^- + Fe(III) \longrightarrow O_2 + Fe(II) \tag{4}$$

The overall chain process is described, in this case, by reactions (1), (2), and (4), and at infinite chain length the net reaction is

$$2H_2O_2 \longrightarrow O_2 + 2H_2O \tag{5}$$

In reality, chain-breaking reactions, such as reaction (6), occur and they account for the conversion of Fe(II) to Fe(III) concomitant with the decomposition of H_2O_2.

$$O_2^- + O_2^- + 2H^+ \longrightarrow O_2 + H_2O_2 \tag{6}$$

Reaction (3) is, in this view, the net reaction of reactions (1) and (4) and is often called the iron-mediated Haber-Weiss reaction.

If present, additional reactants will compete for HO$^\bullet$ with H_2O_2 and Fe(II), and both the stoichiometry and the mechanism of the above-described chemistry can change dramatically. This will be discussed in Sect. 2.2. Additional reading for Sect. 2.1 can be found in Refs. [6–11] and in other chapters of this book.

2.2.   Other Metal Complexes Replacing Fe(II)

The reduced form of complexes of other transition metals such as copper and vanadium can be oxidized by H_2O_2, resulting in production of HO$^\bullet$ [6,8,11–17].

The processes initiated by the reaction of vanadyl (V(IV)) with H_2O_2 in the presence or absence of targets for HO• will be discussed below as an example of the complexity of FLRs and their potential to cause damage.

Upon mixing of V(IV) and H_2O_2, oxygen is produced through the HO• and O_2^--mediated reactions (7), (2), and (8) [12–15].

$$V(IV) + H_2O_2 \longrightarrow V(V) + HO\bullet + HO^- \tag{7}$$

$$V(V) + O_2^- \longrightarrow V(IV) + O_2 \tag{8}$$

Reactions (7) and (8) are analogous to reactions (1) and (4), except for V(IV) replacing Fe(II). This does not mean that the V(IV)-catalyzed Fenton chemistry is completely similar to the Fe(II)-catalyzed process. For example, reaction (8) is thought to be the net of reactions (9) and (10) [13].

$$V(V) + O_2^- \rightleftharpoons V(IV)-O_2 \tag{9}$$

$$V(IV)-O_2 \rightleftharpoons V(IV) + O_2 \tag{10}$$

The intermediate, V(IV)–O_2, is capable of initiating the oxidation of NAD(P)H, which then proceeds *via* O_2^--dependent chain reactions without the involvement of HO• [18].

In the presence of sufficient concentration of targets for HO• (RH) such as NADH, ethanol, or Hepes, the HO• formed in reaction (7) reacts with them [reaction (11)] rather than with H_2O_2 [reaction (2)], whereas in reaction (12) the secondary radical (R•) reacts with oxygen to form O_2^- [13].

In this case, described by reactions (7), (11), and (12), the targets of HO• are being oxidized and consumption of O_2, rather than its release, occurs. In addition, the secondary radicals formed are also capable of causing damage [13].

$$RH + HO\bullet \longrightarrow R\bullet + H_2O \tag{11}$$

$$R\bullet + O_2 \longrightarrow R + O_2^- \tag{12}$$

2.3. Other Peroxides Replacing Hydrogen Peroxide

Fe(II) complexes and the complexes of other transition metals (M) are capable of reducing lipid hydroperoxides (ROOH).

$$M^{n+} + ROOH \longrightarrow M^{(n+1)+} + RO\bullet + HO^- \tag{13}$$

Reaction (13) plays an extremely important role in the process of lipid peroxidation whose mechanism and biological significance are extensively discussed in Chapters 3 and 8 of this book and briefly in other parts of this chapter.

H_2O_2 forms complexes with (V(V)), molybdate, and tungstate, which are now considered to be responsible for a significant part of the pharmacological and toxicological properties of these metal ions, pervanadate being the most studied [19–21]. H_2O_2-decomposing enzymes such as catalase and horseradish peroxidase are practically inactive toward pervanadate [22,23] and it follows that the effective concentration of peroxides will be high in cells accumulating metals capable of forming complexes with H_2O_2.

As far as FLRs are concerned, V(IV) is capable of decomposing diperoxovanadate and O_2 is released in the process [24]. While much remains to be learned about the mechanism of this process, the first reaction should be the reduction of pervanadate by V(IV). In the presence of several organic molecules in addition to V(IV) and pervanadate, oxygen is being consumed, indicating formation of HO• or similarly reactive species [24], as is the case in the process started by the reduction of H_2O_2 by V(IV) described in Sect. 2.2.

2.4. Is HO• the (Only) Reactive Species Generated?

This is a question valid for all FLRs and in some cases, such as the reduction of lipid hydroperoxides by reduced metal complexes [reaction (13)], radicals other than HO• are formed.

Mostly, ongoing discussions concern the classical Fenton reaction [reaction (1)] and the interaction of Cu,Zn superoxide dismutase (SOD) with H_2O_2 [6–9,25–29].

Many years ago, DiGuiseppi and Fridovich compared the ability of several known HO• scavengers to inhibit the damage caused by the Haber-Weiss reaction [30]. They observed that the activity of these scavengers was strongly proportional to their known reactivity toward the free radiolytically generated HO•. It would appear that HO• is the species generated by the Haber-Weiss and consequently the Fenton reaction and causes the effects observed. In other situations, however, the activity of the HO• scavengers is different from the one exerted toward free HO• or uncharacteristic products different from those resulting from free HO• attack are formed [6–9,25–27]. In such cases the effects could be due to high valent transition metal complexes such as Fe(IV), or to "bound HO•," or, to be on the safe side, to species physically different from HO• but having similar reactivity.

Indeed, the mechanism of reaction (1) is probably complex and the reaction proceeds through discrete steps and involves FeO^{2+} as intermediate [6,8,31]. Factors such as pH, ligands, and the reactivity of the targets might determine if HO• or FeO^{2+} will be the dominant oxidizing species [6,9,25–27]. From the viewpoint of a biologist, the significance of this discussion is in the possibility for selective damage caused by FLRs in different situations. For convenience, the active species are referred to as HO• while similar causes of selective damage will be discussed later.

2.5. The Haber-Weiss Reaction: Superoxide Radical Acting as Reductant for Fe(III)

If indeed FLRs are a major factor in the phenomenon of oxidative stress, they must be kept running continuously. Mechanisms should exist that supply Fe(II) and/or reduce Fe(III) to Fe(II) and that regenerate H_2O_2 and other peroxides. The discovery of SOD [33] and the consequent investigations [10,28,30–35] pointed out the extraordinary importance of O_2^- for oxidative stress. Yet the chemist's opinion initially was that O_2^- is poorly reactive [36,37]. How then could O_2 be toxic?

Soon after the discovery of SOD, Beauchamp and Fridovich [38] found that HO• is generated from O_2^- and H_2O_2. They were aware of the studies of Haber and Weiss [3] but not of Barb et al. [4] and proposed that HO• is produced through the iron-independent Haber-Weiss reaction. Furthermore the Haber-Weiss reaction was suggested to be the major toxic pathway in the SOD-based theory of oxygen toxicity [39]. Later it was recognized that iron is required for the O_2^--dependent formation of HO• [40,41] and the iron-dependent Haber-Weiss reaction was promoted as a central mechanism of the toxicity of O_2^-.

This mechanism was proven in a very large number of in vitro experiments [5,6,10,11,27,28,31,34] and was, and still seems to be, widely accepted as the major cause of the toxicity of O_2^- and, by extension, of oxidative stress. A variation of this mechanism is also considered by some authors. In this, O_2^- acting as a reductant has a moderate ability to release iron from ferritin [42,43].

While the above mentioned explanations and observations are

interesting, it is unlikely that O_2^- can trigger the Fenton reaction in particular and FLRs in general by acting as reductant intracellularly, as discussed in Sect. 3 of this chapter.

2.6.   Other Reductants Replacing Superoxide Radical

O_2^- rapidly reduces some iron complexes with rate constants as high as $10^7 \, M^{-1} \, s^{-1}$ [5,44] and others more slowly [45]. However, it is neither unique nor specific in its action. In fact many enzymes and abundant biological reductants such as ascorbate and thiols, efficiently reduce both Fe(III) and Cu(II) and both reduce and release ferritin-bound iron [46–49]. Consequently, in the presence of H_2O_2, or hydroperoxides, and iron, or ferritin, such reductants should catalyze lipid peroxidation or HO• production and indeed this has been reported [46–48].

2.7.   In Vitro Damage Caused by Fenton-like Reactions

We have briefly discussed that the reduction of lipid or metal peroxides creates active species capable of causing further damage. The HO• created when H_2O_2 is the oxidant is with reason regarded [10] as probably the most reactive free radical. Indeed, it reacts extremely rapidly with lipids, sugars, alcohols, and nucleic acids, to name a few [6,8–11, 13,14,46,50]. In some cases, as when attacking DNA, HO• is damaging in several ways, causing strand breaks (HO• is known to cause deoxyribose fragmentation) and DNA base oxidations [9,14,51]. This damage follows a characteristic pattern and this circumstance is useful for discrimination of the toxicity caused by HO• from other pathways both in vitro and in vivo [9,27,51]. The chemical, biochemical, and toxicological aspects of the reactions of HO• and the radicals formed during lipid peroxidation are discussed in several chapters of this book.

2.8.   Site-Specific Fenton-like Reactions and Caged HO•

Will O_2^- or HO• be more toxic in vivo if both radicals are produced at the same rate? It has been suggested, with reason, that the much less discriminately reactive O_2^- will be more toxic because more O_2^- molecules will survive to damage important targets rather than to react with abundant and expendable molecules [52].

However, this hypothesis assumes random distribution of O_2^- and HO• and a lack of selectivity of the actions of the latter. In fact, the damage caused by HO• produced through FLRs, as opposed to that which is radiolytically generated, is often selective [6,9], and this led to creation of terms like "crypto HO•," "caged HO•," "site-specific formation of HO•," etc. This selectivity may have several explanations. HO• or similar reactive species could be formed within a space accessible to some but not all possible targets. Alternatively, Fe or other metal ions could be preferentially bound to some targets, like DNA, and in this situation HO• will be generated adjacent to, and be more likely to react with, those targets. Finally, as was already mentioned, selective damage could be due to difference in the reactivity of the free HO• and species like FeO^{2+}. Those problems are discussed later in this chapter and in other chapters of this book.

2.9. The Ability of Metal-Containing Proteins to Catalyze Fenton-like Reactions

2.9.1. Iron-Sulfur Cluster-Containing Proteins

The mechanism of inactivation by oxidants of a number of [4Fe-4S] cluster-containing dehydratases has been subject to an increasing number of studies and a number of reviews are devoted to this and other aspects concerning iron-sulfur (FeS) cluster-containing proteins and enzymes [31,34,53–58]. For instance, O_2^- reacts rapidly with aconitases of bacteria or mammals, fumarases A and B of *Escherichia coli* (*E. coli*) and other [4Fe-4S] cluster-containing dehydratases [31,34,54,57, 59,60]. The O_2^- oxidizes the clusters, as do other oxidants, resulting in release of Fe(II) [34,55,57,59,61]. This sets the stage for Fenton chemistry and has very important biological implications that are discussed in Sections 3 and 4 of this chapter. Peroxynitrite is another important biological oxidant that reacts rapidly with [4Fe-4S] clusters, leading to inactivation of the cluster-containing enzymes and iron release [60–62], whereas O_2 and NO react much less rapidly with those clusters [57, 59,62,63].

2.9.2. Cu,Zn Superoxide Dismutase

This enzyme has a low level of peroxidase activity and the peroxidation of a number of substrates as well as self-inactivation in the presence of H_2O_2 have been reported [64,65]. The reactions proposed are:

$$H_2O_2 + SOD\text{-}Cu(II) \longrightarrow SOD\text{-}Cu(I) + O_2^- + 2H^+ \tag{14}$$

$$SOD\text{-}Cu(I) + H_2O_2 + H^+ \longrightarrow SOD\text{-}Cu(II) + HO^\bullet + H_2O \tag{15}$$

$$SOD\text{-}Cu(I) + H_2O_2 + H^+ \longrightarrow SOD\text{-}Cu(II)\cdot OH + H_2O \tag{16}$$

Reaction (14) describes the reduction of the enzyme-bound copper by H_2O_2 and is not a matter of debate, whereas serious disagreement exists between the groups led by E. Stadtman [29] and I. Fridovich [28,66] about the nature of the active species formed in reactions (15) or (16) that describe the reduction of H_2O_2 by the SOD-bound Cu(I). Stadtman and colleagues favor the production of free $HO^\bullet$ [reaction (15)], whereas Fridovich [28] and Liochev et al. [66] suggest the formation of enzyme-copper-bound hydroxyl radical or a higher oxidative state of copper [reaction (16)].

    While the arguments of both sides could be read in the cited references, the problem was unequivocally solved by Mason and coworkers [67,68]. They were able to show by EPR-spin trapping that the oxygen atom in the DMPO-OH adduct, produced as a result of the Fenton reaction [reaction (1)], originates from H_2O_2 as can be expected for $HO^\bullet$ being the product of this reaction [67]. On the contrary, a significant portion of the oxygen in the DMPO-OH adduct generated during the interaction of Cu,ZnSOD with H_2O_2 [reactions (14) plus (15) or (16)] comes from water and this is incompatible with free $HO^\bullet$ being formed [68]. Hence, the peroxidative action of Cu,ZnSOD is described by reactions (14) plus (16).

2.9.3 Others

The significance of ferritin has already been noted. O_2^- acts as an oxidant toward Cu(I)-thionein, causing release of copper as Cu(II) [69]; and in the presence of reductants and peroxides copper-dependent FLRs should result. Such reactions of O_2^- might suggest a role for SOD in copper homeostasis in addition to the copper-buffering effect proposed by others [70].

Lipid peroxidation due to destabilization of FeS clusters of electron transfer proteins has been reported [71]. Heme destruction could be another source of free iron [72], whereas the oxidation of metmyoglobin by H_2O_2 produces ferrylmyoglobin, which contains the reactive $Fe(IV)=O$ [73].

2.10. Additional Aspects in Vitro

$HO^\bullet$ can be produced by metal-independent reactions. Thus O_2^- reacts rapidly with HOCl, $HO^\bullet$ is among the products, and this reaction might play an important role during inflammation [6]. Nitric oxide (NO) reacts extremely rapidly with O_2^- forming peroxynitrite [74], which undergoes "spontaneous" decomposition generating $HO^\bullet$ (or similar reactive species) in the process [75,76]. Another scenario for peroxynitrite-triggered $HO^\bullet$ production in biological systems has also been proposed [34]. Since peroxynitrite rapidly inactivates FeS clusters containing enzymes, it is conceivable that intracellularly this will result in an increased "free" iron pool and, consequently, in the presence of H_2O_2, in an increased rate of the Fenton reaction. This proposal [34] was recently confirmed [77].

3. SIGNIFICANCE OF THE FENTON-LIKE REACTIONS IN VIVO

3.1. Existence and Nature of the Cellular "Free" Iron Pools

The existence of intracellular free iron is clear even from studies that do not involve direct measurements. For instance, inactivation of Cu,ZnSOD in endothelial cells [78] sensitizes those cells to H_2O_2 toxicity whereas iron-chelating agents protect. Similarly, E. coli cells deficient in cytoplasmic SODs are much more sensitive to H_2O_2-induced kill than the parental strain and are protected by iron chelators [79].

Direct evidence based on measurements of the intracellular and extracellular iron also exist [77,80,81]. Thus through EPR techniques, the concentration of the free iron in E. coli is estimated to be several micromolar [77,80]. Using similar EPR-based methods, Cooper and colleagues were able to detect a free iron pool in eucaryotic cells [81]. Such pools exist in mitochondria and extracellular fluids [81–84].

It is clearly established that in some situations the concentration of this pool changes. For example, if the regulation of the iron transport in *E. coli* is impaired [85] or the concentration of iron in mitochondria is increased [86], the result is increased toxicity of H_2O_2. Similarly, oxidative stress has been both proposed and established to cause an increase in the intracellular free iron [31,34,77,80], whereas other treatments not necessarily of an oxidative nature can lead to increased extracellular iron concentration [82,87].

The nature of this pool is less understood due to a significant lack of good noninvasive methods [81], but nevertheless some data exist [81, 88] suggesting that, as might be expected, amino acids and nucleotides such as ATP could be among the ligands of the "free" iron.

Thus, despite the highly sophisticated and regulated mechanisms of iron accumulation, transport, storage, and utilization [89–92] free iron pools exist. It is unlikely, of course, that this is just "a nuisance" allowing FLR to occur. In fact, this free iron, also called readily chelatable iron pool and low molecular weight iron pool, is necessary and can be used for the synthesis of iron-containing enzymes and proteins [34,55,77]. It is reasonable to conclude, therefore, that this free iron becomes a nuisance only when its concentration or the rate of production of H_2O_2 and other peroxides increases.

3.2. Intra- and Extracellular Sources of Free Iron, Hydrogen Peroxide, and Superoxide

3.2.1. Hydrogen Peroxide and Superoxide

These two species are being produced by numerous enzymes and during the oxidation of biomolecules by oxygen by mitochondria, microsomes, and other membrane-bound enzymes, xanthine oxidase, glucose oxidase, urate oxidase, monoamine oxidase, and fumarate reductase. They are also produced during autoxidation of reduced flavins, tetrahydropterins, catecholamines, etc. [10,32,34,39,41,50,93]. In some cases, e.g., glucose oxidase and urate oxidase [50,94], oxygen is reduced directly to H_2O_2, whereas in many other cases oxygen is initially converted to O_2^- [10,32,50,95], which then can react with intracellular targets or be dismuted to H_2O_2 and O_2 by SODs.

The steady-state concentrations of O_2^-, H_2O_2, and other peroxides are determined by a balance between all processes that produce and consume them. Part of the consumption is due to reactions with anti-oxidant enzymes such as superoxide dismutases, catalases, and peroxidases, whereas another part of O_2^- and the peroxides will react with targets, exerting toxicity. The ratio between the number of the reactive species consumed by the protective enzymes versus the number of species disappearing through reactions with targets determines the extent of protection of the targets.

For instance, it has been accepted for many years that SODs provide significant protection against the toxicity of O_2^-, but the extent of protection of the targets was unknown and is still unknown in most cases. Recently, a method for intracellular measurement of O_2^- scavenging activity has been developed based on the lucigenin-dependent chemiluminescence [96]. This method established that approximately 95% of the O_2^- produced within *E. coli* cells is scavenged by the normal level of SOD found in the cell [96]. Furthermore, these results are in agreement with earlier estimation of the steady-state concentration of O_2^- in SOD-proficient and deficient *E. coli* cells, including calculations based on the inactivation-reactivation of iron-sulfur cluster-containing enzymes [54,97,98]. Measuring the extent of protection is, in principle, less complicated than estimation of the steady-state concentrations of O_2^- and or H_2O_2. Thus, in the method described above for measuring the SOD activity in vivo [96], it was not necessary to know the rate of production of O_2^-.

Nevertheless, attempts to estimate the intracellular and even intracompartmental concentrations of O_2^- and H_2O_2 have been made [50,54,98,99]. The calculated intracellular concentrations of O_2^-, H_2O_2, and $HO^\bullet$ are low [50,54,98,99]. This actually attests to their reactivity rather than arguing for their unimportance. Many cells, such as neutrophils, are capable of producing O_2^- and H_2O_2 extracellularly together with other reactive species and the rate of production could further be increased by chemical and biological agents [27,100,101].

3.2.2. Sources of Iron

In addition to the previously mentioned potential sources of iron it should be emphasized that the intracellular iron content might increase due to malfunctioning of the complex mechanisms of uptake, storage, and utilization of this metal. For example, impaired regulation of the iron uptake in *E. coli* results in FLR-mediated damage to the cells [85]. In eucaryotic cells the conversion of the cytosolic aconitase into iron regulatory protein-1 (IRP-1, formerly called IRE-BP, etc.) leads to increased iron uptake and more iron is made available due to decreased ferritin biosynthesis [53,55,56,58,89]. Since not only iron deficiency but also oxidative stress triggers the aconitase → active IRP conversion [31,89,90], this might result in enhancement of oxidative stress by FLR-dependent pathways. Many reductants, including O_2^-, are capable of releasing iron from ferritin [42,43,49]. While arguments will be presented later that O_2^- is unlikely to exert toxicity by acting in this way, ferritin undoubtedly serves as source of free iron in some situations. A ferritin-deficient mutant of *Campylobacter jejuni* is more sensitive to H_2O_2 or paraquat toxicities than the parent strain [102], whereas ascorbate mobilizes iron from ferritin in neuroblastoma cells [103].

An increase in the intracellular and extracellular iron pools has been detected in cases such as hemochromatosis and following cancer chemotherapy [81,87]. Such increased extracellular iron is in a form capable of catalyzing free radical-mediated reactions [87] and of reactivating FeS cluster-containing enzymes [82].

It should not be generalized, however, that any increase of the total intracellular iron concentration, especially through supplementation of the extracellular media with iron, must result in toxicity. After all, iron is necessary for the biosynthesis of the active centers of many enzymes. Recently, it was found that supplementation of the growth medium of *sodA sodB E. coli* improves the growth rate of this strain aerobically and does not result in increased kill by H_2O_2 [104]. Clearly, at least in some situations, with the mechanisms for iron uptake and utilization intact, the increased total cell content of iron is not due to increase of free iron concentration capable of catalyzing FLR.

3.3. Evidence of Damage to DNA and Other Biomolecules: Some Questions

There is abundant information about the in vivo damage to DNA, biomolecules, and biomembranes by FLRs or specifically by O_2^- [9,10,14,27,

28,31,33,34,51,77,90,105], including in this book. At this stage it seems reasonable to define some problems that were, or are, unsatisfactorily answered or that seem puzzling, and which the discussion in the remainder of this chapter attempts to resolve.

Thus, O_2^- undoubtedly plays an important role in oxidative stress, yet it is not capable of damaging DNA directly [33,105]. Nevertheless, when its intracellular concentration is increased by increasing its production or by elimination of SODs, severe damage, including that to DNA, occurs [9,28,31,34,51,77,105]. While it is conclusively established that a significant part of the O_2^--dependent damage is mediated by FLRs [9,28,51,105], the mechanism through which O_2^- indirectly triggers FLR in vivo was not clear until recently and is a major topic for discussion late in this chapter.

A large portion of the FLR-mediated damage to DNA in the cell appears not to be due to freely diffusible $HO^{\bullet}$ and probably occurs at the DNA molecule on Fe(II) ions associated with DNA [9]. The question is whether this Fe(II) is always there available for the Fenton reaction or rather appears during oxidative stress.

In addition to the damage to DNA that results in increased mutagenesis and death [9,28,31,34,51,77,106,107], increase in the intracellular O_2^- concentration causes damage to the membranes [108] and growth defects [28,34,109]. Is the damage exerted by O_2^- in all of these cases due to attack of different targets or of one and the same target? Finally, is the mechanism through which O_2^- exerts damage to DNA and other biological molecules and structures specific for O_2^- or can be it be used by other reactive species?

3.4.   Superoxide Radical: Reductant or Oxidant?

As a result of enormous effort following the discovery of SOD about 30 years ago, it became established that a broad range of toxic effects depend on production of $HO^{\bullet}$ by a process that also involves O_2^-, H_2O_2, and Fe ions (complexes) [9,10,27,28,31,33,34,46,47,50,51,54,85,90,106, 109,110,111]. The iron-dependent Haber-Weiss reaction, described in Sect. 2, seemed to be the mechanism of this toxicity both in vitro and in vivo. Indeed, this mechanism was proven numerous times in vitro and it is hard to imagine how $HO^{\bullet}$ could be produced from O_2^-, H_2O_2, and iron in any other way. Yet serious arguments initially expressed by only several groups and scientists [111–113] strongly questioned the Haber-Weiss reaction-based explanation of the toxicity of O_2^-:

1. The concentration of O_2^- intracellularly is too low as compared with that of other reductants for iron; hence, O_2^- cannot contribute significantly to the process of reduction of Fe(III) [113]. Convinced that O_2^- cannot exert its toxicity in vivo by acting as reductant for Fe(III), Czapski et al. [113] hypothesized that O_2^- acts as oxidant for Fe(III) complexes resulting in a higher oxidation state of iron (Fe(IV)), which is the directly damaging agent. This explanation for the toxicity of O_2^- does not provide a role for FLRs.

2. A similar, but not identical, argument was made by Winterbourn [111,112]. Thus, the cellular environment is strongly reducing, even aerobically, and any free iron will normally exist as Fe(II). How could O_2^- act as reductant of Fe(III) then? This is also a good argument against O_2^- acting as an oxidant of free Fe(III), Cu(II), and so forth.

3. Attempts were made to circumvent the above arguments by supposing that O_2^- might trigger FLR by reducing not all available free iron but specifically certain types of iron complexes or proteins outcompeting, in such case, other reductants. Thus release of iron from ferritin by O_2^- serving as reductant was suggested, as a variant of the iron-dependent Haber-Weiss reaction, to provide Fe(II) necessary for the production of HO$^{\bullet}$ in vivo [42,43,114]. There is no evidence, however, that in vivo O_2^- will reduce ferritin more efficiently than other reductants. Therefore, this proposal is subject to the same criticism as that applied to the Haber-Weiss reaction. In *E. coli* the problem has been settled by the finding that none of the two types of ferritin of this organism contributes to the O_2^--induced increase of the intracellular free iron [80].

4. Rowley and Halliwell [115,116] and Imlay and Linn [46] had encountered situations of synergism between O_2^- and other cellular reductants both in vitro and in vivo, but proposed mechanisms based again on O_2^- acting as reductant.

Although unsuccessful in providing a convincing explanation to the ultimate question of how O_2^- triggers FLR in vivo, these investigations and their analyses allow the formulation of criteria to which such an explanation should conform:

1. Intracellularly, O_2^- could not possibly exert toxic action by acting as reductant.
2. It seems likely that O_2^- should act by increasing the intracellular free iron pool rather than interacting with a preexisting pool.
3. O_2^- and cellular reductants might synergize in triggering FLRs.

An explanation that conforms to these criteria and other facts discussed in this chapter is given in Sect. 3.5.

3.5. The Role of Iron-Sulfur Clusters

The nature and the properties of iron-sulfur (FeS) cluster-containing enzymes and the mechanism of their inactivation by O_2^- has been mentioned in Sect. 2.9, studied and reviewed previously [31,34,53–63], and will not be described in detail here. FeS-clusters have been proposed to play a major role in the in vivo production of HO$^\bullet$ through the Fenton reaction and in initiation of FLRs in general [31,34] and the present state of this theory is discussed below.

Several dehydratases of mammalian or microbial origin have been found to be inactivated in vitro or in vivo by hyperbaric oxygen, by O_2^-, by redox cycling agents such as paraquat (PQ), and by peroxynitrite and other oxidants [31,34,54,57,59–63]. Interestingly, all of these enzymes contain [4Fe-4S] centers [34,53,54,55,57,59,60,62].

Most studies concern the inactivation of these enzymes by O_2^- and it is important to note that several growth defects, such as the inability of SOD-deficient E. coli to grow in the absence of branch chain amino acids, are explicable by O_2^--dependent oxidation of the [4Fe-4S] clusters of a target enzyme [34,109,117–119]. The inactivated enzymes can be reactivated in vitro or in vivo by reductive reconstruction of the FeS center [34,54,55,57]. The reactivation of purified enzymes can be achieved non-enzymatically in the presence of thiols and iron [55,57, 120]. However, the in vivo reactivation, which is not well understood, is most likely enzyme-dependent [34,77]. The significance of a number of enzymic and nonenzymic factors has been reviewed [34]; in addition, iron storage proteins such as ferritin are clearly implicated in the reactivation process [77]. It has been proposed that the process of reconstruction itself can be impaired by oxidative stress [34]. In cell extracts from *E. coli*, at least some of the inactivated FeS enzymes undergo "spontaneous" reactivation anaerobically [34,121]. For example, the aerobically inactivated fumarase A can be reactivated anaerobically and EDTA inhibits this reactivation [121]. This and other experiments reviewed previously [34] show that upon inactivation an "easily chelatable" iron pool is created that can be used for reactivation of FeS enzymes or, alternatively, for catalysis of FLRs. Studies on the mechanism of one-electron oxidation of [4Fe-4S] clusters, including by O_2^-, also indicate release of iron from the cluster [55,57,120].

All this provides the basis for an explanation of how O_2^- triggers the Fenton reaction in vivo [31,34], although the authors admit that the idea did not occur to us immediately and we published our initial hypothesis paper [31] with some hesitation. Such is the power of the traditional thinking and the hope that some modified version of the Haber-Weiss reaction could explain the facts!

I will describe this hypothesis by stating the predictions (expectations) it makes [31] and discussing to what extent those predictions are confirmed at present. Thus, increasing the concentration of O_2^- by increasing the rate of production or by decreasing the concentration of SOD should have the following consequences:

1. The activities of the [4Fe-4S]-containing dehydratases will decline. This is rather obvious and, as discussed above, has been observed many times both in vitro and in vivo.

2. The level of free iron will correspondingly increase. This prediction was based on the discussed release of iron from the clusters upon oxidation by O_2^- and the creation of a chelatable iron pool in vitro. It was of extreme importance that this conclusion be confirmed in vivo and this has been done by Keyer et al. [79,80]. Using an EPR method they established an easily chelatable Fe(II) pool in SOD-proficient *E. coli* that was 5–10 times larger in the cells of the *sodA sodB* strain.

3. The rate of production of H_2O_2 will increase as will the rate of production of FeO^{2+} or $HO^•$. Indeed, as already described, Fe(II) released from the cluster should react with H_2O_2 produced by numerous enzymes and nonenzymic processes, in-

cluding by the reduction of O_2^- by the cluster. When one O_2^- acts as an oxidant it yields one H_2O_2, but when it is dismuted by SOD it yields only 0.5 H_2O_2. Thus overproduction of SOD should not result in increased H_2O_2 production. This has been recently confirmed [122]. Recently, experiments using EPR spin trapping methods established that the $HO^\bullet$ formation in *sodA sodB E. coli* is significantly increased, as compared with the control strain [123].

4. The cellular reserve of reductants such as glutathione (GSH) and NAD(P)H will decrease or at least their rate of consumption will increase.

The reactivation of the cluster will require two reducing equivalents because both the oxidation of the cluster by O_2^- and the subsequent oxidation of the released Fe(II) by H_2O_2 are one-electron transfer reactions. Assuming that NADPH is the major although indirect source of reducing equivalents, the net reaction of the O_2^--dependent inactivation and the reductive reactivation of the clusters has been formulated [34] [reaction (17)].

$$\text{NADPH} + O_2^- + 2H^+ \xrightarrow{\text{FeS cluster}} \text{NADP}^+ + HO^\bullet + H_2O \ (17)$$

More reductants should be consumed and more $HO^\bullet$ generated through a cycle consisting of reactions causing reduction of Fe(III) and reaction (1) since reaction (1) and the process of cluster reassembly are in competition for Fe(II).

This prediction has not been strictly confirmed but strong indirect evidence is available. For example, modest inductions of *soxRS*-regulated enzymes [121,124] are detected in *sodA sodB E. coli*, and O_2^- and PQ strongly synergize in inducing the *soxRS*-regulated enzyme fumarase C [121]. The *soxRS* system of *E. coli* and other organisms and the role of the NADPH/NADP ratio to modulate the activity of this system will be discussed in Sect. 3.8.

5. Direct confirmation of the prediction [31,34] that the FeS clusters are the major source of the free iron released by O_2^- and other oxidants is still lacking and is of course extremely difficult to obtain. There are, however, several lines of evidence

that this is the case. Thus, ferritin is not the free iron source in
E. coli whereas overproduction of FeS cluster-containing enzymes had the expected outcome [80]. Remarkably strong coincidence between the time dependence of the intracellular
inactivation of FeS-cluster containing enzymes by peroxynitrite and the time dependence of the appearance of free iron
is seen, whereas similarly the free iron disappears in parallel
with the reactivation of these enzymes [77].

In general, the FeS cluster-based theory of O_2^- toxicity is well supported. Moreover, this theory is applicable in cases when other oxidants
are the toxic agents. Thus, as already mentioned, it could be predicted
that peroxynitrite will increase the intracellular free iron concentration
[34] and this was established recently [77]. Oxygen is also capable of
inactivating FeS cluster-containing enzymes [59,121] and the released
Fe is in a form that could be used in reactivation [121]. It has been predicted that oxygen will also increase the intracellular free iron [31]. The
toxicity of the hyperbaric oxygen, therefore, might be due at least in part
to this direct effect in addition to causing increased production of O_2^-.

Nitric oxide (NO) is another agent that under certain conditions
inactivates FeS cluster-containing enzymes [62,63], although the rate
of this reaction is slow. Nevertheless, at sufficient NO concentration an
increase in the intracellular free iron pool should be expected.

3.6. The Central Role of the Fenton-like Reactions in Oxidative Stress

It is interesting to try to briefly summarize the relative significance of
factors such as O_2^-, FeS cluster-containing enzymes, and FLRs in regard to their role in the intracellular production of HO•, or similarly
reactive radicals, including those produced from lipid peroxides.

(1) O_2^- causes a metabolic slowdown [31,34,109] largely due to
inactivation of FeS cluster-containing dehydratases. The same mechanism is responsible for damage to DNA and other biomolecules and
membranes, contributing to increased mutagenicity and even cell death
because it leads to increased intracellular free iron concentration and
consequently triggers FLRs. Thus, O_2^- is an extremely dangerous, selectively damaging agent. This explains the importance of SODs.

The postulate of O_2^- acting as oxidant might not be true for extracellular fluids [34] and the periplasmic space of microbial cells [79], which are much less reduced environments than the cytoplasm and are not known to contain 4Fe-4S cluster-containing dehydratases. In these compartments $HO^\bullet$ could indeed by generated by the Haber-Weiss reaction or through metal-independent reactions such as the reaction of O_2^- with HOCl.

(2) The significance of the FeS center-containing enzymes for the phenomena of the oxidative stress and $HO^\bullet$ production in particular seems, at least intracellularly, to be broader than that of O_2^-. As already mentioned, a number of other oxidants can attack those clusters, leading to the consequences discussed above.

(3) The importance of FLRs for the production of $HO^\bullet$ is even broader. In fact, FLR could mediate production of $HO^\bullet$ through the FeS center-based mechanism, through the Haber-Weiss reaction, and through increase of intracellular "free metals" by increased metal ion uptake, or release of metal ions from metal storage proteins. True, even FLRs do not represent all cases of $HO^\bullet$ generation.

FLRs have been implicated in the development of an impressive number of diseases and pathological conditions as well as in the toxicity exerted by metals and drugs. The role of FeS cluster-containing enzymes for the intracellular production of $HO^\bullet$, however, began to be appreciated only very recently because the concept itself is very new. Therefore when debating, in Sect. 3.7, the role of FLRs in medical situations the emphasis will be on the recent developments establishing the importance of the FeS center-dependent pathway, whereas only a few other recent findings will be briefly analyzed.

3.7. Selected Examples of Possible Involvement of Fenton-like Reactions In Vivo

3.7.1 MnSOD-Deficient Mice

SOD-deficient organisms have been created [33,107,109] beginning with the pioneering work of Carlioz and Touati [109] who engineered sodA sodB E. coli. Very recently, a mouse entirely lacking functional mitochondrial MnSOD has been constructed and many features of its phenotype were described [107,125]. The analysis of this phenotype supports the conclusion that even in mammals the O_2^- toxicity is likely to be exerted through attack of FeS clusters and a consequent increase of the intracellular free iron pool. Mitochondria are a potent source of O_2^- [50,126] and the mitochondrial aconitase, a [4Fe-4S] cluster-containing enzyme, is O_2^- sensitive [54,127] and therefore the significant reduction of aconitase activity in the MnSOD-deficient mice [107] is not surprising. One of the consequences of the increased iron pool should be development of lipid peroxidation and lipid peroxidation-dependent damage; indeed, convincing evidence of this has been reported in the case of the mitochondrial SOD-deficient mice [107,125]. All of the remaining damage resulting in death of the SOD-deficient mice within a few days to weeks of birth [107,125] could conceivably be a consequence of these basic toxic processes.

It should be noted that the activity of succinate dehydrogenase, which is also an FeS cluster-containing enzyme, is also reduced in the MnSOD-deficient mice [107]. However, the FeS clusters of aconitase and succinate dehydrogenase have different functions. The FeS cluster of succinate dehydrogenase is involved in the electron transport and normally redox cycles, whereas the activity of aconitase is lost upon oxidation of its FeS cluster. Therefore, the reported inactivation of succinate dehydrogenase is probably indirectly due to processes like lipid peroxidation, rather than to direct attack by O_2^-.

3.7.2. Aging

Evidence exists that the mitochondrial production of H_2O_2 and O_2^- increases during aging [126,128]. Thus, while remarkable as an observation, it is not surprising that the mitochondrial aconitase activity in houseflies decreases during aging [128]. Enhancement of the carbonyl content of mitochondrial aconitase and of other proteins and other types of oxidative damage have also been observed during aging [128,129]. Arguably, the O_2^--dependent oxidation of FeS clusters and inactivation of FeS cluster-containing enzymes accompanied with increased intracellular free iron concentration and the consequent FLR play an important role in aging.

3.7.3 Friedreich's Ataxia

Friedreich's ataxia is an autosomal recessive degenerative disease, occurring at a rate of approximately 1/50,000 live births, which is associated with mutations affecting the gene encoding frataxin [86,130,131]. Although the exact function of frataxin in humans is not clear, increased iron content has been reported in the hearts of Friedreich's ataxia patients [131].

The mitochondrial iron content in yeast carrying a deleted frataxin gene counterpart *yfh1* increases dramatically, suggesting that the function of the *yfh1* gene product, and by extension of frataxin, is to control the mitochondrial iron homeostasis [86]. Notably, the $\Delta yfh1$ strain is hypersensitive to H_2O_2, indicating possible involvement of FLR in the pathology of this disease [86]. This argument is supported by the finding that vitamin E deficiency in humans leads to development of a similar ataxia [132].

Another important recent finding is that the activities of aconitase and of the mitochondrial respiratory chain complexes I, II, and III, which also contain FeS clusters, are severely decreased both in humans suffering from this disease and in *yfh1* null yeast [130]. While the decreased aconitase activity could, a priori, be due to different reasons, increase of the mitochondrial steady-state level of O_2^- is strongly indicated [54]. Our working hypothesis is that the increased iron in the mitochondria replaces manganese from the active center of the mitochondrial SOD creating inactive enzyme which results in higher O_2^- concentration. Such competition between Fe and Mn for the active center of the MnSOD has been reported in the case of *E. coli* [133]. For reasons debated in Sect. 3.7.1, it is unlikely that the decreased activity of the respiratory chain complexes is due to direct oxidation of FeS clusters but is likely to be indirectly caused by the increased FLRs following enhancement of the mitochondrial free iron content.

3.7.4. Cancer and Tumor Necrosis Factor Cytotoxicity

The O_2^- concentration in tumor cell mitochondria should be increased because the mitochondrial SOD activity is diminished as compared to the normal cell type from which the tumor arose, whereas the rate of O_2^- production appears unaltered [134]. Hence, O_2^- toxicity leading to enhancement of FLR through attack on FeS clusters might be expected. This might explain the aerobic glycolysis of tumor cells. Indeed, a strong correlation of loss of MnSOD activity and extent of mitochondrial damage is found in cancer cells [51,134]. Elevated serum ferritin levels in cancer patients are used as diagnostic test for various types of tumor development [103]. The intracellular content of ferritin in cancer cells is also increased [103], probably as an adaptation to the increase of the free iron pool.

Tumor necrosis factor (TNF) exerts much higher cytotoxicity when acting on cells having low MnSOD activity, such as cancer cells, or cells in which MnSOD cannot be induced in response to TNF, or when this induction is prevented [135–137]. Despite earlier conclusions to the contrary, it was unequivocally established that TNF does not increase the steady-state level of O_2^- in mitochondria of target cells [138]. How, then, could the protective effect of SOD be explained? The following is a brief description of a published hypothesis [137]. Undeniably, mitochondria are a strong source of O_2^- production [50,126] and sensitive FeS clusters, such as that in aconitase, should constantly suffer oxidation resulting in a certain pool of intramitochondrial free iron [31,34, 137]. This pool will be tolerable or even useful in the absence of a significant level of peroxides. TNF will act to elevate the concentration of hydroperoxides rather than to increase the rate of O_2^- production, and in such a case the existing iron pool will no longer be innocuous. Induction of SOD will decrease this pool by the way of decreasing the rate of oxidation of FeS clusters, and cells in which this adaptation does not occur or that have low SOD activity will be selectively damaged by TNF. Other adaptations to TNF toxicity will be discussed in Sect. 3.8.

3.7.5. Amyotrophic Lateral Sclerosis

Amyotrophic lateral sclerosis (ALS) is a late-onset progressive and fatal motor neuron disease which occurs in both sporadic and familial (SALS and FALS) form. Approximately 15% of the familial cases are linked to point mutations of the cytosolic Cu,ZnSOD. This literature has been reviewed [139,140] including in Chapter 5 of this book; therefore only matters concerning the ability of FALS-associated SOD variants to catalyze FLR are touched on here.

As already explained in Sect. 2.9.2, Cu,ZnSOD has a limited peroxidase activity due to generation, in the presence of H_2O_2, of enzyme-bound reactive species through reactions (14) and (16). Reaction (16) is a typical FLR and the reactive species formed (SOD-Cu(II)-OH) is capable of oxidizing a variety of molecules [64,65]. Initially, results have been presented that the peroxidase activity of the FALS-associated

Cu,ZnSOD variants is increased and this was proposed to be a significant factor in the development of the disease [29,141]. Recently, however, this conclusion has been challenged by several groups that did not observe increased peroxidase activity of the FALS-associated mutant SODs [68,142,143]. The reason for these disagreements is unknown.

One particular variant, H48Q, has been found to have a novel activity and that is an O_2^--dependent peroxidase activity [66]. In this case, the enzyme-bound Cu(II) is reduced by O_2^- through reaction (18) rather than through H_2O_2 as in reaction (14). The second step is the same as in the case of the H_2O_2-only dependent peroxidase activity and is described by reaction (16).

$$SOD(H48Q)Cu(II) + O_2^- \longrightarrow SOD(H48Q)Cu(I) + O_2 \qquad (18)$$

This activity of H48Q must be due to the rate constant for reaction (16) being much higher than it is in the case of normal Cu,ZnSOD [66].

It has been strongly argued in this chapter that O_2^- could not exert effects intracellularly by acting as a reductant because it will not outcompete other reductants; however, any rule can have an exception. O_2^- is the preferred specific reductant for SODs and if the peroxidase activity of normal or mutated SODs is to play any role in vivo, this should likely be O_2^--dependent peroxidase activity [66,143].

3.7.6 Others

The involvement of free radicals and FLRs in particular is demonstrated or strongly suggested in the development of various diseases and other pathological conditions such as reperfusion injury, adult respiratory distress syndrome, etc. [144–146]. If any conclusion should be drawn in those cases in addition to that already reported in the literature, it is that more attention should be paid to the fate of the FeS cluster-containing enzymes and the consequences of the destruction of the clusters.

3.8. Adaptations to Oxidative Stress and Iron Toxicity

Cells and organisms possess elaborate and essential defenses against oxidative stress. This requires the synchronized action of several antioxidant enzymes and nonenzymic antioxidants, and further adjustment is achieved by mechanisms through which the levels or the activity of the coregulated antioxidant proteins change in proportion to the extent of oxidative stress. Therefore, studies of such adaptations provide further clues about the nature of oxidative stress. In this regard the adaptation to oxidative stress in microorganisms and specifically in $E.\ coli$ will be in the focus here because at present it is sufficiently well understood to allow important conclusions to be drawn.

(1) $soxRS$. The activation of SoxR, the sensor protein of the $soxRS$ regulon of $E.\ coli$, induces the transcription of $soxS$. The product of this gene in turn transcriptionally activates the target genes resulting in increased biosynthesis of many enzymes and proteins [124,147–149]. Many of them are identified and include MnSOD, glucose 6-phosphate dehydrogenase, fumarase C, endonuclease IV, etc. [34,147–152]. Clearly, the $soxRS$ system ensures strategic adaptation against oxidative stress. Initially, the biological function of $soxRS$ was thought to be to specifically sense and provide protection from O_2^-, but lately this view has been significantly modified [34,147,150]. The author's view [34] is that the nature of oxidative stress that $soxRS$ senses and combats is the process of oxidative inactivation of FeS cluster-containing enzymes and the resulting increase in the free iron pool and rate of FLR. This is the same toxic mechanism whose existence has been proven and deduced by other means as described in this chapter. Arguments in support of this suggestion have been presented earlier [34] and will be only briefly discussed, although some new developments which seem to provide further insight to the problem are analyzed below.

If O_2^- was the only species that is sensed or combatted by $soxRS$, induction only of SOD would have been sufficient and specific activation of the regulon by O_2^- would have been required. In fact, not only SOD but a number of other enzymes and proteins are induced as well and the analysis of their function provides the main basis of the author's view [34]. The activation of $soxRS$ is certainly not triggered specifically by O_2^- [34,147,150]. In fact, only the oxidized form of SoxR, which is a [2Fe-2S] cluster-containing protein [147,150,151], is transcriptionally active and an unidentified system exists that reduces the oxidized SoxR [34,150,151]. The process of redox regulation of SoxR strongly resembles the process of oxidation-reactivation of [4Fe-4S] clusters and it is very likely that the same or a similar system is involved in both cases [34,153]. Specifically, it has been proposed that the reduction of SoxR is enzyme-dependent, modulated by the NADPH/ NADP ratio, and probably mediated by "doxins" such as flavodoxin or ferredoxin [34,153]. Excitingly, exactly this seems to happen in *Azotobacter vinelandii*, where disruption of the gene for ferredoxin I upregulates its redox partner, the NADPH:ferredoxin reductase [154], which in *E. coli* is controlled by *soxRS* [153]. Moreover, the gene for the NADPH: ferredoxin reductase of *A. vinlandii* has a binding site for SoxS that is approximately 50% identical to the binding site of the corresponding *E. coli* gene [154]. These and other observations of Yannone and Burgess [154] clearly point to the existence of a *soxRS*-like system in *A. vinelandii* that is regulated exactly in the predicted way. In *E. coli*, though, while the prediction about the existence of NADPH-dependent enzymic system for reduction of SoxR still holds, the participation of ferredoxin reductase, flavodoxin, and ferredoxin was not confirmed [149,150]. Of course, it is still likely that another oxidoreductase and/or other doxins are involved in the SoxR reduction.

Until recently, a major puzzle remained unsolved. Nitric oxide is capable of inducing *soxRS* even anaerobically [147], yet two laboratories found independently that it cannot inactivate FeS clustercontaining enzymes [60,61]. What about the theory that *soxRS* senses agents and conditions causing oxidative inactivation of FeS clustercontaining enzymes? This problem was resolved by the finding of two other groups that NO is in fact capable of inactivating FeS clustercontaining enzymes in some situations [62,63]. It appears [63] that those experiments performed at a pH value closer to the physiological pH established inactivation as opposed to those done at higher pH.

NADPH:ferredoxin reductase is among the enzymes of *E. coli* proposed to play a role in repairing the consequences of the inactivation of the FeS clusters [34,153]. Recently, in support of this view, it was reported that *fpr* (the gene of ferredoxin reductase) mutants are hypersensitive to H_2O_2, whereas iron chelators and $HO^{\bullet}$ scavengers significantly diminished this sensitivity [155]. While in most cases oxidative situations that cause inactivation of FeS cluster-containing enzymes should be expected to cause adequate activation of SoxR, the existence of agents more effective in causing destruction of FeS clusters than in activating SoxR, or the other way around, cannot be excluded. The versatility and effectiveness even of such an ingenious adaptive system as *soxRS* must have limits. Even a single agent can trigger adaptations through several regulons. For example, the *soxRS* regulon is only part of the total response to paraquat [156]. In real situations oxidative stress could be caused by the concerted actions of a number of agents that might trigger several adaptive systems. It should be expected that such systems are correlated and interact at different levels in order to provide for maximum adaptation. An example is the fact that the genes coding for many enzymes and proteins playing roles in oxidative stress are often controlled by more than one regulon [147,156–158]. For example MnSOD in *E. coli* is regulated by at least six regulatory systems including *fur*, which regulates iron uptake and homeostasis [10,157]. Aconitases A and B, both [4Fe-4S] cluster-containing enzymes whose activity is inhibited under conditions of oxidative stress, are also under the control of several regulons [158] including *soxRS* in the case of aconitase A [158].

(2)   *oxyR*. This regulon of *E. coli* and of *Salmonella typhimurium* controls the biosynthesis of several enzymes and proteins including catalase, alkyl hydroperoxide reductase, glutathione reductase and glutaredoxin [156,159–161]. Initially it was thought that *oxyR* specifically senses and provides protection from H_2O_2 [156,159,160]. In fact, *oxyR* also senses and protects the cells from other agents such as nitrosothiols [162]. The sensor protein OxyR contains a thiol group whose oxidation activates the regulon [156,159]. These facts suggest that a major role of *oxyR* is actually to protect target thiols containing proteins and enzymes, H_2O_2 being only one of the agents capable of causing oxidations of thiols. Thus, rather than being systems for the protection of different targets from specific oxidants (O_2^- and H_2O_2), *soxRS* and *oxyR* sense and protect specific targets from various oxidants. Clearly, thiols and FeS clusters are among the most sensitive and critical targets and their oxidation is what in fact constitutes, to a great extent, oxidative stress. It should be expected that some cooperation exists between *oxyR* and *soxRS*. For example, H_2O_2 might play a role in the PQ-induced decrease of the activity of the [4Fe-4S] cluster-containing unstable fumarases of *E. coli*, probably inhibiting the reactivation [34,121]. Clearly, activation of the *oxyR* by H_2O_2 should be expected to be important in such a case.

(3)   *Adaptations to oxidative stress in higher organisms*. Our understanding of the mechanisms of adaptation towards oxidative stress in eukaryotes, and especially in mammals, is still limited. Here an

attempt will be made to discuss the possibility that even in the case of higher organisms a significant part of the adaptation to oxidative stress might be due to mechanisms capable of decreasing the intracellular free iron pool and protecting FeS cluster-containing enzymes from inactivation.

As already mentioned, the activation of IRP1 leads to increased cellular uptake of iron and decreased sequestration of iron in ferritin. The biological significance of this adaptation is clear when this activation is triggered by iron deficiency [53,55,56,58]. However, H_2O_2 also causes activation of IRP1 [163,164]. It is desirable in such situations that the level of the intracellular iron pool be decreased. The induction of MnSOD in mammalian cells by H_2O_2 [165] might be one of the mechanisms through which diminution of intracellular iron is achieved. The induction of the mitochondrial MnSOD by TNF also represents an attempt to decrease the intracellular free iron concentration as discussed in Sect. 3.7. In fact, the ferritin heavy chain [166] and ceruloplasmin [167] are also induced by TNF.

4. CONCLUSIONS

The involvement of metal ions and complexes in free radical reactions, and especially the ability of these ions and complexes to catalyze FLRs, has always been a central concern in free radical biology and chemistry. Knowledge in this field has reached a new stage characterized by a deeper understanding of free radical processes taking place in vivo and an ability to explain previously puzzling problems and to predict new phenomena. We can also settle longstanding disputes. Thus, for approximately two decades there was competition between two apparently contradictory views attempting to explain the biological role of SODs.

The majority of the scientists held the view, most actively defended by I. Fridovich [28,31,33,39,93], that the role of SODs is to protect biological targets from the toxicity of O_2^-, due to the ability of these enzymes to rapidly dismute this radical.

The view of the minority, usually represented by J. Fee, was that the ability of SODs to dismute O_2^- is coincidental, biologically irrelevant, and that these enzymes are metal storage proteins or are otherwise involved in metal homeostasis [37,168,169]. The present stage of knowledge allows the unequivocal conclusion that SODs play an important role in iron and probably other metal homeostasis. But this is mostly or even entirely due to the ability of SODs to catalyze dismutation of O_2^-.

Yet the most interesting challenges are still ahead. This should be particularly true with regard to our knowledge about the free radical processes taking place in higher organisms and in pathological conditions. Successful therapeutic and preventive strategies based on O_2^- and other radicals, scavengers, and antioxidants remain to be devised.

ACKNOWLEDGEMENTS

I am thankful to Dr. I. Fridovich for helpful discussions and for critical reading of the manuscript. This work was supported by grants from the Council for Tobacco Research, USA, Inc. (2871AR2), the National Institutes of Health (HL56025-03), and the Amyotrophic Lateral Sclerosis Association.

ABBREVIATIONS

ALS	amyotrophic lateral sclerosis
ATP	adenosine 5'-triphosphate
DMPO	dimethylpyrroline N-oxide
EDTA	ethylenediamine-N,N,N',N'-tetraacetate
EPR	electron paramagnetic resonance
FALS	familial amyotrophic lateral sclerosis
FLR	Fenton-like reaction
GSH	glutathione
HO$^\bullet$	hydroxyl radical
IRP	iron regulatory protein–1
NADH	nicotinamide adenine dinucleotide, reduced form
NADPH	nicotinamide adenine dinucleotide phosphate, reduced form
O_2^-	superoxide radical

PQ	paraquat
SALS	sporadic amyotrophic lateral sclerosis
SOD	superoxide dismutase
sodA and *sodB*	MnSOD and FeSOD genes
soxRS, oxyR	*E. coli* regulons
TNF	tumor necrosis factor
V(IV)	vanadyl
V(V)	vanadate

REFERENCES

1. H. J. H. Fenton, *J. Chem. Soc.*, 65, 899–910 (1894).

2. F. Haber and R. Willstätter, *Ber. Deutsch. Chem. Ges.*, 64, 2844–2856 (1931).

3. F. Haber and J. Weiss, *Proc. Roy. Soc. London*, A147, 332–351 (1934).

4. W. G. Barb, J. H. Baxendale, P. George, and K. R. Hargrave, *Trans. Faraday Soc.*, 47, 462–500 (1951).

5. J. D. Rush and H. J. Bielski, *J. Phys. Chem.*, 89, 5062–5066 (1985).

6. P. Wardman and L. P. Candeias, *Radiat. Res.*, 145, 523–531 (1996).

7. W. H. Koppenol, *Free Rad. Biol. Med.*, 15, 645–651 (1993).

8. S. Goldstein, D. Meyerstein, and G. Czapski, *Free Rad. Biol. Med.*, 15, 435–445 (1993).

9. E. S. Henle and S. Linn, *J. Biol. Chem.*, 272, 19095–19098 (1997).

10. H. M. Hassan, in *Lung Biology in Health and Disease*, Vol. 105, *Oxygen Gene Expression and Cellular Function* (L. B. Clerch and D. J. Massaro, eds.), Marcel Dekker, New York, 1997, pp. 27–47.

11. B. Halliwell and J. M. C. Gutteridge, *Meth. Enzymol.*, 86, 1–85 (1990).

12. H. B. Brooks and F. Sicilio, *Inorg. Chem.*, 10, 2530–2534 (1971).

13. S. I. Liochev and I. Fridovich, *Arch. Biochem. Biophys.*, 291, 379–382 (1991).

14. S. Liochev and E. Ivancheva, *Free Rad. Res. Commun.*, 14, 335–342 (1991).

15. T. Ozawa and A. Hanaki, *Chem. Pharm. Bull.*, *37*, 1407–1409 (1989).

16. G. J. Quinlan, C. Coudray, A. Hubbard, and J. M. C. Gutteridge, *J. Pharm. Sci.*, *81*, 611–614 (1992).

17. O. I. Aruoma, B. Halliwell, E. Gajewski, and M. Dizdaroglu, *Biochem. J.*, *273*, 601–604 (1991).

18. S. I. Liochev and I. Fridovich, *Arch. Biochem. Biophys.*, *279*, 1–7 (1990).

19. J. Li, G. Elberg, D. Gefel, and Y. Shechter, *Biochemistry*, *34*, 6218–6225 (1995).

20. G. Fantus, S. Kadota, G. Deragon, B. Foster, and B. I. Posner, *Biochemistry*, *28*, 8864–8871 (1989).

21. A. Shisheva and Y. Shechter, *Endocrinology*, *133*, 1562–1568 (1993).

22. M. A. Serra, A. Pintar, L. Casella, and E. Sabbioni, *J. Inorg. Biochem.*, *46*, 161–174 (1992).

23. H. N. Ravishankar, A. V. Rao, and T. Ramasarma, *Arch. Biochem. Biophys.*, *321*, 477–484 (1995).

24. H. N. Ravishankar, M. K. Chaudhuri, and T. Ramasarma, *Inorg. Chem.*, *33*, 3788–3793 (1994).

25. E. Luzzato, H. Cohen, C. Stockheim, K. Wieghardt, and D. Meyerstein, *Free Rad. Res.*, *23*, 453–463 (1995).

26. W. K. Pogozelski, T. J. McNeese, and T. D. Tullius, *J. Am. Chem. Soc.*, *117*, 6428–6433 (1995).

27. B. Halliwell and M. C. Gutteridge, *FEBS Lett.*, *307*, 108–112 (1992).

28. I. Fridovich, *Annu. Rev. Biochem.*, *64*, 97–112 (1995).

29. M. B. Yim, J.-H. Kang, H.-S. Yim, H.-S. Kwak, P. B. Chock, and E. R. Stadtman, *Proc. Natl. Acad. Sci. USA*, *93*, 5709–5714 (1996).

30. J. DiGuiseppi and I. Fridovich, *Arch. Biochem. Biophys.*, *205*, 323–329 (1980).

31. S. I. Liochev and I. Fridovich, *Free Rad. Biol. Med.*, *16*, 29–33 (1994).

32. J. M. McCord and I. Fridovich, *J. Biol. Chem.*, *244*, 6049–6755 (1969).

33. I. Fridovich, *J. Biol. Chem.*, *272*, 18515–18517 (1997).

34. S. I. Liochev, *Free Rad. Res.*, 25, 369–384 (1996).

35. I. Fridovich, *J. Biol. Chem.*, 264, 7761–7764 (1989).

36. D. T. Sawyer and J. S. Valentine, *Acc. Chem. Res.*, 14, 393–400 (1981).

37. J. A. Fee, in *Developments in Biochemistry*, Vol. 11B (W. H. Bannister and J. V. Bannister, eds.), Elsevier/North Holland, New York, 1980, pp. 41–48.

38. C. O. Beauchamp and I. Fridovich, *J. Biol. Chem.*, 245, 4641–4646 (1970).

39. J. M. McCord, B. B. Keele, Jr., and I. Fridovich, *Proc. Natl. Acad. Sci. USA*, 68, 1024–1027 (1971).

40. J. M. McCord and E. D. Day, Jr., *FEBS Lett.*, 86, 139–142 (1978).

41. B. Halliwell, *FEBS Lett.*, 92, 321–326 (1978).

42. C. E. Thomas, L. E. Morehouse, and S. D. Aust, *J. Biol. Chem.*, 260, 3275–3280 (1985).

43. D. W. Reif, *Free Rad. Biol. Med.*, 12, 417–427 (1992).

44. G. R. Buettner, T. P. Doherty, and L. K. Pattersen, *FEBS Lett.*, 158, 143–145 (1983).

45. J. M. C. Gutteridge, *J. Trace Elem. Electrolytes Health Dis.*, 5, 271–272 (1991).

46. J. A. Imlay and S. Linn, *Science*, 240, 1302–1309 (1988).

47. C. C. Winterbourn, *Biochem. J.*, 182, 625–628 (1979).

48. G. R. Buettner and B. A. Jurkiewicz, *Radiat. Res.*, 145, 532–541 (1996).

49. H. P. Monteiro, G. F. Vile, and C. C. Winterbourn, *Free Rad. Biol. Med.*, 6, 587–591 (1989).

50. A. Boveris and E. Cadenas, in *Lung Biology in Health and Disease*, Vol. 105, *Oxygen Gene Expression and Cellular Function* (L. B. Clerch and D. J. Massaro, eds.), Marcel Dekker, New York, 1997, pp. 1–25.

51. H. Wiseman and B. Halliwell, *Biochem. J.*, 313, 17–29 (1996).

52. Y. J. Susuki and G. D. Ford, *Free Rad. Biol. Med.*, 16, 63–72 (1994).

53. M. C. Kennedy, in *Metal Ions in Biological Systems*, Vol. 32, *Interactions of Metal Ions with Nucleotides, Nucleic Acids and Their Constituents* (A. Sigel and H. Sigel, eds.), Marcel Dekker, New York, 1996, pp. 579–602.

54. P. R. Gardner, *Biosci. Rep.*, *17*, 33–42 (1997).

55. H. Beinert, M. C. Kennedy, and C. D. Stout, *Chem. Rev.*, *96*, 2335–2373 (1996).

56. T. A. Roualt and R. D. Klausner, *Trends Biochem. Sci.*, *21*, 174–177 (1996).

57. D. H. Flint and R. M. Allen, *Chem. Rev.*, *96*, 2315–2334 (1996).

58. E. Paraskeva and M. W. Hentze, *FEBS Lett.*, *389*, 40–43 (1996).

59. D. H. Flint, J. F. Tuminello, and M. H. Emptage, *J. Biol. Chem.*, *268*, 22369–22376 (1993).

60. A. Hausladen and I. Fridovich, *J. Biol. Chem.*, *269*, 29405–29408 (1994).

61. L. Castro, M. Rodriguez, and R. Radi, *J. Biol. Chem.*, *269*, 29409–29415 (1994).

62. M. C. Kennedy, W. E. Antholine, and H. Beinert, *J. Biol. Chem.*, *272*, 20340–20347 (1997).

63. P. R. Gardner, G. Constantino, C. Szabó, and A. Salzman, *J. Biol. Chem.*, *272*, 25071–25076 (1997)

64. E. K. Hodgson and I. Fridovich, *Biochemistry*, *14*, 5299–5303 (1975).

65. E. K. Hodgson and I. Fridovich, *Biochemistry*, *14*, 5294–5299 (1975).

66. S. I. Liochev, L. L. Chen, R. A. Hallewell, and I. Fridovich, *Arch. Biochem. Biophys.*, *346*, 263–268 (1997).

67. R. V. Lloyd, P. M. Hanna, and R. P. Mason, *Free Rad. Biol. Med.*, *22*, 885–888 (1997).

68. R. J. Singh, H. Karoui, M. Gunther, J. S. Beckman, R. Mason, and B. Kalyanaraman, *Program and Abstracts of the 4th Annual Meeting of the Oxygen Society*, p. 31 (1997).

69. K. Felix, E. Lengfelder, H. J. Hartman, and U. Weser, *Biochim. Biophys. Acta*, *1203*, 104–108 (1993).

70. V. C. Culotta, H. D. Joh, S. J. Lin, K. H. Slekar, and J. Strain, *J. Biol. Chem.*, *270*, 2991–2997 (1995).

71. R. M. Kaschnitz and Y. Hatefi, *Arch. Biochem. Biophys.*, *171*, 292–304 (1975).

72. G. Balla, G. M. Vercellotti, U. Muller-Eberhard, J. Eaton, and H. S. Jacobs, *Lab. Invest.*, *64*, 648–655 (1991).

73. C. Giulivi and E. Cadenas, *Free Rad. Biol. Med.*, *24*, 269–279 (1998).

74. R. E. Huie and S. Padmaja, *Free Rad. Res. Commun.*, *18*, 195–199 (1993).

75. R. Radi, J. S. Beckman, K. M. Bush, and B. A. Freeman, *J. Biol. Chem.*, *266*, 4244–4250 (1991).

76. R. M. Gatti, B. Alvarez, J. Vasquez-Vivar, R. Radi, and O. Augusto, *Arch. Biochem. Biophys.*, *349*, 36–46 (1998).

77. K. Keyer and J. A. Imlay, *J. Biol. Chem.*, *272*, 27652–27659 (1997).

78. H. Hirashi, A. Terano, M. Razandi, T. Sugimoto, T. Hasada, and K. J. Ivey, *J. Biol. Chem.*, *267*, 14812–14817 (1992).

79. K. Keyer, A. Strohmeier-Gort, and J. A. Imlay, *J. Bacteriol.*, *177*, 6782–6790 (1995).

80. K. Keyer and I. Imlay, *Proc. Natl. Acad. Sci. USA*, *93*, 13635–13640 (1996).

81. C. E. Cooper and J. B. Porter, *Biochem. Soc. Trans.*, *25*, 75–80 (1996).

82. J. M. C. Gutteridge, S. Mumby, M. Koizumi, and N. Taniguchi, *Biochem. Biophys. Res. Commun.*, *229*, 806–809 (1996).

83. A. Tangeras, *Biochim. Biophys. Acta*, *843*, 119–207 (1985).

84. A. Tangeras, *Biochim. Biophys. Acta*, *757*, 59–68 (1983).

85. D. Touati, M. Jacques, B. Tardat, L. Bouchard, and S. Despied, *J. Bacteriol.*, *177*, 2305–2314 (1995).

86. M. Babcock, D. deSilva, R. Oaks, S. Davis-Kaplan, S. Jiralerspong, L. Montermini, M. Pandolfo and J. Kaplan, *Science*, *276*, 1709–1712 (1997).

87. T. C. Carmine, P. Evans, G. Bruchelt, R. Evans, R. Handgretinger, D. Niethammer, and B. Halliwell, *Cancer Lett.*, *138*, 33–36 (1995).

88. R. Böhnke and B. F. Matzanke, *Biometals*, *8*, 223–230 (1995).

89. M. W. Hentze and L. C. Kühn, *Proc. Natl. Acad. Sci. USA*, *93*, 8175–8182 (1996).

90. R. Meneghini, *Free Rad. Biol. Med.*, *23*, 783–792 (1997).

91. D. R. Richardson and P. Ponka, *Biochim. Biophys. Acta*, *1331*, 1–40 (1997).

92. J. F. Briat, *J. Gen. Microbiol.*, *138*, 2475–2483 (1992).

93. I. Fridovich, *Arch. Biochem. Biophys.*, *247*, 1–11 (1986).

94. V. Massey, S. Strickland, S. G. Mayhew, L. G. Howell, P. C. Engel, R. G. Mathews, M. Schuman, and P. A. Sullivan, *Biochem. Biophys. Res. Commun.*, *36*, 891–897 (1969)

95. I. Fridovich, *J. Biol. Chem.*, *245*, 4053–4057 (1970).

96. S. I. Liochev and I. Fridovich, *Proc. Natl. Acad. Sci. USA*, *94*, 2891–2896 (1997).

97. J. A. Imlay and I. Fridovich, *J. Biol. Chem.*, *266*, 6957–6965 (1991).

98. P. R. Gardner and I. Fridovich, *J. Biol. Chem.*, *267*, 8757–8763 (1992).

99. B. González-Flecha and B. Demple, *J. Biol. Chem.*, *270*, 13681–13687 (1995).

100. B. M. Babior, *N. Engl. J. Med.*, *298*, 659–668 (1978).

101. C. F. Nathan, *J. Clin. Invest.*, *79*, 319–326 (1987).

102. S. N. Wai, K. Nakayama, K. Umene, T. Moriya, and K. Amako, *Mol. Microbiol.*, *20*, 1127–1134 (1996).

103. S. L. Baader, E. Bill, A. X. Trautwein, G. Bruchelt, and B. F. Matzanke, *FEBS Lett.*, *381*, 131–134 (1996).

104. L. Benov and I. Fridovich, *J. Biol. Chem.*, *273*, 10313–10316 (1998).

105. B. Halliwell and O. I. Aruoma, *FEBS Lett.*, *281*, 9–19 (1991).

106. S. B. Farr, R. D'Ari, and D. Touati, *Proc. Natl. Acad. Sci. USA*, *83*, 8268–8272 (1986).

107. Y. Li, T.-T. Huang, E. J. Carlson, S. Melov, P. C. Ursell, J. L. Oson, L. J. Noble, M. P. Yoshimura, C. Berger, P. H. Chan, D. C. Wallace, and C. Epstein, *Nature Genetics*, *11*, 376–381 (1995).

108. L. Benov, N. M. Kredich, and I. Fridovich, *J. Biol. Chem.*, *35*, 21037–21040 (1996).

109. A. Carlioz and D. Touati, *EMBO J.*, *5*, 623–630 (1986).

110. C. C. Winterbourn, *Toxicol. Lett.*, *82–83*, 969–974 (1995).

111. C. C. Winterbourn, *Free Rad. Biol. Med.*, *14*, 85–90 (1993).

112. C. C. Winterbourn, *Biochem. J.*, *205*, 463 (1982).

113. G. Czapski, S. Goldstein, and D. Meyerstein, *Free Rad. Res. Commun.*, *4*, 231–236 (1988).

114. P. Beimond, A. J. G. Swaak, M. Beindorff, and J. F. Coster, *Biochem. J.*, *239*, 169–173 (1986).

115. D. A. Rowley and B. Halliwell, *FEBS Lett.*, *138*, 33–36 (1982).

116. D. A. Rowley and B. Halliwell, *FEBS Lett.*, *142*, 39–41 (1982).

117. M. J. Pianzzola, U. Soubes, and D. Touati, *J. Bacteriol.*, *178*, 6736–6742 (1996).

118. O. R. Brown and F. Yein, *Biochem. Biophys. Res. Commun.*, *85*, 1219–1224 (1978).

119. O. R. Brown, E. Smyk-Randall, B. Draczynska-Lusiak, and J. A. Fee, *Arch. Biochem. Biophys.*, *319*, 10–22 (1995).

120. M. C. Kennedy, M. H. Emptage, J. L. Dreyer, and H. Beinert, *J. Biol. Chem.*, *258*, 11098–11105 (1983).

121. S. I. Liochev and I. Fridovich, *Arch. Biochem. Biophys.*, *301*, 379–384 (1993).

122. H. D. Teixeira and R. Meneghini, *Program and Abstracts of the 4th Annual Meeting of the Oxygen Society*, p. 62 (1997).

123. M. L. McCormick, G. R. Buettner, and B. E. Britigan, *J. Bacteriol.*, *180*, 622–625 (1998).

124. J. Wu and B. Weiss, *J. Bacteriol.*, *174*, 3915–3920 (1992).

125. R. M. Lebovitz, H. Zhang, H. Vogel, J. Cartwright, Jr., L. Dionne, N. Lu, S. Huang, and M. M. Matzuk, *Proc. Natl. Acad. Sci. USA*, *93*, 9782–9787 (1996).

126. R. S. Sohal, *FASEB. J.*, *11*, 1269–1270 (1997).

127. P. R. Gardner, I. Raineri, L. B. Epstein, and C. W. White, *J. Biol. Chem.*, *270*, 13399–13405 (1995).

128. L.-J. Yan, R. L. Levine, and R. S. Sohal, *Proc. Natl. Acad. Sci. USA*, *94*, 11168–11172 (1997).

129. Y. H. Wei, *Proc. Soc. Exp. Biol. Med.*, *217*, 53–63 (1998).

130. A. Rötig, P. deLonlay, D. Cretien, F. Fouri, M. Koenig, D. Sidi, A. Munnich, and P. Rustin, *Nature Genetics*, *17*, 215–217 (1997).

131. G. Sanchez-Casis, M. Cote, and A. Barbeau, *Can. J. Neurol. Sci.*, *3*, 349–354 (1977).

132. M. Ben Hamida, S. Belal, G. Sirugo, C. Ben Hamida, K. Panayides, P. Ionannou, J. Beckman, T. L. Mandel, F. Hentati, M. Koenig, and L. Middleton, *Neurology*, *43*, 2179–2183 (1993).

133. C. T. Privalle and I. Fridovich, *J. Biol. Chem.*, *267*, 9140–9145 (1992).

134. L. W. Oberley and T. D. Oberley, in *Lung Biology in Health and Disease*, Vol. 105, *Oxygen Gene Expression and Cellular Function* (L. B. Clerch and D. J. Massaro, eds.), Marcel Dekker, New York, 1997, pp. 279–307.

135. G. H. W. Wong, J. H. Elwell, L. W. Oberley, and D. V. Goeddel, *Cell*, *58*, 923–931 (1989).

136. G. H. W. Wong and D. V. Goeddel, *Science*, *24*, 941–944 (1988).

137. S. I. Liochev and I. Fridovich, *Free Rad. Biol. Med.*, *23*, 668–671 (1997).

138. P. R Gardner and C. W. White, *Arch. Biochem. Biophys.*, *334*, 158–162 (1996).

139. E. R. B. McCabe, *Proc. Natl. Acad. Sci. USA*, *92*, 8533–8534 (1995).

140. T. Siddique and H.-X. Denq, *Hum. Mol. Gen.*, *5* (special issue), 1465–1470 (1996).

141. M. Wiedau-Pazos, J. J. Goto, S. Rabizadeh, E. B. Gralla, J. A. Roe, M. K. Lee, J. S. Valentine, and D. E. Bredesen, *Science*, *271*, 515–518 (1996).

142. S. L. Marklund, P. M. Anderson, L. Forsgren, P. Nilsson, P. I. Ohloson, G. Wikander, and A. Oberg, *J. Neurochem.*, *69*, 675–681 (1996).

143. S. I. Liochev, L. L. Chen, R. A. Hallewell, and I. Fridovich, *Arch. Biochem. Biophys.*, *352*, 237–239 (1998).

144. J. J. M. Marx and B. S. van Asbeck, *Acta Haematol.*, *95*, 49–62 (1996).

145. J. M. Gutteridge, *Free Rad. Res. Commun.*, *19*, 141–158 (1993).

146. J. M. McCord, *Clin. Biochem.*, *26*, 351–357 (1993).

147. B. Demple, *Gene*, *179*, 53–57 (1996).

148. I. R. Tsaneva and B. Weiss, *J. Bacteriol.*, *172*, 4197–4205 (1990).

149. P. Gaudu and B. Weiss, *Proc. Natl. Acad. Sci. USA*, *93*, 10094–10098 (1996).

150. P. Gaudu, N. Moon, and B. Weiss, *J. Biol. Chem.*, *272*, 5082–5086 (1997).

151. H. Ding and B. Demple, *Proc. Natl. Acad. Sci. USA*, *94*, 8445–8449 (1997).

152. E. Hidalgo, H. Ding, and B. Demple, *Cell*, *88*, 121–129 (1997).

153. S. I. Liochev, A. Hausladen, W. F. Beyer, Jr., and I. Fridovich, *Proc. Natl. Acad. Sci. USA*, *90*, 993–997 (1993).

154. S. M. Yannone and B. K. Burgess, *J. Biol. Chem.*, *272*, 14454–14458 (1997).

155. A. R. Krapp, V. B. Tognetti, N. Carrillo, and A. Acevedo, *Eur. J. Biochem.*, *249*, 556–563 (1997).

156. S. B. Farr and T. Kogoma, *Microbiol. Rev.*, *55*, 561–585 (1991).

157. I. Compan and D. Touati, *J. Bacteriol.*, *175*, 1687–1696 (1993).

158. L. Cunningham, M. J. Gruer, and J. R. Guest, *Microbiology*, *143*, 3795–3805 (1997).

159. G. Storz, L. A. Tartaglia, and B. N. Ames, *Science*, *248*, 189–194 (1990).

160. B. Demple, *Annu. Rev. Genet.*, *25*, 315–337 (1991).

161. K. Tao, *J. Bacteriol.*, *179*, 5967–5970 (1997).

162. A. Hausladen, C. T. Privalle, T. Keng, J. DeAngelo, and J. Stamler, *Cell*, *86*, 719–729 (1996).

163. E. A. L. Martins, B. L. Robalinho, and R. Meneghini, *Arch. Biochem. Biophys.*, *316*, 128–134 (1995).

164. K. Pantopoulos and M. W. Hentze, *EMBO J.*, *14*, 2917–2924 (1995).

165. S. Borrello, and B. Demple, *Arch. Biochem. Biophys.*, *348*, 289–294 (1997).

166. S. V. Torti, E. L. Kwak, S. C. Miller, L. L. Miller, G. M. Ringold, K. W. Myambo, A. P. Young, and F. Torti, *J. Biol. Chem.*, *263*, 12638–12644 (1988).

167. F. S. Schütze, P. Scheurich, C. Schlüter, U. Ucer, K. Pfizenmaier, and M. Krönke, *J. Immunol.*, *140*, 3000–3005 (1988).

168. J. S. Valentine and M. W. Pantoliano, in *Metal Ions in Biology*, Vol. 3. (T. G. Spiro, ed.), John Wiley and Sons, New York, 1981, pp. 291–358.

169. J. A. Fee, *Mol. Microbiol.*, *5*, 2599–2610 (1991).

2

Reactions of Aliphatic Carbon-Centered and Aliphatic-Peroxyl Radicals with Transition Metal Complexes as a Plausible Source for Biological Damage Induced by Radical Processes

Dan Meyerstein

Chemistry Department, Ben-Gurion University
of the Negev, Beer-Sheva and College of Judea
and Samaria, Ariel, Israel

1. INTRODUCTION

It is commonly accepted that radical reactions play a major role in aging and that such reactions cause a large variety of diseases, e.g., cancer, atherosclerosis, essential hypertension, Alzheimer's disease, amyloidosis, osteoarthritis, etc. [1]. The initiators of the radical-induced deleterious processes are mainly the $^\bullet$OH and $HO_2^\bullet$ radicals. However, very little is known about the detailed mechanisms of reaction of radicals in biological systems. The aim of this chapter is to discuss the plausible contribution of the reactions of aliphatic carbon-centered and aliphatic-peroxyl radicals with transition metal complexes to the biological deleterious processes.

2. SOURCES OF RADICALS IN BIOLOGICAL SYSTEMS

2.1. Hydroxyl Radicals

Hydroxyl radicals ($^\bullet$OH) are formed in biological systems mainly via "Fenton-like" reactions [2,3], via the decomposition of peroxonitrite [4], by the absorption of ionizing radiation [5,6] or of ultraviolet (UV) light, and probably when ozone is introduced to biological systems [7].

The "Fenton reagent" refers to a mixture of hydrogen peroxide and ferrous salts that was shown by Fenton to be an efficient oxidant of many organic compounds [8]. In 1934, Haber and Weiss [9] proposed that hydroxyl radicals formed in reaction (1):

$$Fe(H_2O)_6^{2+} + H_2O_2 + H_3O^+ \longrightarrow Fe(H_2O)_6^{3+} + 2H_2O + {}^\bullet OH \quad (1)$$

are the active intermediates in these oxidations. Since then it is commonly accepted that free hydroxyl radicals are formed in this reaction and in the analogous Fenton-like reactions:

$$M^nL_m + H_2O_2 + H_3O^+ \longrightarrow M^{n+1}L_m + 2H_2O + {}^\bullet OH \quad (1')$$

When H_2O_2 is replaced by ROOH, where R is an alkyl residue, alkoxyl radicals are formed via the analogous reaction (2) [10]:

$$M^nL_m + ROOH + H_3O^+ \longrightarrow M^{n+1}L_m + 2H_2O + RO^\bullet \quad (2)$$

where M^nL_m is a "low-valent" transition metal complex. Peracids react via reaction (3) [10]:

$$M^nL_m + RCO_2OH \longrightarrow M^{n+1}L_m + RCO_2^- + {}^\bullet OH \quad (3a)$$

$$M^nL_m + RCO_2OH \longrightarrow M^{n+1}L_m + OH^- + RCO_2^\bullet \quad (3b)$$

However, the suggestion that ${}^\bullet OH$ radicals are formed in several of these systems was questioned, as comparison of the properties of the intermediates formed in them with those of ${}^\bullet OH$ radicals formed via ionizing radiation indicated that significant differences are often observed [11–13]. Therefore, it was proposed that transition metal complexes in higher oxidation states, e.g., Fe^{IV} and Cu^{III}, are formed in the Fenton-like reactions [11–13].

Recently [14,15] it was shown, based on the Marcus theory [16], that due to thermodynamic considerations, reaction (1') and its analogous processes cannot proceed via the outer-sphere mechanism. Therefore, the first step in Fenton-like reactions has to be one of the following [15]:

$$M^nL_m + ROOH \longrightarrow L_mM^n\text{-}(HOOR) \quad (4a)$$

$$M^nL_m + ROOH \longrightarrow L_{m-1}M^n\text{-}(HOOR) + L \quad (4b)$$

$$M^nL_m + ROOH \longrightarrow L_mM^n\text{-}({}^-OOR) + H_3O^+ \quad (4c)$$

$$M^nL_m + ROOH \longrightarrow L_{m-1}M^n\text{-}({}^-OOR) + H_3O^+ + L \quad (4d)$$

where R = alkyl or H. This step, written for L_mM^n-($^-$OOH) as an example, might be followed by one of the following processes [14], which will cause the oxidation of organic substrates present in the system. Other mechanisms of decomposition of transient complexes of the type L_mM^n-($^-$OOH) are discussed below.

(a) L_mM^n-($^-$OOH) + H_3O^+ $\longrightarrow$ $M^{n+1}L_m$ + OH^- + $^\bullet OH$ + H_2O (5)

i.e., via a process that yields the same products as the mechanism suggested by Haber and Weiss. This reaction will be followed in the presence of a saturated aliphatic compound, RH, by:

$$RH + {}^\bullet OH \longrightarrow R^\bullet + H_2O \tag{6}$$

(b) L_mM^n-($^-$OOH) + H_3O^+ $\longrightarrow$ $L_mM^{n+2}{=}O$ + $2H_2O$ (7)

i.e., via a two-electron oxidation of the central metal cation. This reaction might then be followed by:

$$L_mM^{n+2}{=}O + RH \longrightarrow M^{n+1}L_m + OH^- + R^\bullet \tag{8}$$

or by:

$$L_mM^{n+2}{=}O + RH \longrightarrow M^nL_m + ROH \tag{9}$$

(c) L_mM^n-($^-$OOH) + RH + $2H_3O^+$ $\longrightarrow$ $M^{n+1}L_m$ + $4H_2O$ + $R^\bullet$ (10)

Thus, at least formally, the same products might be formed via the three pathways. The results [14,17–19] indicate that the choice — which of these pathways occurs for a given system — depends on the nature of M^n, of RH, and of L, on the pH, and on the concentration of RH. Therefore, it seems impossible at present to determine whether in a given system in vivo $^\bullet OH$ radicals are indeed formed [15].

It should be noted that in aerated systems that contain a reducing agent, e.g., ascorbate, H_2O_2 might be formed via:

$$M^{n+1}L_m + red \longrightarrow M^nL_m + ox \tag{11}$$

$$M^nL_m + O_2 \longrightarrow M^{n+1}L_m + O_2^{\bullet -} \tag{12}$$

$$2O_2^{\bullet -} + 2H_3O^+ \longrightarrow O_2 + H_2O_2 + 2H_2O \tag{13}$$

which will be followed by the Fenton-like processes. These systems, which were first described by Udenfriend et al. [20], are often referred to as the Udenfriend reagents and serve as models for oxygenases [21].

Finally, it should be pointed out that it was proposed [22] that the reaction of Fe(II) with hypochlorous acid:

$$M^nL_m + ClOH \longrightarrow M^{n+1}L_m + Cl^- + {}^\bullet OH \tag{14}$$

also forms radical products. The results seem to suggest that $^\bullet OH$ radicals are not always formed in this process [22]. It seems probable that, as in the Fenton reaction, the detailed mechanism involves the formation of a transient complex of the type $L_m M^n \cdot ClOH$ and that its mechanism of decomposition depends on the nature of M, L, the specific substrate present and its concentration, and the pH.

Peroxonitrite is formed in biological systems via reaction (15):

$${}^\bullet NO + O_2^{\bullet -} \longrightarrow NO(OO^-) \tag{15}$$

Peroxonitrite is a powerful oxidizing agent that decomposes via reaction (16) [4]:

$$\begin{aligned} NO(OO^-) + H_3O^+ &\rightleftharpoons NO(OOH) \longrightarrow NO_3^- + H_3O^+ \\ NO(OO^-) + H_3O^+ &\rightleftharpoons NO(OOH) \longrightarrow NO_2 + {}^\bullet OH \end{aligned} \tag{16}$$

Approximately 40% of the decomposition yields $^\bullet OH$ radicals.

Ionizing radiation is absorbed homogeneously in the medium producing $^\bullet OH$ radicals as well as $H^\bullet$ atoms and e_{aq}^-. All of these radicals induce deleterious processes in biological systems. However, the $^\bullet OH$ radical seems to cause most of the damage [5,6]. The fact that transition metal complexes are radiosensitizers [23,24], though the $^\bullet OH$ radicals are formed homogeneously, points out that their role in these processes is not limited to the production of the radicals.

2.2. Superoxide Radicals

Superoxide radicals ($O_2^{\bullet -}$) are formed mainly as a side product in the electron transport chains in the mitochondria and endoplasmic reticulum [25] and by some oxidation of heme during oxygen transport [26]. In the electron transport chains dioxygen is reduced to water. However, several electrons always leak to the dioxygen in the early stages of the chain to form $O_2^{\bullet -}$ [25,27]. It has been shown that in *Escherichia coli* this pathway produces 5 μM/s of $O_2^{\bullet -}$ which forms a steady-state concentration of approx. $2 \cdot 10^{-10}$ M due to its superoxide dismutase (SOD)-

induced catalytic decomposition [27]. Hemoglobin and myoglobin carry dioxygen reversibly. However, some oxidation to methemoglobin and ferrimyoglobin forming $O_2^{\bullet -}$ occurs; thus 0.1–0.5% of the hemoglobin is oxidized every hour in the red cells in vivo [26].

Superoxide radicals have been suggested to be formed mainly via oxidation of the reduced form of ubiquinone and cytochrome b-566 [28]. Superoxide radicals are also formed in a variety of enzymatic processes, e.g., in glutathione reductase [27], xanthine oxidase, indoleamine dioxygenase, tryptophan dioxygenase, aldehyde oxidase, and galactose oxidase [29].

Large quantities of superoxide radicals as well as H_2O_2, hypohalous acids, hydroxyl radicals, and singlet oxygen are formed by stimulated neutrophils in order to destroy microorganisms and malignant or senile cells [30].

2.3. Aliphatic Carbon-Centered Radicals

In biological systems aliphatic carbon-centered radicals are formed mainly via hydrogen atom abstraction from saturated aliphatic compounds by $^{\bullet}OH$ radicals:

$$^{\bullet}OH + RH \longrightarrow {}^{\bullet}R + H_2O \tag{17}$$

These reactions are fast for α hydrogens to -OH, -OR, or $-NH_2$ substituents and considerably slower for α hydrogens to $-CO_2^-$ and $-NH_3^+$ substituents [31].

Alternatively, it was proposed that the reaction of $^{\bullet}OH$ radicals with the anionic form of amino acids proceeds via reactions (18) and (19) [32,33]:

$$NH_2CH_2CO_2^- + {}^{\bullet}OH \longrightarrow {}^{+\bullet}NH_2CH_2CO_2^- + OH^- \tag{18}$$

$$^{+\bullet}NH_2CH_2CO_2^- \longrightarrow NH_2CH_2^{\bullet} + CO_2 \tag{19}$$

and not via abstraction of an α hydrogen. The $^{\bullet}OH$ radicals react very quickly with aliphatic unsaturated compounds, and aromatic compounds, via addition to a double bond:

$$^{\bullet}OH + R^1R^2C{=}CR^3R^4 \longrightarrow {}^{\bullet}CR^1R^2C(OH)R^3R^4 \tag{20}$$

Aliphatic radicals can also be produced via oxidations of alkanes by high-valent transition metal complexes of, e.g., Cu(III) [33,34], Fe(IV) or "Fe(V)" [33,35–38], or Co(III) [39]:

$$M^{n+1}L_m + RH \longrightarrow M^nL_m + H_3O^+ + R^\bullet \tag{21}$$

Analogous radicals are obtained via oxidation [40–43] or photooxidation by high-valent metal-carboxylates [44–46]:

$$M^{n+1}\text{-OOCR} \longrightarrow M^n + RCOO^\bullet \tag{22}$$

$$RCOO^\bullet \longrightarrow R^\bullet + CO_2 \tag{23}$$

$$M^{n+1}\text{-OOCR} \xrightarrow{h\nu} M^n + RCOO^\bullet \tag{24}$$

Other mechanisms for the formation of aliphatic radicals in biological systems involve hydrogen abstraction by other radicals present in the system, e.g., by peroxyl radicals [47–50]:

$$R'O_2^\bullet + RH \longrightarrow {}^\bullet R + R'O_2H \tag{25}$$

where $R' =$ alkyl or H, or by leakage of radicals formed in enzymatic processes, e.g., in vitamin B_{12}-catalyzed processes [51,52]:

$$AdoCbl \rightleftharpoons Ado^\bullet + B12_r \tag{26}$$

$$Ado^\bullet + RH \longrightarrow R^\bullet + AdoH \tag{27}$$

or by the tyrosyl radicals present in a variety of enzymes, e.g., ribonucleotide reductase and prostaglandin H synthetase [38].

Another source of aliphatic carbon-centered radicals in biological systems is via the reduction of organic halides, e.g., $CHCl_3$, by low-valent transition metal complexes [53]:

$$M^nL_m + RX \longrightarrow M^{n+1}L_m + RX^{\bullet-} \longrightarrow M^{n+1}L_m + R^\bullet + X^-$$
$$M^nL_m + RX \longrightarrow L_{m-1}M^{n+1}X + R^\bullet + L \tag{28}$$

This type of process might be at least partially the origin of the toxicity of organic halides.

2.4. Thiyl Radicals

In biological systems thiyl radicals are formed mainly via hydrogen abstraction from the $^-$SH groups by $^\bullet$OH [31], H$^\bullet$ [31], RO$_2^\bullet$ [54], or R$^\bullet$ [55,56] radicals:

$$R'SH + {}^\bullet OH/H^\bullet/RO_2^\bullet/R^\bullet \longrightarrow R'S^\bullet + H_2O/H_2/RO_2H/RH \tag{29}$$

The reactions with $^\bullet$OH radicals are very fast, approaching the diffusion-controlled limit [31], whereas those with H$^\bullet$ [31], RO$_2^\bullet$ [54], or R$^\bullet$ [55,56]

are considerably slower. It is of interest to note that the latter reaction is, at least for R = $^\bullet CR^1R^2(OH)$, a reversible process [57,58]:

$$R'SH + {}^\bullet CR^1R^2(OH) \rightleftharpoons R'S^\bullet + HCR^1R^2(OH) \tag{30}$$

This means that although reaction (30) is known as a "repair" reaction in the reverse direction, thiyl radicals might induce biological damage [57]. Thiyl radicals react with $R'S^-$ to form a dimeric anion radical:

$$R'SH + R'S^- \rightleftharpoons R'SSR'^{\bullet -} \tag{31}$$

The same anion radical is formed via the reduction of $R'SSR'$ by e_{aq}^-, $H^\bullet$, and glycine anhydride radical anion [31,55], i.e., this is also a route to the formation of thiyl radicals.

2.5. Aliphatic-Peroxyl Radicals

In biological systems aliphatic peroxyl radicals are formed mainly via the reaction of aliphatic radicals with dioxygen. All of the reactions of aliphatic carbon-centered radicals with dioxygen are very fast, approaching the diffusion-controlled limit [54,55,59]:

$$R^\bullet + O_2 \longrightarrow RO_2^\bullet \tag{32}$$

The same radicals are also formed via the single-electron oxidation of alkyl peroxides by high-valent transition metal complexes [47]:

$$M^{n+1}L_m + ROOH \longrightarrow M^nL_m + RO_2^\bullet + H_3O^+ \tag{33}$$

Alkyl-peroxyl radicals of the types $R^1R^2C(OH)OO^\bullet$, $R^1R^2C(OR)OO^\bullet$, and $R^1R^2C(NR^3R^4)OO^\bullet$ decompose via the formation of $O_2^{\bullet -}$ [54]:

$$R^1R^2C(OH)OO^\bullet \longrightarrow R^1R^2CO + O_2^{\bullet -} + H_3O^+ \tag{34}$$

and analogous processes. These reactions are base-catalyzed [54].

3. REDOX PROPERTIES OF ALKYL AND ALKYL-PEROXYL RADICALS

All of the alkyl and alkyl-peroxyl radicals are relatively strong single-electron oxidizing and reducing agents. Thus their radical-radical reactions often result in disproportionation processes, e.g.:

$$2^{\bullet}CH_2CH_3 \xrightarrow{k_a} C_4H_{10}$$

$$2^{\bullet}CH_2CH_3 \xrightarrow{k_b} C_2H_4 + C_2H_6 \qquad k_b/k_a = 0.35 \text{ [60]} \qquad (35)$$

$$2(CH_3)_2CHOO^{\bullet} \longrightarrow (CH_3)_2CHOH + (CH_3)_2CO + O_2 \text{ [54]} \qquad (36)$$

However, in most processes the alkyl-peroxyl radicals react as oxidizing reagents [54], whereas the redox properties of the aliphatic carbon-centered radicals depend on the substituents on the α carbon. Thus, for example, radicals of the type $^{\bullet}CR^1R^2(OH)$ are relatively strong reducing agents [61]; however, they do oxidize low-valent transition metal complexes, e.g., $Cr(H_2O)_6^{2+}$ [62] and $V(H_2O)_6^{2+}$ [63]. The known redox potentials of many radicals are summarized in two reviews [61, 64] and those of some alkyl-peroxyl radicals in a recent publication [65].

4. THE ROLE OF TRANSITION METAL COMPLEXES IN RADICAL-INDUCED BIOLOGICAL DELETERIOUS PROCESSES

The observation that oxidative stress, which produces the $^{\bullet}OH$ and $O_2^{\bullet-}$ radicals, causes a variety of biological deleterious processes [1–3] led to the suggestion that the major deleterious effects caused by radicals in biological systems are the result of the reactions of hydroxyl radicals. This hypothesis is based on the observation that the $O_2^{\bullet-}$ radical is relatively unreactive toward most biological components [66]. According to this hypothesis, the role of the superoxide free radicals is to reduce a transition metal complex in the system:

$$M^{n+1}L_m + O_2^{\bullet-} \longrightarrow M^nL_m + O_2 \qquad (37)$$

This reaction is followed by the Fenton-like reaction (1′), i.e., the deleterious role of the superoxide radicals is via the metal-catalyzed Haber-Weiss reaction [7], which transforms the "nonreactive" superoxide radicals into the reactive $^{\bullet}OH$ radicals. The role of the transition metal complexes according to this mechanism is mainly via the production of the highly reactive $^{\bullet}OH$ radicals. However, $O_2^{\bullet-}$ was recently shown to be a relatively competent oxidizing agent toward transition metal complexes, e.g., it oxidizes Cu(II)-peptides to the corresponding Cu(III)-peptides [67] and Ni(II)-tetraazamacrocyclic complexes to the corresponding Ni(III) complexes [68,69]. A third mechanism by which the

$O_2^{\bullet-}$ radicals might induce deleterious processes is via their reaction with $^\bullet$NO to form peroxonitrite (reaction (15)) [4].

The hypothesis that the biological deleterious effects of radicals are due mainly to the reactions of the hydroxyl radicals has several drawbacks:

1. It was shown that hydroxyl radical scavengers have a significantly smaller protective effect in biological systems than that predicted from their rates of reaction with the hydroxyl radical. This observation has been attributed to a "site-specific" mechanism, i.e., the metal ion is bound near the biological target and the damage is caused by the hydroxyl radical immediately after its formation [70–72]. The site-specific mechanism is discussed below in more detail.

2. Transition metal complexes, even at relatively low concentrations, were shown to be radiosensitizers, i.e., to enhance the deleterious effect of ionizing radiation [23,24]. As ionizing radiation is absorbed homogeneously in the medium, the site-specific mechanism clearly cannot explain this observation. Furthermore, it is easy to show from the relative rates of reaction of the hydroxyl radicals with the different constituents of the irradiated cells [31] that under experimental conditions nearly all of the radicals are scavenged by the organic constituents of the biological medium and not by the transition metal complexes present at very low concentrations.

Thus it seems reasonable to suggest that the role of transition metal complexes in radiosensitization and probably also in other biological deleterious effects caused by radicals is partially due to the interaction of aliphatic carbon-centered and aliphatic peroxyl radicals with these complexes. Indeed, recent studies indicate that many aliphatic carbon-centered radicals react with first-row transition metal complexes in aqueous solutions [55]:

$$R^\bullet + M^nL_m \longrightarrow \text{products (for details, see below)} \qquad (38)$$

The specific rates of these reactions are usually considerably faster than those of the same radicals with the organic constituents of the biological systems [55]. Therefore, carbon-centered radicals are expected to react selectively with the transition metal complexes present in the biological system.

In most biological systems dioxygen is present, though often at relatively low concentrations. Thus, for example, in muscle cells the partial oxygen pressure is ≤ 5 mm Hg, i.e., its concentration is $\leq 1 \cdot 10^{-5}$ M. Therefore, reaction (38) competes with reaction (32) and the role of the $RO_2^\bullet$ radicals as the source of the biological deleterious processes caused by radicals cannot be neglected. $RO_2^\bullet$ radicals are known to cause peroxidation of cell membranes via the chain reactions (32) and (39) [7]. However, inside the cells reaction (40) is expected to compete effectively with reaction (39), due to their relative rates [54], and has therefore to be considered also as a plausible source for radical-induced biological deleterious processes.

$$RO_2^\bullet + RH \longrightarrow RO_2H + R^\bullet \tag{39}$$

$$RO_2^\bullet + M^nL_m \longrightarrow \text{products (for details, see below)} \tag{40}$$

5. THE "SITE-SPECIFIC" MECHANISM

The site-specific mechanism was proposed [70–72] as an explanation to the observation that in many biological systems hydroxyl radical scavengers are less effective than predicted from known rates of reaction in inhibiting specific deleterious processes. This mechanism might be summarized by the following reactions:

$$M^n(\text{Biol}) + H_2O_2 \longrightarrow M^{n+1}(\text{Biol}) + {}^\bullet OH \tag{41}$$

$$M^{n+1}(\text{Biol}) + {}^\bullet OH \longrightarrow M^{n+1}(\text{Biol}') \tag{42}$$

where "Biol" is the biological target system to which the transition metal cation is bound and "Biol'" is the same molecule damaged by the hydroxyl radical. These reactions might naturally be followed by:

$$M^{n+1}(\text{Biol}') + \text{red} \longrightarrow M^n(\text{Biol}') + \text{ox} \tag{43}$$

where "red" is a reducing species, e.g., the radical adduct of ${}^\bullet OH$ to "Biol," ascorbate, $O_2^{\bullet-}$, and "ox" is the oxidized form of the same species. Thus the process might be a chain process causing repetitive damage to the active site.

However, it was shown [18] that when ${}^\bullet OH$ radicals are formed via reaction (44),

$$Fe^{II}EDTA^{2-} + H_2O_2 + H_3O^+ \longrightarrow Fe^{III}EDTA^- + 2H_2O + {}^{\bullet}OH \ (44)$$

the $^{\bullet}OH$ radicals escape the $Fe^{III}EDTA^-$ though $k(Fe^{III}EDTA^- + {}^{\bullet}OH)$ = $1.5 \cdot 10^9 \ dm^3 \cdot mol^{-1} \cdot s^{-1}$; and 1 mM of salicylate, $k(salicylate + {}^{\bullet}OH)$ = $1.6 \cdot 10^{10} \ dm^3 \cdot mol^{-1} \cdot s^{-1}$, scavenges nearly all of the $^{\bullet}OH$ radicals. A calculation [18] shows that under these conditions the $^{\bullet}OH$ radicals diffuse to a distance of ~100 Å from the site of formation prior to their reaction with the scavenger. Thus cage effects are expected only if:

1. At the site of formation of the $^{\bullet}OH$ radicals a scavenger is present that reacts with it in a reaction that is indeed diffusion-controlled, e.g., if at the site amino acids with aromatic residues or nucleic acids are present. This enables the use of the reaction of tethered $Fe^{II}EDTA^{2-}$ with H_2O_2 for the determination of residues situated nearby on DNA [73].

2. The reaction occurs within a small crevice surrounded by a polymer that inhibits the diffusion out of the formed radical. In this situation, the radical is expected to react with the polymer, even if the reaction is considerably slower than diffusion-controlled but at a variety of sites on the polymer.

An alternative source of the site-specific deleterious processes induced by radical reactions is the preferential reaction of aliphatic peroxyl and carbon-centered radicals with the transition metal complexes present in the system, which directs the damage to the sites surrounding these complexes (see below).

It was reported that transition metal cations, mainly Mn^{2+}, Fe^{2+}, and Cu^{2+}, catalyze the oxidation of amino acids and peptides by H_2O_2 [74–77]. These oxidations often transform primary amino groups to the corresponding carbonyl groups. This process, which is of significant biological importance, was shown to be site-specific. In proteins this oxidation often involves the lysine residues [75]. It was proposed that this specificity is due to the site-specific mechanism and that the $^{\bullet}OH$ radicals formed in the process specifically oxidize the lysine residues bound to the central cation. The above discussion clearly points out that this cannot be true as the rate of reaction of $^{\bullet}OH$ radicals with the acidic form of lysine is only $k(OH + lysine) = 3.5 \cdot 10^8 \ dm^3 \cdot mol^{-1} \cdot s^{-1}$ [31]. Thus, the metal-catalyzed oxidations of proteins have to proceed via reaction (7) or (10) and not via reaction (5) as proposed.

6. REACTIONS OF ALIPHATIC CARBON-CENTERED RADICALS WITH TRANSITION METAL COMPLEXES IN AQUEOUS SOLUTIONS

The kinetics and mechanisms of reaction of many carbon-centered radicals with a large variety of transition metal complexes were studied. The specific rates of most of these reactions are summarized in [55]. In general terms these reactions proceed via one of the following mechanisms:

$$M^nL_m + R^\bullet \longrightarrow M^{n\pm1}L_m + R^\mp \tag{45}$$

$$M^nL_m + R^\bullet \longrightarrow L_mM^{n+1}\text{-}R \text{ or } L_{m-1}M^{n+1}\text{-}R + L \tag{46}$$

$$M^nL_m + R^\bullet \longrightarrow M^{n-1}L_{m-1} + L\text{-}R \text{ or } L^\pm + R^\mp \tag{47}$$

$$M^nL_m + R^\bullet \longrightarrow L_{m-1}M^n\text{-}LR^\bullet \longrightarrow M^{n\pm1}L_{m-1} + LR^\mp \text{ or} \atop + L^\pm + R^\mp \tag{48}$$

$$M^nL_m + R^\bullet \longrightarrow L_{m-1}M^n(L^\pm) + R^\mp \longrightarrow M^{n\pm1}L_m + R^\mp \tag{49}$$

In the following paragraphs these mechanisms and their implications are discussed separately. As the consequences of reaction (46) are of major interest and are most complicated, they will be discussed at the end of this section.

Reaction (45) describes outer-sphere redox processes. The outer-sphere reductions of transition metal complexes by $^\bullet CR^1R^2OH$ or $^\bullet CR^1R^2O^-$ radicals [78–86] are the most abundant of this type of reaction, as these radicals are powerful reducing agents and as the oxidation of these radicals does not require major bond rearrangements. Thus the self-exchange rate for the couple $C(CH_3)_2OH^{+/0}$ has been estimated to be $\sim 10^3$ $dm^3 \cdot mol^{-1} \cdot s^{-1}$ [84]. Some of these reactions produce secondary powerful reducing agents, e.g., Fe(I)deuteroporphyrin [87] and Pb^+_{aq} [88]. It should be noted that $CO_2^{\bullet-}$ is also a powerful reducing agent [61,89], but as it is bent and CO_2 is linear, most of its reactions proceed via the inner-sphere mechanism.

It is more surprising that even unsubstituted α-centered alkyl radicals reduce strongly oxidizing agents via the outer-sphere mechanism. Thus, though many radicals reduce $IrCl_6^{2-}$ via the inner-sphere mechanism (see below), the radicals $^\bullet CH(CH_3)_2$ and $^\bullet C(CH_3)_3$ reduce it partially (50% and 77%, respectively) via the outer-sphere mechanism to yield mainly $H_2C=CHCH_3$ and $H_2C=C(CH_3)_2$, along with some $HOCH(CH_3)_2$ and $HOC(CH_3)_3$ [86]. In addition, the reduction of $Fe(CN)_6^{3-}$ by alkyl radicals seems also to proceed via the outer-sphere mechanism [86]. C_2H_4 is also formed in the reduction of $Fe(phen)_3^{3+}$ and $Fe(bpy)_3^{3+}$ though not in the reduction of the weaker oxidizing agent $Cr(bpy)_3^{3+}$ [90]. The mechanism of these reactions is not clear though they probably proceed via the outer-sphere mechanism [80,90]. Some ethylene is even produced when $^\bullet C_2H_5$ radicals reduce $Co(NH_3)_5X^{2+}$ and $Ru(NH_3)_5X^{2+}$, where X = halide or pseudohalide, indicating that an outer-sphere mechanism also contributes to these processes [90].

Another mechanism, which might be considered outer sphere, is according to reactions (50) and (51) [45]:

$$^\bullet CH_3 + Cl^- \rightleftharpoons CH_3Cl^{\bullet -} \tag{50}$$

$$CH_3Cl^{\bullet -} + Co(NH_3)_5O_2CCH_3^{2+} + 6H_3O^+ \tag{51}$$
$$\rightarrow CH_3Cl + Co_{aq}^{2+} + 5NH_4^+ + HO_2CCH_3$$

Even more surprising is the observation that $^\bullet C(CH_3)_2OH$ radicals oxidize $Ru(NH_3)_6^{2+}$ via the outer-sphere mechanism, though the kinetic data suggest that "an intermediate" ion pair is formed in this process [91]:

$$Ru(NH_3)_6^{2+} + ^\bullet C(CH_3)_2OH \rightleftharpoons \{Ru(NH_3)_6^{3+}, {}^- C(CH_3)_2OH\} \tag{52}$$

$$\{Ru(NH_3)_6^{3+}, {}^-(CH_3)_2OH\} \xrightarrow{H_2O} Ru(NH_3)_6^{3+} + CH(CH_3)_2OH + OH^-$$
$$\tag{53}$$

$$\{Ru(NH_3)_6^{3+}, {}^- C(CH_3)_2OH\} \xrightarrow{H_3O^+} Ru(NH_3)_6^{3+} + CH(CH_3)_2OH + H_2O$$

This mechanism was proposed on the basis of the pH effect on the observed rate of reaction in a pH region where none of the reactants has a pK_a [91]. However, it should be pointed out that a pH effect (in the same pH range), though in the opposite direction, is also observed in the reduction of $Co(NH_3)_6^{3+}$ by $^\bullet CR^1R^2OH$ [79]. Thus a stronger interaction between the radicals and the complexes probably exists and they should not be considered outer-sphere processes.

Processes following the mechanism described in Eq. (47) are mainly observed for halide and analogous complexes as the oxidizing species, e.g. [78,82,86,90]:

$$M(NH_3)_5X^{2+} + R^{\bullet} \longrightarrow RX + M_{aq}^{2+} + 5NH_4^+ \tag{54}$$

Analogous processes in biological systems, if they occur, will produce haloalkyls that are known to be toxic. In principle also the reactions (55) (see for example [92])

$$L_mM^{n+1}\text{-}R + R'^{\bullet} \longrightarrow M^nL_m + RR' \text{ or } RH + R_{-H} \tag{55}$$

(which will be discussed below in detail) follow an analogous mechanism.

Processes following the mechanism described in Eq. (48) are observed for the addition of alkyl [93] and substituted alkyl [90] radicals to an aromatic ligand followed by the reduction of the central cation and proton loss from the aromatic ring, e.g. [90,93]:

$$M(bpy)_3^{3+} + R \longrightarrow (bpy)_2M^{3+} \longrightarrow (bpy)_2M^{2+} + H^+ \tag{56}$$

In principle, the central cation could also be oxidized by its radical ligand. However, such reactions are not known. The addition of $^{\bullet}OH$ radicals and $H^{\bullet}$ atoms to aromatic ligands results in similar processes [81,86,94]. Even the reaction of $^{\bullet}OH$ radicals with methemoglobin, at an unknown site, is followed by intramolecular electron transfer and reduction of Fe(III) to Fe(II) [95,96]. Naturally, such reactions cause structural changes in the biomolecules and might therefore be deleterious.

Processes following the mechanism described in Eq. (49) are abundant, though only processes in which the ligand is reduced are known [97–104]. In principle, the processes in which a radical is reducing an enzyme at a remote site, a process followed by intramolecular electron transfer over long distances, belong also to this class of processes (see, for example, [105–113]).

Processes following the mechanism described in Eq. (46) are the common mechanism of reaction for "low-valent" complexes with ligand exchange rates, or steric structures, which enable a bond formation between the central metal cation and the attacking radical. The products of these reactions are transient or stable complexes with metal-carbon σ bonds. The question of whether the mechanism involves ligand exchange or is accompanied by an increase in the coordination number is often not clear. In some cases the results suggest that an increase in the coordination number occurs at least initially [63,114], whereas in other cases the reaction seems to require a loss of one of the ligands, L [115]. Recently [116] it was shown that the rate-determining step in most of these reactions is the ligand interchange step, i.e., that the number of ligands is not increased in these processes. The specific rates of reactions proceeding via this mechanism were measured for first-row transition metal complexes with different ligands for which M^n = Ti(III) [82,117,118] (though the transient complexes with a Ti^{IV}-C bond were not observed); V(II) [64,119] (though the transient complexes with a V^{III}-C bond were not observed); Cr(II) [62,120–129]; Cr(III) [129]; Mn(II) [130,131]; Mn(III) [131] (though the transient complexes with a Mn^{IV}-C bond were not observed); Fe(II) [130,132–138]; Fe(III) [136, 138,139]; Co(II) [45,92,125,130,140–150]; Ni(I) [132,151–153]; Ni(II) [132, 153–156]; Cu(I) [157–166]; and Cu(II) [159,166–174]. (Analogous reactions were reported for other metal complexes, but as they seem to be without biological interest they are not discussed herein.) Nearly all of the complexes with a metal-carbon σ bond thus formed are unstable under physiological conditions. However, the products of their decomposition, discussed below, might cause biological deleterious processes. Naturally, if stable complexes of the type $L_m M^{n+1}$-R are formed these might have deleterious effects if the complex $M^n L_m$ has a biologically essential role. However, this is the case only for iron-porphyrin complexes [132], which are stable only in anaerobic solutions, and for cobalt complexes.

7. MECHANISMS OF DECOMPOSITION OF THE TRANSIENT COMPLEXES $L_m M^{n+1}$-R

The mechanisms and kinetics of decomposition of the transient complexes $L_m M^{n+1}$-R in aqueous solutions depend on the nature of the central cation, of the ligands L, of the substituents on the aliphatic residue R, on the pH, and on the presence and nature of various substrates S in the medium, e.g., O_2. In the following paragraphs, the major mechanisms observed and their plausible biological consequences are discussed:

(1) The major mechanism observed is the heterolysis of the metal-carbon σ bond:

$$L_mM^{n+1}\text{-}R + H_2O \longrightarrow M^{n+1}L_m + RH + OH^- \tag{57}$$

$$L_mM^{n+1}\text{-}R + H_2O \longrightarrow M^{n-1}L_m + ROH/R_{-H} + H_3O^+ \tag{58}$$

Reaction (57) is observed mainly for complexes M^nL_m which are relatively strong reducing agents and which are not expected to be reduced, e.g., for complexes of Ti(III) [82,117,118]; V(II) [64,119]; Cr(II) [62,121,127–129]; Mn(II) [131,132]; Fe(II) [117,133]; Co(II) [90]; Ni(I) [132,151,152,175]; and Cu(I) [157,158,164]. This reaction is usually acid [64,121,124,128,164] and general base-catalyzed [124,176,177].

Reaction (58) is mainly observed for complexes M^nL_m which can be reduced and are difficult to oxidize, e.g., Fe(III) [139]; Cu(II) [58,76]; and with proper ligands for Co(II) [140,149].

It is of interest to note the mechanisms of decomposition of several complexes of the type $LCu^{III}\text{-}CH_3$:

$$Cu^{III}\text{-}CH_{3aq}^{2+} \longrightarrow Cu_{aq}^+ + CH_3OH \text{ [178]} \tag{59}$$

$$(H_2NCH_2CO_2^-)_2Cu^{III}\text{-}CH_3 \longrightarrow Cu^{III}(H_2NCH_2CO_2^-)_2^+ + CH_4 \text{ [172]} \tag{60}$$

$$(\text{glycylglycylglycine})Cu^{III}\text{-}CH_3^+ \longrightarrow Cu^I(\text{glycylglycylglycine})$$
$$+ CH_3OH \text{ [179]} \tag{61}$$

The shift in mechanism caused by the glycine ligands was attributed [172] to their effect on the redox potential of the $Cu^{III/II}$ couple. However, the glycylglycylglycine ligand stabilizes the Cu^{III} oxidation state even more than the glycines. It has therefore to be concluded that the choice of the mechanism of decomposition of the transient complexes depends on the activation energies of the different plausible mechanisms and not on the total free energy gain. This is plausible as the $^\bullet CH_3$ radicals are both powerful oxidizing and reducing agents [45], and therefore both reaction mechanisms are highly exothermic.

If reaction (57) occurs in a biological system, it is usually not expected to cause deleterious effects because the RH product is usually the source of the $R^\bullet$ radical; see reaction (17). However, the oxidized transition metal complex might in some systems cause deleterious processes, e.g., Cu(III) is a powerful oxidizing agent that oxidizes amino acids [180,181] and peptides [182,183]. On the other hand, reaction (58) is expected to cause a variety of deleterious effects as it changes the nature of the organic substrate. This is especially true for $R = CR^1R^2(OH)$, which in this reaction yields aldehydes and ketones, many of which are toxic. Furthermore, the reduced transition metal complex might initiate deleterious processes, e.g., via reactions (4) and/or (12).

(2)   In many systems studied, the results point out that reaction (46) is an equilibrium process and the mechanism of decomposition of the transient complex $L_{m-1}M^{n+1}$-R involves radical processes:

$$L_m M^{n+1}\text{-}R + L \rightleftharpoons M^n L_m + R^\bullet \qquad (-46')$$

followed by:

$$2R^\bullet \longrightarrow R_2/RH + R_{-H} \qquad (62)$$

or by:

$$R^\bullet + S \longrightarrow \text{products} \qquad (63)$$

This mechanism of decomposition has been observed for a large variety of complexes of Cr(III) [121,128,159]; Mn(III) [130]; Fe(III) [130,132]; Fe(IV) [139]; Co(III) [92,143,144]; Ni(III) [154–156]; and Cu(II) [46,48, 77,159,184].

It should be noted that as reaction (46) proceeds, usually via a mechanism involving a ligand interchange [92,116,144], the measurement of $\Delta H^\#$ of reaction (−46) does not yield the M-C bond strength as was proposed [185–188], though the determination of the bond strengths by this method were performed mainly in aprotic solvents that do not behave as ligands.

The homolytic pathway usually does not contribute to deleterious processes in biological systems as it only "stores" the radicals and prolongs their lifetime. However, for complexes that react with the carbon-centered radicals with rates approaching the diffusion-controlled limit, e.g., several metal(porphyrin) complexes [129,136,138], a cage might be formed in these systems and the radical will therefore attack the biomolecules via a site-specific mechanism and thus might induce specific deleterious processes.

(3)   The mechanism of decomposition of transient complexes of the type $L_m M^{n+1}$-R, which probably is responsible for most of the biological deleterious processes induced by the formation of these complexes, is the β elimination of a good leaving group, X, e.g., $X = -OR, -NR_2$, $-NHC(O)R$, and halides:

$$L_m M^{n+1}\text{-}CR^1R^2CR^3R^4X \longrightarrow M^{n+1}L_m + R^1R^2C=CR^3R^4 + X^- \quad (64)$$

These reactions, which are often acid-catalyzed, were observed for a large variety of such complexes, see, for example, for X = OH [62,121, 126,128,134,140,142,148,157,160,169,189–191]; X = OR [192]; X = NR$_2$ [163,174,193–195]; X = OP(O)(OH)$_2$ [196]; and X = NHC(O)R [197]. All of these reactions might cause severe modifications in a variety of biologically important components, e.g., in peptides, DNA, RNA, sugars, phosphate esters, etc. It should be pointed out that some of these reactions are only catalyzed by the formation of the metal-carbon σ bond, i.e., the reactions also occur in the absence of the transition metal complexes.

(4)   Another mechanism of decomposition that is expected to induce biological deleterious processes is the β elimination of carboxylates:

$$L_m M^{n+1}\text{-}(CR^1R^2CR^3R^4CO_2^-) \longrightarrow M^{n-1}L_m + R^1R^2C=CR^3R^4 + CO_2 \quad (65)$$

This mechanism has been observed only for M^n = Cu(II) [173,174].

(5) β hydride shift reactions are also expected to induce biological deleterious processes:

$$L_m M^{n+1}\text{-}(CR^1R^2CHR^3R^4) \longrightarrow L_m M^{n+1}\text{-}H + R^1R^2C=CR^3R^4 \quad (66)$$

This process has been observed in only a few systems, e.g.:

$$L(H_2O)Co^{III}\text{-}CH(CH_3)OH^{2+} \longrightarrow L(H_2O)Co^{III}\text{-}H^{2+} + CH_3CHO \; [140] \quad (67)$$

$$L(H_2O)Cr^{III}\text{-}CH(CH_3)OH^{2+} \longrightarrow L(H_2O)Cr^{III}\text{-}H^{2+} + CH_3CHO \; [198] \quad (68)$$

$$L(H_2O)Fe^{III}\text{-}(CH_2CH_2NH_3^+) + H_2O \longrightarrow L(H_2O)Fe^{III}\text{-}H + CH_3CHO + NH_4^+ \; [199] \quad (69)$$

where L represents different tetradentate ligands. Also the decomposition of $LCo^{III}\text{-}CH(CH_3)_2$ probably proceeds via this mechanism as $CH_3CH=CH_2$ is the final product [149], though the authors do not discuss the detailed mechanism of this process.

(6)   Another reaction that might contribute significantly as a deleterious process in a variety of biological systems is the reaction of the radical $R^\bullet$ formed in reaction $(-46')$ with dioxygen, i.e., according to reaction (32).

$$R^{\bullet} + O_2 \longrightarrow RO_2^{\bullet} \tag{32}$$

This is expected to be followed in many systems by:

$$M^nL_m + RO_2^{\bullet} \longrightarrow L_mM^{n+1}\text{-OOR} \tag{70}$$

i.e., in the homolytic insertion of dioxygen into the M-C bond [143,155]. The plausible biological consequences of such a process are discussed below.

(7)   The toxicity of halogenated alkanes has been attributed [136, 138] in part to the formation of carbon-centered radicals via:

$$PFe(II) + RX \longrightarrow PFe(III) + X^- + R^{\bullet} \tag{71}$$

where P = a porphyrin and X a halogen atom. This reaction may be followed by reaction (32) and then by lipid peroxidation or by reaction of the peroxyl radical with PFe(III) or PFe(II) (see below).

A large series of alkyl and halogenated alkyl radicals react with PFe(II) [132,136–138] and PFe(III) [136,138], as well as a variety of other metal porphyrins [54,55,81,86,129,131,132,136–138,149,150,153]. Many of the PFe^{III}-R complexes are stable in anaerobic solution but are oxidized by dioxygen to form PFe(III) and oxidized organic products that might be toxic [132].

The complex PFe^{III}-CF$_3$, which is formed via analogous radical processes, decomposes via:

$$PFe^{III}\text{-CF}_3 + PFe(II) \longrightarrow PFe^{III}\text{-CF}_2 + PFe(III) + F^- \tag{72}$$

$$PFe^{III}\text{-CF}_2 + 3H_2O \longrightarrow PFe^{II}\text{-CO} + 2H_3O^+ + 2F^- \tag{73}$$

i.e., via a carbene intermediate [138]. It is of interest to note that the aquation of $(H_2O)_5Cr^{III}\text{-CF}_3^{2+}$ yields $Cr(H_2O)_6^{3+}$, HF, and CO. The data suggested that the Cr-C and the first C-F bond break coherently, i.e., no complex carbene is formed in this reaction [200]. Also, the aquation of $(H_2O)_5Cr^{III}\text{-CHX}_2^{2+}$ (X = Cl; Br; I) yields CO as one of the carbon-containing products, the other being HCO_2H [201].

(8)   In principle, one can expect that the formation of transient complexes of the type L_mM^{n+1}-R might result in the rearrangement of the carbon-skeleton of R in analogy to vitamin B_{12} catalyzed processes [202]. However, only one such reaction was observed until now, probably due to the fact that very simple alkyl radicals are used in most studies. The observed rearrangement [161] is according to reaction (74):

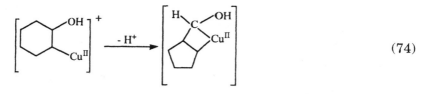

$$(74)$$

This reaction is a key step in the catalytic oxidation of cyclohexene by peroxides [203]. Naturally, if reactions of this type occur in biological systems they are expected to induce deleterious processes.

(9) The transient complexes $L_m M^{n+1}$-R are reducing agents relative to the complexes $M^{n+1}L_m$. Thus they react with a variety of oxidants yielding usually $M^{n+1}L_m$ and various oxidized organic products formed from R [60,128,132,148,150,171,175]. These reactions are not expected to contribute to the deleterious processes induced by radicals as the chances that an oxidizing agent will react with $L_m M^{n+1}$-R prior to its decomposition are very low. An exception to this may be the reaction with dioxygen due to its relatively high concentration in biological systems.

(10) Oxidation of several (porphyrin)Co(III)-alkyl and (porphyrin)-Fe(III)-alkyl complexes was shown to cause intramolecular migration of the alkyl residue to form the corresponding N-alkylmetalloporphyrins, M(II)(N-alkylporphyrin) [204]. This process is reversible and upon reduction of the M(II)(N-alkylporphyrin) complexes the corresponding (porphyrin)M(III)-alkyl complexes are formed [204]. These processes can be induced by a variety of radicals [153]. Especially if alkyl radicals form high-valent (porphyrin)M-alkyl complexes, these reactions are expected to have deleterious biological consequences.

If analogous alkyl migration reactions also occur for complexes with other ligands they will naturally result in a biological deleterious process. Such reactions have not been observed. However, it should be noted that N-benzylglycine is formed quantitatively when an alkaline aqueous solution containing $Cu^{II}(NH_2CH_2CO_2^-)_2$ and $ClCH_2C_6H_5$ is sonolyzed [205]. The mechanism of this process is not known but it was proposed [172] that it proceeds via the formation of $(NH_2CH_2CO_2^-)_2$-Cu^{III}-$CH_2C_6H_5$ followed by the benzyl migration to the glycine ligand.

(11) The transient complexes $L_m M^{n+1}$-R react with alkyl radicals to form various organic products:

$$L_m M^{n+1}\text{-R} + R'^{\bullet} \longrightarrow M^n L_m + (RR' + RH + R_{-H}$$
$$+ R'H + R'_{-H})$$

$$(75)$$

i.e., the formation of carbon-carbon bonds [60,92,130,132,143,148,155, 156,159,172,173] and "disproportionation" of the radicals $R^{\bullet}$ and $R'^{\bullet}$ [60,92,135,148,173]. These reactions are expected to occur only rarely in biological systems as they require that two radicals reach the same site. Thus, if at all, these processes will occur only for stable complexes of the type L_mM^{n+1}-R, e.g., (porphyrin)Fe^{III}-alkyl [136,137], or long-lived complexes of the type L_mM^{n+1}-R, e.g., (glycylglycylglycine)Cu^{III}-CH_3^+, which decomposes with a rate of 0.0035 s^{-1} [179], and (nta)(H_2O)Fe^{III}-CO_2^-, which decomposes via a second-order reaction with a rate of $1.9 \cdot 10^7$ $dm^3 \cdot mol^{-1} \cdot s^{-1}$. Naturally, if such reactions occur in biological system they are expected to have deleterious effects.

(12)   Other mechanisms of decomposition of the transient complexes of the type L_mM^{n+1}-R, e.g., CO insertion into the metal-carbon bond [165] or transalkylation reactions with other transition metal complexes [121], are probably not relevant in biological systems and are therefore not discussed here.

8. REACTIONS OF ALKYL-PEROXYL RADICALS WITH TRANSITION METAL COMPLEXES IN AQUEOUS SOLUTIONS

Alkyl-peroxyl radicals are formed in biological systems mainly via reaction (32) and are therefore expected to be formed in most radical involving processes. In principle the following mechanisms of reaction of these radicals with transition metal complexes might be expected:

$$M^nL_m + ROO^{\bullet} \longrightarrow M^{n \pm 1}L_m + ROO^{\mp} \tag{76}$$

$$M^nL_m + ROO^{\bullet} \longrightarrow L_mM^{n+1}\text{-OOR or } L_{m-1}M^{n+1}\text{-OOR} + L \tag{77}$$

$$M^nL_m + ROO^{\bullet} \longrightarrow M^{n-1}L_{m-1} + L\text{-OOR or } L^{\pm} + ROO^{\mp} \tag{78}$$

$$M^nL_m + ROO^{\bullet} \longrightarrow L_{m-1}M^n\text{-LROO}^{\bullet}$$
$$\longrightarrow M^{n \pm 1}L_{m-1} + LROO^{\mp} \text{ or } + L^{\pm} + ROO^{\mp} \tag{79}$$

$$M^nL_m + ROO^{\bullet} \longrightarrow L_{m-1}M^n(L^{\pm}) + ROO^{\mp}$$
$$\longrightarrow M^{n \pm 1}L_m + ROO^{\mp} \tag{80}$$

However, as the alkyl-peroxyl radicals are poor reducing agents only reactions in which they oxidize transition metal complexes have been reported to date. Furthermore, the results indicate that the self-exchange rate of the couples $ROO^{\bullet}/ROO^{-}$ is very slow, $k \leqslant 10^{-2} \ s^{-1}$ [65]; thus reactions (76) and (80) are not expected to be observed. Indeed, reaction (76) was measured only for $M^nL_m = Fe(CN)_6^{4-}$, and even for this complex, for which reaction (77) cannot occur, the results suggest that "the intimate details of the oxidation reaction include more parameters than those expressed in the basic outer-sphere electron-transfer process" [65]. Oxidations in which the first product observed is an oxidized porphyrin ligand [153,205,206], i.e., processes that seem to proceed via reaction (80), might involve other short-lived intermediates, e.g., they could proceed via reaction (77) or (79). Thus the major route of reaction of alkyl-peroxyl radicals with transition metal complexes is via reaction (77) [42,147,206–210]. As reaction (77) is usually considerably faster than reaction (39) in protic media [54], it will compete with it efficiently in many biological systems. Therefore, many deleterious effects caused by these radicals probably stem from the decomposition of the transient complexes L_mM^{n+1}-OOR.

9. MECHANISMS OF DECOMPOSITION OF THE TRANSIENT COMPLEXES L_mM^{n+1}-OOR

The transient complexes L_mM^{n+1}-OOR are strong, and potentially up to three electron, oxidizing agents. They are therefore plausible intermediates in site-specific deleterious processes in biological systems. It should be noted that the transient complexes L_mM^{n+1}-OOR are also formed as transients in the catalytic oxidations by peroxides [211–214].

$$M^{n+1}L_m + HO_2R \longrightarrow L_mM^{n+1}\text{-OOR} + H^+ \ or \ L_mM^{n+1}(HOOR) \qquad (81)$$

Thus their mechanisms of decomposition are relevant not only in radical processes. The following mechanisms of decomposition of these transient complexes were observed:

(1) Heterolysis of the M-O bond:

$$L_mM^{n+1}\text{-OOR} + H_3O^+ \longrightarrow M^{n+1}L_m + HOOR + H_2O \qquad (82)$$

This mechanism was observed, for example, for $M^{n+1} = Mn(III)$ [42]; Fe(III) [207]; and Co(III) [42]. Most of the alkyl peroxides are unstable and decompose via

$$HOOCH_2R' \longrightarrow R'CHO + H_2O \tag{83}$$

Thus this mechanism will be of deleterious consequences in biological systems due to the toxicity of the aldehydes formed and in some systems also due to the oxidizing properties of the complex $M^{n+1}L_m$ formed. Naturally, if the peroxide HOOR is stable then it will cause deleterious effects as any peroxide.

(2) Heterolysis of the O-O bond:

$$L_mM^{n+1}\text{-}OOR + H_3O^+ \longrightarrow L_mM^{n+3}{=}O + ROH + H_2O \tag{84}$$

followed by oxidations of substrates by $L_mM^{n+3}{=}O$. This mechanism has been proposed as a key step in a variety of enzymatic and non-enzymatic catalytic process (see, for example, [38]). However, to date this reaction has not been observed in aqueous solutions.

(3) Homolysis of the M-O bond:

$$L_mM^{n+1}\text{-}OOR \rightleftharpoons M^nL_m + ROO^\bullet \tag{85}$$

This type of decomposition has been reported for $(H_2O)_5Fe^{III}\text{-}OOR$ [207], $(H_2NCH_2CO_2^-)_2Cu^{III}\text{-}OOR$ [208]; and $(cyclam)(H_2O)Ni^{III}\text{-}OOH$ [69]. It is of interest to note that the equilibrium constants K_{85} for a variety of complexes of the type $(H_2O)_5Fe^{III}\text{-}OOR$ are nearly independent of the nature of R [207] though electron-withdrawing substitutents on R increase the redox potential of the radical $ROO^\bullet$ [65]. This apparent discrepancy was attributed to the observation that electron-withdrawing substitutents on R decrease the pK_a of HOOR [65] and therefore decrease the basicity of the ligand ROO^-, thus decreasing the stabilizing effect of ROO^- on the Fe^{III} central cation [207].

This mechanism of decomposition of the transient complexes naturally causes no biological deleterious effects as it only increases the lifetime of the radicals.

(4) Via oxidation of R, e.g., via:

$$L_mM^{n+1}\text{-}OOCH_2R' + H_3O^+ \longrightarrow M^{n+1}L_m + R'CHO + 2H_2O \tag{86}$$

This is the proposed [215] mechanism of the catalytic oxidation of para-cresol by dioxygen in alkaline media in the presence of $CoCl_2$ (reaction (87)) and the decomposition of the complex $L(H_2O)Co^{III}\text{-}O_2CH_3$, (L = 2,3,9,10-tetramethyl-1,4,8,11-tetraazacyclotetradeca-1,3,8,10-tetraene) [209] as CH_2O is the major product. However, there is no proof that the decomposition does not occur via reactions (82) and (83).

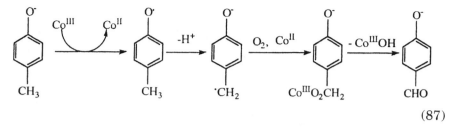

(87)

If this mechanism occurs in biological systems it will clearly induce deleterious effects.

(5) Via oxidation of the ligands L. The transient complexes are strong three-electron oxidants and it is therefore not surprising that in several systems a two-electron oxidation of the ligand is observed. Thus, for example, the decomposition of (glycylglycylglycine)Cu^{III}-O_2CH_3 does not yield CH_2O [208], i.e., it does not occur via reactions (86) or (82) followed by (83), though the ligand oxidation product was still not identified. Similarly, the complex (glycine)$_2Cu^{III}$-O_2CH_3 decomposes to yield equal amounts of CO_2 and CH_2O [208], proving that here also the ligand is oxidized.

Clearly oxidation of the ligands L, via such a mechanism, if it occurs in a biological system, causes deleterious effects.

(6) Via direct oxidation of substrates present in the solution: The transient complexes $L_mM^{(n+1)}$-O_2R are strong and potentially up to three-electron oxidizing agents. Indeed, recent studies suggest that complexes of this type act as intermediates in transition metal-catalyzed oxidations by $HO_2^\bullet$ radicals [216,217]. Such processes will cause deleterious processes in biological systems if the oxidized substrate has an important biological role or if the product of the oxidation is toxic.

(7) Via Fenton-like reactions, i.e., via

$$L_mM^{n+1}\text{-OOR} + M^nL_m + 2H_3O^+ \longrightarrow 2M^{n+1}L_m + RO^\bullet + 3H_2O \quad (88)$$

The transient complexes L_mM^{n+1}-OOR are peroxides and are therefore expected to decompose via Fenton-like reactions. Indeed, the complexes $((H_2O)_5Fe^{III}$-$O_2R)^{2+}$ [207] and the complex $L(H_2O)Co^{III}$-O_2CH_3 [209] decompose partially via this mechanism.

Decomposition via this mechanism produces $RO^\bullet$ radicals that have similar properties to the $^\bullet OH$ radicals and therefore may cause deleterious processes in biological systems.

10. CONCLUDING REMARKS

The purpose of this chapter is to point out that the role of transition metal complexes in the induction of biological deleterious processes is probably dual:

1. The transition metal complexes play a major role in the production of the ·OH radicals, or analogous oxidizing agents, via Fenton-like reactions and probably via catalysis of the Haber-Weiss reaction of superoxide radicals with hydrogen peroxide.

2. The transition metal complexes react preferentially with the secondary, aliphatic carbon-centered, and tertiary, alkyl-peroxyl, radicals formed in the biological systems. These reactions direct the radical-initiated processes to the biological sites where the transition metal complexes are located. Thus they are an alternative explanation to the site-specific effect.

The products of reaction of these radicals with the transition metal complexes, especially those resulting from the decomposition of the transient complexes $L_m M^{n+1}$-R and $L_m M^{n+1}$-OOR, are probably the cause of many biological deleterious effects initiated by radicals.

It should be pointed out that the idea that the reactions of R· and ROO· radicals with transition metal complexes in biological systems are a main source of the radical-induced deleterious processes is not commonly accepted. This hypothesis, outlined herein, has to be checked experimentally in biological systems — a complicated task.

ACKNOWLEDGMENTS

It is a pleasure to thank my many coworkers whose work is cited in this chapter. Without our many discussions and their experimental expertise this chapter could not have been written. Thanks are also due to Mrs. Irene Evens for her ongoing interest and support. Many of the studies described herein were made possible by the financial support of the Israel Science Foundation administered by the Israel Academy of Sciences and Humanities, by the Budgeting and Planning Committee of the Council of Higher Education, by the Israel Atomic Energy Commission, by the German-Israeli Foundation for Scientific Research and

Development (GIF), and by the United States-Israel Binational Science Foundation (BSF), Jerusalem, Israel.

REFERENCES

1. D. Harman, in *Free Radicals, Aging, and Degenerative Diseases* (J. E. Johnson, Jr., R. Walford, D. Harman, and J. Miquel, eds.), Alan R. Liss, New York, 1986, p. 3.
2. B. Halliwell and J. M. C. Gutteridge, *Mol. Aspects Med.*, 8, 89 (1985).
3. B. Halliwell, in *Active Oxygen in Biochemistry* (J. Selverstone Valentine, C. S. Foote, A. Greenberg, and J. L. Liebman, eds.), Blackie, London, 1995, p. 313.
4. G. Merenyi and J. Lind, *Chem. Res. Toxicol.*, 10, 1216 (1997).
5. A. Breccia, C. Rimondi, and G. E. Adams (eds.), *Advanced Topics on Radiosensitizers of Hypoxic Cells*, Plenum Press, New York, 1982.
6. Y. Tabata, *Pulse Radiolysis*, CRC Press, Boca-Raton, 1990.
7. B. Halliwell and J. M. C. Gutteridge, *Free Radicals in Biology and Medicine*, Clarendon Press, Oxford, 1985, pp. 33–36.
8. H. J. Fenton, *J. Chem. Soc.*, 65, 899 (1894).
9. F. Haber and J. J. Weiss, *Proc. Roy. Soc. London (Biol)*, A147, 332 (1934).
10. R. A. Sheldon and J. K. Kochi, *Metal-Catalyzed Oxidation of Organic Compounds*, Academic Press, New York, 1981.
11. W. C. Bray and M. H. Gorin, *J. Am. Chem. Soc.*, 54, 2124 (1934).
12. H. C. Sutton and C. C. Winterbourn, *Free Rad. Biol. Med.*, 6, 53 (1989).
13. S. Goldstein and G. Czapski, *Int. Rev. Exp. Pathol.*, 31, 133 (1990).
14. M. Masarwa, H. Cohen, D. Meyerstein, D. L. Hickman, A. Bakac, and J. H. Espenson, *J. Am. Chem. Soc.*, 110, 4293 (1988).
15. S. Goldstein, D. Meyerstein, and G. Czapski, *Free Rad. Biol. Med.*, 15, 435 (1993).
16. R. A. Marcus and N. Sutin, *Biochim. Biophys. Acta*, 811, 265 (1985).

17. H. Bamnolker, H. Cohen, and D. Meyerstein, *Free Rad. Res.*, *15*, 231 (1991).

18. S. Goldstein, D. Meyerstein, and G. Czapski, in *The Oxygen Paradox* (K. J. A. Davies and F. Ursini, eds.), Cleup University Press, Padova, Italy, 1995, p. 169.

19. E. Luzzatto, H. Cohen, C. Stockheim, K. Wieghardt, and D. Meyerstein, *Free Rad. Res.*, *23*, 453 (1995).

20. S. Udenfriend, C. T. Clark, J. Axelrod, and B. B. Brodie, *J. Biol. Chem.*, *208*, 731 (1954).

21. V. Ulrich and H. Staudinger, Oxygen reactions in model systems, in *Microsomes and Drug Oxidations*, Academic Press, New York, 1969, p. 199.

22. L. K. Folkes, L. P. Candeias, and P. Wardman, *Arch. Biochem. Biophys.*, *323*, 120 (1995).

23. S. Tofigh and K. Frenkel, *Free Rad. Biol. Med.*, *7*, 131 (1989).

24. G. Czapski and S. Goldstein, *Bioelectrochem. Bioenerg.*, *18*, 21 (1987).

25. Ref. 7, p. 113.

26. J. H. Jandl, Blood, in *Textbook of Hematology*, Little, Brown, Boston, 1987, p. 62.

27. J. A. Imlay and I. Fridovich, *J. Biol. Chem.*, *266*, 6957 (1991).

28. H. Nohl, in *Free Radicals, Aging, and Degenerative Diseases* (J. E. Johnson, Jr., R. Walford, D. Harman, and J. Miquel, eds.), Alan R. Liss, New York, 1986, p. 3, and references therein.

29. Ref. 7, p. 108.

30. Ref. 26, p. 458.

31. G. V. Buxton, C. L. Greenstock, W. P. Helman, and A. B. Ross, *J. Phys. Chem. Ref. Data*, *17*, 513 (1988).

32. C. von Sonntag, *The Chemical Basis of Radiation Biology*, Taylor and Francis, London, 1987, p. 394.

33. V. K. Sharma and B. H. J. Bielski, *Inorg. Chem.*, *30*, 4036 (1991).

34. S. Mahapatra, J. A. Halfen, and W. B. Tolman, *J. Am. Chem. Soc.*, *118*, 11575 (1996).

35. S. Mahapatra, V. G. Young, Jr., S. Kaderli, A. D. Zuberbühler, and W. B. Tolman, *Angew. Chem. Int. Ed. Engl.*, *36*, 130 (1997).

36. A. Sen, in *Applied Homogeneous Catalysis with Organometallic*

Compounds (B. Comils and W. A. Herrmann, eds.), VCH, New York, 1996, p. 1081.

37. J. T. Groves, *J. Chem. Ed.*, *62*, 928 (1985).

38. D. Mansuy and P. Battioni, in *Bioinorganic Catalysis* (J. Reedijk, ed.), Marcel Dekker, New York, 1993, p. 395.

39. R. W. Fischer and F. Rohrscheid, in *Applied Homogeneous Catalysis with Organometallic Compounds* (B. Comils and W. A. Herrmann, eds.), VCH, New York, 1996, p. 439.

40. D. Meyerstein, *Inorg. Chem.*, *10*, 2244 (1971).

41. J. Lati and D. Meyerstein, *Inorg. Chem.*, *11*, 2397 (1972).

42. J. Lati and D. Meyerstein, *J. Chem. Soc. Dalton Trans.*, 1105 (1978).

43. S. Goldstein, G. Czapski, H. Cohen, D. Meyerstein, and R. van Eldik, *Inorg. Chem.*, *33*, 3255 (1994).

44. E. R. Kantrowitz, M. Z. Hoffman, and J. F. Endicott, *J. Phys. Chem.*, *75*, 1914 (1971).

45. T. S. Roche and J. F. Endicott, *Inorg. Chem.*, *13*, 1575 (1974).

46. J. C. Scaiano, W. J. Leigh, and G. Ferraudi, *Can. J. Chem.*, *62*, 2355 (1984).

47. C. C. Hobbs, Jr., *Applied Homogeneous Catalysis with Organometallic Compounds* (B. Comils and W. A. Herrmann, eds.), VCH, New York, 1996, p. 521.

48. W. A. Pryor, *ACS Symp. Series*, *277*, 77 (1985).

49. R. L. Rusting, *Sci. Am.*, *267*, 86 (1992).

50. Ref. 7, p. 139.

51. J. M. Pratt, in *Metal Ions in Biological Systems*, Vol. 29 (H. Sigel and A. Sigel, eds.), Marcel Dekker, New York, 1993, p. 229.

52. J. Halpern, *Science*, *227*, 869 (1985).

53. J. K. Kochi, *Organometallic Mechanisms and Catalysis*, Academic Press, New York, 1978, p. 138.

54. P. Neta, R. E. Huie, and A. B. Ross, *J. Phys. Chem. Ref. Data*, *19*, 413 (1990).

55. P. Neta, J. Grodkowski, and A. B. Ross, *J. Phys. Chem. Ref. Data*, *25*, 709 (1996).

56. P. Huston, J. H. Espenson, and A. Bakac, *Inorg. Chem.*, *31*, 721 (1992).

57. C. Schoneich, M. Bonifacic, and K. D. Asmus, *Free Rad. Res. Commun.*, *6*, 393 (1989).

58. P. Huston, J. H. Espenson, and A. Bakac, *J. Am. Chem. Soc.*, *114*, 9510 (1992).

59. A. Marchaj, D. G. Kelley, A. Bakac, and J. H. Espenson, *J. Phys. Chem.*, *95*, 4440 (1991).

60. A. Bakac and J. H. Espenson, *J. Am. Chem. Soc.*, *108*, 719 (1986).

61. P. Wardman, *J. Phys. Chem. Ref. Data*, *18*, 1637 (1989).

62. H. Cohen and D. Meyerstein, *Inorg. Chem.*, *13*, 2434 (1974).

63. J. H. Espenson, A. Bakac, and J. H. Kim, *Inorg. Chem.*, *30*, 4830 (1991).

64. D. M. Stanbury, *Adv. Inorg. Chem.*, *33*, 69 (1989).

65. S. V. Jovanovic, I. Jankovic, and L. Josimovic, *J. Am. Chem. Soc.*, *114*, 9018 (1992) and T. N. Das, T. Dhanasekaran, Z. B. Alfassi, and P. Neta, *J. Phys. Chem.*, *A102*, 280 (1998).

66. B. H. Bielski, D. E. Cabelli, R. L. Arudi, and A. B. Ross, *J. Phys. Chem. Ref. Data*, *14*, 1041 (1985).

67. S. Goldstein, G. Czapski, and D. Meyerstein, *J. Am. Chem. Soc.*, *112*, 6489 (1990).

68. A. Meshulam, H. Cohen, R. van Eldik, and D. Meyerstein, *Inorg. Chem.*, *31*, 2151 (1992).

69. A. Meshulam, H. Cohen, and D. Meyerstein, *Inorg. Chim. Acta* *273*, 266 (1998).

70. G. Czapski, *Israel J. Chem.*, *24*, 29 (1984).

71. R. C. Lynch and I. Fridovich, *Biochim. Biophys. Acta*, *571*, 195 (1979).

72. M. Chevion, *Free Rad. Biol. Med.*, *5*, 27 (1988).

73. R. P. Herzberg and P. B. Dervan, *Biochemistry*, *23*, 2934 (1984).

74. S. D. Aust and B. C. White, *Adv. Free Rad. Biol. Med.*, *1*, 1 (1985).

75. E. R. Stadman, *Free Rad. Biol. Med.*, *9*, 315 (1990).

76. E. R. Stadman and C. N. Oliver, *J. Biol. Chem.*, *266*, 2005 (1991).

77. S. Goldstein and G. Czapski, *Free Rad. Biol. Med.*, *2*, 3 (1986).

78. H. Cohen and D. Meyerstein, *J. Am. Chem. Soc.*, *94*, 6944 (1972).

79. H. Cohen and D. Meyerstein, *J. Chem. Soc. Dalton Trans.*, 2559 (1977).

80. R. M. McHatton and J. H. Espenson, *Inorg. Chem.*, *22*, 784 (1983).

81. J. Grodkowski, P. Neta, C. J. Schlesener, and J. K. Kochi, *J. Phys. Chem.*, *89*, 4373 (1985).

82. A. Bakac, J. H. Espenson, J. Lovric, and M. Orhanovic, *Inorg. Chem.*, *26*, 4096 (1987).

83. A. Bakac, V. Butkovic, J. H. Espenson, J. Lovric, and M. Orhanovic, *Inorg. Chem.*, *28*, 4323 (1989).

84. K. Kusaba, H. Ogino, A. Bakac, and J. H. Espenson, *Inorg. Chem.*, *28*, 970 (1988).

85. A. Bakac, V. Butkovic, J. H. Espenson, R. Marcec, and M. Orhanovic, *Inorg. Chem.*, *25*, 341 (1986).

86. S. Steenken and P. Neta, *J. Am. Chem. Soc.*, *104*, 1244 (1982).

87. D. Brault, R. Santus, E. J. Land, and J. Swallow, *J. Phys. Chem.*, *88*, 5836 (1984).

88. M. Breitenkamp, A. Henglein, and J. Lilie, *Ber. Bunsen Ges. Phys. Chem.*, *80*, 973 (1976).

89. H. A. Schwarz and W. R. Dodson, *J. Phys. Chem.*, *93*, 409 (1989).

90. D. G. Kelley, J. H. Espenson, and A. Bakac, *Inorg. Chem.*, *29*, 4996 (1990).

91. A. Bakac, V. Butkovic, J. H. Espenson, R. Marcec, and M. Orhanovic, *Inorg. Chem.*, *26*, 3249 (1987).

92. D. Meyerstein and H. A. Schwarz, *J. Chem. Soc. Faraday Trans. I*, *84*, 2933 (1988).

93. M. Venturi, S. Emmi, P. G. Fuochi, and G. G. Mulazzani, *J. Phys. Chem.*, *84*, 2160 (1980).

94. M. Z. Hoffman, D. W. Kimmel, and M. G. Simic, *Inorg. Chem.*, *18*, 2479 (1979).

95. J. W. van Leeuwen, J. Tromp, and H. Nauta, *Biochim. Biophys. Acta*, *577*, 394 (1979).

96. W. Bors, J. Wachtveitl, and M. Saran, *Free Rad. Res. Commun.*, *6*, 251 (1989).

97. H. Cohen and D. Meyerstein, *J. Chem. Soc. Dalton Trans.*, 2447 (1975) and H. Cohen and D. Meyerstein, *J. Chem. Soc. Dalton Trans.*, 1976 (1976).

98. K. Wieghardt, H. Cohen, and D. Meyerstein, *Ber. Bunsen Ges. Phys. Chem.*, *82*, 985 (1978) and K. Wieghardt, H. Cohen, and

D. Meyerstein, *Angew. Chem.*, *17*, 608 (1978) and J. V. Beitz, J. R. Miller, H. Cohen, K. Wieghardt, and D. Meyerstein, *Inorg. Chem.*, *19*, 966 (1980).

99.   H. Cohen, M. Nutkovitch, K. Wieghardt, and D. Meyerstein, *J. Chem. Soc. Dalton Trans.*, 943 (1982).

100. A. Bakac, V. Butkovic, J. H. Espenson, R. Marcec, and M. Orhanovic, *Inorg. Chem.*, *30*, 481 (1991).

101. K. D. Whitburn, M. Z. Hoffman, N. V. Brezniak, and M. G. Simic, *Inorg. Chem.*, *25*, 3037 (1986).

102. A. Bakac, V. Butkovic, J. H. Espenson, R. Marcec, and M. Orhanovic, *Inorg. Chem.*, *25*, 2562 (1986).

103. K. Tsukahara and R. G. Wilkins, *Inorg. Chem.*, *28*, 1605 (1989).

104. M. G. Simic, M. Z. Hoffman, and N. V. Brezniak, *J. Am. Chem. Soc.*, *99*, 2166 (1977).

105. S. S. Isied, C. Kuehn, and G. Worosila, *J. Am. Chem. Soc.*, *106*, 1722 (1984).

106. M. P. Jackman, J. McGinnis, R. Powls, G. A. Salmon, and A. G. Sykes, *J. Am. Chem. Soc.*, *110*, 5880 (1988).

107. P. Osvath, G. A. Salmon, and A. G. Sykes, *J. Am. Chem. Soc.*, *110*, 7114 (1988).

108. C. Huuee-Levin, M. Gardes-Albert, K. Benzineb, C. Ferradini, and B. Hickel, *Biochemistry*, *28*, 9848 (1989).

109. O. Farver and I. Pecht, *Inorg. Chem.*, *29*, 4855 (1990).

110. O. Farver and I Pecht, *FEBS Lett.*, *244*, 379 (1989).

111. K. Govindaraju, G. A. Salmon, N. P. Tomkinson, and A. G. Sykes, *J. Chem. Soc. Chem. Commun.*, 1003 (1990).

112. K. Govindaraju, H. E. M. Christensen, E. Lloyd, M. Olsen, G. A. Salmon, N. P. Tomkinson, and A. G. Sykes, *Inorg. Chem.*, *32*, 40 (1993).

113. C. Fenwick, S. Marmor, K. Govindaraju, A. M. English, J. F. Wishart, and J. Sun, *J. Am. Chem. Soc.*, *116*, 3169 (1994).

114. D. Golub, H. Cohen, and D. Meyerstein, *J. Chem. Soc. Dalton Trans.*, 641 (1985).

115. A. Rotman, H. Cohen, and D. Meyerstein, *Inorg. Chem.*, *24*, 4158 (1985).

116. R. van Eldik, H. Cohen, and D. Meyerstein, *Inorg. Chem.*, *33*, 1566 (1994).

117. D. Behar, A. Samuni, and R. W. Fessenden, *J. Phys. Chem.*, *77*, 2055 (1973).

118. J. D. Ellis, M. Green, A. G. Sykes, G. V. Buxton, and R. M. Sellers, *J. Chem. Soc. Dalton Trans.*, 1724 (1973).

119. J. T. Chen and J. H. Espenson, *Inorg. Chem.*, *22*, 1651 (1983).

120. W. A. Mulac, H. Cohen, and D. Meyerstein, *Inorg. Chem.*, *21*, 4016 (1982).

121. J. H. Espenson, in *Advances in Organic and Bioinorganic Mechanisms*, Vol. 1 (A. G. Sykes, ed.), 1982, p. 1.

122. J. H. Espenson, P. Connolly, D. Meyerstein, and H. Cohen, *Inorg. Chem.*, *22*, 1009 (1983).

123. H. Cohen, D. Meyerstein, A. Shusterman, and M. Weiss, *J. Chem. Soc. Chem. Commun.*, 1424 (1985).

124. H. Cohen, W. Gaede, A. Gerhard, D. Meyerstein, and R. van Eldik, *Inorg. Chem.*, *31*, 3805 (1992).

125. A. Bakac and J. H. Espenson, *Inorg. Chem.*, *28*, 3901 (1989).

126. H. Cohen, A. Feldman, R. Ish-Shalom, and D. Meyerstein, *J. Am. Chem. Soc.*, *113*, 5292 (1991).

127. R. van Eldik, W. Gaede, H. Cohen, and D. Meyerstein, *Inorg. Chem.*, *31*, 3695 (1992).

128. J. H. Espenson, *Acc. Chem. Res.*, *25*, 222 (1992).

129. D. M. Guldi, P. Neta, and P. Hambright, *J. Chem. Soc. Faraday Trans.*, *88*, 2337 (1992).

130. H. Cohen and D. Meyerstein, *Inorg. Chem.*, *27*, 3429 (1988).

131. K. M. Morehouse and P. Neta, *J. Phys. Chem.*, *88*, 1575 (1984).

132. D. M. Guldi, M. Kumar, P. Neta, and P. Hambright, *J. Phys. Chem.*, *96*, 9576 (1992).

133. Y. Sorek, H. Cohen, and D. Meyerstein, *J. Chem. Soc. Faraday Trans. I*, *81*, 233 (1985).

134. Y. Sorek, H. Cohen, and D. Meyerstein, *J. Chem. Soc. Faraday Trans. I*, *82*, 3431 (1986).

135. G. Czapsky, S. Goldstein, H. Cohen, and D. Meyerstein, *J. Am. Chem. Soc.*, *110*, 3903 (1988).

136. D. Brault and P. Neta, *J. Am. Chem. Soc.*, *103*, 2705 (1981).

137. D. Brault and P. Neta, *J. Phys. Chem.*, *86*, 3405 (1982).

138. D. Brault and P. Neta, *J. Phys. Chem.*, *91*, 4156 (1987).

139. D. E. Cabelli, J. D. Rush, M. J. Thomas, and B. H. J. Bielski, *J. Phys. Chem.*, *93*, 3579 (1989).

140. H. Elroi and D. Meyerstein, *J. Am. Chem. Soc.*, *100*, 5540 (1978).

141. D. Meyerstein, *Pure Appl. Chem.*, *61*, 885 (1989).

142. Y. Sorek, H. Cohen, and D. Meyerstein, *J. Chem. Soc. Faraday Trans. I*, *85*, 1169 (1989).

143. A. Sauer, H. Cohen, and D. Meyerstein, *Inorg. Chem.*, *28*, 2511 (1989).

144. R. van Eldik, H. Cohen, and D. Meyerstein, *Angew. Chem.*, *30*, 1158 (1991).

145. R. Blackburn, M. Kyaw, G. O. Phillips, and A. J. Swallow, *J. Chem. Soc. Faraday Trans. I*, *71*, 2277 (1975).

146. R. C. McHatton, J. H. Espenson, and A. Bakac, *J. Am. Chem. Soc.*, *108*, 5885 (1986).

147. A. Bakac and J. H. Espenson, *Inorg. Chem.*, *28*, 4319 (1989).

148. S. Lee, J. H. Espenson, and A. Bakac, *Inorg. Chem.*, *29*, 3442 (1990).

149. S. Baral and P. Neta, *J. Phys. Chem.*, *87*, 1502 (1983).

150. M. Kumar, E. Natarajan, and P. Neta, *J. Phys. Chem.*, *98*, 8024 (1994).

151. M. Kelm, J. Lilie, A. Henglein, and E. Janata, *J. Phys. Chem.*, *78*, 882 (1974).

152. M. S. Ram, A. Bakac, and J. H. Espenson, *Inorg. Chem.*, *25*, 3267 (1986).

153. D. M. Guldi, P. Neta, P. Hambright, and R. Rahimi, *Inorg. Chem.*, *31*, 4849 (1992).

154. D. G. Kelley, A. Marchaj, A. Bakac, and J. H. Espenson, *J. Am. Chem. Soc.*, *113*, 7583 (1991).

155. A. Sauer, H. Cohen, and D. Meyerstein, *Inorg. Chem.*, *27*, 4578 (1988).

156. R. van Eldik, H. Cohen, A. Meshulam, and D. Meyerstein, *Inorg. Chem.*, *29*, 4156 (1990).

157. M. Freiberg, W. A. Mulac, K. H. Schmidt, and D. Meyerstein, *J. Chem. Soc. Faraday Soc. I*, *76*, 1838 (1980).

158. H. Cohen and D. Meyerstein, *Inorg. Chem.*, *25*, 1505 (1986).

159. H. Cohen and D. Meyerstein, *Inorg. Chem.*, *26*, 2342 (1987).

160. S. Goldstein, G. Czapski, H. Cohen, and D. Meyerstein, *Inorg. Chem.*, *27*, 4130 (1988).

161. M. Masarwa, H. Cohen, and D. Meyerstein, *Inorg. Chem.*, *30*, 1849 (1991).

162. S. Goldstein, G. Czapski, H. Cohn, and D. Meyerstein, *Inorg. Chim. Acta*, *192*, 87 (1992).

163. S. Goldstein, G. Czapski, H. Cohen, D. Meyerstein, J. K. Cho, and S. Shaik, *Inorg. Chem.*, *31*, 798 (1992).

164. N. Navon, G. Golub, H. Cohen, and D. Meyerstein, *Organometallics*, *14*, 5670 (1995).

165. A. Szulc, D. Meyerstein, and H. Cohen, *Inorg. Chim. Acta*, *270*, 440 (1998).

166. G. V. Buxton and J. C. Green, *J. Chem. Soc. Faraday Trans. I*, *74*, 697 (1978).

167. M. Freiberg and D. Meyerstein, *J. Chem. Soc. Chem. Commun.*, 127 (1977).

168. W. A. Mulac and D. Meyerstein, *J. Chem. Soc. Chem. Commun.*, 893 (1979).

169. M. Freiberg and D. Meyerstein, *J. Chem. Soc. Faraday Soc. I*, *76*, 1825 (1980).

170. L. J. Kirschenbaum and D. Meyerstein, *Inorg. Chem.*, *19*, 1373 (1980).

171. M. Masarwa, H. Cohen, and D. Meyerstein, *Inorg. Chem.*, *25*, 4897 (1986).

172. M. Masarwa, H. Cohen, R. Glaser, and D. Meyerstein, *Inorg. Chem.*, *29*, 5031 (1990).

173. M. Masarwa, H. Cohen, J. Saar, and D. Meyerstein, *Israel J. Chem.*, *30*, 361 (1990).

174. S. Goldstein, G. Czapski, H. Cohen, and D. Meyerstein, *Inorg. Chem.*, *31*, 2439 (1992).

175. A. Bakac and J. H. Espenson, *J. Am. Chem. Soc.*, *108*, 713 (1986).

176. H. Cohen and D. Meyerstein, *Inorg. Chem.*, *23*, 84 (1984).

177. W. Gaede, R. van Eldik, H. Cohen, and D. Meyerstein, *Inorg. Chem.*, *32*, 1997 (1993).

178. G. Ferraudi, *Inorg. Chem.*, *17*, 2506 (1978).

179. C. Mansano-Weiss, H. Cohen, and D. Meyerstein, *J. Inorg. Biochem.*, *47*, 54 (1992).

180. D. Meyerstein, *Inorg. Chem.*, *10*, 2244 (1971).

181. J. Lati and D. Meyerstein, *J. Chem. Soc. Dalton Trans.*, 1105 (1978).

182. S. Goldstein, G. Czapski, H. Cohen, D. Meyerstein, and R. van Eldik, *Inorg. Chem.*, *33*, 3255 (1994).

183. D. Margerum, W. M. Scheper, M. R. McDonald, F. C. Fredericks, L. Wang, and H. D. Lee, in *Bioinorganic Chemistry of Copper* (K. D. Karlin and Z. Tyeklar, eds.), Chapman and Hall, New York, 1993, p. 213.

184. A. Mayouf, H. Lemmetyinen, I. Sychttchikowa, and J. Koskikallio, *Int. J. Chem. Kinet.*, *24*, 579 (1992).

185. J. Halpern, *Pure Appl. Chem.*, *51*, 2171 (1979).

186. J. Halpern, *Pure Appl. Chem.*, *55*, 1059 (1983).

187. R. G. Finke, D. A. Schiraldi, and B. Mayer, *Coord. Chem. Rev.*, *54*, 1 (1984).

188. D. C. Woska and B. B. Wayland, *Inorg. Chim. Acta*, *270*, 197 (1998).

189. H. Cohen and D. Meyerstein, *J. Chem. Soc. Faraday Trans. I*, *84*, 4157 (1988).

190. Y. Sorek, H. Cohen, W. A. Mulac, K. H. Schmidt, and D. Meyerstein, *Inorg. Chem.*, *22*, 3040 (1983).

191. W. A. Mulac and D. Meyerstein, *J. Am. Chem. Soc.*, *104*, 4124 (1982).

192. H. Cohen and D. Meyerstein, *Angew. Chem. Int. Ed.*, *34*, 779 (1985).

193. S. Goldstein, G. Czapski, H. Cohen, D. Meyerstein, and S. Shaik, *J. Chem. Soc. Faraday Trans.*, *89*, 4045 (1993).

194. S. Goldstein, G. Czapski, H. Cohen, and D. Meyerstein, *Inorg. Chim. Acta*, *192*, 87 (1992).

195. H. Cohen, R. van Eldik, W. Gaede, A. Gerhard, S. Goldstein, G. Czapski, and D. Meyerstein, *Inorg. Chim. Acta*, *227*, 57 (1994).

196. S. Goldstein, G. Czapski, H. Cohen, and D. Meyerstein, *Free Rad. Biol. Med.*, *9*, 371 (1990).

197. S. Goldstein, G. Czapski, H. Cohen, and D. Meyerstein, *Free Rad. Biol. Med.*, *17*, 11 (1994).

198. P. Huston, J. H. Espenson, and A. Bakac, *Inorg. Chem.*, *30*, 4826 (1991).

199. E. Luzzatto, H. Cohen, and D. Meyerstein, *Abstracts of the International Conference on Bioinorganic Chemistry*, La Jolla, CA (1993), *J. Inorg. Biochem.*, *51*, 225 (1993).

200. S. K. Malik, W. Schmidt, and L. O. Spreer, *Inorg. Chem.*, *13*, 2986 (1974).

201. M. J. Akhtar and L. O. Spreer, *Inorg. Chem.*, *18*, 3327 (1979).

202. L. G. Marzilli, in *Bioinorganic Catalysis* (J. Reedijk, ed.), Marcel Dekker, New York, 1993, p. 227.

203. C. Arroldi, A. Citero, and F. Miniiski, *J. Chem. Soc. Perkin Trans. II*, 531 (1983).

204. D. K. Lavallee, *The Chemistry and Biochemistry of N-Substituted Porphyrins*, VCH, New York, 1987.

205. D. Brault and P. Neta, *J. Phys. Chem.*, *88*, 2857 (1984).

206. D. M. Guldi, J. Field, J. Grodkowski, P. Neta, and E. Vogel, *J. Phys. Chem.*, *100*, 13609 (1996).

207. C. Mansano-Weiss, H. Cohen, and D. Meyerstein, *J. Inorg. Biochem.*, *59*, 393 (1995).

208. C. Mansano-Weiss, H. Cohen, and D. Meyerstein, International Meeting on Reaction Mechanism, Canterbury, England, 1996.

209. E. Solomon, H. Cohen, and D. Meyerstein, 24th IUPAC Conference on Solution Chemistry, Lisbon, Portugal, 1995.

210. G. I. Khaikin, Z. B. Alfassi, R. E. Huie, and P. Neta, *J. Phys. Chem.*, *100*, 7072 (1996).

211. J. O. Edwards and R. Curci, in *Catalytic Oxidations with Hydrogen Peroxide as Oxidant* (G. Strukul, ed.), Kluwer Academic, Netherlands, 1992, p. 97.

212. B. Meunier, in Ref. 211, p. 153.

213. G. Gtrukul, in Ref. 211, p. 177.

214. D. T. Sawyer, *Oxygen Chemistry*, Oxford University Press, New York, 1991.

215. R. A. Sheldon and J. K. Kochi, *Metal-Catalyzed Oxidations of Organic Compounds*, Academic Press, New York, 1981.

216. B. H. J. Bielski, *Trans. Roy. Soc. London*, *B311*, 473 (1985).

217. D. E. Cabelli, *Free Rad. Biol. Med.*, *6*, 171 (1989).

3

Free Radicals as a Result of Dioxygen Metabolism

Bruce P. Branchaud

Department of Chemistry, University of Oregon,
Eugene, OR 97403-1253, USA

1. INTRODUCTION

The controlled combustion of food using dioxygen provides the energy necessary to support life. Carbohydrates, proteins, and fats all have a high energy content (caloric content), indicating that their combustion is a very exothermic process. Nevertheless, in the absence of a catalyst, the powerful oxidizing agent dioxygen is effectively inert to reactions with organic biomolecules. This is fortunate because it allows energy to be captured and stored in biomolecules through photosynthesis. The stored energy can then be used by living things in a controlled fashion through biological enzyme-catalyzed combustion of food biomolecules.

Dioxygen is kinetically inert in the absence of a catalyst because dioxygen exists in its ground state as a triplet diradical, $^3O_2$, with two electrons having unpaired spins. Essentially all organic biomolecules exist as ground state singlets, with all electrons spin-paired. The direct reaction between ground state triplet dioxygen and an organic molecule in the singlet ground state must give a product in a high-energy triplet excited state in order to conserve the spin in the system. This fact makes most direct reactions of ground state triplet dioxygen with organic molecules "uphill" thermodynamically and thus unfavorable under normal conditions.

Dioxygen becomes very reactive when it participates in one-electron reactions. These include radical chain reactions, such as autoxidation of lipids and other organic molecules (Secs. 3.2 and 4.1). These also include electron transfer reactions. One-electron reduction of dioxygen generates superoxide radical anion with an unpaired electron on oxygen. The unpaired electron can relax its spin, thus avoiding the singlet-triplet problem found in the direct reaction of ground state triplet dioxygen with organic biomolecules. One-electron reduction of dioxygen to produce superoxide and further reduction reactions produce highly reactive radicals and radical-derived products known collectively as *reactive oxygen species* (ROS). ROS are implicated in numerous degenerative processes and diseases, including aging [1–5], cancer [6], inflammation [7], amyotrophic lateral sclerosis, Down syndrome [8], and Alzheimer's disease [9,10]. Some of these topics are reviewed in other chapters in this volume. General discussions can be found in several recent monographs and reviews on ROS [11–17].

Dioxygen metabolism provides nature with a paradoxical dilemma. Biological catalysts of dioxygen metabolism must use one-electron processes to overcome the singlet-triplet problem discussed above. This means that ROS must be generated, albeit usually very transiently and in the protected environment of enzyme active sites or at redox centers in multienzyme complexes. Apparently none of these systems are infallible. For example, it has been estimated that 1–6% of the dioxygen that enters the mitochondrial respiratory chain is not completely reduced to water but instead escapes as one-electron-reduced superoxide radical anion [18–20].

The production of superoxide is the starting point for a series of reactions that lead to the formation of ROS. The common ROS include superoxide radical anion ($O_2^{\cdot-}$), hydrogen peroxide (H_2O_2), alkyl hydroperoxides (ROOH), peroxyl radicals (ROO$^\cdot$), alkoxyl radicals (RO$^\cdot$), and hydroxyl radical (HO$^\cdot$). Hydrogen peroxide is produced by monooxygenase enzymes and by the disproportionation of superoxide by the enzyme superoxide dismutase (reviewed in Chapter 5 in this volume). Alkyl hydroperoxides are produced by radical chain autoxidation of organic molecules, especially lipids (Secs. 3.2 and 4.1). Hydroxyl radical and alkoxyl radicals are produced by the Fenton reaction and Fenton-type chemistry (Secs. 3.3 and 5). When ROS get into the general milieu of biological systems, reactions can occur to cause degenerative and sometimes irreparable damage. Thus the redox processes essential for life also produce the ROS that are harmful to life.

The number of original research articles, review articles, book chapters, and monographs published annually on ROS is staggering. The experienced reader in this field might ask: Do we really need another review on ROS? This chapter attempts to be different from other recent reviews by focusing on the chemical aspects of ROS rather than the biochemical and biomedical aspects, which have been covered extensively elsewhere. A treatment of the fundamental chemistry should provide a framework for a better understanding of ROS.

2. REDOX CHEMISTRY, ACID-BASE CHEMISTRY, AND BOND ENERGIES OF DIOXYGEN AND REACTIVE OXYGEN SPECIES

This section will highlight some key facts necessary to understand the radical reactions of dioxygen and ROS. Readers interested in greater detail on pH, solvent, and other effects should consult the recent monograph by Sawyer [21].

Figure 1 shows the products that can be formed from one-electron redox reactions of dioxygen [21,22]. Based on that information, the major species that will exist under physiological conditions at pH $\sim$7 are dioxygen ($^3O_2$), superoxide radical anion ($O_2^{\bullet-}$), hydrogen peroxide (HOOH), hydroxyl radical (HO$^{\bullet}$), and neutral water (H_2O).

Figure 2 shows the relevant redox reactions that interconvert these species, along with the standard reduction potentials versus the normal hydrogen electrode at pH 7 [21]. Based on those data, it can be seen that the most difficult reduction is the first one, i.e., the conversion of dioxygen to superoxide radical anion. If a reductant is available that is strong enough to perform that step, then it should be possible to form all of the other ROS shown in Fig. 2.

Table 1 shows selected bond dissociation energies for oxygen-containing molecules and related molecules [21,23]. Some important observations relevant to this article can be made using the data in Table 1. First, oxygen-oxygen bonds are weaker than analogous carbon-oxygen or carbon-carbon bonds. Reactions that break oxygen-oxygen bonds and form carbon-oxygen bonds should be exothermic. This helps explain why oxygen is such a good oxidant. Second, oxygen-hydrogen

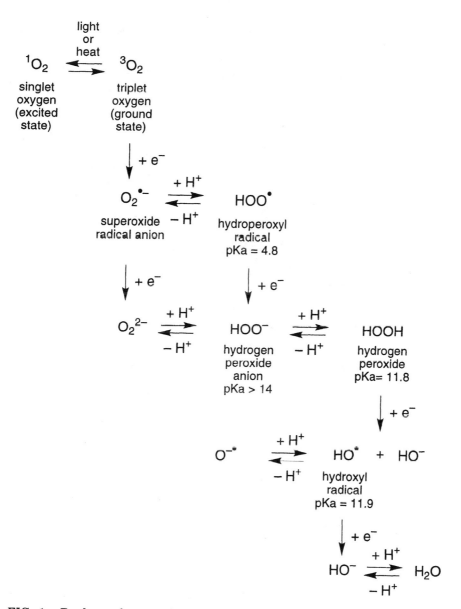

FIG. 1. Products that can be formed from one-electron redox reactions of dioxygen.

$$E^0 \text{ (volts)}$$
vs NHE at pH 7

$$^3O_2 \xrightarrow{\; + e^- \;} O_2^{\bullet-} \qquad\qquad -0.33$$

$$O_2^{\bullet-} \xrightarrow{\; + e^-, + 2\,H^+ \;} HOOH \qquad\qquad +0.89$$

$$HOOH \xrightarrow{\; + e^-, + H^+ \;} HO^\bullet + H_2O \qquad +0.38$$

$$HO^\bullet \xrightarrow{\; + e^-, + H^+ \;} H_2O \qquad\qquad +2.31$$

FIG. 2. Reduction potential versus the normal hydrogen electrode (NHE) at pH 7 for dioxygen and major reactive oxygen species at that pH.

TABLE 1

Bond Dissociation Energies (D^0_{298}) (kcal/mol)
for Selected Oxygen-Containing Molecules
and Related Molecules

Bond	D^0_{298}	Bond	D^0_{298}
$O{=}O$	119	$H{-}OH$	119
$H_2C{=}O$	175	$H{-}CH_3$	105
$H_2C{=}CH_2$	163		
		$H{-}OCH_3$	104
$HO{-}OH$	51	$H{-}CH_2CH_3$	101
$H_3C{-}OH$	91		
$H_3C{-}CH_3$	88	$H{-}OCH_2CH_3$	105
		$H{-}CH_2CH_2CH_3$	100
$H_3CO{-}OCH_3$	38		
$t\text{-}BuO{-}OBu\text{-}t$	38	$H{-}OC(CH_3)_3$	105
$H_3C{-}OCH_3$	80	$H{-}CH_2C(CH_3)_3$	100
$H_3C{-}CH_2CH_3$	85		
		$H{-}OOH$	88
		$H{-}OOCH_3$	88
		$H{-}OOBu\text{-}t$	89

bonds are stronger than analogous carbon-hydrogen bonds. Reactions that break carbon-hydrogen bonds and form oxygen-hydrogen bonds should be exothermic. This also helps explain why oxygen is such a good oxidant. Third, oxygen-hydrogen bonds in peroxides are much weaker than other types of oxygen-hydrogen bonds and the oxygen-hydrogen bonds in peroxides are weaker than many types of carbon-hydrogen bonds. This point will be important in explaining the reactivity and selectivity in autoxidation reactions (Secs. 3.2 and 4.1).

The data presented in this section can be interpreted to lead to two significant general conclusions. First, it is easy to form radicals and other ROS by one-electron reduction of dioxygen and species derived from it. Second, bond energy considerations indicate that many reactions of ROS with organic molecules should be thermodynamically favorable.

3. MAJOR RADICAL REACTIONS OF DIOXYGEN AND REACTIVE OXYGEN SPECIES

3.1. Hydrogen Atom Abstraction Reactions Using Oxy Radicals

Table 2 shows the calculated enthalpy changes for the reactions of superoxide radical anion ($O_2^{\bullet -}$), hydroperoxyl radical (HOO$^{\bullet}$), peroxyl radical (ROO$^{\bullet}$), alkoxyl radical (RO$^{\bullet}$), and hydroxyl radical (HO$^{\bullet}$) with the C-H bonds of some common organic molecules that have been chosen to represent the common types of C-H bonds in biomolecules [21,23]. Analysis of the data in the table brings out several important points.

Resonance delocalization greatly stabilizes radicals, so that cleavage of C-H bonds that generate resonance-stabilized radicals will be easier than the cleavage of C-H bonds that do not generate resonance-stabilized radicals. For example, the H-CH$_2$CH$_2$CH$_3$ bond in propane is quite strong with a bond dissociation energy of 100 kcal/mol compared to the much weaker H$-$CH$_2$CH$=$CH$_2$ in propene with a bond dissociation energy of 86 kcal/mol. The H$-$CH(CH$=$CH$_2$)$_2$ bond in 1,4-pentadiene is exceptionally weak for a C-H bond, with a bond dissociation energy of only 76 kcal/mol, because cleavage of the C-H bond produces a radical that is stabilized by conjugation with two double bonds.

Superoxide radical anion is not reactive with C-H bonds. All reac-

TABLE 2

Enthalpy Change (ΔH) (kcal/mol) for the Reaction
of Common Types of C–H Bonds of Oxy Radicals
R–H + oxy radical ⟶ R• + oxy–H

Bond	D^0_{298} (kcal/mol)	ΔH $O_2^{•-}$	ΔH HOO•	ΔH ROO•	ΔH RO•	ΔH HO•
$H-C_6H_5$	111	+48	+23	+22	+7	−8
$H-CH=CH_2$	111	+48	+23	+22	+7	−8
$H-CH_3$	105	+42	+17	+16	+1	−14
$H-CH_2CH_2CH_3$	100	+37	+12	+11	−4	−19
$H-CH_2COCH_3$	98	+35	+10	+9	−6	−21
$H-C(CH_3)_3$	96	+33	+8	+7	−8	−23
$H-CH_2OCH_3$	93	+30	+5	+4	−11	−26
$H-C(CH_3)_2OH$	91	+28	+3	+2	−13	−28
$H-CH_2C_6H_5$	88	+25	0	−1	−16	−31
$H-COCH_2CH_3$	87	+24	−1	−2	−17	−32
$H-CH_2CH=CH_2$	86	+23	−2	−3	−18	−33
$H-CH(C_6H_5)_2$	81	+18	−7	−8	−23	−38
$H-C(CH_3)_2CH=CH_2$	77	+14	−11	−12	−27	−42
$H-CH(CH=CH_2)_2$	76	+13	−12	−13	−28	−43

Note: D^0_{298} in kcal/mol used for calculations are $H-OO^- = 63$, $H-OOH = 88$, $H-OOR = 89$, $H-OR = 104$, and $H-OH = 119$.

tions of superoxide with C-H bonds are endothermic (positive ΔH for reaction) and thus are thermodynamically "uphill" and unfavorable. This may seem surprising at first because superoxide is regarded as a very potent member of the class of ROS. The important point here is that superoxide does not cause oxidative damage to biomolecules by direct reaction with them but through electron transfer and radical-radical reactions, which produce other compounds that can cause oxidative damage to biomolecules.

Hydroperoxyl radical (HOO•) and related peroxyl radicals (ROO•) have similar reactivities. These radicals do not react with strong C-H bonds because the ΔH is very positive for those reactions. Instead they will react with weak C-H bonds that produce resonance-stabilized radicals in exothermic (negative ΔH) reactions that are thermodynamically "downhill" and favorable. For example, the reactions of $H-CH(CH=CH_2)_2$ with $HOO^\bullet$ is exothermic by 12 kcal/mol and with $ROO^\bullet$ the reaction is exothermic by 13 kcal/mol. The data in Table 2 on $HOO^\bullet$ and $ROO^\bullet$ explain the selectivity seen in lipid peroxidation by autoxidation (Secs. 3.2 and 4.1).

Hydroxyl radical ($HO^\bullet$) is the most reactive of all oxy radicals. The reaction of $HO^\bullet$ with any C-H bond found in biomolecules is exothermic due to the formation of the exceptionally strong H-OH bond, with a bond dissociation energy of 119 kcal/mol. The hydroxyl radical is capable of reacting with any C-H bond that it encounters, to abstract a hydrogen atom and form a radical on the molecule that it attacked: $HO^\bullet$ + R-H $\longrightarrow$ H_2O + $R^\bullet$.

3.2. Radical Chain Autoxidation of Carbon-Hydrogen Bonds by Dioxygen

Autoxidation is the oxidation of carbon-hydrogen bonds in organic molecules by dioxygen. Autoxidation causes considerable degradation in biological systems and in anything that contains oxidizable carbon-hydrogen bonds, i.e., in almost all things made of organic compounds. A general mechanism for radical chain autoxidation of carbon-hydrogen bonds is shown in Fig. 3. Initiation can be caused by a variety of radical species that are produced either adventitiously in biological systems or from intentionally added radical precursors in autoxidation reactions in vitro. Chain propagation will be efficient if the R-H bond is sufficiently weak that the $ROO^\bullet$ can break the R-H bond (more will be said about the energetics of this type of reaction in Sec. 4.1). The other chain propagation step, the reaction of $R^\bullet$ with O_2, will be facile for all common $R^\bullet$ because it is essentially a very efficient radical-radical combination reaction of $R^\bullet$ with the triplet radical dioxygen ground state $^\bullet O\text{-}O^\bullet$ to form a C-O bond and produce $ROO^\bullet$.

3.3. Fenton and Fenton-like Reactions

In 1894, Fenton reported that H_2O_2 was a strong oxidant in the presence of Fe(II) [24]. In 1934, Haber and Weiss proposed that the reaction of Fe(II) with H_2O_2 produced the hydroxyl radical, $HO^\bullet$, as shown in

Initiation

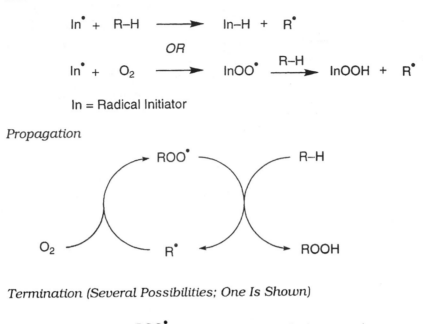

In = Radical Initiator

Propagation

Termination (Several Possibilities; One Is Shown)

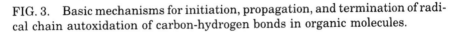

FIG. 3.   Basic mechanisms for initiation, propagation, and termination of radical chain autoxidation of carbon-hydrogen bonds in organic molecules.

Fig. 4. Several low-valent metals can also generate a strong oxidant through metal-mediated reduction of H_2O_2, including Cu(I) [25]. The Fenton reaction alone might not be very important in biological systems but the Haber-Weiss reaction, in which superoxide is used to reduce Fe(III) to Fe(II), allows a single metal center to undergo numerous catalytic cycles of reaction. The overall stoichiometry of the Haber-Weiss reaction is shown in Fig. 4, in which iron is used as a catalyst for the reduction of H_2O_2 by $O_2^{\bullet-}$ to produce O_2, HO^-, and $HO^{\bullet}$.

    For a long time the hydroxyl radical was believed to be the strong oxidant produced in the Fenton reaction. Over the past decade or so there has been significant debate on this point. The key issues are

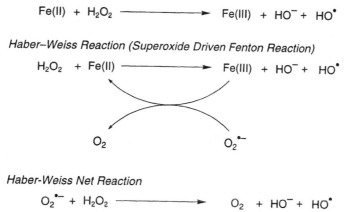

Fenton Reaction

$$Fe(II) + H_2O_2 \longrightarrow Fe(III) + HO^- + HO^{\bullet}$$

Haber–Weiss Reaction (Superoxide Driven Fenton Reaction)

$$H_2O_2 + Fe(II) \longrightarrow Fe(III) + HO^- + HO^{\bullet}$$

$$O_2 \qquad\qquad O_2^{\bullet-}$$

Haber-Weiss Net Reaction

$$O_2^{\bullet-} + H_2O_2 \longrightarrow O_2 + HO^- + HO^{\bullet}$$

FIG. 4.   Basic mechanisms for the Fenton reaction and the Haber-Weiss reaction.

illustrated and summarized in Fig. 5. The classic Fenton reaction, shown at the left of Fig. 5, uses Fe(II) only as an electron donor to H_2O_2. Such a reaction is an outer-sphere electron transfer reaction, with no direct bonding interactions between electron donor and acceptor [26]. Other possible inner-sphere electron transfer mechanisms do involve direct bonding between iron and hydrogen peroxide, shown at the right of Fig. 5. Direct interaction of Fe(II) with HOO^- as a ligand could produce Fe(II)OOH. The Fe(II)OOH could react further to generate hydroxyl radical or to generate Fe(IV)=O. The central question is what is the principal oxidant under a given set of conditions.

Hydroxyl radical is produced from water by high-energy ionizing radiation. The properties and reactions of hydroxyl radical produced by ionizing radiation have been studied extensively and are well known [25].

Hydroxyl radical is also known to be produced by the reaction of Fe(II) with H_2O_2 under strongly acidic conditions [27]. The kinetics of the reaction under acidic conditions are purely bimolecular, first order in Fe(II) and H_2O_2 [28], and are consistent with the classic outer sphere mechanism for the Fenton reaction.

Many reactions between Fe(II) and H_2O_2 under neutral conditions

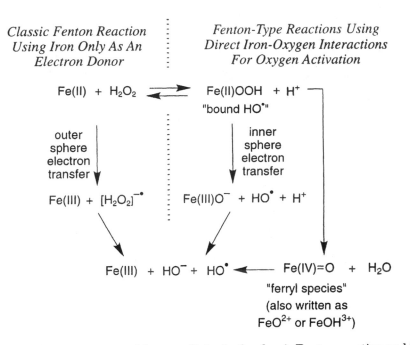

FIG. 5. Basic reactions and intermediates in the classic Fenton reaction and in metal-centered Fenton-like reactions.

have been interpreted as standard Fenton reactions, generating hydroxyl radical as the active oxidant [29]. Other more recent work has come to different conclusions [27,30–33].

Under neutral conditions, H_2O_2 (pK_a 11.8) is partially ionized. The HOO^- can add to Fe(II) to produce Fe(II)OOH. It has been argued that the strong Lewis basicity of HOO^- drives its complexation with Fe(II) [30]. The Fe(II)OOH has been called "bound $HO^\bullet$" and has been proposed to react directly with oxidizable substrates to generate products that are the same as or similar to those that would be generated by the reaction of free $HO^\bullet$ [30]. Differences in product distributions and kinetic isotope effects have been interpreted to distinguish between "bound $HO^\bullet$" (Fe(II)OOH or a product derived from it) and "free $HO^\bullet$" [31].

Although the "bound $HO^\bullet$" model using Fe(II)OOH as the active oxidant under neutral conditions has one proponent, there are several advocates of Fe(IV)=O (ferryl radical, FeO^{2+}, or $FeOH^{3+}$). The issue of such species was raised and discounted in a review of the first extensive mechanistic work on the Fenton reaction, work that concluded that the Fenton reaction under neutral conditions generated hydroxyl radical [29].

It was reported in 1986 that scavenging of the oxidant formed by the reaction of Fe(II)EDTA with H_2O_2 showed a different pattern of scavenging efficiencies than would be expected for $HO^{\bullet}$ [32]. These data were interpreted to favor a ferryl (FeO^{2+})-EDTA complex as the likely oxidant and not $HO^{\bullet}$. Subsequent similar studies came to the same conclusion [27,33].

A recent comprehensive review on the role of free radicals and catalytic metal ions in human disease [34] came to the conclusion that "most scientists seem to agree that when iron ions are added as iron-EDTA complexes to H_2O_2 or to systems generating both O^{2-} and H_2O_2, $^{\bullet}OH$ is formed." In support of this conclusion, extensive evidence on spin-trapping studies, aromatic hydroxylation studies, and studies with several other hydroxyl radical scavengers were cited. To explain the results of the 1986 study of Rush and Koppenol [32] discussed above, it was proposed that an intermediate (undefined, possibly a "ferryl species") could form that could act as an oxidant but that could also lead to the production of hydroxyl radical under the appropriate conditions (see reactions of Fe(II)OOH and Fe(IV)=O in Fig. 5). Consistent with this interpretation, recent electron spin resonance (ESR) spin-trapping studies indicate that Fenton systems can produce a ferryl-like oxidizing species and/or hydroxyl radical (spin-trapped as an adduct to 5,5-dimethyl-1-pyrroline N-oxide, DMPO, a standard spin trap) in varying amounts as a function of iron ligand and the concentration of iron [35]. Interestingly, nonphysiological "high" concentrations of iron (~0.1 mM), typically used for mechanistic studies of Fenton chemistry, favor the formation of the "ferryl species" [35], whereas lower, more physiological concentrations of iron (<1 μM) lead to quantitative formation of $HO^{\bullet}$ (characterized as the DMPO spin adduct) [36].

What can be concluded from the foregoing? First, the reactants in the Fenton reaction can produce different oxidants under different conditions. The key factors appear to be iron ligand, pH, and concentra-

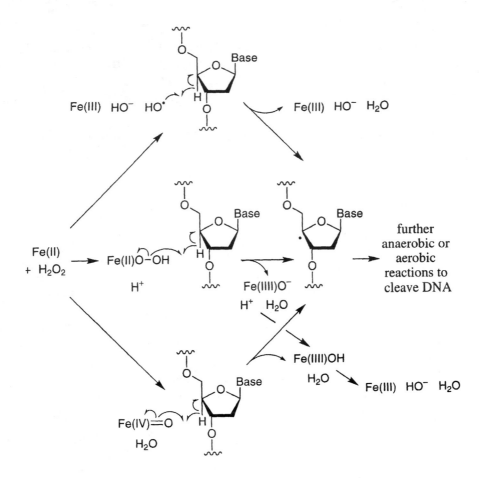

Overall Reaction For All Three Mechanisms

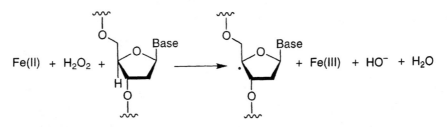

FIG. 6.   Three possible mechanisms for abstraction of a hydrogen atom using the classic Fenton reaction and metal-centered Fenton-like reactions. The common abstraction of the hydrogen from the 4 position of deoxyribose in DNA is used as an example.

tions of reactants. Second, despite considerable study, it is still not possible to predict with certainty what the active oxidant species will be under a given set of conditions, especially under different types of physiological conditions. Third, for biochemically and biomedically oriented scientists who are interested in the consequences of the radical reactions of Fenton chemistry, the mechanistic nuances may not be as important as the products of the reactions. In that respect, all Fenton-type chemistry, regardless of the active oxidant, gives similar products. This point is illustrated in Figs. 6 and 7. In Fig. 6 mechanisms for the cleavage of DNA by formation of a radical at C-4, a common mode of radical-induced DNA cleavage, are shown for HO•, Fe(II)OOH ("bound HO•") and for Fe(IV)=O ("ferryl species"). In each case, plausible mechanisms can be written to form the C-4 radical that can lead to DNA cleavage (for further mechanistic discussion of radical-mediated DNA cleavage, see the cited references) [37–39]. In Fig. 7 mechanisms for the formation of 8-hydroxyguanosine, the most common oxidative reaction of a DNA base, are shown for HO•, Fe(II)OOH ("bound HO•") and for Fe(IV)=O ("ferryl species"). In each case plausible mechanisms can be written to form 8-hydroxyguanosine (for further discussion of oxidative modification of DNA bases, see the cited references) [14,39,40].

4. LIPID PEROXIDATION, AN ESPECIALLY FACILE AND MAJOR TYPE OF RADICAL CHAIN AUTOXIDATION

4.1. Thermochemistry of Lipid Peroxidation Explains Why It Is So Facile

Lipid peroxidation is autoxidation of the hydrocarbon portion of a fatty acid incorporated as an ester into a lipid [41,42]. The reactions occurs almost exclusively for polyunsaturated fatty acid esters that have a CH_2 group with alkenes on either side of it, i.e., the CH bonds are doubly activated by two neighboring alkenes. The energetic consequence of this double activation was discussed in Sec. 3.1. Recalling that discussion and again considering the data in Table 2, it is easy to understand why the representative peroxidation (autoxidation) of the arachidonic acid ester shown in Fig. 8 proceeds essentially exclusively as shown. Of all the C-H bonds available in arachidonic acid for reaction

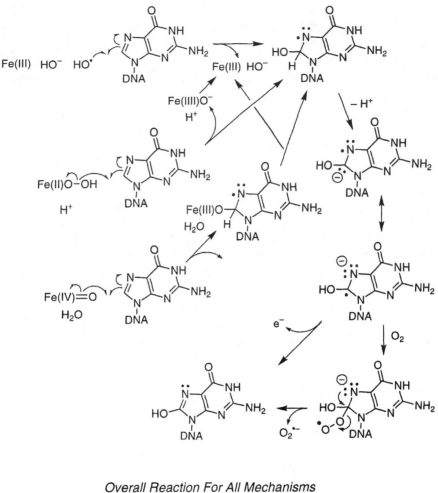

Overall Reaction For All Mechanisms

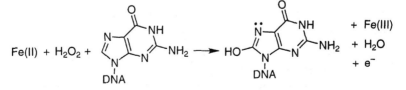

FIG. 7.   Possible mechanisms for hydroxylation of guanosine in the 8 position using the classic Fenton reaction and metal-centered Fenton-like reactions. Other similar mechanisms are possible, leading to the same overall reaction.

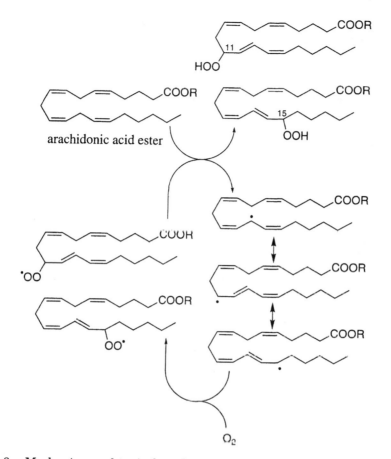

FIG. 8. Mechanism and typical products of lipid peroxidation (autoxidation) using an arachidonic acid ester as an example.

with ROO• those of the CH_2 groups doubly activated by adjacent al-kenes will be the most reactive. The pentadienyl radical intermediates formed in lipid peroxidation have several resonance structures as shown. The different sites of reactivity of pentadienyl radicals can lead to different products, in the case of the arachidonic acid ester in Fig. 8 to react with O_2 at the 11 position and the 15 position (reactions at positions 5 and 9 of the other diene system are also possible). This is a typical type of reactivity, to react at the ends of the conjugated system to leave the remaining double bonds in conjugation.

4.2. Lipid Hydroperoxides as Primary Products of Lipid Peroxidation

The primary products of lipid peroxidation are lipid hydroperoxides such as those shown in Fig. 8. Lipid hydroperoxides are not stable final products. They can undergo further reactions as discussed in the next section.

5. LIPID HYDROPEROXIDES AS A MAJOR SOURCE OF RADICALS

Lipid hydroperoxides are transiently stable but will eventually undergo further reactions. One reaction is simple reduction of the O-O bond to produce lipid alcohol products. This is a benign reaction that serves to eliminate the radical-generating potential of lipid hydroperoxides. In terms of radical chemistry, a very serious reaction is shown in Fig. 9. As with the O-O bond in H_2O_2, the O-O bond in lipid hydroperoxides can undergo Fenton chemistry. The reaction could generate HO$^•$ and lipid alcohols, or it could generate HO$^-$ and alkoxy radicals as shown in Fig. 9. Alkoxy radicals are unstable, highly reactive intermediates that can undergo several possible reactions. As shown in Table 1, the H-O bond in an alcohol (~104–105 kcal/mol) is not as strong as the H-O bond in water (119 kcal/mol), but the H-O bond in an alcohol is still sufficiently strong that alkoxy radicals are reasonably reactive in hydrogen atom abstraction reactions from C-H bonds (consider bond energies and calculations in Tables 1 and 2). Thus alkoxy radicals might be expected to promote further rounds of radical reactions, including autoxidation/lipid peroxidation. However, alkoxy radicals are also prone to facile radical fragmentation reactions, as shown in Fig. 9, to generate aldehyde products and alkyl radicals. The alkyl radical can engage in further autoxidation/lipid peroxidation reactions [43]. The aldehyde products and other products derived from them have a rich biochemistry as biochemical messenger molecules [44].

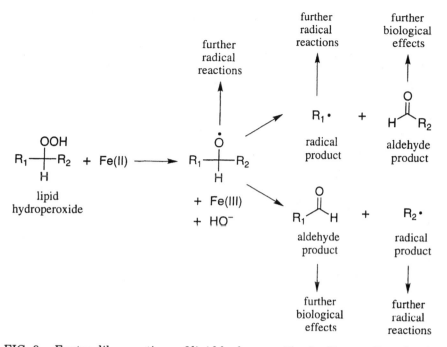

FIG. 9. Fenton-like reactions of lipid hydroperoxides leading to alkoxyl radicals that can undergo further radical reactions, including carbon chain fragmentation, to generate aldehyde products and carbon-centered reactive radicals.

6. PRODUCTION OF REACTIVE OXYGEN SPECIES BY RESPIRATORY BURST OF PHAGOCYTES

Phagocyte cells in the immune system engulf foreign entities by phagocytosis, then attack them with ROS and other toxic molecules. Much is known about the chemistry, biochemistry, and medical consequences of these processes [45–50]. Only the basic chemical reactions that involve radical reactions of oxygen species will be presented here.

The principal enzymes utilized by phagocytes to generate ROS and other toxic molecules are NADPH oxidase and myeloperoxidase. NADPH oxidase produces $O_2^{\bullet-}$. Myeloperoxidase produces HOCl (bleach).

NADPH oxidase catalyzes the reduction of O_2 to $O_2^{\bullet-}$ with the following stoichiometry: $2O_2 + NADPH \longrightarrow 2O_2^{\bullet -} + NADP^+ + H^+$. This reaction, which leads to significant oxygen consumption, is called the "respiratory burst." The production of $O_2^{\bullet -}$ by this reaction can then be followed by the standard cascade of other reactive oxygen species reactions discussed previously in this chapter.

Myeloperoxidase catalyzes the reaction of hydrogen peroxide with halide ions with the following stoichiometry: $X^- + H_2O_2 + H^+ \longrightarrow HOX + H_2O$, where $X = Cl^-$, Br^-, or I^-. Subsequent reactions can generate singlet oxygen, through the reaction of HOCl with H_2O_2: $HOCl + H_2O_2 \longrightarrow {}^1O_2 + H_2O + HCl$. The reaction of HOCl with amines can produce toxic chloramines. A recent report provides evidence that hydroxyl radical can be produced from $O_2^{\bullet -}$ and HOCl in a myeloperoxidase-mediated reaction that does not utilize soluble iron or other metals (i.e., it is a non-Fenton reaction that produces Fenton-like products): $O_2^{\bullet -} + HOCl \longrightarrow HO^{\bullet} + O_2 + Cl^-$ [51].

Despite considerable study, the mechanistic details of the respiratory burst are not fully understood. It is clear that a toxic mix of ROS, bleach, and toxic metabolites from them are all produced to have a strong cytotoxic effect.

7. CONCLUSIONS

Much is known about radicals derived from dioxygen metabolism. The key players, the ROS, have been known for decades. Literally dozens of papers are published weekly on ROS, mostly in biochemically and medically oriented journals. Despite this wealth of knowledge, much fundamental knowledge remains obscure about the detailed reaction chemistry of ROS. It is hoped that the chemical facts and analysis presented in this chapter provide a chemical and mechanistic perspective to aid in thinking about ROS chemistry.

ACKNOWLEDGMENTS

Research by the author on oxygen enzymology and other aspects of oxygen biochemistry has been, and continues to be, supported by the U.S. National Science Foundation.

ABBREVIATIONS

DMPO	5,5-dimethyl-1-pyrroline N-oxide
EDTA	ethylenediamine-N,N,N',N'-tetraacetate
HO$^{\bullet}$	hydroxyl radical
NADPH	nicotinamide adenine dinucleotide phosphate (reduced)
NHE	normal hydrogen electrode
$O_2^{\bullet-}$	superoxide radical anion
$^1O_2$	singlet oxygen
$^3O_2$	triplet dioxygen
ROOH	alkyl hydroperoxide
RO$^{\bullet}$	alkoxy or alkoxyl radical
ROO$^{\bullet}$	peroxy or peroxyl radical
ROS	reactive oxygen species

REFERENCES

1. D. Harman, *Gerontology, 11*, 298–300 (1956).

2. D. Harman, *Mutat. Res., 275*, 257–266 (1992).

3. B. N. Ames, M. K. Shigenaga, and T. M. Hagen, *Proc. Natl. Acad. Sci. USA, 90*, 7915–7922 (1993).

4. H. Nohl, *Br. Med. Bull., 49*, 653–667 (1993).

5. Y. H. Wei, *Proc. Soc. Exp. Biol. Med., 217*, 53–63 (1998).

6. S. Cerda and S. A. Weitzman, *Mutat. Res., 386*, 141 (1997).

7. V. R. Winrow, P. G. Winyard, C. J. Morris, and D. R. Blake, *Br. Med. Bull., 49*, 506–522 (1993).

8. J. B. deHaan, E. J. Wolvetang, F. Cristiano, R. Iannello, C. Bladier, M. J. Kelner, and I. Kola, *Reactive oxygen species and their contribution to pathology in Down syndrome*, in *Antioxidants in Disease Mechanisms* (H. Sies, ed.), Academic Press, San Diego, 1997, pp. 379–402.

9. G. Multhaup, T. Ruppert, A. Schlicksupp, L. Hesse, D. Beher, C. L. Masters, and K. Beyreuther, *Biochem. Pharmacol., 54*, 533–539 (1997).

10. W. R. Markesbery, *Free Rad. Biol. Med., 23*, 134–147 (1997).

11. S. Hippeli and E. E. F. Elstner, *Z. Naturforsch. C, 52*, 555–563 (1997).

12. S. Ahmad (ed.), *Oxidative Stress and Antioxidant Defenses in Biology*, Chapman and Hall, New York, 1995.

13. C. Rice-Evans, B. Halliwell, and G. G. Lunt, *Free Radicals and Oxidative Stress: Environment, Drugs and Food Additives*, Portland Press, London, 1995.

14. B. Halliwell and O. I. Aruoma, *DNA and Free Radicals*, Ellis Horwood, New York, 1993.

15. L. Packer and A. N. Glazer, *Oxygen Radicals in Biological Systems, Part B, Oxygen Radical and Antioxidants*, Academic Press, New York, 1990.

16. E. Cadenas, *Annu. Rev. Biochem.*, *58*, 79–110 (1989).

17. L. Packer, *Oxygen Radicals in Biological Systems*, Academic Press, New York, 1984.

18. K. U. Ingold, T. Paul, M. J. Young, and L. Doiron, *J. Am. Chem. Soc.*, *119*, 12364 (1997).

19. J. F. Turrens, *Biosci. Rep.*, *17*, 3–8 (1997).

20. S.-S. Liu, *Biosci. Rep.*, *17*, 259–272 (1997).

21. D. T. Sawyer, *Oxygen Chemistry*, Oxford University Press, New York, 1991.

22. M. J. Green and A. O. Hill, *Chemistry of dioxygen*, in *Oxygen Radicals in Biological Systems* (L. Packer, ed.), Academic Press, New York, 1984, pp. 3–22.

23. *Bond strengths in polyatomic molecules*, in *CRC Handbook of Chemistry and Physics* (D. R. Lide and H. P. R. Frederikse, eds.), CRC Press, Boca Raton, 1996, pp. 9-64–9-69.

24. H. J. H. Fenton, *J. Chem. Soc. Trans.*, *65*, 899–910 (1894).

25. B. Halliwell and J. M. C. Gutteridge, *Role of free radicals and catalytic metal ions in human disease: an overview*, in *Oxygen Radicals in Biological Systems, Part B, Oxygen Radical and Antioxidants* (L. Packer and A. N. Glazer, eds.), Academic Press, New York, 1990, pp. 1–85.

26. R. D. Cannon, *Electron Transfer Reactions*, Butterworths, Boston, 1980.

27. S. Rahhal and H. W. Richter, *J. Am. Chem. Soc.*, *110*, 3126–3133 (1988).

28. H. N. Po and N. Sutin, *Inorg. Chem.*, *7*, 621–624 (1968).

29. C. Walling, *Acc. Chem. Res.*, *8*, 125–131 (1975).

30. D. T. Sawyer, A. Sobkowiak, and T. Matsushita, *Acc. Chem. Res.*, *29*, 409–416 (1996).

31. D. T. Sawyer, C. Kang, A. Llobet, and C. Redman, *J. Am. Chem. Soc.*, *115*, 5817–5818 (1993).

32. J. D. Rush and W. H. Koppenol, *J. Biol. Chem.*, *261*, 6730–6733 (1986).

33. S. Rahhal and H. W. Richter, *Radiat. Phys. Chem.*, *32*, 129–135 (1988).

34. B. Halliwell and J. M. C. Gutteridge, *Role of free radicals and catalytic metal ions in human disease: an overview*, in *Oxygen Radicals in Biological Systems, Part B, Oxygen Radical and Antioxidants* (L. Packer and A. N. Glazer, eds.), Academic Press, New York, 1990, pp. 1–85, see discussion on pp. 15–19.

35. I. Yamazaki and L. H. Piette, *J. Am. Chem. Soc.*, *113*, 7588–7593 (1991).

36. I. Yamazaki and L. H. Piette, *J. Biol. Chem.*, *265*, 13589–13594 (1990).

37. J. Stubbe and J. W. Kozarich, *Chem. Rev.*, *87*, 1107–1136 (1987).

38. G. Pratviel, J. Bernadou, and B. Meunier, *Angew. Chem. Int. Ed. Engl.*, *34*, 746–769 (1995).

39. A. P. Breen and J. A. Murphy, *Free Rad. Biol. Med.*, *18*, 1033–1077 (1995).

40. H. Kasai, *Mutat. Res.*, *387*, 147–163 (1997).

41. K. H. Cheeseman, *Lipid peroxidation and cancer*, in *DNA and Free Radicals* (B. Halliwell and O. I. Aruoma, eds.), Ellis Horwood, New York, 1993, pp. 109–144.

42. N. A. Porter, Chemistry of lipid peroxidation, in *Oxygen Radicals in Biological Systems* (L. Packer, ed.), Academic Press, New York, 1984, pp. 273–282.

43. H. Esterbauer, R. J. Schaur, and H. Zollner, *Free Rad. Biol. Med.*, *11*, 81–128 (1991).

44. A. Hammer, M. Ferro, H. M. Tillian, F. Tatzber, H. Zollner, E. Schauenstein, and R. J. Schaur, *Free Rad. Biol. Med.*, *23*, 26–33 (1997).

45. J. M. Robinson and J. A. Badwey, *Histochemistry*, *103*, 163–180 (1995).

46. M. J. Karnovsky, *Histochemistry*, *102*, 15–27 (1994).

47.   A. W. Segal, *Protoplasma*, *184*, 86–103 (1995).

48.   A. J. Thrasher, N. H. Keep, F. Wientjes, and A. W. Segal, *Biochim. Biophys. Acta*, *1227*, 1–24 (1994).

49.   R. Seifert and G. Schultz, *Physiology, Biochemistry and Pharmacology*, *117*, 1–338 (1991).

50.   J. A. Badwey and M. L. Karnovsky, *Annu. Rev. Biochem.*, *49*, 695–726 (1980).

51.   C. L. Ramos, S. Pou, B. E. Britigan, M. S. Cohen, and G. M. Rosen, *J. Biol. Chem.*, *267*, 8307–8312 (1992).

4

Free Radicals as a Source of Uncommon Oxidation States of Transition Metals

George V. Buxton[1] and Quinto G. Mulazzani[2]

[1]Centre for Joint Honours in Science,
University of Leeds, Leeds LS2 9JT, UK

[2]Istituto di Fotochimica e Radiazioni d'Alta Energia
del C.N.R., 40129 Bologna, Italy

1. INTRODUCTION

1.1. Nature of the Species

Free radicals and transition metal ions are ubiquitous in biological systems. The variable valency of transition metal ions plays a key role in the catalysis of redox processes in which the oxidation states of the metal are inherently unstable and/or have only a transient existence during the reaction sequence. In many cases where the overall reaction involves multiple electron transfer, the detailed mechanism frequently comprises a sequence of one-electron steps. Free radicals, being atoms or molecules with an unpaired electron, are the archetypal one-electron redox reagents, so that the elementary step in the interaction of free radicals with transition metal ions is a one-electron process. The elucidation of the complex mechanisms that take place in nature can be advanced to a great extent if one is able to measure in isolation these individual steps. To achieve this goal use is made of the fast time-resolved methods of pulse radiolysis and flash photolysis to generate the desired free radicals and to measure their rates of reaction with the species of interest. Often these species are model compounds rather than the biological systems themselves.

Previous surveys of uncommon oxidation states of metal ions can be found in the literature [1–3].

1.2.   Interaction of Free Radicals with Transition Metal Ions

A characteristic property of the transition metals in solution is that they are coordinated to ligands, so that one must consider the complete entity — metal center and ligands — in its reaction with free radicals. The possible reactions are of two main types: outer-sphere electron transfer, in which an electron is transferred between the free radical and the metal center with the ligands of the latter remaining intact; and inner-sphere electron transfer, whereby the ligands are involved. In the latter case the reaction may occur first at a ligand, turning it into a radical, followed by electron transfer to or from the metal center, or the free radical may itself become bonded to the metal ion, either by exchanging with one of the ligands or by increasing the coordination number of the metal ion by one, before electron transfer takes place. In some cases the free radical remains bound to the metal center and the oxidation number of the metal is not well defined. Examples of each of these kinds of reaction are described in Secs. 5 and 6.

2.   PROPERTIES OF FREE RADICALS

2.3.   Acid-Base Forms and Reduction Potentials

In aqueous media it is important to recognize that free radicals with a dissociable proton exist in acidic and basic forms that can have quite different reactivities. As a general rule, the basic form of the radical has a lower reduction potential, making it a weaker oxidant or stronger reductant. Of particular interest here are the hydroxyl ($^{\bullet}$OH) and hydroperoxyl ($HO_2^{\bullet}$) radicals, which have pK_a values of 11.9 and 4.7, respectively [4].

$$^{\bullet}OH \rightleftharpoons O^{\bullet -} + H^+ \tag{1}$$
$$HO_2 \rightleftharpoons O_2^{\bullet -} + H^+ \tag{2}$$

In its neutral form the hydroxyl radical is a strong oxidant, the $^{\bullet}$OH/OH$^-$ couple having a standard reduction potential, E^0, of 1.8 V in neutral solution and 2.7 V at low pH where the extra energy is derived from the neutralization of OH$^-$ by H$^+$. The basic form of the radical,

$O^{\bullet-}$, is important only in strongly alkaline solution (pH $\geqslant$ 13) and with $E^0 = 1.6$ V it is a weaker oxidant than the neutral form. A more significant difference in the two forms is the fact that $^{\bullet}OH$ is an electrophile whereas $O^{\bullet-}$ is a nucleophile. Thus in their reactions with organic ligands, the sites of attack can be different, with $^{\bullet}OH$ adding at centers of unsaturation such as double bonds and aromatic rings, and $O^{\bullet-}$ abstracting H from alkyl groups.

$HO_2^{\bullet}$ has a reduction potential of 1.5 V, making it a reasonably strong oxidizing radical, whereas $O_2^{\bullet-}$ is both an oxidant ($E^0 = 1.0$ V) and a reductant ($E^0 = -0.33$ V, referred to O_2 at 1 atm).

The hydrated electron, e_{aq}^-, is a powerful reductant with $E^0 = -2.9$ V; this makes it suitable to generate hyperreduced forms of the simple aquo ions of the first-row transition metal ions such as Mn^+, Fe^+, Co^+, Ni^+, and Zn^+. Its acidic form is the hydrogen atom formed by reaction (3) with $k_3 = 2.3 \times 10^{10}$ dm^3 mol^{-1} s^{-1}:

$$e_{aq}^- + H^+ \longrightarrow H^{\bullet} \tag{3}$$

The reverse transformation is achieved through reaction (4) with $k_4 = 2.2 \times 10^7$ dm^3 mol^{-1} s^{-1}, so that $H^{\bullet}$ has a formal $pK_a = 9.6$:

$$H^{\bullet} + OH^- \longrightarrow e_{aq}^- \tag{4}$$

The hydrogen atom is generally not nearly so reactive as the hydrated electron with respect to metal ions. In its reactions with organic ligands it behaves like the hydroxyl radical but, again, is less reactive. Compilations of the reduction potentials of inorganic and organic free radicals commonly used as one-electron redox reagents in aqueous solution have been published [5,6].

3. GENERATION OF FREE RADICALS

In order to study the kinetics and mechanisms of the reactions of free radicals, fast time-resolved methods are required because the rates of these processes often approach the diffusion-controlled limit. Two particularly suitable methods are pulse radiolysis and flash photolysis, with the former being more widely used than the latter for reactions in aqueous solution.

3.1. Radiolysis of Water

The interaction of ionizing radiation, such as ^{60}Co γ-rays or electrons with energies of the order of a few MeV, with matter is nonspecific; it is governed by the number of electrons in the system. Thus for aqueous media interaction is mainly with the water component whose radiation chemistry is well understood. It is summarized by reaction (5) at 10^{-12} s after the ionization event has occurred:

$$0.47H_2O \xrightarrow{\quad} 0.47e_{aq}^- \mid 0.47^\bullet OII \mid 0.47II^+ \tag{5}$$

where the numbers are the radiation chemical yields (G values) in units of $\mu mol\ J^{-1}$ for radiation of low linear energy transfer (LET) [4] such as ^{60}Co γ-rays and energetic electrons. Because the ionization events occur in clusters, the products of reaction (5) do not have a homogeneous distribution. As time elapses they diffuse and react, so that a fraction of them combine to form molecular products and the remainder escape into the bulk solution. Thus by 10^{-7} s after the ionization event the situation in neutral water is represented by reaction (6):

$$0.41H_2O \xrightarrow{\quad} 0.27e_{aq}^- \mid 0.06H^\bullet + 0.27^\bullet OH + 0.27H^+$$
$$+ 0.04H_2 + 0.07H_2O_2 \tag{6}$$

For a radical that reacts at the diffusion-controlled rate, i.e., the rate constant k is $\sim 10^{10}\ dm^3\ mol^{-1}s^{-1}$, the yields in reaction (6) will apply for solute concentrations [S] up to $10^{-3}\ mol\ dm^{-3}$; at higher values of $k[S]$ the yields of available radicals increase by about 0.04 for each 10-fold increase in the value of $k[S]$ because scavenging of the radicals by solute increasingly competes with their interreactions to form the molecular products.

3.2. Photolysis

In this method a suitable solute has to be chosen to generate the desired free radicals. The more commonly used solutes are those that on photo-excitation undergo homolytic fission of a bond to give a pair of identical radicals. Hydrogen peroxide, peroxydisulfate ion, and dithionate ion are typical examples:

$$H_2O_2 + h\nu \longrightarrow 2\,{}^\bullet OH \qquad\qquad (7)$$

$$S_2O_8^{2-} + h\nu \longrightarrow 2SO_4^{\bullet-} \qquad\qquad (8)$$

$$S_2O_6^{2-} + h\nu \longrightarrow 2SO_3^{\bullet-} \qquad\qquad (9)$$

Photoionization of inorganic ions can be used to generate hydrated electrons:

$$Fe(CN)_6^{4-} + h\nu \longrightarrow Fe(CN)_6^{3-} + e_{aq}^- \qquad\qquad (10)$$

$$I^- + h\nu \longrightarrow I^\bullet + e_{aq}^- \qquad\qquad (11)$$

The radiolysis and photolysis methods both have advantages and disadvantages. The main advantage of radiolysis for aqueous systems is that the solvent is the source of the radicals, so that dilute solutions can be easily studied. A potential disadvantage is that the primary products comprise approximately equal numbers of strongly reducing radicals (e_{aq}^-, $H^\bullet$) and oxidizing radicals (${}^\bullet OH$), but this is avoidable by interconversion of the different primary radicals to a single kind or to secondary radicals having the desired properties, either reducing or oxidizing. An advantage of the photolysis method is that a single type of radical can be generated, as exemplified by reactions (7)–(9). Problems can arise, however, because of interfering thermal reactions of the photolyte with the other solutes and light absorption by the solutes of interest. More generally, the concentration of radicals that can be generated is greater with pulse radiolysis than with flash photolysis, so that the need for signal averaging is avoided in the former method. Nevertheless, the two methods together provide complementary data that can be invaluable in resolving ambiguities.

3.3. Generation of Secondary Radicals in Water Radiolysis

It is usually desirable to have totally oxidizing or totally reducing conditions, rather than the mixed conditions produced by the primary radicals of water radiolysis, and these are achieved by adding appropriate solutes. For effectively oxidizing conditions saturation of the solution with nitrous oxide is almost invariably the chosen method because of reactions (12) and (13). These reactions, at pH < 11, are complete in $5 \cdot 10^{-8}$ s [4]:

$$e_{aq}^- + N_2O \longrightarrow N_2 + O^{\bullet -} \tag{12}$$

$$O^{\bullet -} + H_2O \longrightarrow {}^{\bullet}OH + OH^- \tag{13}$$

The H$^{\bullet}$ remaining from reaction (6) represents only 10% of the total radicals and often can be neglected without seriously affecting the interpretation of the chemistry.

For reducing conditions an organic scavenger RH for $^{\bullet}$OH and H$^{\bullet}$ is used to generate a radical R$^{\bullet}$, which may be reducing to complement e_{aq}^-, or relatively unreactive to leave e_{aq}^- as the sole reductant. Formate ion provides an example of the first condition:

$${}^{\bullet}OH/H^{\bullet} + HCO_2^- \longrightarrow CO_2^{\bullet -} + H_2O/H_2 \tag{14}$$

and 2-methyl-2-propanol (*tert*-butanol) is generally used to achieve the second condition because the product radical is relatively unreactive:

$${}^{\bullet}OH + (CH_3)_3COH \longrightarrow {}^{\bullet}CH_2(CH_3)_2COH + H_2O \tag{15}$$

This latter system is used in conjunction with peroxydisulfate to generate the powerful oxidant $SO_4^{\bullet -}$ through reaction (16):

$$e_{aq}^- + S_2O_8^{2-} \longrightarrow SO_4^{\bullet -} + SO_4^{2-} \tag{16}$$

N$_2$O-saturated solutions can be used to generate a variety of secondary radicals, either oxidizing or reducing, by suitable choice of the scavenger for $^{\bullet}$OH, and further radical conversion can also be achieved:

$$CO_2^{\bullet -} + O_2 \longrightarrow O_2^{\bullet -} + CO_2 \tag{17}$$

4.   MEASUREMENT OF REACTION RATES

The general principle is to generate the desired radicals, as described above, using a short pulse of energetic electrons from an electron accelerator, or light flash from a rare gas/halogen laser, so that the radicals are produced in a time that is much shorter than their reaction time. Most commonly, optical absorption measurements are used to monitor the rate of change of their concentration as a function of the solute concentration under pseudo-first-order conditions to obtain their reaction rate constants. In the case of pulse radiolysis, conductivity measurements can also be made. These mainly provide information on the uptake or release of protons during and/or after the radical reaction;

this is particularly useful for reactions involving transition metal ions where the state of hydrolysis may change, or release of ligands may occur, as a result of the redox process.

Rate constants for the reactions of free radicals described above are available in the following compilations: (a) A. B. Ross, W. G. Mallard, W. P. Helman, G. V. Buxton, R. E. Huie, and P. Neta, *NDRL-NIST Solution Kinetics Database, Version 2*, NIST Standard Reference Data, Gaithersburg, MD, 1994; (b) a database is also available on the worldwide web site of the Notre Dame Radiation Chemistry Data Center at http//allen.rad.nd.edu., which includes rate constants for reactions of transients from metal ions and metal complexes in aqueous solution.

5. INTERACTION OF FREE RADICALS WITH SIMPLE TRANSITION METAL IONS

Here simple ions are defined as those transition metals having as ligands H_2O, OH^-, or O^{2-}, i.e., the aquo and hydrolyzed aquo complexes.

The hyperreduced states are most readily obtained by reaction with the hydrated electron:

$$e_{aq}^- + M^{n+} \longrightarrow M^{(n-1)+} \tag{18}$$

and reaction (18) is, of course, an outer-sphere electron transfer process. The hyperreduced ions Co^+, Ni^+, and Cd^+, for example, are themselves strong reducing agents and, because of their positive charge, have been substituted for the hydrated electron in studies of the redox chemistry of metalloproteins [7] when these molecules carry a high negative charge on the surface.

Hyperoxidized ions are conveniently produced by reaction (19):

$$\cdot OH + M^{n+} \longrightarrow MOH^{n+} \tag{19}$$

There are several examples to show that reaction (19) occurs as written rather than by outer-sphere electron transfer. Thus, it has been shown [8] by optical, conductivity, and polarographic detection that when M^{n+} is Tl^+, Ag^+, Cu^{2+}, and Sn^{2+}, the products are the corresponding OH adducts. Outer-sphere electron transfer is considered unlikely in any case because of the large reorganization energy involved in going from $\cdot OH$ to OH^- in water. In the case where M^{n+} is the substitution-inert ion $Cr(H_2O)_6^{3+}$, there is kinetic evidence for the formation of a seven-coordinate precursor complex with $^{\bullet}OH$ before electron transfer takes place to form $Cr(IV)$ [9].

Other hyperoxidized metal ions are exemplified by $Fe(IV)$ and $Fe(V)$ which, together with $Fe(VI)$, are of great interest because these oxidation states are involved in catalysis in many industrial and biological processes. Studies of the simple hypervalent states of $Fe(IV)$ and $Fe(V)$ in aqueous solution have been reviewed by Bielski [10]. They are conveniently produced by pulse radiolysis:

$$Fe(OH)_4^- + {}^{\bullet}OH \longrightarrow FeO(OH)_n^{2-n} + H_2O/OH^- \tag{20}$$

$$Fe(VI) + e_{aq}^- \longrightarrow Fe(V) \tag{21}$$

In alkaline solution, pH > 11, reaction (21) can be written as:

$$FeO_4^{2-} + e_{aq}^- \longrightarrow FeO_4^{3-} \tag{22}$$

At lower pH mono-, di-, and triprotonated forms are involved. The general trend in the reactivity of hypervalent iron is that $Fe(IV)$ and $Fe(V)$ react several orders of magnitude faster than $Fe(VI)$, which may be due to the free radical character of the Fe-O bonds in $Fe(IV)$ and $Fe(V)$ [10].

6. INTERACTION OF FREE RADICALS WITH COMPLEXED TRANSITION METAL IONS

6.1. Reduction of the Metal Center

Direct reduction at the metal center of transition metal complexes, represented as M-L, is effected by reducing radicals, $R^{\bullet}$, such as e_{aq}^-, $H^{\bullet}$, $CO_2^{\bullet-}$, ${}^{\bullet}CH_2OH$, $(CH_3)_2COH$, and $O_2^{\bullet-}$:

$$M\text{-}L + R^{\bullet} \longrightarrow M^-\text{-}L + R^+ \tag{23}$$

A large number of complexes with ligands such as NH_3, cyclam (1,4,8,11-tetraazacyclotetradecane), ethylenediamine, CN^-, halide ions, etc., react with e_{aq}^- to undergo simple reduction at the metal center.

The hydrated electron reacts by electron transfer, which means that there must be available a vacant orbital at a suitable energy level so that the transfer process is isoenergetic. This means that some reorganization of the nuclear framework of the donor and acceptor molecules may have to precede the transfer step. The details of this kind of reaction have been worked out by Marcus [11] and others following him, but suffice it to say here that if the separation in energy levels of the ground states of the donor and acceptor is large then the rate of transfer is low. In such a case, transfer into a higher energy orbital to create an electronically excited state of the acceptor may be more probable.

A classic example of the electron transfer product being formed in an excited state is provided by the reaction of e_{aq}^- with $Ru(bpy)_3^{3+}$ (bpy = 2,2'-bipyridine):

$$e_{aq}^- + Ru(bpy)_3^{3+} \longrightarrow Ru(bpy)_3^{2+*} \qquad (24)$$

where the product emits light. In fact, two excited-state products have been proposed [12], the second one not showing luminescence. The overall free energy change for reaction (24) is −4 eV, which makes it highly exergonic so that capture of the electron into a higher energy orbital of the acceptor reduces the energy change required.

The reduction of $Co(bpy)_3^{3+}$ by e_{aq}^- was suggested [13] to result in the cobalt(II) ion being formed in the low-spin $t_{2g}{}^6e_g$ excited state followed by slow relaxation to the high-spin ground state, as might be expected for a change in spin multiplicity:

$$e_{aq}^- + Co(bpy)_3^{3+} \longrightarrow Co(bpy)_3^{2+} \qquad (25)$$

An alternative idea put forward [14] was that reaction (25) proceeds via electron addition to form the ligand radical anion followed, apparently, by reaction with another $Co(bpy)_3^{3+}$ to form the Co^{II} product. A third suggestion [15] based on data obtained at low concentrations of $Co(bpy)_3^{3+}$, which revealed a new intense absorption band not detected in the earlier work, was that the transient species represents neither low-spin Co^{II} nor the simple ligand radical anion coordinated to Co^{III}. Further investigation using $CO_2^{\bullet-}$, $^{\bullet}CH_2OH$, and $(CH_3)_2COH$, as well as e_{aq}^-, provided unequivocal evidence for the rapid formation ($k = 10^8$–10^{10} dm^3 mol^{-1} s^{-1}) of high-spin $Co(bpy)_3^{2+}$, which then slowly equilibrates with loss of a ligand [16]:

$$Co(bpy)_3^{2+} + 2H_2O \longrightarrow Co(bpy)_2(H_2O)_2^{2+} + bpy \qquad (26)$$

It was concluded that any precursor of the high-spin product must have a lifetime of <0.5 μs and that the previous erroneous assignments [13–15] were due to inappropriate experimental conditions.

Reduction of the metal center by the hydrogen atom is efficient for certain complexes. Examples include $Fe(CN)_6^{3-}$ and pentammine-cobalt(III) complexes where the sixth coordination position is occupied by a simple anion such as halide, NCS^-, N_3^-, NO_2^-, CN^-, PO_4^{3-}, CH_3COO^-, $CF_3CO_2^-$, or H_2O. Although it is clear that reduction occurs at the metal center, it has been proposed that the likely mechanism is an inner-sphere (bridged) process [17] because of the large reorganization energy involved in converting the hydrophobic H atom to the strongly solvated H_3O^+:

$$(NH_3)_5CoX^{2+} + H^\bullet \longrightarrow [(NH_3)_5CoXH]^{2+}$$
$$\longrightarrow Co^{2+} + 5NH_3 + X^- + H^+ \qquad (27)$$

The H atom reduces the metal center in $Ru(bpy)_3^{3+}$:

$$H^\bullet + Ru(bpy)_3^{3+} \longrightarrow H^+ + Ru(bpy)_3^{2+} \qquad (28)$$

but no luminescence is observed in this case [12] even though the re-duction potential of $H^\bullet$ is only 0.3 V less than that of e_{aq}^-. This suggests that an inner-sphere mechanism is involved. Similar reduction occurs with Cd^+ and the *tert*-butanol radical $^\bullet CH_2(CH_3)COH$.

The carboxyl radical ion, $CO_2^{\bullet-}$, is an effective reductant and has the advantage that all the primary radicals produced by water radio-lysis can be converted to it, making the system simpler for study. Some examples of reduction of the metal center by reaction with $CO_2^{\bullet-}$ are the following:

$$CO_2^{\bullet-} + Ni(CN)_4^{2-} \longrightarrow Ni(CN)_4^{3-} + CO_2 \qquad (29)$$

Confirmation of this electron transfer mechanism [18] is provided by the fact that the same product, identified by its absorption spectrum, is obtained when e_{aq}^- is the reductant. $Ni(CN)_4^{3-}$ represents the inter-mediate stage in the two-electron reduction of $Ni(CN)_4^{2-}$ to $Ni_2(CN)_6^{4-}$ which is brought about electrochemically or by the action of sodium or potassium amalgam. The rate constants for the intermediate steps in the reduction mechanism have been determined using pulse radiolysis and are $k_{30} = (8.1 \pm 1.0) \times 10^3 \ s^{-1}$ and $k_{31} = (7.4 \pm 1.0) \times 10^7 \ dm^3 \ mol^{-1} \ s^{-1}$ [18]:

$$Ni(CN)_4^{3-} \longrightarrow Ni(CN)_3^{2-} + CN^- \qquad (30)$$
$$2Ni(CN)_3^{2-} \longrightarrow Ni_2(CN)_6^{4-} \qquad (31)$$

A second example is the reduction of copper(II) complexed with tetra-aza macrocyclic ligands (L) [19], which occurs by the following mechanism:

$$Cu^{II}L^{2+}(planar) + CO_2^{\bullet-} \longrightarrow Cu^{I}L^{+}(planar) + CO_2 \qquad (32)$$

$$Cu^{I}L^{+}(planar) + H^{+} \rightleftharpoons Cu^{III}L(H)^{2+} \qquad (33)$$

$$Cu^{I}L^{+}(planar) \rightleftharpoons Cu^{I}L^{+}(tetrahedral) \qquad (34)$$

$$Cu^{III}L(H)^{2+} \rightleftharpoons Cu^{I}L^{+}(tetrahedral) + H^{+} \qquad (35)$$

$$Cu^{III}L(H)^{2+} \{or\ Cu^{I}L^{+}(tetrahedral)\} + H^{+} \longrightarrow Cu^{I} + H_2L^{+} \quad (36)$$

in which rearrangement of the geometry of the coordination shell takes place followed by ligand loss. The rates of rearrangement and ligand loss are reported to increase with increasing size of the macrocycle but the mechanism remains unchanged [19].

Reduction of Cu(II)-peptide complexes serves as a model for studying the mechanism of electron transfer in copper proteins, although it must be recognized that there may be differences in the precise environment of the copper ion in the proteins and the model complexes. When $CO_2^{\bullet-}$ is used as the reducing agent for Cu(II)-triglycine, the reduction occurs directly at the copper(II) ion [20]:

$$CO_2^{\bullet-} + Cu^{II}\text{-}(gly)_3 \longrightarrow Cu^{I}\text{-}(gly)_3 + CO_2 \qquad (37)$$

whereas with the hydrated electron 50% of the reduction of the Cu(II) is direct and 50% via electron addition to the carbonyl group of the ligand followed by intramolecular electron transfer to the copper:

$$e_{aq}^{-} + Cu^{II}\text{-}(gly)_3 \longrightarrow Cu^{II}\text{-}(gly)_3^{-} \longrightarrow Cu^{I}\text{-}(gly)_3 \qquad (38)$$

Analogous behavior is exhibited by Cu(II)-glutathione, which contains a disulfide bridge. In this case the electron can add to the disulfide bridge, to form $RSSR^{-}$, as well as to the carbonyl group. The reaction with $CO_2^{\bullet-}$ produces 90% reduction directly at the metal center and only 10% at the disulfide bridge [20] even though $CO_2^{\bullet-}$ reacts efficiently with free glutathione. A suggested explanation for this result is that in the complex the copper is more exposed than the disulfide bridge [20]. As with $Cu^{II}\text{-}(gly)_3$, reduction of Cu(II)-glutathione by e_{aq}^{-} is a multistep process involving addition to the $-S-S-$ and $C=O$ moieties of the ligand followed by intramolecular electron transfer, with the possibility that this transfer proceeds from the carbonyl group to the copper via the disulfide bridge.

6.2.   Oxidation of the Metal Center

Reactions of oxidizing radicals such as $^{\bullet}OH$, $CO_3^{\bullet-}$, $Cl_2^{\bullet-}$, $Br_2^{\bullet-}$, $(SCN)_2^{\bullet-}$, and $SO_4^{\bullet-}$ with metal complexes have been investigated in some detail. Particular attention has been paid to the oxidation of nickel(II) and copper(II) complexes.

In the case of nickel(II) complexed with tetra-aza macrocyclic ligands there is clear evidence [21,22] for oxidation of the metal center to Ni(III). The nickel(III) complexes are sufficiently long-lived for their reactions with solutes such as halide ions, metal ions, and hydrogen peroxide to be measured [21].

The mechanism for oxidation by $^{\bullet}OH$ is thought to be via abstraction of H from a ligand followed by proton-assisted repair in which an electron is transferred from the metal to the ligand and the resultant ligand radical anion is protonated. In the case of radical anions, direct oxidation of the metal center occurs by outer-sphere electron transfer at close to diffusion-controlled rates. The two types of reaction are summarized as:

$$^{\bullet}OH + Ni^{II}L \longrightarrow Ni^{II}L'^{\bullet}CH + H_2O \tag{39}$$

$$Ni^{II}L'^{\bullet}CH + H^+ \longrightarrow Ni^{III}L \tag{40}$$

$$X_2^{\bullet-} + Ni^{II}L \longrightarrow Ni^{III}L + 2\ X^- \tag{41}$$

The precise behavior of the Ni(III) complexes varies with the composition of the macrocyclic ligand. Ligands that contain NH groups are converted to nitrogen-centered radical complexes, $Ni^{II}L^{\bullet}$, at pH >3 [21]:

$$Ni^{III}L \longrightarrow Ni^{II}L^{\bullet} \tag{42}$$

$Ni^{II}L^{\bullet}$ may also be formed from $Ni^{II}L'^{\bullet}CH$ [21]. The stability of the Ni^{III}-macrocycle complexes is enhanced considerably when the counterion is SO_4^{2-} instead of ClO_4^-. For example, the half-life of $[Ni(cyclam)]^{3+}$ is increased from ~ 10 h in 0.3 mol dm^{-3} ClO_4^- to 5 d in 0.1 mol dm^{-3} SO_4^{2-} at pH 1.6. The stabilizing effect is attributed [22] to kinetic factors, perhaps resulting from conformational changes in the complexes, rather than thermodynamic factors.

The influence of kinetic factors is also apparent in a study of the oxidation of first-row bivalent transition metal complexes containing ethylenediamine tetraacetate (EDTA) or nitrilotriacetate (NTA) ligands (L) [23]. Oxidation of the metal center by $^\bullet OH$ is observed for Ni^{II}, Cu^{II}, and Fe^{II} but not for Mn^{II}. For $Co^{II}(EDTA)^{2-}$ only a small fraction of $^\bullet OH$ oxidize the metal center, the remainder reacting with the ligand by abstracting a hydrogen atom, and for $Co^{II}(NTA)$ reaction is exclusively with the ligand [23]. There is no evidence for metal to ligand electron transfer following reaction of $^\bullet OH$ with the ligand even though $Mn^{III}(L)$ and $Co^{III}(L)$ are stable and the corresponding $M^{III}(L)/M^{II}(L)$ couples have lower reduction potentials than those for nickel and copper. A plausible explanation put forward [23] for these observations is that $^\bullet OH$ reacts via a short-lived precursor complex in which it interacts strongly with the CO group of the carboxylate moiety prior to the electron transfer step. Then, the rate of this transfer resulting in oxidation of the metal center depends on the magnitude of the reorganization energy involved. This is expected to be large for the cobalt and manganese complexes, but considerably smaller for nickel, copper, and iron [23].

The kinetics of oxidation of the Fe^{II}, Mn^{II}, and Co^{II} complexes by other radicals such as $Br_2^{\bullet-}$, $(SCN)_2^{\bullet-}$, and $^\bullet O_2CH_2C(CH_3)_2OH$ indicate an inner-sphere mechanism in these cases [23]. Oxidation of $Ni^{II}(EDTA)$ by $CO_3^{\bullet-}$ is also suggested [24] to proceed via an inner-sphere mechanism leading ultimately to the formation of $Ni^{III}(EDTA)^-$; the latter is believed to be a five-coordinate species that adds water to the sixth position with $k = 1.1 \times 10^3$ s^{-1} [24]:

$$Ni^{II}(EDTA)(H_2O)^{2-} + CO_3^{\bullet-} \longrightarrow H_2O + [Ni^{II}(EDTA)CO_3^{\bullet-}]^{3-}$$
$$\longrightarrow Ni^{III}(EDTA)^- + CO_3^{2-} \tag{43}$$

$$Ni^{III}(EDTA)^- + H_2O \longrightarrow Ni^{III}(EDTA)(H_2O)^- \tag{44}$$

Interactions of free radicals with copper complexes are of considerable interest because of the part that copper plays in biological systems. Oxidation of copper(II) to copper(III) by free radicals has been reported for complexes containing NH_3, EDTA, amines, amino acids, etc. [25], as well as macrocyclic ligands [22,26]. The copper(III) complexes are generally less stable than their nickel(III) counterparts due, it is suggested [26], to Cu^{III} being more strongly oxidizing so that ligand-to-metal electron transfer readily occurs to produce the Cu^{II}-ligand radical complexes. Indeed, because of the similarity of the absorption spectra of many of the oxidized products, which show a single broad band with a maximum near 300 nm, it has been argued [26] that the observed product is not a Cu^{III} species but rather is the Cu^{II}-ligand radical complex. Somewhat different conclusions were drawn from product measurements in a γ-radiolysis study of Cu^{II}-amino acid complexes [27]. The data were interpreted in terms of a mechanism in which ~ 65% of •OH react with the ligand to form the coordinated radical, and the remainder react directly with the metal center to form a copper(III) species in which •OH is bonded to Cu^{II} rather than forming a true Cu^{III} state.

On the other hand, Meyerstein et al. [22] did obtain a spectrum for the product of the reaction of •OH with [Cu(cyclam)]$^{2+}$ that shows a maximum near 400 nm which is similar to that observed for Cu^{III}-macrocycle complexes in acetonitrile.

Reaction of $O_2^{\bullet-}$ with copper(I) and copper(II) complexes is of interest because of the role of copper ions in the dismutation of this radical. In the case of superoxide dismutase (SOD) the reactions are:

$$\text{Enzyme-}Cu^{2+} + O_2^{\bullet-} \longrightarrow \text{enzyme-}Cu^+ + O_2 \tag{45}$$

$$\text{Enzyme-}Cu^+ + O_2^{\bullet-} + 2H^+ \longrightarrow \text{enzyme-}Cu^{2+} + H_2O_2 \tag{46}$$

A number of copper compounds have been shown to have almost the same catalytic activity as SOD, including the aquo complex and complexes with some amino acids or salicylates as ligands. The mechanistic details are illustrated by copper(II)-1,10-phenanthroline, $Cu(phen)_2^{2+}$, as the catalyst [28]. The relevant reactions are:

$$Cu(phen)_2^{2+} + O_2^{\bullet-} \longrightarrow Cu(phen)_2^+ + O_2 \tag{47}$$

$$Cu(phen)_2^+ + O_2^{\bullet-} + 2H^+ \longrightarrow Cu(phen)_2^{2+} + H_2O_2 \tag{48}$$

with $k_{47} = 1.9 \times 10^9$ and $k_{48} = 3 \times 10^8$ dm^3 mol^{-1} s^{-1}. These data yield a turnover rate constant $k_{cat} = (5.1 \pm 0.9) \times 10^8$ dm^3 mol^{-1} s^{-1}, which is only six times lower than that of SOD. Note that in reaction (48) $O_2^{\bullet-}$ is an oxidant, which is unusual for its reactions with metal ions. The suggestion is that here reduction of $O_2^{\bullet-}$ proceeds via a bridged, inner-sphere mechanism in which incompletely coordinated Cu^I binds $O_2^{\bullet-}$ prior to electron transfer [28].

Another example where $O_2^{\bullet-}$ acts as an oxidant is in its reaction with cobalt(II)-sepulchrate, Co(sep)$^{2+}$ (sepulchrate = 1,3,6,8,10,13,16,19-octaazabicyclo[6.6.6]eicosane) where it is proposed [29] that hydrogen atom abstraction from a secondary amine group (N-H) in the ligand is followed by electron transfer to, and protonation of, the ligand:

$$Co(sep)^{2+} + O_2^{\bullet-} \longrightarrow Co(sep - H)^{2+} + HO_2^- \qquad (49)$$

$$Co(sep - H)^{2+} + H^+ \longrightarrow Co(sep)^{3+} \qquad (50)$$

6.3. Addition to the Metal Center

Reactions of $^\bullet OH$ with complexes having inert ligands have already been noted as proceeding by addition to the metal center to produce a hydrolyzed or aquated higher oxidation state of the metal ion. Several other reactions have also been investigated in which carbon centered radicals in particular bind to the metal center, and these are detailed in Chapter 2 of this volume. Two other interesting addition reactions involve the binding of $CO_2^{\bullet-}$ and $H^\bullet$ to N-racemic and N-meso stereoisomers of the cobalt(II) macrocycle CoL^{2+} (L = 5,7,7,12,14,14-hexamethyl-1,4,8,11-tetraazacyclotetradeca-4,11-diene). The products are regarded as six-coordinate cobalt(III) complexes with CO_2^{2-} and H^- bound as ligands [30].

Analogous reactions have been reported [31] for reduction of Ni-(cyclam)$^{2+}$ by $CO_2^{\bullet-}$ and $H^\bullet$. The rates of these reactions approach the diffusion-controlled limit and there is direct evidence for an inner-sphere mechanism for both radicals. The resulting products Ni(cyclam)(CO$_2$)$^+$ and Ni(cyclam)(H)$^{2+}$ have dissociation constants, with respect to release of CO_2 and H^+, respectively, which are several orders of magnitude larger than their CoL^{2+} counterparts referred to above.

6.4. Reaction at the Ligand

6.4.1. Electron Transfer

As well as oxidation and reduction occurring at the metal center, as described in the preceding sections, these processes can be confined to the ligand without change in the oxidation number of the metal ion. This latter type of redox chemistry is exhibited by some water-soluble metalloporphyrins (MP), with M = Zn^{II}, Pd^{II}, Ag^{II}, Cd^{II}, Cu^{II}, Sn^{IV}, and Pb^{II}, and P = tetrakis(X)porphine where X is 4-sulfophenyl, 4-pyridyl, N-methylpyridinium, or N-propylsulfonatopyridinium. These compounds have been shown to react with e_{aq}^- and $(CH_3)_2\dot{C}OH/(CH_3)_2\dot{C}O^-$ to yield π-radical anions [32]. These species are thought to decay by disproportionation, and the central metal ion exerts a strong influence on their stability. Thus the Sn^{IV} complexes have lifetimes greater than 10 s whereas for others the second-order rate constants are in the range 4×10^6 to 10^9 dm^3 mol^{-1} s^{-1}. Reaction of the parent compounds with the oxidizing radical $Cl_2^{\bullet-}$, $Br_2^{\bullet-}$, or $SO_4^{\bullet-}$ generates the corresponding π-radical cations. These are themselves powerful oxidants and are unstable in water, but the decay route remains obscure [32].

The reduction of $Ru^{II}(bpy)_3^{2+}$ by e_{aq} and other reducing radicals [14,33] also occurs at the ligand, giving a species that contains a coordinated radical anion [14], i.e., $[Ru^{II}(bpy)_2(bpy^{\bullet-})]^+$. Analogs of $Ru(bpy)_3^{2+}$ that contain the ligands 2,2'-bipyrazine, 2,2'-bipyrimidine, and 2-(2'-pyridyl)pyrimidine behave in the same way. However, when the complexes contain different ligands, the added electron is localized on the one that is more easily reduced. The reduced complexes, $[Ru^{II}(L^{\bullet-})]^+$, that contain at least one protonatable nitrogen atom undergo acid-base equilibria, and the pK_a of the acid forms $[Ru^{II}(LH^\bullet)]^{2+}$ have been determined, as have the rate constants for protonation of $[Ru^{II}(L^{\bullet-})]^+$ and deprotonation of $[Ru^{II}(LH^\bullet)]^{2+}$, respectively [34,35].

By considering that $[Ru^{II}(L^{\bullet-})]^+$ and $[Ru^{II}(LH^\bullet)]^{2+}$ are orbital analogs of $[Ru^{III}(L^{\bullet-})]^{2+*}$ and $[Ru^{III}(LH^\bullet)]^{3+*}$, i.e., the unprotonated and protonated excited states of $Ru^{II}L^{2+}$, respectively, correlations between these species have been made and the empirical equation $pK_a[Ru^{II}(LH^\bullet)]^{2+} - pK_a[Ru^{III}(LH^\bullet)]^{3+*} = 4.8$ has been derived [36,37]. The use of this relationship, together with the knowledge of $pK_a[Ru(bpy)_2(dppzH^\bullet)^{2+}] = 10$, (dppz = dipyrido[3,2-$a$:2',3'-$c$]phenazine) [38], has helped to clarify the nature of the so-called light-switch mechanism of $Ru(phen)_2(dppz)^{2+}$ [39], i.e., one of the species used as DNA intercalator (see, for example, Chapter 7 of this volume).

6.4.2. Addition

When the metal complex has unsaturated organic ligands, free radicals may add to the center of unsaturation with the oxidation state of the metal center remaining unchanged. An example of this is the reaction of $^\bullet OH$ with $Ru(bpy)_3^{2+}$ where the radical adds to one of the aromatic rings of the bipyridine ligand [40–42]:

$$\bullet OH + Ru(bpy)_3^{2+} \longrightarrow Ru(bpy)_2(bpyOH^\bullet)^{2+} \qquad (51)$$

The H atom also adds to a bipyridine ligand [14], whereas $Cl_2^{\bullet-}$ and $SO_4^{\bullet-}$ oxidize the metal center [41,43].

Reaction (51) has been exploited [40] to produce polymer molecules with covalently bound $Ru(bpy)_3^{2+}$ residues by choosing conditions such that reaction (52) competes with (51) and then the product radicals combine:

$$\bullet OH + polymer \longrightarrow (polymer)^\bullet \qquad (52)$$

Addition of $\bullet OH$ to the bipyridine ligand has also been observed with $Co(bpy)_3^{3+}$, $Fe(bpy)_3^{2+}$, $Fe(Me_2bpy)^{2+}$, and $Os(bpy)_3^{2+}$ [44]. Except for $Co(bpy)_3^{3+}$, oxidation of the metal center in each case is feasible on thermodynamic grounds, so that the barrier to electron transfer from metal to ligand is attributed to kinetic factors [44]. Thus, although the $(bpyOH)^\bullet$ ligands are potential precursors to electron transfer, more facile channels for the decay of the ligand radical moiety must be predominant.

6.4.3.   Hydrogen Atom Abstraction

Interaction of $\bullet OH$ with saturated organic ligands generally results in abstraction of a hydrogen atom to leave a radical coordinated to the metal center. The same kinetic and mechanistic considerations apply as in the case of radical addition, and examples have already been given in earlier parts of Sec. 6.

7.   CONCLUSIONS

Use of the time-resolved methods of pulse radiolysis and flash photolysis over the past 30 years or so has produced a wealth of information on the kinetics and mechanisms of the interactions of free radicals with transition metal complexes to produce unusual oxidation states of the metal ions. The general principle that there is no direct link between the rate of a reaction and its thermodynamic feasibility is much in evidence, making predictions somewhat insecure substitutes for actual measurements.

ABBREVIATIONS

bpy	2,2'-bipyridine
dppz	dipyrido[3,2-a:2',3'-c]phenazine
EDTA	ethylenediamine-N,N,N',N'-tetraacetate
gly	glycine
NTA	nitrilotriacetate
phen	1,10-phenanthroline
sep	sepulchrate (= 1,3,6,8,10,13,16,19-octaazobicyclo[6.6.6]-eicosane)
SOD	superoxide dismutase

REFERENCES

1. G. V. Buxton and R. M. Sellers, *Coord. Chem. Rev.*, 22, 195–274 (1977).

2. (a) M. Z. Hoffman, in *Inorganic Reactions and Methods*, Vol. 15 (J. J. Zuckerman, ed.), VCH, New York, 1986, pp. 272–297. (b) G. V. Buxton, Q. G. Mulazzani, and A. B. Ross, *J. Phys. Chem. Ref. Data*, 24, 1055–1349 (1995).

3. B. G. Ershov, *Russ. Chem. Rev.*, 66, 93–105 (1997).

4. G. V. Buxton, in *Radiation Chemistry: Principles and Applications* (Farhataziz and M. A. J. Rodgers, eds.), VCII, New York, 1987, p. 321 ff.

5. D. M. Stanbury, *Adv. Inorg. Chem.*, 33, 69–138 (1989).

6. P. Wardman, *J. Phys. Chem. Ref. Data*, 18, 1637–1755 (1989).

7. K. Govindaraju, G. A. Salmon, N. P. Tompkinson, and A. G. Sykes, *J. Chem. Soc., Chem. Commun.*, 1003–1004 (1990).

8. K.-D. Asmus, M. Bonifacic, P. Toffel, P. O'Neill, D. Schulte-Frohlinde, and S. Steenken, *J. Chem. Soc., Faraday Trans. I*, 74, 1820–1826 (1978).

9. G. V. Buxton, F. Djouider, D. A. Lynch, and T. N. Malone, *J. Chem. Soc., Faraday Trans.*, 93, 4265–4268 (1997).

10. B. H. J. Bielski, *Free Rad. Res. Commun.*, 12–13, 469–477 (1991).

11. R. A. Marcus, *J. Chem. Phys.*, 24, 966–978 (1956).

12. C. D. Jonah, M. S. Matheson, and D. Meisel, *J. Am. Chem. Soc.*, *100*, 1449–1456 (1978).

13. W. L. Waltz and R. G. Pearson, *J. Phys. Chem.*, *73*, 1941–1952 (1969).

14. J. H. Baxendale and M. Fiti, *J. Chem. Soc., Dalton Trans.*, 1995–1998 (1972).

15. M. Z. Hoffman and M. Simic, *J. Chem. Soc., Chem. Commun.*, 640–641 (1973).

16. M. G. Simic, M. Z. Hoffman, R. P. Cheney, and Q. G. Mulazzani, *J. Phys. Chem.*, *83*, 439–443 (1979).

17. J. Halpern and J. Rabani, *J. Am. Chem. Soc.*, *88*, 699–704 (1966).

18. Q. G. Mulazzani, M. D. Ward, G. Semerano, S. S. Emmi, and P. Giordani, *Int. J. Radiat. Phys. Chem.*, *6*, 187–201 (1974).

19. M. Freiberg, D. Meyerstein, and Y. Yamamoto, *J. Chem. Soc., Dalton Trans.*, 1137–1141 (1982).

20. M. Faraggi and J. G. Leopold, *Radiat. Res.*, *65*, 238–249 (1976).

21. K. D. Whitburn and G. S. Laurence, *J. Chem. Soc., Dalton Trans.*, 139–148 (1979).

22. E. Zeigerson, G. Ginsburg, D. Meyerstein, and L. J. Kirschenbaum, *J. Chem. Soc., Dalton Trans.*, 1243–1247 (1980).

23. J. Lati and D. Meyerstein, *J. Chem. Soc., Dalton Trans.*, 1105–1118 (1978).

24. P. C. Mandal, D. K. Bardhan, S. Sarkar, and S. N. Bhattacharyya, *J. Chem. Soc., Dalton Trans.*, 1457–1461 (1991).

25. D. Meyerstein, *Inorg. Chem.*, *10*, 2244–2249 (1971).

26. K. D. Whitburn and G. S. Laurence, *J. Chem. Soc., Dalton Trans.*, 334–337 (1979).

27. G. R. A. Johnson, N. B. Nazhat, and R. A. Saadalla-Nazhat, *J. Chem. Soc., Faraday Trans. I*, *85*, 677–689 (1989).

28. S. Goldstein and G. Czapski, *J. Am. Chem. Soc.*, *105*, 7276–7280 (1983).

29. A. Bakac, J. H. Espenso, I. I. Creaser, and A. M. Sargeson, *J. Am. Chem. Soc.*, *105*, 7624–7628 (1983).

30. C. Creutz, H. A. Schwarz, J. F. Wishart, E. Fujita, and N. Sutin, *J. Am. Chem. Soc.*, *113*, 3361–3371 (1991).

31. C. A. Kelly, Q. G. Mulazzani, M. Venturi, E. L. Blinn, and M. A. J. Rodgers, *J. Am. Chem. Soc.*, *117*, 4911–4919 (1995).

32. A. Harriman, M. C. Richoux, and P. Neta, *J. Phys. Chem.*, *87*, 4957–4965 (1983).

33. Q. G. Mulazzani, S. Emmi, P. G. Fuochi, M. Z. Hoffman, and M. Venturi, *J. Am. Chem. Soc.*, *100*, 981–983 (1978).

34. M. D'Angelantonio, Q. G. Mulazzani, M. Venturi, M. Ciano, and M. Z. Hoffman, *J. Phys. Chem.*, *95*, 5121–5129 (1991).

35. F. Casalboni, Q. G. Mulazzani, C. D. Clark, M. Z. Hoffman, P. L. Orizondo, M. W. Perkovic, and D. P. Rillema, *Inorg. Chem.*, *36*, 2252–2257 (1997).

36. H. Sun and M. Z. Hoffman, *J. Phys. Chem.*, *97*, 5014–5018 (1993).

37. H. Sun, M. Z. Hoffman, and Q. G. Mulazzani, *Res. Chem. Intermed.*, *20*, 735–754 (1994).

38. Q. G. Mulazzani, M. D'Angelantonio, M. Venturi, M.-L. Boillot, J.-C. Chambron, and E. Amouyal, *N. J. Chem.*, *13*, 441–447 (1989).

39. E. J. C. Olson, D. Hu, A. Hormann, A. M. Jonkman, M. R. Arkin, E. D. A. Stemp, J. K. Barton, and P. F. Barbara, *J. Am. Chem. Soc.*, *119*, 11458–11467 (1997).

40. P. Neta, J. Silverman, V. Markovic, and J. Rabani, *J. Phys. Chem.*, *90*, 703–707 (1986).

41. Q. G. Mulazzani, M. Venturi, F. Bolletta, and V. Balzani, *Inorg. Chim. Acta*, *113*, L1–L2 (1986).

42. C. Creutz and N. Sutin, *Proc. Nat. Acad. Sci. USA*, *72*, 2858–2862 (1975).

43. H. Gorner, C. Stradowski, and D. Schulte-Frohlinde, *Photochem. Photobiol.*, *47*, 15–29 (1988).

44. A. C. Maliyackel, W. L. Waltz, J. Lilie, and R. J. Woods, *Inorg. Chem.*, *29*, 340–348 (1990).

5

Biological Chemistry of Copper-Zinc Superoxide Dismutase and Its Link to Amyotrophic Lateral Sclerosis

Thomas J. Lyons,* Edith Butler Gralla,* and Joan Selverstone Valentine

Department of Chemistry and Biochemistry,
University of California Los Angeles,
Los Angeles, CA 90095-1569, USA

---

*These two authors contributed equally to the preparation of this review.

1. INTRODUCTION

The evolution of photosynthesis on earth infused the atmosphere with
increasing amounts of O_2. Beginning around 550 million years ago,
sufficient O_2 was generated to change the nature of the atmosphere
from reducing to oxidizing and thus to change the nature of life itself
[1]. Indeed, this change may have fueled the evolutionary explosion that

characterized the Cambrian era. Most of the organisms that survived the transition and adapted to (indeed, thrived on) an oxidizing atmosphere such as we have today were those that "learned" how to use reduction of dioxygen to water ($O_2 + 4e^- + 4H^+ \rightarrow 2H_2O$) to produce energy in a form that could be stored. However, like most good things, oxygen utilization has its down side: partially reduced species derived from dioxygen can be highly reactive and toxic. So, concurrent with the use of O_2, organisms evolved mechanisms to protect themselves from the byproducts of oxygen utilization.

Earthly aerobic life forms have evolved several systems that work together to defend against the deleterious side effects of oxygen metabolism. These systems include (1) enzymes that react directly with the intermediate reduction products to remove them, (2) systems that shelter redox active metal ions from reactions with cellular components, (3) small antioxidant molecules that intercept radicals, and (4) elaborate repair and degradation systems that identify and repair or recycle damaged macromolecular components.

This chapter focuses on the first type of system, in fact on a single enzyme—the antioxidant enzyme copper-zinc superoxide dismutase (CuZnSOD). Superoxide dismutases (SODs) use metal ions to catalyze the disproportionation of O_2^- (superoxide) via reactions (1) and (2).

$$O_2^- + M^{n+} \longrightarrow O_2 + M^{(n-1)+} \tag{1}$$

$$O_2^- + M^{(n-1)+} + 2H^+ \longrightarrow H_2O_2 + M^{n+} \tag{2}$$

SODs constitute part of the primary line of defense against what are known as reactive oxygen species (ROS) such as superoxide (O_2^-, the one-electron reduction product of dioxygen), hydrogen peroxide (H_2O_2, the two-electron reduction product of dioxygen), and hydroxyl radical (•OH, resulting from one-electron reduction of hydrogen peroxide followed by O-O bond cleavage to yield •OH and OH$^-$). Sometimes other reactive compounds derived from these three are also included under the label ROS, such as peroxynitrite (ONOO$^-$) and lipid hydroperoxides. SODs, together with catalases and peroxidases (which catalyze removal of H_2O_2), severely limit production of all these products by removing O_2^- and H_2O_2.

Erythrocuprein, isolated from bovine blood, was found to possess SOD activity in 1969 and was therefore renamed CuZnSOD [2]. Since that time many superoxide dismutases have been isolated, and some form of SOD has been found in every aerobic and aerotolerant organism.

There are two major forms of CuZnSOD: the dimeric intracellular form present in abundant quantities in the cytosol and nucleus of almost all eukaryotic organisms, and a higher molecular weight, tetrameric, extracellular homolog of much lower abundance (ECSOD) [3]. The glycosylated ECSOD is located in the interstitial spaces of tissue as well as plasma and is bound mostly to heparin sulfate proteoglycans. To date all eukaryotes have been found to contain cytosolic CuZnSOD, with the notable exception of some marine arthropods which seem to have replaced it with a cytosolic manganese SOD [4].

Plants contain cytosolic and chloroplast-specific (plastidic) CuZnSODs and usually contain multiple isozymes [5]. Several reports have demonstrated a peroxisomal localization for CuZnSODs as well. Sunflower cotyledons provide an interesting curiosity in that they do not seem to have a mitochondrial MnSOD but instead have CuZnSOD localized to this compartment [6].

Procaryotic homologs of CuZnSOD have also been found in the periplasm of many bacteria, indicating a remarkable degree of conservation of this enzyme throughout evolution [7]. There have even been published reports of viral CuZnSOD genes, supporting the hypothesis that this gene is critical for aerobic life [8].

Another, evolutionarily unrelated, family of SODs uses manganese or iron as the metal cofactor (MnSOD and FeSOD). MnSOD is found as a tetramer in the matrix of mitochondria, and both MnSOD and FeSOD (dimeric or tetrameric) are widely distributed in prokaryotes. These SODs are also well studied, both biologically and physically, but the details are beyond the scope of this chapter (for access to this literature, see [9–13]). Another, more recently discovered SOD family uses nickel as the catalytic metal ion (NiSOD) [14–16].

It can be truthfully stated that CuZnSOD is one of the most thoroughly characterized proteins in the scientific literature, and thousands of research articles describe properties ranging from physical characterization to biological activity. Any chapter reviewing CuZnSOD must therefore of necessity be selective in its coverage, and in this review we have chosen to present an overview of the biophysical properties of CuZnSOD followed by a review of some of the more recent biological results, particularly those relating to biological function, including a discussion of the role of this enzyme in the human neurodegenerative disease amyotrophic lateral sclerosis (ALS). In the sec-

tions addressing the biological function of CuZnSOD, we have somewhat broadened the scope to include other SODs, particularly the
MnSOD, since the SODs are closely linked by a common function.

A large number of excellent reviews of various aspects of SOD
structure and function have been published over the years, to which the
reader is referred for more comprehensive coverage. For biophysical
characterization, see [17–23]. For discussion of SOD mechanisms, see
[24]. For more information on basic biological function, see [25–29]; for
SOD in yeast, see [30,31]; for a review of recent developments in biological damage due to reactive oxygen species, see [32]; for ALS and the
role of SOD, see [33,34] and Sec. 5 of this chapter.

2. BIOPHYSICAL PROPERTIES OF WILD-TYPE SUPEROXIDE DISMUTASE

A wealth of knowledge regarding structure-function relationships in
CuZnSOD lies buried in stacks of journals dating back to the original
isolation of erythrocuprein in 1939 [35]. Yet, as researchers periodically
discover, the textbooks must be rewritten and dogma put to the test.
The recent discovery that mutations in CuZnSOD cause familial amyotrophic lateral sclerosis (FALS) [36,37] has revitalized research routes
down what was previously thought by many to be a too-well-trodden
path. At first glance it seems reasonable that mutations in CuZnSOD
might be disease causing, based on knowledge of its function in protecting against damage in vivo by ROS. But this initial conclusion must be
reconsidered when it is confronted by the evidence that the ALS phenotype is due to a gain of function rather than a loss of SOD activity
[33,38–40]. This new evidence has become the impetus to explore ave-
nues of fundamental research on CuZnSOD that were not pursued
before in order to fill in the chemical details and provide an explanation
for the disease causing properties.

2.1. Protein Structure

While ECSOD is tetrameric [41] and the periplasmic CuZnSOD from
Escherichia coli is monomeric [42], cytoplasmic CuZnSOD is invariably

a dimer of identical subunits. (The only exception reported to date is the monomeric isozyme IV from *Oryzae sativa* (rice) [43].) Each subunit of cytoplasmic CuZnSOD is approximately 16 kD, contains around 150 amino acid residues, and binds one zinc ion and one catalytic copper ion [22]. We will use the amino acid numbering for the human CuZnSOD, which has 153 amino acids, throughout this chapter (see Fig. 1 for an alignment of human, bovine, and yeast CuZnSOD sequences).

The overall tertiary fold is an immunoglobulin-like β barrel, consisting of eight antiparallel β sheets with a conserved disulfide bond [44,45]. The dimer is held together by hydrophobic contacts that cover approximately 50 Å^2 [45]. High-resolution crystal structures are available for the human [46,47], yeast [48–51], bovine [45,52–58], frog [56,59,60], and spinach [61] enzymes. They demonstrate the conservation of protein fold as well as amino acid sequence [62,63]. The wealth of crystallographic data has been recently reviewed by Bertini et al. [17] and the structure of the human wild type CuZnSOD dimer is shown in Fig. 2.

2.2. Nature of the Dimer Interaction

While the intersubunit contacts are quite strong, there is evidence that, at least in vitro, the subunits can be separated. The bovine, yeast, and swordfish enzymes were shown to be monomerized in 8 M urea [64,65], and the wheat germ subunits were reported to dissociate in sodium dodecyl sulfate (SDS) [66,67]. Dissociation was proposed to be dependent on the metallation state of the protein [68,69], although these results were later questioned [70]. It has also been reported that simple dilution is enough to significantly effect the monomer/dimer equilibrium [71]. Guanidinium hydrochloride [72,73] and SDS also influence the monomer/dimer equilibrium, as do succinylation [74], point mutations at sites not in the dimer interface [75], and temperature [71]. Introduction of charged residues into the hydrophobic interface using site-directed mutagenesis [76–79] was found to be sufficient to prevent dimerization of the subunits of CuZnSOD, but the monomeric mutant enzyme was found to have very low SOD activity.

Related to the monomer/dimer equilibrium is the phenomenon of intersubunit communication or cooperativity, which is supported by

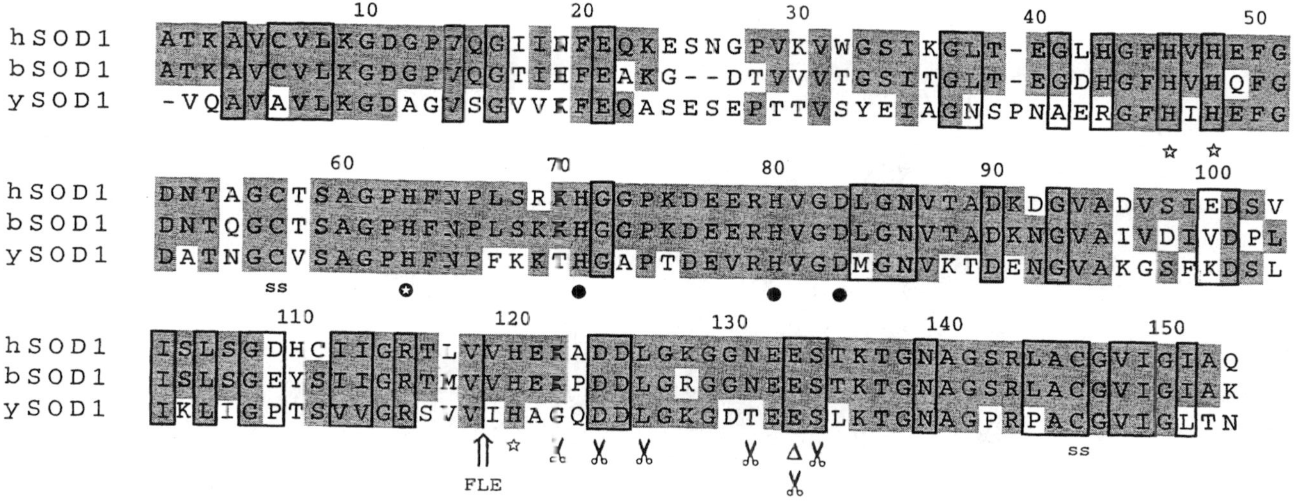

FIG. 1. Sequence alignment of human, bovine, and yeast CuZnSOD. Shaded areas indicate amino acid identity in at least two sequences. Numbering is according to the human sequence. Boxed areas indicate positions at which ALS-causing missense mutations have been found in the human gene. (●) Amino acids involved in zinc binding. (☆) copper binding amino acids. (Note that His63 is a ligand for both copper and zinc.) (ss) Position of the two cysteines involved in the disulfide bridge. (✀) Position of ALS mutations that result in a premature stop and truncation of the protein. (⇑) An ALS mutation that results in an insertion of the tripeptide FLE. (△) An ALS mutation that results in the deletion of one amino acid.

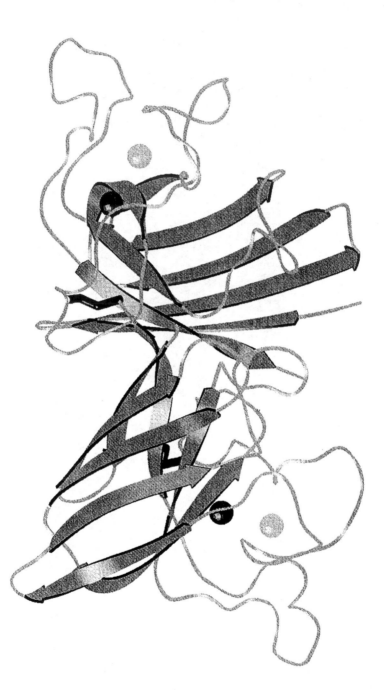

several lines of evidence, at least in the case of the bovine, yeast, and human enzymes. Studies using nuclear magnetic resonance (NMR) [80], differential scanning colorimetry (DSC) [81], pulse radiolysis [82], and metal titration [83–85] suggested that the two subunits communicated with each other and that structural changes in one subunit could affect the conformation of the other subunit.

Computational studies have predicted that an inherent subunit asymmetry should exist in dimeric CuZnSOD [86,87]. This theoretical prediction is challenged by the published crystal structures, which show equivalent subunits for all species. However, it is now believed that the yeast enzyme possesses significant asymmetry when stripped of its metals [88,89], and one human FALS-causing mutant was shown both in solution and in the crystalline state to exhibit significant differences between the subunits [90].

The teleological reason that CuZnSOD forms a dimer is still a matter of debate. The fact that the monomer prepared by site-directed mutagenesis is not SOD-active may indicate that dimer contact is crucial for proper formation of the active site. However, the *E. coli* enzyme is monomeric and active [91,92]. One recent publication regarding the structure and stability of *E. coli* CuZnSOD suggests that the β-barrel fold is enough to stabilize the structure of CuZnSOD, but the additional stability toward heat and pH denaturation is conferred by dimerization [93]. It is also possible that the CuZnSOD protects its amino and carboxy termini from proteolysis by burying them in the dimer interface.

The importance of dimerization in generating the final electrostatic field that attracts superoxide to the active site is also unknown. Dimerization may maximize the positive surface of the protein, ensur-

---

FIG. 2.   Three dimensional structure of the human wild-type CuZnSOD. Note that the basic fold is an 8-stranded β barrel. β strands are indicated by broad arrows and loops by ropes. The lighter sphere is the zinc ion; the darker one is the copper ion. The view we have chosen shows that the copper ion is located deeper in the protein and that the zinc ion is more peripherally located. However, the copper ion has one coordination site accessible to small anions through the narrow active site channel while the zinc ion has no open coordination sites. It serves to stabilize the neighboring loop structure. The disulfide bond, which also serves to stabilize a loop, is shown in black.

ing optimal collisions with its negative substrate. That the monomeric point mutant is inactive despite an apparently identical structure (determined by NMR) supports the last idea [78], but the existence of a fully active monomeric protein from *E. coli* tends to refute it [42].

Overall, CuZnSOD is a remarkably stable protein. The melting temperature, as gauged by DSC, can be as high as 100°C, in the case of the bovine enzyme, and the protein is denatured by neither 8 M urea nor 1% SDS [94]. Recombinant human CuZnSOD is not proteolytically degraded by trypsin, aminopeptidase M, or serum [95]. The structural attributes that result in this unusual stability are unknown.

2.3. Structural Heterogeneity

It has been known for some time that CuZnSOD from a variety of sources produces a pattern of multiple bands on nondenaturing polyacrylamide gels [96,97]. Little is known about the nature of these electromorphs, but there are many theories as to their origins. In some instances, further purification of some of these individual bands, followed by another round of electrophoresis, yielded the same pattern of multiple bands, leading to the suggestion that the band contained forms capable of equilibrating with each other [98,99].

An early suggestion of labile sulfur bound to CuZnSOD was proposed because some preparations of CuZnSOD displayed a 320-nm UV band indicative of thiolate charge transfer [100]. Another thought was that the different forms represented covalently modified derivatives of the enzyme, perhaps the products of oxidative modification or glycation [101–107]. Such conclusions were based on observations of the increased heterogeneity when CuZnSOD was exposed to conditions that increase such phenomena. It is also possible that the multiple bands represent different metallation states of the protein, i.e., apo- versus holo-protein [85,108,109]. However, other evidence suggests that they are the result of folding isomers of the enzyme [98,99]. The discrepancy between total and electron paramagnetic resonance (EPR)-detectable copper [110,111] and the presence of conformational sub-states detected using EPR [112] support this last theory. Based on the large amount of evidence of heterogeneity, it seems possible that CuZnSOD has several quasi-stable structures, each differing from the others minutely.

2.4. Cysteines

The disulfide bond, which is conserved in all eukaryotic CuZnSODs, links Cys57 and Cys146 (see Figs. 1 and 2). The mammalian CuZnSODs contain a third cysteine, Cys6, that is buried in the dimer interface, while the human protein possesses yet another cysteine, Cys111, that is exposed on the surface of the protein [62,63]. The buried Cys6 in the human and bovine proteins has been mutagenized to an alanine in order to create a more thermostable enzyme. The human enzyme is also stabilized by mutagenesis of the surface Cys111 to serine [57,113,114].

The fact that CuZnSOD possesses a disulfide bond is interesting in itself [115]. Such bonds stabilize protein structures, and they are common in secreted proteins but are not commonly found in intracellular proteins. One possible explanation is that it plays a role in the folding pathway of CuZnSOD.

The role of the other cysteines in the CuZnSOD sequence is equally intriguing. While free, nondisulfide cysteines are common as ligands in metalloproteins, there is no evidence to support the notion that they function in this capacity in CuZnSOD. Also of questionable relevance is the fact that some proteins contain redox-active thiol groups that have catalytic functions. Surface-exposed thiols have even been proposed to possess a novel function—the reversible binding of nitric oxide [116]. Since human CuZnSOD possesses both a buried and an exposed thiol, it will be interesting to see whether new discoveries will be made about their functions.

2.5. Metal Binding Sites

The Zn^{2+} ion occupies a tetrahedral binding site in a loop consisting of residues 63–83 (human CuZnSOD numbering). Zinc is ligated to three nitrogens from histidyl imidazole rings (His63, His71, His80) and one carboxylic oxygen from aspartate (Asp83) (Fig. 3). In the oxidized form of the enzyme, His63 forms an imidazolate bridge to the bound Cu^{2+} ion, which is tetragonally ligated to four histidines (His46, His48, His63, His120). In addition to the direct imidazolate bridge between the Cu^{2+} and Zn^{2+} ions, there is a bridge formed by hydrogen bonds that

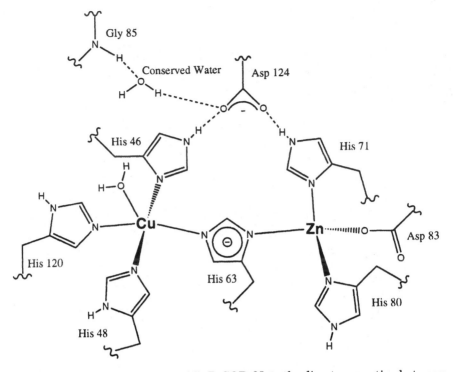

FIG. 3.   The active site region of CuZnSOD. Note the direct connection between
the two metal ions through the imidazolate bridge formed by His63, as well as
an indirect connection via the secondary bridge formed by Asp124 which connects two histidine ligands, one for the zinc and one for the copper ion. The
water coordinated to the copper ion is replaced by superoxide anion during
catalysis. Hydrogen bonds are indicated by dotted lines, and connections to the
protein backbone by squiggles. See text for more discussion.

links the two metal binding sites. This hydrogen bonding network
connects His71 of the zinc site with His46 of the copper site via the
carboxylate group on Asp124 [45]. This secondary bridge seems to be
more vital to the integrity of the zinc binding site because site-directed
mutants modified at position 124 are zinc-deficient but copper-replete
[117]. In solution and in one reported crystal structure, reduction of
Cu^{2+} to Cu^+ results in breakage of the imidazolate bridge [51,118–123]
with the Cu^+ ion moving away from His63 [51].

The zinc in wild-type CuZnSOD can be replaced by Co^{2+} with neither loss in activity [124–126] nor significant perturbation of the protein structure [55]. It can similarly be replaced, with retention of some or all SOD activity, by Cd^{2+} [127], Cu^{2+} [128], Hg^{2+} [129], Ni^{2+} [130], and VO^{2+} [131], providing a wealth of structural probes for the zinc site. The copper can be replaced with Co^{2+} [132], Zn^{2+} [80], Cd^{2+} [133], Ni^{2+} [134], or Ag^{+} [135], although none of these copper-substituted CuZnSOD derivatives is SOD-active.

2.6. In Vivo Metallation State

CuZnSOD isolated from yeast exposed to nontoxic levels of Ag^{+} has been found to contain silver, indicating that the metallation state of CuZnSOD can be altered in vivo [136]. Other studies in yeast demonstrated that CuZnSOD can buffer toxic levels of copper [137] and that the copper-binding protein, metallothionein, can functionally substitute for CuZnSOD [138]. Many studies on dietary copper and zinc deficiency in whole organisms have shown significant alterations in the SOD activity of intracellular CuZnSOD [129,139–142]. But the question has always remained whether this effect is due to alterations in metallation states as opposed to CuZnSOD protein levels.

Several studies have concluded that CuZnSOD exists in vivo as both a metallated pool and a metal-deficient pool. In the yeast *Saccharomyces cerevisiae* grown under anaerobic conditions, it was shown that a significant portion of the CuZnSOD is inactive and that it can be reactivated by the addition of copper ions to cell extracts [143]. Similar experiments in human K562 cells led to the conclusion that the same is true in this mammalian system if the cells are undifferentiated [144, 145]. More compelling data were provided by a group studying human lymphoblasts who showed that ^{64}Cu incorporation into an apoprotein pool of CuZnSOD could be achieved by addition of the copper isotope to intact cells and that this incorporation was accompanied by an increase in SOD activity [146]. Similar results were obtained in the presence of cycloheximide, demonstrating that it was not de novo synthesis that accounted for the increased copper binding and SOD activity. A separate study of CuZnSOD isolation under conditions of copper deficiency led to the conclusion that the isolated protein was both copper- and zinc-deficient, suggesting that metal ions may be inserted into the apoprotein in a cooperative fashion. It was also observed in the same study that the SOD activity in lysates from copper-deficient cells was very low [108].

This line of research has recently gained momentum due to the discovery of copper chaperones that play an important role in facilitating proper metal ion insertion into CuZnSOD apoprotein within yeast and human cells. In the case of *S. cerevisiae*, the relevant gene is *LYS7* and its human homolog is *ccs1* (Copper Chaperone for SOD) [147]. Lys7 and CCS are multidomain proteins with one domain that is homologous to the CuZnSOD monomer. It seems likely that this domain recognizes apoCuZnSOD and, perhaps, like a classic protein chaperone [148], modifies the protein structure to make it more amenable to metal binding.

3. ENZYMATIC ACTIVITY

3.1. Redox Reactivity

CuZnSOD is remarkably specific with respect to its reactivity, and the rates of reaction of both the oxidized, cupric form and the reduced, cuprous form with O_2^- appear to have been optimized, with rate constants for both approaching the diffusion limits [149]; see reactions (3) and (5). At the same time, rates of reactions with other potential substrates such as H_2O_2 [150,151], NADH, and ascorbate [152] appear to have been minimized.

CuZnSOD activity is high and nearly constant over a wide pH range (pH = 5–9). Removal of Zn^{2+} makes the enzyme markedly pH-dependent although it does not alter the maximal activity [153]. The zinc-deficient enzyme also reacts more readily with peroxynitrite than does the native enzyme [154]. These data demand a rethinking of the role of zinc in this protein, because it is clearly not only a structural ion that confers stability but also a regulatory ion that modulates the reactivity of copper in the enzyme. Ironically, the belated awareness of this role for zinc is in keeping with the fact that zinc was not even known to be a cofactor in CuZnSOD until 31 years after the protein's initial isolation [35,155].

The products of the SOD reaction, O_2, and H_2O_2, can also react with CuZnSOD, although the reactions are much slower than the reactions of the enzyme with O_2^-. Thus O_2 reacts slowly with reduced, cuprous CuZnSOD forming the oxidized, cupric enzyme and superoxide; see reactions (3) and (4) [156].

$$E\text{-}Cu^{2+} + O_2^- \longrightarrow E\text{-}Cu^+ + O_2 \quad k = 2 \times 10^9 \text{ M}^{-1}\text{s}^{-1} \quad (3)$$
$$E\text{-}Cu^+ + O_2 \longrightarrow E\text{-}Cu^{2+} + O_2^- \quad k = 1.5 \text{ M}^{-1}\text{s}^{-1} [156] \quad (4)$$

Hydrogen peroxide can function both as a reductant and as an oxidant of CuZnSOD. The oxidized enzyme is reduced by the addition of H_2O_2 [157,158], and the resulting reduced CuZnSOD can then react further with H_2O_2, albeit slowly, to form hydroxyl radical or a hydroxyl-like species [159,160] that is capable of hydroxylating substrates and of irreversibly inactivating the enzyme [161] by oxidizing a single histidyl residue [162]; see reactions (5)–(7).

$$E\text{-}Cu^+ + O_2^- + 2H^+ \longrightarrow E\text{-}Cu^{2+} + H_2O_2 \quad k = 2 \times 10^9 \text{ M}^{-1}\text{s}^{-1} \quad (5)$$
$$E\text{-}Cu^{2+} + H_2O_2 \longrightarrow E\text{-}Cu^+ + O_2^- + 2H^+ \quad k = 0.56 \text{ M}^{-1}\text{s}^{-1} [158] \quad (6)$$
$$E\text{-}Cu^+ + HO_2^- + H^+ \longrightarrow E\text{-}Cu^{2+} + OH^- + {}^\bullet OH$$
$$k = 2.6 \times 10^3 \text{ M}^{-1}\text{s}^{-1} [160] \quad (7)$$

CuZnSOD also reacts with peroxynitrite resulting in nitrosylation of tyrosine residues [163,164]. As stated above, this activity can be increased by the removal of zinc from the protein.

$$E\text{-}Cu^{2+} + OONO^- + H^+ \longrightarrow E\text{-}Cu^{2+}(OH^-) + NO_2^+ \quad (8)$$

Hydrosulfide anion, HS^- [165,166], is oxidized by CuZnSOD producing hydrosulfide radical ($HS^\bullet$). $HS^\bullet$ can then recombine with another $HS^\bullet$ or with HS^- to form elemental sulfur (S^0) or polysulfide anion (S_n^{2-}). The oxidized enzyme is then regenerated by reaction with O_2 [165,166]. Reactions of CuZnSOD with NO to give NO^- have also been reported [167]; see reactions (9)–(11).

$$E\text{-}Cu^{2+} + HS^- \longrightarrow E\text{-}Cu^+ + HS^\bullet \quad (9)$$
$$E\text{-}Cu^+ + NO \longrightarrow E\text{-}Cu^{2+} + NO^- \quad (10)$$

3.2. Dismutase Mechanism

CuZnSOD scavenges superoxide anion in a nearly pH-independent manner over a wide range of pH. The catalytic rate constant has been determined to be $2 \times 10^9 \, M^{-1} \, s^{-1}$, which is extremely high for an enzyme and approaches the diffusion-controlled limit. The protein surface surrounding the active site channel is highly positively charged [168,169]. It is this positively charged electrostatic field that is postulated to steer the superoxide substrate into the active site. Altering the electrostatic field by covalent modification [170], ionic strength manipulation [171], or site-directed mutagenesis can diminish [172] and, remarkably, even enhance [173] the activity of the enzyme. Even the reaction with peroxide, which reacts in the form of the HO_2^- anion, is believed to be electrostatically guided [174].

The superoxide anion is believed to bind to the axial position on the roughly tetragonal Cu^{2+}, replacing a weakly coordinated water molecule, forming a hydrogen bond with Arg143. Electron transfer to the Cu^{2+} ion results in breaking of the imidazolate bridge and release of dioxygen. Another superoxide then reacts with the trigonal Cu^+,

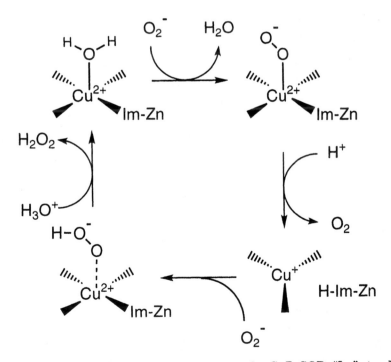

FIG. 4.   Mechanism of superoxide dismutation by CuZnSOD. "Im" stands for the bridging imidazolate (His63). See text for more detailed explanations.

oxidizing it to form Cu^{2+}. The imidazolate bridge between copper and zinc is reformed and peroxide is released. The details of the timing of the proton transfers are unknown, but evidence suggests that the role of the imidazolate bridge (and consequently the zinc ion as well) is to ensure that the peroxide ligand is only weakly bound in an axial position, ensuring rapid release of the peroxide product [175,176] (see Fig. 4). Under normal conditions of low superoxide concentrations, the imidazolate bridge between the zinc and copper sites is broken upon reduction. This step is relatively slow, however, and requires significant atomic rearrangement; it is not expected to occur in the event of saturating substrate concentrations [122]. In the latter case the imidazolate bridge is expected to remain intact during catalysis; however, this situation is unlikely to occur in vivo because CuZnSOD exists in vast excess of its substrate inside of a normal cell.

4.   EVIDENCE CONCERNING BIOLOGICAL FUNCTION

4.1.   Historical Background

Almost from the moment that the SOD activity of CuZnSOD was first announced by McCord and Fridovich in 1909 [2], this enzyme has been widely recognized as an important antioxidant enzyme, and its high SOD activity and its ubiquitous presence in aerobic organisms strongly support this conclusion. Nevertheless, an early controversy in the field arose from the concerns of some chemists (including one of the present authors), who were reluctant to assign to superoxide an extremely deleterious role in causing generalized damage without knowledge of the specific chemical reactions that cause superoxide to be toxic in vivo [177]. Much of this concern arose as detailed information concerning the chemical reactivity of superoxide accumulated. In particular (1) that the rate of spontaneous disproportionation of superoxide is fast and it therefore will not accumulate to high concentrations, even in the absence of an SOD; (2) that cupric ion itself is an excellent catalyst of superoxide disproportionation, albeit only at low pH, suggesting the possibility that the SOD activity of CuZnSOD might be artifactual; and (3) that superoxide is usually not capable of fast oxidations of substrates because a proton is needed before it can accept an electron from a substrate (reaction (11)), and it is not strongly basic in water [24].

$$O_2^- + e^- + HX \longrightarrow HO_2^- + X^- \qquad (11)$$

Moreover, the explanation of superoxide toxicity, widely accepted by some at that time, that it acted as a reductant of redox metal ions in the Fenton reaction (reactions (12) and (13)) seemed improbable to others because other reducing agents such as ascorbate could also play that role and yet were not considered to be highly toxic.

$$M^{n+} + O_2^- \longrightarrow M^{(n-1)+} + O_2 \qquad (12)$$

$$M^{(n-1)+} + H_2O_2 \longrightarrow M^{n+} + OH^- + \cdot OH \qquad (13)$$

Time has provided much experimental evidence to support the hypothesis originally put forward by the enzymologists that CuZnSOD is indeed a central antioxidant enzyme. The corollary that superoxide is an important endogenously generated toxin also is by and large true. But the chemists were also right: Superoxide does not cause random damage but reacts with certain vulnerable targets, particularly exposed 4Fe-4S clusters, which are nonetheless extremely important for survival. Superoxide may also participate in Fenton-type reactions, but it is likely that the damage here is also site-specific, with hydroxyl radical, $\cdot$OH, being generated and reacting in the close vicinity of metal ions bound more or less specifically to biological molecules [178]. These conclusions are based on much work from many laboratories working with different organisms, some of which we will review in the following sections.

Although this chapter focuses primarily on CuZnSOD, it is important in any discussion of biological function that the other SODs be considered as well because they play a vital role in concert with CuZnSOD in vivo, MnSOD, found in the mitochondria of eukaryotes, will therefore be covered in some detail in the following section. The prokaryotic FeSOD and MnSOD, and the CuZnSOD found in chloroplasts of plants, will not be thoroughly discussed here, although their importance to aerobic life is unquestionable.

4.2. Genetic Manipulation of SOD Levels in Whole Organisms

4.2.1. Knockouts

SOD gene knockouts were made first in bacteria [179,180] and then in yeast [181–185]. *Escherichia coli* lacking SOD were found to grow slowly in air, to have an air-dependent auxotropy for branched chain amino acids, and to be very sensitive to redox cycling drugs such as paraquat or phenazine methosulfate. The phenotype of *Saccharomyces cerevisiae* that lack CuZnSOD (*sod1Δ*) was found to be similar though not identical. While the gene is not strictly essential, the strains grow poorly in air, are extremely sensitive to redox-cycling drugs, have air-dependent auxotrophies for cysteine or methionine and for lysine, and will not grow in 100% oxygen. *S. cerevisiae* lacking the mitochondrial MnSOD (*sod2Δ*), on the other hand, grow normally in air, but are sensitive to redox cycling drugs, will not grow in 100% oxygen, and have difficulty growing in nonfermentable carbon sources (that require respiration for their utilization). (Yeast are normally grown in medium containing glucose, the substrate for glycolysis, as the carbon source.)

Indicators of oxidative damage tend to increase in older individuals or populations, implying that oxidative damage plays a role in the aging process, and it has been suggested that it is the main cause of aging [186]. While this hypothesis remains to be proven, it seems likely that oxidative damage is one of several (or many) causes of aging. To study the role of antioxidant proteins in this process, "knockout" mutants and "overexpressors" have been made and characterized in yeast as well as a few higher organisms. The results from knockout experiments have been clear, the overexpressor experimental results were less so.

In yeast, a stationary phase model system has been developed in our laboratory for the purpose of studying oxidative damage associated with aging. Yeast that run out of food enter stationary phase, a nongrowing state in which they live on stored nutrients, and can survive in water for weeks to months. Mutants that lack either CuZnSOD or MnSOD show dramatic decreases in this "chronological life span" and evidence of oxidative damage to cellular components [187]. CuZnSOD mutants, for example, show increased rates of nuclear mutation, and MnSOD mutants show lowered aconitase and succinate dehydrogenase activities and a precipitous drop in oxygen consumption, particularly as they enter into stationary phase, indicating damage to mitochondrial components resulting in decreased energy production [187,188]. Mutants lacking CuZnSOD in *Drosophila melanogaster* were found to have a shortened life span, as well as decreased lifetime activity [189].

Recently, transgenic mice lacking CuZnSOD or MnSOD have been constructed [39,190–192]. Rather surprisingly, the CuZnSOD-minus mice grew and developed normally. In fact, they were little affected, unless injury occurred. Even traditional indicators of oxidative damage (protein carbonyls, lipid peroxidation, and GSSG/GSH ratio) were unaffected, at least in brain. Because these investigators were focused on determining whether lack of CuZnSOD activity could be a cause of the paralytic disease ALS, they assayed motor neuron function and found that axonal regrowth following injury was markedly impaired [39]. Another group, working with a different strain of *sod1-* mice, reported that the main effect of the removal of active CuZnSOD was a drastic decrease in female fertility. In *sod1* −/− mothers, embryos died shortly after implantation; in heterozygous (−/+) or wild-type mothers, this effect was not observed [190], indicating it is a maternal effect.

Mice lacking MnSOD, on the other hand, were severely impaired. They developed normally in utero but died shortly after birth. Two groups have constructed such mice. At birth, the mice made by Epstein and coworkers [191] appeared almost normal, except they were pale, hypothermic (cooler than normal), and tired easily. However, their condition deteriorated rapidly following birth and they all died within 10 days. At autopsy, enlarged hearts and lipid deposits in muscle and liver were observed. Mitochondria were ultrastructurally normal, but the activities of aconitase and succinate dehydrogenase were quite low. The second group constructed MnSOD knockout mice in a different strain background and found a somewhat different phenotype [192]. Their mice survived for 3 weeks, and exhibited severe anemia and neurological symptoms (degeneration of large CNS neurons, weakness, and circling behavior). At later stages, mitochondrial structural damage was evident. Interestingly, in these mice CuZnSOD levels were increased by about 25%. Overall, the affected cell types are those with obligatory requirements for high levels of oxidative metabolism — cardiac myocytes, neurons, hepatocytes, and hematopoietic cells. When the mice from the Epstein laboratory were treated with SOD mimics, the early death was somewhat delayed, and neurological symptoms and spongiform degeneration of the cortex and specific brain stem nuclei were observed [193].

Mice lacking ECSOD have also been constructed and the effect of its absence was mild. Homozygotes developed normally, and no changes were observed in a wide range of other antioxidant enzymes. The ECSOD null mutant mice were more sensitive to 99% oxygen, accumulating lung damage faster and dying sooner than their wild-type counterparts [194].

It is interesting to contrast the effects of SOD knockouts in yeast with those of mammals. MnSOD is apparently much more important in mice, while the CuZnSOD is more important in yeast. The small magnitude of the effect of the absence of MnSOD in yeast may be an artifact of the somewhat artificial conditions in which most lab yeast are grown — they always have plenty of glucose and other nutrients and thus do not need to respire, enter stationary phase, or sporulate, all processes that absolutely require mitochondrial respiration. Nevertheless, it is still true that the CuZnSOD is far more important to the survival of yeast, possibly because each yeast cell in real life could be directly exposed to the air at any time, while murine cells are mostly well protected by physical barriers from direct contact with air. Interestingly, when fetal fibroblast cells are cultured from knockout mice, the phenotype is more like that observed in yeast. $sod1\Delta$ cells are difficult to culture (most die soon after plating), and far more sensitive to the redox cycling drug paraquat, while $sod2\Delta$ yeast are easier to culture and not as sensitive to paraquat, although they are more sensitive than wild-type cells [195]. Mice heterozygous for MnSOD deletion are grossly normal but were found to be more susceptible to cerebral infarction following focal cerebral ischemia [196].

In cell culture models it was shown that downregulation of CuZn-SOD by antisense RNA in neurons resulted in increased apoptotic death after serum withdrawal. Also microinjection of CuZnSOD protein was protective in the same system. Other studies showed that Cu-ZnSOD overexpression in another neuronal cell culture system resulted in protection against this same type of induced apoptosis [197].

4.2.2. Overexpressors

There has been interest in the possible consequences of too much Cu-ZnSOD ever since CuZnSOD was found to be located on human chromosome 21, an extra copy (or partial copy) of which causes Down syndrome. CuZnSOD levels in many Down patients are approximately 150% of normal, as might be expected [198]. It was reported around 10 years ago that transgenic mice overexpressing human CuZnSOD had altered neuromuscular junctions in the tongue, similar to those seen in Down patients [199]. Decreased blood levels of serotonin were also observed in the mice, as in patients, and attributed to more active uptake of serotonin by platelets; reuptake of neurotransmitters in the brain may be similarly affected [200]. More recently, ultrastructural abnormalities in the thymus similar to those found in Down patients have been observed in transgenic mice overexpressing human CuZnSOD. It was suggested that these abnormalities may account for some of the immune system defects in Down syndrome [201]. It is clear, however, that an altered CuZnSOD level is not the whole picture. There are many other genes on chromosome 21 that are also duplicated and that may have effects, including a gene implicated in familial Alzheimer's disease. In addition, CuZnSOD levels are not always elevated in patients. Down syndrome can be caused by partial duplications of chromosome 21, and often these cases show normal CuZnSOD levels [202]. Thus, while CuZnSOD overexpression may account for some of the features of Down syndrome, we must look elsewhere for the whole story.

A longstanding theory, which nevertheless remains to be definitively proven or disproven, holds that accumulation of oxidative damage is a primary cause of aging [186,203,204]. Workers have investigated the possibility that alterations in CuZnSOD can alter aging in several different organisms. It is clear that a lack of CuZnSOD can have drastic consequences on life span, in, e.g., *Drosophila* [189], and yeast in stationary phase (chronological life span) [187]. Mice without MnSOD die within a few days of birth (see above), but an effect on life span of CuZnSOD-lacking mice has not been reported.

What is less clear is whether overexpression of CuZnSOD and/or other antioxidant enzymes can enhance life span or even improve resistance to oxidative stress. Reveillaud et al. [205] reported that overexpression of bovine CuZnSOD in *Drosophila* led to increased resistance to paraquat and to a slight life span extension. Orr and Sohol reported that overexpression of CuZnSOD only led to increased life span if catalase was also overexpressed [206]. More recently, Sun and Tower [207] found that turning on extra CuZnSOD activity in adult *Drosophila* (as opposed to having it present during the larval stages as well) can extend mean life span up to 20% [208]. Overexpression of MnSOD in mice was reported to protect against adriamycin-mediated cardiac damage [208].

On the other hand, there is some evidence that overexpression of SOD by itself is slightly deleterious, resulting in increased susceptibility to oxidative damage, and that simultaneous overexpression of either catalase or glutathione peroxidase (enzymes that degrade hydrogen peroxide) can prevent this effect. The first observation that extra SOD activity could be harmful was made in *E. coli* overexpressing MnSOD [209]. In transgenic tobacco plants, overexpression of chloroplastic CuZnSOD caused increased resistance to oxidative stress. However, ascorbate peroxidase was also elevated in these strains, indicating that the increased protection may have resulted from the double overexpression, rather than just overexpression of the SOD [210,211]. A similar observation was made in cultured mammalian cells overexpressing CuZnSOD. It was noted that the degree of resistance to oxidative stress correlated with the balance between CuZnSOD and glutathione peroxidase (either cotransfected or endogenously induced), rather than with the absolute levels of CuZnSOD [212]. Careful examination of mammalian cells transfected with human SOD showed that the transfection caused a variety of changes in different clones in the expression of other antioxidant enzymes, particularly glutathione peroxidase, and some enzymes not normally associated with oxygen metabolism [213]. On the other hand, in transfected cell lines without elevated Gpx, the resistance to paraquat correlated directly with the SOD level [214]. Thus it is evident that it is not sufficient to consider a single antioxidant enzyme in isolation. The absolute levels and the balance between the different activities are probably both important.

4.3. CuZnSOD as a Component of Antioxidant Systems

Normal metabolic processes involving oxygen necessarily generate O_2^- and/or H_2O_2; if redox-active metal ions are present, then they can catalyze the production of the extremely reactive hydroxyl radical ($^\bullet OH$) or other reactive species from these reactants. Even a process as efficient in its use of oxygen as mitochondrial respiration leaks reactive oxygen species at a measurable rate, which has been estimated (from in vitro experiments) at from 1% to 4% of the oxygen utilized. Other cellular reactions also generate superoxide of H_2O_2 as product. Antioxidant enzymes as well as small-molecule antioxidants are strategically located to deal with these leaks.

CuZnSOD and MnSOD are structurally unrelated enzymes that nevertheless catalyze exactly the same reaction. MnSOD is located in the mitochondrial matrix and CuZnSOD in the cytoplasm and nucleus;

thus these two SODs cover most of the aqueous compartments in the cell. CuZnSOD has also been reported to be found in the intermembrane space of mitochondria [9] (see also [215]). Because hydrogen peroxide is one of the products of superoxide dismutation, enzymes that degrade H_2O_2 are usually located in close proximity: catalase and/or glutathione peroxidase in the cytoplasm, glutathione peroxidase (or cytochrome c peroxidase in yeast) in the intermembrane space of mitochondria. In addition, large amounts of catalase are found in peroxisomes, where heavy generation of H_2O_2 occurs without intermediate production of O_2^-.

There are no known antioxidant enzymes in the lipid compartments; protection is provided by small-molecule antioxidants — primarily vitamin E (or α-tocopherol) and the reduced form of coenzyme Q (QH_2 or ubiquinol). In mammalian systems these are kept reduced by the ascorbic acid cycle. Membranes are quite vulnerable to oxidative damage, particularly if they contain, as most membranes do, polyunsaturated fatty acids. Small-molecule antioxidants in the aqueous compartments are ascorbic acid, glutathione (GSH), and uric acid (in plasma). Ascorbic acid and GSH have enzyme systems that keep them reduced. GSH reductase reduces the oxidized form of glutathione (GSSG) to two molecules of GSH at the expense of NADPH. Thus the system that produces NADPH is important [216].

4.4. Connections Between Metal Metabolism and CuZnSOD

In yeast, lack of CuZnSOD (*sod1Δ*) causes a severe phenotype, which is described above. Perhaps surprisingly, however, strains of yeast lacking both SODs can be constructed. This presents a useful system for finding out what other cellular components can be involved in antioxidant protection. A number of genetic suppressors of the defects observed in yeast lacking SOD have been isolated and are summarized in Table 1 (see also [30,31]). The gene products involved are located in various cellular compartments and organelles, but they have in common an involvement in metal metabolism, particularly copper or manganese metabolism. Simply adding either copper or manganese to the culture medium significantly improves the growth of these strains, and

TABLE 1

Mutations that Rescue the Phenotype of Yeast Lacking SOD

Gene	Rescue by	Mechanism	Reference
CUP1	Overexpr.	Gene for copper metallothionein (MT); required for rescue of *sod1Δ* yeast by excess copper in the medium; MT with bound copper has SOD-like, or superoxide scavenging, activity in vitro, and may have similar activity in vivo as well.	138
bsd2	Deletion	Located in the endoplamic reticulum, regulates translocation to the cell surface of *SMF1* and *SMF2* metal transporters; deletion results in increased intracellular copper.	217
pmr1	Deletion	P-type ATPase in Golgi involved in Mn transport; deletion results in increased Mn accumulation in cytoplasm, which is required for rescue.	220
ATX1	Overexpr.	Copper chaperone in cytoplasm carries copper from *CTR1* copper transporter at plasma membrane to *CCC2* copper transporter into secretory pathway; not clear why overexpression rescues *sod1Δ* yeast, may deliver copper to other site(s) as well.	218
ATX2	Overexpr.	Golgi membrane protein involved in Mn metabolism; overexpression results in increased Mn accumulation, which is required for rescue.	219
coq3, *atp2*, other *pet* mutants	Deletion	Nuclear petite (respiration-deficient) mutations; lack of respiration decreases generation of superoxide.	187

these suppressors appear to cause similar changes in intracellular metal ion levels.

Out of this work has also come the discovery of copper chaperones — small soluble proteins that accept copper at the plasma membrane and deliver it to various cellular targets (reviewed in [221]). The yeast work led directly to the isolation of similar human proteins, indicating that these functions are widespread if not universal. Of particular relevance to the topic of this review was the identification of the gene for a protein that delivers copper to CuZnSOD [147]. In yeast, this gene is *LYS7*, and the phenotype of *lys7* mutants is due to their lack of CuZnSOD activity. Thus, even a protein such as CuZnSOD, which is soluble and cytoplasmic and easy to metallate in vitro, requires special delivery of copper in order to be active. This discovery may also have relevance for the familial forms of the neurodegenerative disease ALS (discussed in Sec. 5).

4.5. Targets of Superoxide Damage

O_2^- is a key player in hydroxyl radical generation in vivo because its dismutation is the primary source of cellular H_2O_2; H_2O_2 in the presence of redox-active metal ions (iron or copper) and a reducing agent can then be further reduced to form the dangerously reactive hydroxyl radical (the Fenton reaction). While O_2^- may act as a reducing agent for metal ions in vivo, many other cellular reducing agents can also perform this role, leaving one to wonder exactly why superoxide itself appeared so toxic. In addition, superoxide on a chemical level is a rather sluggish reactant.

Until recently, there was little hard evidence for toxicity due to direct reactions of O_2^- itself. For one thing, because of the spontaneous dismutation of superoxide to H_2O_2 and O_2, it is difficult to experimentally separate the effects of O_2^- from those of H_2O_2. However, in the past few years, specific targets for superoxide-mediated damage have been identified. Certain iron-sulfur cluster-containing enzymes with exposed 4Fe-4S clusters have been shown to be directly inactivated by superoxide in vivo as well as in vitro. Superoxide partially disassembles the 4Fe-4S cluster, inactivating the enzyme and releasing iron. These targets include the tricarboxylic acid cycle enzyme aconitase and dihydroxyacid dehydratase, an enzyme involved in branched chain amino acid synthesis in *E. coli* [222,223]. The sensitivity of aconitase to inactivation by superoxide, as well as its ubiquitous presence in respiration-competent organisms, has led to its use as a sensor for superoxide in vivo [224].

Elevated levels of oxidative stress have long been known to result in oxidation of DNA, and recent results suggest that free intracellular iron is involved in this oxidation [225,226]. A widely accepted theory is that "free" iron may bind loosely to DNA, where it can act as a catalyst for the generation, from hydrogen peroxide, of a very reactive species that then reacts with and modifies DNA in the immediate vicinity. Among the species suggested to attack DNA are hydroxyl radical, an iron ferryl radical, or an iron-bound hydroxyl radical [227]. Increases in cellular iron have recently been reported to cause iron deposition in the nucleus [225]. Cellular reductants such as superoxide, ascorbate, and NADH may then reduce the bound iron to the ferrous state where Fenton chemistry may then occur in close proximity to DNA bases and sugars [226]. This type of chemistry is also possible for copper ions adventitiously bound to DNA, which has been observed in vitro with NADH as the reductant [228] and would be expected to result in modified bases and/or single-strand breaks in the local DNA. That such accumulation of iron is dangerous is evidenced by the phenotype of yeast *yfh1* mutants. These mutants lack the yeast equivalent of the human protein responsible for Friedreich's ataxia (frataxin) and accumulate iron in their mitochondria. They invariably lose mitochondrial function due to damage to the mitochondrial DNA and are hypersensitive to added hydrogen peroxide [229].

The astute reader will have noticed that another effect of the inactivation of iron-sulfur cluster enzymes is the release of "free" iron into the cellular fluids. Because this free iron can then participate in the Fenton reaction, leading to further damage at sites close to the iron ion, it can be very dangerous. Such a mechanism has been proposed and discussed by Liochev and Fridovich [230]. (See also Chaps. 1 and 3 of this volume.)

In conclusion, it now appears that the toxicity of superoxide is due primarily to two factors: (1) its role as the main precursor of mitochondrially generated H_2O_2 and (2) its ability to react with exposed 4Fe-4S clusters. The second factor may be critically important as it leads to release of free iron as well as inactivation of certain key enzymes.

4.6.   Role of SOD in Apoptosis and Cell Death

There has recently been a great surge of interest in programmed cell death, also known as apoptosis. In response to various kinds of signals, both natural and damage-induced, most if not all kinds of cells can enter a pathway in which they essentially self-destruct in a manner designed to minimize damage to surrounding cells and to promote easy cleanup of the debris. Apoptosis is important during normal development of multicellular organisms, when certain cells are programmed to die during remodeling of various tissues, as well as being a defense mechanism against cancer. Undesired apoptosis has been implicated in some neurological diseases and other pathological states. There is a large and fascinating literature on this subject, which is, unfortunately, somewhat distant from the focus of this chapter. The reader is referred to some recent reviews for further information [231–233].

In recent years, it has become apparent that oxidative signals can cause apoptosis and that SOD levels can influence the degree to which these signals are effective. For example, an early result showed that overexpression of wild-type CuZnSOD in a neural cell line caused increased resistance to induction of apoptosis by two different signals — either withdrawal of growth factors or treatment with a calcium ionophore [234]. It has also been shown that mitochondrial changes are intimately involved in the pathway; one of the early steps in the committed pathway to apoptosis is the release of some mitochondrial contents into the cytoplasm — specifically cytochrome c and calcium (reviewed in [235,236]).

Interestingly, a death pathway may be a very early feature of eukaryotes. There are indications that even as simple a eukaryote as *S. cerevisiae* has primitive features of such a pathway, although death in this organism does not show all the defining hallmarks of apoptosis as seen in mammalian cells. As is described above, yeast mutants that lack CuZnSOD die quickly after entering stationary phase. The human antiapoptotic protein Bcl-2 [237] is partially able to prevent this death [238], indicating some relationship between the human and yeast pathways. However, it should be noted in passing that yeast do not contain an obvious homolog of Bcl-2. The evidence for a death program in yeast has been augmented by other investigators. Two different groups [239,240] have shown that some of the typical morphological changes can be observed in dying yeast, and a third group [241] showed that cytochrome c release and decreased cytochrome oxidase activity was modulated in yeast by Bcl-2 family members, in a manner parallel to that seen in mammalian cells.

5. ROLE OF MUTANT HUMAN CuZnSUPEROXIDE DISMUTASE IN AMYOTROPHIC LATERAL SCLEROSIS

5.1. Amyotrophic Lateral Sclerosis

Amyotrophic lateral sclerosis (ALS, Lou Gehrig's disease) is a neuro-degenerative disease, with a mean age of onset of 55 years, characterized by the slow loss of large motor neurons in the spinal cord and brain [242]. In the vast majority of cases, the disease is sporadic and has no known cause (sporadic ALS, or SALS). There are, however, some encouraging data that suggest the involvement of defectively spliced glutamate transporters and glutamate excitotoxicity [243]. In approximately 10% of cases, ALS is inherited and approximately one-fifth of those familial ALS (FALS) cases are associated with mutations in sod1, the gene that encodes human CuZnSOD [244]. There is also evidence that a small percentage of SALS cases may also be the result of somatic or inherited mutations in sod1 [245]. In the latter case, it may be poorly documented family histories that lead to the classification as SALS and not FALS. While most FALS-causing sod1 mutations are dominant, requiring only one copy of the mutant gene for the ALS phenotype, two mutations have been reported in homozygous patients that apparently leave heterozygotes unaffected [246,247].

5.2. Gain-of-Function Mechanism for FALS Mutant CuZnSODs

Several studies support the conclusion that FALS mutations in CuZnSOD are dominant and exert their effects due to a gain of function: (1) Several of the FALS mutant human CuZnSOD genes (A4V, L38V, G93A, G93C, G37R, G41D, and G85R, but not H46R and H48Q which have no SOD activity) have been found to rescue the oxygen-sensitive

phenotype of *sod1Δ* yeast (see Sec. 4 above), leading to the conclusion
that most FALS mutant CuZnSODs are capable of full or nearly full
functionality in yeast [234,248]. (2) FALS mutant CuZnSOD genes
(A4V, G93A, L38V, G93C, G37R) introduced into cultured neuronal cells
promoted apoptosis where the wild-type human CuZnSOD gene was
antiapoptotic, despite higher-than-normal total SOD activity in all
cases [248]. (3) Constitutive expression of the FALS mutant G93A but
not wild-type human CuZnSOD in a human neuroblastoma cell line
induced a loss of mitochondrial membrane potential and an increase in
cytosolic calcium concentration [249]. (4) Transgenic mice overexpressing FALS mutant CuZnSODs (A4V, G93A, G85R, G37R) developed a
motor neuron degenerative syndrome despite normal or above-normal
SOD activities [250–253], whereas transgenic mice overexpressing
wild-type human CuZnSOD [250,252,253] as well as mice with no Cu-
ZnSOD did not [39].

5.3.   Structural Properties of FALS Mutant CuZnSODs

Since the initial study that reported mutations in *sod1*, the number of
different mutations has risen from 11 to almost 60 [34,37,244,245,247,
254–256]. When mapped onto the three-dimensional structure of human CuZnSOD these mutations tend to cluster at the dimer interface
and in loop regions at the ends of β strands [36,107,257]; however, there
is no real pattern to the distribution of mutations. Figure 1 illustrates
the positions of FALS-mutated residues in the primary structure of human CuZnSOD. The alignment of human, bovine, and yeast sequences
shows that not all of the positions are conserved from species to species.
Six of the reported mutations cause premature stop codons that result
in proteins that are truncated by 20–33 amino acids. Since all regions of
the protein contain ALS-causing mutations, one can speculate that the
structure of CuZnSOD is exquisitely sensitive to even small perturbations in the protein fold.

5.3.1.   X-Ray Structure of Human G37R CuZnSOD

To date only one crystal structure for a mutant CuZnSOD has been
published. In addition to high-resolution crystallography, human G37R

has been studied extensively both in vivo and in vitro. Mice that are transgenic for G37R develop motor neuron disease [258]. The mutant protein has been reported to have full specific activity but a twofold reduction in biological half-life relative to wild-type [38]. G37R is also one of the mutations that was shown to enhance apoptosis in a dominant fashion in neural cell culture [234].

The x-ray structure of the human G37R was determined and analyzed to 1.9 Å resolution [90]. The structure of G37R CuZnSOD shows typical β-barrel topology consistent with known CuZnSOD structures and there are no gross deviations from the wild-type and thermostable mutant human CuZnSOD protein coordinates in the Protein Data Bank (pdb 1 spd, pdb 1 sos) [36,47]. While human CuZnSOD expressed in *E. coli* is not N-acetylated as it is in human cells, human G37R CuZnSOD is properly modified by the yeast expression system [40,259]. The replacement of glycine by arginine causes remarkably little rearrangement of the protein backbone relative to wild type.

The only major change in the structure is that the two subunits in the G37R dimer have distinct environments and are different in structure at their copper binding sites. In one subunit, the bridging imidazolate (His63) coordinates both metal ions and the metal ion in the copper site has a four-coordinate ligand geometry suggesting an oxidized copper site. The other subunit shows a distorted trigonal planar geometry at the copper site and no electron density between the copper ion and His63. This is consistent with a reduced copper site. The subunit asymmetry seen in this crystal structure may indicate that mutations can affect subunit communication and the symmetrical nature of the human dimer.

5.3.2. Truncation Mutations

While the majority of FALS-causing mutations are single substitutions of one amino acid residue for another, six of the reported mutations result in proteins truncated by as many as 33 amino acids. These proteins are missing part or all of the last β strand, which is involved in dimer contact; it is difficult to imagine that they are long-lived enough to cause disease. Indeed, several studies have failed to find evidence that truncated protein is even synthesized [260–262]. How these mutations result in the same gain of function as the others is enigmatic, and it is precisely this question that seems to derail otherwise elegant theories as to why *sod1* mutations cause this disease. It may well be that these mutations hold the key to uncovering the pathogenicity of FALS mutants.

5.4. Hypotheses for Toxicity of Mutant CuZnSOD

5.4.1. Protein Aggregation and Neurofilament Abnormalities

Protein aggregation has been implicated in CuZnSOD-associated FALS in several studies. Human FALS mutant CuZnSODs, but not wild-type human CuZnSOD, was observed to aggregate when expressed in cultured spinal motor neurons [263]. In a yeast two-hybrid assay, FALS mutant CuZnSOD was found to bind to two proteins, lysyl-tRNA synthetase and translocon-associated protein delta, whereas wild-type did not [264]. It is also interesting to note that expression of many of the FALS mutant human CuZnSODs in *E. coli* led to formation of inclusion bodies, whereas expression of the wild type did not [288]. Aggregation might also explain the toxicity of the unstable truncation mutations.

Swelling of the axon of the motor neuron and neurofilament abnormalities are frequently associated with ALS and have been observed in the transgenic mice expressing FALS mutant CuZnSODs as well [265–267]. The hypothesis that the deleterious effect of the FALS mutant CuZnSODs cause neurofilament abnormalities is further supported by the recent demonstration of impaired axonal transport in the ventral roots of human G93A CuZnSOD transgenic mice and the appearance of neurofilament inclusions and vacuoles in vulnerable motor neurons [268].

5.4.2. Alterations in Metal Binding Properties

Since CuZnSOD is an abundant component of eukaryotic cells and it is both a copper- and a zinc-binding protein, one cannot rule out its involvement in metal metabolism. Both copper and zinc have been implicated in neurotoxicity [269,270], and alterations in their intracellular levels or partitioning could potentially mediate motor neuron death in ALS. Several ALS mutations were found to alter the characteristic metal ion binding properties of wild-type CuZnSODs in the case of both yeast FALS analogs [248,271] and the mutant human enzymes

[271–273]. Also, non-protein-bound copper was found to elute close to mutant SOD protein during purification from ALS erythrocytes. This suggested that at least for G37R and H46R copper could leak out during the purification process [260].

Refuting this idea is a study that demonstrated copper replete protein isolated from patients with the D90A mutation [274]. Total blood and plasma copper and zinc levels were shown to be virtually the same in FALS patients and nonneurological controls, and no increase in "free" copper was seen in FALS erythrocytes [275]. Although this rules out gross changes in organismal metal metabolism, it should be noted that blood is not cerebrospinal fluid and erythrocytes are not neurons, so these data do not rule out localized changes in the nervous system or changes in intracellular localization or state of chelation for copper and zinc. It is also possible that relatively small changes in the intracellular levels of these metals could result in toxicity and that this might go unnoticed due to the insensitivity of the experimental method.

5.4.3. Catalysis of Oxidations by Hydrogen Peroxide

Copper is strongly implicated in the pathogenic mechanism of FALS mutant CuZnSODs. Copper chelators provide a modicum of protection from FALS mutant-induced death in both a cell culture model of FALS [234] and in the FALS SOD-expressing transgenic mice [276]. As was stated before in Sec. 3, CuZnSOD is remarkably specific for superoxide. It is possible that these mutants have lost this specificity and are able to react with other favorable substrates to a greater extent than is the wild-type protein. This theory fits with the dominant gain of function that is the working hypothesis for disease initiation.

Wild-type CuZnSOD can react with hydrogen peroxide to generate hydroxyl radical ($^{\bullet}$OH) or a hydroxyl radical-like species, which is decidedly capable of damaging cellular components. Fortunately, nature has designed CuZnSOD to minimize this thermodynamically favorable but biologically undesirable reaction [159,277]. This reaction can be monitored using the spin trap 5,5'-dimethyl-1-pyrroline N-oxide as an acceptor for the hydroxyl radical. In this fashion it was shown that two human FALS mutants, A4V and G93A, possessed higher peroxidative activities than did wild-type human CuZnSOD [40,278,279], while in vitro this mechanism uses peroxide as a reductant of the oxidized enzyme — a reaction that is known to be quite slow [158]. In vivo, other cellular reductants are likely to play a part as well, especially considering the fact that the actual levels of intracellular hydrogen peroxide are likely to be quite low. A recent study observed a superoxide-dependent peroxidase activity for the H48Q FALS CuZnSOD mutant enzyme. The same activity was not observed for the G93A, G93R, or E100G mutants [280]. The possibility that ascorbate could act as a cellular reductant for FALS mutants is also intriguing [271].

While this theory is not proven it must be noted that there is an increasing body of evidence for excess oxidative damage in tissues of ALS patients. Oxidatively damaged nucleic acids, proteins, and lipids have recently been reported in tissue samples obtained from both sporadic ALS and SOD-associated FALS patients [281]. Moreover, vitamin E was shown to have a beneficial effect on transgenic mice expressing G93A mutant CuZnSOD [282].

5.4.4. Catalysis of Tyrosine Nitration

As previously mentioned, CuZnSOD has the ability to react with peroxynitrite to catalyze the nitration of tyrosine residues. Peroxynitrite is formed by the rapid reaction of superoxide with nitric oxide in vivo and it is possible that the toxicity of mutant CuZnSODs is mediated through the further reaction of this powerful oxidant with the enzyme to nitrate protein-bound tyrosine residues [283].

Supporting this hypothesis is evidence from G37R transgenic mice that exhibit a two- to threefold elevation in free nitrotyrosine levels in spinal cord tissue relative to normal mice or mice expressing high levels of wild-type human enzyme [284]. This is also supported by a study that demonstrated increased 3-nitrotyrosine immunoreactivity in the motor neurons of both sporadic and familial ALS patients [285,286]. The peroxynitrite-dependent nitration activity is also enhanced in FALS mutants that have been depleted of zinc [272]. The same group also showed that CuZnSOD was capable of nitrating neurofilament L, providing a likely target within motor neurons [154].

5.5. Conclusions

While there is no consensus regarding the factor or factors that contribute to the pathogenicity of mutant CuZnSODs, clear progress has been made in revealing the more subtle characteristics of this well-studied enzyme. Biophysical characterization of mutant and wild-type CuZnSODs continues to shed light on the structural requirements for proper function just as biological studies have continued to expand the definition of proper function. An illustration of the complexity of the problem is the transgenic mouse model. To date four lines of mice with four different mutations have been studied. What has been determined is that each mutation seems to have a different pathological course with the same end-result—death by motor neuron disease. It is possible that there are multiple properties of the mutant enzymes that contribute to the toxicity, including SOD, peroxidative, and nitration activities, as well as protein stability. The inability to correlate properties of individual mutations with disease severity coupled with heterogeneity within FALS families suggests the possibility of other as-yet-unidentified factors that contribute to the progression of the disease. One study advanced the idea that apolipoprotein E genotyping may play a role [287] in SALS. This line of research may prove beneficial for the FALS community as well.

ACKNOWLEDGMENTS

Research support from the ALS Association and the USPHS (GM28222 and DK46828) is gratefully acknowledged. We also thank Hongbin Liu for providing Fig. 2.

ABBREVIATIONS

ALS	amyotrophic lateral sclerosis
CNS	central nervous system
CuZnSOD	copper, zinc superoxide dismutase
DSC	differential scanning calorimetry
ECSOD	extracellular superoxide dismutase
EPR	electron paramagnetic resonance
FALS	familial amyotrophic lateral sclerosis (inherited form of the disease)
GSH	reduced glutathione

GSSG	oxidized glutathione
MnSOD	manganese superoxide dismutase
NADH	nicotinamide adenine dinucleotide (reduced)
NADPH	nicotinamide adenine dinucleotide phosphate (reduced)
NMR	nuclear magnetic resonance
QH_2	reduced coenzyme Q, or ubiquinol
ROS	reactive oxygen species
SALS	sporadic amyotrophic lateral sclerosis
SDS	sodium dodecyl sulfate
SOD	superoxide dismutase
sod1Δ	gene deletion of CuZnSOD
sod2Δ	gene deletion of MnSOD

REFERENCES

1. M. Dole, *J. Gen. Physiol.*, *49*, 5–27 (1965).
2. J. M. McCord and I. Fridovich, *J. Biol. Chem.*, *244*, 6049–6055 (1969).
3. S. Marklund, *Acta Chem. Scand.*, *27*, 1458–1460 (1973).
4. M. Brouwer, T. H. Brouwer, W. Grater, J. J. Enghild, and I. B. Thogersen, *Biochemistry*, *36*, 13381–13388 (1997).
5. J. G. Scandalios, in *Oxidative Stress and the Molecular Biology of Antioxidant Defenses* (J. G. Scandalios, ed.), Cold Spring Harbor Laboratory Press, Cold Spring Harbor, NY, 1997, pp. 527–568.
6. F. J. Corpas, L. M. Sandalios, L. A. Del Rio, and R. N. Trelease, *New Phytol.*, *138*, 307–314 (1998).
7. J. S. Kroll, P. R. Langford, K. E. Wilks, and A. D. Keil, *Microbiology*, *141*, 2271–2279 (1995).
8. M. D. Tomalski, R. Eldridge, and L. K. Miller, *Virology*, *184*, 149–161 (1991).
9. A. F. Miller and D. L. Sorkin, *Comments Mol. Cell. Biophys.*, *9*, 1–48 (1997).
10. M. M. Whittaker and J. W. Whittaker, *Biochemistry*, *36*, 8923–8931 (1997).
11. M. M. Whittaker and J. W. Whittaker, *Biochemistry*, *35*, 6762–6770 (1996).

12. D. Touati, in *Oxidative Stress and the Molecular Biology of Antioxidant Defenses* (J. G. Scandalios, ed.), Cold Spring Harbor Laboratory Press, Cold Spring Harbor, NY, 1997, pp. 447–493.

13. J. A. Fee, in *Superoxide and Superoxide Dismutases* (A. M. Michelson, J. M. McCord, and I. Fridovich, eds.), Academic Press, New York, 1977, pp. 173–192.

14. H. D. Youn, H. Youn, J. W. Lee, Y. I. Yim, J. K. Lee, Y. C. Hah, and S. O. Kang, *Arch. Biochem. Biophys.*, *334*, 341–348 (1996).

15. H. D. Youn, E. J. Kim, J. H. Roe, Y. C. Hah, and S. O. Kang, *Biochem. J.*, *318*, 889–896 (1996).

16. E. J. Kim, H. P. Kim, Y. C. Hah, and J. H. Roe, *Eur. J. Biochem.*, *241*, 178–185 (1996).

17. I. Bertini, S. Mangani, and M. S. Viezzoli, *Adv. Inorg. Chem.*, *45*, 127–250 (1998).

18. I. Bertini, L. Banci, and M. Piccioli, *Coord. Chem. Rev.*, *100*, 67–103 (1990).

19. J. V. Bannister, W. H. Bannister, and G. Rotilio, *CRC Crit. Rev. Biochem.*, *22*, 110–180 (1987).

20. G. Rotilio, L. Calabrese, A. Rigo, and E. M. Fielden, *Adv. Exp. Med. Biol.*, *148*, 155–168 (1982).

21. H. M. Steinman, in *Superoxide Dismutase*, Vol. 1 (L. W. Oberley, ed.), CRC Press, Boca Raton, 1982, pp. 11–68.

22. J. S. Valentine and M. W. Pantoliano, in *Copper Proteins* (T. G. Spiro, ed.), John Wiley and Sons, New York, 1981, pp. 292–358.

23. J. S. Valentine, in *Bioinorganic Chemistry* (I. Bertini, H. B. Gray, S. J. Lippard, and J. S. Valentine, eds.), University Science Books, Mill Valley, CA, 1994, pp. 253–313.

24. D. E. Cabelli, D. Riley, J. A. Rodriguez, J. S. Valentine, and H. Zhu, in *Biomimetic Oxidations Catalyzed by Transition Metal Complexes* (B. Meunier, ed.), Imperial College Press (in press).

25. I. Fridovich, *J. Biol. Chem.*, *272*, 18515–18517 (1997).

26. I. Fridovich, *Annu. Rev. Biochem.*, *64*, 97–112 (1995).

27. I. Fridovich, *Adv. Enzymol. Rel. Areas Mol. Biol.*, *58*, 61–97 (1986).

28. I. Fridovich, *Adv. Exp. Biol. Med.*, *74*, 530–539 (1976).

29. I. Fridovich, *Annu. Rev. Biochem.*, *44*, 147–159 (1975).

30. E. B. Gralla and D. J. Kosman, *Adv. Genet.*, *30*, 251–319 (1992).

31. E. B. Gralla, in *Oxidative Stress and the Molecular Biology of Antioxidant Defenses* (J. G. Scandalios, ed.), Cold Spring Harbor Laboratory Press, Cold Spring Harbor, NY, 1997, pp. 495–525.

32. J. S. Valentine, D. L. Wertz, T. J. Lyons, L.-L. Liou, J. J. Goto, and E. B. Gralla, *Curr. Op. Chem. Biol.*, 2, 253–262 (1998).

33. R. H. Brown, Jr., *Curr. Opin. Neurobiol.*, 5, 841–846 (1995).

34. T. Siddique, D. Nijhawan, and A. Hentati, *J. Neural Transm. Suppl.*, 49, 219–233 (1997).

35. T. Mann and D. Keilin, *Proc. Roy. Soc.*, B126, 303 (1939).

36. H.-X. Deng, A. Hentati, J. H. Tainer, Z. Iqbal, A. Cayabyab, W.-Y. Hung, E. D. Getzoff, P. Hu, B. Herzfeldt, R. P. Roos, C. Warner, G. Deng, E. Soriano, C. Smyth, H. E. Parge, A. Ahmed, A. D. Roses, R. A. Hallewell, M. Pericak-Vance, and T. Siddique, *Science*, 261, 1047–1051 (1993).

37. D. R. Rosen, T. Siddique, D. Patterson, D. A. Figlewicz, P. Sapp, A. Hentati, D. Donaldson, J. Goto, J. P. O'Regan, H.-X. Deng, Z. Rahmani, A. Krizus, D. McKenna-Yasek, A. Cayabyab, S. M. Gaston, R. Berger, R. E. Tanzi, J. J. Halperin, B. Herzfeldt, R. Van den Bergh, W.-Y. Hung, T. Bird, G. Deng, D. W. Mulder, C. Smyth, N. G. Laing, E. Soriano, M. A. Pericak-Vance, J. Haines, G. A. Rouleau, J. S. Gusella, H. R. Horvitz, and R. H. Brown, Jr., *Nature*, 362, 59–62 (1993).

38. D. R. Borchelt, M. K. Lee, H. S. Slunt, M. Guarnieri, Z. S. Xu, P. C. Wong, R. H. Brown, D. L. Price, S. S. Sisodia, and D. W. Cleveland, *Proc. Natl. Acad. Sci. USA*, 91, 8292–8296 (1994).

39. A. G. Reaume, J. L. Elliott, E. K. Hoffman, N. W. Kowall, R. J. Ferrante, D. F. Siwek, H. M. Wilcox, D. G. Flood, M. F. Beal, R. H. J. Brown, R. W. Scott, and W. D. Snider, *Nature Genet.*, 13, 43–47 (1996).

40. M. Wiedau-Pazos, J. J. Goto, S. Rabizadeh, E. B. Gralla, J. A. Roe, M. K. Lee, J. S. Valentine, and D. E. Bredesen, *Science*, 271, 515–518 (1996).

41. T. D. Oury, J. D. Crapo, Z. Valnickova, and J. J. Enghild, *Biochem. J.*, 317, 51–57 (1996).

42. A. Battistoni and G. Rotilio, *FEBS Lett.*, 374, 199–202 (1995).

43. S. Kanematsu and K. Asada, *Plant Cell Physiol.*, 30, 381–391 (1989).

44. J. S. Richardson, D. C. Richardson, K. A. Thomas, E. W. Silverton, and D. R. Davies, *J. Mol. Biol.*, *102*, 221–235 (1976).

45. J. A. Tainer, E. D. Getzoff, K. M. Beem, J. S. Richardson, and D. C. Richardson, *J. Mol. Biol.*, *160*, 181–217 (1982).

46. H. E. Parge, E. D. Getzoff, C. S. Scandella, R. A. Hallewell, and J. A. Tainer, *J. Biol. Chem.*, *261*, 16215–16218 (1986).

47. H. E. Parge, R. A. Hallewell, and J. A. Tainer, *Proc. Natl. Acad. Sci. USA*, *89*, 6109– 6113 (1992).

48. K. Djinovic, G. Gatti, A. Coda, L. Antolini, G. Pelosi, A. Desideri, M. Falconi, F. Marmocchi, G. Rotilio, and M. Bolognesi, *J. Mol. Biol.*, *225*, 791–809 (1992).

49. K. Djinovic, G. Gatti, A. Coda, L. Antolini, G. Pelosi, A. Desideri, M. Falconi, F. Marmocchi, G. Rotilio, and M. Bolognesi, *Acta Crystallogr.*, *B47*, 918–927 (1991).

50. F. Frigerio, M. Falconi, G. Gatti, M. Bolognesi, A. Desideri, F. Marmocchi, and G. Rotilio, *Biochem. Biophys. Res. Commun.*, *160*, 677–681 (1989).

51. N. L. Ogihara, H. E. Parge, P. J. Hart, M. S. Weiss, J. J. Goto, B. R. Crane, J. Tsang, K. Slater, J. A. Roe, J. S. Valentine, D. Eisenberg, and J. A. Tainer, *Biochemistry*, *35*, 2316–2321 (1996).

52. W. R. Rypniewski, S. Mangani, B. Bruni, P. L. Orioli, M. Casati, and K. S. Wilson, *J. Mol. Biol.*, *251*, 282–296 (1995).

53. K. A. Thomas, B. H. Rubin, C. J. Bier, J. S. Richardson, and D. C. Richardson, *J. Biol. Chem.*, *249*, 5677–5683 (1974).

54. J. S. Richardson, K. A. Thomas, B. H. Rubin, and D. C. Richardson, *Proc. Natl. Acad. Sci. USA*, *72*, 1349–1353 (1975).

55. K. Djinovic, A. Coda, L. Antolini, G. Pelosi, A. Desideri, M. Falconi, G. Rotilio, and M. Bolognesi, *J. Mol. Biol.*, *226*, 227–238 (1992).

56. K. Djinovic-Carugo, F. Polticelli, A. Desideri, G. Rotilio, K. S. Wilson, and M. Bolognesi, *J. Mol. Biol.*, *240*, 179–183 (1994).

57. D. E. McRee, S. M. Redford, E. D. Getzoff, J. R. Lepock, R. A. Hallewell, and J. A. Tainer, *J. Biol. Chem.*, *265*, 14234–14241 (1990).

58. C. D. Smith, M. Carson, M. van der Woerd, J. Chen, H. Ischiropoulos, and J. S. Beckman, *Arch. Biochem. Biophys.*, *299*, 350–355 (1992).

59. K. Djinovic-Carugo, C. Collyer, A. Coda, M. T. Carri, A. Battistoni, G. Bottaro, F. Polticelli, A. Desideri, and M. Bolognesi, *Biochem. Biophys. Res. Commun.*, *194*, 1008–1011 (1993).

60. K. Djinovic-Carugo, A. Battistoni, M. T. Carri, F. Polticelli, A. Desideri, G. Rotilio, A. Coda, K. S. Wilson, and M. Bolognesi, *Acta Crystallogr., Sect. D: Biol. Crystallogr.*, *1*, 176–188 (1996).

61. Y. Kitagawa, N. Tanaka, Y. Hata, M. Kusunoki, G. Lee, Y. Katsube, K. Asada, S. Aibara, and Y. Morita, *J. Biochem.*, *109*, 477–485 (1991).

62. W. H. Bannister, J. V. Bannister, D. Barra, J. Bond, and F. Bossa, *Free Rad. Res. Commun.*, *12–13*, 349–361 (1991).

63. D. Bordo, K. Djinovic, and M. Bolognesi, *J. Mol. Biol.*, *238*, 366–386 (1994).

64. J. V. Bannister, A. Anastasi, and W. H. Bannister, *Biochem. Biophys. Res. Commun.*, *81*, 469–472 (1978).

65. F. Marmocchi, G. Venardi, F. Bossa, A. Rigo, and G. Rotilio, *FEBS Lett.*, *94*, 109–111 (1978).

66. A. Rigo, F. Marmocchi, D. Cocco, P. Viglino, and G. Rotilio, *Biochemistry*, *17*, 534–537 (1978).

67. F. Marmocchi, G. Venardi, G. Caulini, and G. Rotilio, *FEBS Lett.*, *44*, 337–339 (1974).

68. F. Marmocchi, G. Caulini, G. Venardi, D. Cocco, L. Calabrese, and G. Rotilio, *Physiol. Chem. Phys.*, *7*, 465–471 (1975).

69. G. Caulini, F. Marmocchi, G. Venardi, and G. Rotilio, *Boll. Soc. Ital. Biol. Sper.*, *50*, 1091–1094 (1974).

70. D. P. Malinkowski and I. Fridovich, *Biochemistry*, *18*, 5055–5060 (1979).

71. K. Inouye, A. Osaki, and B. Tonomura, *J. Biochem.*, *115*, 507–515 (1994).

72. G. Mei, N. Rosato, N. Silva, Jr., R. Rusch, E. Gratton, I. Savini, and A. Finazzi-Agro, *Biochemistry*, *31*, 7224–7230 (1992).

73. N. Silva, Jr., E. Gratton, G. Mei, N. Rosato, R. Rusch, and A. Finazzi-Agro, *Biophys. Chem.*, *48*, 171–182 (1993).

74. F. Marmocchi, I. Mavelli, A. Rigo, R. Stevanato, F. Bossa, and G. Rotilio, *Biochemistry*, *21*, 2853–2856 (1982).

75. S. Folcarelli, A. Battistoni, M. T. Carri, F. Polticelli, M. Falconi,

L. Nicolini, L. Stella, N. Rosato, G. Rotilio, and A. Desideri, *Protein Eng.*, *9*, 323–325 (1996).

76. I. Bertini, M. Piccioli, M. S. Viezzoli, C. Y. Chiu, and G. T. Mullenbach, *Eur. Biophys. J.*, *23*, 167–176 (1994).

77. L. Banci, I. Bertini, C. Y. Chiu, G. T. Mullenbach, and M. S. Viezzoli, *Eur. J. Biochem.*, *234*, 855–860 (1995).

78. L. Banci, M. Benedetto, I. Bertini, R. Del Conte, M. Piccioli, T. Richert, and M. S. Viezzoli, *Magnet. Reson. Chem.*, *35*, 845–853 (1997).

79. L. Banci, L. Bertini, M. S. Viezzoli, E. Argese, E. F. Orsega, C. Y. Chiu, and G. T. Mullenbach, *J. Biol. Inorg. Chem.*, *2*, 295–301 (1997).

80. S. J. Lippard, A. R. Burger, K. Ugurbil, J. S. Valentine, and M. W. Pantoliano, in *Advances in Chemistry Series, No. 162, Bioinorganic Chemistry II* (K. Raymond, ed.), American Chemical Society, Washington, D.C., 1977, pp. 251–262.

81. J. A. Roe, A. Butler, D. M. Scholler, J. S. Valentine, L. Marky, and K. Breslauer, *Biochemistry*, *27*, 950–958 (1988).

82. E. M. Fielden, P. B. Roberts, R. C. Bray, D. J. Lowe, G. N. Mautner, G. Rotilio, and L. Calabrese, *Biochem. J.*, *139*, 49–60 (1974).

83. A. Rigo, P. Viglino, L. Calabrese, D. Cocco, and G. Rotilio, *Biochem. J.*, *161*, 27–30 (1977).

84. M. S. Viezzoli and Y. Wang, *Inorg. Chim. Acta*, *153*, 189–191 (1988).

85. J. Hirose, S. Toida, H. Kuno, S. Ozaki, and Y. Kidani, *Chem. Pharm. Bull. (Tokyo)*, *36*, 2103–2108 (1988).

86. M. Falconi, R. Gallimbeni, and E. Paci, *J. Comput.-Aided Mol. Des.*, *10*, 490–498 (1996).

87. G. Chillemi, M. Falconi, A. Amadei, G. Zimatore, A. Desideri, and A. DiNola, *Biophys. J.*, *73*, 1007–1018 (1997).

88. J. C. Dunbar, B. Holmquist, and J. T. Johansen, *Biochemistry*, *23*, 4330–4335 (1984).

89. T. Lyons, A. Nersissian, J. J. Goto, H. Zhu, E. B. Gralla, and J. S. Valentine, *J. Biol. Inorg. Chem.* (in press).

90. P. J. Hart, H. Liu, M. Pellegrini, A. M. Nersissian, E. B. Gralla, J. S. Valentine, and D. Eisenberg, *Prot. Sci.*, *7*, 545–555 (1998).

91. L. Benov, H. Sage, and I. Fridovich, *Arch. Biochem. Biophys.*, *340*, 305–310 (1997).

92. L. T. Benov and I. Fridovich, *J. Biol. Chem.*, *269*, 25310–25314 (1994).

93. A. Battistoni, S. Folcarelli, L. Cervoni, F. Polizio, A. Desideri, A. Giartosio, and G. Rotilio, *J. Biol. Chem.*, *273*, 5655–5661 (1998).

94. J. R. Lepock, L. D. Arnold, B. H. Torrie, B. Andrews, and J. Kruuv, *Arch. Biochem. Biophys.*, *241*, 243–251 (1985).

95. Y. Senoo, K. Katoh, Y. Nakai, Y. Hashimoto, K. Bando, and S. Teramoto, *Acta Medica Okayama*, *42*, 169–174 (1988).

96. J. Kajihara, M. Enomoto, K. Seya, Y. Sukenaga, and K. Katoh, *J. Biochem.*, *104*, 638–642 (1988).

97. W. H. Bannister and E. J. Wood, *Life Sci.*, *9*, 229–233 (1970).

98. N. Crosti, *Biochem. Genet.*, *16*, 739–742 (1978).

99. E. Wenisch, K. Vorauer, A. Jungbauer, H. Katinger, and P. G. Righetti, *Electrophoresis*, *15*, 647–653 (1994).

100. L. Calabrese, G. Federici, W. H. Bannister, J. V. Bannister, G. Rotilio, and A. Finazzi-Agro, *Eur. J. Biochem.*, *56*, 305–309 (1975).

101. B. P. Sharonov and I. V. Churilova, *Biochem. Biophys. Res. Commun.*, *189*, 1129–1135 (1992).

102. I. Mavelli, M. R. Ciriolo, and G. Rotilio, *Biochem. Biophys. Res. Commun.*, *117*, 677–681 (1983).

103. S. L. Jewett, *Inorg. Chim. Acta*, *79*, 144–145 (1983).

104. K. Arai, S. Iizuka, Y. Tada, K. Oikawa, and N. Taniguchi, *Biochim. Biophys. Acta*, *924*, 292–296 (1987).

105. N. Taniguchi, K. Arai, and N. Kinoshita, *Meth. Enzymol.*, *179*, 570–581 (1989).

106. N. Kawamura, T. Ookawara, K. Suzuki, K. Konishi, M. Mino, and N. Taniguchi, *J. Clin. Endocrinol. Metab.*, *74*, 1352–1354 (1992).

107. D. E. Bredesen, L. M. Ellerby, P. J. Hart, M. Wiedau-Pazos, and J. S. Valentine, *Ann. Neurol.*, *42*, 135–137 (1997).

108. L. Rossi, E. Marchese, A. De Martino, G. Rotilio, and M. R. Ciriolo, *Biometals*, *10*, 257–262 (1997).

109. S. L. Jewett, G. S. Latrenta, and C. M. Beck, *Arch. Biochem. Biophys.*, *215*, 116–128 (1982).

110. S. A. Goscin and I. Fridovich, *Biochim. Biophys. Acta*, *289*, 276–283 (1972).

111. S. A. Cockle and R. C. Bray, in *Superoxide and Superoxide Dismutases* (A. M. Michelson, J. M. McCord, and I. Fridovich, eds.), Academic Press, New York, 1977, pp. 215–216.

112. F. Polizio, A. R. Bizzarri, S. Cannistraro, and A. Desideri, *Eur. Biophys. J. Biophys. Lett.*, *26*, 291–297 (1997).

113. R. A. Hallewell, K. C. Imlay, P. Lee, N. M. Fong, C. Gallegos, E. D. Getzoff, J. A. Tainer, D. E. Cabelli, P. Tekamp-Olson, G. T. Mullenbach, and L. S. Cousens, *Biochem. Biophys. Res. Commun.*, *181*, 474–480 (1991).

114. J. R. Lepock, H. E. Frey, and R. A. Hallewell, *J. Biol. Chem.*, *265*, 21612 21618 (1990).

115. G. E. Schulz and R. H. Schirmer, *Principles of Protein Structure*, Springer-Verlag, New York, 1979.

116. L. Jia, C. Bonaventura, J. Bonaventura, and J. S. Stamler, *Nature*, *380*, 221–226 (1996).

117. L. Banci, I. Bertini, D. E. Cabelli, R. A. Hallewall, J. W. Tung, and M. S. Viezzoli, *Eur. J. Biochem.*, *196*, 123–128 (1991).

118. A. Merli, G. Rossi, K. Djinovic-Carugo, M. Bolognesi, A. Desideri, and G. Rotilio, *Biochem. Biophys. Res. Commun.*, *210*, 1040–1044 (1995).

119. I. Ascone, R. Castaner, C. Tarricone, M. Bolognesi, M. E. Stroppolo, and A. Desideri, *Biochem. Biophys. Res. Commun.*, *241*, 119–121 (1997).

120. I. Bertini, C. Luchinat, and R. Monnanni, *J. Am. Chem. Soc.*, *107*, 2178–2179 (1985).

121. S. Hashimoto, S. Ohsaka, H. Takeuchi, and I. Harada, *J. Am. Chem. Soc.*, *111*, 8926–8928 (1989).

122. L. M. Murphy, R. W. Strange, and S. S. Hasnain, *Structure*, *5*, 371–379 (1997).

123. M. E. McAdam, E. M. Fielden, F. Lavelle, L. Calabrese, D. Cocco, and G. Rotilio, *Biochem. J.*, *167*, 271–274 (1977).

124. P. O'Neill, E. M. Fielden, D. Cocco, G. Rotilio, and L. Calabrese, *Biochem. J.*, *205*, 181–187 (1982).

125. L. Calabrese, G. Rotilio, and B. Mondovi, *Biochim. Biophys. Acta*, *263*, 827–829 (1972).

126. L. Calabrese, D. Cocco, L. Morpurgo, B. Mondovi, and G. Rotilio, *Eur. J. Biochem.*, *64*, 465–470 (1976).

127. K. M. Beem, W. E. Rich, and K. V. Rajagopalan, *J. Biol. Chem.*, *249*, 7298–7305 (1974).

128. J. A. Fee and R. G. Briggs, *Biochim. Biophys. Acta*, *400*, 439–450 (1975).

129. H. J. Forman and I. Fridovich, *J. Biol. Chem.*, *248*, 2645–2649 (1973).

130. L.-J. Ming and J. S. Valentine, *J. Am. Chem. Soc.*, *109*, 4426–4428 (1987).

131. J. Kajihara, M. Enomoto, K. Katoh, K. Mitsuta, and M. Kohno, *Agric. Biol. Chem.*, *54*, 495–499 (1990).

132. L. Calabrese, D. Cocco, and A. Desideri, *FEBS Lett.*, *106*, 142–144 (1979).

133. P. Kofod, R. Bauer, E. Danielsen, E. Larsen, and M. J. Bjerrum, *Eur. J. Biochem.*, *198*, 607–611 (1991).

134. L.-J. Ming and J. S. Valentine, *J. Am. Chem. Soc.*, *112*, 6374–6383 (1990).

135. K. M. Beem, D. C. Richardson, and K. V. Rajagopalan, *Biochemistry*, *16*, 1930–1936 (1977).

136. M. R. Ciriolo, P. Civitareale, M. T. Carri, A. De Martino, F. Galiazzo, and G. Rotilio, *J. Biol. Chem.*, *269*, 25783–25787 (1994).

137. V. C. Culotta, H. D. Joh, S. J. Lin, K. H. Slekar, and J. Strain, *J. Biol. Chem.*, *270*, 29991–29997 (1995).

138. K. T. Tamai, E. B. Gralla, L. M. Ellerby, J. S. Valentine, and D. J. Thiele, *Proc. Natl. Acad. Sci. USA*, *90*, 8013–8017 (1993).

139. C. T. Dameron and E. D. Harris, *Biochem. J.*, *248*, 663–668 (1987).

140. R. A. DiSilvestro, *Arch. Biochem. Biophys.*, *274*, 298–303 (1989).

141. A. Levieux, D. Levieux, and C. Lab, *Free Radic. Biol. Med.*, *11*, 589–595 (1991).

142. K. L. Olin, M. S. Golub, M. E. Gershwin, A. G. Hendrickx, B. Lonnerdal, and C. L. Keen, *Am. J. Clin. Nutr.*, *61*, 1263–1267 (1995).

143. F. Galiazzo, M. R. Ciriolo, M. T. Carri, P. Civitareale, L. Marcocci, F. Marmocchi, and G. Rotilio, *Eur. J. Biochem.*, *196*, 545–549 (1991).

144. C. Steinkuhler, M. T. Carri, G. Micheli, L. Knoepfel, U. Weser, and G. Rotilio, *Biochem. J.*, *302*, 687–694 (1994).

145. C. Steinkuhler, O. Sapora, M. T. Carri, W. Nagel, L. Marcocci, M. R. Ciriolo, U. Weser, and G. Rotilio, *J. Biol. Chem.*, *266*, 24580–24587 (1991).

146. N. Petrovic, A. Comi, and M. J. Ettinger, *J. Biol. Chem.*, *271*, 28331–28334 (1996).

147. V. C. Culotta, L. W. Klomp, J. Strain, R. L. Casareno, B. Krems, and J. D. Gitlin, *J. Biol. Chem.*, *272*, 23469–23472 (1997).

148. J. R. Ellis, *Science*, *250*, 954–959 (1990).

149. G. Rotilio, R. C. Bray, and E. M. Fielden, *Biochim. Biophys. Acta*, *268*, 605–609 (1972).

150. E. K. Hodgson and I. Fridovich, *Biochem. Biophys. Res. Commun.*, *54*, 270–274 (1973).

151. M. A. Symonyan and R. M. Nalbandyan, *FEBS Lett.*, *28*, 22–24 (1972).

152. J. A. Fee and P. E. DiCorleto, *Biochemistry*, *12*, 4893–4899 (1973).

153. M. W. Pantoliano, J. S. Valentine, A. R. Burger, and S. J. Lippard, *J. Inorg. Biochem.*, *17*, 325–341 (1982).

154. J. P. Crow, Y. Z. Ye, M. Strong, M. Kirk, S. Barnes, and J. S. Beckman, *J. Neurochem.*, *69*, 1945–1953 (1997).

155. R. J. Carrico and H. F. Deutsch, *J. Biol. Chem.*, *245*, 723–727 (1970).

156. P. Viglino, M. Scarpa, F. Coin, G. Rotilio, and A. Rigo, *Biochem. J.*, *237*, 305–308 (1986).

157. P. Viglino, M. Scarpa, D. Cocco, and A. Rigo, *Biochem. J.*, *229*, 87–90 (1985).

158. A. Rigo, P. Viglino, M. Scarpa, and G. Rotilio, in *Superoxide and Superoxide Dismutase in Chemistry, Biology and Medicine* (G. Rotilio, ed.), Elsevier, Amsterdam, 1986, pp. 184–188.

159. M. B. Yim, P. B. Chock, and E. R. Stadtman, *Proc. Natl. Acad. Sci. USA*, *87*, 5006–5010 (1990).

160. D. E. Cabelli, D. Allen, B. J. Bielski, and J. Holcman, *J. Biol. Chem.*, *264*, 9967–9971 (1989).

161. E. K. Hodgson and I. Fridovich, *Biochemistry*, *14*, 5294–5298 (1975).

162. K. Uchida and S. Kawakishi, *J. Biol. Chem.*, *269*, 2405–2410 (1994).

163. J. S. Beckman, H. Ischiropoulos, L. Zhu, M. van der Woerd, C. Smith, J. Chen, J. Harrison, J. C. Martin, and M. Tsai, *Arch. Biochem. Biophys.*, *298*, 438–445 (1992).

164. J. S. Beckman, J. Chen, J. P. Crow, and Y. Z. Ye, *Prog. Brain Res.*, *103*, 371–380 (1994).

165. D. G. Searcy, J. P. Whitehead, and M. J. Maroney, *Arch. Biochem. Biophys.*, *318*, 251–263 (1995).

166. D. G. Searcy, *Arch. Biochem. Biophys.*, *334*, 50–58 (1996).

167. D. Deters and U. Weser, *Biometals*, *8*, 25–29 (1995).

168. W. H. Koppenol, in *Oxygen and Oxy-Radicals in Chemistry and Biology* (M. A. J. Rogers and E. L. Powers, eds.), Academic Press, New York, 1981, pp. 671–674.

169. E. D. Getzoff, J. A. Tainer, P. K. Weiner, P. A. Kollman, J. S. Richardson, and D. C. Richardson, *Nature*, *306*, 287–290 (1983).

170. D. P. Malinkowski and I. Fridovich, in *Chemical and Biochemical Aspects of Superoxide and Superoxide Dismutase* (J. V. Bannister and H. A. O. Hill, eds.), Elsevier, New York, 1980, pp. 299–317.

171. A. Cudd and I. Fridovich, *J. Biol. Chem.*, *257*, 11443–11447 (1982).

172. W. F. Beyer, I. Fridovich, G. T. Mullenbach, and R. Hallewell, *J. Biol. Chem.*, *262*, 11182–11187 (1987).

173. E. D. Getzoff, D. E. Cabelli, C. L. Fisher, H. E. Parge, M. S. Viezzoli, L. Banci, and R. A. Hallewall, *Nature*, *358*, 347–351 (1992).

174. P. Viglino, M. Scarpa, G. Rotilio, and A. Rigo, *Biochim. Biophys. Acta*, *952*, 77–82 (1988).

175. J. A. Graden, L. M. Ellerby, J. A. Roe, and J. S. Valentine, *J. Am. Chem. Soc.*, *116*, 9743–9744 (1994).

176. L. M. Ellerby, D. E. Cabelli, J. A. Graden, and J. S. Valentine, *J. Am. Chem. Soc.*, *118*, 6556–6561 (1996).

177. D. T. Sawyer and J. S. Valentine, *Acc. Chem. Res.*, *14*, 393–400 (1981).

178. P. Saltman, *Semin. Hematol.*, *26*, 249–256 (1989).

179. D. Touati, *Free Rad. Res. Commun.*, *8*, 1–9 (1989).

180. A. Carlioz and D. Touati, *EMBO J.*, *5*, 623–630 (1986).

181. X. F. Liu, I. Elashvili, E. B. Gralla, J. S. Valentine, P. Lapinskas, and V. C. Culotta, *J. Biol. Chem.*, *267*, 18298–18302 (1992).

182. E. B. Gralla and J. S. Valentine, *J. Bacteriol.*, *173*, 5918–5920 (1991).

183. A. P. van Loon, B. Pesold-Hurt, and G. Schatz, *Proc. Natl. Acad. Sci. USA*, *83*, 3820–3824 (1986).

184. C. A. Marres, A. P. Van Loon, P. Oudshoorn, H. Van Steeg, L. A. Grivell, and E. C. Slater, *Eur. J. Biochem.*, *147*, 153–161 (1985).

185. T. Bilinski, Z. Krawiec, A. Liczmanski, and J. Litwinska, *Biochem. Biophys. Res. Commun.*, *130*, 533–539 (1985).

186. R. S. Sohal and R. Weindruch, *Science*, *273*, 59–63 (1996).

187. V. D. Longo, E. B. Gralla, and J. S. Valentine, *J. Biol. Chem.*, *271*, 12275–12280 (1996).

188. V. L. Longo, L. L. Liou, J. S. Valentine, and E. B. Gralla (submitted).

189. J. P. Phillips, S. D. Campbell, D. Michaud, M. Charbonneau, and A. J. Hilliker, *Proc. Natl. Acad. Sci. USA*, *86*, 2761–2765 (1989).

190. Y.-S. Ho, M. Gargano, J. Cao, R. T. Bronson, I. Heimler, and R. J. Hutz, *J. Biol. Chem.*, *273*, 7765–7769 (1998).

191. Y. Li, T. T. Huang, E. J. Carlson, S. Melov, P. C. Ursell, J. L. Olson, L. J. Noble, M. P. Yoshimura, C. Berger, P. H. Chan, D. C. Wallace, and C. J. Epstein, *Nature Genet.*, *11*, 376–381 (1995).

192. R. M. Lebovitz, H. Zhang, H. Vogel, J. Cartwright, Jr., L. Dionne, N. Lu, S. Huang, and M. M. Matzuk, *Proc. Natl. Acad. Sci. USA*, *93*, 9782–9787 (1996).

193. S. Melov, J. A. Schneider, B. J. Day, D. Hinerfeld, P. Coskun, S. S. Mirra, J. D. Crapo, and D. C. Wallace, *Nature Genet.*, *18*, 159–163 (1998).

194. L. M. Carlsson, F. Jonsson, T. Edlund, and S. L. Marklund, *Proc. Natl. Acad. Sci. USA*, *92*, 6264–6268 (1995).

195. T.-T. Huang, M. Yasunami, E. J. Carlson, A. M. Gillespie, A. G. Reaume, E. K. Hoffman, P. H. Chan, R. W. Scott, and C. J. Epstein, *Arch. Biochem. Biophys.*, *344*, 424–432 (1997).

196. K. Murakami, T. Kondo, M. Kawase, Y. Li, S. Sato, S. F. Chen, and P. H. Chan, *J. Neurosci.*, *18*, 205–213 (1998).

197. L. J. Greenlund, T. L. Deckwerth, and E. M. J. Johnson, *Neuron*, *14*, 303–315 (1995).

198. P. M. Sinet, F. Lavelle, A. M. Michelson, and H. Jerome, *Biochem. Biophys. Res. Commun.*, *67*, 904–909 (1975).

199. K. B. Avraham, M. Schickler, D. Sapoznikov, R. Yarom, and Y. Groner, *Cell*, *54*, 823–829 (1988).

200. M. Schickler, H. Knobler, K. B. Avraham, O. Elroy-Stein, and Y. Groner, *EMBO J.*, *8*, 1385–1392 (1989).

201. B. Nabarra, M. Casanova, D. Paris, E. Paly, K. Toyoma, I. Ceballos, and J. London, *Mech. Ageing Dev.*, *96*, 59–73 (1997).

202. R. De La Torre, A. Casado, E. Lopez-Fernandez, D. Carrascosa, V. Ramirez, and J. Saez, *Experientia*, *52*, 871–873 (1996).

203. K. B. Beckman and B. N. Ames, in *Oxidative Stress and the Molecular Biology of Antioxidant Defenses* (J. Scandalios, ed.), Cold Spring Harbor Laboratory Press, Cold Spring Harbor, NY, 1997, pp. 201– 246.

204. K. Z. Guyton, M. Gorospe, and N. J. Holbrook, in *Oxidative Stress and the Molecular Biology of Antioxidant Defenses* (J. G. Scandalios, ed.), Cold Spring Harbor Laboratory Press, Cold Spring Harbor, NY, 1997, pp. 247–272.

205. I. Reveillaud, A. Niedzwiecki, K. G. Bensch, and J. E. Fleming, *Mol. Cell. Biol.*, *11*, 632–640 (1991).

206. W. C. Orr and R. S. Sohal, *Science*, *263*, 1128–1130 (1994).

207. J. Sun and J. Tower (in preparation).

208. H. C. Yen, T. D. Oberley, S. Vichitbandha, Y. S. Ho, and D. K. St. Clair, *J. Clin. Invest.*, *98*, 1253–1260 (1996).

209. M. D. Scott, S. R. Meshnick, and J. W. Eaton, *J. Biol. Chem.*, *262*, 3640–3645 (1987).

210. A. S. Gupta, J. L. Heinen, A. S. Holaday, J. J. Burke, and R. D. Allen, *Proc. Natl. Acad. Sci. USA*, *90*, 1629–1633 (1993).

211. A. S. Gupta, R. P. Webb, A. S. Holaday, and R. D. Allen, *Plant Physiol.*, *103*, 1067–1073 (1993).

212. P. Amstad, R. Moret, and P. Cerutti, *J. Biol. Chem.*, *269*, 1606–1609 (1994).

213. M. J. Kelner, R. Bagnell, M. Montoya, L. Estes, S. F. Uglik, and P. Cerutti, *Free Rad. Biol. Med.*, *18*, 497–506 (1995).

214. T.-T. Huang, E. J. Carlson, S. A. Leadon, and C. J. Epstein, *FASEB J.*, *6*, 903–910 (1992).

215. D. M. Glerum and A. Tzagoloff, personal communication.

216. K. H. Slekar, D. J. Kosman, and V. C. Culotta, *J. Biol. Chem.*, *271*, 28831–28836 (1996).
217. X. F. Liu, F. Supek, N. Nelson, and V. C. Culotta, *J. Biol. Chem.*, *272*, 11763–11769 (1997).
218. S. J. Lin, R. A. Pufahl, A. Dancis, T. V. O'Halloran, and V. C. Culotta, *J. Biol. Chem.*, *272*, 9215–9220 (1997).
219. S. J. Lin and V. C. Culotta, *Mol. Cell. Biol.*, *16*, 6303–6312 (1996).
220. P. J. Lapinskas, K. W. Cunningham, X. F. Liu, G. R. Fink, and V. C. Culotta, *Mol. Cell. Biol.*, *15*, 1382–1388 (1995).
221. J. S. Valentine and E. B. Gralla, *Science*, *278*, 817–818 (1997).
222. P. R. Gardner, *Biosci. Rep.*, *17*, 33–42 (1997).
223. D. H. Flint and R. M. Allen, *Chem. Rev.*, *96*, 2315–2334 (1996).
224. P. R. Gardner, I. Raineri, L. B. Epstein, and C. W. White, *J. Biol. Chem.*, *270*, 13399–13405 (1995).
225. K. Keyer and J. A. Imlay, *Proc. Natl. Acad. Sci. USA*, *93*, 13635–13640 (1996).
226. R. Meneghini, *Free Rad. Biol. Med.*, *23*, 783–792 (1997).
227. E. S. Henle and S. Linn, *J. Biol. Chem.*, *272*, 19095–19098 (1997).
228. S. Oikawa and S. Kawanishi, *Biochemistry*, *35*, 4584–4590 (1996).
229. M. Babcock, D. de Silva, R. Oaks, S. Davis-Kaplan, S. Jiralerspong, L. Montermini, M. Pandolfo, and J. Kaplan, *Science*, *276*, 1709–1712 (1997).
230. S. I. Liochev and I. Fridovich, *Free Rad. Biol. Med.*, *16*, 29–33 (1994).
231. S. W. Hetts, *JAMA*, *279*, 300–307 (1998).
232. M. E. Peter, A. E. Heufelder, and M. O. Hengartner, *Proc. Natl. Acad. Sci. USA*, *94*, 12736–12737 (1997).
233. J. L. Rinkenberger and S. J. Korsmeyer, *Curr. Opin. Genet. Dev.*, *7*, 589–596 (1997).
234. S. Rabizadeh, E. B. Gralla, D. R. Borchelt, R. Gwinn, J. S. Valentine, S. Sisodia, P. Wong, M. Lee, H. Hahn, and D. E. Bredesen, *Proc. Natl. Acad. Sci. USA*, *92*, 3024–3028 (1995).
235. S. Papa and V. P. Skulachev, *Mol. Cell. Biochem.*, *174*, 305–319 (1997).
236. N. Zamzami, T. Hirsch, B. Dallaporta, P. X. Petit, and G. Kroemer, *J. Bioenerg. Biomembr.*, *29*, 185–193 (1997).

237. J. C. Reed, *Semin. Hematol.*, *34*, 9–19 (1997).

238. V. D. Longo, L. M. Ellerby, D. E. Bredesen, J. S. Valentine, and E. B. Gralla, *J. Cell Biol.*, *137*, 1581–1588 (1997).

239. F. Madeo, E. Frohlich, and K. U. Frohlich, *J. Cell Biol.*, *139*, 729–734 (1997).

240. B. Ink, M. Zornig, B. Baum, N. Hajibagheri, C. James, T. Chittenden, and G. Evan, *Mol. Cell. Biol.*, *17*, 2468–2474 (1997).

241. S. Manon, B. Chaudhuri, and M. Guerin, *FEBS Lett.*, *415*, 29–32 (1997).

242. R. H. Brown, Jr., *Arch. Neurol.*, *54*, 1246–1250 (1997).

243. L. A. Bristol and J. D. Rothstein, *Ann. Neurol.*, *39*, 676–678 (1996).

244. M. E. Cudkowicz and R. H. Brown, Jr., *J. Neurol. Sci.*, *139 Suppl.*, 10–15 (1996).

245. M. Jackson, A. Al-Chalabi, Z. E. Enayat, B. Chioza, P. N. Leigh, and K. E. Morrison, *Ann. Neurol.*, *42*, 803–807 (1997).

246. P. M. Andersen, L. Forsgren, M. Binzer, P. Nilsson, V. Ala-Hurula, M.-L. Karenen, L. Bergmark, A. Saarinen, T. Haltia, I. Tarvainen, E. Kinnunen, B. Udd, and S. L. Marklund, *Brain*, *119*, 1153–1172 (1996).

247. C. Hayward, D. J. H. Brock, R. A. Minns, and R. J. Swingler, *J. Med. Genet.*, *35*, 174 (1998).

248. C. Nishida, E. B. Gralla, and J. S. Valentine, *Proc. Natl. Acad. Sci. USA*, *91*, 9906–9910 (1994).

249. M. T. Carri, A. Ferri, A. Battistoni, L. Famhy, R. Gabbianelli, F. Poccia, and G. Rotilio, *FEBS Lett.*, *414*, 365–368 (1997).

250. M. C. Dal Canto and M. E. Gurney, *Acta Neuropathol. (Berl)*, *93*, 537–550 (1997).

251. L. I. Bruijn, M. W. Becher, M. K. Lee, K. L. Anderson, N. A. Jenkins, N. G. Copeland, S. S. Sisodia, J. D. Rothstein, D. R. Borchelt, D. L. Price, and D. W. Cleveland, *Neuron*, *18*, 327–338 (1997).

252. M. E. Gurney, H. Pu, A. Y. Chiu, M. C. Dal Canto, C. Y. Polchow, D. D. Alexander, J. Caliendo, A. Hentati, Y. W. Kwon, H. X. Deng, W. Chen, P. Zhai, R. L. Sufit, and T. Siddique, *Science*, *264*, 1772–1775 (1994).

253. M. E. Gurney, *J. Neurol. Sci.*, *152 Suppl. 1*, S67–S73 (1997).

254. T. Siddique, *Cold Spring Harbor Symp. Quant. Biol.*, *61*, 699–708 (1996).

255. R. W. Orrell, J. J. Habgood, D. I. Shepherd, D. Donnai, and J. de Belleroche, *Eur. J. Neurol.*, *4*, 48–51 (1997).

256. M. Kostrzewa, M. S. Damian, and U. Mueller, *Hum. Genet.*, *98*, 48–50 (1996).

257. J. S. Valentine, P. J. Hart, and E. B. Gralla, in *Copper Transport and Its Disorders: Molecular and Cellular Aspects* (A. Leone and J. F. B. Mercer, eds.), Plenum Press (in press).

258. D. W. Cleveland, L. I. Bruijn, P. C. Wong, J. R. Marszalek, J. D. Vechio, M. K. Lee, X. S. Xu, D. R. Borchelt, S. S. Sisodia, and D. L. Price, *Neurology, 47 Suppl. 2*, S54–S62 (1996).

259. D. O. Natvig, K. Imlay, D. Touati, and R. A. Hallewell, *J. Biol. Chem.*, *262*, 14697–14701 (1987).

260. Y. Ogawa, H. Kosaka, T. Nakanishi, A. Shimizu, N. Ohoi, H. Shouji, T. Yanagihara, and S. Sakoda, *Biochem. Biophys. Res. Commun.*, *241*, 251–257 (1997).

261. Y. Watanabe, N. Kuno, Y. Kono, E. Nanba, E. Ohama, K. Nakashima, and K. Takahashi, *Acta Neurol. Scand.*, *95*, 167–172 (1997).

262. Y. Watanabe, Y. Kono, E. Nanba, E. Ohama, and K. Nakashima, *FEBS Lett.*, *400*, 108–112 (1997).

263. H. D. Durham, J. Roy, L. Dong, and D. A. Figlewicz, *J. Neuropathol. Exptl. Neurol.*, *56*, 523–530 (1997).

264. C. B. Kunst, E. Mezey, M. J. Brownstein, and D. Patterson, *Nature Genet.*, *15*, 91–94 (1997).

265. T. L. Williamson, J. R. Marszalek, J. D. Vechio, L. I. Bruijn, M. K. Lee, Z. Xu, R. H. Brown, Jr., and D. W. Cleveland, *Cold Spring Harbor Symp. Quant. Biol.*, *61*, 709–723 (1996).

266. P. H. Tu, M. E. Gurney, J. P. Julien, V. M. Lee, and J. Q. Trojanowski, *Lab. Invest.*, *76*, 441–456 (1997).

267. L. I. Bruijn and D. W. Cleveland, *Neuropathol. Appl. Neurobiol.*, *22*, 373–387 (1996).

268. B. Zhang, P. Tu, F. Abtahian, J. Q. Trojanowski, and V. M. Lee, *J. Cell Biol.*, *139*, 1307–1315 (1997).

269. J. Y. Koh, S. W. Suh, B. J. Gwag, Y. Y. He, C. Y. Hsu, and D. W. Choi, *Science*, *272*, 1013–1016 (1996).

270. Z. L. Harris and J. D. Gitlin, *Am. J. Clin. Nutr.*, *63*, 836S–841S (1996).

271. T. J. Lyons, H. Liu, J. J. Goto, A. M. Nersissian, J. A. Roe, J. A. Graden, C. Café, L. M. Ellerby, D. E. Bredesen, E. B. Gralla, and J. S. Valentine, *Proc. Natl. Acad. Sci. USA*, *93*, 12240–12244 (1996).

272. J. P. Crow, J. B. Sampson, Y. Zhuang, J. A. Thompson, and J. S. Beckman, *J. Neurochem.*, *69*, 1936–1944 (1997).

273. M. T. Carri, A. Battistoni, F. Polizio, A. Desideri, and G. Rotilio, *FEBS Lett.*, *356*, 314–316 (1994).

274. S. L. Marklund, P. M. Andersen, L. Forsgren, P. Nilsson, P. I. Ohlsson, G. Wikander, and A. Öberg, *J. Neurochem.*, *69*, 675–681 (1997).

275. A. Radunovic, H. T. Delves, W. Robberecht, P. Tilkin, Z. E. Enayat, C. E. Shaw, Z. Stevic, S. Apostolski, J. F. Powell, and P. N. Leigh, *Ann. Neurol.*, *42*, 130–131 (1997).

276. A. F. Hottinger, E. G. Fine, M. E. Gurney, A. D. Zurn, and P. Aebischer, *Eur. J. Neurosci.*, *9*, 1548–1551 (1997).

277. M. B. Yim, P. B. Chock, and E. R. Stadtman, *J. Biol. Chem.*, *268*, 4099–4105 (1993).

278. H. S. Yim, J. H. Kang, P. B. Chock, E. R. Stadtman, and M. B. Yim, *J. Biol. Chem.*, *272*, 8861–8863 (1997).

279. M. B. Yim, J. H. Kang, H. S. Yim, H. S. Kwak, P. B. Chock, and E. R. Stadtman, *Proc. Natl. Acad. Sci. USA*, *93*, 5709–5714 (1996).

280. S. I. Liochev, L. L. Chen, R. A. Hallewell, and I. Fridovich, *Arch. Biochem. Biophys.*, *346*, 263–268 (1997).

281. R. J. Ferrante, S. E. Browne, L. A. Shinobu, A. C. Bowling, M. J. Baik, U. MacGarvey, N. W. Kowall, R. H. Brown, Jr., and M. F. Beal, *J. Neurochem.*, *69*, 2064–2074 (1997).

282. M. E. Gurney, F. B. Cutting, P. Zhai, A. Doble, C. P. Taylor, P. K. Andrus, and E. D. Hall, *Ann. Neurol.*, *39*, 147–157 (1996).

283. J. S. Beckman, M. Carson, C. D. Smith, and W. H. Koppenol, *Nature*, *364*, 584 (1993).

284. L. I. Bruijn, M. F. Beal, M. W. Becher, J. B. Schulz, P. C. Wong, D. L. Price, and D. W. Cleveland, *Proc. Natl. Acad. Sci. USA*, *94*, 7606–7611 (1997).

285. M. F. Beal, R. J. Ferrante, S. E. Browne, R. T. Matthews, N. W. Kowall, and R. H. Brown, Jr., *Ann. Neurol.*, *42*, 644–654 (1997).

286. R. J. Ferrante, L. A. Shinobu, J. B. Schulz, R. T. Matthews, C. E. Thomas, N. W. Kowall, M. E. Gurney, and M. F. Beal, *Ann. Neurol.*, *42*, 326–334 (1997).

287. B. Moulard, A. Sefiani, A. Laamri, A. Malafosse, and W. Camu, *J. Neurol. Sci.*, *139 Suppl.*, 34–37 (1996).

6

DNA Damage Mediated by Metal Ions with Special Reference to Copper and Iron

José-Luis Sagripanti

Molecular Biology Branch, CDRH, Food and Drug
Administration, 12709 Twinbrook Parkway,
Rockville, MD 20852, USA

1. INTRODUCTION

Our interest on metal ions started with the discovery that low levels of microwaves can damage DNA in the presence of copper ions [1]. Subsequent discovery of base-specific DNA damage mediated by Cu(II) [2] made the study of metal-mediated DNA damage a main effort of my laboratory. The need to identify binding sites that could explain the observed sequence-specific DNA damage promoted a parallel study of metal-DNA interactions. Our interest on copper broadened to iron primarily through the contribution of Shinya Toyokuni, who was awarded a postdoctoral fellowship to work with us. Pressing needs at the U.S. Food and Drug Administration to improve disinfectants, combined with our experience in virology, soon brought several metal ions into focus in microbiology projects. Because of this previous experience, the main thrust of this chapter will relate to our work with copper and iron. I believe that these metals are good choices for a chapter on metal-mediated DNA damage because copper and iron are two major transition metals in the biological environment that can catalyze extensive DNA damage in vitro and in vivo [2–6]. Furthermore, discussion of DNA lesions created by these metals can be related to clinical conditions, since it is well established that overload of copper or iron in humans or rodents leads to specific diseases associated with oxidative DNA damage and an increased incidence of cancer [5,6].

　　We consider metal-mediated DNA damage as an intermediate step after metal-DNA binding or interaction and before the onset of any related biological consequences. Thus, our examination of metal-mediated DNA damage will be complemented by discussion of noncovalent interactions or binding between metal ions and DNA, as well as by attempts to associate the DNA damage observed in vitro with lesions measured in relevant biological models.

2. BINDING OF METAL IONS TO DNA

Early studies showed that nucleic acids isolated from a variety of sources contain large amounts of firmly bound metals, including iron, zinc, copper, chromium, manganese, nickel, and aluminum [7–9]. The binding to DNA of Be(II), Co(II), Mn(II), Cu(II), Cr(III), Ni(II), and, to a lesser extent, Pb(II), Cd(II) or Zn(II) alters DNA polymerase activity and increases the incorporation of genetic errors [10,11]. Infidelity in DNA synthesis by many metal ions is likely associated with interference in the binding of Mg(II), which is an essential ion for proper interaction between DNA and DNA polymerase [12]. Cadmium or nickel can substitute zinc in zinc finger proteins, altering the binding of transcription factors, a process that has been associated with metal-induced DNA damage and carcinogenesis [13–15].

Promising techniques to elucidate interactions of DNA with metal ions (particularly for Zn(II)) have included the search for metal binding domains by nucleic acid hybridization and functionally genetic approaches [16]. We studied the binding of Cu(II) by determining (1) the number and affinity of DNA sites by equilibrium and chemical binding analysis, and (2) the metal specificity of DNA sites, through competition studies with other divalent cations.

2.1 Binding Sites in DNA by Equilibrium Analysis

At least two kinds of binding sites for Cu(II) [17] were identified in the DNA double helix by incubating the metal ion with DNA and determining the amounts of copper remaining free in solution and bound to DNA after the sample reached equilibrium [18]. One DNA site for Cu(II) is saturable and present on the average once every base pair. These binding sites in high abundance had a stability constant (K_a) of 2.4×10^4 M^{-1}, a value comparable to those previously obtained by spectroscopic analysis [19]. The frequency of appearance of the high-abundance site measured by equilibrium analysis is similar to that estimated by measuring the release of Cu(II) during DNA renaturation [20]. However, we do not consider this copper binding site simply as counterion for neutralization of phosphate charges. Since the stability constant reported for the interaction of Cu(II) and phosphate is 7×10^2

M^{-1} [21], binding to phosphate alone could not explain the higher affinity we found for Cu(II)-DNA. The low competition observed for other divalent cations (discussed below) also indicates a more specific interaction between copper and DNA.

A comparison of the concentration of high abundance Cu(II) binding sites in double-stranded DNA (dsDNA) [17] and the amount of copper tightly bound per mg of DNA in chromatin [8] indicates that only 1 in 3700 copper sites are actually occupied in DNA in vivo. These unsaturated low affinity binding sites present in every base pair could be ideal candidates to envision a regulatory role of copper at the DNA level.

The other Cu(II) binding site in DNA is present on the average once every four nucleotides, has a high affinity, and presents marked deviation from linearity with a positive slope at low occupancy of DNA sites (Fig. 1), suggesting a cooperative effect in copper binding. The abundance obtained for this site agrees with earlier reported stoichiometry (one complex per four bases) of DNA base-copper complexes [22]. The shape of the plot obtained for Cu(II) resembles that reported for carcinogenic divalent metals such as Hg(II), Cd(II), and Ni(II) [23–25].

2.2. Chemical Studies

Studies on modified bases and nucleotide crystals have shown that Cu(II) binds covalently to the N-7 of purine bases, assuming either square-pyramidal or octahedral coordination geometries [26]. Analysis of the electrostatic potentials at the surface of DNA shows that the most electronegative region in DNA corresponds to a sequence containing two or more guanosines (≈ -85 e/Å potential) [27]. The carbonyl oxygen, O6, is a unique feature of guanine that contributes to the high electrostatic potential of polyguanosine (poly G) sequences in DNA. Poly G sequences present the most intense electronegative target in DNA for $Cu(H_2O)_6^{2+}$. Adenine shows a considerably smaller (≈ -43 e/Å) electronegative potential than guanine due in part to lack of a carbonyl oxygen. Cytosine and thymine present much less electronegative regions than guanine and adenine. Hence, electrostatic considerations suggest that interaction of copper with cytosine and thymine should be expected to occur after guanine and adenine binding sites have been occupied.

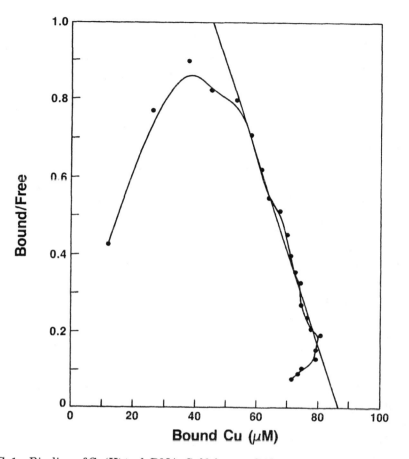

FIG. 1   Binding of Cu(II) to dsDNA. Calf thymus DNA (56 µg) was incubated in
a 1-mL volume at 20°C for 1 h with various concentrations of Cu(II) (10–400
µM). Bound and free Cu(II) were separated by DNA precipitation and measured
by flame atomic absorption spectrophotometry. Each point represents the average of seven independent experiments. (Reproduced with permission from [17].)

Computer-assisted modeling of the three-dimensional interaction between copper and DNA predicted eight possible copper-binding structures [27]. After calculating the number and geometry of coordination
sites, as well as the strain energy and the distortion they would exert
over the DNA helix, modeling analysis also indicated that interaction of
copper with sites involving guanine is most likely.

Recently, elucidation of the crystal structure of the Z-DNA sequence d(CGCGCG), soaked with Cu(II) chloride solution after crystallization, indicated that Cu(II) ions do not bind to phosphate moieties
but instead form a covalent coordinate bond to N-7 of each guanine [28].
Soaking with copper did not perturb crystallized DNA [28]; however,
studies using laser Raman spectroscopy have shown that copper has a
pronounced structural effect when DNA is in solution [29], raising
questions about the relevance of crystallographic data of preformed
crystals soaked with metal salts.

2.3. Metal Specificity of DNA Interaction

Before a metal ion can interact with DNA in a living organism, it may
compete for the target site with a variety of other essential cations. It
is well known that some metal ions can antagonize the effects of others
both in vitro and in vivo. For example, cadmium binding to DNA in vitro
was antagonized by zinc, magnesium, and calcium [24]. Magnesium,
manganese, and calcium reduced nickel binding to DNA in vitro [25].
The relative capacities of these metals to antagonize binding of cadmium and nickel to DNA correlates remarkably with their abilities to
prevent cadmium and nickel carcinogenesis [24,25]. Other studies have
shown that zinc administration attenuates copper toxicity in Wilson's
disease [30,31].
    The capacity of nine metals to antagonize Cu(II) binding to DNA is
shown in Fig. 2. The displacement of Cu(II) was approximately a linear
function of the log antagonist concentration in the range tested. The
relative efficacy of several divalent cations to antagonize Cu(II) binding
was Ni $\simeq$ Cd $\simeq$ Mg $\gg$ Zn $\simeq$ Hg > Ca > Pb $\gg$ Mn, whereas Cr(VI)
enhanced Cu(II) binding to DNA. The distorted octahedral coordination
sphere proposed for Cu(II) and the strong tendency to coordinate donor
atoms equatorially [32] may be responsible for the specific interaction of
Cu(II) with DNA.
    The value for the apparent stability constant of 2.4×10^4 M^{-1}
obtained for Cu(II) binding sites in high abundance (see previous section) is lower than the reported values for cadmium (4×10^4 M^{-1}; [24])
and for nickel (19×10^4 M^{-1}; [25]). The fact that metals with higher
affinity constants for DNA, such as cadmium and nickel, displace Cu(II)

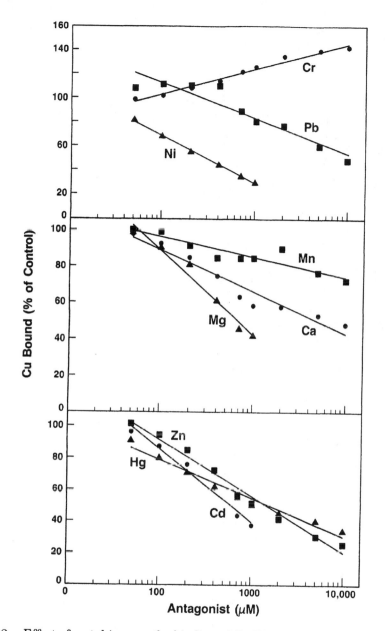

FIG. 2. Effect of metal ions on the binding of Cu(II) to DNA. Cu(II) (150 μM)
was mixed with various concentrations of competing metal ion (50–10 000 μM)
before incubation with dsDNA. Cu(II) bound to DNA was separated from free
Cu(II) by precipitation and measured by atomic absorption. The binding of
Cu(II) in absence of competing ions was considered as 100% and each point in
the graph represents the average of five to nine independent determinations.
(Reproduced with permission from [17].)

SAGRIPANTI

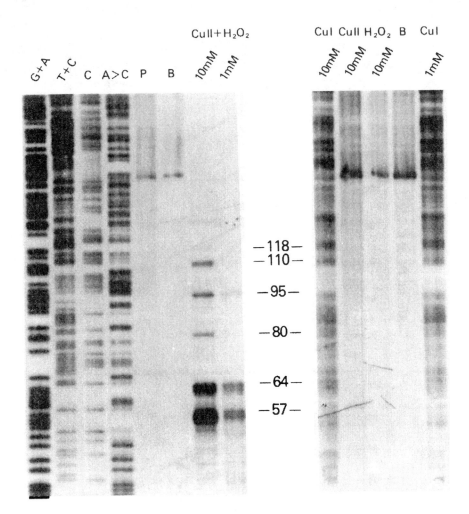

FIG. 3. Sequence specificity of DNA damage produced by Cu(II) plus H_2O_2. Columns 7 and 8 (from the left) are gel electrophoresis of ^{32}P 3'-end-labeled fragments of plasmid pZ189 DNA (partial sequence at bottom) treated with Cu(II) plus H_2O_2 for 30 min at 23°C followed by treatment with piperidine at

minimally leads us to speculate that different metal ions prefer different binding sites in the DNA double helix. Furthermore, the low antagonism of Zn(II) suggests that the mitigating effect of zinc in anticopper therapy for Wilson's disease [30,31] is not mediated by displacement of Cu(II) from DNA binding sites.

2.4. Specificity of DNA Damage

During earlier studies, we investigated whether preferential interaction of copper with a particular DNA base or DNA sequence could focus energy into a narrow region of DNA and explain the DNA damage produced by low levels of microwaves [1]. Enzymatic sequencing analysis after treating DNA with Cu(II) did not reveal alkali-labile sites, UV or T4-endonuclease sensitivity, or lesions that block DNA polymerization [2]. However, incubation with hot piperidine (as required in chemical sequencing) revealed lesions at polyguanosine regions (poly G) produced by Cu(II) and H_2O_2 (Fig. 3). This selectivity for poly G was independently confirmed shortly thereafter by another group in Japan [33].

Catalase prevented Cu(II)-mediated DNA damage, but superoxide dismutase (SOD) had little effect. Cu(I) damage lacked the sequence specificity observed for Cu(II) and was about 20 times less effective in producing strand breaks than a similar concentration of Cu(II) and H_2O_2 [2]. The poly G selectivity by Cu(II) in plasmid DNA correlated with inactivation of plasmid transforming ability, demonstrating an association between DNA damage and biological viability. These results suggested that DNA damage mediated by copper ions occurred after a key step favoring base sequence specificity. This step could be the binding of Cu(II) (but not Cu(I)) to an electronegative region involving at least two contiguous guanosines.

90°C for 30 min. The bands reflect the cutting frequency produced by the treatment, and the arrows point to specific cutting at sites of two or more adjacent guanosine residues. The effect of Cu(I), Cu(II), or H_2O_2 alone followed by hot piperidine (right panel) indicates either less specificity or no damage. First four columns (on the left) correspond to treatment of DNA with Maxam and Gilbert sequencing reagents followed by hot piperidine treatment. (Reproduced with permission from [2].)

3. DNA LESIONS

Numerous metal-mediated oxidative lesions produced in DNA have been chemically characterized by several specialized groups, whose findings will not be repeated here [34–36]. The following discussion will focus on efforts to accurately measure DNA breakage and to relate in vitro studies with DNA damage produced by metal ions in vivo.

3.1. DNA Strand Breaks

Chromium, nickel, cobalt, iron, cadmium, lead, and copper compounds have been shown to produce DNA strand breaks [37]. Velocity sedimentation of DNA was the first major technique used for studying strand breaks produced by metal-containing systems as well as by other DNA-damaging reactions. However, since DNA is very sensitive to shear forces, the reproducibility and sensitivity of these analyses were largely affected by the methods of DNA isolation [38]. More recently, DNA unwinding techniques have been used for the determination of DNA single-strand breaks (SSBs) [39]. Results obtained by these methods are semiquantitative and obscured by protein-DNA crosslinking [40–42].

We studied the production of strand breaks by metal ions in vitro on supercoiled plasmid DNA. The assay, based on the relaxation and linearization of supercoiled DNA, is simple yet sensitive and quantitative (Fig. 4). Careful production of plasmid DNA results in largely intact supercoiled DNA (form I) with small amounts of the other forms [1,3]. A single-strand break (SSB, nick) allows the DNA molecule to uncoil by rotating around the unbroken strand, resulting in the relaxation of the supercoiled DNA structure into a circular form (form II). A double-strand break (DSB) cuts the DNA and opens the molecule into a linear form (form III). The detection sensitivity after electrophoresis in 0.8–1.0% agarose gel stained with ethidium bromide is 1 ng of DNA when gel photographs are analyzed by densitometry [43]. Densitometry of stained gels allows one to measure the amount of intact (supercoiled) DNA remaining after metal treatment in comparison to that present in untreated controls, and to quantify the production of circular and linear DNA molecules.

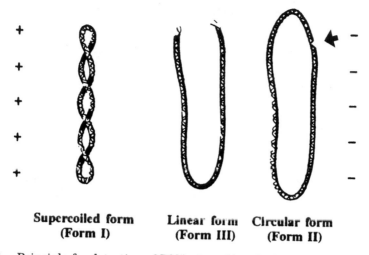

Supercoiled form   **Linear form**   **Circular form**
(Form I)   **(Form III)**   **(Form II)**

FIG. 4. Principle for detection of DNA strand breaks by relaxation of super-coiled plasmid DNA. Careful preparation of plasmid DNA results mainly in supercoiled form (form I). A single-strand break (pointed by the arrow) relaxes form I to circular form (form II). A double-strand break opens form I or II to linear form (form III). Forms I, II, and III can be separated by agarose gel electrophoresis. (For pZ189 plasmid, relative mobility of form I > form III > form II). (Reproduced with permission from [3].)

3.1.1. Single-Strand Breaks

Copper(II) and iron(III) mediated the production of SSBs and DSBs in supercoiled DNA in the presence of redox agents (e.g., ascorbate, hydro-gen peroxide, cysteine), making these metal ions good models for our study. A linear relationship was obtained when DNA damage [as log (% remaining supercoiled DNA)] was plotted against the concentration of, or incubation time with, systems containing a metal ion and either H_2O_2, ascorbate, or cysteine [44]. This linear relationship indicated that the production of DNA SSBs follows a single-hit kinetics (exponential kinetics). Derivation and substitution in the single-hit kinetic equation provides a useful formula relating the number of SSBs and % remaining supercoiled DNA:

$$SSB = n(\ln 10)[2 - \log(S)] \tag{1}$$

where n is the number of plasmid molecules in the system, calculated considering Avogadro's number and nanomoles of untreated plasmid DNA sampled as a control in the gels, and S is the % supercoiled DNA remaining after treatment, as measured in the agarose gels after electrophoresis [44].

3.1.2. Double-Strand Breaks

DNA DSBs have been considered events more lethal than SSBs [45] and responsible for chromosome aberrations and oncogenic transformation [46,47]. The induction of DSBs in all of the systems that we studied was dependent on the levels of SSBs. Regression analysis of strand breaks in several Fe(III)-containing systems revealed a linear relationship between the production of SSBs and DSBs after a threshold value (Fig. 5, inset) that followed the equation:

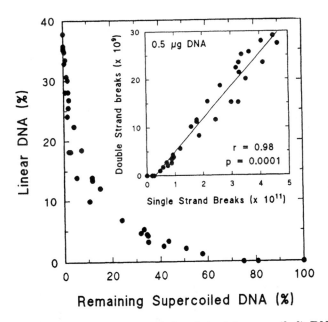

FIG. 5.   Relationship between remaining intact (supercoiled) DNA, single-strand breaks, and double-strands breaks produced by 0.1 mM Fe(III) combined with 0.4 mM citrate and 0.1 mM ascorbate for different periods in pZ189 DNA samples analyzed after treatment by quantitative agarose gel electrophoresis. (Reproduced with permission from [44].)

$$DSB = (a \times 10^{-2}) \times SSB - (b \times 10^9) \qquad (2)$$

where a (slope) and b (Y-axis intersect) are constants [44]. Equation (2) indicates that no DSB is detected before a certain threshold is reached in the number of SSBs. In Fe(III)-containing systems, this threshold corresponded to 35% remaining supercoiled plasmid DNA form. Above this threshold, several additional damaging hits (28 on the average) are necessary to yield one DSB [44]. These observations suggests that DSB could be the result of a secondary event at (or near) an SSB site.

3.1.3. Protection of DNA Breakage

Metal chelators, free radical scavengers, and enzymes of the oxygen reduction pathway can modulate the DNA damage mediated by metal ions. The supercoiled plasmid relaxation assay was useful for quantitatively determining the protection afforded by those substances to DNA damage mediated by copper or iron ions [2,3,44]. The protection conferred by different substances can be calculated by the equation:

$$\text{Protection (\%)} = 100 \times [1 - [(S_c - S_{m+a})/(S_c - S_m)]]$$
$$\text{also Protection (\%)} = 100 \times (S_{m+a} - S_m)/(S_c - S_m) \qquad (3)$$

where S_c is the supercoiled plasmid DNA remaining in the untreated control, S_{m+a} is the supercoiled DNA remaining after treatment with metal ion plus reductant in the presence of scavenger, chelator, etc., and S_m is the supercoiled DNA remaining after treatment with metal ion plus reductant in the absence of any protection [2].

Ethylenediaminetetraacetic acid (EDTA) and bathocuproine inhibited DNA damage mediated by copper ions, and deferoxamine (DF) and adenosine diphosphate inhibited DNA damage by ferric ions. In contrast, nitrilotriacetic acid (NTA), EDTA, diethylenetriaminepentaacetic acid (DTPA), and citrate increased the DNA damage mediated by Fe(III) [3].

Free radical scavengers such as sodium azide, dimethylsulfoxide, t-butyl alcohol, mannitol, and ethanol completely inhibited iron-mediated DNA damage [44]. These results indicate the participation of free radicals in the mechanism of DNA damage mediated by Fe(III). Only sodium azide and t-butyl alcohol completely protected the DNA damage mediated by Cu(II) [2], suggesting a lesser involvement of free radicals or generation of radicals with a short radii of action.

The role assigned to superoxide in the generation of damaging species by metals has been controversial. Although considered to play a major role by some authors, others have found its participation to be relatively minor [48]. In our systems, SOD inhibited DNA damage only when H_2O_2 was used as reductant with Fe(III). In the presence of Fe(II) alone or Fe(III) with either cysteine or ascorbate, SOD failed to protect DNA. These findings suggest that physiologically important reductants such as ascorbate and cysteine could participate in iron-mediated DNA damage without apparent $O_2^{\bullet-}$ involvement. Native or denatured SOD conferred little or no protection to DNA exposed to Cu(II) and H_2O_2 [2], suggesting that participation of $O_2^{\bullet-}$ in DNA damage mediated by metal ions may not be as common as originally assumed [49–51].

Catalase efficiently protected DNA from iron- or copper-mediated damage in all of the systems studied. The protection was independent of the reducing agent used or the presence of chelator. The effect of catalase indicates that H_2O_2 is a key intermediary in the DNA-damaging reaction mediated by iron or copper ions. The synergistic increase in iron-mediated DNA damage produced by H_2O_2 in vitro correlates with the huge increase (50–90 times compared to only 5–10 times the risk of each factor alone) in the prevalence of lung cancer seen when asbestos inhalation (a source of iron) is combined with smoking (a source of H_2O_2) [52].

These effects of free radical scavengers and enzymes on iron- or copper-containing systems agree with previous observations suggesting a cyclic reduction and oxidation of metal ions and indicating that H_2O_2 and, finally, $^{\bullet}OH$ are common important intermediates in the DNA damaging reactions [50–54].

3.1.4. Alkali-Labile Sites

Alkali-labile sites in DNA have been associated with exposure to certain metal ions, particularly Cr(VI), Cu(II), and Ni(II) [55]. Alkali-labile sites are a consequence of DNA depurination and modification of the sugar moiety, which results in chemical instability and breakage of the phosphodiester backbone after treatment with alkali [36]. It was suggested that the deoxy residue at an apurinic site is no longer fixed in

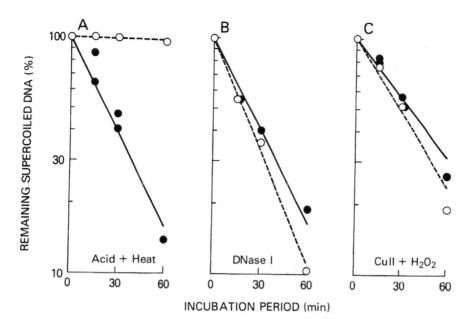

FIG. 6.   Study of alkali-labile sites in plasmid DNA. DNA was treated with
(A) acid and heat (30 mM citrate buffer, pH 4.25, at 70°C) to create alkali-labile
apurinic sites, (B) DNase I (83 ng/mL at 21°C) to produce strand breaks without
formation of alkali-labile sites, and (C) Cu(II) 10 μM plus H_2O_2 100 μM at 21°C.
The percent supercoiled plasmid remaining after treated samples were incubated for 2 h at 21°C in neutral (broken line) or alkali buffer (2 M NaOH, glycine,
pH 12.6) was determined after agarose gel electrophoresis. (Reproduced with
permission from [2].)

the furanose form but instead is in equilibrium with an unstable open
form (with a free aldehyde group) susceptible in alkaline conditions to β
elimination and consequent chain cleavage [56].

The kinetics of DNA breakage after alkaline or neutral treatment was studied after incubation with acid and heat, DNase I, or
Cu(II) plus H_2O_2 (Fig. 6). Of these, acid-plus-heat treatment is known
to induce alkali-labile sites whereas digestion with DNase I breaks
DNA without producing this lesion. The breakage of DNA by alkali
treatment after incubation with Cu(II) plus H_2O_2 (solid line in the
figure) may suggest the formation of alkali-labile sites. However, com-

parison with positive (acid plus heat) and negative (DNAse I) controls demonstrate that Cu(II) plus H_2O_2 does not result in a substantial number of alkali labile sites [2].

Some systems reported to produce alkali-labile sites may not survive objective comparison with adequate controls. Before reporting the presence of alkali-labile sites, two conditions should be met: (1) more breakage occurs after alkali treatment than after similar incubation at neutral pH, and (2) positive and negative controls for alkali-labile sites are included for comparison.

3.2. Base Modifications

Although we previously discussed the utility of the supercoiled plasmid DNA system to measure total DNA breakage, SSBs, and DSBs, as well as the effect of chelators, enzymes and scavengers, analysis of plasmid DNA is generally limited to in vitro testing. Difficulties in preserving intact, large DNA molecules and in obtaining reproducible and sensitive measurement of damage in genomic DNA have slowed advances in the quantitative study of DNA strand breakage in vivo.

In contrast to genomic DNA breakage, oxidative DNA base modifications have been precisely measured and extensively characterized [34–36,57]. A number of DNA base modifications have been identified after DNA damage caused by nickel, cobalt, iron, and copper ions . One of the best characterized base modifications is 8-hydroxy-2'-deoxyguanosine (8-OHdG), which induces G:C to T:A transversions during DNA replication [58–62]. 8-OHdG was proposed as a marker for carcinogenesis after increased levels of 8-OHdG were reported following exposure to salts of Cr(VI), nickel acetate, ferric-NTA, and a variety of other carcinogens including ionizing radiation, $KBrO_3$, 2-nitropropane, ciprofibrate, choline deficiency, asbestos, 12-o-tetradecanoylphorbol-13-acetate, and 4-nitroquinoline 1-oxide [60,63].

8-OHdG has been measured in DNA samples hydrolyzed to deoxyribonucleosides (with a mixture of nucleases and alkaline phosphatase) either by gas chromatography with mass spectrometry or by high-performance liquid chromatography with electrochemical detection [59,60,64]. The amount of 8-OHdG can be precisely measured after damage of DNA in vitro (on plasmid DNA) or in genomic DNA from animals or patients.

Could DNA breaks and 8-OHdG produced by metal ions arise by a common chemical reaction? If so, would it be possible to estimate DNA breaks by measuring 8-OHdG?

We simultaneously measured the number of DNA SSBs and 8-OHdG molecules by treating supercoiled plasmid DNA pZ189 with Cu(II) or Fe(III) in the presence of different reducing agents. The number of DNA SSBs observed was linearly related to the number of 8-OHdG present (Fig. 7) by the equation:

$$SSB = c + d \, (8\text{-OHdG}) \tag{4}$$

where c is the intersection with the y axis, and d is the slope. In all systems studied, c was small (typically <0.05), ruling out a threshold effect in the production of either lesion [43]. The slope was similar in systems containing Cu(II) or Fe(III) and various agents associated with metal reactivity in vivo, such as H_2O_2, ascorbate, or cysteine. The aver-

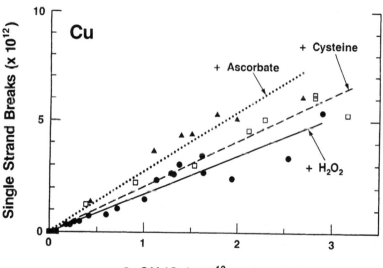

FIG. 7.   Production of single-strand breaks as a function of 8-OHdG formation. Plasmid DNA was treated for 30 min at 37°C with 20 μM Cu(II) in the presence of variable concentration (20 μM–10 mM) of either H_2O_2, L-cysteine, or L-ascorbate. The number of strand breaks was determined after gel electrophoresis analysis, and 8-OHdG was measured by high-performance liquid chromatography with electrochemical detection. (Reproduced with permission from [43].)

age number (1.7) of SSBs per 8-OHdG produced in the presence of H_2O_2 was identical for Cu(II) or Fe(III) [43]. Although still not definitively established, H_2O_2 is believed to play a major role in the production of oxidative DNA damage in vivo due to cellular production and diffusion through membranes [65]. Thus, SSB levels in vivo could be estimated by measuring the number of 8-OHdG molecules and multiplying this amount by 1.7.

The strong correlation obtained for all of the data plots (correlation coefficients ranging from 0.92 to 0.99) demonstrated that production of SSBs and 8-OHdG must be produced via a common or closely interrelated chemical mechanism. Under this perspective, the oxidative base modification and strand breakage produced by metal ions in DNA could be viewed as different endpoints of a common chemical process.

4. BIOLOGICAL IMPLICATIONS OF METAL-DNA DAMAGE

Functional roles of metal ions in the regulation of gene expression, immunity, the endocrine system, the brain, cell protection and growth, and in genetic disturbances, neoplasia, renal or cardiovascular failure, and other diseases have been the center of research previously reviewed [66].

Breaks in the phosphodiester backbone of DNA constitute one of the main lesions induced in vivo by ionizing and microwave radiation [1,67], and by several therapeutic drugs including the bleomycin and neocarzinostatin classes of antitumor antibiotics [68,69]. The association of copper to the production of DNA strand breaks by the antitumor drugs camptothecin and streptonigrin is well established [70,71]. Oxidative DNA strand breakage associated with free radical formation is also involved in aging [72], light sensitization [73], mutagenesis and carcinogenesis [74]. The DNA in numerous congenital human disorders shows reduced stability and increased sensitivity to the induction of damage [75]. The present discussion will focus on some of the biological roles of metal ions in relation to metal-mediated DNA damage.

4.1. Deleterious Effects

Compounds of chromium(VI), nickel, cadmium, antimony, beryllium, cobalt, lead, titanium, and arsenic have been classified as carcinogenic to humans or experimental mammals [12,76]. In addition, compounds of selenium, manganese, copper, iron, molybdenum, platinum, silver, zinc, aluminum, and mercury have been mutagenic in bacteria or phages, or have produced transformation of hamster embryo cells [76,77]. The function of metal ions in genetic regulation and their role in development of tumors, leukemia, and other malignant diseases has been reviewed in another volume of this series [78].

Wilson's disease and genetic (idiopathic, hereditary, or primary) hemochromatosis (GH) are human conditions that proceed with disruption of copper or iron homeostasis, respectively, and that present (both diseases) an increased hepatic copper or iron concentration and reduced stability of DNA [5,68,79]. Large human prospective studies suggest that excess stores of body iron are associated with carcinogenesis (primary hepatocellular carcinoma, esophagus cancer, colon cancer, lung cancer, bladder cancer) [80,81] and neoplastic cell growth (acute lymphocytic leukemia, neuroblastoma, Hodgkin's disease) [82–84]. The risk of primary hepatocellular carcinoma is over 200 times greater in GH patients than in the matched control population [85,86].

It was demonstrated that "free" iron in the serum of GH patients exists largely as complexes with citrate [87]. A concentration of Fe-citrate, two orders of magnitude smaller than the serum free iron concentration observed in severe GH patients, produced DNA strand breaks in vitro when physiological concentrations of biological antioxidants like ascorbate or cysteine were present [44]. Ascorbate at concentrations similar to those in normal serum induced strong iron-mediated DNA breakage, while deferoxamine (DF) inhibited this damage. The effect of ascorbate and DF observed in vitro correlates with the clinical observation that vitamin C (L-ascorbate) can be harmful to iron-overloaded patients unless DF is administered simultaneously [88,89]. Albumin, the major serum and hepatic protein, did not inhibit the DNA damage mediated by Fe-citrate [44].

Injection of Fe(III) chelated to NTA induces adenocarcinomas localized to the proximal tubules of the kidney in rodents [90–92]. Maximal Fe-NTA-mediated DNA breakage in vitro was induced under conditions of neutral pH, low ionic strength, presence of reducing agent, and absence of albumin. These conditions are present exclusively in the cortical proximal tubules of the kidney, the only location where toxicity and carcinogenicity of Fe-NTA has been observed. Thus, localized DNA breakage seems to explain the anatomic site preferred by Fe-NTA-induced carcinogenesis.

The effect of iron combined with other substances of biological relevance like citrate, ascorbate, and metal chelators (e.g., DF and NTA) agrees with clinical observations and suggests that the DSBs and SSBs observed in vitro play an important role in the carcinogenesis associated with hemochromatosis.

Copper is another transition metal with potent catalytic activity for the production of DNA damage [2,33,54,93]. Mutagenic activity of cupric ions has recently been reported using plasmid DNA [94], and high incidence of hepatocellular carcinoma has been observed in Long-Evans cinnamon (LEC) rats, a congenital animal model for copper toxicosis, where copper accumulates in the liver [95].

We implanted male Wistar rats with osmotic minipumps that continuously administered either saline, $CuCl_2$, or a copper chelate (cupric-NTA) at a rate that maintained serum copper concentrations at levels higher than in untreated controls [64]. A comparable level of copper intake could be reached during use of copper-containing intrauterine devices (IUDs), where the amount of copper administered in our experiments could be released in the uterus of women after a few months of continued IUD use [96–98].

At different times post-implantation, we measured the levels of 8-OHdG in DNA of kidney, liver, and tissue surrounding the pump implant, since production of 8-OHdG has been associated with mutagenesis and carcinogenesis (see Sec. 3.2 of this chapter). Hepatic and renal levels of 8-OHdG in $CuCl_2$ or Cu-NTA-treated animals were significantly higher than in control animals [64]. In contrast, histopathological changes in kidneys and livers of rats exposed to $CuCl_2$ and Cu-NTA were limited to mild changes involving hepatic focal necrosis and slightly increased mitotic activity in the renal proximal tubules. These observations demonstrate that DNA damage (as estimated by levels of 8-OHdG) is easily detected before histopathological changes, suggesting that DNA lesions could be early markers of metal toxicity.

Finding increased levels of a carcinogenic marker like 8-OHdG in kidney, liver, and tissues surrounding the site of delivery suggests that these organs and tissues may be a target for carcinogenesis by copper and agree with reports of primary hepatocellular carcinoma in humans as a complication of Wilson's disease (in which copper accumulates in liver and kidney) [99]. DNA lesions produced by copper in liver and kidney also agree with reports of liver cirrhosis, and of the production of renal adenocarcinoma in Wistar rats receiving copper-NTA [100].

4.2. Beneficial Effects

Therapeutic administration of compounds containing copper, zinc, magnesium, iron, platinum, gallium, gold, and lithium in a variety of diseases has been recently reviewed [66]. The antitumor or carcinostatic properties of complexes of platinum, copper, and ruthenium have been reviewed in another volume of this series [78]. These and several additional metal ions complexed to alkylating agents increased the survival of leukemic mice [101]. The role of metal ions (primarily copper, but also iron, cobalt, nickel, zinc, and aluminum) in the DNA damage and antitumor effect of bleomycin, tallysomycin, phleomycin, streptonigrin, daunomycin, adriamycin, chromycin, olivomycin, and mithramycin was reviewed in a previous volume of this series [102]. Analgesic and antiinflammatory activities of copper complexes have been reviewed elsewhere [103,104].

Our experience producing DNA damage in vitro and studying DNA lesions in vivo was employed to develop metal-based microbicidal agents. Either copper or iron ions by themselves inactivated several viruses with an efficacy similar to that of some substances used in hospital disinfection [105,106]. The microbicidal effect of metals, however, was enhanced further by the same reducing agents employed to produce SSBs, DSBs, and 8-OHdG. Cupric ascorbate was particularly effective in the inactivation of a variety of pathogenic bacteria, including *Pseudomonas aeruginosa, Staphylococcus aureus, Escherichia coli* 0157:H7, *Salmonella typhimurium, Yersinia enterocolitica, Shigella sonnei,* and *Vibrios cholerae, vulnificus,* and *parahemolyticus* [107].

Even bacterial spores, considered among the most resistant living organisms, were killed by adequate combinations of metal ion and reducing agent [108,109].

Human immunodeficiency virus (HIV-1) was inactivated by either cupric or ferric ions when the virus was free in solution [110]. Since HIV can be transmitted by cell-to-cell contact, we evaluated the ability of Cu(II) and Fe(III) ions to inactivate intracellular virus. Exposure to these metal ions once, and for a relatively short time, sufficed to prevent syncytia formation and synthesis of virus-specific p24 antigen in cells infected with HIV, while still preserving host cell viability [110].

Cu(II) has been previously shown to inactivate in vitro almost stoichiometrically a purified and cloned HIV protease [111], but we believe that protease inhibition is not the main cause of HIV inactivation by copper ions. The virus appears to be more resistant than the protease, and, in contrast to protease inhibition, HIV inactivation was not stoichiometric but required approximately 10^{11} copper atoms per infectious virus particle. Furthermore, the sensitivity shown by HIV for cupric and ferric ions does not appear to be specific for the HIV protease but instead was similar to that reported for Junin virus, herpes simplex virus (HSV), and bacteriophages $\phi6$, T7, and ϕX174 [106].

Cupric ions inactivated HSV under a variety of conditions. The addition of reducing agents allowed reduction of Cu(II) to a concentration similar to that in blood, with low cytotoxicity and retaining HSV killing activity [112]. Dose-response as well as kinetic experiments demonstrated that the enhancing virucidal effect of several substances of biological significance followed the order ascorbate $\gg$ hydrogen peroxide $>$ cysteine. The relative order of enhancement of these substances in Cu(II)-mediated HSV killing was identical to that previously reported for the DNA damage mediated by ferric ions [3,4]. A window of effective concentration observed for cysteine and lack of such a window for peroxide and ascorbic acid also resembled the results found for those compounds in enhancing the DNA damage by Fe(III) ions [4]. HSV killing in the presence of albumin demonstrates that Cu(II) preserves its virucidal activity even in the presence of known copper-complexing proteins. The relative enhancing power of reducing agent and the effects, or lack thereof, of metal chelators, catalase, SOD, and free radical scavengers on copper-mediated HSV inactivation were similar to those previously observed in DNA damage mediated by Cu(II) and Fe(III) [2–4]. This suggests that HSV (and likely other viruses) is inactivated by a mechanism paralleling that observed during DNA damage produced by copper ions in vitro.

The concentration of copper ascorbate that inactivated half of the HSV infectivity present (ID_{50}) in VERO cells was similar to the ID_{50} of acyclovir for HSV types 1 and 2 in the same cell line [113–115] and also similar to the concentrations of ganciclovir or vidavirine that inhibit most strains of HSV [116]. This comparison demonstrates that the antiviral potency of cupric ascorbate compares to that of drugs used clinically in antiviral therapy. The doses with virucidal activity are similar to the concentration of copper in normal serum and below the doses administered in the treatment of many patients with rheumatoid arthritis [117].

Cupric ascorbate left no HSV available to infect untreated cells, stopping HSV replication irreversibly. This was demonstrated by the absence of virus growth after subculturing infected cells treated with cupric ascorbate onto untreated cell monolayers where the biosynthetic machinery was intact [112]. Cell recovery after treatment with cupric ascorbate suggests that a potential advantage of metal-mediated viral killing over specific antiviral drugs resides in a rather broad spectrum of molecular damage that the host cell can repair much more quickly for itself than for the virus.

5. CONCLUSION

The DNA in many human disorders, including xeroderma pigmentosum, ataxia telangiectasia, Hutchinson-Gilford progeria, retinoblastoma, Fanconi's anemia, Bloom's, Cockayne's, and Werner's syndromes, and possibly Down syndrome, show increased sensitivity to the induction of damage [75]. DNA damage also develops as a secondary effect in the course of other diseases such as rheumatoid arthritis and various inflammatory conditions. The participation of metal ions in DNA damage associated with carcinogenesis, metabolic imbalances, and several other pathologies is well documented (see Sec. 4.1). In particular, we reported that those biochemical conditions optimal to produce SSBs by Fe-NTA in vitro are identical to the conditions in the only anatomic location where tumors appear after Fe-NTA treatment in vivo.

We presented equations to quantitatively determine the number of SSBs, DSBs, the protection of chelators, free radical scavengers, and other substances on the DNA damage mediated by metal ions. We also discussed the association between DNA strand breakage and formation of 8-OHdG. Direct correlation between SSB and 8-OHdG could facilitate reliable estimation of DNA breakage arising in various disease states. Precise estimation of SSB could be clinically important in those diseases where symptomatology and damage of DNA have been established.

Lesions mediated by metal ions do not seem to appear at random through the DNA sequence. Instead, the evidence suggests that each metal shows preference for particular sites in DNA. We envisioned metal interaction leading to DNA damage as a dynamic process whereby (1) a particular metal ion, either associated only with hydration solvent molecules or tightly bound to chelators, enzymes, or other substances, arrives at the neighborhood of DNA; (2) electrostatic forces first guide the ion to particular regions of the DNA helix; (3) at closer range, the ion favors its residence time to specific DNA sites that more closely complement the electronic orbital signature of the metal; and (4) in the relative intimacy of the bondage, electrons flow between metal and DNA, resulting in a variety of products including free radicals, SSBs, DSBs, and base modifications like 8-OHdG.

Risk and deleterious consequences of DNA damage has been a main arena for establishing the role of metal ions in biological systems. Although less understood, we believe that beneficial effects recently associated with metal ions in cancer and microbial diseases could open a new era in metallotherapy.

ABBREVIATIONS

8-OHdG	8-hydroxy-2'-deoxyguanosine
DF	deferoxamine
DSB	double-strand break
dsDNA	double-stranded DNA
DTPA	diethylenetriaminepentaacetic acid
EDTA	ethylenediamine-N,N,N',N'-tetraacetic acid

GH	genetic hemochromatosis
HIV	human immunodeficiency virus
HSV	herpes simplex virus
ID_{50}	dose of antimicrobial agent that reduces infectivity to half of that in untreated controls
IUD	intrauterine device
NTA	nitrilotriacetic acid
SOD	superoxide dismutase
SSB	single-strand break

REFERENCES

1. J. L. Sagripanti, M. L. Swicord, and C. C. Davis, *Radiat. Res.*, *110*, 219–231 (1987).

2. J. L. Sagripanti and K. H. Kraemer, *J. Biol. Chem.*, *264*, 1729–1734 (1989).

3. S. Toyokuni and J. L. Sagripanti, *J. Inorg. Biochem.*, *47*, 241–248 (1992).

4. S. Toyokuni and J. L. Sagripanti, *Carcinogenesis*, *14*, 223–227 (1993).

5. B. R. Bacon and R. S. Britton, *Hepatology*, *11*, 127–137 (1990).

6. M. Mori, A. Hattori, N. Sawaki, N. Tsuzuki, N. Sawada, M. Oyamada, N. Sugawara, and A. Enomoto, *Am. J. Pathol.*, *144*, 200–204 (1994).

7. W. E. C. Wacker and B. L. Vallee, *J. Biol. Chem.*, *234*, 3257–3262 (1959).

8. S. E. Bryan, D. L. Vizard, D. A. Beary, R. A. Labiche, and K. J. Hardy, *Nucleic Acid Res.*, *9*, 5811–5823 (1981).

9. A. Jendryczko, M. Drozdz, J. Tomala, and K. Magner, *Neoplasma*, *33*, 239–244 (1986).

10. M. A. Sirover and L. A. Loeb, *Science*, *194*, 1434–1436 (1976).

11. L. K. Tkeshelashvili, C. W. Shearman, R. A. Zakour, and L. A. Loeb, *Cancer Res.*, *40*, 2455–2460 (1980).

12. D. Beyersmann and A. Hartwig, *Arch. Toxicol.*, *Suppl. 16*, 192–198 (1994).

13. B. Sarkar, *Nutrition*, *11*, 646–649 (1995).

14. M. Nagaoka, J. Kuwahara, and Y. Sugiura, *Biochem. Biophys. Res. Commun.*, *194*, 1515–1520 (1993).

15. F. W. Sunderman, Jr. and A. M. Barber, *Ann. Clin. Lab. Sci.*, *18*, 267–288 (1988).

16. J. M. Berg, in *Metal-DNA Chemistry* (T. D. Tullius, ed.), American Chemical Society, Washington, D.C., 1989, pp. 90–96.

17. J. L. Sagripanti, P. L. Goering, and A. Lamanna, *Toxicol. Applied Pharmacol.*, *110*, 477–485 (1991).

18. G. Scatchard, *Ann. N. Y. Acad. Sci.*, *51*, 660–672 (1949).

19. G. N. Sayenko, A. P. Babii, and I. A. Bagaveyev, *Biophysics*, *31*, 451–456 (1986).

20. H. Richard, J. P. Schreiber, and M. Duane, *Biopolymers*, *12*, 1–10 (1973).

21. H. Sigel, in *Metal-DNA Chemistry* (T. D. Tullius, ed.), American Chemical Society, Washington, D.C., 1989, pp. 159–204.

22. L. E. Michenkova and V. I. Ivanov, *Biopolymers*, *5*, 615–625 (1967).

23. R. B. Simpson, *J. Amer. Chem. Soc.*, *86*, 2059–2065 (1964).

24. M. P. Waalkes and L. A. Poirier, *Toxicol. Applied Pharmacol.*, *75*, 539–546 (1984).

25. K. S. Kasprzak, M. P. Waalkes, and L. A. Poirier, *Toxicol. Applied Pharmacol.*, *82*, 336–343 (1986).

26. V. Swaminathan, and M. Sundaralingam, *Crit. Rev. Biochem.*, *6*, 245–336 (1979).

27. J. L. Sagripanti and L. Schroeder, *J. Inorg. Biochem.*, *47*, 58 (1992). (Abstract)

28. T. F. Kagawa, B. H. Geierstanger, A. H. J. Wang, and P. S. Ho, *J. Biol. Chem.*, *266*, 20175–20184 (1991).

29. H. A. Tajmir, M. Langlais, and R. Savoie, *Nucl. Acid Res.*, *16*, 751–762 (1988).

30. G. J. Brewer, V. Yuzbasiyan-Gurkan, D. Y. Lee, and H. Appelman, *J. Lab. Clin. Med.*, *114*, 633–638 (1989).

31. M. L. Schilsky, R. R. Blank, M. J. Czaja, M. A. Zern, I. H. Scheinberg, R. J. Stockert, and I. Sternlieb, *J. Clin. Invest.*, *84*, 1562–1568 (1989).

32. H. Sigel and B. R. Martin, *Chem. Rev.*, *82*, 385–426 (1982).

33. K. Yamamoto and S. Kawanishi, *J. Biol. Chem.*, *264*, 15435–15440 (1989).

34. J. Cadet, M. Berger, and A. Shaw, in *Mechanisms of DNA Damage and Repair* (M. G. Simic, L. Grossman, and A. C. Upton, eds.), Plenum Press, New York, 1986, pp. 69–74.

35. M. Dizdaroglu, *Mutat. Res.*, *275*, 331–342 (1992).

36. A. P. Breen and J. A. Murphy, *Free Radic. Biol. Med.*, *18*, 1033–1077 (1995).

37. A. Hartwig, *Biometals*, *8*, 3–11 (1995).

38. G. Ahnstrom and K. Erixon, *Int. J. Radiat. Biol.*, *23*, 285–289 (1973).

39. G. Ahnstrom, and K. Erixon, in *DNA Repair: A Laboratory Manual of Research Procedures* (E. C. Friedberg and P. C. Hanawalt, eds.), Marcel Dekker, New York, 1981, pp. 403–418.

40. H. C. Birnboim and J. J. Jevcak, *Cancer Res.*, *41*, 1889–1892 (1981).

41. K. W. Kohn and R. A. Grimek-Ewig, *Cancer Res.*, *33*, 1849–1853 (1973).

42. F. Hutchinson, *Radiat. Res.*, *120*, 182–186 (1989).

43. S. Toyokuni and J. L. Sagripanti, *Free Radic. Biol. Med.*, *20*, 859–864 (1006).

44. S. Toyokuni and J. L. Sagripanti, *Free Rad. Biol. Med.*, *15*, 117–123 (1993).

45. J. F. Ward, W. F. Blakely, and E. I. Joner, *Radiat. Res.*, *103*, 383–392 (1985).

46. P. E. Bryant, *Int. J. Rad. Biol.*, *46*, 57–65 (1984).

47. P. E. Bryant and A. C. Riches, *Br. J. Cancer*, *60*, 852–854 (1989).

48. B. Halliwell, *Free Rad. Res. Commun.*, *9*, 1 (1990).

49. J. M. Gutteridge, *Free Rad. Res. Commun.*, *9*, 119–125 (1990).

50. S. F. Wong, B. Halliwell, R. Richmond, and W. R. Skowroneck, *J. Inorg. Biochem.*, *14*, 127–134 (1981).

51. M. S. Baker and J. M. Gebicki, *Arch. Biochem. Biophys.*, *246*, 581–588 (1986).

52. J. H. Jackson, I. U. Schraufstatter, P. A. Hyslop, K. Vosbeck, R. Sauerheber, S. A. Weitzman, and C. G. Cochrane, *J. Clin. Invest.*, *80*, 1090–1095 (1987).

53. J. E. Schneider, M. M. Browning, and R. A. Floyd, *Free Rad. Biol. Med.*, *5*, 287–295 (1988).

54. S. H. Chiou, *J. Biochem.*, *96*, 1307–1310 (1984).

55. K. S. Kasprzak, *Cancer Invest.*, *13*, 411–430 (1995).

56. T. Lindahl and S. Ljungquist, in *Molecular Mechanisms of DNA Repair* (P. C. Hanawalt and R. B. Setlow, eds.), Plenum Press, New York, 1975, pp. 31–38.

57. M. Dizdaroglu, G. Rao, B. Halliwell, and E. Gajewski, *Arch. Biochem. Biophys.*, *285*, 317–324 (1991).

58. H. Kasai, H. and S. Nishimura, *Nucleic Acid Res.*, *12*, 2137–2145 (1984).

59. M. Dizdaroglu, *Free Rad. Biol. Med.*, *10*, 225–242 (1991).

60. R. A. Floyd, *Carcinogenesis*, *11*, 1447–1450 (1990).

61. Y. Kuchino, F. Mori, H. Kasai, H. Inoue, S. Iwai, K. Miura, E. Ohtsuka, and S. Nishimura, *Nature*, *327*, 77–79 (1987).

62. K. C. Cheng, D. S. Cahill, H. Kasai, S. Nishimura, and L. A. Loeb, *J. Biol. Chem.*, *267*, 166–172 (1992).

63. K. S. Kasprzak, B. Diwan, N. Konishi, M. Misra, and J. M. Rice, *Carcinogenesis*, *11*, 647–652 (1990).

64. S. Toyokuni and J. L. Sagripanti, *Toxicol. Applied Pharmacol.*, *126*, 91–97 (1994).

65. B. Halliwell and J. M. Gutteridge, in *Free Radicals in Biology and Medicine*, Vol. 2, Clarendon Press, Oxford, 1989, pp. 79–81.

66. G. Berthon (ed.), *Handbook of Metal-Ligand Interactions in Biological Fluids*, Marcel Dekker, New York, 1995.

67. W. D. Henner, L. O. Rodriguez, S. M. Hecht, and W. A. Haseltine, *J. Biol. Chem.*, *258*, 711–713 (1983).

68. L. S. Kappen and I. H. Goldberg, *Biochemistry*, *22*, 4872–4878 (1983).

69. Y. Sugiura, T. Takita, and H. Umezawa, in *Metal Ions in Biological Systems*, Vol. 19, *Antibiotics and Their Complexes* (H. Sigel, ed.), Marcel Dekker, New York, 1985, pp. 81–108.

70. J. Kuwahara, T. Suzuki, K. Funakoshi, and Y. Sugiura, *Biochemistry*, *25*, 1216–1221 (1986).

71. Y. Sugiura, J. Kuwahara, and T. Suzuki, *Biochim. Biophys. Acta*, *782*, 254–261 (1984).

72. C. Bernstein, in *DNA and Free Radicals* (B. Halliwell and O. I. Aruoma, eds.), Ellis Horwood, Chichester, UK, 1993, pp. 193–210.

73. B. Epe, in *DNA and Free Radicals* (B. Halliwell and O. I. Aruoma, eds.), Ellis Horwood, Chichester, UK, 1993, pp. 41–65.

74. B. Halliwell and J. M. Gutteridge, in *Free Radicals in Biology and Medicine*, Vol. 2, Clarendon Press, Oxford, UK, 1989, pp. 466–508.

75. C. F. Arlet and A. R. Lehmann, *Annu. Rev. Genet.*, *12*, 95–115 (1978).

76. C. P. Flessel, A. Furst, and S. B. Radding, in *Metal Ions in Biological Systems*, Vol. 10, *Carcinogenicity and Metal Ions* (H. Sigel, ed.), Marcel Dekker, New York, 1980, pp. 23–54.

77. E. C. Foulkes, *Metal Carcinogenesis*. Biological effects of heavy metals series, Vol. 2, CRC Press, Boca Raton, 1990.

78. H. Sigel (ed.), *Metal Ions in Biological Systems*, Vol. 10, *Carcinogenicity and Metal Ions*, Marcel Dekker, New York, 1980.

79. I. H. Scheinberg and I. Sternlieb, *Wilson's Disease*, W.B. Saunders, Philadelphia, 1984.

80. R. G. Stevens, D. Y. Jones, M. S. Micozzi, and P. R. Taylor, *N. Engl. J. Med.*, *319*, 1047–1052 (1988).

81. J. V. Selby and G. D. Friedman, *Int. J. Cancer*, *41*, 677–682 (1988).

82. D. Potaznic, S. Groshen, D. Miller, R. Bagin, R. Bhalla, M. Schwartz, and M. deSousa, *Am. J. Pediatr. Hematol./Oncol.*, *9*, 350–355 (1987).

83. A. E. Evans, G. J. D'angio, K. Propert, J. Anderson, and H. W. Hann, *Cancer*, *59*, 1853–1859 (1987).

84. H. W. Hann, B. Lange, M. W. Stahlhut, and K. A. McGlynn, *Cancer*, *66*, 313–316 (1990).

85. R. A. Bradbear, C. Bain, V. Siskind, F. D. Schofield, S. Webb, E. M. Azelsen, J. W. Halliday, M. L. Bassett, and L. W. Powell, *J. Natl. Cancer Inst.*, *75*, 81–84 (1985).

86. C. Niderau, R. Fisher, A. Sonnenberg, W. Stremmel, H. J. Trampish, and G. Strohmer, *N. Engl. J. Med.*, *313*, 1256–1262 (1985).

87. M. Grootveld, J. D. Bell, B. Halliwell, O. I. Aruoma, A. Bomford, and P. J. Sadler, *J. Biol. Chem.*, *264*, 4417–4422 (1989).

88. J. A. Murray, *Br. Med. J.*, *284*, 1401– (1982).

89. B. Halliwell, *Br. Med. J.*, *285*, 296 (1982).

90. J. L. Li, S. Okada, S. Hamazaki, Y. Ebina, and O. Midorikawa, *Cancer Res.*, *47*, 1867–1869 (1987).

91. Y. Ebina, S. Okada, S. Hamazaki, F. Ogino, J. L. Li, and O. Midorikawa, *J. Natl. Cancer Inst.*, *76*, 107–113 (1986).

92. S. Hamazaki, S. Okada, Y. Ebina, M. Fujioka, and O. Midorikawa, *Am. J. Pathol.*, *123*, 343–350 (1986).

93. O. I. Aruoma, B. Halliwell, E. Gajewski, and M. Dizdaroglu, *Biochem. J.*, *273*, 601–604 (1991).

94. L. K. Tkeshelashvili, T. McBride, K. Spence, and L. A. Loeb, *J. Biol. Chem.*, *266*, 6401–6406 (1998).

95. Y. Li, Y. Togashi, S. Sato, T. Emoto, J. H. Kang, N. Takeichi, H. Kobayashi, Y. Kojima, Y. Une, and J. Uchino, *Jpn. J. Cancer Res.*, *82*, 490–492 (1991).

96. A. J. Moo-Young, H. J. Tatum, L. S. Wan, and M. Lane, in *Analysis of Intrauterine Contraception* (F. Hefnawi and S. J. Segal, eds.), Elsevier, New York, 1975, pp. 439–457.

97. T. H. Goh and S. L. Tong, *Contraception*, *33*, 411–420 (1986).

98. Y. Y. Tsong and H. A. Nash, *Contraception*, *44*, 385–391 (1991).

99. J. Polio, R. E. Enriquez, A. Chow, W. M. Wood, and C. E. Atterbury, *J. Clin. Gastroenterol.*, *11*, 220–224 (1989).

100. S. Toyokuni, S. Okada, S. Hamazaki, M. Fujioka, J. L. Li, and O. Midorikawa, *Am. J. Pathol.*, *134*, 1263–1274 (1989).

101. M. D. Joesten, in *Metal Ions in Biological Systems*, Vol. 11, *Metal Complexes as Anticancer Agents* (H. Sigel, ed.), Marcel Dekker, New York, 1980, pp. 285–304.

102. J. C. Dabrowiak, in *Metal Ions in Biological Systems*, Vol. 11, *Metal Complexes as Anticancer Agents* (H. Sigel, ed.), Marcel Dekker, New York, 1980, pp. 305–336.

103. J. R. J. Sorenson, *Biology of Copper Complexes*, Humana Press, Clifton, NJ, 1987.

104. J. R. J. Sorenson, *Inflammatory Diseases and Copper*, Humana Press, Clifton, NJ, 1982.

105. J. L. Sagripanti, *Appl. Environm. Microbiol.*, *58*, 3157–3162 (1992).

106. J. L. Sagripanti, L. B. Routson, and C. D. Lytle, *Appl. Environ. Microbiol.*, *59*, 4374–4376 (1993).

107. J. L. Sagripanti, C. A. Eklund, P. A. Trost, K. C. Jinneman, C. Abeyta, C. A. Kaysner, and W. E. Hill, *Am. J. Infection Control*, *25*, 335–339 (1997).

108. J. L. Sagripanti and A. Bonifacino, *Appl. Environ. Microbiol.*, *62*, 545–551 (1996).

109. J. L. Sagripanti and A. Bonifacino, *Am. J. Hosp. Infect.*, *24*, 364–371 (1996).

110. J. L. Sagripanti and M. M. Lightfoote, *AIDS and Human Retroviruses*, *12*, 333–336 (1996).

111. A. R. Kalstrom and R. L. Levine, *Proc. Natl. Acad. Sci. USA*, *88*, 5552–5556 (1991).

112. J. L. Sagripanti, L. B. Routson, A. Bonifacino, and C. D. Lytle, *Antimicrob. Ag. Chemother.*, *41*, 812–817 (1997).

113. R. Arky, *Physicians' Desk Reference*, 49th Edition, Medical Economics Data Production, Montvale, NJ, USA, 1995, pp. 827.

114. D. W. Barry and B. Nusinoff, in *Herpes Virus and Virus Chemotherapy* (R. Kono, and A. Nakajima, eds.), Excerpta Medica, New York, 1985, pp. 269–270.

115. J. J. O'Brien and D. M. Campoli-Richards, *Drugs*, *37*, 233–309 (1989).

116. G. R. Douglas, in *The Pharmacological Basis of Therapeutics* (A. Goodman-Gilman, T. W. Rall, A. S. Nies, and P. Taylor, eds.), Pergamon Press, Elmsford, NY, 1990, pp. 1182–1201.

117. J. Aaseth and T. Norseth, in *Handbook of Toxicology of Metals*, Vol. 2 (L. Friberg, G. Nordberg, and V. Vouk, eds.), Elsevier, New York, 1986, pp. 233–254.

7

Radical Migration Through the DNA Helix: Chemistry at a Distance

Shana O. Kelley and Jacqueline K. Barton

Division of Chemistry and Chemical Engineering,
Beckman Institute, California Institute of Technology,
Pasadena, CA 91125, USA

1. INTRODUCTION

The reaction of the DNA bases with radical species generated by radiation, carcinogens, or oxidative stress can lead to mutagenic damage [1]. The efficiency and dynamics of radical transport through the DNA helix therefore hold profound biological implications. Intriguing questions concerning charge migration through DNA arise that can now begin to be addressed using well-defined chemical experiments. Does radical migration through DNA occur over long molecular distances? How is it modulated by DNA sequence and the structural variations in DNA? Is it physiologically important? How general is this phenomenon? These are issues that need to be addressed in the context of delineating mechanisms of DNA damage and repair.

    The presence of an extended, π-stacked arrangement of base pairs at the core of the essential genetic material is useful to consider. Structurally similar solid-state materials are conductive when oxidatively doped [2]. However, double-helical nucleic acids may be a unique example of a *molecular* π-stacked array. Although electron transfer reactions through the σ-bonded matrices found in proteins have been extensively investigated both experimentally [3] and theoretically [4], this phenomenon has not been characterized in detail for π-stacked media. Hence, the DNA double helix, structurally well characterized and synthetically accessible, may represent a valuable medium in which to study the migration of radicals from a fundamental perspective.

Studies of electron transfer within DNA at the molecular level have provided conflicting assessments concerning the ability of the DNA base stack to facilitate charge transport [5–38]. Here we focus on experiments employing probes intercalated or stacked within the double helix of DNA [18–33,38]. These studies, with well-stacked donors and acceptors, now comprise a significant body of work with consistent conclusions. Efficient long-range charge transport over remarkable molecular distances has been elucidated in diverse systems utilizing spectroscopy, electrochemistry, and the visualization of chemical reactions (Table 1). Stacking, both among the DNA bases and between reactants and the DNA bases, has been identified as a key parameter dictating the efficiency of this phenomenon. As such, the role of reactant and base stacking is emphasized in this chapter where we describe evidence from our laboratory that reveals long-range radical migration within the DNA helix.

2. CHARGE TRANSPORT IN DNA: CHEMICAL AND BIOLOGICAL HISTORY

For over 30 years, given the relevance of charge migration in nucleic acids to DNA damage [39], chemists and biologists have directed considerable effort toward understanding this phenomenon. A significant number of studies employing spectroscopic techniques [7–23], pulse radiolysis methods [34–37], and theoretical approaches [15–17] have been carried out. However, many seemingly contradictory results have been obtained, and these have impeded the formation of a definitive picture with respect to the efficiency of DNA-mediated charge transport.

One family of experiments that has been conducted involve monitoring the migration of radical species generated by ionizing radiation within the DNA double helix [34–37]. Based on these pulse radiolysis

TABLE 1

DNA Assemblies and Experiments
to Probe Long-Range Electron Transfer in DNA

Assembly	Techniques	Result
Rh ... **Ru**	Fluorescence quenching	Efficient fluorescence quenching over 40 Å
Rh ... **Et**	Fluorescence quenching	Fast electron transfer ($k > 10^{10}$ s^{-1}) over 17–36 Å with shallow distance dependence ($\beta < 0.1$ Å^{-1})
Et ... **Z**	Fluorescence quenching	Distance dependence (0.2–0.4 Å^{-1}) is sensitive to stacking of reactants
	Electrochemistry	Efficient electron transfer over 40 Å, detection of base mismatches based on charge transmission through DNA
Rh ... **G G**	Gel electrophoresis	Long-range oxidation of guanine doublets (17–34 Å), reaction is sensitive to integrity of base stack
Ru^{ox} ... **G G**	Gel electrophoresis Transient absorption spectroscopy	Guanine doublets oxidized from > 30 Å, GA mismatches also susceptible to damage
Rh ... **T⬦T**	HPLC	Thymine dimer repaired by long-range electron transfer through the DNA helix

studies, which rely on radicals generated either within DNA directly or in aqueous solution (e.g., $OH^{\bullet}$, hydrated electrons, $SO_4^{\bullet-}$, and $COO^{\bullet-}$), estimates have been made that radicals can travel anywhere from 1 to 200 base pairs within DNA. A variety of experimental conditions have been employed, and it is not clear as to how best to determine the range of distance limits obtained. However, the majority of the studies employing intercalators as spin traps have produced data consistent with charge migration over a distance of more than 25 Å [35]. In addition, electron spin resonance measurements specifically monitoring the reduced form of the intercalating antitumor agent daunomycin provided evidence for the disproportionation of this radical over 100 base pairs [37].

Theoretical studies have also provided differing assessments of the DNA base stack as a medium for charge transport. Quantum mechanical models have predicted that charges could move from base to base on the femtosecond time scale, which would allow long-range migration (>100 Å) in picoseconds, albeit in discrete steps [15]. Redfield theory has also been used to explain observations of efficient electron transfer in DNA by describing a "delocalized" band within the stacked bases [16]. However, other calculations have predicted a much steeper dependence of charge transport on distance, with a value for β, the decay of electronic coupling with distance [4], similar to that observed in σ-dominated protein systems (1 Å^{-1}) [3]. From the conclusions of these calculations, it would be expected that electron transfer would be prohibitively slow over long distances (>20 Å) where extremely efficient reactions had been observed experimentally (vide infra).

Most recently, chemists have focused their efforts on studying discrete photoinduced electron transfer events between donors and acceptors bound to DNA (Fig. 1A) [8–14,18–26]. Small molecules can bind to DNA in a variety of ways depending on their structural properties; hence these studies have taken advantage of molecules that bind to DNA either through electrostatic association with the anionic phosphate backbone, hydrophobic interaction with the grooves of the helix, or via intercalation of a planar, aromatic, heterocyclic moiety between the DNA bases that entails electrostatic, hydrophobic, and dipolar interactions. Many DNA-binding molecules both are positively charged and contain hydrophobic or aromatic functionalities, so that a mixture of these binding modes may also prevail under a given set of conditions.

One of the first studies of photoinduced electron transfer through DNA employed ethidium (Et), a classical organic intercalator, as an electron donor, and acridine antitumor drugs as acceptors [8,9]. Efficient quenching of the fluorescent singlet excited state of ethidium was observed, presumably due to reduction of the excited chromophore by the acceptors. This reaction appeared to take place on a time scale that

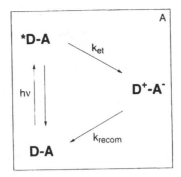

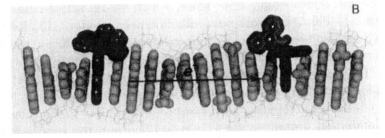

FIG. 1.   (A) Photoinduced electron transfer cycle. (B) Schematic depiction of metallointercalator electron donors and acceptors bound to DNA.

precluded diffusional contact and gave perhaps the first indication that intercalated species were afforded strong electronic coupling through the base stack. In a different study where the groove-binding acceptor methyl viologen was utilized in conjunction with this same donor, rates of electron transfer representing an enhancement over those in the absence of DNA were obtained, but the time scale of this quenching reaction was slower than that observed with intercalating acridine acceptors, indicating that diffusional contact was involved in the quenching reaction with the more mobile viologen acceptor [10].

At about this time, our laboratory was investigating the relationship between the structural features and the DNA binding properties of small transition metal complexes of the formula $M(phen)_3^{2+}$ (M = Ru, Co, Rh, Cr). In the course of these studies, it was discovered that the presence of DNA greatly accelerated quenching reactions between these complexes [18,19]. However, since the architecture of these molecules allows both intercalation and binding to the surface of the DNA grooves, there were also diffusional components in the quenching profiles. In order to selectively assess the role of the DNA helix in modulating electronic coupling between intercalated species, a new family of probes was required.

The metallointercalators $Ru(phen)_2dppz^{2+}$ and $Rh(phi)_2bpy^{3+}$ represented the next generation of coordination complexes employed to study the DNA base stack as a medium for electron transfer (Fig. 1B). These molecules exhibit large DNA binding constants ($K \geqslant 10^6$ M^{-1}) and highly favor intercalative binding [26,41]. Moreover, the metal-to-ligand charge transfer (MLCT) excited state of $Ru(phen)_2dppz^{2+}$, localized on the intercalated dppz ligand, is energetically suited (E^0 ($Ru^{3+}/*Ru^{2+}$) ▪ -0.6 V) to undergo electron transfer with $Rh(phi)_2bpy^{3+}$, which features a low potential reduction (E^0 (Rh^{3+}/Rh^{2+}) = $+0.03$ V) localized on the intercalated 9,10-phenanthrene quinone diimine (phi) ligand [20]. The spectroscopic properties of this pair and the donor analog $Os(phen)_2dppz^{2+}$ also provide handles for the characterization of charge-separated intermediates during the electron transfer and the monitoring of reaction dynamics via transient absorption [22,23].

Indeed, remarkably fast electron-transfer kinetics ($k_{et} \geqslant 10^{10}$ s^{-1}) at dilute donor-acceptor loadings (1 ($Ru(II)/Rh(III)$)/50 bp) have been observed for this reactant pair in the presence of calf thymus DNA [23]. In addition, this reaction is extremely sensitive to the interaction of these molecules with the DNA base stack. Much lower efficiencies were measured for the Λ enantiomers of these complexes, presumably due to poorer intercalation for these complexes with chiralities that are not complementary to right-handed DNA. Surprisingly, the reaction kinetics monitored via transient absorption were not affected by variations in the loading of the intercalators and hence appear to be remarkably insensitive to distance. It was subsequently suggested that this insensitivity to loading does not reflect the ability of the DNA base stack to mediate long-range electron transfer but instead reflects a cooperative clustering between the intercalators [42,43]. Although theoretical modeling of the quenching data in terms of cooperative binding was put forth and putative experimental support for this mechanism was obtained by monitoring perturbations in the circular dichroism spectra of the intercalated complexes, the sensitivity of the reaction kinetics for electron transfer between $Ru(phen)_2dppz^{2+}$ and $Rh(phi)_2bpy^{3+}$ to sequence and donor-acceptor chirality [22] was still inconsistent with this model. Moreover, a recent reinvestigation of these circular dichroism experiments and an NMR study of the intercalated complexes revealed no indication of cooperative binding and instead confirmed the reasonable expectation that the preferred arrangement of these cationic metal complexes on a DNA helix was one in which the complexes were spatially separated [43]. It appears then that the ultrafast electron transfer kinetics observed in this system actually reflect efficient charge transport through the DNA base stack over long molecular distances.

During the course of these investigations, electron transfer across a fixed distance was also studied in a 15-mer DNA duplex with derivatives of the metallointercalators $Ru(phen)_2dppz^{2+}$ and $Rh(phi)_2phen^{3+}$ attached to the 5' termini of complementary strands [21]. In this system, the luminescence of $Ru(phen)_2dppz^{2+}$ was quantitatively quenched on a subnanosecond time scale in the presence of the acceptor located 40 Å down the helix. Here, where the locations of the reactants were indisputable, results consistent with long-range electron transfer were again found.

Substantially different conclusions about DNA as an electron transfer medium have been drawn from other systems that would appear to be quite similar in design. For instance, where the intercalator ethidium was used as a photoreductant in the presence of DAP (DAP = N,N'-dimethyl-2,7-diazapyrenium dichloride), electron transfer rates of $\sim 10^6$–10^8 s^{-1} were measured at high intercalator loadings where donor-acceptor distances were estimated from a nonrandom binding distribution as 10–17 Å [11]. Although the reactant separations were not well known, a value for β was reported as 1 Å^{-1}. This value indicates a significantly greater sensitivity to distance than the intercalator systems described above. Since ethidium was previously observed to undergo much more efficient electron transfer with the intercalated acridine acceptors mentioned previously, it is possible that the binding mode of the DAP acceptor used in this study, which may be a combination of weak intercalation and groove binding, precludes fast-reaction dynamics because of poor intercalation into the base stack.

Slower electron transfer kinetics and steeper distance dependencies have also been observed in other systems with reactants bound at fixed positions. In a series of stilbene-bridged DNA hairpins, the electron-transfer kinetics for the oxidation of guanine by this chromophore ranged from 10^{12} s^{-1} at a donor-acceptor separation of 3.4 Å to 10^{8} s^{-1} at a distance of 17 Å, and a value of β = 0.6 Å^{-1} was calculated [12]. In addition, in an octamer duplex modified with two transition metal complexes through σ-linkages, an electron transfer rate constant of 10^{6} s^{-1} was measured over a ~20 Å through-space separation, indicating a much steeper dependence on distance than was observed in any of the previously described systems with intercalated probes [13].

Although all of the experiments described have sought to under-stand charge transport through the same medium, the DNA helix, widely varying dynamics and distance dependencies have been re-ported. In many cases, however, the reactants employed (solution-borne radicals, nonintercalated probes) do not interact strongly with the base stack, and it is these systems that show the largest sensitivity to dis-tance and provide the most conservative estimates of the efficiency of radical migration within DNA. On the other hand, in the majority of systems where intercalators are employed as reactants or radical scav-engers, reactions occurring over extended distances are observed on fast time scales. These trends indicate that the interactions between reactants and the DNA base stack cannot be neglected in interpreting experimental data and that π stacking may strongly influence the kinetics and distance dependences of DNA-mediated electron transfer reactions.

Our most recent work, as described here, includes an important focus on the role of stacking in modulating the efficiency of charge transport in DNA. Taken together, these studies not only may provide a cohesive picture of the nature of DNA as an electron-transfer medium but also may help to reconcile the controversy surrounding previous disparate results. These studies serve to demonstrate that stacking within the DNA bridge is essential for efficient, long-range charge transfer.

3. DNA AS A BRIDGE FOR LONG-RANGE CHARGE TRANSPORT

Our first studies of electron transfer between intercalating probes within the DNA base stack provided promising indications that exceptionally efficient electron transfer can be mediated by this macro-molecular structure. However, to explore the effects of distance and structural perturbations within the DNA helix in an unequivocal fashion, we have turned our attention almost exclusively to studies of reactants covalently attached to synthetic DNA duplexes, where electron transfer can be monitored within structurally well-characterized assemblies. In order to expand the range of methodologies employed in studying DNA-mediated electron transfer and to provide a route to practical applications, electrochemical methods have most recently been used to study long-range redox processes at gold surfaces modified with DNA duplexes. These studies have provided important complementary information concerning this medium as a bridge for electron transfer.

3.1. Photoinduced Electron Transfer in DNA

3.1.1. Electron Transfer Between Intercalators: Ethidium and Rh(phi)₂bpy³⁺

3.1.1. Electron Transfer Between Intercalators: Ethidium and $Rh(phi)_2bpy^{3+}$

Efficient electron transfer with ultrafast dynamics was first characterized in assemblies containing metallointercalators. To better understand the role of reactant structure on this chemistry and to explore the scope of this chemistry, the organic intercalator ethidium was studied as an excited-state electron donor in studies with $Rh(phi)_2bpy^{3+}$ as an acceptor [24]. The intercalative interaction of ethidium as well as the fluorescence properties of this molecule both in the absence and presence of DNA have been extensively characterized. Furthermore, ethidium is amenable to derivatization and can be covalently attached to DNA to allow the systematic study of donor-acceptor separation.

When noncovalently bound to DNA, photoexcited ethidium is efficiently quenched by the metallointercalator $Rh(phi)_2bpy^{3+}$. This quenching reaction takes place on a subnanosecond time scale even when reactants are separated by >30 Å (1 Et/2 Rh/50 base pairs), assuming a random distribution of intercalators along the helix. Ethidium binds DNA from the minor groove side of the helix and differs significantly in structure from the major-groove-binding metallointercalator $Ru(phen)_2dppz^{2+}$. Hence, clustering interactions of the organic donor with $Rh(phi)_2bpy^{3+}$ become improbable when this reactant pair is noncovalently bound to DNA and cannot also account for this fast reaction. Thus, the observation of ultrafast electron transfer between the Et/Rh(III) pair strongly implicates the DNA base stack in the mediation of this efficient reaction. However, it is worth noting that when ethidium is utilized as the donor, quenching yields are lower at given loadings of acceptor in comparison to the reaction for the Δ-*Ru(II) donor, but higher than those observed with Λ-*Ru(II). This trend may reflect differences in the integrity of the intercalation for ethidium as compared to the ruthenium enantiomers.

3.1.2. Distance Dependence

In order to study the reaction between ethidium and $Rh(phi)_2bpy^{3+}$ as a function of donor-acceptor separation, these intercalators were derivatized and attached to the termini of DNA oligonucleotides via flexible linkers [24]. Duplexes ranging from 10 to 14 base pairs in length were synthesized, leading to a range of separations for intercalated Et-Rh(III) of ~24–38 Å (Fig. 2A). These assemblies were extensively characterized by photophysical and biochemical methods, and were found to behave as well-formed duplexes with the same structural characteristics as the unmodified analogs. For instance, fluorescence polarization measurements allowed the calculation of duplex volumes for Et and Et/Rh-modified duplexes, and the values obtained from experiment were in good agreement with those calculated based on the anticipated cylindrical dimensions of the modified helices. Moreover, thermal denaturation measurements revealed that the duplexes with tethered intercalators exhibited similar amounts of hypochromicity as unmodified duplexes but exhibited higher melting temperatures, indicating that well-hybridized assemblies are formed and stabilized by the presence of the intercalators.

Fluorescence quenching measurements demonstrated that photoinduced electron transfer between these tethered intercalators proceeded over extended distances up to ~40 Å. The efficiency of the reaction was seen to depend only weakly on distance, implying that β must be quite small, possibly ≤ 0.1 Å^{-1}. Moreover, when the electron-transfer kinetics were monitored, it was found that the reaction occurred on a remarkably fast time scale over the entire range of donor-acceptor distances studied ($k_{et} \geq 10^{10}$ s^{-1}). Thus, instead of increased donor-

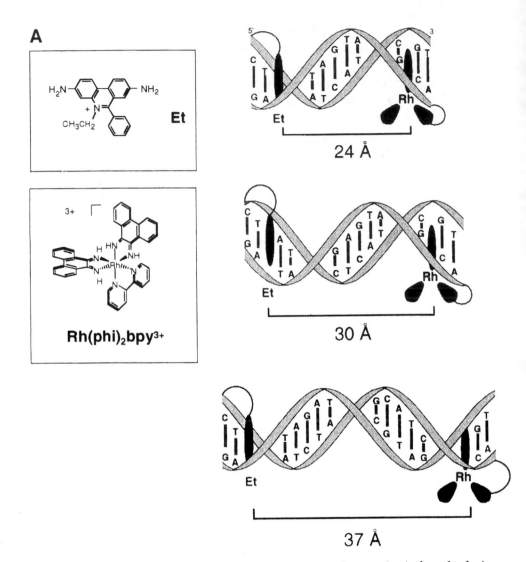

FIG. 2. (A) DNA duplexes, ranging from 10 to 14 base pairs in length, derivatized with the intercalating electron donor, ethidium, and the intercalating electron acceptor, Rh(phi)$_2$bpy^{3+}. (B) Et/Rh(III)-modified duplexes containing Watson-Crick (TA) and mismatched base pairs (CA). In the paired, well-stacked duplex, efficient electron transfer is observed, whereas when the base stacking within the duplex is perturbed due to the CA mismatch, the yield of reaction is significantly attenuated. (Adapted from [24].)

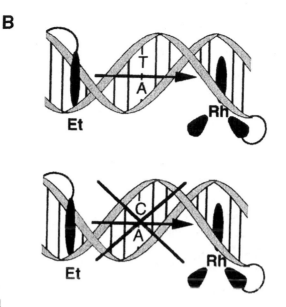

FIG 2. Continued

acceptor distance causing the reaction to slow down, it appeared that increasing the reactant separation changed the *yield* of very fast electron transfer.

This behavior seemed to indicate that the reaction might be "gated" by processes slower than the electron transfer. It was then necessary to consider the dynamics of the DNA helix in interpreting the observed distance dependences. DNA has a range of internal motions that occur on the time scales of nanoseconds to seconds. It was therefore proposed that these dynamics might introduce defects in the base-stack-mediated pathway that prohibited the very strong electronic coupling that must facilitate the ultrafast reaction. The probability of having a fully stacked pathway, then, would depend on the number of base pairs separating the donor and acceptor, and could lead to an exponential distance dependence in the yield of quenching. If this hypothesis is correct, this first study of the distance dependence of a DNA-mediated electron transfer reaction between tethered intercalators indicates that the DNA helix mediates fast, long-range electron transfer reactions exceptionally well. However, eventually the efficiency of this process is limited by the dynamic nature of this molecular assembly in solution.

3.1.3. Sensitivity to Stacking

The proposal that disruptions in base stacking could attenuate the efficiency of electron transfer processes proceeding through the double helix prompted us to explore further the effect of perturbations in stacking [24]. Thermal denaturation experiments showed a striking correlation between the loss of electron transfer in ethidium/rhodium-modified duplexes and the loss of hypochromicity during the melting of the duplex. Since hypochromicity provides a measure of base stacking, this correlation pointed to a direct relationship between this parameter and the efficiency of electron transfer.

To engineer systematically perturbations in stacking into the electron-transfer pathway in these ethidium/rhodium(III)-modified duplexes, single-base changes were made in the duplexes, leading to mismatched base pairs (Fig. 2B). The mismatches incorporated, CA and GA, have been shown to possess very distinct structural properties. The CA mismatches exhibit poor hydrogen bonding and do not stack well in duplex DNA, creating short-range disruptions in the double helix. In contrast, GA mismatches are actually very well stacked due to the increased aromatic surface area for this purine-purine pair. In fact, the efficiency of electron transfer in duplexes we observe is quite consistent with the predicted changes in stacking. While significant electron transfer is evident between ethidium and rhodium in the fully Watson-Crick base-paired duplex as well as in the duplex containing the GA mismatch, the efficiency is drastically attenuated in the presence of an intervening CA mismatch. These results highlight the sensitivity of long-range electron transfer to the integrity of stacking along the base-stack-mediated pathway. These results also provided the first direct evidence of an electron transfer through DNA being *mediated* by the DNA base pair stack.

3.2. Photoinduced Oxidation of Modified Bases

Having observed that photoinduced electron transfer between inter-calating reactants could proceed over extended distances in DNA, we then sought to examine whether a DNA base could also participate as a reactant in these same ultrafast reactions. Using a DNA base as a reactant had the added advantage of providing the opportunity to probe still more directly how the stacking of a reactant influences the efficiency of electron transfer.

3.2.1. Photoinduced Oxidation of 7-Deazaguanine

An analog of the natural base guanine, 7-deazaguanine (dz-G/Z), can be used to probe electron transport processes in DNA [25]. This modified base differs by only one atom from the natural form, but the C to N substitution is manifested in an oxidation potential that is ~300 mV lower for deazaguanine relative to guanine. This lowered potential allows the selective oxidation of this base within a heterogeneous sequence of DNA by appropriate reactants. Indeed, in DNA duplexes containing deazaguanine, significantly lowered fluorescence quantum yields had been observed for noncovalently bound ethidium relative to guanine-containing duplexes, and these observations provided a direct indication that the excited state of this intercalator can oxidize the unnatural base. A variety of other intercalators, including Ru(phen)$_2$dppz^{2+} and acridine orange, upon photoexcitation also appear to react with the modified base in luminescence quenching experiments. However, Os(phen)$_2$dppz^{2+}, which is isostructural to Ru(phen)$_2$dppz^{2+} but has a lower excited-state reduction potential, does not react with deazaguanine. Presumably, this latter reaction is prohibited due to insufficient driving force.

3.2.2. Distance and Sequence Dependence of Direct Base Oxidation

The effects of distance and sequence on the photoinduced oxidation of deazaguanine were monitored in ethidium-modified duplexes containing this base (Fig. 3) [25]. A series of duplexes containing tethered ethidium and either guanine or deazaguanine were synthesized featuring a range of ethidium/dz-G separations spanning 6–24 Å. Using fluorescence quenching measurements to assess the efficiency of the base oxidation as a function of distance, it was again observed that photoinduced electron transfer in DNA is only weakly sensitive to distance. Moreover, as in the studies with tethered intercalators, the reaction kinetics as monitored by time-resolved single-photon counting pro-

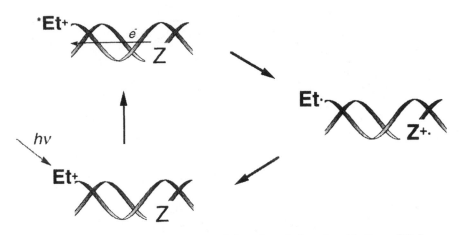

FIG. 3.   Schematic representation of the photoinduced oxidation of 7-deaza-
guanine by ethidium. In the excited state, ethidium is sufficiently oxidizing to
react selectively with 7-deazaguanine, allowing the intervening natural bases
to be probed as a medium for electron transfer.

ceeded on a fast time scale and did not appear to be greatly attenuated
by distance. Here, too, the results obtained were consistent with the
idea that the dynamics of the DNA bases may dictate the yield of electron transfer at a given donor-acceptor distance.

The efficiency of the photooxidation of deazaguanine was also
monitored in assemblies where the base sequence flanking deazaguanine was varied. Although changing the sequence to the 5' side of
the reactant had little effect on the reaction efficiency, the mutation of
the base to the 3' side resulted in up to twofold changes in the amount
of fluorescence quenching observed. The variation in quenching as a
function of flanking sequence, 5'-dz-GG > 5'-dz-GT $\geq$ 5'-dz-GA > 5'-dz-
GC, was correlated with calculated ionization potentials [5] for stacked
base pairs that increase in the order 5'-GG < 5'-GA $\leq$ 5'-GT < 5'-GC.
Thus, it appears that changes in efficiency in this case might result
from fluctuations in thermodynamic driving force.

The effect of base mismatches on long-range electron transport
through the DNA base stack was also monitored in these assemblies.
In this system, the importance of stacking interactions between reactants and the DNA bases could be evaluated through the incorporation
of deazaguanine into mispairs. Interestingly, when deazaguanine was

paired with any base other than its natural mate C, the efficiency of the photooxidation reaction was drastically attenuated. These results may indicate that the modified base is not coupled to the base stack as strongly when part of a mispair. Hence, it appears to be crucial not only for the DNA bases themselves to be well stacked for the facilitation of long-range electron transfer, but for the reactants to interact strongly with the base stack for efficient charge transport.

The overall distance dependence of the reaction between ethidium and deazaguanine was examined as a function of flanking sequence [25]. Here the remarkable finding was made that in addition to the reaction efficiency being dependent on the environment of the modified base, the distance dependence was also influenced by this factor. When deazaguanine was flanked by two purine bases, a more shallow distance dependence (0.2 Å^{-1}) was observed compared to that measured when the base was surrounded by one pyrimidine and one purine base (0.4 Å^{-1}) (Fig. 4). A possible explanation for this trend may be based on the strength of stacking interactions within this site. For the purine-purine site, the greater overall aromatic surface area would certainly enhance the stacking of deazaguanine relative to the purine-pyrimidine site. These results may provide the first experimental demonstration that stacking interactions between the bases and DNA-bound reactants modulate the overall efficiency of DNA-mediated electron transfer reactions.

Given these many observations, it is now instructive to revisit the seemingly contradictory body of evidence collected by various researchers for and against long-range charge transport mediated by the DNA helix. From the results just described, it appears that the integrity of stacking interactions between reactants and the DNA base stack may dictate the efficiency of electron transport. The various results obtained by other researchers also can be reconciled in this context. When electron transfer through DNA is studied with reactants tethered to the backbone of DNA or with probes that are not stacked through strong intercalative binding modes, much slower kinetics and steeper distance dependencies have been reported [10–14]. These seemingly conflicting results may then just be an even more dramatic demonstration of the importance of stacking interactions. The DNA helix is structurally unique and is almost completely dominated by an ensemble of noncovalent interactions. Stacking between the DNA bases contributes to the thermodynamic stabilization of the double-helical structure of this biopolymer. Perhaps, then, it should not be surprising that

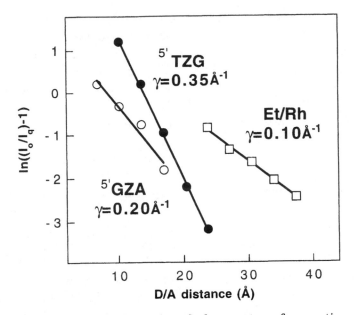

FIG. 4.   Dependence of DNA-mediated electron transfer reactions on distance. Data shown correspond to fluorescence quenching data obtained as a function of distance for photoinduced electron transfer between ethidium and $Rh(phi)_2bpy^{3+}$ (empty squares), and the photooxidation of 7-deazaguanine by ethidium (deazaguanine flanking sequence: 5'-GZA (empty circles), 5'-TZG (filled circles)). Remarkably, although all of these reactions take place through the same medium, the DNA base stack, significantly different distance dependences are observed. These trends may be influenced by the differing energetics of these reactions, but as the two 7-deazaguanine reactions have almost identical driving forces, it is more likely that the distance dependence is modulated by the integrity of the stacking interactions that couple the reactants into the DNA bridge. (Adapted from [24] and [25].)

stacking is essential to facilitate electron-transfer processes within this medium.

3.3.   Electron Transfer at DNA-Modified Electrodes

To probe more directly the transport of electrons through the DNA helix, we have initiated electrochemical studies [31–33]. This work, involving the fabrication of DNA-modified electrodes, has allowed us to explore different aspects of DNA-mediated charge transport. Moreover, these studies have led us to exploit the unique characteristics of DNA-mediated electron transfer chemistry on a practical level in developing novel biosensing applications.

In fact, studies of electron transfer at DNA films corroborate the conclusions of spectroscopic studies. At the DNA-modified surfaces, electron transfer mediated by the base stack proceeds over long molecular distances and appears to occur on a fast time scale compared to σ-bonded systems. Furthermore, the efficiency of charge transport is exquisitely sensitive to perturbations in stacking caused by mismatches. Since this sensitivity can be measured by simply quantitating a current, this effect provides the basis for the design of DNA-based biosensors capable of detecting point mutations within DNA [33].

3.3.1. DNA-Modified Surfaces

The discovery that thiol-containing compounds spontaneously assemble on gold surfaces has provided a new class of materials that have proven to be exceptionally useful for the study of electron-transfer processes [45]. In particular, the development of well-ordered self-assembled monolayers formed by aliphatic alkanethiols of variable lengths on gold surfaces has allowed the application of electrochemical techniques to the study of electron transfer through σ-bonded media. Because the construction of these monolayers permits a redox-active species to be spatially separated by a well-defined distance from an electrode surface, the kinetics of electron transfer through the resulting films can be monitored as a function of distance. These studies of electron transfer through σ-bonded monolayers have confirmed evaluations of the electronic coupling ($\beta \sim 1$ Å^{-1}) in this medium obtained spectroscopically.

The modification of surfaces with DNA appended to an alkanethiol tether provides another new class of materials that are potentially useful for the study of electron-transfer processes, structural characterization of nucleic acids, and biosensing applications. In our laboratory, we have fabricated and extensively characterized DNA-modified surfaces covered with small (15 base pair) duplexes (Fig. 5) [32]. For the investigation of electron transfer through these films, it is crucial that the orientation and packing of the duplexes on the surface is established. By exploiting conditions similar to those used to crystallize

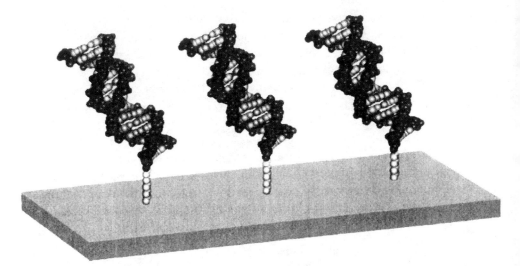

FIG. 5.   Illustration of thiol-modified duplexes immobilized on a gold surface.
Atomic force microscopy studies indicate that the DNA helices are oriented at
an angle of ~45° from the surface at open circuit and are attracted to the
electrode at positive potentials and repelled from the surface at negative potentials. (Adapted from [32].)

DNA, namely, using high concentrations of divalent metal ion, a high
surface density of DNA duplexes was immobilized on the gold surface;
spectroscopic and biochemical characterizations support this dense
packing. The incorporation of a radioactive label on the deposited helices, for example, allowed the direct quantitation of surface coverage,
which appeared to approach the theoretical limit for close-packed duplexes oriented upright with respect to the surface [31].

Atomic force microscopy (AFM) has allowed the surface morphol-
ogy of these DNA films to be probed directly [32]. These measurements
revealed that the surfaces are densely covered with a monolayer 45 Å
in thickness, indicating that the DNA helices are oriented at an angle of
~45° from the surface. We also explored the effect of applied potential on
the orientation of the DNA monolayer to examine the effect of electrostatic forces on the monolayer. In these experiments, where the surface
was modified at lower coverages, we found that when the electrode was
positively charged the polyanionic DNA appeared to be attracted to the

metal surface, thereby compressing the monolayer thickness (~20 Å). In contrast, when a negative charge was induced, the monolayer thickness increased significantly (~50 Å). These potential-dependent changes were furthermore and perhaps not surprisingly found to be reversible, suggesting the basis for the design of a nanoscale "switch."

These results also provide important information to consider in the selection of reactants for the study of electron transfer at the DNA-modified surfaces. Based on our photophysical studies, it was clear that to study electron transfer mediated by the base stack it was imperative to utilize intercalators as redox probes. Moreover, for unambiguous results to be obtained from electrochemical experiments, it is also important that the DNA duplexes assume an upright orientation with respect to the surface; otherwise more direct reaction of the intercalator with the electrode, not mediated by the DNA base stack, might "short-circuit" the DNA-mediated reaction. Thus, given the characterization of surfaces by AFM, reactants were chosen such that negative potentials were employed and a through-helix electron transfer pathway could be directly probed.

3.3.2. Redox Reactions of Intercalators at DNA-Modified Electrodes

Our initial studies of electron transport through DNA films centered around the electrochemical properties of reporter molecules bound noncovalently to the DNA-modified surfaces through intercalation. The cyclic voltammetry of methylene blue (MB), a three-ringed organic heterocycle that binds to DNA via intercalation, was probed first at the DNA-modified electrode [31]. A pronounced electrochemical response was observed at low concentrations of MB (<1 μM), providing strong evidence that MB bound tightly and was electronically well coupled to the modified surface. Moreover, coulometric titrations of MB at surfaces modified with DNA duplexes yielded well-behaved Langmuir isotherms, and an association constant was determined that agreed well with literature values for MB binding to DNA in solution. These measurements demonstrated the utility of DNA-modified surfaces for the study of binding phenomena between small molecules and DNA.

Studies of the electron-transfer kinetics of MB at the modified surface indicated that fast charge transport was required for the reduction of this intercalator through the DNA films. Significantly, the measured rate constant for the reduction of MB at DNA-modified gold was of the same order of magnitude as those measured for electroactive species attached directly to electrodes by aliphatic tethers of lengths similar to our thiol-terminated linker. However, the DNA duplex is approximately four times longer than the aliphatic linker, suggesting that the electron transfer rate in these films might be limited only by transport through the σ system rather than through the much longer π-stacked DNA. These results therefore also pointed to exceptionally efficient charge transport mediated by the DNA base stack, but since the binding site of MB in this system cannot be unambiguously ascertained, a more well-defined system was required to draw definitive conclusions. We therefore turned our attention to developing systems in which the intercalator binding sites could be precisely defined by covalently linking intercalated probes to the immobilized DNA helices.

Daunomycin (DM), a redox-active antitumor agent, was an ideal probe for further investigation of electron transfer through the DNA monolayers [33]. This intercalator displays a reversible reduction at negative potentials and, importantly, can be attached to specific sites on a DNA duplex by covalent crosslinking to the exocyclic amine (N2) of guanine [46]; DM bound in this fashion to a DNA duplex had been crystallographically characterized [47]. Thus, a series of thiol-derivatized duplexes with a single GC base step were synthesized to address unambiguously the effect of distance on electron transfer through the DNA helix using electrochemical techniques (Fig. 6).

Daunomycin-labeled DNA duplexes derivatized with the alkythiol linker were prepared, characterized using UV-vis spectroscopy, duplex-melting assays, and mass spectrometry, and then deposited onto the gold surface [33]. The modified DNA duplexes contained DM crosslinked a range of positions on the DNA helix. High surface coverages were also apparent in these films, and AFM studies revealed densely packed monolayers with heights greater than 45 Å at open circuit. Thus, electron transfer through the DNA helix could be assessed systematically as a function of distance.

Remarkably, no qualitative differences were detected for the electrochemical responses of covalently crosslinked daunomycin even when the intercalator binding sites varied in position by 40 Å. Cyclic voltammograms measured as a function of scan rate were essentially

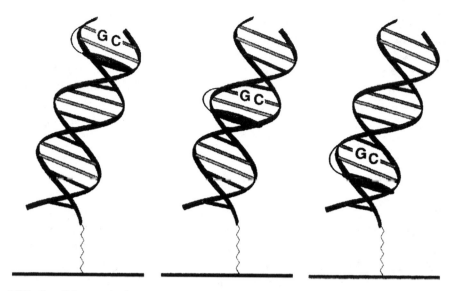

FIG. 6. Schematic diagram depicting DNA duplexes immobilized on gold surfaces used to study the distance-dependent reduction of daunomycin. Regardless of the position of the intercalator along the duplex, comparable electron transfer kinetics were observed, indicating that the DNA base stack is remarkably efficient at mediating long-range charge transport. (Adapted from |33|.)

invariant for the entire series of DM-modified duplexes, indicating that within the resolution of this experiment, increasing the through-helix DM/gold separation does not substantially affect the rate of electron transfer. If, indeed, the pathway for reduction proceeds through the helix, these results strongly suggested that the distance dependence of electron transfer through DNA is exceptionally shallow. For comparison, the rate constant would be expected to drop by more than 15 orders of magnitude in moving the daunomycin binding site 40 Å, given a σ-only tunneling matrix ($\beta = 1\,\text{Å}^{-1}$). But might the pathway for electron transfer in fact be "short-circuited" in the DNA films?

A CA mismatch was incorporated into the region of the helix intervening between daunomycin and the electrode surface as a means to disrupt stacking within the proposed base pair-mediated pathway. Upon introduction of this perturbation in stacking, the electrochemical response observed with fully paired duplexes was essentially switched off (Fig. 7). Apparently, the long-range reduction of daunomycin requires a fully stacked DNA duplex; long-range electron transfer cannot proceed without proper coupling between the DNA bases. These data confirmed that the pathway must indeed proceed through the stacked bases and confirmed further that the lack of a detectable distance dependence for the long-range reduction of daunomycin through the DNA helix must reflect the remarkable ability of the base-stacked medium to facilitate charge transport.

Perhaps more importantly, the attenuation of the long-range reduction of daunomycin in the presence of a CA mismatch revealed that a *single-point mutation within a given sequence could be accurately detected electrochemically*. The achievement of such a sensitive means to detect a small change in DNA sequence has been a high priority in the development of DNA-based diagnostics. Thus, we sought to explore further the versatility and applicability of this approach based on long-range charge transduction through the DNA helix. A variety of mismatches were tested and it was found that mispairs of varying compositions could be detected. Since our approach utilized exclusively double-stranded DNA duplexes, the generation of a single-stranded surface capable of assaying test sequences in solution was another important feature to confirm the practicality of this approach. Indeed, electrodes originally containing DNA duplexes could be stripped of the complementary strands by heat denaturation, and, after adding the test strand and dipping the electrode into a DM solution, a CA mismatch could be detected in the new DNA films duplexes formed in situ. Therefore, the sensitivity of charge transduction through these DNA films to stacking perturbations such as DNA mismatches can indeed be powerfully exploited. Charge transduction through DNA provides the basis for the design of DNA-based biosensors capable of detecting genetic mutations.

Thus, it has been demonstrated both spectroscopically and electrochemically that long-range electron transfer for species intercalated within the base stack proceeds over exceptionally long distances and can occur on ultrafast time scales. The DNA bases provide a bridge for electron transport that allows very strong coupling between reactants. Our results indicate also that π-stacked media may provide more efficient pathways for electron transfer than σ-bonded systems. Importantly, the properties of charge transport through the base stack are

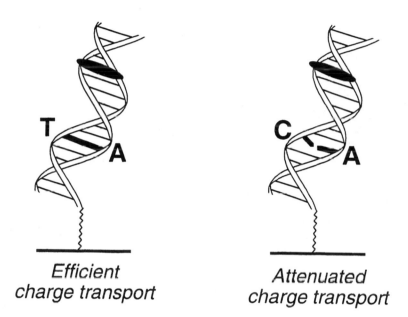

Efficient
charge transport

Attenuated
charge transport

FIG. 7.   Schematic illustration of the effect of a CA mismatch on the electrochemical response of redox-active intercalators bound to DNA films.

highly sensitive to the perturbation of stacking interactions. Base-base interactions are essential to facilitate long-range electron transfer. The sensitivity of this reaction to stacking may also be important in reconciling the less efficient electron transfer observed by others in studies of DNA-mediated electron transfer using reactants that did not interact closely with the base stack. Indeed, efficient charge transport clearly depends on and requires that feature which may be unique to a DNA helix: the π-stacked array of aromatic heterocyclic base pairs.

4. DNA AS A REACTANT IN LONG-RANGE CHARGE TRANSFER REACTIONS

Electron transfer reactions involving DNA that are proposed to occur in the biological milieu result in the formation of irreversible base damage [39]. In addition, a particularly prevalent lesion, the thymine dimer, is not only formed but can be repaired by electron transfer in vivo [40]. Given the demonstrations that the transport of electrons can proceed over long molecular distances, as described (vide supra), an intriguing question becomes whether the damage or repair of DNA lesions can be promoted from a remote position along the DNA helix by electron transfer. This question offers profound implications for molecular mechanisms of carcinogenesis and mutagenesis.

4.1. Oxidative Damage at 5'-GG-3' Sites in DNA

Guanine is the most easily oxidized DNA base ($E^0 \sim +1.3$ V versus NHE) and is presumed to be the primary target for oxidative damage in the cell [39]. Many investigations of base damage in vitro have confirmed this expectation [48–50]. In fact, both theoretical [51] and empirical studies have uncovered the fact that 5'-GG-3' doublets are preferentially oxidized, with damage concentrated at the 5'-G of the pair. The oxidation of guanine in DNA leads to piperidine-sensitive lesions, sites where after treatment with piperidine strand breaks result. Although some of the earlier studies sought to assign this preferential reactivity to specific binding, theoretical studies [51] have provided a more reasonable explanation for this trend by demonstrating that the ionization potentials for stacked guanines are the lowest of any dinucleotide, and that the HOMO for this pair is in fact localized on the 5'-G. Therefore, the specific formation of piperidine-sensitive strand breaks at the 5'-G within a 5'-GG-3' doublet can be considered to be a signature of electron transfer-induced damage. Other types of oxidative damage (e.g., singlet oxygen) do not appear to be sensitive to these small stacking-dependent changes in thermodynamic driving force.

4.1.1. Oxidative Damage at 5'-GG-3' Sites by Rh(phi)$_2$bpy^{3+}

Complexes of Rh(III) containing phi ligands have been employed extensively in our laboratory as structural probes of nucleic acids and have been utilized in designing strategies for the recognition of specific base sequences [52]. These complexes intercalate from the major groove of the double helix using the phi ligand and display very high affinity for B-form DNA. The position and extent of binding can be easily assayed for these complexes, as their photochemical properties lead to strand scission upon irradiation at wavelengths in the near-UV region. These strand breaks have been attributed to direct hydrogen abstraction by the photoexcited phi complex at sites of intercalation [53]. Perhaps of greater interest in this context, Rh(phi)$_2$bpy^{3+} has been shown not only to promote direct strand scission upon photolysis at 313 nm but also to oxidize guanine bases when irradiated at 365 nm. This rhodium complex is a potent photooxidant, with an excited state reduction potential (E(Rh^{3+}*/Rh^{2+})) of ~2.0 V versus NHE [54]. Because both position of intercalation and site of oxidation can be probed in discrete photochemical reactions using this complex, this intercalator could be harnessed as an ideal probe of long-range base oxidation in DNA.

Indeed, the irradiation at 365 nm of Rh(phi)$_2$bpy^{3+} noncovalently bound to DNA induced piperidine-sensitive lesions specifically at 5'-GG-3' (underlined G is site of preferential cleavage) doublets [27]. Irradiation at higher energy (313 nm) of this complex bound to DNA produced random strand breaks, illustrating that this molecule binds nonspecifically to the DNA helix. Thus, the localization of oxidative damage at 5'-GG-3' sites, initiated by irradiation at 365 nm, provided the first indications that damage generated all along the DNA duplex might migrate to these thermodynamically favored sites.

To determine the contribution of long-range charge migration through the DNA helix to the formation of irreversible lesions at 5'-GG-3' sites, an assembly was therefore constructed that contained an intercalating Rh(III) photooxidant covalently tethered to one end of a DNA duplex and 5'-GG-3' doublets in the center and near the other end of the duplex. In this manner, by spatially separating the bound intercalator from the site(s) of preferential oxidation, we could determine if an electron hole, injected into the base stack, could migrate to the remote 5'-GG-3' sites and yield oxidative damage to DNA from a distance.

The results are illustrated schematically in Fig. 8. In the first assembly prepared, the intercalator, constrained by the tether, was spatially separated from the two guanine doublets by 17 and 34 Å of intervening base stack. With irradiation of the assembly at 313 nm, direct strand breaks were evident only near the end of the duplex. However, upon 365 nm irradiation, damage of both 5'-GG-3' sites, revealed as piperidine-sensitive strand breaks, was observed [27]. These data indicated for the first time in a chemically well-defined system that the migration of an injected hole can promote oxidative damage to DNA from a distance. Interestingly, no attenuation in the yield of damage was observed for the distal relative to the proximal guanine doublet, suggesting that the hole migration through DNA is not dramatically affected by distance. Moreover, the primary oxidative lesion, established by digestion without piperidine treatment, was found to be 8-oxo-guanine, a common oxidative lesion to DNA in vivo [1]. Thus a physiologically relevant and important oxidative lesion in DNA can be initiated by the facile transport of electrons through the DNA base stack over long molecular distances. Therefore, DNA damage may result not only from localized events but may also be promoted from remote sites along the DNA helix.

The results obtained in this system confirm that components of the DNA base stack can participate in electron transfer as reactants, but here again, the integrity of the DNA bridge is still crucial [28].

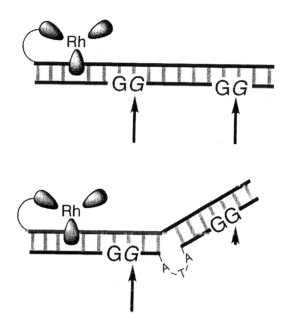

FIG. 8. Rhodium(III)-modified assemblies for the investigation of long-range guanine oxidation. With the rhodium photooxidant tethered to the end of a 16-base-pair DNA duplex, damage occurs with equal intensity at guanine doublets located 14 and 37 Å down the helix. However, in the presence of an ATA bulge intervening between the two sites, damage at the distal guanine doublet is significantly diminished. Again, base stacking appears to be essential for efficient, long-range charge transport through the DNA base stack. (Adapted from [28].)

When bulged regions were introduced to disrupt the base stacking between the Rh(III) intercalator and guanine doublets, the yield of guanine damage at the distal site was significantly diminished (Fig. 8). This observation again implicates the extent of stacking between the DNA bases as an important parameter that modulates the efficiency of DNA-mediated charge transport. Moreover, the sensitivity of this reaction to stacking may provide the basis for a convenient assay of DNA structure. Studies are currently in progress to monitor the effects of protein-induced DNA bending or base flipping; these effects are difficult to detect without very sophisticated structural determinations, and thus the electronics of the base stack may provide a useful handle for the identification of these structurally subtle events.

The actual distance over which an electron hole can migrate becomes an intriguing question. If no attenuation is observed over 40 Å, can damage migrate hundreds of Ångstroms? This question is currently also under investigation in our laboratory. We have already found significant damage to DNA resulting from oxidation by an intercalator bound 200 Å from the lesion [55]. In fact, in a family of assemblies where the distal 5'-GG-3' doublet was systematically moved away from the proximal site, little diminution in oxidative damage was seen over 75 Å. However if a 5'-TATA-3' sequence, one that is thought to have particular flexibility, was inserted within this assembly, a substantial decrease of oxidative damage at long range was observed. If charge transport can proceed over such distances in DNA within the cell, the biological implications become increasingly interesting. The long-range transmission of chemical information, modulated by DNA sequence, structure, and perhaps by protein binding, could certainly be an attractive goal in vivo. This efficiency of electron transport through the DNA helix might provide a viable mechanism for the execution of remote signaling.

4.1.2.   Oxidative Damage at 5'-GG-3' Sites and Base Mismatches by Ru(phen)$_2$dppz^{3+}

Ruthenium intercalators can also be utilized to probe long-range oxidative damage in DNA. Here, a flash-quench approach (Fig. 9) is utilized

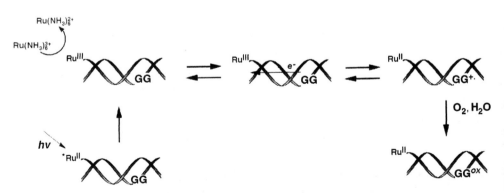

FIG. 9.   Flash-quench approach for long-range guanine oxidation through the base stack of DNA. This technique, which produces high yields of oxidized product, allowed the spectroscopic characterization of the photochemically generated guanine radical cation within duplex DNA. (Adapted from [29].)

where DNA-bound Ru(phen)$_2$dppz^{2+} is excited with visible light, and oxidatively quenched with a diffusible quencher to generate Ru(III) (E^0 (Ru$^{(III/II)}$) ~ +1.6 V). This ground state oxidant then can react with guanine [55]. Moreover, the reaction involving Ru(phen)$_2$dppz^{2+} has allowed the spectroscopic characterization of the guanine radical intermediate formed in the DNA duplex by transient absorption spectroscopy.

DNA duplexes covalently modified with a Ru(II) intercalator have also been used in establishing oxidative damage to 5'-GG-3' sites through charge migration over appreciable distances [29]. Again, piperidine-sensitive lesions are generated at 5'-GG-3' sites. The quantum yield for this reaction is at least three orders of magnitude higher than for the direct Rh(III) reaction, consistent with the more efficient photochemistry of ruthenium compared to rhodium. Thus damage from a distance can be generated with potent ground-state intercalating oxidants as well. With these ruthenium-modified DNA assemblies, it was illustrated that where the damage is concentrated on the DNA helix depends on thermodynamic parameters. In a ruthenium-modified duplex containing a single 5'-GG-3' site, only the 5'-G was modified. However, when the 3'-G of the 5'-GG-3' was mutated to a C, with no other changes to the duplex, oxidative damage was apparent, and almost equally, at all of the single G bases. In addition, guanines incorporated within GA mismatches were identified as sites particularly susceptible to oxidative damage, whereas those within GT mismatches were not affected. These mispairs may differ in redox potential or solvent accessibility; the susceptibility of the GA mispair to oxidative damage is certainly intriguing from a biological standpoint.

4.2.   Oxidative Repair of Thymine Dimers

Another attractive target for DNA-mediated electron transfer chemistry is the repair of a base lesion. The primary photochemical lesion in DNA is the cyclobutyl thymine dimer, which results from a [2+2] photocycloaddition between adjacent thymine bases on the same polynucleotide strand. In *E. coli*, the thymine dimer is repaired by photolyase in an electron transfer reaction involving a flavin cofactor [40]. This reaction repairs the dimer through a reductive event, but model studies have revealed that this lesion can also be repaired oxidatively (E^0 (T<>T) ~

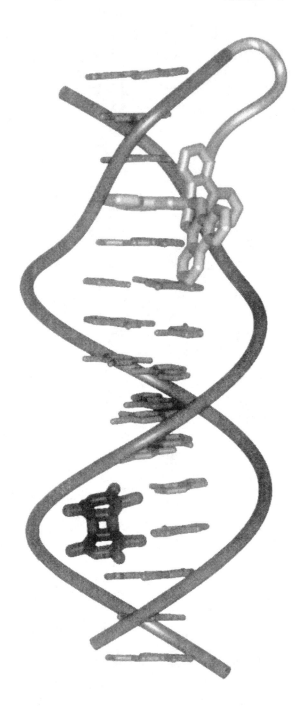

+2 V). Hence the repair of a thymine dimer represents a physiologically important reaction that can be triggered cleanly by electron or hole transfer and involves no subsequent atom transfer reactions. In our laboratory, the metallointercalator $Rh(phi)_2bpy^{3+}$ was an attractive reactant to employ for the study of oxidative thymine dimer repair triggered by an electron transfer reaction mediated by the DNA base stack over long distances.

The Rh(III) intercalator was tethered to the end of a DNA duplex containing a thymine dimer, which had been specifically incorporated in the DNA duplex at a site over 25 Å from the site of rhodium intercalation [30] (Fig. 10). In a high-temperature, high performance liquid chromatography assay, the efficient repair of the dimer was monitored as a function of irradiation of the bound rhodium with 400-nm light. Again, the products of a chemical reaction occurring as a result of charge migration over an extraordinary distance were detected. This experiment clearly depended on and underscored the fact that the DNA base stack can be remarkably efficient in transporting charge. As we had observed in tests of oxidative damage to DNA from a distance, here too we found the efficiency of DNA repair to be relatively insensitive to distance. In fact, the repair yield increased slightly when the dimer was moved down the helix away from the intercalator. Additionally, and paralleling our other studies of long-range electron transfer, the stacking of the DNA bases was found to be an important factor in modulating this reaction. In the presence of bulged regions within the DNA base stack intervening between the intercalator and the dimer, the yield of repair is drastically diminished, suggesting that the pathway for this reaction has been significantly affected by the perturbation in base stacking.

Given that the chemical reactivity of guanine doublets and thymine dimers within the DNA π stack were elucidated, experiments could also be designed to probe competitively the effect of the differing thermodynamics of these sites on their reactivities. To evaluate the

FIG. 10. Repair of a thymine dimer by an intercalating Rh(III) photooxidant. In a 15-base-pair DNA duplex covalently modified with $Rh(phi)_2bpy^{3+}$, this physiologically deleterious lesion is repaired via long-range charge transfer over >30 Å. (Adapted from [30].)

reactivity of the thymine dimer in the presence of guanine doublets, and vice versa, duplexes were synthesized that contained one or both of these reactants and a covalently tethered oxidant (either $Rh(phi)_2bpy^{3+}$ or $Ru(phen)_2dppz^{2+}$ derivatives) [38]. Thymine dimers are more difficult to oxidize than guanine by at least 0.7 V, and are not thermodynamically accessible with the Ru(III) oxidant. Thus, these assemblies represent an interesting opportunity to see whether these thermodynamic or perhaps kinetic parameters dominated the photochemical reactions. In a duplex containing a Rh(III) oxidant, a proximal and distal guanine doublet, and a thymine dimer, the dimer is still repaired, although the efficiency is decreased by 20–40%, compared to sequences without GG sites. The guanine doublets also undergo twofold diminished oxidation in the presence of the dimer compared to sequences lacking this lesion. These results indicate that although repair of the thymine dimer is not thermodynamically favored, kinetic factors must drive this reaction. Interestingly, the Ru(III) oxidant does not repair the thymine dimer, but it does oxidize guanines on either side of the dimer.

5. CONCLUSIONS

In reactions monitored spectroscopically, electrochemically, and chemically, we have now demonstrated that a whole series of intercalated reactants with different structural properties can undergo remarkably efficient electron transfer over long molecular distances through the DNA helix. The base stack therefore serves as a bridging medium that is particularly favorable for facilitating long-range electronic coupling. Moreover, this stack of aromatic heterocycles participates in chemical reactions triggered from remote locations along the DNA helix, and these reactions are pertinent to biological base damage mechanisms. As DNA is a molecular π stack, stacking interactions both between the DNA bases and between reactants and these bases appear to be critical to facilitate long-range reactions. This observation may help to explain the results indicating much poorer electron transfer efficiency through DNA in systems employing nonintercalating reactants.

Now that the ability of the DNA base stack to serve as a medium for long-range charge transport has been demonstrated to be a general phenomenon in these chemical systems, important biological issues need to be considered. Since the oxidative damage that is generated may be the result of radical migration over long distances that can concentrate in thermodynamically favored sites, might cellular damage arise similarly?

There are many issues remaining to be explored with respect to the fundamental characteristics of electron transfer reactions through the DNA double helix. It remains to be seen over what maximum distance charge transport can proceed; to address this question, we will require methodologies to study this phenomenon in biochemically generated DNA sequences. In addition, although it has been established that photoinduced electron transfer reactions facilitated by the DNA base stack occur within hundreds of picoseconds, more detailed measurements on time scales requiring sophisticated instrumentation will be required to measure the actual effect of distance on the rate of electron transfer for reactions involving intercalators. Additionally, since we have identified stacking as an important parameter that modulates the efficiency of DNA-mediated charge transport, the development of experimental methods to quantitate this effect will facilitate this correlation, and only approaches that include this parameter might correctly explain this phenomenon from a theoretical standpoint. We need also to unravel the relationship between sequence and stacking and/or the dynamics of the base stack. Efforts are currently underway in our laboratory to examine electron transfer between modified bases in DNA. These studies will allow us to focus more specifically on sequence and stacking, as well as answering the question of whether electron transfer proceeds preferentially down one strand. Perhaps systematic studies of DNA-mediated electron transfer as a function of sequence will provide a new and fuller description of the sequence-dependent structure of DNA.

Finally, since the established biological function of DNA involves the transmission of information based on the structural characteristics of the DNA bases within the double helix, we might now imagine that this function might also be influenced by the electronic properties of these stacked bases. Could nature access the pathway for electron transport through DNA for the communication of information? Moreover, just as we can so accurately detect the presence of base mismatches by the attenuation of electron transfer, might mutagenic changes in the genomic code be identified in this manner in vivo? The questions now presented by the finding that efficient electron transport proceeds through the interior of one of the most essential biological materials are endless.

ACKNOWLEDGMENTS

We are grateful to the National Institutes of Health and the National Foundation for Cancer Research for their financial support of this work. In addition, we thank our colleagues and collaborators, as referenced throughout the chapter, for their efforts and their ideas.

ABBREVIATIONS

AFM	atomic force microscopy
bpy	2,2'-bipyridine
DAP	N,N'-dimethyl-2,7-diazapyrenium dichloride
DM	daunomycin
dppz	dipyridophenazine
dz-G or Z	7-deazaguanine
Et	ethidium
HOMO	highest occupied molecular orbital
MB	methylene blue
MLCT	metal-to-ligand charge transfer
NHE	normal hydrogen electrode
phen	1,10-phenanthroline
phi	9,10-phenanthrene-quinone diimine
UV-Vis	ultraviolet-visible

REFERENCES

1. J. Cadet, in *DNA Adducts: Identification and Significance* (K. Hemminki, A. Dipple, D. E. G. Shuker, F. F. Kadlubar, D. Segerback, and H. Bartsch, eds.), IARC Publications, Lyon, 1994.

2. T. J. Marks, *Science, 227*, 881 (1985).

3. B. E. Bowler, A. L. Raphael, and H. B. Gray, *Prog. Inorg. Chem.*, *38*, 259 (1990).

4. R. A. Marcus and N. Sutin, *Biochim. Biophys. Acta*, *811*, 265 (1985).

5. R. S. Snart, *Biopolymers*, *6*, 293 (1968).

6. C. Hélène and M. Charlier, *Photochem., Photobiol.*, *25*, 429 (1977).

7. J. D. Magan, W. Blau, D. T. Croke, D. J. McConnell, and J. M. Kelly, *Chem. Phys. Lett.*, *141*, 489 (1987).

8. L. M. Davis, J. D. Harvey, and B. C. Baguley, *Chem.-Biol. Interact.*, *62*, 45 (1987).

9. B. C. Baguley and M. LeBret, *Biochemistry*, *23*, 937 (1994).

10. P. Fromherz and B. Rieger, *J. Am. Chem. Soc.*, *108*, 5361 (1986).

11. A. M. Brun and A. Harriman, *J. Am. Chem. Soc.*, *114*, 3656 (1992).

12. F. D. Lewis, T. Wu, Y. Zhang, R. L. Letsinger, S. R. Greenfield, and M. R. Wasielewski, *Science*, *277*, 673 (1997).

13. T. J. Meade and J. F. Kayyem, *Angew. Chem. Int. Ed. Engl.*, *34*, 352 (1995).

14. K. Fukui and K. Tanaka, *Angew. Chem. Int. Ed. Engl.*, *37*, 158 (1998).

15. D. Dee and M. E. Baur, *J. Chem. Phys.*, *60*, 541 (1974).

16. A. K. Felts, W. T. Pollard, and R. A. Freisner, *J. Phys. Chem.*, *99*, 2929 (1995).

17. S. Priyadarshy, S. M. Risser, and D. N. Beratan, *J. Phys. Chem.*, *100*, 17678 (1996).

18. J. K. Barton, C. V. Kumar, and N. J. Turro, *J. Am. Chem. Soc.*, *108*, 6391 (1986).

19. D. Purugganan, C. V. Kumar, N. J. Turro, and J. K. Barton, *Science*, *241*, 1645 (1986).

20. C. J. Murphy, M. R. Arkin, N. D. Ghatlia, S. Bossman, N. J. Turro, and J. K. Barton, *Proc. Natl. Acad. Sci. USA*, *91*, 5315 (1994).

21. C. J. Murphy, M. A. Arkin, Y. Jenkins, N. D. Ghatlia, S. Bossman, N. J. Turro, and J. K Barton, *Science*, *262*, 1025 (1993).

22. M. R. Arkin, E. D. A. Stemp, R. E. Holmlin, J. K. Barton, A. Hörmann, E. J. C. Olson, and P. A. Barbara, *Science*, *273*, 475 (1996).

23. R. E. Holmlin, E. D. A. Stemp, and J. K. Barton, *J. Am. Chem. Soc.*, *118*, 5236 (1996).

24. S. O. Kelley, R. E. Holmlin, E. D. A. Stemp, and J. K. Barton, *J. Am. Chem. Soc.*, *119*, 9861 (1997).

25. S. O. Kelley and J. K. Barton, *Chem. Biol.*, *5*, 413 (1998).

26. R. E. Holmlin, P. J. Dandliker, and J. K. Barton, *Angew. Chem. Int. Ed.*, *36*, 2714–2730 (1997).

27. D. B. Hall, R. E. Holmlin, and J. K. Barton, *Nature*, *382*, 731 (1996).

28. D. B. Hall and J. K. Barton, *J. Am. Chem. Soc.*, *119*, 5045 (1997).

29. M. R. Arkin, E. D. A. Stemp, and J. K. Barton, *Chem. Biol.*, *4*, 389 (1997).

30. P. J. Dandliker, R. E. Holmlin, and J. K. Barton, *Science*, *275*, 1465 (1997).

31. S. O. Kelley, J. K. Barton, N. M. Jackson, and M. G. Hill, *Bioconj. Chem.*, *8*, 31 (1997).

32. S. O. Kelley, J. K. Barton, N. M. Jackson, L. McPherson, A. Potter, E. M. Spain, M. J. Allen, and M. G. Hill, *Langmuir* (in press).

33. S. O. Kelley, N. M. Jackson, M. G. Hill, and J. K. Barton, (submitted).

34. E. M. Fielden, S. C. Lillicrap, and A. B. Robins, *Radiat. Res.*, *48*, 421 (1971).

35. P. M. Cullius, J. D. McClymont, and M. C. R. Symons, *J. Chem. Soc. Faraday Trans.*, *86*, 591 (1990).

36. J. H. Miller, D. L. Frasco, C. E. Swanberg, and A. Rupprecht, in *Radiation Research: A Twentieth Century Perspective*, Vol. 2 (J. D. Chapman, W. C. Dewey, and G. F. Witmore, eds.), Academic Press, San Diego, 1992.

37. C. Houee-Levin, M. Gardes-Albert, A. Rouscilles, C. Ferradini, and B. Hickel, *Biochemistry*, *30*, 8216 (1991).

38. P. J. Dandliker, M. E. Nunez, and J. K. Barton, *Biochemistry*, *37*, 6491 (1998).

39. S. Steenken, *Chem. Rev.*, *89*, 503 (1989).

40. A. Sancar and G. B. Sancar, *Annu. Rev. Biochem.*, *57*, 29 (1988).

41. C. M. Dupureur and J. K. Barton, in *Comprehensive Supramolecular Chemistry*, Vol. 5 (J. M. Lehn, ed.), Pergamon Press, Oxford, 1995.

42. E. J. C. Olson, D. Hu, A. Hörmann, and P. F. Barbara, *J. Phys. Chem.*, *101*, 299 (1997).

43. P. Lincoln, E. Tuite, and B. Norden, *J. Am. Chem. Soc.*, *119*, 1454 (1997).

44. S. J. Franklin, C. R. Treadway, and J. K. Barton, *Inorg. Chem.*, *37*, 5198 (1998).

45. H. O. Finklea, *Electroanal. Chem.*, *19*, 109 (1996).

46. F. Leng, R. Savkur, I. Fokt, T. Przewloka, W. Priebe, and J. B. Chaires, *J. Am. Chem. Soc.*, *118*, 4732 (1996).

47. A. H.-J. Wang, Y.-G. Gao, Y.-C. Liaw, and Y.-K. Li, *Biochemistry*, *30*, 3812 (1991).

48. H. Kasai, Z. Yamaizumi, M. Berger, and J. Cadet, *J. Am. Chem. Soc.*, *114*, 9692 (1992).

49. I. Saito, M. Takayama, H. Sugiyama, K. Nakatani, A. Tsuchida, and M. Yamamoto *J. Am. Chem. Soc.*, *117*, 6406 (1995).

50. D. T. Breslin and G. B. Schuster, *J. Am. Chem. Soc.*, *119*, 5043 (1997).

51. H. Sugiyama and I. Saito, *J. Am. Chem. Soc.*, *118*, 7063 (1996).

52. T. W. Johann and J. K. Barton, *Phil. Trans. Roy. Society*, *354*, 299–324 (1996).

53. A. Sitlani, E. C. Long, A. M. Pyle, and J. K. Barton, *J. Am. Chem. Soc.*, *114*, 2303–2312 (1992).

54. C. Turro, A. Evenzahav, S. Bossmann, J. K. Barton, and N. J. Turro, *Inorg. Chim. Acta*, *243*, 101 108 (1996).

55. M. Nunez, D. Hall, and J. K. Barton, unpublished results.

56. E. D. A. Stemp, M. R. Arkin, and J. K. Barton *J. Am. Chem. Soc.*, *119*, 2921 (1997).

8

Involvement of Metal Ions in Lipid Peroxidation: Biological Implications

Odile Sergent, Isabelle Morel, and Josiane Cillard

Laboratoire de Biologie Cellulaire et Végétale,
INSERM U456, Faculté de Pharmacie,
F-35043 Rennes Cedex, France

1.   INTRODUCTION

Metals ranging from a few nontransition (lead, cadmium, mercury) to many transition elements (vanadium, chromium, iron, nickel, copper, silver) have been reported as capable of inducing oxidative stress and especially lipid peroxidation. The severity of oxidative stress and the types of pathologies induced by metal ions have been extensively reviewed by Stohs and Bagchi [1]. Oxidative stress results in the inability of the antioxidant defense system to compensate the increase in oxygen-derived free radicals. Thus, oxidative damage can occur in cellular membrane lipids. In general, the term *lipid peroxidation* is used to describe the nonenzymatic oxidative degradation of fats.

Lipid peroxidation is now recognized to be a critical stage in toxicological processes. Peroxidation of polyunsaturated fatty acids in lipid bilayer membranes causes loss of fluidity, a fall in membrane potential, increased permeability to protons and calcium ions, and, eventually, breakdown of cell membranes due to cellular deformities. The structural and functional integrity of the cell membranes is necessary for signal transduction, molecular recognition and transport, cellular metabolism, etc. Hence it is important to understand the mechanisms of lipid peroxidation in order to prevent it.

This chapter focuses only on processes occurring in biological systems. Moreover, only lipid peroxidation induced by metal ions will be described and catalysis by heme-containing proteins and nonheme iron enzymes will be excluded. Differences in the toxicities of metal ions can also be related to differences in solubility and transport in the body, but these topics will not be dealt with in this chapter.

In this chapter, the possible involvement of metal ions during the whole process of lipid peroxidation will be developed and consequences on the evaluation of lipid peroxidation and on tissue damage will be considered.

2. OUTLINE OF THE MECHANISM OF LIPID PEROXIDATION

During lipid peroxidation molecular oxygen is incorporated into unsaturated lipids (LH) to form lipid hydroperoxides (LOOH). Lipid peroxidation proceeds through three phases: initiation, propagation, and termination (Fig. 1) [2–4].

The initiation reaction consists of allylic hydrogen atom abstraction from a polyunsaturated fatty acid chain of a phospholipid molecule by an oxidant of sufficient chemical reactivity and formation of a carbon-centered lipid alkyl radical (L$^{\bullet}$). The resulting lipid radical undergoes diene conjugation. Despite intensive work, the nature of the primary oxidant is still unknown. One of the most potent oxidants that can be formed from oxygen is the hydroxyl radical [4], which is for this reason one of the most popular candidates for the initiation of lipid peroxidation. It is extremely reactive and can attack many cell constituents including lipids, proteins, and nucleic acids. However, polyunsaturated fatty acids (PUFAs) are particularly susceptible to oxidation because of the weakness of the double bonds [5]. Then oxygen will add there rapidly to form the peroxyl radical (LOO$^{\bullet}$).

The next stage is the propagation process. Whereas oxidants are required to initiate lipid peroxidation, once peroxyl radicals are formed,

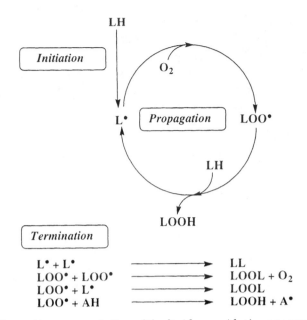

FIG. 1.   Schematic representation of the lipid peroxidation process. LH, unsaturated fatty acid; LOOH, lipid hydroperoxide; LO•, alkoxyl radical; LOO•, peroxyl radical; AH, antioxidant.

the reaction progresses alone to form lipid hydroperoxides; peroxyl radicals continue the oxidative chain reaction by abstracting a hydrogen atom from another nearby lipid leading to the formation of hydroperoxides and, hence, propagating the oxidative process. From the initial oxidation of a single PUFA, it is estimated that in vivo 10–15 cycles will occur before the termination step [6].

Ultimately, the oxidative reactions will stop by various termination reactions, i.e., by bimolecular reactions of two lipid alkyl radicals and two lipid peroxyl radicals to form nonradical products such as ketones and secondary alcohols. If the system contains antioxidants, termination can occur by reaction of the alkyl or peroxyl radicals with physiological antioxidants such as α-tocopherol, glutathione, etc. In this case, the initial rate of oxidation, called lag phase, is slow because these antioxidants first scavenge peroxyl radicals, then the antioxidants are consumed and the rate of propagation increases.

Breakdown of alkoxyl and peroxyl radicals can lead to the formation of many compounds. For example, by β cleavage of alkoxyl radicals, hydrocarbon gases are formed [7]; cyclization and cleavage of peroxyl radicals can lead to malondialdehyde formation; lipid radical decomposition gives 2-alkenal derivatives and low molecular weight alcohols, esters, and hydrocarbons [4].

3. INVOLVEMENT OF METAL IONS IN VARIOUS STAGES OF THE LIPID PEROXIDATION PROCESS

The initiation stage is the key event for the development of lipid peroxidation. To inhibit it would lead to the prevention of possible tissue damage. However, in biological systems this stage is the most controversial of the lipid peroxidation process promoted by metal ions and many assumptions can be discussed.

3.1. Initiation

3.1.1. Redox Cycling of Metals

Transition metals (iron, copper, chromium, and vanadium) are easily reduced or oxidized in biological systems [1,8]. At trace levels, they can catalyze the production of free radicals when these metals are available in a redox-active form. Two principal mechanisms have been described:

(a) The primary or first-chain lipid peroxidation initiation (hydroperoxide-independent lipid peroxidation). Transition metals can catalyze the production of hydroxyl radicals ($\cdot$OH) through both the Fenton and Haber-Weiss reactions (see also Chapter 1 of this book) [9,10]. Then the hydroxyl radical can abstract a hydrogen atom from a polyunsaturated fatty acid of a phospholipid molecule. These reactions are especially well described for iron and copper.

In the Fenton reaction, the hydroxyl radical is produced from hydrogen peroxide in the presence of Fe^{2+}. Nickel oxides containing Ni(III) [11], nickel stabilized by chelation to peptides containing a glycylgylcyl L-histidine [12], and cobalt [13] have also been reported to induce Fenton-like reactions. However, using electron spin resonance (ESR) techniques, Kadiiska et al. were unable to demonstrate the effect of cobalt [14].

In the Haber-Weiss reaction, superoxide anion (O_2^-) reduces metal ions before the Fenton reaction occurs:

$$O_2^- + M^{(n+1)+} \longrightarrow O_2 + M^{n+} \text{ (reduction of metal)}$$

$$H_2O_2 + M^{n+} \longrightarrow {}^{\bullet}OH + OH^- + M^{(n+1)+} \text{ (Fenton reaction)}$$

$$O_2^- + H_2O_2 \longrightarrow O_2 + {}^{\bullet}OH + OH^- \text{ (Haber-Weiss reaction)}$$

It should be noted that an important step in this process is the reduction of the metal. It is now well recognized that ascorbate, NADH, and NADPH could also be reductants instead of the superoxide anion [8].

(b) *The secondary lipid peroxidation initiation (hydroperoxide-dependent lipid peroxidation).* The other type of reaction promoted by iron, copper, cobalt, and manganese is the decomposition of *preformed* lipid hydroperoxides [15–17], which can be obtained by lipoxygenase- or radical-mediated pathways.

Ferric or cupric ions can react with preformed lipid hydroperoxides to produce peroxyl radicals:

$$LOOH + M^{(n+1)+} \longrightarrow LOO^{\bullet} + H^+ + M^{n+}$$

Then LOO${}^{\bullet}$ can propagate lipid peroxidation by abstracting a hydrogen atom from another nearby lipid, leading to the formation of hydroperoxides.

Similarly, ferrous and cuprous ions can react with lipid hydroperoxides to form alkoxyl radicals:

$$LOOH + M^{n+} \longrightarrow LO^{\bullet} + OH^- + M^{(n+1)+}$$

Subsequently, LO${}^{\bullet}$ can generate L${}^{\bullet}$ through a β scission and oxygen will add rapidly to form LOO${}^{\bullet}$ and thus, hydroperoxides by abstraction of a hydrogen atom from another lipid.

Ferrous ions can decompose lipid peroxides more rapidly than ferric ions, thereby generating larger amounts of free radicals [18]. This could explain why Ramanathan et al. observed an increase of lipid peroxidation with $FeSO_4$ compared to $FeCl_3$ in fish phospholipid liposomes [19].

Chromium(III) [20], vanadate(V), vanadyl(IV) [21], and nickel oxides containing Ni(II) [11] have also been described as inducing lipid peroxidation via hydroperoxide decomposition.

For both types of initiation reactions (a) and (b), to be catalytically active, iron must be "free," i.e., not stored in proteins like hemoglobin, myoglobin, ferritin, hemosiderin, or transferrin (see Chapter 1 of this book). However, based on the low solubility of metal ions at physiological pH, iron must be chelated to be solubilized. The identity of the physiological iron-chelating species is still unknown. It has been suggested that the catalytically active iron pool exists in a low molecular weight form chelated to citrate, amino acids, nucleotides, etc. [22,23]. Likewise, copper would be chelated to glutathione [24]. Reducing agents such as nitric oxide [25], superoxide anion [26,27], and NADH [28] are known to facilitate the release of iron from ferritin, an iron storage protein. By producing reducing agents, various xenobiotics such as alcohol [29], paraquat, diquat, nitrofurantoin, adriamycin, daunomycin, and diaziquone [27,30] can also release iron from ferritin. Likewise, iron can also be released from transferrin when iron-chelating agents are present and from hemoglobin and myoglobin when heme is damaged [23].

Selenium is another metal ion that can interfere with redox cycling processes described above for transition metal ions. Selenium action is based on its capacity to facilitate elimination of H_2O_2 in biological systems. Selenium-dependent enzymes such as glutathione peroxidase are able to eliminate H_2O_2 by catalyzing the transfer of a hydrogen atom from reduced glutathione to H_2O_2. The protective effect of selenium against peroxidation of lipid membranes exposed to oxidizing conditions has been well demonstrated [31].

3.1.2. Formation of Metal-Oxygen Complexes

The nature of the primary oxidant responsible for the first-chain initiation of lipid peroxidation by iron in biological systems is a very controversial topic, especially with the observations of Aust et al. [9,32,33]. They proposed species other than hydroxyl radicals to initiate lipid peroxidation; lipid oxidation in the presence of H_2O_2 can be promoted by ferryl or perferryl iron ($Fe^{3+}O_2^-$) and especially mixed metal-O_2 complexes [Fe^{2+}-O_2-Fe^{3+}]. Cobalt and manganese can prevent the formation of such iron-oxygen species [34]. The generation of these iron--

oxygen complexes occurs if the Fe^{2+}/Fe^{3+} ratio approaches unity [33]. The autoxidation of Fe^{2+} with H_2O_2 or a chelator and the reduction of Fe^{3+} by superoxide anion or reductants like ascorbate, NADPH, cysteine, dithiothreitol, and glutathione may represent an alternate mean to generate Fe^{3+} or Fe^{2+}, respectively, in order to obtain the adequate ratio. However, excess Fe^{2+} oxidation or Fe^{3+} reduction will inhibit lipid peroxidation by changing the Fe^{2+}/Fe^{3+} ratio [33]. For example, at low concentrations, ascorbate (25 μM) can promote lipid peroxidation, whereas at higher concentrations (50 μM) it inhibits lipid peroxidation [33].

Vanadium has also been reported to initiate lipid peroxidation via the formation of a peroxy-vanadyl [V(IV)-OO•] complex [21]. However, only vanadyl but not vanadate could initiate lipid peroxidation because vanadate first has to be reduced to vanadyl by a flavoprotein [21].

3.1.3. Depletion of Glutathione and Other Antioxidants

Many metal ions such as lead and cadmium do not undergo redox cycling and depletion of glutathione is one of their mechanisms of promoting lipid peroxidation. Glutathione, the tripeptide L-γ-glutamyl-L-cyteinylglycine, present in millimolar concentration in all cells, functions both as a direct scavenger of reactive oxygen species (ROS) and a cofactor in their metabolic detoxification [35,36]. For instance, the activity of glutathione peroxidase (GPx) depends on glutathione (GSH) to reduce H_2O_2 or lipid hydroperoxides (ROOH) in water or stable alcohols, respectively, as shown below:

$$ROOH + 2GSH \xrightarrow{GPx} ROH + H_2O + GSSG$$

Then regeneration of GSH is in turn dependent on glutathione reductase (Gr) activity:

$$GSSG + NADPH \xrightarrow{Gr} 2GSH + NADP^+$$

Glutathione transferase, another glutathione-dependent enzyme, detoxifies a variety of electrophilic compounds to less toxic forms by conjugation with thiol groups such as glutathione. Another property of this enzyme is to reduce lipid peroxides; therefore, it is called selenium-independent glutathione peroxidase.

Mercury, copper, and cadmium possess a high affinity for sulf-hydryl groups and depletion of glutathione simultaneously with lipid peroxidation was observed in the liver [37–41], in the kidney [37,42,43], and in the heart [37]. Studies from Strubelt et al. suggested that the decrease was due to metal binding to glutathione and subsequent elimination [39]. Sugiyama reported in his review many investigations about the formation of chromium-glutathione complexes; the decreased levels of glutathione in hepatocytes, thymocytes, or erythrocytes; and the involvement of glutathione depletion in chromium-induced lipid peroxidation [44]. Loading 3T3 cells with nickel also leads to the depletion of cytoskeletal protein sulfhydryls as well as glutathione [45]. Donaldson and Knowles reported the formation of a complex with lead and glutathione due to the high affinity of lead for sulfhydryl groups [46]. Moreover, supplementation of Chinese hamster ovary (CHO) cells with N-acetylcysteine, known to enhance glutathione levels, inhibits lipid peroxidation induced by lead [47]. Arsenic may also induce lipid peroxidation via glutathione depletion only in the rat liver; yet the mechanism of action remains unknown [48]. However, some metals such as zinc and selenium can prevent lipid peroxidation by inhibiting glutathione depletion. Thus, zinc has been reported to induce resistance to lipid peroxidation induced by ultraviolet radiation [49]. One of the possible mechanisms is the protection of thiol groups of molecules such as glutathione from oxidation [50]. Likewise, the protective effect of selenium toward lipid peroxidation induced in rat liver was also demonstrated to be due to a direct action on glutathione levels [51]. Glutathione depletion by metals results in a decrease of glutathione peroxidase activity. This is observed in heart, liver, and kidney of rats treated with cadmium for 30 days [38]. In the same manner, in rat kidney cortex, acute intoxication with chromium induces a marked decrease of glutathione peroxidase activity, which promotes lipid peroxidation [52]. Lead has also been reported to interact with glutathione peroxidase by decreasing its activity [42,53].

Moreover, other antioxidant enzyme activities are also inhibited by metals such as superoxide dismutase (SOD), which is responsible for the catalytic dismutation of the potentially toxic superoxide anion radical to H_2O_2 [54]; catalase (CAT) then eliminates H_2O_2 [55]. A decrease of SOD and CAT activities is observed in acute intoxication with chromium [52] or lead [42]. In the same way, in kidney and liver, MnSOD and CAT activities were also found to be decreased in acute [56,57] and chronic intoxication with cadmium for 30 days [38,56]. Hussain et al. assumed that the inhibition of SOD activity could be due to a direct interaction of cadmium with the enzyme, where cadmium replaces zinc to form Cu,CdSOD [56].

Nevertheless, except for glutathione depletion due to the high affinity of some metals to sulfhydryl groups, it is difficult to ascertain whether the decrease in antioxidants is a cause or a consequence of oxidative damage (see also Sect. 4.3).

3.1.4. Activation of Phagocytes

Cadmium displays a weak redox potential under conditions prevailing in biological systems, suggesting that free radicals must be produced indirectly. Manca et al. hypothesized that cadmium may induce lipid peroxidation in the lung by activating macrophages [58]. In vitro, Amoruso et al. showed that cadmium activated rat and human phagocytes to produce superoxide anion [59].

Chromium has also been reported to activate peritoneal macrophages [60]. It is well known that these cells produce large quantities of O_2^- via NADPH oxidase and subsequently hydroxyl radicals.

3.1.5. Modification of Fatty Acid Composition of Membranes

Lead can induce lipid peroxidation by increasing the relative tissue concentrations of arachidonic acid as a percentage of total fatty acids [46,61,62]. Arachidonic acid is a fatty acid with four double bonds, which makes it very susceptible to oxidation when compared to, say, linoleic acid, which has two double bonds.

Nickel induces an increase of free unsaturated fatty acids through the cleavage of membrane phospholipids via activation of phospholipase A_2 or direct interaction with membranes [63,64].

In the membrane, zinc is linked mainly to phospholipids protecting them from possible damage [65]. Indeed, zinc deficiency can also lead to an increase in unsaturated fatty acids, thereby supplying substrates for lipid peroxidation [66].

3.1.6. Mitochondrial Dysfunction

For cadmium, mitochondrial dysfunction has also been reported [67] in rat hepatocytes with a consequent enhancement of lipid peroxidation. Likewise, a dysfunction of the mitochondrial electron transport chain has been observed in kidney mitochondria of rats treated with mercury leading to an increase of H_2O_2 levels and of lipid peroxidation [43]. The ubiquinone-cytochrome b region is the principal site of H_2O_2 formation in kidney mitochondria [43].

3.1.7. Increased Iron or Copper Levels

Recently, it has been suggested that cadmium can induce lipid peroxidation by releasing iron from ferritin [68]. Nickel has also been reported to significantly increase liver iron levels and consequently to induce lipid peroxidation via iron [69]. Nevertheless, Stinson et al. showed that desferrioxamine, a chelator of iron, was unable to inhibit lipid peroxidation induced by nickel in liver homogenates [70]. Nickel may also induce lipid peroxidation in the liver by depleting plasma ceruloplasmin and hence increasing plasma copper levels [71].

3.1.8. Activation of Microsomal Enzymes

Chromium can be metabolized by the electron transporter cytochrome P450 [72]. This could lead to the efflux of superoxide anion from the electron transport chain.

3.2. Propagation

Redox-active metals rapidly promote the breakdown of lipid hydroperoxides to form a broad spectrum of low molecular weight products such as free radicals, epoxides, aldehydes, and hydrocarbon gases. By formation of free radicals, transition metals can enhance the propagation of lipid peroxidation (see Sect. 3.1.1). Ferrous ions can cleave hydroperoxides to highly reactive alkoxyl ($LO^\bullet$) radicals which in turn abstract hydrogen from lipids to form new lipid alkyl radicals.

Selenium can also interfere with the propagation stage because selenium-dependent systems such as glutathione peroxidase [73] or seleno amino acids (selenocystine, selenocysteine, selenocystamine, and selenomethionine) [74] can eliminate lipid hydroperoxides (LOOH) to form nontoxic alcohol derivatives (LOH):

$$LOOH + 2GSH \longrightarrow LOH + H_2O + GSSG$$

The protective effect of selenium against lipid peroxidation has been well documented in leukemia cells exposed to *tert*-butyl hydroperoxide [31], in rats after physical exercise [75], in phenylketonuric children [76], or in rats receiving autotransfusion [77].

At high concentration (nearly 5×10^{-6} M), copper has been reported to act as a chain-breaking antioxidant in photosensitized lipid peroxidation [78,79]. To a lesser extent, manganese and cobalt can do the same [79,80].

$$LOO^{\bullet} + M^{n+} + H^+ \longrightarrow LOOH + M^{(n+1)+}$$

3.3. Termination

It has been reported that increasing concentrations of copper and manganese (e.g., $>10^3$ pmol/100 g fish and 10^5 pmol/100 g fish, respectively) caused a diminution in the extent of lipid peroxidation in cooked food [81]. It is likely that these metal ions at high concentrations inhibit lipid peroxidation through chain termination [82].

4. BIOLOGICAL EXAMINATIONS OF THE EFFECT OF METAL IONS ON LIPID PEROXIDATION

In biological systems, the involvement of metal ions in lipid peroxidation is difficult to define. Indeed, cells can express cellular self-defense response and very often cells and tissues are exposed to more than one metal ion. At last, the evaluation of lipid peroxidation in biological systems is questionable because there are many sources of misinterpretation, especially when metal ions are present.

4.1. Adaptive Response

In cells, the adaptation in response to a primary oxidative insult by metals is manifested by increased antioxidant levels. For instance, under chronic treatment with chromium, the glutathione content is significantly increased together with an increased activity of glutathione peroxidase and glutathione reductase; therefore, a decrease of lipid peroxidation occurs compared to acute intoxication [52]. Moreover, in this model the activities of SOD and CAT were also enhanced. Adaptive processes are particularly well described for cadmium. Li et al. showed that 3T3 cells exposed to cadmium exhibit a biphasic response with an initial decrease of GSH after 2 h of incubation, whereas after 8 h an overshooting is noted [45]. Likewise, SOD and GSH transferase activities were increased in rat heart, liver, and kidney after 1 day of treatment with cadmium with a consequent decrease of lipid peroxidation, which had first increased in the early stage [37]. A resistance to cadmium-induced lipid peroxidation due to the elevation of glutathione content has also been observed in V79 Chinese hamster fibroblasts [83], in rat glomerular mesangial cells [84], in liver of rats intoxicated with cadmium for 1 year [85], and in red blood cells of rats treated with cadmium for 30 days [86]. In the last mentioned model, antioxidant enzyme activities such as SOD, CAT, glutathione peroxidase, glutathione transferase, and glutathione reductase were also elevated when compared to control rats [86].

For lead, adaptive processes have been less described. Nevertheless, after chronic administration of lead, a diminution in lipid peroxidation has been correlated with an elevation of glutathione levels and of SOD, CAT, and glutathione transferase activities [46,53].

Defensive adaptation of cells can explain the opposite results obtained with the same metal ion during chronic intoxication. The results depend on the period of observation, i.e., either during initial oxidative insult, or later, during adaptive response, or even later, during secondary insult when antioxidants are suppressed. In microvessels, treatment for 30 days with cadmium enhances SOD and GPx activities without inducing lipid peroxidation, whereas treatment of rats for 90 days results in lipid peroxidation simultaneously with a decrease of SOD,

CAT, glutathione peroxidase, and transferase activities as well as of vitamin E, glutathione, ascorbic acid, and ceruloplasmin [87]. The authors suggest that the absence of lipid peroxidation and the elevation of antioxidant enzyme activities may be due to an initial induction of oxidative stress by cadmium. Then excessive production of free radicals by cadmium leads to a secondary insult suppressing the antioxidant activities. In this connection, opposite results can be observed with lead. Treatment with lead for 72 h gives an adaptive response with an elevation of antioxidant activities and a lack of lipid peroxidation, whereas intoxication for a longer period, such as 8 weeks, leads to a strong decrease of antioxidant activities and an induction of lipid peroxidation [88,89].

4.2. Effect of Metal Ion Combination

Combined exposure to two or more metals can have opposite effects on lipid peroxidation.

Aluminum and metals chemically and physically related to aluminum (scandium, gallium, indium, yttrium, and beryllium) are capable of stimulating iron-induced lipid peroxidation in a time and dose manner [90–93] but cannot initiate lipid peroxidation [94]. For these metals without redox capacity, many mechanisms have been proposed. First, they can act by affecting the physical properties of the membranes, producing subtle changes in the rearrangement of lipids that could increase the susceptibility of fatty acids to free radical attack and hence facilitate the initiation or propagation of lipid peroxidation [92,93,95]. According to Xie and Yokel, this would be the more likely mechanism [94]. Second, aluminum can induce disruption of the iron metabolism. Aluminum has been shown to bind to ferritin in the brain and to facilitate the release of catalytically active iron [96]. It also increases transferrin receptor and reduces ferritin synthesis [97], leading to increased iron uptake and decreased storage in a protected site. Catalytically active iron could then participate in the production of oxygen free radicals causing lipid peroxidation. Moreover, Al^{3+} and also Pb^{2+} may stimulate iron-induced lipid peroxidation by displacing iron from certain inhibitory sites on the membrane [98]. For other metal ion combinations with iron, no mechanism has yet been proposed to explain the stimulation of iron-induced lipid peroxidation by these metals. For instance, vanadium enhanced lipid peroxidation induced by iron in liver, erythrocytes [99], and kidney [100].

On the contrary, cobalt [101], manganese [102], zinc [103–105], and aluminum [92], the latter two being redox-inactive metals, can exert an antioxidant action by replacing iron at cellular membrane binding sites. This is what Chevion has called the "push" antioxidant mechanism [103]. Other combinations of metal ions have been reported to be protective toward lipid peroxidation. Zinc can inhibit copper-induced lipid peroxidation in diluted human plasma by competing with copper [106]. Zinc, cadmium, and silver inhibit nickel-induced lipid peroxidation [107] and the authors suggest that the mechanism of antioxidant protection works by an increase of Zn-, Cd-, and Ag-metallothionein levels. Metallothioneins are now considered to be anti-oxidant proteins [108].

4.3. Possible Misinterpretations in Biological Systems Containing Metal Ions

Many methods are available to study lipid peroxidation, including those that use thiobarbituric acid–reactive substances (TBARS) and conjugated dienes. The first reaction is based on the formation of a fluorescent red complex from malondialdehyde (MDA), an end-product of lipid hydroperoxide decomposition, and thiobarbituric acid (TBA) when samples are heated in acid medium [109]. Evaluation of the complex is performed either by spectrophotometry at 532 nm or fluorimetry at 553 nm. When redox cycling metals are present in the medium, results can be artificially modified leading to overestimated values [110]. Indeed metal ions (see Sect. 3.2) can decompose preexisting hydroperoxides into free radicals [17,111], which can stimulate further lipid peroxidation during the TBA reaction process. Moreover, other molecules such as amino acids, biliary pigments, sugars, and nucleic acids can be transformed into aldehydes able to react with TBA and can be responsible for other errors [109,112]. For instance, in erythrocyte membranes, TBARS appear first after 1 h of incubation with iron whereas polyunsaturated fatty acids and conjugated dienes remain unchanged. After 2 h of incubation, simultaneously the conjugated dienes increase and the polyunsaturated fatty acids disappear [113]. In fact, in this experiment, lipid peroxidation is induced only after 2 h, whereas after 1 h of incubation sugars and amino acids present on the membrane surface are responsible for the reaction with TBA.

The method of evaluating conjugated dienes [114,115] is based on the formation of a double-bond resonance that leads to ultraviolet absorption. Conjugated dienes correspond to hydroperoxides and hydroxylated derivatives or other decomposition products. This method measures the presence of conjugated dienes at a precise time and does not take into account factors that can change their production or degradation such as metal ions [116]. Since redox cycling metal ions decompose hydroperoxides to MDA, the conjugated dienes do not increase in the same manner and MDA may cause misinterpretations [117].

For these reasons, many authors suggest the use of at least two lipid peroxidation indices in order to obtain a careful evaluation of lipid peroxidation [7,109,118,119].

Experimental conditions should also be taken into consideration, i.e., the presence of contaminating metals in reagents and metal ligation by buffers such as phosphate or Tris [120]. Moreover, the effect of mechanical tissue disruption in order to obtain homogenates should also be considered. This process can lead to the release of metal ions from their storage sites and from metalloproteins that are hydrolyzed by proteases released from damaged lysosomes [7,121]. These metal ions can decompose preexisting lipid hydroperoxides to MDA and stimulate further lipid peroxidation.

Another point raised by Schaich and Borg should better be considered by researchers [10], namely, that little attention is given to reaction conditions that could change the binding site of metals to the membrane modifying the catalytic activity of metal ions. Results from Driomina et al. are in agreement herewith [122]. They found that the kinetics and efficiency of iron-induced lipid peroxidation in membranes depend on the concentration of iron bound to the surface. Moreover, Shertzer et al. showed that the toxic effect of free iron and membrane-bound iron is different: only free iron is able to induce lipid peroxidation whereas only bound iron is able to catalyze hydroperoxide toxicity [34]. Membrane surface charges have also been reported to affect the ability of iron to initiate lipid peroxidation [123–126].

Finally, chelators and complexing agents are critical determinants of the catalytic mode and effectiveness of metals. Chelators vary in their metal affinities, their charge and solubility in lipids [10], as well as in the redox potentials of their complexes so that they can change the ability of ferrous ion to act as an oxidizing agent [8]. Effects and properties of chelates on lipid peroxidation induced by transition metal ions have been well described by Aust et al. [8]. Chelate-to-metal ratio can also have an impact on the extent of the propagation stage [8].

5. CONSIDERATIONS ON TISSUE DAMAGE

The occurrence of lipid peroxidation induced by metals has been recognized in many experimental models. Nevertheless, except with iron and copper [119,127–135], for which a growing body of evidence exhibits its involvement in metal ion toxicity, it is not really established for the other metal ions (Table 1). The involvement of lipid peroxidation in metal toxicity can be demonstrated by showing that treatment with an antioxidant prevents cell damage or at least that lipid peroxidation occurs simultaneously with cell damage. Few papers realized this (Table 1); most studies only show the effect of antioxidants on lipid peroxidation and the consequences of metal ion injury on endogenous antioxidant contents.

However, two pathologies have been shown to be well connected with lipid peroxidation: fibrogenesis and atherosclerosis.

5.1. Fibrogenesis

Chronic insult of organs like liver, kidney, or lung with metal overload such as iron or copper can lead to fibrosis, i.e., cell damage and/or necrosis followed by excessive deposition of extracellular matrix components [128,134]. Byproducts of lipid peroxidation, especially aldehydes (malondialdehyde, hydroxynonenal), have been reported to activate hepatic quiescent stellate cells, which leads to an increase in collagen production and procollagen mRNA levels [135–137]. Moreover α-tocopherol, a chain-breaking antioxidant, was capable of inhibiting the activation and the proliferation of hepatic stellate cells and to stop iron-induced

TABLE 1

Possible Relationship Between Lipid Peroxidation and Toxicity Induced
by Metal Ions Excluding Iron and Copper (Partial List)[a]

Metal	Acute or chronic intoxication	Model for lipid peroxidation investigation	Simultaneous toxicity	Toxicity reversed by antioxidants	Ref.
Arsenic	Acute	Rat liver, kidney and heart homogenates	n.d.	n.d.	[48]
Cadmium	Acute	Rat liver mitochondria and microsomes	Inhibition of enzyme activities	No	[68]
	Acute	Rat hepatocytes	Enzyme leakage, loss of intracellular potassium	Yes	[167]
	Acute	Rat lung, brain, liver, kidney, testes, and heart homogenates	Enzyme leakage	n.d.	[168]
	Acute	Rat heart, kidney, and liver homogenates	n.d.	n.d.	[37]
	Acute	Rat liver homogenates	n.d.	n.d.	[169]
	Acute	Rat hepatocytes	Enzyme leakage	n.d.	[170]
	Acute	Rat hepatocytes	Enzyme leakage, loss of intracellular potassium	n.d.	[171]
	Acute	Rat liver mitochondria and microsomes, urine collection	n.d.	n.d.	[40]

	Type	Model	Effect		Ref.
	Acute	Rat hepatocytes and mitochondria	Mitochondria dysfunction	No	[67]
	Acute	Lung homogenates	Inflammatory response	n.d.	[58]
	Acute	Rat liver perfusate	Enzyme leakage	Yes	[39]
			Decrease of biliary flow	No	
			Mitochondrial dysfunction	No	
	Acute and chronic	Human fetal hepatic cell line (WRL-68 cells)	Trypan blue exclusion, ultrastructural alterations, enzyme leakage	n.d.	[172]
	Chronic	Rat blood and gastric mucosa	Decrease of mucin and PGE$_2$ content of the gastric mucosal barrier	n.d.	[173]
	Chronic	Brain microvessels	Increase of blood–brain barrier permeability	n.d.	[87]
	Chronic	Rat kidney homogenates	n.d.	n.d.	[174]
	Chronic	Rat liver homogenates	n.d.	n.d.	[51]
Chromium	Acute	Rat kidney homogenates	n.d.	n.d.	[52]
	Acute	Rat hepatocytes	Enzyme leakage	No	[175]
	Acute	Rat heart homogenates	n.d.	n.d.	[176]
	Acute	Rat liver mitochondria and microsomes, urine collection	n.d.	n.d.	[60]
	Chronic	Rat liver mitochondria and microsomes, urine collection	n.d.	n.d.	[177]

TABLE 1

Continued

Metal	Acute or chronic intoxication	Model for lipid peroxidation investigation	Simultaneous toxicity	Toxicity reversed by antioxidants	Ref.
Lead	Chronic	Chick liver homogenates and microsomes	n.d.	n.d.	[61]
	Chronic	Rat brain homogenates	n.d.	n.d.	[178]
	Chronic	Fish liver homogenates	Inhibition of hepatic enzyme activity	n.d.	[179]
	Chronic	Rat brain homogenates	Inhibition of acetylcholin-esterase activity	n.d.	[88]
	Chronic	Rat serum	Triglycerides, total cholesterol, HDL blood content, histological alteration of aorta	n.d.	[180]
	Chronic	Rat liver homogenates	n.d.	n.d.	[89]
	Acute	Rat brain homogenates	n.d.	n.d.	[179]
	Acute	CHO cells	Inhibition of colony formation	Yes	[47]
	Acute	Liver, brain, and heart from chick embryos	n.d.	n.d.	[53]

Metal	Exposure	System/Tissue	Effect	Result	Ref.
	Acute	Liver mitochondria and microsomes from chick embryos	Inhibition of hepatic enzyme activities, decrease of mitochondrial cytochrome and cytochrome P450 content	n.d.	[181]
Mercury	Acute	Rat kidney homogenates	Blood urea nitrogen, histological alteration of kidney	Yes	[182]
	Acute	Rat kidney mitochondria	Mitochondrial dysfunction	n.d.	[43]
	Acute	Rat tissues (liver, kidney, lung, testes, and serum)	Hepatic enzyme leakage, blood urea nitrogen, serum creatinine	n.d.	[183]
	Acute	Rat hepatocytes	Enzyme leakage, loss of intracellular potassium	n.d.	[171]
	Acute	Rat liver perfusate	Enzyme leakage / Decrease of biliary flow / Mitochondrial dysfunction	No / No / No	[39]
	Acute and chronic	Human fetal hepatic cell line (WRL-68 cells)	Trypan blue exclusion, ultrastructural alterations, enzyme leakage	n.d.	[172]
Nickel	Chronic	Rat liver homogenates	n.d.	n.d.	[51]
	Acute	Rat liver homogenates	n.d.	n.d.	[69]
	Acute	Rat liver homogenates and nuclei	n.d.	n.d.	[70]

TABLE 1

Continued

Metal	Acute or chronic intoxication	Model for lipid peroxidation investigation	Simultaneous toxicity	Toxicity reversed by antioxidants	Ref.
Nickel (continued)					
	Acute	CHO cells	n.d.	n.d.	[184]
	Acute	Mice testis homogenates	Fertility test	n.d.	[185]
	Acute	Mice lung homogenates	Lung enzyme leakage, inhibition of lung enzyme activity	n.d.	[94]
	Acute	Mice liver homogenates	Animal mortality	Yes	[107]
	Acute	Mice liver homogenates	Animal mortality	Yes	[71]

	Acute	Alveolar macrophages from guinea pigs	Release of proinflammatory products	n.d.	[64]
	Acute	Human lung cells form L132 cell line	Release of proinflammatory products	n.d.	[63]
	Chronic	Guinea pig skin areas	Histological alterations of skin	n.d.	[186]
Vanadium	Acute	Liver homogenates and perfusate	Inhibition of enzyme activities, enzyme leakage	Yes	[187]
	Acute	Rat hepatocytes	Enzyme leakage, loss of intracellular potassium	Yes	[188]
	Chronic	Liver homogenates	n.d.	n.d.	[99]

[a]The effects of metal ion combination is not taken into consideration in this table. Almost all of these studies used TBARS as indices of lipid peroxidation. n.d., not demonstrated.

fibrogenesis in the gerbil [138]. However, for Pietrangelo et al., lipid peroxidation of hepatic stellate cells is not responsible for this process
because there is no collocation of collagen gene expression and aldehyde adduct staining [139,140].

5.2. Atherogenesis

Biochemical modifications of low-density lipoprotein (LDL), the major
cholesterol transport protein in the circulation, significantly enhance
atherogenicity of cholesterol [141]. Oxidation of LDL has been linked to
the development of atherosclerotic lesions. Compositional properties of
LDL relevant for its susceptibility to oxidation, kinetics of oxidation,
and consequences for the formation of atherosclerotic plaques have
been extensively studied and reviewed [119,132,133,142–144]. One
should remember that breakdown products of lipid hydroperoxides
modify the structural and functional integrity of apolipoprotein B and
that once modified, LDL are preferentially taken up by macrophages,
giving lipid-laden "foam cells." These foam cells are one of the main
components of atherosclerotic plaques. Moreover, oxidized LDL can
release highly cytotoxic lipid peroxidation products, which leads to
endothelial cell death, platelet aggregation, accumulation of inflammatory cells, and release of biologically active factors.
      Metal ions are involved at both levels in LDL oxidation either
directly or via cell oxidation. First, transition metal ions can modify
LDL by inducing peroxidation of lipids carried in LDL. Most researchers
have observed that cuprous ions were better and faster oxidants than
ferrous or ferric ions in obtaining lipoprotein modifications leading to
the recognition by macrophages [119]. However, in some instances, iron
ions can become more efficacious to induce lipid peroxidation of LDL,
namely, at mild acidic pH or in the presence of thiol reductants such as
cysteine, mercaptoethanol, and glutathione [121]. Autoxidation of thiols
generates superoxide anion, which can induce lipid peroxidation of
LDL when metal ions are present [132]. Lynch and Frei showed that
copper but not iron is capable of inducing LDL lipid peroxidation
even in the absence of superoxide anion, suggesting that copper acts by
breaking down preformed lipid hydroperoxide yielding alkoxyl and peroxyl radical species, whereas iron would require superoxide anion and

the formation of perferryl or ferryl species [145]. Indeed, LDL has been reported to contain preformed hydroperoxides of phospholipids or cholesteryl esters [146,147]. Moreover, Esterbauer et al. suggested that the weak ability of iron to decompose hydroperoxide compared to copper could explain a lower recognition by macrophages of iron-oxidized lipoproteins because the production of hydroperoxide byproducts is necessary to obtain appropriate LDL modifications for this recognition [119]. Other differences between iron and copper comprise the better ability of LDL to reduce Cu(II) over Fe(III), the first step in the initiation of lipid peroxidation [148,149], and the better accessibility of copper to reactive regions of the LDL particle [149]. Indeed, only with copper can the LDL core be attacked.

On the other hand, metal ions can catalyze oxidation of LDL by cells contained in atherosclerotic lesions (endothelial cells, smooth muscle cells, and monocyte-macrophages) [150–153]. Superoxide anion produced by these cells may be involved in LDL oxidation [132,154]. Another process has been described by Fuhrman et al. [155]. Iron can induce lipid peroxidation in macrophage membranes, which can subsequently cause oxidation of extracellular LDL lipids. The subsequent oxidation is independent of the presence of transition metal ions in the medium and would be related to lipid peroxide release to the medium by macrophages.

The role of vitamin E (α-tocopherol), a highly effective chain-breaking antioxidant that is present in LDL, is a matter of controversy. For Kontush et al., α-tocopherol in lipoprotein is a reductant of Cu(II) to Cu(I), which leads to the initiation of lipid peroxidation induced by Cu(II) [156]. Moreover, this reaction would abolish the antioxidative property of α-tocopherol toward oxidation of LDL. However, dietary vitamin E has been shown to increase the resistance of LDL to oxidative stress in vitro [157]. As last, for Maiorino et al., α-tocopherol can neither be considered a prooxidant nor an antioxidant in lipid peroxidation of LDL [158]. Recently, in the absence of α-tocopherol, preformed hydroperoxides have been shown capable of reducing Cu(II) to Cu(I), promoting the first stage of initiation of LDL lipid peroxidation [159].

Although metal ions can induce lipid peroxidation in LDL, their involvement in vivo is still controversial. Many amino acids or proteins (transferrin, ceruloplasmin, albumin, bilirubin, etc.) can scavenge them in plasma. Yet advanced atherosclerotic lesions contain significant amounts of catalytically active copper and iron that can catalyze the oxidation of LDL by macrophages [160,161]. But Swain and Gutteridge claim that chelatable forms of copper and iron found in freshly taken atherosclerotic material were not sufficient to stimulate lipid peroxidation [162].

High-density lipoprotein (HDL) is believed to have a protective role in atherosclerosis. Indeed, HDL can inhibit LDL peroxidation induced by copper and iron [163–165]. However, in conditions of oxidative stress for HDL, the results are controversial. For some researchers, oxidized HDL has no significant effect on LDL peroxidation [164], whereas for Tribble et al. oxidized HDL still inhibits LDL oxidation and even LDL could be capable of inhibiting peroxidative stress within the HDL particle [165]. Many possible mechanisms have been suggested [165]. One group hypothesized the metal ion sequestration by HDL-associated proteins [166].

6. CONCLUSIONS

An extensive literature has been accumulated concerning the increase of lipid peroxidation with many metal ions. Therefore, it is logical that many authors suggest the involvement of lipid peroxidation in the toxicity associated with these metals. However, it should be noted that very often at best only a linear correlation between the increase of lipid peroxidation and the toxicity has been established. Multiple mechanisms are also involved for a given metal. Further research is needed to delineate the specific roles of lipid peroxidation in the pathogenesis of metal ion toxicity and to determine the possible treatment by various antioxidants of individuals who are intoxicated with these metals.

ACKNOWLEDGMENT

The authors thank the Langlois Foundation (Rennes, France), which has supported our lab financially for many years.

ABBREVIATIONS

AH	antioxidant
CAT	catalase
GPx	glutathione peroxidase
Gr	glutathione reductase
GSH	glutathione, reduced form
GSSG	glutathione, oxidized form
HDL	high-density lipoprotein
L·	lipid alkyl radical
LDL	low-density lipoprotein
LH	lipid
LO·	lipid alkoxyl radical
LOO·	lipid peroxyl radical
LOOH	lipid hydroperoxide
MDA	malondialdehyde
NADH	nicotinamide adenine dinucleotide, reduced form
NADP$^+$	nicotinamide adenine dinucleotide, oxidized form
NADPH	nicotinamide adenine dinucleotide phosphate, reduced form
·OH	hydroxyl radical
PGE$_2$	prostaglandin E$_2$
PUFA	polyunsaturated fatty acid
SOD	superoxide dismutase
TBA	thiobarbituric acid
TBARS	thiobarbituric acid-reactive substances

REFERENCES

1. S. J. Stohs and D. Bagchi, *Free Rad. Biol. Med.*, *18*, 321–336 (1995).
2. A. W. Girotti, *J. Free Rad. Biol. Med.*, *1*, 87–95 (1985).
3. T. A. Dix and J. Aikens, *Chem. Res. Toxicol.*, *6*, 2–18 (1993).

4.   R. Buccala, *Redox Rep.*, *2*, 291–307 (1996).

5.   H. W. Gardner, *Free Rad. Biol. Med.*, *7*, 65–86 (1989).

6.   J. M. McCord, *Nutr. Rev.*, *54*, 85–88 (1996).

7.   B. Halliwell and J. M. C. Gutteridge, in *Free Radicals in Biology and Medicine*, Clarendon Press, New York, 1989, pp. 188–276.

8.   S. D. Aust, L. A. Morehouse, and C. E. Thomas, *J. Free Rad. Biol. Med.*, *1*, 3–25 (1985).

9.   G. Minotti and S. D. Aust, *Chem. Biol. Interact.*, *71*, 1–19 (1989).

10. K. M. Schaich and D. C. Borg, *Lipids*, *27*, 209–218 (1992).

11. P. O'Brien and H. J. Salacinski, *Arch. Toxicol.*, *70*, 787–800 (1996).

12. J. Torreilles and M. C. Guerin, *FEBS Lett.*, *272*, 58–60 (1990).

13. C. P. Moorhouse, B. Halliwell, M. Grootveld, and J. M. C. Gutteridge, *Biochim. Biophys. Acta*, *843*, 261–268 (1985).

14. M. B. Kadiiska, K. R. Maples, and R. P. Mason, *Arch. Biochem. Biophys.*, *275*, 98–111 (1990).

15. W. A. Waters, *J. Am. Chem. Soc.*, *48*, 427–433 (1971).

16. J. F. Black, *J. Am. Chem. Soc.*, *100*, 527–535 (1978).

17. E. N. Frankel, W. E. Neff, E. Selke, and D. D. Brooks, *Lipids*, *22*, 322–327 (1987).

18. B. A. Svingen, J. A. Buege, F. O. O'Neal, and S. D. Aust, *J. Biol. Chem.*, *254*, 5892–5899 (1979).

19. L. Ramanathan, N. P. Das, and Q.-T. Li, *Biol. Trace Elem. Res.*, *40*, 59–70 (1994).

20. X. Shi, N. S. Dalal, and K. S. Kasprzak, *Arch. Biochem. Biophys.*, *302*, 294–299 (1993).

21. J. Z. Byczkowski and A. P. Kulkarni, *Biochim. Biophys. Acta*, *1125*, 134–141 (1992).

22. A. Jacobs, *Blood*, *50*, 433–437 (1977).

23. B. Halliwell and J. M. C. Gutteridge, in *Free Radicals in Biology and Medicine*, Clarendon Press, New York, 1989, pp. 34–36.

24. J. H. Freedman, M. R. Ciriolo, and J. Peisach, *J. Biol. Chem.*, *264*, 5598–5605 (1989).

25. D. W. Reif and R. D. Simmons, *Arch. Biochem. Biophys.*, *283*, 537–541 (1990).

26. P. Biemond, A. J. G. Swaak, H. G. Van Eijk, and J. F. Koster, *Free Rad. Biol. Med.*, *4*, 185– 198 (1988).

27. D. W. Reif, *Free Rad. Biol. Med.*, *12*, 417–427 (1992).

28. R. Topham, M. Goger, K. Pearce, and P. Schultz, *Biochem. J.*, *261*, 137–143 (1989).

29. S. Shaw, E. Jayatilleke, and C. S. Lieber, *Alcohol*, *5*, 135–140 (1987).

30. V. M. Samokyszyn, C. E. Thomas, D. W. Reif, M. Saito, and S. D. Aust, *Drug Metab. Rev.*, *19*, 283–303 (1988).

31. P. G. Geiger, F. Lin, and A. W. Girotti, *Free Rad. Biol. Med.*, *14*, 251–266 (1993).

32. V. M. Samokyszyn, D. M. Miller, D. W. Reif, and S. D. Aust, *J. Biol. Chem.*, *264*, 21–26 (1989).

33. G. Minotti and S. D. Aust, *Lipids*, *27*, 219–226 (1992).

34. H. G. Shertzer, G. L. Bannenberg, and P. Moldeus, *Biochem. Pharmacol.*, *44*, 1367–1373 (1992).

35. B. Ketterer, B. Coles, and D. J. Meyer, *Environ. Health Perspect.*, *49*, 59–69 (1983).

36. B. Halliwell and J. M. C. Gutteridge, in *Free Radicals in Biology and Medicine*, Clarendon Press, New York, 1989, pp. 126–130.

37. S. Sarkar, P. Yadav, R. Trivedi, A. K. Bansal, and D. Bhatnagar, *J. Trace Elements Md. Biol.*, *9*, 144–149 (1995).

38. B. Ognjanovic, R. V. Zikic, A. Stajn, Z. S. Saicic, M. M. Kostic, and V. M. Petrovic, *Physiol. Res.*, *44*, 293 300 (1995).

39. O. Strubelt, J. Kremer, A. Tilse, J. Keogh, R. Pentz, and M. Younes, *J. Toxicol. Environ. Health*, *47*, 267–283 (1996).

40. D. Bagchi, M. Bagchi, E. A. Hassoun, and S. J. Stohs, *Biol. Trace Elem. Res.*, *52*, 143–154 (1996).

41. H. Zheng, J. Liu, Y. H. Liu, and C. D. Klaassen, *Toxicol. Lett.*, *87*, 139–145 (1996).

42. G. Gstraunthaler, W. Pfaller, and P. Kotanko, *Biochem. Pharmacol.*, *32*, 2969–2972 (1983).

43. B. O. Lund, D. M. Miller, and J. S. Wood, *Biochem. Pharmacol.*, *45*, 2017–2024 (1993).

44. M. Sugiyama, *Free Rad. Biol. Med.*, *12*, 397–407 (1992).

45. W. Li, Y. Zhao, and I. N. Chou, *Toxicology*, 77, 65–79 (1993).

46. W. E. Donaldson and S. O. Knowles, *Comp. Biochem. Physiol.*, *104C*, 377–379 (1993).

47. N. Ercal, P. Treeratphan, P. Lutz, T. C. Hammond, and R. H. Matthews, *Toxicology*, *108*, 57–64 (1996).

48. O. Ramos, L. Carrizales, L. Yanez, J. Mejia, L. Batres, D. Ortiz, and F. Diaz-Barriga, *Environ. Health Perspect.*, *103*, Suppl. 1, 85–88 (1995).

49. M.-J. Richard, P. Guiraud, M. Leccia, J. Beani, and A. Favier, *Biol. Trace Elem. Res.*, *37*, 187–199 (1993).

50. P. N. B. Gibbs, M. G. Gore, and P. M. Jordan, *Biochem. J.*, *225*, 573–580 (1985).

51. S. V. S. Rana and P. R. Boora, *Bull. Environ. Contam. Toxicol.*, *48*, 120–124 (1992).

52. T. Sengupta, D. Chattopadhyay, N. Ghosh, G. Maulik, and G. C. Chatterjee, *Ind. J. Biochem. Biophys.*, *29*, 287–290 (1992).

53. B. V. Somashekaraiah, K. Padmaja, and A. R. K. Prasad, *Free Rad. Biol. Med.*, *13*, 107–114 (1992).

54. I. Fridovich, *Annu. Rev. Biochem.*, *44*, 147–159 (1975).

55. B. Halliwell and J. M. C. Gutteridge, in *Free Radicals in Biology and Medicine*, Clarendon Press, New York, 1989, pp. 86–92.

56. T. Hussain, G. S. Shukla, and S. F. Chandra, *Pharmacol. Toxicol.*, *60*, 355–358 (1987).

57. T. Koizumi and Z. G. Li, *J. Toxicol. Environ. Health*, *37*, 25–36 (1992).

58. D. Manca, A. C. Ricard, B. Trottier, and G. Chevalier, *Arch. Toxicol.*, *68*, 364–369 (1994).

59. M. A. Amoruso, G. Witz, and B. D. Goldstein, *Toxicol. Lett.*, *10*, 133–138 (1982).

60. D. Bagchi, E. A. Hassoun, M. Bagchi, and S. J. Stohs, *Comp. Biochem. Physiol.*, *110C*, 177–187 (1995).

61. L. J. Lawton and W. E. Donaldson, *Biol. Trace Elem. Res.*, *28*, 83–97 (1991).

62. S. J. Yiin and T. H. Lin, *Biol. Trace Elem. Res.*, *50*, 167–172 (1995).

63. P. Shirali, E. Teissier, T. Marez, H. F. Hildebrand, and J. M. Haguenoer, *Carcinogenesis*, *15*, 759–762 (1994).

64. E. Teissier, P. Shirali, M. H. Hannothiaux, T. Marez, and J. M. Haguenoer, *J. Appl. Toxicol.*, *14*, 167–171 (1994).

65. M. Chvapil, D. Montgomery, J. C. Ludwig, and C. F. Zukoshi, *Proc. Soc. Exp. Biol. Med.*, *162*, 480–484 (1979).

66. J. F. Sullivan, M. M. Jetton, H. K. J. Hahn, and R. E. Burch, *Am. J. Clin. Nutr.*, *33*, 51–56 (1980).

67. L. Muller, *Toxicology*, *40*, 285–295 (1986).

68. E. Casalino, C. Sblano, and C. Landriscina, *Arch. Biochem. Biophys.*, *346*, 171–179 (1997).

69. M. Athar, S. K. Hasan, and S. K. Srivastava, *Biochem. Biophys. Res. Commun.*, *147*, 1276–1281 (1987).

70. T. J. Stinson, S. Jaw, E. H. Jeffery, and M. J. Plewa, *Toxicol. Appl. Pharmacol.*, *117*, 98–103 (1992).

71. R. C. Srivastava, M. M. Husain, S. K. Srivastava, S. K. Hasan, and A. Lai, *Bull. Environ. Contam. Toxicol.*, *54*, 751–759 (1995).

72. J. D. Garcia and K. W. Jenette, *J. Inorg. Chem.*, *14*, 281–295 (1981).

73. W. G. Hoekstra, *Fed. Proc.*, *34*, 2083 (1975).

74. K. Yasuda, H. Watanabe, S. Yamazaki, and S. Toda, *Biochem. Biophys. Res. Commun.*, *96*, 243–249 (1980).

75. V. Reddy, C. Kumar, M. Prasad, and P. Reddana, *Biochem. Int.*, *26*, 863–871 (1992).

76. B. C. Wilke, M. Vidailhet, A. Favier, C. Guillemin, V. Ducros, J. Arnaud, and M.-J. Richard, *Clin. Chim. Acta*, *207*, 137 142 (1992).

77. J. Simoni, G. Simoni, E. L. Garcia, S. D. Prien, R. M. Tran, M. Feola, and G. T. Shires, *Art. Cells Blood Subst. Immob. Biotech.*, *23*, 469–486 (1995).

78. L. L. Costanzo, G. De Guidi, S. Giuffrida, S. Sortino, G. Condorelli, and G. Pappalardo, *J. Inorg. Biochem.*, *57*, 115–125 (1995).

79. L. L. Costanzo, G. De Guidi, S. Giuffrida, S. Sortino, and G. Condorelli, *J. Inorg. Biochem.*, *59*, 1–13 (1995).

80. S. Giuffrida, G. De Guidi, P. Milano, S. Sortino, G. Condorelli, and L. L. Costanzo, *J. Inorg. Biochem.*, *63*, 253–263 (1996).

81. L. Ramanathan and N. P. Das, *Biol. Trace Elem. Res.*, *34*, 35–44 (1992).

82. K. Sato and G. R. Hegarty, *J. Food Sci.*, *36*, 1098–1102 (1971).

83. L. S. Chubatsu, M. Gennari, and R. Meneghini, *Chem. Biol. Interact.*, *82*, 99–110 (1992).

84. T. A. Chin and D. M. Templeton, *Toxicology*, *77*, 145–156 (1993).

85. T. Kamiyama, H. Miyakawa, J. P. Li, T. Akiba, J. Liu, J.-H. Liu, F. Marumo, and C. Sato, *Res. Commun. Mol. Pathol. Pharmacol.*, *88*, 177–186 (1995).

86. M. M. Kostic, B. Ognjanovic, S. Dimitrijevic, R. V. Zikic, A. Stajn, G. L. Rosic, and R. V. Zivkovic, *Eur. J. Haematol.*, *51*, 86–92 (1993).

87. A. Shukla, G. S. Shukla, and R. C. Srimal, *Hum. Exp. Toxicol.*, *15*, 400–405 (1996).

88. R. Sandhir, D. Julka, and K. D. Gill, *Pharmacol. Toxicol.*, *74*, 66–71 (1994).

89. R. Sandhir and K. D. Gill, *Biol. Trace Elem. Res.*, *48*, 91–97 (1995).

90. J. M. C. Gutteridge, G. J. Quinlan, I. Clark, and B. Halliwell, *Biochim. Biophys. Acta*, *835*, 441–447 (1985).

91. S. V. Verstraeten and P. I. Oteiza, *Arch. Biochem. Biophys.*, *322*, 284–290 (1995).

92. P. I. Oteiza, C. G. Fraga, and C. L. Keen, *Arch. Biochem. Biophys.*, *300*, 517–521 (1993).

93. P. I. Oteiza, *Arch. Biochem. Biophys.*, *308*, 374–379 (1994).

94. C. X. Xie and R. A. Yokel, *Arch. Biochem. Biophys.*, *327*, 222–226 (1996).

95. T. Ohyashiki, T. Karino, and K. Matsui, *Biochim. Biophys. Acta*, *1170*, 182–188 (1993).

96. H. B. Johnston, S. M. Thomas, and C. K. Atterwill, *Toxicol. In Vitro*, *7*, 229–233 (1993).

97. K. Abreo, J. Glass, S. Jain, and M. L. Sella, *Kidney Int.*, *45*, 636–641 (1994).

98. O. I. Aruoma, B. Halliwell, M. J. Laughton, G. J. Quinlan, and J. M. C. Gutteridge, *Biochem. J.*, *258*, 617–620 (1989).

99. K. H. Thompson and J. H. McNeill, *Res. Commun. Chem. Pathol. Pharmacol.*, *80*, 187–200 (1993).

100. E. Russanov, H. Zaporowska, E. Ivancheva, M. Kirkova, and S. Konstantinova, *Comp. Biochem. Physiol.*, *107C*, 415–421 (1994).

101. Y. Tampo and M. Yonaha, *Arch. Biochem. Biophys.*, *289*, 26–32 (1991).

102. Y. Tampo and M. Yonaha, *Free Rad. Biol. Med.*, *13*, 115–120 (1992).

103. M. Chevrion, *Free Rad. Res. Commun.*, *13*, 691–696 (1991).

104. A. J. F. Searle and A. Tomasi, *J. Inorg. Biochem.*, *17*, 161–166 (1982).

105. A. W. Girotti, J. P. Thomas, and J. E. Jordan, *J. Free Rad. Biol. Med.*, *1*, 395–401 (1985).

106. P. M. Filipe, A. C. Fernandes, and C. F. Manso, *Biol. Trace Elem. Res.*, *47*, 51–56 (1995).

107. R. C. Srivastava, S. K. Hasan, J. Gupta, and S. Gupta, *Biochem. Mol. Biol. Int.*, *30*, 261–270 (1993).

108. M. Sato and I. Bremner, *Free Rad. Biol. Med.*, *14*, 325–337 (1993).

109. D. R. Janero, *Free Rad. Biol. Med.*, *9*, 515–540 (1990).

110. M. E. Götz, A. Dirr, A. Freyberger, R. Burger, and P. Riederer, *Neurochem. Int.*, *22*, 255–262 (1993).

111. M. J. Davies, in *Free Radicals, Metal Ions and Biopolymers* (P. C. Beaumont, D. J. Deeble, B. J. Parsons, and C. Rice-Evans, eds.), Richelieu Press, London, 1989, pp. 303–311.

112. R. P. Bird and H. H. Draper, in *Methods in Enzymology: Oxygen Radicals in Biological Systems*, Vol. 105 (L. Packer, ed.), Academic Press, Orlando, 1984, pp. 299–319.

113. C. Tallineau, L. Barrier, B. Fauconneau, A. Guettier, and A. Piriou, *Biol. Trace Elem. Res.*, *47*, 3–7 (1995).

114. R. O. Recknagel and E. A. Glende, in *Methods in Enzymology: Oxygen Radicals in Biological Systems*, Vol. 105 (L. Packer, ed.), Academic Press, Orlando, 1984, pp. 331–337.

115. F. P. Corongiu, G. Poli, M. U. Dianzani, K. H. Cheeseman, and T. F. Slater, *Chem. Biol. Interact.*, *44*, 289–297 (1983).

116. M. U. Dianzani, *Alcohol Alcoholism*, *20*, 161–173 (1985).

117. O. Sergent, I. Morel, P. Cogrel, M. Chevanne, N. Pasdeloup, P. Brissot, G. Lescoat, P. Cillard, and J. Cillard, *Chem. Phys. Lipids*, *65*, 133–139 (1993).

118. T. F. Slater, in *Methods in Enzymology: Oxygen Radicals in Bio-*

logical Systems, Vol. 105 (L. Packer, ed.), Academic Press, Orlando, 1984, pp. 283–293.

119. H. Esterbauer, J. Gebicki, H. Puhl, and G. Jürgens, *Free Rad. Biol. Med.*, *13*, 341–390 (1992).

120. D. M. Miller, G. R. Buettner, and S. D. Aust, *Free Rad. Biol. Med.*, *8*, 95–108 (1990).

121. C. Rice-Evans, D. Leake, K. R. Bruckdorfer, and A. T. Diplock, *Free Rad. Res.*, *25*, 285–311 (1996).

122. E. S. Driomina, V. S. Sharov, and Y. A. Vladimirov, *Free Rad. Biol. Med.*, *15*, 239–247 (1993).

123. K. Kogure, K. Fukuzawa, H. Kawano, and H. Terada, *Free Rad. Biol. Med.*, *14*, 501–507 (1993).

124. K. Fukuzawa and T. Fujii, *Lipids*, *27*, 227–233 (1992).

125. K. Fukazawa, T. Seko, K. Minami, and J. Terao, *Lipids*, *28*, 497–503 (1993).

126. Y. Tampo and M. Yonaha, *Lipids*, *31*, 1029–1038 (1996).

127. I. Morel, G. Lescoat, J. Cillard, N. Pasdeloup, P. Brissot, and P. Cillard, *Biochem. Pharmacol.*, *39*, 1647–1655 (1990).

128. A. Pietrangelo, *Semin. Liver Dis.*, *16*, 13–30 (1996).

129. R. S. Britton, *Semin. Liver Dis.*, *16*, 3–12 (1996).

130. R. J. Sokol, *Semin. Liver Dis.*, *16*, 39–46 (1996).

131. T. P. Ryan and S. D. Aust, *Crit. Rev. Toxicol.*, *22*, 119–141 (1992).

132. J. A. Berliner and J. W. Heinecke, *Free Rad. Biol. Med.*, *20*, 707–727 (1996).

133. H. Esterbauer and P. Ramos, *Rev. Physiol. Biochem. Pharmacol.*, *127*, 31–64 (1995).

134. G. Poli and M. Parola, *Free Rad. Biol. Med.*, *22*, 287–305 (1997).

135. M. Parola, M. Pinzani, A. Casini, E. Albano, G. Poli, A. Gentilini, P. Gentilini, and M. U. Dianzani, *Biochem. Biophys. Res. Commun.*, *194*, 1044–1050 (1993).

136. G. A. Ramm, L. Li, R. S. Britton, R. O'Neil, Y. Koayashi, and B. R. Bacon, *Am. J. Physiol.*, *268*, G451–G458 (1995).

137. K. S. Lee, M. Buck, K. Houglum, and M. Choijkier, *J. Clin. Invest.*, *96*, 2461–2468 (1995).

138. A. Pietrangelo, R. Gualdi, G. Casalgrandi, G. Montosi, and E. Ventura, *J. Clin. Invest.*, *95*, 1824–1831 (1995).

139. A. Pietrangelo, R. Gualdi, and G. Montosi, *Hepatology*, *20*, 290A (1994).

140. A. Pietrangelo, G. Montosi, and C. Garuti, *J. Hepatol.*, *23* (Suppl. 1), 77 (1995).

141. D. Steinberg, S. Parthasarathy, T. E. Carew, J. D. Khoo, and J. L. Witztum, *N. Engl. J. Med.*, *320*, 915–924 (1989).

142. J. L. Witztum and D. Steinberg, *J. Clin. Invest.*, *88*, 1785–1792 (1991).

143. A. Kontush, C. Hübner, B. Finckh, A. Kohlschütter, and U. Beisiegel, *Free Rad. Res.*, *24*, 135–147 (1996).

144. K. D. Croft, P. Williams, S. Dimmitt, R. Abu-Amsha, and L. J. Beilin, *Biochim. Biophys. Acta*, *1254*, 250–256 (1995).

145. S. M. Lynch and B. Frei, *J. Lipids Res.*, *34*, 1745–1753 (1993).

146. W. Sattler, D. Mohr, and R. Stocker, *Meth. Enzymol.*, *233*, 469–489 (1994).

147. J. P. Thomas, B. Kalyanaraman, and A. W. Girotti, *Arch. Biochem. Biophys.*, *315*, 244–254 (1994).

148. S. M. Lynch and B. Frei, *J. Biol. Chem.*, *270*, 5158–5163 (1995).

149. D. L. Tribble, B. M. Chu, G. A. Levine, R. M. Krauss, and E. L. Gong, *Arterioscler. Thromb. Vasc. Biol.*, *16*, 1580–1587 (1996).

150. U. P. Streinbrecher, *Biochim. Biophys. Acta*, *959*, 20–30 (1988).

151. J. W. Heinecke, H. Rosen, and A. Chait, *J. Clin. Invest.*, *74*, 1890–1894 (1984).

152. K. Hiramatsu, H. Rosen, and J. W. Heinecke, *Arteriosclerosis*, *7*, 55–60 (1987).

153. G. M. Wilklins and D. S. Leake, *Biochem. Soc. Trans.*, *18*, 1170–1171 (1990).

154. J. W. Heinecke, M. Kawamur, L. Suzuki, and A. Chait, *J. Lipid Res.*, *34*, 2051–2061 (1993).

155. B. Fuhrman, J. Oiknine, and M. Aviram, *Atherosclerosis*, *111*, 65–78 (1994).

156. A. Kontush, S. Meyer, B. Finckh, A. Kohlschütter, and U. Beisiegel, *J. Biol. Chem.*, *271*, 11106–11112 (1996).

157. S. A. Wiseman, M. A. P. Van Den Boom, N. J. De Fouw, M. Groot Wassink, J. A. F. Op Den Kamp, and L. B. M. Tuburg, *Free Rad. Biol. Med.*, *19*, 617–626 (1995).

158. M. Maiorino, A. Zamburlini, A. Roveri, and F. Ursini, *Free Rad. Biol. Med.*, *18*, 67–74 (1995).

159. R. P. Patel, D. Svistunenko, M. T. Wilson, and D. M. Darley-Usmar, *Biochem. J.*, *322*, 425–433 (1997).

160. C. Smith, M. J. Mitchinson, O. I. Aruoma, and B. Halliwell, *Biochem. J.*, *286*, 901–905 (1992).

161. D. J. Lamb, M. J. Mitchinson, and D. S. Leake, *FEBS Lett.*, *374*, 12–16 (1995).

162. J. Swain and J. M. C. Gutteridge, *FEBS Lett.*, *368*, 513–515 (1995).

163. S. Parthasarathy, J. Barnett, and L. G. Fong, *Biochim. Biophys. Acta*, *1044*, 275–283 (1990).

164. M. Hahn and M. T. Ravi Subbiah, *Biochem. Mol. Biol. Int.*, *33*, 699–704 (1994).

165. D. L. Tribble, B. M. Chu, E. L. Gong, F. Van Venrooij, and A. V. Nichols, *J. Lipid Res.*, *36*, 2580–2589 (1995).

166. S. T. Kunitake, M. R. Jarvis, R. L. Hamilton, and J. P. Kane, *Proc. Natl. Acad. Sci. USA*, *89*, 6993–6997 (1992).

167. M. W. Fariss, *Toxicology*, *69*, 63–77 (1991).

168. D. Manca, A. C. Ricard, B. Trottier, and G. Chevalier, *Toxicology*, *67*, 303–323 (1991).

169. R. Sumathi, G. Baskaran, and P. Varalakshmi, *Jpn. J. Med. Sci. Biol.*, *49*, 39–48 (1996).

170. H. Zheng, J. Liu, Y. Liu, and C. D. Klaassen, *Toxicol. Lett.*, *87*, 138–145 (1996).

171. N. H. Stacey and C. D. Klaassen, *J. Toxicol. Environ. Health*, *7*, 139–147 (1981).

172. L. Bucio, V. Souza, A. Albores, A. Sierra, E. Chavez, A. Carabez, and M. C. Gutierrez-Ruiz, *Toxicology*, *102*, 285–299 (1995).

173. G. Öner, V. N. Izgut-Uysal, and Ü. K. Sentürk, *Fd. Chem. Toxicol.*, *32*, 799–804 (1994).

174. G. Öner, Ü. K. Sentürk, and N. Izgüt-Uysal, *Biol. Trace Elem. Res.*, *48*, 111–117 (1995).

175. S. Ueno, N. Susa, Y. Furukawa, K. Aikawa, and I. Itagaki, *Jpn. J. Vet. Sci.*, *51*, 137–145 (1989).

176. C. Coudray, P. Faure, S. Rachidi, A. Jeunet, M.-J. Richard, A.-M.

Roussel, and A. Favier, *Biol. Trace Elem. Res.*, *32*, 161–170 (1992).

177. D. Bagchi, E. A. Hassoun, M. Bagchi, D. F. Muldoon, and S. J. Stohs, *Comp. Biochem. Physiol.*, *110C*, 281–287 (1995).

178. Shafiq-Ur-Rehman, *Toxicol. Lett.*, *21*, 333–337 (1984).

179. S. S. Chaurasia, P. Gupta, A. Kar, and P. K. Maiti, *Bull. Environ. Contam. Toxicol.*, *56*, 649–654 (1996).

180. A. Skoczynska, R. Smolik, and M. Jelen, *Arch. Toxicol.*, *67*, 200–204 (1993).

181. B. V. Somashekaraiah, K. Padmaja, and A. R. K. Prasad, *Biochem. Int.*, *27*, 803–809 (1992).

182. G. Girardi and M. M. Elias, *Free Rad. Biol. Med.*, *18*, 61–66 (1995).

183. Y. L. Huang, S. L. Cheng, and T. H. Lin, *Biol. Trace Elem. Res.*, *52*, 193–206 (1996).

184. X. Huang, Z. Zhuang, K. Frenkel, C. B. Klein, and M. Costa, *Environ. Health Persp.*, *102* (Suppl. 3), 281–84 (1994).

185. J. Xie, T. Funakoshi, H. Shimada, and S. Kojima, *Toxicology*, *103*, 147–155 (1995).

186. A. K. Mathur, B. N. Gupta, S. Singh, S. Narang, and R. Shanker, *Bull. Environ, Contam. Toxicol.*, *49*, 871–878 (1992).

187. M. Younes and O. Strubelt, *Toxicology*, *66*, 63–74 (1991).

188. N. H. Stacey and C. D. Klaassen, *Toxicol. Appl. Pharmacol.*, *58*, 8–18 (1981).

9

Formation of Methemoglobin and Free Radicals in Erythrocytes

Hans Nohl and Klaus Stolze

Institute of Pharmacology and Toxicology,
Veterinary University of Vienna, A-1210 Vienna, Austria

1. INTRODUCTION

Liganding of dioxygen to hemoglobin is associated with the dislocation of an electron from the iron atom orbital to the molecular orbital of O_2. The small percentage of methemoglobin (MetHb) formation that is always present in healthy individuals ($\sim 3\%$) may be due to destabilization of the oxo-iron bond as a result of the changed electron delocalization. The heme-bound oxygen molecule, which inserts this electron into the antibonding molecular orbital ($\pi_2 p$), is released as superoxide anion whereas ferrous heme iron is oxidized to the ferric state, forming methemoglobin. Erythrocytes are adjusted to this special case. The high SOD activity (together with catalase) protects these cells from the threat of oxygen radicals while MetHb reductase recycles the nonphysiological form of the O_2 carrier back to the physiologically active form. It seems that even slight modifications in the heme pocket allowing access of small anions are sufficient to destabilize O_2 binding to heme iron with the risk of both a decreased capacity of O_2 transportation and the development of oxidative stress. This interrelation is impressively demonstrated by an inborn defect of the protein moiety of hemoglobin (β-thalassemia) leading to a flood of oxygen radicals in erythrocytes of these individuals. Oxidants such as chlorates, perchlorates, $K_3(Fe(CN)_6)$, or some transition metals cause MetHb generation by the formation of redox couples with the heme center and remove the electron from the heme-iron without the formation of free radicals. A much more complicated chemical mechanism causing MetHb formation was found with some reducing MetHb generators. Formation of MetHb by these compounds requires an interaction with oxyhemoglobin (HbO_2). Therefore, oxygen activation possibly resulting in the formation of radicals has to be taken into account.

The ability of reducing xenobiotics such as hydroxylamine derivatives or phenolic compounds to oxidize oxyhemoglobin to the ferric form has been reported by several authors [1,2]. A correct stoichiometry for this type of "cooxidation" reaction can only be obtained when one assumes the bound dioxygen molecule as the active oxidant that oxidizes both the ferrous heme center of hemoglobin and the reducing xenobiotic (R-H):

$$Hb^{2+}O_2 + R-H \longrightarrow [MetHb^{3+}-O-O^{2-}] + H^+ + R^\bullet \qquad (1)$$

The fate of the MetHb^{3+}-O-O^{2-} complex, which is only stable at well below room temperature [3], depends on the nature of the reducing xenobiotic, R-H. With aromatic amines partial decomposition to hydrogen peroxide has been reported by Eyer et al. [1], whereas hydroxylamines [4–6], hydroxamic acids [7], and phenolic compounds [8,9] do not release hydrogen peroxide. Instead, the observation of low-level chemiluminescence [10–12] suggests the transient formation of the compound I ferryl heme species (eq. (2)), which in the case of hemoglobin as well as myoglobin consists of an Fe(IV) heme iron and a protein radical most likely centered at a tyrosine residue [13–15].

$$[TyrH/MetHb^{3+}-O-O^{2-}] + H^+ \longrightarrow [Tyr^\bullet/Hb^{4+}=O^{2-}] + H_2O^- \qquad (2)$$

As long as excess oxyhemoglobin [16] or reducing xenobiotic [17] is present, the compound I species is rapidly reduced to MetHb^{3+} (eqs. (3) and (4)).

$$[Tyr^\bullet/Hb^{4+}=O^{2-}] + 2Hb^{2+}O_2 + 3H^+$$
$$\longrightarrow 3[Tyr-H/MetHb^{3+}] + 2\,O_2 + H_2O \qquad (3)$$

$$[Tyr^\bullet/Hb^{4+}=O^{2-}] + 2\,R-H + H^+$$
$$\longrightarrow [Tyr-H/MetHb^{3+}] + 2\,R^\bullet + H_2O \qquad (4)$$

When butylated hydroxyanisole (BHA) was used as MetHb generator, secondary reactions including the formation of Hb-based thiyl radicals on position β 93 [18], and the addition of the BHA-derived t-butyl-4-methoxyphenoxyl radical to the SH group of Hb were observed [9]. The existence of these reactive intermediates as well as the secondary reaction products was inferred from experiments with isolated oxyhemoglobin. The establishment of this reaction sequence in intact erythrocytes requires permeability of the respective MetHb generators through the erythrocytic membrane in order to access Hb in the cytosol. The appearance of the reactive metabolites in the erythrocytes may have harmful consequences. Considering the high antioxidant defense capacity in erythrocytes, one can assume that relatively high levels of

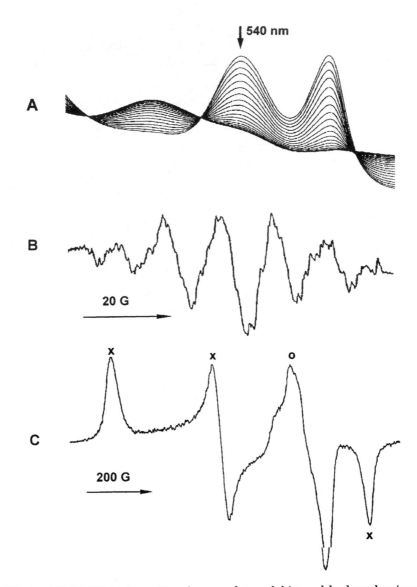

FIG. 1.   (A) MetHb^{3+} formation from oxyhemoglobin and hydroxylamine: 20
repetitive scans of a system consisting of Hb^{2+}O$_2$ (0.1 mM) and NH$_2$OH (0.3
mM) in pH 7.4 phosphate buffer in the visible range between 450 nm and 650
nm, with three isosbestic points at 473, 524, and 587 nm. (B) ESR spectrum of
the NH$_2$O$^\bullet$ radical obtained by rapid mixing of (1) Hb^{2+}O$_2$ (6.38 mM), DETAPAC
(1 mM) and (2) NH$_2$OH (100 mM, pH 7.0), DETAPAC (1 mM) at a total flow rate
of 20 mL/min. (C) The Hb^{2+}NO complex (o) and the MetHb^{3+}-NH$_2$OH adduct
(x), recorded at low temperature (110 K).

radicals will be tolerated. A possible consequence of an imbalanced development of prooxidant formation will be irreversible degradation of hemoglobin and other proteins (Schiff base formation, Heinz bodies), a decrease of the thiol redox status, and oxidative membrane alterations. The formation of $MetHb^{3+}$ can easily be followed spectrophotometrically at the characteristic absorption band of oxyhemoglobin [19] at 540 nm where little interference with the absorption of the xenobiotics and their usually colorless metabolites can be expected. When hydroxylamine (0.3 mM) is added to a phosphate-buffered oxyhemoglobin solution (0.1 mM, pH 7.4), three distinct isosbestic points at 473, 524, and 587 nm appeared (Fig. 1A), indicating that reaction intermediates other than HbO_2 and $MetHb^{3+}$ are either colorless or present at very low concentrations. Additional species that can be expected are simple nitrogen species such as N_2 or N_2O [20], and the one-e⁻ oxidation product $NH_2O^{\bullet}$, which is a free radical detectable by ESR spectroscopy [4]. In addition, one may expect the existence of nitric oxide, which remains firmly bound to hemoglobin in the form of its Hb^{2+}-NO complex [4], and the compound I and compound II type of ferryl hemoglobin species that have characteristic absorption bands in the UV-VIS spectra and, in the case of compound I, a protein radical detectable by ESR spectroscopy [14].

2.   ESR DETECTION OF FREE RADICAL INTERMEDIATES

2.1.   Hydroxylamine Compounds (RR′N-OH)

When isolated $Hb^{2+}O_2$ was rapidly mixed with NH_2OH in a continuous flow mixing cell, the ESR spectrum shown in Fig. 1B appeared. Identification of the radical as the one-e⁻ oxidation product of NH_2OH (the $NH_2O^{\bullet}$ radical) was done by comparing the characteristic ESR parameters with data from the literature [21]. When the mixture was rapidly frozen, two additional paramagnetic species were detected at 77 K: $MetHb^{3+}$-NH_2OH and Hb^{2+}-NO (Fig. 1C). All ESR parameters are listed in Table 1.

Rapid mixing of an erythrocyte suspension with hydroxylamine hydrochloride also resulted in the formation of the $NH_2O^{\bullet}$ species (see above, Fig. 1B). However, the maximum obtainable signal intensity was significantly lower and the mixing flow rate between hydroxylamine and erythrocytes had to be decreased considerably when compared to

TABLE 1

ESR Parameters of Free Radicals Formed from Oxyhemoglobin
and Various Hydroxylamine and Phenolic Compounds

Compound	Species	g	Coupling constants
NH_2OH	$NH_2O^{\bullet}$	2.0062	12.6 G ($H_{(2)}$,N)
	Hb-NO	2.060/1.986 (shoulder/min)	
	MetHb-NH_2OH	2.46/2.20/1.91	
CH_3NHOH	$CH_3NHO^{\bullet}$	2.0055	14.8 G ($H_{(4)}$,N)
	MetHb-CH_3NHOH	2.38/2.17/1.93	
$(CH_3)_2NOH$	$(CH_3)_2NO^{\bullet}$	2.0054	14.8 G ($H_{(6)}$) 17.0 G (N)
	MetHb-$(CH_3)_2NOH$	2.49/2.22/1.89	
HyUr	H_2N-CO-$NHO^{\bullet}$	2.0063	11.7 G (H) 8.05 G (N)
	MetHb-HyUr	2.52/2.24/1.86	
p-HyAn	CH_3O-C_6H_4-$O^{\bullet}$	2.0044	5.05 G ($H_{(2)}$) 2.05 G ($H_{(3)}$)
BHA	BHA$^{\bullet}$		5.3 G ($H_{(1)}$) 1.85 G ($H_{(3)}$)
BHT	BHT$^{\bullet}$		0.65 G ($H_{(1)}$) 11.65 G ($H_{(3)}$) 1.6 G ($H_{(2)}$)

Data compiled from [4–6,8,9]

the homogeneous system, indicating that penetration of hydroxylamine through the erythrocyte membrane has become the rate-limiting
step of the reaction.

   Similar experiments were performed with substituted hydroxylamine compounds. The best results were obtained with the flow system CH_3NHOH (0.9 M)/HbO_2 (4.74 mM), and the stationary systems
$(CH_3)_2NOH$ (0.9 M)/HbO_2 (4.74 mM) and H_2N-CO-NHOH (200 mM)/
HbO_2 (4.74 mM). At room temperature, the respective nitroxyl radical

species were observed; after freezing the incubation mixture, additional low-spin iron species were detected at low temperature (between 77 and 110 K). All spectral parameters are listed in Table 1.

Similar results were obtained in incubation mixtures with intact erythrocytes with the compounds N,N-dimethylhydroxylamine and N-hydroxyurea. With N-methylhydroxylamine only a composite spectra of the original nitroxyl radical and a paramagnetic dimerization product could be obtained.

2.2. Phenolic Compounds (p-Hydroxyanisole, BHA, BHT)

Phenolic compounds are also able to undergo one-electron oxidation in the presence of oxyhemoglobin, thereby forming transient phenoxyl radicals. Important compounds are p-hydroxyanisole, which has been tested as a candidate for the treatment of malignant melanoma, and the antioxidants butylated hydroxyanisole (BHA) and butylated hydroxytoluene (BHT), which are used as food additives. In contrast to the hydroxylamine compounds, phenolic substances are poorly water-soluble at neutral pH, which renders the ESR detection of their oxidation products, the very short-lived phenoxyl radicals, very difficult.

Detection of the p-hydroxyanisole-derived phenoxyl radical (the p-methoxyphenoxyl radical) was possible using a flow system of a concentrated oxyhemoglobin solution in aqueous phosphate buffer, pH 7.4 (see Table 1). The BHA-derived phenoxyl radical was successfully detected in a weakly buffered system containing 10% ethanol in order to increase the solubility of BHA (Fig. 2 and Table 1). With BHT, the respective BHT-derived 2,6-di-*tert*-butyl-4-methylphenoxyl radical could not be detected under these conditions. However, incubation of BHT with a MetHb^{3+}/H$_2$O$_2$ model system resulted in an intensive ESR signal of the compound I-derived protein radical superimposed by weak lines of the BHT-derived phenoxyl radical. After computer subtraction of the interfering compound I signal, the remaining signal could be identified as the BHT phenoxyl radical (Table 1).

No ESR signals were detected in incubations of intact erythrocytes with the phenols p-hydroxyanisole, BHA, or BHT. In addition to the poor solubility of these compounds, further limitations exist in erythro-

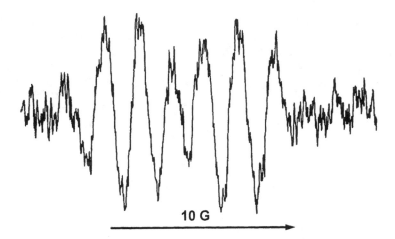

FIG. 2.   Formation of the BHA-derived phenoxyl radical. $Hb^{2+}O_2$ (3 mM) was mixed with BHA (10 mM) in 22.7 mM phosphate buffer containing 10% ethanol.

cyte suspensions such as the diffusion barrier across the erythrocyte membrane and interaction with a high concentration of antioxidants.

3.   CHEMILUMINESCENCE DETECTION OF THE FERRYL HEMOGLOBIN PROTEIN RADICAL

Although the presence of three isosbestic points in our spectrophotometric investigations have shown that more than 99% of hemoglobin must be present in the form of either oxyhemoglobin or methemoglobin, the transient formation of the highly reactive ferryl hemoglobin cannot be excluded. The high redox potential (+0.99 V) of this species both in its compound I and its compound II form renders these species highly toxic [22]. From our ESR investigations it can be concluded that one-electron transfer reactions take place, which makes the involvement of these higher oxidation states of the heme iron highly probable. During the electron transfer steps the intermediate existence of excited electronic states such as singlet oxygen ($^1O_2$) or excited carbonyls [23] can be expected. A sensitive method for their detection in biological systems is the measurement of low-level chemiluminescence.

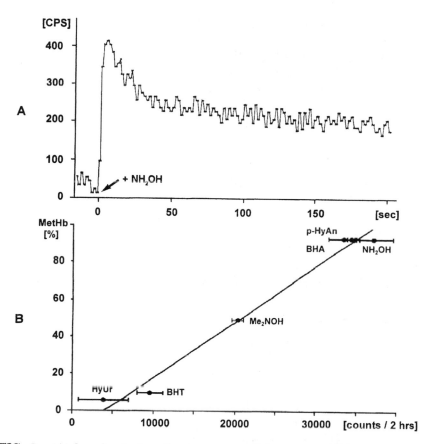

FIG. 3. (A) Low-level chemiluminescence of the system NH_2OH (4.0 mM)/ $Hb^{2+}O_2$ (1.05 mM). (B) Correlation of MetHb formation and integrated chemiluminescence of erythrocytes incubated for 2 h with hydroxylamines (NH_2OH, Me_2NOH, HyUr) or phenols (p-HyAn, BHA, BHT).

In Fig. 3A, a time scan of chemiluminescence developed by the $Hb^{2+}O_2/NH_2OH$ mixture is shown, having a maximum of only about 400 cps immediately after mixing. The chemiluminescence intensities of the other xenobiotics were even lower, corresponding roughly with their MetHb formation rates [10,11]. These low intensities render specific investigations such as the effect of selective quenchers or spectral resolution using monochromators practically impossible. On the other hand, much higher light intensities can be achieved when the species responsible for light emission is produced in a model system, by direct mixing of $MetHb^{3+}$ with H_2O_2. For this reason, all experiments for further characterization were performed with this system. The emitted light is independent of the presence of catalase once the compound I ferryl hemoglobin has been formed. A good correlation exists between the compound I concentration, as indicated by its ESR intensity, and the intensity of light emission [10]. This proves that the light emission process is strongly related to compound I formation. Compound II is not a candidate for the light emitting species because its half-life (>20 min) exceeds the time of light emission by more than two orders of magnitude. Reactive oxygen species have also been excluded as indicated by the effect of different selective quenchers and enhancers of light emission listed in Table 2.

Superoxide and hydroxyl radicals could easily be excluded as SOD and •OH quenchers did not show any effect. A discrepancy existed between the strong decrease of light emission observed in experiments using a 600-nm short-wave pass cutoff filter (which blocks the 634 nm/701 nm emission of $^1O_2$) and the absence of an intensity increase in

TABLE 2

Effect of Various Selective Quenchers and Enhancers
of Chemiluminescence on the $MetHb^{3+}/H_2O_2$
and $MetMyo^{3+}/H_2O_2$ Systems

Conditions	Chemiluminescence (% of control)		Detectable species
	MetHb	MetMyo	
DABCO	159.6	180.9	$^1O_2$
600-nm filter	8.3	30.7	$^1O_2$
D_2O	103.0	105.2	$^1O_2$
SOD	67.8	96.7	$O_2^{•-}$
Mannitol	101.9	99.7	•OH
NEM	48.3	106.2	SH groups
Ascorbate	26.5	5.0	Prooxidants
Uric acid	9.3	33.2	Prooxidants
Desferal	11.8	14.7	Prooxidants

Data compiled from [11].

D_2O buffer (which enhances singlet oxygen emission by a factor of 5–10). Therefore, a set of experiments with the specific singlet oxygen trap 9,10-anthracenedipropionic acid (ADPA) was performed using an HPLC system with fluorescence detection for product identification [11], which clearly showed that no singlet oxygen is produced in the $MetHb/H_2O_2$ system.

Our observation that no light emission was observed in the hematin/H_2O_2 system unless bovine serum albumin (BSA) was added clearly proves the involvement of amino acid residues of the globin moiety in the light emitting step [10,11]. Further characterization involved partial spectral resolution of the emitted light [24] where only a small contribution of wavelengths between 400 and 600 nm was observed, indicating that excited carbonyls do not play a major role.

When whole erythrocytes were incubated with NH_2OH, p-HyAn, or BHA, MetHb formation was virtually complete after 2 h, whereas only 50% MetHb was formed in the incubation mixture containing Me_2NOH, and very little with BHT and HyUr. All incubation mixtures showed light emission that correlated well with the MetHb formation when the emitted light was integrated and corrected for background noise (see Fig. 3B).

Due to the low intensity of the emitted light, a complete spectral resolution of the erythrocyte/NH_2OH system was not possible. However, the use of a 600-nm cutoff filter reduced light intensity by at least 70%, indicating that most of the emitted light stems from the same reaction pathway found in hemoglobin solution [10,24], i.e., compound I decomposition via excited porphyrin states [25].

4. SPIN LABELING OF THE ERYTHROCYTE MEMBRANE

Virtually no correlation was observed between hemolysis and MetHb formation rates, indicating that the formation of MetHb is not directly linked to the process leading to hemolysis. In order to clarify the mechanism leading to complete hemolysis in the presence of BHA, we recorded the membrane fluidity changes upon the addition of xenobiotics to erythrocytes or Hb-free erythrocyte ghosts measuring the linewidth changes of the spin-label 5-doxylstearic acid (4 µg/mL) [26]. In the case of BHA, a higher mobility of the erythrocyte membrane lipids was observed both in whole erythrocytes (+8%) and in hemoglobin-free ghosts (+12.5%). This effect can be explained in terms of a direct hydrophobic interaction with the membrane bilayer. For NH_2OH two opposite effects coexist: (1) a decrease in membrane fluidity when measured in whole erythrocytes (−10%), probably due to oxidative changes in membrane constituents such as crosslinking, and (2) a small but significant fluidity increase measured with membrane ghosts in the absence of HbO_2 (+4%), which indicates a direct membrane effect. All other investigated compounds showed only minor effects on the membrane fluidity.

5. IRON LIBERATION

Iron release has previously been described [27] as a consequence of porphyrin ring opening. In the presence of reactive oxygen species such as superoxide or hydrogen peroxide, iron ions play a significant role as prooxidant catalysts. Therefore, the ability of xenobiotics to release iron from erythrocytes is an important aspect of their toxicity. With most investigated xenobiotics (Me_2NOH, HyUr, p-HyAn, BHA, and BHT), the observed iron release was not significantly different from the respective control value (0.40 μM), except for hydroxylamine (NH_2OH) where approximately 18.3 μM iron (i.e., almost 1% of total heme iron) was released from an erythrocyte suspension containing 2 mM HbO_2 [26]. Approximately 11.2 μM remained in the cells whereas 7.1 μM was detected outside of the erythrocytes, measured as the proportion that permeated a dialysis membrane during the 2-h incubation. Similar results were obtained in homogeneous HbO_2 solution (14.4 μM total iron released).

6. REACTIONS INVOLVING THIOL GROUPS (R-SH)

Thiol groups, which are particularly sensitive to oxidative stress, indicate the formation of prooxidants at an early stage before other compounds are affected. In the case of reduced glutathione, SH groups also play a vital role in protecting membrane lipids and proteins from oxidative degradation. Prooxidants, which might be formed during xenobiotic MetHb formation, such as xenobiotic-derived free radicals, globin-

TABLE 3

Xenobiotic-Induced Thiol Level Changes[a]

Compound	Thiol level change relative to control (%)
NH_2OH	−43
HyUr	−14
p-HyAn	−14
BHA	−6
BHT	+5

[a]Erythrocytes (2mM Hb) were incubated with the respective xenobiotic (2 mM) for 2 h at 37°C. For ESR measurements, aliquots (0.2 mM) were mixed with the biradical RSSR (0.2 mM) and incubated for 30 min at 25°C. Data compiled from [26].

derived radicals, ferryl heme compounds, or iron ions released from hemoglobin, are reduced by the GSH/GSH-peroxidase system or react directly with GSH, so that accumulation of these dangerous species does not take place in the early phase of oxidative stress. Thiol redox status was recorded using a modification of the biradical technique according to Weiner et al. [28,29] (Table 3). Since thiols spontaneously autoxidize even in the absence of xenobiotics, values are expressed as differences versus the respective controls (erythrocytes (2mM Hb), 2h/37°C, no xenobiotic). With all xenobiotics a significant decrease in thiol levels was observed, except for BHT where the reduced thiol groups remained approximately 5% above the respective control value, indicating that the antioxidant properties of BHT are predominant.

7. ELECTROPHORESIS OF MEMBRANES WITH SDS-PAGE

After incubation of erythrocytes with NH_2OH, ghosts were prepared and subjected to SDS-PAGE [30]. The formation of a new band around 30 kD (Fig. 4A, marked "Hb_2") can be seen, possibly an Hb dimer linked

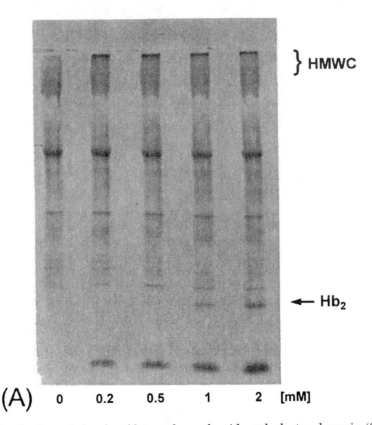

(A)   0   0.2   0.5   1   2   [mM]

FIG. 4. Sodium dodecyl sulfate polyacrylamide gel electrophoresis (SDS-PAGE) of ghosts prepared from xenobiotic-preincubated erythrocytes (1 mM Hb). (A) Preincubation with NH_2OH (0.2–2 mM) for 4 h at ambient temperature. The gel was run under nonreducing conditions. The formation of high molecular weight compounds above 250 kD (HMWCs) and the appearance of a new band around 30 kD (Hb dimerization product, indicated "Hb_2") are indicated. (B) Preincubation with t-butylhydroquinone (0.1–2 mM) for 4 h at ambient temperature. The gel was run under reducing conditions (addition of dithiothreitol). The formation of HMWCs above 250 kD is indicated (molecular weight markers at 30, 43, 67, and 94 kD; see column at left).

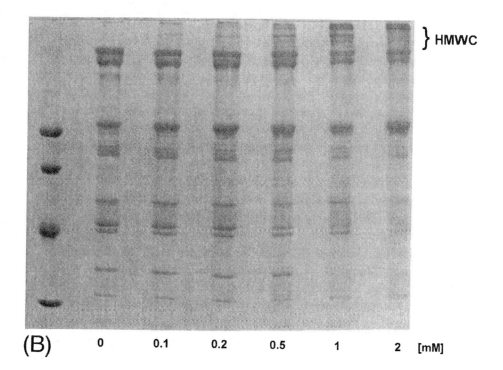

(B)   0   0.1   0.2   0.5   1   2   [mM]

via S-S bridges formed by thiol group oxidation, since the band became considerably weaker when the gel was run in the presence of the thiol compound dithiothreitol (not shown), which reduces S-S bridges to free SH groups. BHA seems to react mainly via its contaminant and metabolite *t*-butylhydroquinone, which has a strong effect on the formation of high molecular weight compounds above bands 1 and 2 of spectrin (Fig. 4B, marked "HMWC").

8. CONCLUSION

Hydroxylamine and phenol compounds are reducing xenobiotics that are long known to induce methemoglobin formation from oxyhemoglobin in a cooxidation-type reaction [4–9]. The first step of the complex

reaction mechanism involves a one-electron transfer from the xenobiotic to the heme-bound dioxygen, resulting in the formation of ESR-
detectable nitroxide or phenoxyl radicals, respectively. The concomitant
reduction of the heme-bound dioxygen leads to an unstable peroxide
complex that cannot be observed at room temperature [3]. Release
of hydrogen peroxide from this peroxide complex was not observed.
Instead, formation of a compound I-type ferryl hemoglobin species
was detected, a process accompanied by low-level chemiluminescence,
which could also be observed when hydrogen peroxide was added to a
methemoglobin solution. Formation of ferryl hemoglobin cannot be detected spectrophotometrically in the oxyhemoglobin/hydroxylamine (or
phenol compound) system because intermediately formed ferryl hemoglobin is rapidly reduced to methemoglobin by excess oxyhemoglobin
[16] or reducing xenobiotic [17]. Evidence on the existence of compound
I ferryl hemoglobin as a transient reaction intermediate came from
the spectral characteristics of the emitted light, which was similar to
chemiluminescence spectra observed in the methemoglobin/hydrogen
peroxide system. Furthermore, the intensity of the emitted light correlated well with the methemoglobin formation rate of the system. When
intact erythrocytes were used instead of solubilized oxyhemoglobin,
methemoglobin formation could also be detected following the addition
of the xenobiotics although the reaction rate was decreased. Detection
of the resulting free radicals was more difficult due to the presence of
antioxidants in the erythrocytes. To obtain well-resolved ESR spectra
the concentration of (readily water-soluble) hydroxylamines under
study had to be increased. With the poorly water-soluble phenolic compounds ESR spectra could not be obtained. However, the observed
chemiluminescence and MetHb formation indicated that phenolic compounds also react with oxyhemoglobin incorporated in intact erythrocytes. In addition, a significant decrease of thiol levels even with those
xenobiotics having a low MetHb formation rate indicated the formation
of oxidizing reaction intermediates in red blood cells. Xenobiotic-
induced iron release from hemoglobin was significant only with unsubstituted hydroxylamine, possibly due to the high affinity of its ultimate
oxidation product, nitric oxide (NO), to iron ions. Membrane destruction was especially evident after exposure to the BHA metabolite
t-butylhydroquinone, where high concentrations of polymerized membrane fragments were detectable by SDS-PAGE.

In conclusion, the main features of MetHb generator-induced toxicity are as follows:

A decrease of oxygen transport capacity due to the conversion of ferrous hemoglobin to the ferric methemoglobin, which cannot bind molecular oxygen, or due to chemical alteration of the globin moiety (formation of Hb dimers as shown by SDS-PAGE).

Alterations of thiol-dependent enzymatic activities as a consequence of SH group oxidation.

Alteration of the erythrocyte membrane (fluidity change, formation of high molecular weight degradation products), which may effect metabolic exchange activities, decrease the deformability of erythrocytes (which is a requisite condition for the passage of capillaries), and decrease the membrane resistance to hemolysis.

Taken together, these effects may result in anemia and harmful ischemia of vital organs.

ACKNOWLEDGMENTS

The authors thank G. Marik and P. Martinek for their skillful technical assistance and the Oesterreichische Fonds zur Förderung der wissenschaftlichen Forschung for financial support (project numbers P07150-ME and P09684-ME).

ABBREVIATIONS

ADPA	anthracene-9,10-dipropionic acid
BHA	butylated hydroxyanisole
BHT	butylated hydroxytoluene
BSA	bovine serum albumin
DABCO	diazabicyclo[2.2.2]octane
DETAPAC	diethylenetriaminepentaaceteic acid
ESR	electron spin resonance
GSH	glutathione, reduced form

Hb   hemoglobin
HPLC   high-performance liquid chromatography
p-HyAn   p-hydroxyanisole
HyUr   hydroxyurea
MetHb   methemoglobin
MetMyo   metmyoglobin
NEM   N-ethylmaleimide
NO   nitric oxide
SDS-PAGE   sodium dodecyl sulfate polyacrylamide gel electro-
            phoresis
SOD   superoxide dismutase

REFERENCES

1. P. Eyer, H. Hertle, M. Kiese, and G. Klein, *Mol. Pharmacol.*, *11*, 326–334 (1975).

2. P. A. Riley, in *Hydroxyanisole: Recent Advances in Anti-Melanoma Therapy* (P. A. Riley, ed.), IRL Press, Oxford, 1984, p. 25 ff.

3. Z. Gasyna, *FEBS Lett.*, *106*, 213–218 (1979).

4. K. Stolze and H. Nohl, *Biochem. Pharmacol.*, *38*, 3055–3059 (1989).

5. K. Stolze and H. Nohl, *Free Rad. Res. Commun.*, *8*, 123–131 (1990).

6. K. Stolze and H. Nohl, *Biochem. Pharmacol.*, *40*, 799–802 (1990).

7. R. Heilmair, W. Lenk, and H. Sterzl, *Biochem. Pharmacol.*, *36*, 2963–2972 (1987).

8. K. Stolze and H. Nohl, *Free Rad. Res. Commun.*, *11*, 321–327 (1991).

9. K. Stolze and H. Nohl, *Free Rad. Res. Commun.*, *16*, 159–166 (1992).

10. H. Nohl and K. Stolze, *Free Rad. Biol. Med.*, *15*, 257–263 (1993).

11. K. Stolze, Y. Liu, and H. Nohl, *Photochem. Photobiol.*, *60*, 91–95 (1994).

12. Y. Liu and H. Nohl, *Photochem. Photobiol.*, *62*, 433–438 (1995).

13. D. Tew and P. R. Ortiz de Montellano, *J. Biol. Chem.*, *263*, 17880–17886 (1988).

14. K. M. McArthur and M. J. Davies, *Biochim. Biophys. Acta, 1202,* 173–181 (1993).

15. D. J. Kelman, J. A. DeGray, and R. P. Mason, *J. Biol. Chem., 269,* 7458–7463 (1994).

16. C. Giulivi and K. J. A. Davies, *J. Biol. Chem., 265,* 19453–19460 (1990).

17. T. Shiga and K. Imaizumi, *Arch. Biochem. Biophys., 167,* 469–479 (1975).

18. K. R. Maples, P. Eyer, and R. P. Mason, *Mol. Pharmacol., 37,* 311–318 (1990).

19. A. Grisk, *Praktikum der Pharmakologie und Toxikologie,* VEB Gustav Fischer Verlag, Jena, 1969.

20. D. A. Bazylinski, R. A. Arkowitz, and T. C. Hollocher, *Arch. Biochem. Biophys., 259,* 520–526 (1987).

21. C. J. W. Gutch and W. A. Waters, *J. Chem. Soc., 1965,* 751–755 (1965).

22. W. H. Koppenol and J. F. Liebman, *J. Phys. Chem., 88,* 99–101 (1984).

23. J. R. Kanofsky, *Chem.-Biol. Interact., 70,* 1–28 (1989).

24. K. Stolze and H. Nohl, *Biochem. Pharmacol., 49,* 1261–1267 (1995).

25. G. Vasváry, S. Elzemzam, and D. Gál, *Biochem. Biophys. Res. Commun., 197,* 1536–1542 (1993).

26. K. Stolze, A. Dadak, Y. Liu, and H. Nohl, *Biochem. Pharmacol., 52,* 1821–1830 (1996).

27. M. Ferrali, C. Signorini, L. Ciccoli, and M. Comporti, *Biochem. J., 285,* 295–301 (1992).

28. L. M. Weiner, H. Hu, and H. M. Swartz, *FEBS Lett., 290,* 243–246 (1991).

29. H. Nohl, K. Stolze, and L. M. Weiner, in *Methods in Enzymology, Vol. 251, Sect. III: Monothiols: Measurement in Organs, Cells, Organelles, and Body Fluids* (L. Packer, ed.), Academic Press, San Diego, 1995, p. 191 ff.

30. K. Stolze and H. Nohl, Electrophoretic detection of erythrocyte membrane alterations during xenobiotic-induced methemoglobin formation (in preparation).

10

Role of Free Radicals and Metal Ions in the Pathogenesis of Alzheimer's Disease

Craig S. Atwood, Xudong Huang, Robert D. Moir, Rudolph E. Tanzi, and Ashley I. Bush

Genetics and Aging Unit, Departments of Psychiatry and Neurology, Massachusetts General Hospital, Harvard Medical School, Boston, MA 02114, USA

1. INTRODUCTION

There is a growing body of evidence indicating that alterations in cerebral metal ion and oxidative metabolism play a fundamental role in many neurological diseases prevalent in society. Such diseases include Alzheimer's disease (AD) [1,2], vascular dementia [3], Parkinson's disease (PD) [4,5], Huntington's disease (HD) [6,7], and prion diseases (scrapie, bovine spongiform encephalopathy, and Creutzfeldt-Jakob disease) [8,9], which collectively we estimate account for up to 70% of sporadic neurodegenerative diseases (see [10]). A common neuropathological feature of these diseases is the presence of focal (and often global) markers of oxidative stress associated with proteinaceous deposits in specific regions of the brain.

In common with these diseases are a number of cuproproteins including amyloid protein precursor (APP), prion protein (PrP), and monoamine oxidases that are implicated in the pathogenesis of AD and vascular dementia, prion disease and PD, respectively. Abnormalities in brain zinc and iron metabolism also have been reported for a variety of neurodegenerative disorders, including AD, PD, neuronal dystrophy, and HD. Furthermore, familial diseases such as Wilson's disease, Menkes' disease, hereditary hemochromatosis, and familial amyotrophic lateral sclerosis that result from mutations in essential transition metal ion genes (ATP7A, ATP7B, HFE, and SOD-1, respectively) emphasize the importance of normal transition metal ion homeostasis in brain function [11–13]. The neurodegenerative processes underlying these diseases appear to be a result of increased oxidative damage resulting from alterations in neuronal metal ion homeostasis. Both neuronal iron and copper have been implicated in oxidative mechanisms of injury and cell death since free $Fe(II)/Cu(I)$ promote reactive oxygen species (ROS) generation [14]. This chapter reviews transition metal ion dysregulation and oxidative stress in the pathogenesis of AD.

2. NORMAL CEREBRAL TRANSITION METAL ION HOMEOSTASIS AND OXIDATIVE METABOLISM

While transition metal ion metabolism in a number of tissues has been studied extensively, the transport, uptake, and cellular regulation and metabolism of metal ions in the brain remains poorly understood. The importance of metal ions in the brain is indicated by the fact that the brain contains high concentrations of metal ions relative to other tissues of the body. Although the brain is relatively tolerant to these large concentrations of metal ions, the concentrations of zinc, and both redox metals ions, copper, and iron, are tightly regulated because of their potential to impact on toxic oxygen radical formation.

2.1. Zinc

Zinc is a redox-inert metal ion that is crucial to the normal functioning of over 200 proteins and enzymes [15,16] in their structural, catalytic, and regulatory roles [17,18]. Zinc is a small ion (0.65 Å) that binds mostly to nitrogen and sulfur donors and is easily exchanged between ligands [18] and is thereby exchangeable between cell compartments. Excellent reviews on the neurobiology of zinc have recently been published [19,20]. Below is a summary of the salient aspects of normal neuronal zinc metabolism.

2.1.1. Neuronal Zinc Metabolism

Zinc is most concentrated in the neocortical and hippocampal areas of the brain [21]. The concentration of zinc in the neocortex is between 150 and 200 μM [22,23] (Table 1), which is one order of magnitude greater than that in blood (11–23 μM) [24]) and three orders of magnitude greater than that in cerebrospinal fluid (CSF) (0.15 μM) [23]. The excitatory nerve terminals of the hippocampus and fascia dentata, critical regions for memory function, contain the highest concentrations of zinc in the body (see [23,25]). It has been estimated that the concentration of zinc in the vesicles of the mossy fibers of the terminal boutons is as high as 220–300 μM [23].

TABLE 1
Microanalysis of Metal Ion Concentrations in Plaque and Neuropil[a]

Factor	Copper (μg/g) (μM)	Iron (μg/g) (μM)	Zinc (μg/g) (μM)
Total senile plaque	25.0 (393)	52.5 (940)	69.0 (1055)
AD neuropil	19.2 (304)	38.8 (695)	51.4 (786)
Control neuropil	4.4 (69)	18.9 (338)	22.6 (346)

[a]For purposes of comparison, these values have been converted to molar concentrations (in brackets) assuming a sample density of 1 g/cm³.
Source: Derived from Lovell et al. [70].

There are three main cellular pools of zinc in the brain: (1) a vesicular pool localized in the synaptic vesicles (SV) of nerve terminals; (2) a membrane-bound metalloprotein, or protein-metal complex pool involved in both metabolic reactions and nonmetabolic reactions (i.e., structural functions); and (3) an ionic pool of free or loosely bound ions in the cytoplasm [23]. Zinc-enriched neurons (ZENs), those that release zinc during synaptic activity and have chelatable zinc pools in their terminal boutons, have been identified in mammalian brain (see [23,26]). Chelatable, or labile, zinc has mainly been detected in the telencephalon, with particularly high concentrations in the vesicular compartments of the hippocampal mossy fiber terminals [27].

2.1.2. Zinc Transport, Cellular Uptake, and Storage

Plasma zinc is mainly bound to plasma proteins (98%) such as albumin with the remainder as free Zn(II) [28] and its uptake from the blood to the brain is homeostatically controlled [29]. Zinc is actively taken up [30] and stored in SVs in nerve terminals throughout the telencephalon [31,32]. Howell and coworkers [32] have described a high- and low-affinity uptake mechanism for zinc into neurons with K_m values of approximately 15 and 361 μM, respectively, while Wensink and coworkers [33] have reported a saturable and higher affinity uptake for zinc ($K_m = 0.25$ μM). Zn(II) entry into neurons appears to overlap with routes of Ca(II) entry. Recently, Choi and coworkers [34] identified four specific routes of zinc entry into neurons: (1) entry through voltage-gated Ca(II) channels, (2) transport-mediated exchange with intracellular Na(I), (3) NMDA receptor-gated channel Ca(II) channels, and (4) Ca(II)-permeable channels gated by certain subtypes of AMPA or kainate receptors.

Zinc transporters involved in the sequestration (ZnT-2/ZnT-3) and extrusion (ZnT-1) of zinc have been cloned [35–37]. ZnT-1 is localized mostly to the plasma membrane and appears to be expressed ubiquitously [37]. ZnT-2 is hardly expressed in the brain but may be involved in vesicular zinc transport [37,38]. ZnT-3 is expressed in the brain and testis, another zinc-rich organ, and is detected particularly within neurons of the entorhinal cortex, amygdala, and hippocampal dentate granule cells and pyramidal cells of the CA1 and CA3 regions, regions of the brain that are rich in vesicular zinc [36]. ZnT-3 is localized to the membranes of all clear, small, round SVs in the mossy fiber boutons of both mouse and monkey [38]. Since up to 60–80% of these SVs contain Timm's-stainable zinc, it has been suggested that ZnT-3 is responsible for the transport of zinc into SVs, and hence for the ability of these neurons to release zinc upon excitation [36,38]. There is conflicting evidence at present as to whether this zinc transporter operates via an energy-dependent mechanism or acts as a channel or facilitated transporter [35,39].

Metallothioneins (MTs) are a storage pool for Zn(II) (and Cu(II)) in the brain. Four metalloproteins that bind zinc have been identified in the mouse, two of which are expressed in most organs (MT-I and MT-II) and are induced by a variety of agents including metals, hormones, and xenobiotics [41] and two (MT-III and MT-IV) that are expressed mainly in brain [42,43] and squamous epithelia [40]. MT-III is expressed predominantly in neurons and choroid plexus [44]. MT-I, II, and III bind zinc and copper with nearly identical stoichiometry and affinity [45] and confer similar resistance to zinc and cadmium toxicity [46]. MT-III is expressed predominantly in neurons that sequester zinc in SVs, i.e., in the dentate gyrus and hippocampal structures [44]. MT-III protects against excitotoxicity, preventing kainate-induced toxicity in glutaminergic neurons compared to normal mice [38]. Glutaminergic neurons release both glutamate and zinc during transmission and seizures. Thus, MT-III may play an important role in protecting cells from zinc toxicity resulting from excitotoxicity and oxidative damage (see Sec.

2.1.4). MT-III does not appear to influence steady-state content of zinc in SVs [47] but may play a role in recycling zinc by transporting zinc from the plasma membrane to SVs following synaptic transmission [32,38].

2.1.3. Synaptic Zinc Release

Zinc, like glutamate, is released during paroxysmal activity [23,32,48]. Zinc is released in a calcium-dependent manner from the hippocampal mossy fiber terminals during spontaneous activity [49], after stimulation evoked electrically [32], after administration of K(I) [48,49], and after administration of kainic acid [48,49]. Endogenously released zinc may reach extracellular peak concentrations of 300 μM in the hippocampus during neurotransmission [32,48]. Even higher concentrations might be expected at the synaptic cleft; it is not known if released zinc is in free ionic form or in a low-affinity complex with small or large organic compounds.

The physiological purpose of such high zinc concentrations in the hippocampus is unclear but some workers have proposed that this large transsynaptic movement of zinc may have a normal signaling function [23,26] and be involved in long-term potentiation, i.e., in the processing of memory formation [50]. Changes in behavior of multiple channels and receptors [51,52], in particular inhibition of NMDA receptors [53], have been reported. Because zinc can be rapidly neurotoxic at such high concentrations (see Sec. 3.1.1), active transport is likely to occur to remove the zinc from the interstitial space and to maintain homeostasis. An energy-dependent transport system has been described for the reuptake of zinc into vesicles following synaptic transmission [23,33]. Chelation of zinc to apothionein or other proteins and amino acids is likely to help maintain homeostatic control [22,35–37,43].

2.1.4. Zinc Neurotoxicity

Exogenous zinc has been shown to induce both necrotic and apoptotic death in neurons in vivo and in vitro [19,20]. Necrotic mechanisms (cell edema and lysis) of zinc-induced death have been observed in neuronal cells when exposed to extracellular concentrations of 50–1000 μM (e.g., [54,55]). Zinc-induced cell death in nonneuronal cells generally occurs at lower extracellular concentrations (7.5–200 μM) [56]. Rat primary cortical cell cultures made zinc-deficient by the addition of the cell-permeable zinc chelator N,N,N',N'-tetrakis(2-pyridylmethyl)-ethylenediamine (TPEN) die rapidly with features typical of apoptosis; membrane blebbing, chromatin condensation, and increased DNA fragmentation (Atwood et al., unpublished results). Thus, neurons have a narrow window of zinc tolerability compared to other cell types, further emphasizing the importance of zinc as a regulator of neuronal survival. Interestingly, zinc accumulation in neurons of the hippocampal hilus and CA1, as well as in the cerebral cortex, thalamus, striatum, and amygdala, has been observed in degenerating neurons after periods of cerebral ischemia (e.g., [57]) or seizure activity [23]. Neuronal death as a result of cerebral ischemia is thought to be via apoptosis (see [19]). A reduction in ischemic-induced neuronal death following intraventricular ethylenediamine N,N,N',N'-tetraacetate (EDTA) injection suggests zinc uptake by neurons is critical for apoptotic death induced by ischemia [57].

The exact intracellular mechanisms of zinc-induced toxicity in neuronal cells are unknown but may involve zinc interactions with receptors (which may induce excitotoxic injury) and interactions with the activities of critical enzymes (such as those involved in respiration), transport proteins, structural proteins, and intracellular calcium-binding proteins (dysregulation of intracellular calcium concentrations) (see [19] for a review). Zinc and copper are biologically antagonistic; alterations in zinc metabolism directly impact copper metabolism and vice versa [58]. Therefore, zinc toxicity may be mediated through copper; zinc competition for cuproproteins may inhibit critical metabolic functions such as the antioxidant protective effects of Cu/Zn-SOD or other cuproproteins leading to increased ROS and cellular damage. Mechanisms of apoptotic death as a result of zinc exposure or deprivation also are unclear. Under normal zinc concentrations, zinc has been proposed as an antioxidant against ROS and free radicals [59]. However, zinc deficiency may result in decreased antioxidant activity by Cu/Zn-SOD and increased oxidative stress, leading to other cellular changes observed in apoptosis (reviewed in [19]).

The importance of zinc in the brain is shown by the fact that zinc stores are preserved in the brain, so that zinc malnutrition can be life threatening without significantly depleting cerebral zinc stores [60,61].

The consequences of clinical zinc deficiency include impaired mentation and memory functions.

2.2. Copper

Copper is a redox-active essential trace metal ion required for a number of cellular enzymes including cytochrome c oxidase, Cu/Zn-SOD, lysyl oxidase, and dopamine-β-monoxygenase [62,63]. Copper also plays a critical role in the assimilation of iron into both microbial and mammalian cells [64,65] and modulates a number of regulatory responses in cells, including transcriptional activation or repression, changes in protein stability and the modulation of protein trafficking [66,67].

2.2.1. Neuronal Copper Metabolism

The brain contains the highest cellular concentrations of copper in the body next to the liver. Copper is most concentrated in the gray matter (60–110 μM) being consistently higher than white matter (25–79 μM; Table 1) [62,68–70]. These concentrations are approximately one-half order of magnitude greater than in blood (11–24 μM) [24] and about one order of magnitude greater than in CSF (1–8 μM) [08]. Nerve terminals and mitochondria contain the highest copper concentration per unit protein [71].

The locus ceruleus (110–400 μM) and substantia nigra (80–120 μM) are characterized by the highest copper concentrations in the human brain [72–74], where copper is localized to nerve terminals and secretory vesicles, primarily within the soluble matrix of the latter, where it may reach concentrations in the hundreds of micromolar [75,76,315]. Copper also is distributed to mitochondria of the granular layers of cerebellum and throughout ependymal cells and epithelial cells of blood vessels [74].

At least 10 acidic copper-binding proteins have been identified in brain (see [69]). Cytochrome oxidase (20%), copper/zinc superoxide dismutase (Cu/Zn-SOD) (25%), and neurocuprein, whose brain content may be twice that of Cu/Zn-SOD [77,78], account for much of the total copper content of the brain. Other neuronal copper-containing proteins include albocupreins I and II, dopamine-β-monooxygenase, amine oxi-dases, and copper bound to low molecular weight compounds such as peptides and amino acids [69]. Unlike in plasma, ceruloplasmin (CP) is not a major cuproprotein in the brain; copper bound to CP has been estimated to account for ≤1% of brain copper [79]. We have recently shown that the Aβ protein, which deposits in AD, binds copper with a stoichiometry of 1:3 [80] (see Sec. 5.2.2) and may act as a copper-binding protein in vivo. Another protein that deposits in the brain in prion diseases, PrP, also is a major copper binding/transporting protein [9].

2.2.2. Copper Transport, Cellular Uptake, and Storage

The inherent toxicity of highly redox-reactive copper has required that cells evolve special ways of transporting copper required for biological functions essential for life. Elaborate cellular machinery are involved in recruiting, trafficking, compartmentalizing, and, ultimately, inserting copper into appropriate proteins. These include the binding of copper to peptides/proteins (chaperones) thereby allowing for its safe transport across membranes and adequate sequestration within the cell and pre-venting damage to lipids, DNA/RNA, proteins, and carbohydrates. Re-views on copper transport have been published by Linder and Hazegh-Azam [63] and Valentine and Gralla [67].

Although CP (~65–70%), albumin (18%), and transcuprein (12%) contains most of the bound copper within plasma, low molecular weight amino acid complexes of copper [62,63] are the primary components of the exchangeable plasma copper pool responsible for the transmem-brane transport of copper into the brain [81,82].

Copper uptake by rat brain (hypothalamic tissue slices) is via a carrier-mediated facilitated diffusion process that is analogous to that of neutral amino acids (system L) and occurs via a two-ligand-dependent saturable process: a high-affinity (6 μM), low-capacity (23 pmol/min/mg) process and a low-affinity (40 μM), high-capacity (425 pmol/min/mg) process [82,83] that likely operate in a positive cooperative manner to transport complexed copper to brain tissue. Copper transport by the low-affinity process is inhibitable by excess ligand (histidine), suggest-ing a common recognition site for the ligand and the copper-ligand complex [82]. The physicochemical properties of the copper complex are an important factor determining copper uptake by brain tissue and dissociation of copper-amino acid complexes appears to occur at the membrane [83].

As with other tissues [62], CP can transport copper to the brain [84,85], although this is not essential for copper transport to tissues since aceruloplasminemic patients have a defect in iron homeostasis rather than copper metabolism [86,87]. The mechanism of copper transport to the brain by CP is unclear. CP has been reported not to cross the blood-brain barrier (BBB) [88] and nonhepatic cells are known to be capable of disengaging copper from CP without absorbing the holo-CP protein into the cytosol [89]. In contrast to these studies, specific receptors have been found on many tissues for CP and radiolabeled CP is taken up by rat brain [90]. CP also has been detected intracellularly in human neurons and astrocytes [79,91] and in rat brain and primary rat astrocyte culture [92]. Whether this CP is from extracellular or intracellular sources remains unclear; however, interleukin-1β has been shown to increase CP expression in primary rat astrocytes, indicating that neurons are capable of synthesizing CP [92].

Copper bound to CP is not readily dissociable [69,90]. The transport of copper from CP across the membrane involves a dissociative reduction via sulfhydryl groups probably located on the membrane [93, 94]. Interestingly, it has recently been shown that amyloid protein precursor (APP), a neuronal transmembrane protein, binds and reduces Cu(II) resulting in a corresponding oxidation of cysteines 144 and 158 in APP that involves an intramolecular reaction leading to a new disulfide bridge [95]. This reaction, specific for the reduction of Cu(II), may be a means of transporting extracellular copper to neurons. Recently, a number of proteins involved in the transport of copper across the cell plasma membrane to soluble cytoplasmic copper transporters/chaperones and further across intracellular membranes have been identified in yeast (and have mammalian homologs) (reviewed in [11,67]).

GSH (reduced glutathione) and MTs mediate cellular copper storage/detoxification (see Sec. 2.2.4). GSH is the most abundant non-protein thiol in mammalian cells, and is able to chelate and detoxify metals soon after entering the cell [96,97]. GSH has been implicated in the incorporation of Cu(I) in MTs [98] as well as in copper donation to both intra- and extracellular proteins like Cu/Zn-SOD [99] and hemocyanin [100].

2.2.3. Synaptic Copper Release

The presence of extracellular copper has been shown to modulate the secretory function of peptidergic neurons and plays an important modulatory role in the CNS. Copper is released in a calcium-dependent manner from vesicles of peptidergic neurons following K(I)-induced depolarization of hypothalamic tissues in vitro [82,101]. It has been calculated that the extracellular copper released by depolarization may reach concentrations as high as 15 μM [82]. Copper released from vesicles following synaptic depolarization is rapidly reuptaken into vesicles, possibly via the low-affinity copper uptake mechanism (see Sec. 2.2.2). The release of copper (bound to an intracellular chelator) during depolarization is thought to oxidize thiols of the luteinizing hormone-releasing hormone (LHRH) granule in an oxygen-dependent manner thereby altering the permeability of granule membranes and mediating the release of LHRH and α-melanotropin from hypothalamic granules of explants of the median eminence area (MEA) [102–104]. The K_m of copper for LHRH release (4 μM) is within the concentration range for copper found in the synapse (~15 μM) [104]. These alterations in membrane permeability are specific for complexed copper and, to a lesser extent, zinc [102,103]. Extracellular nucleotide phosphates facilitate copper uptake by hypothalamic slices (via an interaction with a purinergic receptor) and hence copper stimulation of the release of LHRH from the MEA of rats [105].

A role for PrP binding to copper for normal synaptic transmission has been implicated from studies showing abnormal electrophysiological responses in the presence of excess copper in Purkinje cells from PrP gene-ablated mice [9]. Thus, copper may normally bind PrP^c and implicates PrP^c in copper sequestration following synaptic transmission. The half-maximal binding of PrP^c for copper is 5.9 μM and indicates that PrP^c would be able to compete for synaptic copper and may therefore control the activity of other membrane-associated copper-binding proteins.

2.2.4. Copper Neurotoxicity

Although copper is essential for a number of enzymes and iron transport, excess copper accumulation is highly toxic due to its proclivity to engage in redox reactions that result in the formation of hydroxyl radicals, a reactive species that can cause extensive damage to nucleic acids, proteins, and lipids [14]. Redox-active metal ions are intimately involved in enhancing free radical generation via Fenton and Haber-Weiss chemistry as outlined in the following reactions [106]; reduced Fe(II)/Cu(I) reacts with molecular oxygen to generate the superoxide anion (O_2^-):

$$M^{n+} + O_2 \rightarrow M^{(n+1)+} + O_2^-$$

The O_2^- generated undergoes dismutation to hydrogen peroxide (H_2O_2) either catalyzed by SOD or spontaneously.

$$O_2^- + O_2^- + 2H^+ \rightarrow H_2O_2 + O_2$$

The reaction of reduced metals with H_2O_2 generates the highly reactive hydroxyl radical (OH$^\bullet$) by the Fenton reaction. Cu(I) catalyzes this reaction at a rate constant magnitudes higher than that for Fe(II) [14].

$$M^{n+} + H_2O_2 \rightarrow M^{(n+1)+} + OH^\bullet + OH^-$$

Additionally, the Haber-Weiss reaction can form OH$^\bullet$ in a reaction catalyzed by $M^{(n+1)+}/M^{n+}$:

$$O_2^- + H_2O_2 \rightarrow OH^\bullet + OH^- + O_2$$

The generation of free radicals (including reduced metal ions) and ROS, if not scavenged, can result in lipid peroxidation, localized amino acid oxidation, and modifications to DNA [14].

Copper also may be toxic by its inappropriate incorporation into proteins that normally bind other metal ligands. For example, copper and zinc inhibit lysosomal and cytoplasmic proteases in human cerebral cortex and may be neurotoxic via their ability to shut down the normal proteolytic degradation mechanisms within neurons [107]. Thus, cells must be able to sense appropriate amounts of copper in which to accumulate for cellular requirements, without accumulating toxic levels. The failure to establish and maintain copper homeostasis results in at least two genetic disorders, Wilson's disease and Menkes disease (see [63,108]), and chronic copper exposure slightly above cellular requirements, exceeding the detoxifying cellular capacity, may be responsible for many other neurological diseases (see Sec. 3).

Mechanisms of copper detoxification include the upregulation of copper-binding proteins, such as MT and S100b, cysteine-rich cytoplasmic proteins that sequester intracellular copper thereby lowering free copper to less toxic levels [38,109,110], and Cu/Zn-SOD [111] to help protect cells from oxidative damage. The binding of Cu(I) to transcription factors provides a direct link between a toxic copper sensor and the activation of detoxification genes (see [66,112,113]).

Dietary copper deficiency in animals (or chelation therapy) causes enzootic ataxia, which is characterized by neurological abnormalities and pathologically by focal spongy edematous necrosis (reviewed in [69]). The neurological problems appear to arise from altered myelin synthesis, decreased catecholamine (dopamine to norepinephrine) levels as a result of decreased tyrosine hydroxylase and dopamine-β-monooxygenase activity, decreased cytochrome oxidase activity (which may lead to impaired lipid biosynthesis), and vascular defects arising from a failure of elastin and collagen crosslinking likely due to the decreased activity of the cuproprotein lysyl oxidase, a key enzyme in the crosslinking process. All of these impaired processes may result from the downregulation of proteins at the level of transcription or mRNA turnover. The copper deficiency syndrome in humans, Menkes' disease, also is characterized by severe neurodegeneration and decreased myelination based on low sulfatide and proteolipid protein concentrations perhaps as a result of decreased cuproprotein activity.

2.3. Iron

2.3.1. Neuronal Iron Metabolism

Next to the liver, the brain contains the highest cellular concentrations of iron in the body. Like copper and zinc, iron is not distributed equally throughout the brain. The highest nonheme iron concentrations are found in the globus pallidus, substantia nigra, putamen, caudate nucleus, dentate nucleus, and intrapeduncular nucleus (800–2000 μM), and significant concentrations also are found in the hippocampus and cerebral cortex (400–600 μM) [314]. These concentrations are approximately one to two orders of magnitude greater than that in the blood (9–32 μM) [24]. Iron is an important constituent in brain, playing a role in electron transfer and as a cofactor for certain enzymes, including those involved in catecholamine and myelin synthesis.

2.3.2. Iron Transport, Cellular Uptake, and Storage

Cells have developed two approaches for utilizing iron that is essentially insoluble in neutral solutions: ligand secretion and ferric reduction, ferrous ions being much more soluble at physiological pH. Like copper, ferric iron is sequestered by peptides/proteins and in plasma most is bound to the glycoprotein transferrin (Tf) [114]. In contrast to CP, Tf's functional properties extend only to its ability to bind and release iron; its metal cofactors do not act in a catalytic role. Reviews on iron transport have recently been published [114,115].

The major portion of iron transported across the BBB is via its binding to Tf (Fe_2-Tf) [116], which binds to high-affinity cell surface Tf receptors (TfR) and triggers endocytosis of the Tf/TfR complex (reviewed in [117,118]). Neurons have a high-affinity receptor for Tf [119]. Iron crosses the luminal membrane of the capillary endothelium by receptor-mediated endocytosis of ferric Tf at an average rate of 80 nmol/kg/h [115]. Within the endothelium, exposure of the complex to the acidic pH of the endosome is thought to release iron from Tf, which then enters an intracellular pool for incorporation into cellular proteins or to be stored as ferritin [120], or is absorbed from the vesicular system into cytoplasm and transported across the abluminal plasma membrane into interstitial fluid as one or more low molecular weight complexes. Apo-Tf remains bound to TfR at the low pH of the acidic vesicle (pH ≤ 6.0), and the apo-Tf/TfR complex is then recycled to the cell surface where apo-Tf dissociates at the pH of the blood (~pH 7.4) [118,121]. Within interstitial fluid, transported iron will bind with any unsaturated Tf synthesized or transported to the brain-CSF system. From interstitial fluid, ferric Tf is taken up by neurons and glial cells by receptor-mediated endocytosis [115].

Iron also may be transported rapidly cross the complete BBB through Tf-independent pathways utilizing plasma membrane-based transport systems [114,116]. Transport mediated by both Tf-dependent and -independent import mechanisms involves at least two functional activities: a ferrireductase that converts Fe(III) to Fe(II) and a carrier mechanism that subsequently translocates Fe(II) across membrane lipid bilayers [65,122,123]. Most iron entering the brain across the capillary endothelium finally leaves the brain with the bulk outflow of CSF through arachnoid villi and other channels [115]. This explains the inability to increase brain nonheme iron level above baseline values even after long-term iron therapy [124].

Regulation of cellular iron transport is strictly regulated but also is affected by intracellular copper concentrations (reviewed in [64, 125]). Dietary copper deficiency leads to microcytic hypochromic anemia [126,127]; this apparent iron deficiency results from loss of oxidase activity by the copper requiring CP [87]. This is supported by studies showing that patients with hereditary aceruloplasminemia (a mutation within CP) also cannot mobilize iron from tissues [87,128,129] and may have increased basal ganglia iron levels [87]. Also, in Wilson's disease severe CP deficiency is associated with clinical signs of iron deficiency [130]. Defects in CP lead to neurological abnormalities associated with the deposition of iron in cells in the CNS, indicating that the ferroxidase acitivity of CP is required for normal iron homeostasis in the CNS.

Ferritin, the main intracellular iron storage protein in the brain, is a large molecule composed of L- and H-chain subunits (19 and 21 kD, respectively) capable of binding up to 4500 atoms of iron, but also binds aluminum [131] and zinc [132]. L-Ferritin is normally localized to microglia and oligodendrocytes, with little being present in astrocytes, neurons, or extracellular fluid [133–135]. Although ferritin cannot cross the BBB [136], L-chain ferritin has been detected in the CSF of human and rat brains, particularly during inflammation (e.g., meningitis [137]) or following cerebrovascular stroke (e.g., [138]). It is not clear as to how ferritin is passaged into CSF, but its presence may indicate an impairment in the BBB.

2.3.3. Synaptic Iron Release

Iron is thought to play a crucial role in the maintenance of dopaminergic and GABAergic neurotransmission (see [139,140]). Fe(II) (and Cu(II) and Mn(II)), but not Fe(III), also modulates the binding of dopamine and serotonin to "serotonin binding proteins" that are present in soluble extracts from calf brain [141,142]. These metal ions promote binding by virtue of their ability to enhance the oxidation of dopamine into dopamine-O-quinone, a derivative known to undergo covalent association with sulfhydryl groups of proteins [141,142]. The function of such interactions is unknown.

2.3.4. Iron Neurotoxicity

Although the redox potential of iron is lower than that of copper, the redox chemistry of iron toxicity is similar to that of copper (see Sec. 2.2.4) and may partake in the generation of ROS deleterious to cellular components. The role of iron in dopaminergic-induced neurodegeneration has been reviewed by Ben-Shachar et al. [143].

Iron deficiency has a profound affect on the BBB and brain function, selectively altering the integrity of the BBB for insulin, glucose, and valine transport [143]. Nutritional iron deficiency also induces significant reductions in the level of brain nonheme iron in rats and is accompanied by selective reduction of dopamine D2 receptor Bmax.

3. ALTERED CEREBRAL METAL ION AND OXIDATIVE METABOLISM IN ALZHEIMER'S DISEASE

3.1. Metal Ions

3.1.1. Zinc

Zinc concentration remains relatively constant in brain and CSF throughout adult life [144]. However, several observations indicate that zinc metabolism is altered in AD (Table 1). While differences in the direction of zinc changes have been reported by different investigators for the AD brain compared to control brain, the general consensus from larger more controlled studies using bulk- and micro-analyses indicates the concentration of zinc is significantly elevated in the hippocampus, inferior perietal lobule [145,146], amygdala and olfactory region [146–148], cerebellum, and frontal and temporal poles [146] of the AD brain. Likewise, a recent study using microparticle-induced x-ray emission analyses that enable metal ion determination in microlocalities of the cortical and accessory basal nuclei of the amygdala demonstrated that levels of zinc are elevated within AD neuropil compared with control neuropil, and further concentrated within the core and periphery of plaque deposits [70]. Some investigators [149,150] did not observe a change in the zinc concentration in the hippocampus, although Wenstrup et al. [150] did show a significantly elevated zinc concentration in nuclei-enriched subcellular fractions from AD brain compared with control brain. In contrast, decreased temporal lobe zinc concentrations using bulk analyses have been reported [151–153], although one of these studies used formalin-fixed tissue and others had only a small number of subjects.

Other observations indicating altered zinc metabolism in the AD brain include increased hepatic total zinc with reduced zinc bound to MT [154], an increase in extracellular $Zn(II)$-metalloproteinase activities in AD hippocampus [155], decreased levels of the zinc-chelating protein MT III [42], decreased plasma zinc [156], and possibly increased zinc in CSF [157]. NMDA-mediated excitotoxicity, believed to play a role in AD pathology, also is accompanied by raised free zinc levels in the hippocampus, which serves as a homeostatic attempt to inhibit the NMDA receptor [158]. Also, there is a pervasive defect of systemic zinc metabolism in Down syndrome (DS), so that zinc supplementation, even in the absence of low serum zinc levels, has been advocated as a means of improving developmental, immunological, and endocrine defects in the disorder [159,160].

Collectively, these reports indicate that there may be an abnormality in the uptake or distribution of zinc in the AD brain causing high extracellular and low intracellular concentrations in the brain. AD pathology is localized to those regions of the brain with abnormal zinc metabolism. If there is a failure of the stringent homeostatic mechanisms that regulate interstitial zinc levels in the brain, then changes in the intra- and extracerebral zinc concentration could lead to neuronal death. Recent and comprehensive reviews on the involvement of zinc in AD have been published [161,162].

3.1.2. Copper

Early studies using bulk analysis techniques for the measurement of copper in brain found no significant differences between AD and control brain [151] or a decrease in copper in AD brain (amygdala and hippocampus) [145,163]. The decreases observed in these areas were attributable to decreases in copper-containing enzymes such as CP [164] (not seen by Loeffler et al. [79]) and cytochrome oxidase [165] as a result of neuronal loss in these areas of the AD brain. These studies have been superseded by analyses at the microprobe level that clearly demonstrate elevated copper levels within AD neuropil of the cortical and accessory basal nuclei of the amygdala compared with control neuropil, and a further significant elevation within the core and periphery of plaque deposits (Table 1) [70]. We also have found that the extraction of Aβ deposits from brain tissue into aqueous buffers is increased in the presence of chelators of Cu(II) and Cu(I) [166], providing further evidence that copper accumulates in amyloid plaques and participates in the deposition of Aβ in amyloid plaques. A 2.2-fold increase in the concentration of copper, and an accompanying increase in CP [167] in the CSF of AD patients [168] suggests that the reuptake of copper into neurons may be compromised in AD [82].

Concurrent increases in CP in AD hippocampus, entorhinal cortex, frontal cortex, and putamen [79] (not seen in the superior temporal cortex [164]) have been reported. Marked accumulation of CP also is observed in AD hippocampus within neurons, astrocytes, and neuritic plaques [79]. High local concentrations of CP within plaques may contribute to the increased concentration of copper found in plaques. The main component of amyloid, the Aβ protein, which we have recently shown binds one copper ion when soluble, but three copper ions in the aggregated state [80] (see Sec. 5.2.2) also likely contributes to the increased copper load of plaques.

3.1.3. Iron

Most investigations using bulk analysis techniques report an increase in the concentration of iron in neocortical gray matter [169], frontal pole [146], temporal pole [146], inferior parietal [146], hippocampus [145, 146], amygdala [145–147], and the olfactory pathway [148] (not seen by Cornett et al. [146]). Microprobe iron determination in the cortical and accessory basal nuclei of the amygdala also demonstrated that levels of iron are elevated in AD neuropil compared with control neuropil [70]. Techniques that enable a more localized measurement of iron indicate discrete areas of iron accumulation; histochemical analyses indicate that iron is localized to deposits in the cytoplasm of neurofibrillary tangles (NFTs) bearing neurons, oligodendroglia, walls of capillaries, and senile plaque (SP) cores and cells associated with SP [70,134,135, 170–175], whereas microprobe analyses show elevations of iron in NFT-bearing neurons and SP [70,174].

Like the copper-binding proteins, iron transport and storage proteins also are elevated in concentration and deposit around SP in AD. Ferritin is elevated in concentration in AD brains compared with age-matched controls [131,176], and is thought to be attributable to the increase in ferritin-rich activated microglia in AD brain, which cluster around amyloid plaques [134,173,177,178]. Ferritin has been found in SP in AD [171]; ferritin-containing microglia are present at the center of some mature plaques, whereas diffuse plaques which likely represent the earliest stages of plaque formation are relatively free of ferritin-rich cells [179]. Thus, the microglial response is probably not an initiating event in amyloid formation but rather a response to inflammation and neurodegeneration. Secreted forms of ferritin in the CSF of patients with suspected AD are increased [180], indicating that ferritin either is being released from injured or dying cells, or is released from cells in response to increased extracellular iron concentrations. Either possibility indicates that ferritin is important as an antioxidant, binding potentially reactive free iron. Transferrin also is homogeneously distributed around SP in AD [134] and in astrocytes in cerebral white matter in AD rather than in oligodendroglia, where it is present in normal individuals [133].

3.1.4. Aluminum

Aluminum was one of the first metal ions linked to AD. However, the influence of aluminum on the pathogenesis of AD remains controversial. The proposed connection between aluminum and AD has been reviewed by Markesbery and Ehmann [181] and Xu et al. [182]. At the bulk analysis level alterations in the concentration of aluminum have been observed (e.g. [151,182–184]), and these studies have been superseded by analyses at the microprobe level indicating that highly localized accumulations of aluminum occur within hippocampal NFTs (e.g., [174,182]) and/or SPs [185,186]. Other researchers have been unable to find differences between control and AD brain (e.g., [176]). Tokutake et al. [186] recently suggested that accumulation of aluminum as aluminosilicate complexes in SP and NFTs depends on their lipofuscin content, a variable that may help to explain differences between previous studies. More work is required to clarify any role aluminum might have in AD pathology.

3.2. Oxidative Metabolism

A number of markers of increased oxidative stress/mitochondrial dysfunction from autopsy tissue have been found. The first evidence that oxidative stress may play a role in the pathogenesis of AD came from energy metabolism studies showing that glucose-6-phosphate dehydrogenase (G6PDH) and 6-phosphogluconate dehydrogenase activities [187] were increased in AD brain. Heme oxygenase-1 (HO1) levels also are increased [188]. Since G6PDH main function is to supply reducing equivalents in the form of NADPH, it was postulated that the increase in G6PDH may be a secondary response to an increased demand for NADPH by the GSH-dependent peroxide detoxifying system [187]. This postulate is consistent with the fact that glycolytic enzymes and the rate of glucose utilization are markedly depressed in the AD brain [189].

Other global markers of oxidative stress in the AD brain that involve oxygen radical mediated attack include (1) increased protein oxidation (increased protein carbonyl formation, in the inferior parietal lobule and hippocampus [190,191]); (2) peroxynitrite-mediated protein nitration [192,193]; (3) lipid peroxidation [189,194 197]; (4) increased 4-hydroxynonenal, an aldehyde product of lipid peroxidation [194, 197]; (5) both nuclear and mitochondrial DNA oxidation (formation of 8-hydroxy-2'-deoxyguanosine) [198]; (6) changes in lipofuscin particle size in the frontal cortex (lipofuscin formation is thought to be free radical-mediated) [199] and advanced Maillard reaction products [200, 201]. Other changes include a reduction in glutamine synthetase [190,191,202], cytochrome c oxidase [165], and catalase activity [203], and a marked (80%) decrease in creatine kinase activity [190,191,202]. Recently, redox-active iron also was found in the SP of AD brain [175].

While these global studies provide evidence for oxidative stress in the AD brain, the most convincing evidence for increased oxidative stress comes from immunohistochemical studies in which staining by a large number of antibodies of oxidative stress markers is observed in neuritic plaques and NFTs. Localized increases in Cu/Zn-SOD [204], Mn-SOD [204,205], catalase [204], ferritin and/or Tf [134,173] and HO-1 [188], have been observed in neuritic plaques and/or NFTs. Other markers of oxidative stress in these areas include the identification of redox-active iron [175], pyrraline and pentosidine [206], 3-nitrotyrosine [193], AGE-modified tau [200], and neurofilament-related protein carbonyls (e.g., [190]). An increase in receptors for advanced glycation end products (AGE) in neurons and microglia near SPs and in the vasculature [207] also indicates localized oxidative stress in these areas.

Further evidence to indicate that the AD brain is under oxidative stress can be implied from the observation that there is a strong spatial correlation between antioxidant enzyme activity and markers of lipid peroxidation and the areas of the brain particularly affected by AD lesions, suggesting that antioxidant activity is increased in a compensatory manner for the increased free radical generation [208]. Circulating plasma antioxidant levels appear to be decreased, while tissue levels of most intracellular antioxidants and antioxidant enzymes are variable, with some being increased and others decreased (reviewed in [2]). Although cellular systems may be unable to cope with the increased ROS burden, treatment of individuals with the antioxidants vitamin E and Selegiline [209], or antiinflammatory medications [210], or drugs that target the inflammatory response of microglia [211,212] has been reported to delay the progression of clinical AD. The administration of the metal ion chelator desferrioxamine has been shown to delay the progression of dementia in AD [213].

3.3. Prion Diseases

PrP^c is the normal cellular form of prion protein that is the precursor to the pathogenic protease-resistant forms (PrP^{Sc}) believed to cause scrapie, bovine spongiform encephalopathy (BSE), and Creutzfeldt-Jakob disease [8]. PrP^{Sc} can be distinguished from PrP^c by its relative resistance to proteinase K digestion [214,215]. PrP^c is thought to be converted to PrP^{Sc} during the disease process either by a posttranslational chemical modification or by a conformational change (e.g., [215]).

The N-terminal region of mammalian PrP contains a highly conserved sequence of five to six tandem repeats, with the octapeptide (PHGGGWGQ) being found in the last four repeats. PHGGGWGQ motif shares a common sequence (PHG) with the histidine-rich glycoprotein thought to be involved in plasma copper transport [216]. Like Aβ, the N-terminal repeat domain of PrP^c (residues 23–98) has been shown to bind multiple copper ions (5–6) via histidine residues in a positive cooperative manner [9,217–219]. Specific copper binding sites also have recently been identified on recombinant Syrian hamster prion protein containing residues 29–231 [316]. The dissociation constant for the Octa$_4$ peptides corresponding to tetra repeats of the octapeptide of mammalian PrP is 6.7 μM [219] and a similar value of 5.9 μM was obtained for the half-maximal binding of copper to the 23 98 fragment of PrP^c [9].

PrP^c gene-ablated mice also exhibit severe reductions in the copper content, but not iron or zinc, of membrane-enriched brain extracts and a similar reduction in synaptosomal and endosome-enriched subcellular fractions [9]. An alteration in phenotype of this mouse includes a reduction in Cu/Zn-SOD activity [220] and altered electrophysiological responses in the presence of excess copper, indicating that PrP^{Sc} is a cuproprotein [9]. These studies indicate that PrP^c is a principal copper-binding protein in brain membrane fractions that may be an important copper chaperone for other proteins.

Like Aβ, the secondary structure of PrP^c is predominantly α-helical whereas PrP^{Sc} is β-sheet [217]. Interestingly, binding of Cu(II) promotes a change in conformation from α-helix to β-sheet structure [219, 221,222,316]. The role of such alterations on the pathogenicity of PrP and Aβ remains to be fully determined, although both fibrillar forms of the protein do show increased resistance to proteolysis.

4. CEREBRAL AMYLOID DEPOSITION IN NEUROLOGICAL DISEASES AND AGING

Amyloid plaques seen in AD are dynamic entities whose composition changes with the progression of the disease [223]. Amyloid plaques are primarily composed of the Aβ protein, but a host of other compounds have been colocalized with Aβ, including amyloid P component, ubiquitin, $α_1$-antichymotrypsin, cytokines, complement proteins, growth factors, cholinesterases, $α_2$-macroglobulin, non-Aβ component (synuclein) and precursor, apolipoprotein E, and specific proteoglycans. The co-

localization of these components with Aβ in specific locales in the AD
(and DS) brain indicates a possible interaction between these molecules
that may be important in amyloidosis. While some components appear
to have a direct role in the deposition of Aβ in amyloid (e.g., proteoglycans and apolipoprotein E), others may simply be attracted to the
hydrophobic amyloid mass. Stabilization of the amyloid may result
from chemical modifications to Aβ or other components. Neither the
percentage of these components within amyloid plaques nor the rate of
accumulation of each of these components over the progression of the
disease has been determined. Recent reviews on the formation of neuritic plaques and NFT have been published [161,162,223–225]. This
section will focus on the roles of metal ions in amyloidogenesis.

4.1. APP/Aβ Processing and Metabolism

APP is a family of alternatively spliced proteins that are ubiquitously
expressed and whose function remains unclear. As is the case with
other amyloid proteins, Aβ originates from the normal cellular metabolism of a larger precursor protein. Aβ is a normally soluble and constitutive protein found in biological fluids and tissue (see [80] and
references therein). Soluble Aβ is produced by constitutive cleavage of
its transmembrane parent molecule, the 695 to 770-amino-acid precursor, amyloid protein precursor as a mixture of polypeptides manifesting
carboxyl-terminal and amino-terminal heterogeneity — the products of
the combined action of proteolytic events, possibly proteases, whose
identities are still unknown (see [161] for a review).

Aβ aggregates to form diffuse amorphous deposits in AD, as well
as following head injury [226] and in healthy aged individuals [227].
Combined with other inflammatory proteins such as proteoglycans,
amyloid P component, and apolipoprotein E, Aβ is found in the brains
of individuals affected by AD and DS as dense, focal, extracellular
deposits of twisted β-pleated sheet fibrils in the neuropil (SPs) and in
cerebral blood vessels (amyloid congophilic angiopathy; see reviews
above). The deposition of Aβ, however, is not confined to the brain
parenchyma, having been detected in amyloid diseases of the muscle
[228], blood vessels [229], and in the kidneys, lungs, skin, subcutaneous
tissue, and intestine of AD patients [230,231].

The discovery of pathogenic mutations of APP close to or within the Aβ domain [232] indicates that the metabolism of Aβ and APP is intimately involved with the pathophysiology of this predominantly sporadic disease.

4.2. Amyloid Structure

Amyloids are a class of noncrystalline, water-insoluble, yet ordered protein aggregates [233]. In situ proteins of gray matter exist predominantly in an α-helical and/or unordered conformation, whereas within amyloid deposits a β-sheet structure predominates [234]. Studies using x-ray diffraction patterns, solid-state NMR, and infrared employed to characterize the secondary structure and conformation of synthetic Aβ and Aβ fragments suggest that the major conformation in amyloid plaques is a repeating, oligomeric, antiparallel cross β-pleated sheet structure (see [235]). The ability of Aβ to form amyloid fibrils is directly correlated with the content of β-sheet formation [236], which in turn is correlated with the neurotoxicity of Aβ [237]. Interestingly, most of the sedimentable Aβ in AD brain does not appear to be amyloid (Cherny et al., personal communication).

5. NEUROCHEMICAL FACTORS THAT PRECIPITATE ALZHEIMER Aβ PROTEIN

There are several lines of evidence indicating that overproduction of Aβ species is unlikely to account for the initiation of cerebral amyloid in sporadic and familial AD cases. While $Aβ_{1-42}$ overexpression occurs throughout life in individuals with a genetic cause of AD [238] and in amyloid-bearing APP transgenic mice [239], the onset of disease is usually after the fourth decade of life or 9 months, respectively, indicating that age related stochastic events initiate amyloidosis. Were elevated cortical Aβ concentrations to be solely responsible for the initiation of amyloid, then it would be difficult to explain why the amyloid deposits are focal (related to synapses and the cerebrovascular lamina media) and not uniform in their distribution. Furthermore, while Aβ is expressed as a constitutive protein in all tissues, its deposition is only found in the brain. Finally, to attribute amyloid initiation to the presence of $A\beta_{1-42}$ alone is problematic because the peptide is a normal component of healthy CSF. From these sets of observations, it seems highly likely that pathogenic mechanisms apart from overproduction initiate $A\beta$ amyloid deposition in sporadic and familial AD.

Therefore, an abnormal milieu of the diseased Alzheimer brain impacts on $A\beta$ solubility-specific neuronal environment that promotes $A\beta$ deposition and toxicity. $A\beta$ accumulates in the brain in AD and DS in forms that have limited solubility in water [240,241] but also in forms that can be extracted under harsher conditions and exhibit apparent sodium dodecyl sulfate (SDS)-resistant polymerization on polyacrylamide gel electrophoreses (PAGE) [240,242]. $A\beta_{1-40}$ is the predominant soluble species in biological fluids, while $A\beta_{1-42}$, a minor species of $A\beta$ that is more insoluble in vitro, is the predominant species found in plaques and deposits associated with AD and DS (e.g., [240–242]). $A\beta_{1-40}$ has been reported as the major $A\beta$ peptide in cerebral cortical extracts from AD brain [243], whereas diffuse SPs have been reported as containing primarily $A\beta_{17-42}$ [244].

The physicochemical properties of $A\beta$ have been reviewed [235]. A significant conformational transition is essential for the transformation of soluble $A\beta$ into a fibrillar form. The kinetics of aggregation are dependent on the relative ratios of secondary structures of the $A\beta$ peptides as determined by solvent hydrophobicity, $A\beta$ fragment length, pH, concentration, seeding, ionic strength, conformation, temperature, and metal ions. These different solubilities are dependent on the conformation that the $A\beta$ peptide adopts, which in turn is determined largely by the above conditions (see [245]). A nucleation-dependent polymerization model for amyloid formation has been proposed (see [246]). The following section will assess candidate neurochemical environments for their potential to induce β-sheet formation, aggregation, and chemical modifications of Alzheimer $A\beta$.

5.1. Metal Ions

APP contains multiple cysteine and carboxylic acid residues capable of binding metal ions in a nonspecific manner. The modulation of $A\beta$ aggregation is commonly regulated though the affinities of many neurochemical factors such as zinc, copper, transthyretin, apolipoprotein E, and pH for the His residues of Aβ [80,245,247–249]. At least two classes of metal binding site on Aβ exist; a Cu(II)/Ni(II) site that is responsible for significant Aβ assembly only under mildly acidic (pH 6.6–7.0) conditions, and a Zn(II)/Co(II) site that mediates significant Aβ assembly at pH 7.4 [80].

5.1.1. Zinc

An interaction between APP and metal ions was first recognized when APP was copurified with a zinc-modulated proteolytic mechanism from human plasma [250]. Biochemical dissection of the metabolic associations between zinc and APP subsequently identified a novel, specific, and saturable zinc binding site (K_d = 764 nM) on the ectodomain (residues 181–188) of the protein, within exon 5, at the end of the cysteine-rich region [251].

Zinc also specifically and saturably binds Aβ itself, manifesting high-affinity binding (K_{assoc} = 107 nM) with a 1:1 (zinc:Aβ) stoichiometry, and low-affinity binding (K_{assoc} = 3.2–5.2 μM) with a 2:1 stoichiometry [247,252]. Zinc binding is histidine-mediated [80] and has been mapped to a stretch of contiguous residues between positions 6 and 28 of the Aβ sequence [247]. Zinc binding to Aβ, which straddles the lysine 16 position of α-secretase cleavage [253], specifically inhibits the cleavage of APP by α-secretase [254] and inhibits α-secretase type (tryptic) cleavage of Aβ. Thus, zinc may influence the generation of Aβ from APP and may increase the biological half-life of Aβ by protecting the peptide from proteolytic attack [247].

Aβ solubility is dramatically lowered by the presence of zinc [80, 247,248,255–260]. Although there has been considerable controversy as to the concentration of zinc required for Aβ aggregation and therefore its physiological plausibility as a catalyst for Aβ aggregation [161], it has now been shown that this reaction also occurs at physiological concentrations of Aβ (low nM such as found in CSF [247], Moir et al., unpublished data; Atwood et al., unpublished data) in phosphate buffered saline and in CSF [259]. Rat Aβ$_{1-40}$ (which has three amino acid substitutions) binds zinc less avidly (K_{assoc} = 3.8 μM, with 1:1 stoichiometry) and is unaffected by zinc below 25 μM [255,258], perhaps explaining the scarcity with which these animals form cerebral Aβ amyloid [261,262]. In the absence of zinc, the solubilities of the rat and the human Aβ species are indistinguishable [255].

5.1.2. Copper

APP has a high-affinity copper binding site (10 nM at pH 7.5) located within residues 135–155 of the cysteine-rich domain of APP695 that is present in all APP splice isoforms [263]. The two essential histidines in the copper binding site of APP are conserved in the related protein APLP2. Copper binding was shown to inhibit homophilic APP binding. This reaction is very specific because APP does not bind or reduce other metals such as Fe(III), Ni(II), Co(II), or Mg(II). This copper-binding domain is encoded by exon 4 and contains a consensus motif for type II copper-binding proteins. Subsequently, it was shown that APP can reduce Cu(II) to Cu(I) [95]. The copper ion-mediated redox reaction leads to the oxidation of cysteines 144 and 158 in APP that results in an intramolecular reaction leading to disulfide bond formation from free sulfhydryl groups in APP [95]. Reaction kinetics were unaffected by either O_2^- or H_2O_2 and the reaction was specific for copper ions.

Evidence for an interaction between Cu(II) and $A\beta_{1-40}$ was first observed by the stabilization of an apparent $A\beta_{1-40}$ dimer by Cu(II) on gel chromatography [247] and by the displacement of ^{65}Zn(II) from Aβ when coincubated with excess Cu(II) [252]. Subsequently, aggregated Aβ was shown to exhibit multiple specific binding sites for Cu(II) presented as a chelate, with both low- and high-binding-affinity sites on $A\beta_{1-40}$ (0.025–1 nM and 16–40 nM, respectively; Atwood et al., unpublished observations).

Cu(II)-induced Aβ aggregation at pH 7.4 is considerably less than that induced by Zn(II) [247,255,256]. However, unlike other biometals, marked Cu(II)-induced aggregation of physiological concentrations (low nM) of $A\beta_{1-40}$ and $A\beta_{1-42}$ emerges as the solution pH is lowered from 7.4 to 6.8 [80]. The stoichiometry of Cu(II) binding to soluble Aβ was ⩽1, but it increased upon aggregation to 3 at both pH 7.4 and 6.6. Cu(II)-induced precipitation at pH 6.6, which is specific for certain proteins [40], is completely reversible, with either chelation or alkalinization returning the pH to 7.4. The concentration of Cu(II) required for Aβ aggregation (⩾5 μM) [80] is well within the concentration range for copper found in the synapse (~15 μM) [104].

Cerebral $A\beta$ deposition is not a feature of aged rats [261], although soluble $A\beta_{1-40}$ is produced by rat neuronal tissue [264]. Rat $A\beta$ contains three amino acid substitutions (Arg→Gly, Tyr→Phe, and His→Arg at positions 5, 10, and 13, respectively [261]) that appear to alter the physicochemical properties of the peptide, preventing it from precipitating in the neocortex. Rat $A\beta_{1-40}$ and histidine-modified human $A\beta_{1-40}$ were not aggregated by Cu(II) (or Zn(II) or Fe(III)), indicating that histidine residues are essential for metal-mediated $A\beta$ assembly. Thus, Cu(II) coordination to histidine at position 13 and tyrosine at position 10 may coordinate the aggregation of $A\beta_{1-40}$. Unlike Zn(II), access of Cu(II) (and Fe(III)) to these residues is limited above pH 7.0. These results indicate that H^+-induced conformational changes may unmask a metal binding site on $A\beta$ that mediates reversible assembly of the peptide.

5.1.3. Iron

Low micromolar concentrations of Fe(III) induce considerably less $A\beta$ aggregation than Zn(II) [247,255,256], and the ability of Fe(III) to aggregate $A\beta$ is modulated to a lesser extent by pH compared with Cu(II) [80]. Fe(III)-induced $A\beta$ aggregation is mediated via Fe(III) coordination to histidine residues [80], suggesting that the association of redox-active iron with SP from AD brains recently reported by Smith et al. [175] may be due to the metal ion binding directly to $A\beta$ as well as iron binding proteins such as CP, ferritin, and Tf present in the plaque vicinity. Iron facilitates $A\beta$ toxicity to cultured cells [265].

5.1.4. Aluminum

Although aluminum salts have been shown to precipitate soluble $A\beta$ in vitro, the concentrations of aluminum required to precipitate $A\beta$ are far above those that are physiologically possible [247,257]. However, precipitation of $A\beta$ onto column glass that contains aluminosilicate is enhanced by zinc, and kaolin, a hydrated aluminum silicate suspension, does accelerate the precipitation of $A\beta$ by zinc [247]. Furthermore, circular dichroism studies have shown that physiologically relevant concentrations of aluminum disrupt the partially helical conformation that $A\beta$ adopts in a membrane-mimicking solvent [266]. Therefore, although Al(III) may not be able to precipitate $A\beta$ directly, the possibility that it may destabilize Aβ conformation thereby promoting its aggregation warrants further investigation.

5.2. Post-translational Modifications of Aβ

The identification of racemized, isomerized, sulfoxated, pyroglutamate and SDS-resistant polymer species of Aβ from AD and DS brains indicates a large number of post-translational modifications occur in Aβ deposits [e.g., 317]. These conformational and/or chemical modifications may promote Aβ deposition and/or the maturation of diffuse Aβ deposits into hard core amyloid plaques. Synthetic $Aβ_{1-40}$ has a molecular mass on SDS-PAGE of ~4 kd, with a major isoelectric point at 7.1 ± 0.2 [267]. Both $Aβ_{1-40}$ and $Aβ_{1-42}$ form apparent higher molecular weight SDS-resistant oligomers [268–272]. Dyrks et al. [270] has shown that the formation of these species of Aβ or the COOH-terminal 100 residue requires metal-catalyzed oxidation: H_2O_2 in conjunction with metal-catalyzed oxidation systems (Fe-hemin, Fe-hemoglobin, or Fe-EDTA).

We have found, like Dyrks et al. [270], that Fe(III) induces little chemical modification to $Aβ_{1-40}$ or $Aβ_{1-42}$ in vitro, except in the presence of unphysiologically high concentrations of H_2O_2 or ascorbic acid [172, 270]. However, in the absence of a metal-catalyzed oxidation system, copper alone induces significant chemical modification to Aβ as determined by SDS-PAGE, in the order $Aβ_{1-42} \gg Aβ_{1-40} \gg$ rat $Aβ_{1-40}$, concordant with the reductive capacity of each peptide [272] (Sec. 6.1). The mechanism of Aβ modification appears to involve the oxidation of histidine, tyrosine, and methionine residues [271], residues associated with metal ion binding to Aβ [80], and available for local attack by reduced metal ions or ROS. In addition to these changes, we recently found that Cu(II) can induce the formation of pyroglutamate and racemized species of Aβ (Atwood et al., unpublished observations). These Cu(II)-induced reactions may therefore explain many of the post-translational modifications observed in AD amyloid.

Oxidation of Aβ by radical damage has been suggested as a causative agent for Aβ neurotoxicity [273–275]. It is not known if oxidation or racemization is required for nucleation of amyloid formation or whether oxidized or racemized species Aβ are more neurotoxic.

5.3. Conditions That Promote the Release and Sequestration of Metal Ions by Proteins

The binding of metal ions to metalloproteins is reversibly modulated by pH and the redox state of the system, which alter the stability constant for that protein-metal ion complex. The release of Cu(I) [276] and Fe(II) (e.g. [277,278]) from metalloproteins is promoted by mildly acidic environments.

Other factors that promote metal ion release from ligands include free radicals and ROS (such as HOCl, H_2O_2, and $O_2^{\bullet-}$), ascorbate, catecholamines, and amino acids such as L-glutamate, L-aspartate, and L-cysteic acid (e.g., [279,280]). GSH, and other biological protein disulfides also are effective in releasing metal ions from MT [281].

Thus, mildly acidotic conditions and the generation of GSH disulfide and ROS by cells under oxidative stress may lead to the release of metal ions from their ligands and increased "free" metal ions. Metal ions dissociated in this fashion may serve to modulate cellular responses in certain disease states but may in themselves be toxic.

6. SOURCES OF NEURONAL FREE RADICALS

The large body of evidence indicating oxidative stress in the AD brain raises the question of the origin of this damage. Sources of oxidative stress in AD-affected tissue could include (1) mitochondrial dysfunction leading to defective mitochondrial electron transport systems [282]; (2) AGE products from Maillard chemistry, which in the presence of transition metals can undergo redox cycling with consequent ROS production inducing lipid peroxidation [188,200,283,284]. AGE and Aβ also may activate specific receptors (receptor for advanced glycation end-products and class A scavenger-receptor) thereby increasing ROS production [207]; (3) reactive microglia [285] (see also below); (4) redox metals, such as copper and iron [175,181], and aluminum, which also accumulates in NFT-containing neurons and SPs (see Sec. 3.4) and stimulates iron induced lipid peroxidation [286]; and (5) Aβ itself, the main component of amyloid [191,195,287–294].

6.1. Generation of Free Radicals by Aβ

Synthetic Aβ peptides have been shown to induce lipid peroxidation of synaptosomes [289], and to exert neurotoxicity [191,287] or vascular endothelial toxicity through a mechanism that involves the generation of cellular O_2^-/H_2O_2 and is abolished by the presence of SOD [295] or catalytic synthetic O_2^-/H_2O_2 scavengers [203]. Antioxidant vitamin E and the spin-trap compound phenyl-α-*tert*-butylnitrone have been shown to protect against Aβ-mediated neurotoxicity in vitro [290,296]. Furthermore, neurons cultured from subjects with Down syndrome, a condition complicated by the invariable premature deposition of cerebral Aβ [297] and the overexpression of soluble $Aβ_{1-42}$ in early life [298], exhibit lipid peroxidation and apoptotic cell death caused by increased generation of H_2O_2 [299].

Support for the Aβ hypothesis comes from the findings that Aβ generates oxygen radicals through a putative Aβ peptide fragmentation mechanism that is O_2-dependent, metal-independent, and involves the sulfoxation of the methionine at Aβ residue 35 [289–291]. The $Aβ_{25-35}$ peptide, although not a natural peptide, has been reported to exhibit H_2O_2-like reactivity toward aqueous Fe(II), nitroxide spin probes, and synaptosomal membrane proteins [290]. $Aβ_{1-40}$ also has been reported to generate the OH• [293].

Recently, we have found that Aβ also can generate ROS via metal ion-dependent pathways [294]. Human Aβ peptides in vitro reduce Fe(III) and Cu(II) to Fe(II) and Cu(I), respectively, and also facilitate the subsequent reduction of molecular oxygen into O_2^- and its dismutation to H_2O_2. The highly reactive OH• also is produced by Aβ peptides in the presence of Fe(III) or Cu(II) (1 μM). The amounts of reduced metal ions and ROS were both greatest when generated by $Aβ_{1-42}$ > $Aβ_{1-40}$ ≫ rat $Aβ_{1-40}$ = $Aβ_{1-28}$, a chemical relationship that correlates with the relative neurotoxicity of these peptides. These findings [294] and others elaborated on above strongly suggest, in contrast to previous studies [289–291], that metal ions are likely to be involved in the generation of ROS by Aβ in vivo. Thus, it is possible that both metal-dependent and -independent mechanisms of free radical generation operate in vivo. Together these findings indicate that the accumulation of Aβ in vivo may contribute directly to oxidative stress in brain tissue and may be a major source of free radical damage in AD.

Aβ itself has been shown to alter neuronal metabolism, inactivating glutamine synthetase and creatine kinase enzymes in brain cytosolic extract and cell free incubates [195]. An Aβ fragment also has been shown to induce in a time- and dose-dependent manner decreases in catalase activity, and increases in Cu/Zn-SOD and Mn-SOD activities and cellular peroxides in neuronal cell cultures — a pattern similar to that found in the AD brain [203]. This divergent shift in antioxidant enzymes may contribute to the cascade of neuronal injury.

Other amyloidoses occur in humans, and although these amyloids are composed of different proteins, they are all characterized by the antiparallel β-sheet conformation [300]. Amyloidogenic proteins may share a common cytotoxic mechanism, since three other amyloids — amylin, calcitonin, and atrial natriuretic peptide — are all toxic to clonal and primary neurons and increase the intracellular H_2O_2 level [296, 301]. Thus, the cytotoxic action of these peptides may be mediated through a free radical pathway indistinguishable from that of Aβ [301].

6.2.   Intracellular Free Radical Generation

Defects in mitochondrial electron transport systems as a result of age-related changes in mitochondrial function may be accompanied by free radical leakiness, which could be an initiating source of ROS and may promote subsequent damage [282]. The generation of reducing equivalents in neurons may be diminished as a result of lowered energy metabolism and cytochrome c oxidase levels that have been described in the AD brain (reviewed in [2]).

6.3.   Reactive Astrocytes and Microglia

Astrocytes tend to surround amyloid deposits but are absent from the congophilic zone of AD amyloid plaques in human brain (e.g., [302]). Microglia/macrophages, on the other hand, are intimately associated with the fibrillar amyloid deposits, and, at the ultrastructural level, amyloid fibrils are found in the cytoplasm of these cells [302,303], possibly indicating an attempt at removal of the amyloid (e.g., by phagocytosis).

Activated microglia, that cluster around neuritic but not diffuse amyloid plaques [304] generate ROS that may interact with redox metal ions within Aβ deposits [175]. Furthermore, the NO and O_2^- generated by activated microglia [305] react to form peroxynitrite, which can nitrosylate tyrosine residues that are detected in SPs [192, 193]. Thus, iron-rich ferritin secreted from activated microglia may interact with factors in the extracellular environment such as ascorbate, catecholamines, or acidic pH, resulting in the liberation of iron from ferritin, which may then impact on Aβ precipitation. In support of this hypothesis, secreted ferritin is increased in the CSF of AD, vascular dementia, and PD patients [173,177,178].

7. NEUROCHEMICAL ENVIRONMENTS LEADING TO AMYLOID DEPOSITION AND NEURONAL DEATH IN ALZHEIMER'S DISEASE

Alzheimer's disease is complicated by impaired cerebral energy metabolism and mild acidosis, which may be due to mitochondrial dysfunction [2,306]. Such impaired neuronal energy metabolism may impact on Aβ deposition in a number of ways. Energy-dependent mechanisms act rapidly to reassimilate metal ions released during synaptic transmission [23]. Therefore, damage to neuronal energy metabolism may inhibit the reuptake of zinc/copper into the neuron, with the consequent fall in intraneuronal zinc/copper levels and transiently raised extracellular metal ions. Alterations in energy metabolism also lead to acidosis and oxidative stress (increased glutathione disulfide and ROS),

---

FIG. 1   Model of metal ion mediated amyloidogenesis. Soluble Aβ is recruited into a diffuse amyloid deposit together with metal ions, such as following head injury where Aβ:metal ion interactions may be physiologically purposive in maintaining the structural integrity of a region while promoting other wound-healing processes. However, in conditions where the regional pH falls, such as during inflammation, copper may compete for metal ion binding sites on Aβ, triggering Aβ-induced copper reduction and the generation of reactive oxygen species. The generation of reactive oxygen species would promote the oxidative modification (carbonylation and polymerization) of Aβ, and the maturation of a diffuse deposit into a hard core amyloid plaque.

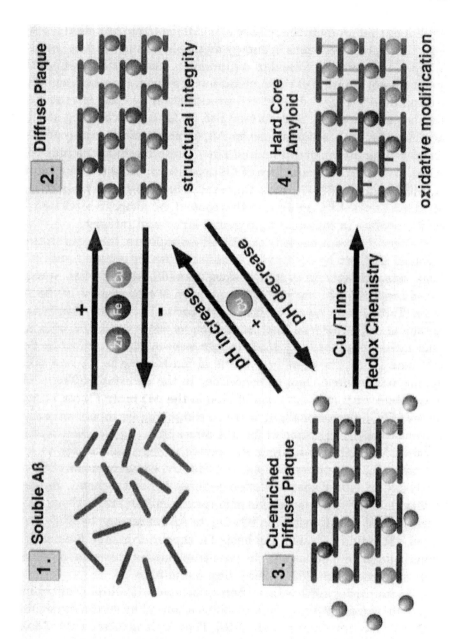

conditions that promote the release of metal ions from proteins (see Sec. 5.5). Therefore, alterations in energy metabolism, such as those seen in AD [307], trauma, or vascular compromise, may not only affect the compartmentalization of these metal ions but liberate metal ions that precipitate Aβ (Fig. 1). Indeed, Aβ precipitation by Cu(II) in this case may be potentiated by acidosis (see Sec. 5). Given the binding affinity stoichiometry of these metal ions for Aβ, Aβ accumulations may act as a sink for trapping copper, iron, and zinc. These mechanisms may contribute to the rapid appearance of Aβ deposits reported to occur following head injury [226,227], while the reversible aggregation properties of Aβ at local sites of injury may, in this context, be compatible with a role for the peptide in maintaining regional structural integrity.

Given the redox capacity of Aβ:Cu/Fe complexes, the sequestration of copper and iron by Aβ may be undesirable under certain conditions. Since zinc competes for copper binding sites [68, Atwood et al., unpublished observations], the high concentration of zinc present in the AD brain (Table 1) may be a protective response to limit the inappropriate uptake of copper or iron into areas such as those deposited with Aβ. Although aggregated $Aβ_{1-40}$ binds copper more avidly than zinc (see Sec. 5.2.1 and 5.2.2), the equilibria of metal ion binding to Aβ and other ligands will be dependent on conditions in the neuronal milieu of the brain. Such high concentration of zinc in the AD brain (Table 1) may, however, lead to neuronal cell death by competing for copper on critical enzymes (see Sec. 2). Support for this comes from studies showing that cultured hippocampal neurons are protected from Aβ toxicity at low, but not high, zinc concentrations and that the administration of metal ion chelators like desferrioxamine reduces hypoxic/ischemic damage [308] and delays the progression of dementia in AD [213]. Although the physiological role of metal ion binding to Aβ remains to be fully elucidated, the data do explain the basis for copper, iron, and zinc enrichment in amyloid plaques and the post-translational chemical modifications of amyloid Aβ [80,247,255] (Sec. 3 and 5.2).

Systemic amyloidoses are often associated with chronic inflammation [309]. An example is the amyloidosis caused by serum amyloid A, an acute phase reactant protein [310]. That Aβ is involved with inflammation is indicated by the fact that the aspartate residue 7 of Aβ is critical for classical complement pathway activation [311]. Inflammation is characterized by mild acidosis and the mobilization of metal ions to inflamed sites. For example, serum copper and zinc levels increase during inflammation, associated with increases in CP, that transfers Cu(II) to enzymes active in processes of basic metabolism and wound healing such as cytochrome oxidase and lysyl oxidase [312]. Therefore, similar mechanisms as described above may operate to precipitate Aβ during inflammation. Such an exchange of Cu(II) at low pH has been described as mediating the binding of serum amyloid P component, an acute phase reactant that also forms an amyloid, to the cell wall polysaccharide zymosan [313].

8. CONCLUSIONS

There is now excellent evidence to indicate that metal ion homeostasis is altered in AD. A number of metal ions, in particular Zn(II), Fe(III), Cu(II), and Al(III), accumulate in the neuropil of the AD brain and are further enriched within amyloid deposits where their concentration may reach millimolar levels. Aβ, the major component of amyloid deposits in the AD brain, avidly binds these metal ions and may explain their enrichment in plaque pathology. Together with the global oxidative stress observed in the AD brain, focal markers of oxidative stress have been localized to amyloid plaque deposits. Evidence indicates that this oxidative stress could be due to the accumulation of Aβ, which can generate ROS in the presence of redox-active metal ions. Alterations in cellular metal ion and hydrogen ion homeostasis, as a result of impaired energy metabolism, provide a plausible neurochemical environment for the deposition of Aβ in AD and DS, and following head injury and vascular compromise. These oxidative properties of Aβ likely promote the chemical modification of Aβ that have been identified in amyloid plaques, and the maturation of diffuse Aβ deposits to "hard"-core amyloid plaques. Both antioxidants and metal ion chelators, which inhibit the redox activity of Aβ in the presence of metal ions, may be of therapeutic use.

ACKNOWLEDGMENTS

The authors thank Richard Scarpa for his help in compiling the manuscript.

ABBREVIATIONS

AD	Alzheimer's disease
AGEs	advanced glycation end products
AMPA	α-amino-3-hydroxy-5-methylisoxazole-4-propionic acid
APP	amyloid protein precursor
BBB	blood-brain barrier
BSE	bovine spongiform encephalopathy
CNS	central nervous system
CP	ceruloplasmin
CSF	cerebrospinal fluid
DS	Down syndrome
EDTA	ethylenediamine-N,N,N',N'-tetraacetate (N,N'-1,2-ethanediylbis[N-(carboxymethyl)glycine])
G6PDH	glucose-6-phosphate dehydrogenase
GABA	γ-aminobutyric acid
GSH	glutathione, reduced form
HD	Huntington's disease
HO-1	heme oxygenase-1
LHRH	luteinizing hormone-releasing hormone
MEA	median eminence area
MT	metallothionein
NADPH	nicotinamide adenine dinucleotide phosphate (reduced)
NFT	neurofibrillary tangle
NMDA	N-methyl-D-aspartate
NMR	nuclear magnetic resonance
PD	Parkinson's disease
PrP	prion protein
PrP^c	prion protein, normal form
PrP^{Sc}	prion protein, protease-resistant form
ROS	reactive oxygen species
SDS-PAGE	sodium dodecyl sulfate polyacrylamide gel electrophoresis
SOD	superoxide dismutase
SP	senile plaque
SV	synaptic vesicle
Tf	transferrin

TfR	transferrin receptor complex
TPEN	N,N,N',N'-tetrakis(2-pyridylmethyl) ethylenediamine
ZEN	zinc-enriched neuron
ZnT	zinc transporters

REFERENCES

1. G. Benzi and A. Moretti, *Neurobiol. Aging*, *16*, 661–674 (1995).

2. L. A. Shinobu and M. F. Beal, in *Metals and Oxidative Damage in Neurological Disorders* (J. R. Connor, ed.), Plenum Publishing, New York, 1997, 237–275.

3. J. H. Morris, in *The Neuropathology of Dementia* (M. M. Esiri and J. H. Morris, eds.), Cambridge University Press, Cambridge, UK, 1997, 137–171.

4. A. D. Owen, A. H. Schapira, P. Jenner, and C. D. Marsden, *J. Neural Transm. Suppl.*, *51*, 167–173 (1997).

5. C. D. Ciccone, *Phys. Ther.*, *78*, 313–319 (1998).

6. A. H. Schapira, *Curr. Opin. Neurol.*, *9*, 260–264 (1996).

7. C. V. Borlongan, K. Kanning, S. G. Poulos, T. B. Freeman, D. W. Cahill, and P. R. Sanberg, *J. Fla. Med. Assoc.*, *83*, 335–341 (1996).

8. S. B. Prusiner, *Science*, *278*, 245–251 (1997).

9. D. R. Brown, K. Qin, J. W. Herms, A. Madlung, J. Manson, R. Strome, P. E. Fraser, T. Kruck, A. von Bohlen, W. Schulz-Schaeffer, A. Giese, D. Westaway, and H. Kretzschmar, *Nature*, *390*, 684–687 (1997).

10. M. M. Esiri and J. H. Morris (eds.), *The Neuropathology of Dementia*, Cambridge University Press, Cambridge, UK, 1997.

11. M. DiDonato and B. Sarkar, *Biochim. Biophys. Acta*, *1360*, 3–16 (1997).

12. R. S. Britton and K. E. Brown, *Hepatology*, *21*, 1195–1197 (1995).

13. B. A. Hosler and R. H. Brown, *Adv. Neurol.*, *68*, 41–46 (1995).

14. B. Halliwell and J. M. C. Gutteridge, *Meth. Enzymol.*, *186*, 1–85 (1990).

15. B. L. Vallee and K. H. Falchuk, *Physiol. Rev.*, *73*, 79–117 (1993).

16. A. S. Prasad, *Ann. Intern. Med.*, *125*, 142–144 (1996).

17. B. L. Vallee, *Zinc Enzymes*, *27*, 23–33 (1983).

18. R. J. P. Willams, in *Zinc in Human Biology* (C. F. Mills, ed.), 1989, pp. 15–31.

19. M. P. Cuajungco and G. J. Lees, *Neurobiol. Dis.*, *4*, 137–169 (1997).

20. M. P. Cuajungco and G. J. Lees, *Brain Res. Rev.*, *23*, 219–236 (1997).

21. W. D. Ehmann, W. R. Markesbery, M. Allaudin, T. I. Hossain, and E. H. Brubaker, *Neurotoxicology*, *7*, 195–206 (1986).

22. M. Ebadi, *Meth. Enzymol.*, *205*, 363–387 (1991).

23. C. J. Frederickson, *Int. Rev. Neurobiol.*, *31*, 145–238 (1989).

24. R. A. Jacob, in *Fundamentals of Clinical Chemistry* (N. W. Tietz, ed.), W. B. Saunders, Philadelphia, 1987, pp. 517–532.

25. C. J. Fredrickson and G. Danscher, *Prog. Brain Res.*, *83*, 71–84 (1990).

26. C. J. Frederickson and D. W. Moncrieff, *Biol. Signals*, *3*, 127–139 (1994).

27. F. M. S. Haug, *Histochemie*, *8*, 355–368 (1967).

28. M. L. Failla, M. van de Veerdonk, W. T. Morgan, and J. C. Smith, *J. Lab. Clin. Med.*, *100*, 943–953 (1982).

29. J. R. Blair-West, D. A. Denton, A. P. Gibson, and M. J. McKinely, *Brain Res.*, *507*, 6–10 (1990).

30. G. Wolf, M. Schutte, and W. Romhild, *Neurosci. Lett.*, *51*, 277–280 (1984).

31. Y. Ibata and N. Otsuka, *J. Histochem. Cytochem.*, *17*, 171–175 (1969).

32. G. A. Howell, M. G. Welch, and C. J. Frederickson, *Nature*, *308*, 736–738 (1984).

33. J. Wensink, A. J. Molenaar, U. D. Woroniecka, and C. J. A Van den Hamer, *J. Neurochem.*, *50*, 782–789 (1988).

34. S. L. Sensi, L. M. Canzoniero, S. P. Yu, H. S. Ying, J. Y. Koh, G. A. Kershner, and D. W. Choi, *J. Neurosci.*, *17*, 9554–9564 (1997).

35. R. D. Palmiter and S. D. Findley, *EMBO J.*, *14*, 639–649 (1995).

36. R. D. Palmiter, T. B. Cole, and S. D. Findley, *EMBO J.*, *15*, 1784–1791 (1996).

37. R. D. Palmiter, T. B. Cole, C. J. Quaife, and S. D. Findley, *Proc. Natl. Acad. Sci. USA*, *93*, 14934–14939 (1996).

38. M. Aschner, M. G. Cherian, C. D. Klaassen, R. D. Palmiter, J. C. Erickson, and A. I. Bush, *Toxicol. Appl. Pharmacol.*, *142*, 229–242 (1997).

39. V. S. K. R. Vepachedu and P. M. Mohan, *Biochemistry*, *35*, 9301, (1996).

40. C. J. Quaife, S. D. Findley, J. C. Erickson, G. J. Froelick, E. J. Kelly, B. P. Zambrowicz, and R. D. Palmiter, *Biochemistry*, *33*, 7250–7259 (1994).

41. R. D. Palmiter, *Experientia (Suppl.)*, *52*, 63–80 (1987).

42. Y. Uchida, K. Takio, K. Titani, Y. Ihara, and M. Tomonaga, *Neuron*, *7*, 337–347 (1991).

43. R. D. Palmiter, S. D. Findley, T. E. Whitmore, and D. M. Durnam, *Proc. Natl. Acad. Sci. USA*, *89*, 6333–6337 (1992).

44. B. A. Masters, C. A. Quaife, J. C. Erickson, E. J. Kelly, G. J. Froelick, B. P. Zambrowicz, R. L. Brinster, and R. D. Palmiter, *J. Neurosci.*, *14*, 5844–5857 (1994).

45. A. K. Sewell, L. T. Jensen, J. C. Erickson, R. D. Palmiter, and D. R. Winge, *Biochemistry*, *34*, 4740–4747 (1995).

46. R. D. Palmiter, *Toxicol. Appl. Pharmacol.*, *135*, 139–146 (1995).

47. J. C. Erickson, G. Hollopeter, S. A. Thomas, G. J. Froelick, and R. D. Palmiter, *J. Neurosci.*, *17*, 1271–1281 (1997).

48. S. Y. Assaf and S. H. Chung, *Nature*, *308*, 734–736 (1984).

49. G. Charton, C. Rovira, Y. Ben-Ari, and V. Leviel, *Exp. Brain Res.*, *58*, 202–205 (1985).

50. J. H. Weiss, J. Koh, C. W. Christine, and D. W. Choi, *Nature*, *338*, 212 (1989).

51. N. L. Harrison and S. J. Gibbons, *Neuropharmacology*, *33*, 935–952 (1994).

52. T. G. Smart, X. Xie, and B. J. Krishek, *Prog. Neurobiol.*, *42*, 393–441 (1994).

53. S. Peters, J. Y. I. Koh, and D. W. Choi, *Science*, *236*, 589–593 (1987).

54. M. Yokoyama, J. Koh, and D. W. Choi, *Neurosci. Lett.*, *71*, 351–355 (1986).

55. D. W. Choi, M. Yokoyama, and J. Koh, *Neuroscience*, *24*, 67–79 (1988).

56. M. Provinciali, G. Di Stefano, and N. Fabris, *Int. J. Immunopharmacol.*, *17*, 735–744 (1995).

57. J. Y. Koh, S. W. Suh, B. J. Gwag, Y. Y. He, C. Y. Hsu, and D. W. Choi, *Science*, *272*, 1013–1016 (1996).

58. C. A. Owen, *Biological Aspects of Copper*, Noyes, Park Ridge, NJ, 1982, pp. 117–129.

59. T. M. Bray and W. J. Bettger, *Free Rad. Biol. Med.*, *8*, 281–291 (1990).

60. R. M. O'Neal, G. W. Pla, M. R. Fox, F. S. Gibson, and B. E. Fry, Jr, *J. Nutr.*, *100*, 491–497 (1970).

61. J. C. Wallwork, D. B. Milne, R. L. Sims, and H. H. Sandstead, *J. Nutr.*, *113*, 1895–1905 (1983).

62. M. C. Linder, *The Biochemistry of Copper*, New York, Plenum Press, 1991.

63. M. C. Linder and M. Hazegh-Azam, *Am. J. Clin. Nutr.*, *63*, 797–811 (1996).

64. C. C. Askwith, D. de Silva, and J. Kaplan, *Mol. Microbiol.*, *20*, 27–34 (1996).

65. J. Kaplan and T. V. O'Halloran, *Science*, *271*, 1510–1512 (1996).

66. D. J. Thiele, *Nucleic Acids Res.*, *20*, 1183–1191 (1992).

67. J. S. Valentine and E. B. Gralla, *Science*, *278*, 817–818 (1997).

68. C. A. Owen, *Biological Aspects of Copper*, Noyes, Park Ridge, NJ, 1982, pp. 161–189.

69. R. M. Nalbandyan, *Neurochem. Res.*, *8*, 1211–1232 (1983).

70. M. A. Lovell, J. D. Robertson, W. J. Teesdale, J. L. Campbell, and W. R. Markesbery, *J. Neurol. Sci.*, *158*, 47–52 (1998).

71. Y. Matsuba and Y. Takahashi, *Anal. Biochem.*, *36*, 182–190 (1970).

72. P. J. Warren, C. J. Earl, and R. H. S. Thompson, *Brain*, *83*, 709–717 (1960).

73. R. M. Smith, in *Trace Element Neurobiology and Deficiencies: Neurobiology of the Trace Elements*, Humana Press, Clifton, NJ, 1983.

74. M. Sato, K. Ohtomo, T. Daimon, T. Sugiyama, and K. Iijima, *J. Histochem. Cytochem.*, *42*, 1585–1591 (1994).

75. R. W. Colburn and J. W. Mass, *Nature*, *208*, 37 (1965).

76. R. E. Mains and B. A. Eipper, *Endocrinology*, *115*, 1683–1690 (1984).

77. S. G. A. Sharoyan, A. A. Shaljian, R. M. Nalbandyan, and H. C Buniatian, *Biochim. Biophys. Acta*, *493*, 478–487 (1977).
78. J. R. Prohaska, *Physiol. Rev.*, *67*, 858–901 (1987).
79. D. A. Loeffler, P. A. LeWitt, P. L. Juneau, A. A. F. Sima, H.-U. Nguyen, A. J. deMaggio, C. M. Brickman, G. J. Brewer, R. D. Dick, M. D. Troyer, and L. Kanaley, *Brain Res.*, *738*, 265–274 (1996).
80. C. S. Atwood, R. D. Moir, X. Huang, R. C. Scarpa, N. M. E. Bacarra, D. M. Romano, M. A. Hartshorn, R. E. Tanzi, and A. I. Bush, *J. Biol. Chem.*, *273*, 12817–12826 (1998).
81. J. G. Chutkow, *Proc. Soc. Exp. Biol. Med.*, *158*, 113–116 (1978).
82. D. E. Hartter and A. Barnea, *J. Biol. Chem.*, *263*, 799–805 (1988).
83. B. M. Katz and A. Barnea, *J. Biol. Chem.*, *265*, 2017–2021 (1990).
84. T. Terao and C. A. Owen, *Am. J. Physiol.*, *224*, 682–686 (1973).
85. T. Terao and C. A. Owen, *Mayo Clin. Proc.*, *49*, 376–381 (1974).
86. J. I. Logan, K. B. Harveyson, G. B. Wisdom, A. E. Hughes, and G. P. R. Archbold, *Quart. J. Med.*, *87*, 663–670 (1994).
87. Z. L. Harris, Y. Takahashi, H. Miyajima, M. Serizawa, R. T. A. MacGillvray, and J. D. Gitlin, *Proc. Natl. Acad. Sci. USA*, *92*, 2539–2543 (1995).
88. K. A. Moshkov, S. Lakatos, J. Hajdu, P. Zavodsky, and S. A. Neifakh, *Eur. J. Biochem.*, *94*, 127–134 (1979).
89. S. J. Orena, C. A. Goode, and M. C. Linder, *Biochem. Biophys. Res. Commun.*, *139*, 822–829 (1986).
90. M. C. Linder and J. R. Moor, *Biochim. Biophys. Acta*, *499*, 329–336 (1977).
91. R. J. Seitz and W. Wechsler, *Acta Neuropathol.*, *73*, 145–152 (1987).
92. B. N. C. Patel and S. David, *Proc. Soc. Neurosci.*, *21*, 564 (1995).
93. S. S. Percival and E. D. Harris, *J. Nutr.*, *119*, 779–784 (1989).
94. E. D. Harris, *Prog. Clin. Biol. Res.*, *380*, 163–179 (1993).
95. G. Multhaup, A. Schlicksupp, L. Hesse, D. Beher, T. Ruppert, C. L. Masters, and K. Beyreuther, *Science*, *271*, 1406–1409 (1996).
96. H. Fukino, M. Hirai, Y. M. Hsueh, S. Moriyasu, and Y. Yamane, *J. Toxicol. Environ. Health*, *19*, 75–89 (1986).
97. G. K. Andrews, K. R. Gallant, and M. G. Cherian, *Eur. J. Biochem.*, *166*, 527–531 (1987).

98. J. H. Freedman, M. R. Ciriolo, and J. Peisach, *J. Biol. Chem.*, *264*, 5598–5605 (1989).

99. M. R. Ciriolo, A. Desideri, M. Paci, and G. Rotilio, *J. Biol. Chem.*, *265*, 11030–11034 (1990).

100. M. Brouwer and T. Brouwer-Hoexum, *Biochemistry*, *31*, 4096–4102 (1992).

101. D. E. Hartter and A. Barnea, *Synapse*, *2*, 412–415 (1988).

102. G. H. Burrows and A. Barnea, *Endocrinology*, *110*, 1456–1458 (1982).

103. G. E. Rice and A. Barnea, *J. Neurochem.*, *41*, 1672–1679 (1983).

104. A. Barnea and G. Cho, *Endocrinology*, *115*, 936–943 (1984).

105. A. Barnea, G. Cho, and B. M. Katz, *Brain Res.*, *541*, 93–97 (1991).

106. W. R. Markesbery, *Free Rad. Biol. Med.*, *23*, 134–147 (1997).

107. G. Falkous, J. B. Harris, and D. Mantle, *Clin. Chim. Acta*, *238*, 125–135 (1995).

108. R. E. Tanzi, K. Petrukhin, I. Chernov, J. L. Pellequer, W. Wasco, B. Ross, D. M. Romano, E. Parano, L. Pavone, L. M. Brzustowicz, M. Devoto, J. Peppercorn, A. I. Bush, I. Stemlieb, M. Piratsu, J. F. Gusella, O. Evgrafov, G. K. Penchaszadeh, B. Honig, I. S. Edelman, M. B. Soares, I. H. Scheinberg, and T. C. Gilliam, *Nature Genet.*, *5*, 344–350 (1993).

109. E. J. Kelly and R. D. Palmiter, *Nature Genet.*, *13*, 219–222 (1996).

110. T. Nishikawa, I. S. Lee, N. Shiraishi, T. Ishikawa, Y. Ohta, and M. J. Nishikimi, *J. Biol. Chem.*, *272*, 23037–23041 (1997).

111. E. B. Gralla, D. J. Thiele, P. Silar, and J. S. Valentine, *Proc. Natl. Acad. Sci. USA*, *88*, 8558–8562 (1991).

112. P. Zhou and D. J. Thiele, *Proc. Natl. Acad. Sci. USA*, *88*, 6112–6116 (1991).

113. J. A. Graden, M. C. Posewitz, J. R. Simon, G. N. George, I. J. Pickering, and D. R. Winge, *Biochemistry*, *35*, 14583–14589 (1996).

114. J. A. Gutierrez and M. Wessling-Resnick, *Crit. Rev. Eukaryot. Gene Expr.*, *6*, 1–14 (1996).

115. M. W. Bradbury, *J. Neurochem.*, *69*, 443–454 (1997).

116. F. Ueda, K. B. Raja, R. J. Simpson, I. S. Trowbridge, and M. W. Bradbury, *J. Neurochem.*, *60*, 106–113 (1993).

117. H. A. Huebers and C. A. Finch, *Physiol. Rev.*, 67, 520–582 (1987).

118. D. R. Richardson and P. Ponka, *Biochim. Biophys. Acta*, 1331, 1–40 (1997).

119. A. J. Roskams and J. R. Connor, *Proc. Natl. Acad. Sci. USA*, 87, 9024–9027 (1990).

120. E. C. Theil, *Annu. Rev. Biochem.*, 56, 289–315 (1987).

121. R. Roberts, A. Sandra, G. C. Siek, J. J. Lucas, and R. E. Fine, *Ann. Neurol.*, 32, S43–S50 (1992).

122. D. de Silva, C. Askwith, D. J. Eide, and J. Kaplan, *J. Biol. Chem.*, 270, 1098–1101 (1995).

123. I. Jordan and J. Kaplan, *Biochem. J.*, 302, 875–879 (1994).

124. D. Ben-Shachar, S. Yehuda, J. P. Finberg, I. Spanier, and M. B. Youdim, *J. Neurochem.*, 50, 1434–1437 (1988).

125. J. J. Winzerling and J. H. Law, *Annu. Rev. Nutr.*, 17, 501–526 (1997).

126. G. R. Lee, S. Nacht, J. N. Lukens, and G. E. Cartwright, *J. Clin. Invest.*, 47, 2058–2069 (1968).

127. C. A. Owen, Jr., *Am. J. Physiol.*, 224, 514–518 (1973).

128. Z. L. Harris and J. D. Gitlin, *Am. J. Clin. Nutr.*, 63, 836S–841S (1996).

129. H. Morita and S. Ikeda, K. Yamamoto, S. Morita, K. Yoshida, S. Nomoto, M. Kato, and N. Yanagisawa, *Ann. Neurol.*, 37, 646–656 (1995).

130. H. P. Roeser, G. R. Lee, S. Nacht, and G. E. Cartwright, *J. Clin. Invest.*, 49, 2408–2417 (1970).

131. J. G. Joshi, M. Dhar, M. Clauberg, and V. Chauthaiwale, *Environ. Health Persp.*, 102S, 207–213 (1994).

132. D. Price and J. G. Joshi, *Proc. Natl. Acad. Sci. USA*, 79, 3116–3119 (1982).

133. J. R. Connor, S. L. Menzies, S. M. St. Martin, and E. J. Mufson, *J. Neurosci. Res.*, 27, 595–611 (1990).

134. J. R. Connor, S. L. Menzies, S. M. St. Martin, and E. J. Mufson, *J. Neurosci. Res.*, 31, 75–83 (1992).

135. J. R. Connor, B. S. Snyder, J. L. Beard, R. E. Fine, and E. J. Mufson, *J. Neurosci. Res.*, 31, 327–335 (1992).

136. J. Xu and E. A. Ling, *J. Anat.*, 184, 227–237 (1994).

137. E. Zappone, V. Bellotti, M. Cazzola, M. Ceroni, F. Meloni, P. Pedrazzoli, and V. Perfetti, *Haematologica*, *71*, 103–107 (1986).

138. R. Hallgren, A. Terent, L. Wide, K. Bergstrom, and G. Birgegard, *Acta Neurol. Scand.*, *61*, 384–392 (1980).

139. R. Ashkenazi, D. Ben-Shachar, and M. B. Youdim, *Pharmacol. Biochem. Behav.*, *17*, 43–47 (1982).

140. M. B. Youdim, D. Ben-Shachar, S. Yehuda, and P. Riederer, *Adv. Neurol.*, *53*, 155–162 (1990).

141. M. Jimenez Del Rio, C. Velez Pardo, J. Pinxteren, W. De Potter, G. Ebinger, and G. Vauquelin, *Eur. J. Pharmacol.*, *247*, 11–21 (1993).

142. C. Velez-Pardo, M. Jimenez del Rio, G. Ebinger, and G. Vauquelin, *Neurochem. Int.*, *26*, 615–622 (1995).

143. D. Ben-Shachar, G. Eshel, P. Riederer, and M. B. Youdim, *Ann. Neurol.*, *32*, S105–S110 (1992).

144. W. R. Markesbery, W. D. Ehmann, T. I. Hossain, M. Alauddin, and D. T. Goodin, *Ann. Neurol.*, *10*, 511–516 (1981).

145. M. A. Deibel, W. D. Ehmann, and W. R. Markesbery, *J. Neurol. Sci.*, *143*, 137–142 (1996).

146. C. R. Cornett, W. R. Markesbery, and W. D. Ehmann, *Neurotoxicology*, *19*, 339–345 (1998).

147. C. M. Thompson, W. R. Markesbery, W. D. Ehmann, Y. X. Mao, and D. E. Vance, *Neurotoxicology*, *9*, 1–7 (1988).

148. D. L. Samudralwar, C. C. Diprete, B. F. Ni, W. D. Ehmann, and W. R. Markesbery, *J. Neurol. Sci.*, *130*, 139–145 (1995).

149. L. A. Hershey, C. O. Hershey, and A. W. Varnes, *Neurology*, *34*, 1197–1201 (1984).

150. D. Wenstrup, W. D. Ehmann, and W. R. Markesbery, *Brain Res.*, *533*, 125–131 (1990).

151. N. I. Ward and J. A. Mason, *J. Radioanal. Nucl. Chem.*, *113*, 515–526 (1987).

152. F. M. Corrigan, G. P. Reynolds, and N. I. Ward, *Biometals*, *6*, 149–154 (1993).

153. Q. S. Deng, G. C. Turk, D. R. Brady, and Q. R. Smith, *Neurobiol. Aging*, *15*(Suppl. 1), S113 (1994).

154. E. Lui, M. Fisman, C. Wong, and F. J. Diaz, *Am. Geriatr. Soc.*, *38*, 633–639 (1990).

155. J. R. Backstrom, C. A. Miller, and Z. A. Tokes, *J. Neurochem.*, *58*, 983–992 (1992).

156. C. Jeandel, M. B. Nicolas, F. Dubois, F. Nabet-Belleville, F. Penin, and G. Cuny, *Gerontology*, *35*, 275–282 (1989).

157. C. O. Hershey, L. A. Hershey, A. Varnes, S. D. Vibhakar, P. Lavin, and W. H. Strain, *Neurology*, *33*, 1350–1353 (1983).

158. D. W. Choi, *Stroke*, *21*, SIII20–SIII22 (1990).

159. B. Bjorksten, O. Back, K. H. Gustavson, G. Hallmans, B. Hagglof, and A. Tarnvik, *Acta Paediatr. Scand.*, *69*, 183–187 (1980).

160. M. Purice, C. Maximilian, I. Dumitriu, and D. Ioan, *Endocrinologie*, *26*, 113–117 (1988).

161. C. S. Atwood, R. D. Moir, X. Huang, R. E. Tanzi, and A. I. Bush, in *Molecular Mechanisms of Dementia* (R. E. Tanzi and W. Wasco, eds.), Humana Press, Totowa, NJ, 1997, pp. 225–237.

162. C. S. Atwood, X. Huang, R. D. Moir, R. E. Tanzi, and A. I. Bush, in *Alzheimer's Disease: Biology, Diagnosis and Therapeutics* (K. Iqbal, B. Winblad, T. Nishimura, M. Takeda, and H. M. Wisniewski, eds.), John Wiley and Sons, Chichester, 1997, pp. 329–337.

163. L.-O. Plantin, Y. Lysing-Tunell, and K. Kristensson, *Biol. Trace Elem. Res.*, *13*, 69–75 (1987).

164. J. R. Connor, P. Tucker, M. Johnson, and B. Snyder, *Neurosci. Lett.*, *159*, 88–90 (1993).

165. W. D. Parker, Jr., and J. K. Parks, *Neurology*, *45*, 482–486 (1995).

166. R. Cherny, C. L. Masters, K. Beyreuther, D. Fairlie, R. E. Tanzi and A. I. Bush, *Soc. Neurosci.*, *23*, 534 (1997).

167. D. A. Loeffler, A. J. DeMaggio, P. L. Juneau, C. M. Brickman, G. A. Mashour, J. H. Finkelman, N. Pomara, and P. A. LeWitt, *Alzheimer Dis. Assoc. Disord.*, *8*, 190–197 (1994).

168. H. Basun, L. G. Forssell, L. Wetterberg, and B. Winbald, *J. Neural Transm. Parkinson's Dis. Dementia Sect.*, *4*, 231–258 (1991).

169. W. D. Ehmann, W. R. Markesbery, M. Alauddin, T. I. M. Hossain, and E. H. Brubaker, *Neurotoxicology*, *7*, 197–206 (1986).

170. L. Goodman, *J. Nerv. Ment. Dis.*, *118*, 97–130 (1953).

171. R. C. Switzer, T. L. Martin, S. K. Campbell, J. C. Parker, and E. D. Caldwell, *Soc. Neurosci. Abstr.*, *12*, 100 (1986).

172. J. Fleming and J. G. Joshi, *Proc. Natl. Acad. Sci. USA*, *84*, 7866–7870 (1987).

173. I. Grundke-Iqbal, J. Fleming, Y. C. Tung, H. Lassmann, K. Iqbal, and J. G. Joshi, *Acta Neuropathol.*, *81*, 105–110 (1990).

174. P. F. Good, D. P. Perl, L. M. Bierer, and J. Schmeidler, *Ann. Neurol.*, *31*, 286–292 (1992).

175. M. A. Smith, P. L. Harris, L. M. Sayre, and G. Perry, *Proc. Natl. Acad. Sci. USA*, *94*, 9866–9868 (1997).

176. D. J. Dedman, M. Trefry, J. M. Candy, G. A. A. Taylor, C. M. Morris, C. A. Bloxham, R. H. Perry, J. A. Edwardson, and P. M. Harrison, *Biochem. J.*, *287*, 509–514 (1992).

177. S. R. Robinson, D. F. Noone, J. Kril, and G. Halliday, *Alzheimer's Res.*, *1*, 191–196 (1995).

178. C. I. Batton, B. S. O'Dowd, D. F. Noone, J. Kril, and S. R. Robinson, *Alzheimer's Res.*, *3*, 45–50 (1997).

179. T. Ohgami, T. Kitamoto, and J. Tateishi, *J. Neurochem.*, *61*, 1553–1556 (1993).

180. M. A. Kuiper, C. Mulder, G. J. van Kamp, P. Scheltens, and E. C. Wolters, *J. Neural Transm. Park. Dis. Dement. Sect.*, *7*, 109–114 (1994).

181. W. R. Markesbery and W. D. Ehmann, in *Alzheimer's Disease* (R. D. Terry, R. Katzman, and K. L. Bick, eds.), 1994, pp. 353–367.

182. N. Xu, V. Majidi, W. D. Ehmann, and W. R. Markesbery, *J. Anat. Atom. Spectrom.*, *7*, 749–751 (1992).

183. D. R. Crapper, S. S. Kirshnam, and A. J. Dalton, *Science*, *180*, 511–513 (1973).

184. D. R. Crapper, S. S. Kirshnam, and S. Quittkat, *Brain*, *99*, 67–80 (1976).

185. J. M. Candy, J. Klinowski, R. H. Perry, E. K. Perry, A. Fairbairn, A. E. Oakley, T. A. Carpenter, J. R. Atack, G. Blessed, and J. A. Edwardson, *Lancet*, *1*, 354–357 (1986).

186. S. Tokutake, H. Nagase, S. Morisake, and S. Oyanagi, *Neurosci. Lett.*, *185*, 99–102 (1995).

187. R. N. Martins, C. G. Harper, G. B. Stokes, and C. L. Masters, *J. Neurochem.*, *46*, 1042–1045 (1986).

188. M. A. Smith, R. K. Kutty, P. L. Richey, S. D. Yan, D. Stern, G. J. Chader, B. Wiggert, R. B. Petersen, and G. Perry, *Am. J. Pathol.*, *145*, 42–47 (1994).

189. L. Balzacs and M. Leon, *Neurochem. Res.*, *19*, 1131–1137 (1994).

190. C. D. Smith, J. M. Carney, P. E. Starke-Reed, C. N. Oliver, E. R. Stadtman, R. A. Floyd, and W. R. Markesbery, *Proc. Natl. Acad. Sci. USA*, *88*, 10540–10543 (1991).

191. K. Hensley, N. Hall, R. Subramaniam, P. Cole, M. Harris, M. Aksenov, M. Aksenova, S. P. Gabbita, J. F. Wu, J. M. Carney, M. Lovell, W. R. Markesbery, and D. A. Butterfield, *J. Neurochem.*, *65*, 2146–2156 (1995).

192. P. F. Good, P. Werner, A. Hsu, C. W. Olanow, and D. P. Perl, *Am. J. Pathol.*, *149*, 21–28 (1996).

193. M. A. Smith, P. L. Richey Harris, L. M. Sayre, J. S. Beckman, and G. Perry, *J. Neurosci.*, *17*, 2653–2657 (1997).

194. M. A. Lovell, W. D. Ehmann, M. P. Mattson, and W. R. Markesbery, *Neurobiol. Aging*, *18*, 457–461 (1997).

195. K. Hensley, J. M. Carney, M. P. Mattson, M. Aksenova, M. Harris, J. F. Wu, R. A. Floyd, and D. A. Butterfield, *Proc. Natl. Acad. Sci. USA*, *91*, 3270–3274 (1994).

196. K. V. Subbarao, J. S. Richardson, and L. C. Ang, *J. Neurochem.*, *55*, 342–345 (1990).

197. L. M. Sayer, D. A. Zelasko, P. L. Harris, G. Perry, R. G. Salomon, and M. A. Smith, *J. Neurochem.*, *68*, 2092–2097 (1997).

198. P. Mecocci, U. MacGarvey, and M. F. Beal, *Ann. Neurol.*, *36*, 747–751, (1994).

199. J. H. Dowson, C. Q. Mountjoy, M. R. Cairns, and H. Wilton-Cox, *Neurobiol. Aging*, *13*, 493–500 (1992).

200. M. A. Smith, S. Taneda, P. L. Richey, S. Miyata, S. D. Yan, D. Stern, L. M. Sayre, V. M. Monnier, and G. Perry, *Proc. Natl. Acad. Sci. USA*, *91*, 5710–5714 (1994).

201. M. A. Smith, P. L. Richey, R. K. Kutty, B. Wiggert, and G. Perry, *Mol. Chem. Neurol.*, *24*, 227–230 (1995).

202. J. M. Carney and A. M. Carney, *Life Sci.*, *55*, 2097–2103 (1994).

203. W. Gsell, R. Conrad, M. Hickethier, E. Sofic, L. Frolich, I. Wichart, K. Jellinger, G. Moll, G. Ransmayr, H. Beckmann, and P. Riederer, *J. Neurochem.*, *64*, 1216–1223 (1995).

204. M. A. Pappolla, R. A. Omar, K. S. Kim, and N. K. Robakis, *Am. J. Pathol.*, *140*, 621–628 (1992).

205. A. Furuta, D. L. Price, C. A. Pardo, J. C. Troncoso, Z. S. Xu, N. Taniguchi, and L. J. Martin, *Am. J. Pathol.*, *146*, 357–367 (1995).

206. M. A. Smith, P. L. Richey, S. Taneda, R. K. Kutty, L. M. Sayre, V. M. Monnier, and G. Perry, *Ann. NY Acad. Sci.*, *738*, 447–454 (1994).

207. S. D. Yan, X. Chen, J. Fu, M. Chen, H. Zhu, A. Roher, T. Slattery, L. Zhao, M. Nagashima, J. Morser, A. Migheli, P. Nawroth, D. Stern, and A. M. Schmidt, *Nature*, *382*, 685–691 (1996).

208. M. A. Lovell, W. D. Ehmann, S. M. Butler, and W. R. Markesbery, *Neurology*, *45*, 1594–1601 (1995).

209. M. Sano, C. Ernesto, R. G. Thomas, M. R. Klauber, K. Schafer, M. Grundman, P. Woodbury, J. Growdon, C. W. Cotman, E. Pfeiffer, L. S. Schneider, and L. J. N. Thal, *N. Engl. J. Med.*, *336*, 1216–1222 (1997).

210. J. C. Breitner, *Annu. Rev. Med.*, *47*, 401–411 (1996).

211. M. Rother, B. Kittner, K. Rudolphi, M. Rossner, and K. H. Labs, *Ann. NY Acad. Sci.*, *777*, 404–409 (1996).

212. P. Schubert, T. Ogata, C. Marchini, S. Ferroni, and K. Rudolphi, *Ann. NY Acad. Sci.*, *825*, 1–10 (1997).

213. D. R. Crapper-McLachlan, A. J. Dalton, T. P. A. Kruck, M. Y. Bell, W. L. Smith, W. Kalow, and D. F. Andrews, *Lancet*, *337*, 1304 (1991).

214. D. C. Bolton, M. P. McKinley and S. B. Prusiner, *Science*, *218*, 1309–1311 (1982).

215. A. L. Horwich and J. S. Weissman, *Cell*, *89*, 499–510 (1997).

216. T. W. Hutchens, R. W. Nelson, M. H. Allen, C. M. Li, and T. T. Yip, *Mass. Spectrom.*, *21*, 151–159 (1992).

217. K.-M. Pan, M. Baldwin, J. Nguyen, M. Gasset, A. Serban, D. Groth, I. Mehlhorn, Z. Huang, R. J. Fletterick, F. E. Cohen, and S. B. Prusiner, *Proc. Natl. Acad. Sci. USA*, *90*, 10962–10966 (1993).

218. M. P. Hornshaw, J. R. McDermott, and J. M. Candy, *Biochem. Biophys. Res. Commun.*, *207*, 621–629 (1995).

219. M. P. Hornshaw, J. R. McDermott, J. M. Candy, and J. H. Lakey, *Biochem. Biophys. Res. Commun.*, *214*, 993–999 (1995).

220. D. R. Brown, W. J. Schulz-Schaeffer, B. Schmidt, and H. A. Kretzschmar, *Exp. Neurol.*, *146*, 104–112 (1997).

221. R. Riek, S. Hornemann, G. Wider, R. Glockshuber, and K. Wüthrich, *FEBS Lett.*, *413*, 282–386 (1997).

222. T. Miura, A. Hori-i, and H. Takeuchi, *FEBS Lett.*, *396*, 248–252 (1996).

223. C. W. Cotman and A. J. Anderson, *Mol. Neurobiol.*, *10*, 19–45 (1995).

224. F. M. LaFerla, B. T. Tinkle, C. J. Bieberch, C. C. Haudenschild, and G. Jay, *Nature Genet.*, *9*, 21–29 (1995).

225. C. P. Maury, *Lab. Invest.*, *72*, 4–16 (1995).

226. G. W. Roberts, S. M. Gentleman, A. Lynch, and D. I. Graham, *Lancet*, *338*, 1422–1423 (1991).

227. I. Mackenzie, *J. Neuropathol. Exp. Neurol.*, *156*, 437–442 (1993).

228. V. Askanas, R. B. Alvarez, and W. K. Engel, *Ann. Neurol.*, *34*, 551 (1993).

229. G. G. Glenner and C. W. Wong, *Biochem. Biophys. Res. Commun.*, *120*, 885–890 (1984).

230. G. Skodras, J. H. Peng, J. C. Parker, Jr., and P. J. Kragel, *Ann. Clin. Lab. Sci.*, *23*, 275–280 (1993).

231. C. L. Joachim, H. Mori, and D. J. Selkoe, *Nature*, *341*, 226–230 (1989).

232. C. van Broeckhoven, J. Haan, E. Bakker, J. A. Hardy, W. van Hul, A. Wehnert, M. Vegter-van der Vlis, and R. A. C. Roos, *Science*, *248*, 1120–1122 (1990).

233. P. T. Lansbury, Jr, P. R. Costa, J. M. Griffiths, E. J. Simon, M. Auger, K. J. Halverson, D. A. Kocisko, Z. S. Hendsch, T. T. Ashburn, R. G. Spencer, B. Tidor, and R. G. Griffin, *Nature Struct. Biol.*, *2*, 990–998 (1995).

234. L. P. Choo, D. L. Wetzel, W. C. Halliday, M. Jackson, S. M. LeVine, and H. H. Mantsch, *Biophys. J.*, *71*, 1672–1679 (1996).

235. J. E. Maggio and P. W. Mantyh, *Brain Pathol.*, *6*, 147–162 (1996).

236. K. Halverson, P. E. Fraser, D. A. Kirschner, and P. T. Lansbury, Jr., *Biochemistry*, *29*, 2639–2644 (1990).

237. C. J. Pike, A. J. Walencewicz-Wasserman, J. Kosmoski, D. H. Cribbs, C. G. Glabe, and C. W. Cotman, *J. Neurochem.*, *64*, 253–265 (1995).

238. D. Scheuner, C. Eckman, M. Jensen, X. Song, M. Citron, N. Suzuki, T. D. Bird, J. Hardy, M. Hutton, W. Kukull, E. Larson, E. Levy-Lahad, M. Viitanen, E. Peskind, P. Poorkaj, G. Schellenberg, R. Tanzi, W. Wasco, L. Lannfelt, D. Selkoe, and S. Younkin, *Nature Med.*, *2*, 864–870 (1996).

239. K. Hsiao, P. Chapman, S. Nilsen, C. Eckman, Y. Harigaya, S. Younkin, F. Yang, and G. Cole, *Science*, *274*, 99–102 (1996).

240. Y. M. Kuo, M. R. Emmerling, C. Vigopelfrey, T. C. Kasunic, J. B. Kirkpatrick, G. H. Murdoch, M. J. Ball, and A. E. Roher, *J. Biol. Chem.*, *271*, 4077–4081 (1996).

241. A. E. Roher, M. O. Chaney, Y.-M Kuo, S. D. Webster, W. Blaine Stine, L. J. Haverkamp, A. S. Woods, R. J. Cotter, J. M. Tuohy, G. A. Krafft, B. S. Bonnell, and M. R. Emmerling, *J. Biol. Chem.*, *271*, 20631–20635 (1996).

242. C. L. Masters, G. Simms, N. A. Weinman, G. Multhaup, B. L. McDonald, and K. Beyreuther, *Proc. Natl. Acad. Sci. USA*, *82*, 4245–4249 (1985).

243. H. Mori, K. Takio, M. Ogawara, and D. J. Selkoe, *J. Biol. Chem.*, *267*, 17082–17086 (1992).

244. E. Gowing, A. E. Roher, A. S. Woods, R. J. Cotter, M. Chaney, S. P. Little, and M. J. Ball, *J. Biol. Chem.*, *269*, 10987–10990 (1994).

245. C. Soto, M. C. Branes, J. Alvarez, and N. C. Inestrosa, *J. Neurochem.*, *63*, 1191–1198 (1994).

246. J. T. Jarrett and P. T. Lansbury, Jr., *Cell*, *73*, 1055–1058 (1993).

247. A. I. Bush, W. H. Pettingell, M. Paradis, Jr., and R. E. Tanzi, *J. Biol. Chem.*, *269*, 12152–12158 (1994).

248. X. Huang, C. S. Atwood, R. D. Moir, M. A. Hartshorn, J.-P. Vonsattel, R. E. Tanzi, and A. I. Bush, *J. Biol. Chem.*, *272*, 26464–26470 (1997).

249. A. L. Schwarzman, L. Gregori, M. P. Vitek, S. Lyubski, W. J. Strittmatter, J. J. Enghilde, R. Bhasin, J. Silverman, K. H. Weisgraber, P. K. Coyle, M. G. Zagorski, J. Talafous, M. Eisenberg, A. M. Saunders, A. D. Roses, and D. Goldgraber, *Proc. Natl. Acad. Sci. USA*, *91*, 8368 (1994).

250. A. I. Bush, K. Beyreuther, and C. L. Masters, *Pharmacol. Ther.*, *56*, 97–117 (1992).

251. A. I. Bush, G. Multhaup, R. D. Moir, T. G. Williamson, D. H. Small, B. Rumble, P. Pollwein, K. Beyreuther, and C. L. Masters, *J. Biol. Chem.*, *268*, 16109–16112 (1993).

252. A. Clements, D. Allsop, D. M. Walsh, and C. H. Williams, *J. Neurochem.*, *66*, 740–747 (1996).

253. F. S. Esch, P. S. Keim, E. C. Beattie, R. W. Blacher, A. R. Culwell, T. Oltersdorf, D. McClure, and P. J. Ward, *Science*, *248*, 1122–1124 (1990).

254. G. W. Roberts, S. M. Gentleman, A. Lynch, L. Murray, M. Lan-

don, and D. I. Graham, *J. Neurol. Neurosurg. Psychiatry, 57*, 419–425 (1994).

255. A. I. Bush, W. H. Pettingell, G. Multhaup, M. Paradis, Jr., J.-P. Vonsattel, J. F. Gusella, K. Beyreuther, C. L. Masters, and R. E. Tanzi, *Science, 265*, 1464–1467 (1994).

256. A. I. Bush, R. D. Moir, K. M. Rosenkranz, and R. E. Tanzi, *Science, 268*, 1921–1922 (1995).

257. P. W. Mantyh, J. R. Ghilardi, S. Rogers, E. DeMaster, C. J. Allen, E. R. Stimson, and J. E. Maggio, *J. Neurochem., 61*, 1171–1174 (1993).

258. W. P. Esler, E. R. Stimson, J. M. Jennings, J. R. Ghilardi, P. W. Mantyh, and J. E. Maggio, *J. Neurochem., 66*, 723–732 (1996).

259. A. M. Brown, D. M. Tummolo, K. J. Rhodes, J. R. Hofmann, J. S. Jacobsen, and J. J. Sonnenberg-Reines, *Neurochemistry, 69*, 1204–1212 (1997).

260. M. Kawahara, N. Arispe, Y. Kuroda, and E. Rojas, *Biophys. J., 73*, 67–75 (1997).

261. E. M. Johnstone, M. O. Chaney, F. H. Norris, R. Pascual, and S. P. Little, *Mol. Brain Res., 10*, 299–305 (1991).

262. B. D. Shivers, C. Hilbich, G. Multhaup, M. Salbaum, K. Beyreuther, and P. H. Seeburg, *EMBO J., 7*, 1365–1370 (1988).

263. L. Hesse, D. Beher, C. L. Masters, and G. Multhaup, *FEBS Lett., 49*, 109–116 (1994).

264. J. Busciglio, D. H. Gabuzda, P. Matsudaira, and B. A. Yankner, *Proc. Natl. Acad. Sci. USA, 90*, 2092–2096 (1993).

265. D. Schubert and M. Chevion, *Biochem. Biophys. Res. Commun., 216*, 702–707 (1995).

266. C. Exley, N. C. Price, S. M. Kelly, and J. D. Birchall, *FEBS Lett., 324*, 293–295 (1993).

267. S. J. Tomski and R. M. Murphy, *Arch. Biochem. Biophys., 294*, 630–638 (1992).

268. C. Hilbich, B. Kisters-Woike, J. Reed, C. L. Masters, and K. Beyreuther, *J. Mol. Biol., 218*, 149–163 (1991).

269. D. Burdick, B. Soreghan, M. Kwon, J. Kosmoski, M. Knauer, A. Henschen, J. Yates, C. Cotman, and C. Glabe, *J. Biol. Chem., 267*, 546–554 (1992).

270. T. Dyrks, E. Dyrks, T. Hartmann, C. L. Masters, and K. Beyreuther, *J. Biol. Chem.*, *267*, 18210–18217 (1992).

271. T. Dyrks, E. Dyrks, C. L. Masters, and K. Beyreuther, *FEBS Lett.*, *324*, 231–236 (1993).

272. C. S. Atwood, X. Huang, R. D. Moir, R. C. Scarpa, N. M. E. Bacarra, M. A. Hartshorn, L. E. Goldstein, D. M. Romano, R. E. Tanzi, and A. I. Bush, *Soc. Neurosci.*, *27*, 731 (1997).

273. C. Behl, J. Davis, G. M. Cole, and D. Schubert, *Biochem. Biophys. Res. Commun.*, *186*, 944–950 (1992).

274. H. Fabian, G. I. Szendrei, H. H. Mantsch, B. D. Greenberg, and L. Otvos, Jr., *Eur. J. Biochem.*, *221*, 959–964 (1994).

275. K. Hensley, M. Aksenova, J. M. Carney, M. Harris, and D. A. Butterfield, *Neuroreport*, *6*, 493–496 (1995).

276. C. A. Owen, Jr., *Proc. Soc. Exp. Biol. Med.*, *149*, 681–682 (1975).

277. J. K. Brieland and J. C. Fantone, *Arch. Biochem. Biophys.*, *284*, 78–83 (1991).

278. D. J. Lamb and D. S. Leake, *FEBS Lett.*, *338*, 122–126 (1994).

279. H. Fliss and M. Menard, *Arch. Biochem. Biophys.*, *287*, 175–179 (1991).

280. K. N. Allen, A. Lavie, A. Glasfeld, T. N. Tanada, D. P. Gerrity, S. C. Carlson, G. K. Farber, G. A. Petsko, and D. Ringe, *Biochemistry*, *33*, 1488–1494 (1994).

281. W. Maret, *Proc. Natl. Acad. Sci. USA*, *91*, 237–241 (1994).

282. W. D. Parker Jr., J. Parks, C. M. Filley, and B. K. Kleinschmidt-DeMasters, *Neurology*, *44*, 1090–1096 (1994).

283. J. W. Baynes, *Diabetes*, *40*, 405–412 (1991).

284. S. D. Yan, S. F. Yan, X. Chen, J. Fu, P. Kuppusamy, M. A. Smith, G. Perry, G. C. Godman, P. Nawroth, J. L. Zweier, and D. Stern, *Nature Med.*, *1*, 693–699 (1995).

285. C. A. Colton, J. Snell, O. Chernyshev, and D. L. Gilbert, *Ann. NY Acad. Sci.*, *738*, 54–63 (1994).

286. J. M. Gutteridge, G. J. Quinlan, I. Clark, and B. Halliwell, *Biochim. Biophys. Acta*, *835*, 441–447 (1985).

287. C. Behl, J. B. Davis, R. Lesley, and D. Schubert, *Cell*, *77*, 817–827 (1994).

288. Y. Goodman and M. P. Mattson, *Exp. Neurol.*, *128*, 1–12 (1994).

289. D. A. Butterfield, K. Hensley, M. Harris, M. Mattson, and J. Carney, *Biochem. Biophys. Res. Commun.*, *200*, 710–715 (1994).

290. D. A. Butterfield, L. Martin, J. M. Carney, and K. Hensley, *Life Sci.*, *58*, 217–228 (1996).

291. K. Hensley, D. A. Butterfield, N. Hall, P. Cole, R. Subramaniam, R. Mark, M. P. Mattson, W. R. Markesbery, M. E. Harris, M. Aksenov, M. Aksenova, J. F. Wu, and J. M. Carney, *Ann. NY Acad. Sci.*, *786*, 120–134 (1996).

292. R. Subramanim, B. J. Howard, K. Hensley, M. Aksenova, J. M. Carney, and D. A. Butterfield, *Alzheimer's Res.*, *1*, 141 (1996).

293. T. Tomiyama, A. Shoji, K. Kataoka, Y. Suwa, S. Asano, H. Kaneko, and N. Endo, *J. Biol. Chem.*, *271*, 6839–6844 (1996).

294. X. Huang, C. S. Atwood, L. E. Goldstein, M. A. Hartshorn, R. D. Moir, G. Multhaup, R. E. Tanzi, and A. I. Bush, *Soc. Neurosci.*, *27*, 648 (1997).

295. T. Thomas, G. Thomas, C. Mclendon, T. Sutton, and M. Mullan, *Nature*, *380*, 168–171 (1996).

296. M. P. Mattson and Y. Goodman, *Brain Res.*, *676*, 219–224 (1995).

297. B. Rumble, R. Retallack, C. Hilbich, G. Simms, G. Multhaup, R. Martins, A. Hockey, P. Montgomery, K. Beyreuther, and C. L. Masters, *N. Engl. J. Med.*, *320*, 1446–1452 (1989).

298. J. K. Teller, C. Russo, L. M. DeBusk, G. Angelini, D. Zaccheo, F. Dagna-Bricarelli, P. Scartezzini, S. Bertolini, D. M. A. Mann, M. Tabaton, and P. Gambetti, *Nature Med.*, *2*, 93–95 (1996).

299. J. Busciglio and B. A. Yankner, *Nature*, *378*, 776–779 (1995).

300. J. D. Sipe, *Annu. Rev. Biochem.*, *61*, 947–975 (1992).

301. D. Schubert, C. Behl, R. Lesley, A. Brack, R. Dargusch, Y. Sagara, and H. Kimura, *Proc. Natl. Acad. Sci. USA*, *92*, 1989–1993 (1995).

302. A. D. Snow, R. T. Sekiguchi, D. Nochlin, R. N. Kalaria, and K. Kimata, *Am. J. Pathol.*, *144*, 337–347 (1994).

303. H. M. Wisniewski and J. Weigel, *Acta Neuropathol.*, *85*, 586–595 (1993).

304. P. Cras, M. Kawai, S. Siedlak, P. Mulvihill, P. Gambetti, D. Lowery, P. Gonzalez-DeWhitt, B. Greenberg, and G. Perry, *Am. J. Pathol.*, *137*, 241–246 (1990).

305. C. A. Colton and D. L. Gilbert, *FEBS Lett.*, *223*, 284–288 (1987).

306. C. M. Yates, J. Butterworth, M. C. Tennant, and A. Gordon, *J. Neurochem.*, *55*, 1624–1630 (1990).

307. C. Messier and M. Gagnon, *Behav. Brain Res.*, *75*, 1–11 (1996).

308. E. Karwatowska-Prokopczuk, E. Czarnowska, and A. Beresewicz, *Cardiovasc. Res.*, *26*, 58–66 (1992).

309. R. Kisilevsky, *Lab. Invest.*, *49*, 381–390 (1983).

310. P. D. Gorevic, A. B. Cleveland, and E. C. Franklin, *Ann. NY Acad. Sci.*, *389*, 380–393 (1982).

311. P. Velazquez, D. H. Cribbs, T. L. Poulos, and A. J. Tenner, *Nature Med.*, *3*, 77–79 (1997).

312. V. Giampaolo, F. Luigina, A. Conforti, and M. Roberto, in *Inflammatory Diseases and Copper* (J. R. J. Sorenson, ed.), Humana Press, Totowa, NJ, 1980, pp. 329–345.

313. L. A. Potempa, B. M. Kubak, and H. Gewurz, *J. Biol. Chem.*, *260*, 12142–12147 (1985).

314. P. Riederer, E. Sofic, W. D. Rausch, B. Schmidt, G. P. Reynolds, K. Jellinger, and M. B. Youdim, *J. Neurochem.*, *52*, 515–520 (1989).

315. K. S. Rajan, R. W. Colburn, and J. M. Davis, *Life Sci.*, *18*, 423–432 (1976).

316. J. Stockel, J. Safar, A. C. Wallace, F. E. Cohen, and S. B. Prusiner, *Biochemistry*, *37*, 7185–7193 (1998).

317. A. E. Roher, J. D. Lowenson, S. Clarke, C. Wolkow, R. Wang, R. J. Cotter, I. M. Reardon, H. A. Zürcher-Neely, R. L. Heinrickson, M. J. Ball, and B. D. Greenberg, *J. Biol. Chem.*, *268*, 3072–3083 (1993).

11

Metal Binding and Radical Generation of Proteins in Human Neurological Diseases and Aging

[1]Gerd Multhaup and [2]Colin L. Masters

[1]ZMBH-Center for Molecular Biology Heidelberg,
University of Heidelberg, D-69120 Heidelberg, Germany

[2]Department of Pathology, University of Melbourne,
and Neuropathology Laboratory, Mental Health
Research Institute of Victoria, Parkville,
Victoria, 3052, Australia

1. INTRODUCTION: TRANSITION METAL IONS IN BIOLOGICAL SYSTEMS

Organisms take great care in the handling of the transition metals iron and copper that can undergo oxidation-reduction reactions. Systems operating with high specificity use transport and storage proteins to minimize reactivity and availability and thus safeguard the cell against toxicity [1]. An adult human contains about 4 g of iron and 80–100 mg copper with both levels tightly regulated because of their essential yet toxic nature [2,3].

The major iron-binding proteins are hemoglobin in plasma, myoglobin, and the transport protein transferrin, which is only about 30% loaded with iron on average [4]. The body takes up iron from the diet by Nramp2, the first mammalian protein known to mediate iron uptake across cell membranes [5]. Iron is then stored intracellularly in an inactivated form bound to ferritin and hemosiderin. This mechanism requires a ferroxidase activity that is provided by ceruloplasmin, the synthesis of which was found to be upregulated by iron-deficient cells [6]. Thus, ceruloplasmin as the major copper-binding protein in plasma has been shown to be implicated in iron metabolism [6] that also has amine oxidase and superoxide dismutase (SOD) activities [2]. Recently, mutations in the ceruloplasmin gene causing a failure of iron export from various tissues [7] unraveled the iron/copper-linked metabolism. Copper deficiency resulted in a secondary iron deficiency and copper turned out to be the missing cofactor of ceruloplasmin.

An excellent model system for studies of metal ion import has been provided by yeast cells. Multicopper oxidases that are homologous to ceruloplasmin were identified in yeast such as Fet3p in *Saccharomyces cerevisiae* [8–10] that facilitates high-affinity iron uptake. In contrast to its human homolog ceruloplasmin, Fet3p has a single hydrophobic domain and is localized to the plasma membrane [9]. To explain transport across cell membranes a model has been proposed in which Fet3 and a permease, Ftr1p, constitute a molecular complex that is responsible for translocating iron either out of the cell or across organelle membranes [11]. A P-type ATPase encoded by the CCC2 gene is required for copper loading of the apoprotein form of Fet3p in a post-Golgi compartment [8,12].

Structural and functional homologs of Ccc2p P-type ATPases are associated with two inherited disorders of human copper metabolism: Menkes' syndrome [13–15] and Wilson's disease [16–18]. Both disorders have been linked to a defect in the export of copper from the cytosol [15,19] and a failure to deliver copper to ceruloplasmin in Wilson's disease. Progressive neurodegeneration and connective tissue disturbances are the main manifestations of Menkes' disease. The candidate genes for Wilson's disease and Menkes' syndrome encode membrane proteins that are homologous to bacterial membrane Cu(II) and Cd(II) transporters [19] and possess special mechanisms to regulate the cellular export of this transition metal [20,21]. Thus, in the area of iron transport and copper export the molecular mechanisms of iron and copper homeostasis in mammals are tightly linked and have started to unfold.

Nevertheless, the molecular basis for copper uptake in mammalian cells remains an enigma. At least, there have been some candidate copper uptake genes identified, hCTR1 and hCTR2 [22]. Since copper uptake must be carefully controlled, multiple uptake pathways may exist. Copper binding and reduction of the amyloid precursor protein of Alzheimer's disease [23,24], intracellular Cu trafficking as Cu(I), and internalization of amyloid precursor protein (APP) as the mode by which copper could be delivered into the cell are compatible with the increasing knowledge of copper transport [25]. Likewise, the prion protein can exist in a Cu-metalloprotein form in vivo. Prion protein gene-ablated mice have altered cellular phenotypes and show a reduction in the intracellular Cu/Zn-SOD activity [26]. This suggests that prion proteins might also be either directly or indirectly involved in copper uptake and/or intracellular copper transport.

Yeast cells acquire copper as Cu(I) through the action of the high-affinity copper uptake genes CTR1, CTR3, and FRE1, each of them involved in high-affinity copper transport in yeast [27]. Most interestingly, Ctr1p is a stable copper transport protein and exposure of yeast cells to copper concentrations of about 10 μM triggers degradation of cell surface Ctr1p. The CTR1 transcript level is regulated by copper and internalization of Ctr1p delivers copper into the cell [28], analogous to the uptake of the transferrin receptor and iron-bound transferrin from plasma [29].

2.   FREE RADICALS AND HUMAN DISEASES

Oxidative stress, a condition describing the production of oxygen radicals beyond a threshold for proper antioxidant neutralization, has been implicated in the pathogenesis of several neurodegenerative diseases, such as amyotrophic lateral sclerosis (ALS), Parkinson's disease, Alzheimer's disease (AD), and Huntington's disease, and in chronic manganese-induced neurotoxicity [30]. In transmissible animal and human prion diseases such as bovine spongiform encephalopathy (BSE), scrapie, and Creutzfeldt-Jakob disease, neuronal death and vacuolation are characteristics of central nervous system degeneration.

Much of the interest in the association of neurodegenerative disease with oxidative damage has emerged from the observation that transition metals normally present in the body may play a role in disease pathogenesis through free radical formation. Metal-induced oxidant stress can damage critical biological molecules and initiate a cascade of events including mitochondrial dysfunction, excitotoxicity, and a rise in cytosolic free calcium, leading to cell death.

The mode of neuronal cell death after damage by excitotoxins or free radicals, however, has remained controversial. One of the potentially neurotoxic events is the activation of nitric oxide synthase followed by the subsequent production of nitric oxide (NO) [31]. A mechanism has been proposed by which NO reacts with superoxide anion O_2^- to generate peroxynitrite ($ONOO^-$), resulting in dose-dependent neuronal damage [32,33]. Superoxide radical and hydrogen peroxide are produced in vivo and catalytic metal ions are available although in limited amounts to produce hydroxyl radicals in many normal biochemical reactions [34]. In general, when the balance between the production of these oxygen-derived species (such as O_2^-, H_2O_2, and $^\bullet OH$)

and antioxidant is disturbed, oxidative stress results [35]. Endogenous sources appear to account for most of the free radicals produced in the cellular aerobic metabolism [32]. Hydrogen peroxide and peroxynitrite, although not themselves free radicals, are important contributors to the cellular redox state. For instance, about 10^{12} O_2 molecules are processed by each rat cell daily. The leakage of partially reduced oxygen molecules from various components of the cellular electron transport chains is about 2%, yielding about 2×10^{10} O_2^- and H_2O_2 molecules per cell per day. Superoxide radicals are formed in almost all aerobic cells by a variety of cytosolic and membrane-bound enzymes, including xanthine oxidase, the cytochrome P450 complex, phospholipase A_2, and are produced during the respiratory burst of phagocytic cells [36]. It is well established that O_2^- is produced in human cells and is the precursor of H_2O_2 that is formed by SOD-catalyzed dismutation [37]. Hydrogen peroxide has a limited reactivity, but it can cross biological membranes, whereas O_2^- species move very slowly unless there is an anion channel through which they can travel [38,39]. Since superoxide species and H_2O_2 at physiological levels are poorly reactive (reviewed in [40]), the toxicity of O_2^- and H_2O_2 was proposed to derive from a conversion to the much more reactive ${}^\bullet OH$. Accordingly, an interaction between O_2^- and nitric oxide (${}^\bullet NO$) leading to ${}^\bullet OH$ could be demonstrated in vitro [32].

Copper(I) can catalyze the same reaction to form ${}^\bullet OH$ in the metal ion-catalyzed Haber-Weiss reaction [37] ($O_2^- + H_2O_2 \xrightarrow{Fe/Cu} {}^\bullet OH + OH^- + O_2$), or in the involvement of iron by the so-called superoxide-driven Fenton reaction, (Fe(II) + $H_2O_2 \rightarrow$ Fe(III) + ${}^\bullet OH$ + OH^-). The rate constant for a reaction of Cu(I) with H_2O_2 (Cu(I) + $H_2O_2 \rightarrow$ Cu(II) + ${}^\bullet OH$ + OH^-) is magnitudes higher than that for Fe(II) [37]. A Cu(III) (cupryl) intermediate has been suggested to be formed as well as, or instead of, ${}^\bullet OH$ and there is an ongoing controversy over the physiological significance of copper-dependent radical production in vivo (reviewed in [37]). Nevertheless, ${}^\bullet OH$ formation requires the presence of catalyzing metal ions. Damage might occur when such radicals react with the binding molecule and are not accessible to added scavengers. More generally, radicals are thought to be involved in lipid peroxidation of the cell membrane, leading to increased membrane fluidity and disturbance of divalent ion homeostasis [37].

3. SUPEROXIDE DISMUTASE IN FAMILIAL
   AMYOTROPHIC LATERAL SCLEROSIS: A GAIN
   OF FUNCTION

The most convincing link so far between neurological disorders and
oxygen radical formation is the strong association observed between
familial amyotrophic lateral sclerosis (FALS) and mutations in the Cu/
Zn-SOD gene. The human motor neuron disease ALS is characterized
by the degeneration of large motor neurons of the spinal cord, brain-
stem, and motor cortex. ALS occurs in sporadic and familial forms that
are clinically and pathologically similar [41]. Some pedigrees of autosomal dominant FALS have missense point mutations in the gene located
on chromosome 21, encoding cytosolic Cu/Zn superoxide dismutase-1
(SOD1) [42]. Currently, much evidence argues strongly that the disease
arises not from loss of SOD1 function but rather from an adverse or
novel property of the mutant SOD1 molecule [43,44]. Transgenic mice
that overexpress the human SOD1 gene develop a clinically analogous
form of motor neuron disease similar to human FALS [45]. An insight
into the mechanism was gained after it had been discovered that SOD is
inactivated by H_2O_2 that rapidly reduced Cu(II) at the active site [46,
47] and Cu/Zn-SOD was found to generate free $^{\bullet}OH$ radicals from H_2O_2
[48]. The most important findings were published by Wiedau-Pazos and
colleagues who reported that in an in vitro system FALS-associated
mutant SOD1 enzyme catalyzes the reduction of H_2O_2, thereby acting
as a peroxidase. Thus, Cu/Zn-SOD has a peroxidative function that
utilizes its own dismutation product, H_2O_2, as a substrate. This was
found to occur more rapidly with mutant than wild-type SOD1, with the
mutants at least twice as reactive to that of the wild-type enzyme.
These findings were confirmed by Yim et al. [48] who showed that the
same FALS mutant G93A of Cu/Zn-SOD, overexpressed in insect cells,
generates more free radicals during the peroxidase reaction. Using the
spin-trapping method it was found that the enhanced free radical-
generating function of the mutant is due to a decrease in K_m for hydrogen peroxide [48]. A proposed model by Wiedau-Pazos et al. [49] results
from the following two equations: $SOD\text{-}Cu^{2+} + H_2O_2 \rightarrow SOD\text{-}Cu^+ + O_2^-$
$+ 2H^+$ and $SOD\text{-}Cu^+ + H_2O_2 \rightarrow SOD\text{-}Cu^{2+}(^{\bullet}OH) + OH^-$. This suggests
that in the critical step H_2O_2 is reduced and accepts an electron from

Cu^+. In this process the Cu ion is bound by imidazole groups on neighboring histidine residues within the SOD molecule, as has been described for the amyloid precursor protein of Alzheimer's disease earlier [50]. This suggests that oxygen radicals might be responsible for the selective degeneration of motor neurons occurring in this fatal disease [51]. One specific example of a Cu-dependent gain of function would be disruption in intracellular Cu trafficking. At least three genes have been identified that genetically interact with SOD and influence the intracellular trafficking of Cu (for a review, see [44]). Their gene products are clearly involved in metal homeostasis and SOD mutations adversely affect Cu homeostasis, thus leading to dysfunction and death of neurons.

Both APP and SOD might be involved in Cu(II)/Cu(I) homeostasis. An interplay of APP and SOD could affect the metabolism of Cu(II), as has been suggested for mutant SOD in FALS. In agreement with this, neurons in trisomy 21 exhibit a three- to fourfold increase in intracellular reactive oxygen species [52]. Therefore, both SOD and APP genes that are overexpressed in Down syndrome may be responsible for the premature onset of AD [53,54].

4. RADICAL GENERATION OF AMYLOID PRECURSOR PROTEIN IN ALZHEIMER'S DISEASE

4.1. Metal Binding of APP

Of particular importance to AD is the binding of metal ions that are potentially neurotoxic. The amyloid precursor protein (APP) interacts specifically with Zn(II) and Cu(II) at two distinct sites in the ectodomain [50,55]. The two other known members of the APP gene family, APLP1 and APLP2 [56–59], have conserved binding sites for Zn(II) [55] whereas the Cu(II) site is only present in APLP2 [50]. Zn(II) and Cu(II) binding can influence APP conformation, stability, and homophilic interactions [23,50,55]. Since transmembrane APP binds copper(II), our working hypothesis is that APP could be involved in Cu homeostasis. The accumulation of APP in neurites, as it occurs in AD [60–62], may lead to disruption of Cu compartmentalization and hence to Cu toxicity [63]. APP not only binds copper ions but has the intrinsic activity to reduce specifically bound copper(II) to copper(I) [23].

Studies on metal ion binding to APP were performed to investigate an association between metals and APP metabolism. A novel Zn(II) binding motif was discovered in the cysteine-rich amino-terminal region of APP between residues 181–200 within exon 5 [55] and is distinct from the Zn(II) binding sites in the Aβ region [64]. Zinc(II) binding has been shown to modulate the functional properties of APP, possibly by enhancing its macromolecular conformation. Incubation of APP with Zn(II) increases binding of APP to heparin [65,66] and potentiates the inhibition of coagulation factor XIa by APP-KPI$^+$ isoforms [67,68]. Zn(II) binding also influences the cleavage of the APP molecule, potentially via α-secretase activity [55,69]. The in vivo influences of zinc on APP metabolism were tested in rats given a dietary Zn(II) supply. This caused an increase in membrane-associated forms and a reduction of soluble forms of APP [70]. The latter may be due to either altered APP trafficking or processing. The active transport of neuronal Zn(II) along cell processes or from cell to cell has been hypothesized [71].

4.2. APP Reduces Cu(II) to Cu(I)

Whereas Zn(II) is assumed to play a purely structural role, we found that APP binds Cu(II) with a dissociation constant of 10 nM at pH 7.5 and reduces it to Cu(I) [23,50]. Since Zn(II) exists exclusively in one oxidation state, only APP-Cu(II) complexes are sensitive to redox reactions. The APP copper binding motif corresponds to type II copper-binding proteins and is encompassed by residues APP 135–155 within exon 4. Incubation of human APP, bacterial fusion proteins of APP (TP-APP$_{770}$, TP-APP$_{N262}$), or a synthetic peptide that represents the APP copper binding site (APP 135–156) with physiological concentrations of Cu(II) in the presence of the Cu(I) indicator molecule bathocuproine disulfonate (BC) [23,50] led to a reduction of Cu(II) to Cu(I) as indicated by formation of a peak with maximal absorbance at 480 nm, characteristic of the BC-Cu(I) complex.

To map the expected cystine formed during the reduction of Cu(II) to Cu(I) in APP, proteolytic fragments of the purified fusion protein APP (APP$_{N262}$) were used containing the previously identified Cu(II) binding site of APP residing within APP residues 135–155

[23,63]. A digestion of the copper-oxidized and carboxymethylated APP with endoprotease Asp-N yielded one major peak followed by a minor one after affinity chromatography on copper(II)-charged chelating Sepharose. To confirm the participation of Cys-144 and Cys-158 in the redox reaction and cystine formation [23], electrospray ionization mass spectrometry (ESI-MS) was employed to obtain a complete analysis of the peptides eluting from the affinity column. These data indicated that one high-performance liquid chromatography (HPLC) fraction indeed contained a single oxidized peptide (APP residues 142–166) and thus permitted the assignment of a disulfide linkage between Cys-144 and Cys-158, differing in two mass units from the peptide in its reduced form [72].

Thus, the reduction of Cu(II) to Cu(I) by APP resulted in a corresponding oxidation of cysteines 144 and 158 in APP that involved an intramolecular reaction leading to a new disulfide bridge. This reaction was found to be very specific, as APP neither bound nor reduced Fe(III) and did not bind other metals such as Ni(II), Co(II), or Mg(II). Thus, Cu(II) binding leads to oxidative modification of APP resulting in cystine and Cu(I) formation, which indicates that APP has an in vitro function in electron transfer to Cu(II). APP thus also participates in electron transfer reactions. This transfer is Cu ion-mediated, has features of a redox reaction, and leads to intramolecular disulfide bond formation of APP, indicating that free sulfhydryl groups of APP are involved. The formation of APP-Cu(I) complexes suggests a fast electron transfer mechanism following direct binding of Cu(II) to APP. Since the oxidation of two cysteines forming one cystine liberates two electrons and since only one electron is required for Cu(II) reduction to Cu(I), this remaining electron enhances the production of hydroxyl radicals, which could then attack sites near the location of the metal. The implication of oxygen radical-induced neuronal damage raises the possibility that Cu-mediated toxicity contributes to neurodegeneration in Alzheimer's disease.

4.3. APP, Copper Transport, and Neuronal Degeneration

Copper is an essential trace element in the brain that is required for a number of enzyme activities including cytochrome c oxidase, Cu/Zn-SOD, and dopamine-β-hydroxylase (a key player in the catecholamine biosynthetic pathway in the nervous system) [73,74]. Zinc and copper ions are needed for thermally stable native Cu/Zn-SOD and to restore full catalytic activity. Whereas zinc can be substituted with cadmium, mercury, and cobalt ions [75], no metal can replace copper in restoring catalytic function.

Despite the essential role in electron transfer, copper is highly reactive and potentially toxic. Thus specialized pathways have evolved for the trafficking of this metal within cells [76]. The following data provide strong evidence that APP may play a role in copper transport. In neurons, APP is first delivered from the cell body to the axonal surface and then to the dendritic plasma membrane [71,77]. The function of APP transport in epithelial and neuronal cells is not known, but we propose that APP has an important physiological role in the cellular transport of the metal ions Zn(II) and Cu(II)/(I) in the periphery and in the central nervous system. Indirect evidence indicates that APP may also function in copper uptake during internalization. APP molecules are normally reinternalized from the cell surface via endocytosis signals in the cytoplasmic tail [78,79] and then processed to Aβ. The safe sequestration of transition metal ions is probably an important antioxidant defense in its own right. This is compatible with the presence of a single Cu(II)/Cu(I) binding site in APP [23] and the fact that organisms take great care in the handling of transition metal ions to minimize availability and hence reactivity.

APP-mediated reduction of bound Cu(II) to Cu(I) suggests that APP is acting as an extracellular copper reductase. Therefore, an ion-mediated redox reaction leads to disulfide bridge formation via free sulfhydryl groups in APP [23]. This is compatible with our knowledge of copper uptake and copper chaperone proteins that act as soluble cytoplasmic Cu(I) receptors in eukaryotic cells, suggesting that copper is taken up and transported as Cu(I) rather than Cu(II) [80,81]. Furthermore, sulfhydryl-modifying reagents inhibit the uptake of copper, implicating sulfur amino acids as required for the transport process in vivo [2,82]. A regulatory mechanism for copper uptake has been discovered in the copper-dependent turnover of Ctr1p protein [28]. When cells are grown in low concentrations of copper, Ctr1p is a stable protein. An increase in copper concentrations up to 10 μM induces cleavage of Ctr1p at the cell surface [28]. This cleavage is believed to release the extracellular domain of Ctr1p that contains the copper binding site and thus inhibits further copper uptake. Since this may represent a general mechanism for other plasma membrane proteins to regulate the uptake of ligands, the same function may be attributed to APP shedding by α-secretase cleavage of APP, which occurs on or close to the cell surface [83] to release a large amino-terminal fragment of APP and precludes formation of Aβ. This proteolytic pathway may inhibit copper uptake since the N-terminal domain of APP is extracellular and binds copper. The existence of an alternative and nonamyloidogenic α-secretase pathway for APP may indicate that APP could function as "mammalian Ctr1p" in brain and represent a rate-limiting component responsible for the delivery of copper across the membrane to the cytosolic copper chaperone proteins of mammalian cells [22]. In contrast, if cells are grown in low concentrations of copper the internalization of cell surface APP could be favored and the intracellular processing would be shifted to β- and γ-secretase cleavage of APP, i.e., the production of Aβ in Alzheimer's disease.

As mentioned before, two human inherited disorders illustrate the importance of copper in recessive disorders of copper metabolism, Wilson's disease and Menkes' disease. The defective genes encode membrane Cu-transporting P-type ATPases. A failure to express functional proteins can result in decreased copper efflux, whereas overexpression of the Menkes' gene will increase copper efflux and confer a stronger resistance to the toxic effects of copper [20]. Neuronal degeneration is a discrete clinical finding of Menkes' disease and also of familial amyotrophic lateral sclerosis.

In Alzheimer's disease, APP-Cu(I) complexes on the surface of neurons may be particularly vulnerable to peroxides generated by extracellular forms of Cu/Zn-SOD. Such complexes are spontaneously formed because APP itself reduces bound Cu(II) to Cu(I). The formation of APP-Cu(I) complexes occurs even in the absence of hydrogen peroxide [23]. Thus, sporadic AD could arise from a perturbation of free radical homeostasis and resulting neuronal toxicity by reactive oxygen species. This model is consistent with the slow onset of AD: younger persons may have greater antioxidant capacity and can withstand free radical stress. Aging coupled to environmental insults or genetic defects could exacerbate the consequences of APP fragmentation. Even the most prevalent risk factor associated with late onset AD has been shown to be linked to cytotoxicity modulated by the varying antioxidant activity of the apoE isoforms [84]. The E4 allele has a higher frequency in AD patients than in age-matched controls and was found to possess the lowest activity in protecting cells from hydrogen peroxide cytotoxicity [84].

5. INVOLVEMENT OF OXYGEN RADICALS IN Aβ AMYLOID FORMATION

When we investigated the site specificity of metal ion-catalyzed modification of cysteine residues in APP of Alzheimer's disease, the utilization of hydrogen peroxide led to C-terminal fragmentation of APP in a "peroxidative" reaction of APP-Cu(I) complexes. Thus, a possible mechanism for the relationship between the oxidation site and the higher order structure of APP could be proposed. Consequently, the results suggest that a cytotoxic gain of function of APP-Cu(I) complexes in the redox reaction with H_2O_2 might result in a perturbation of free radical homeostasis and/or lead to an accumulation of oxidized protein that is known to be associated with a number of diseases, including Alzheimer's [85,86].

ESI-MS and sequencing revealed that an intrachain disulfide bridge was formed between Cys-144 and Cys-158 according to an intrinsic activity of the primary sequence during the reduction of Cu(II) to Cu(I) bound to APP [72]. Fragmentations of APP-Cu(I) complexes were seen upon exposure of the protein to hydrogen peroxide. Electron paramagnetic resonance (EPR) studies showed that the copper binding site of oxidized APP contained EPR-silent Cu(I) which is, however, nullified in the presence $H_2O_2^-$. Thus, APP catalyzed the reduction of H_2O_2 in a "peroxidative" reaction. This indicated that Cu(I) in oxidized APP rapidly reoxidized to Cu(II) with H_2O_2. Therefore, the following reaction process may be proposed for the Cu(I)-catalyzed Fenton-like reaction of APP as the first step (APP-Cu(I) + H_2O_2 → APP-Cu(II)•OH + OH^-). The reaction scheme implies that a bound hydroxyl radical with Cu(II) (i.e., Cu(II)-•OH) but not free •OH was generated upon interaction of APP-Cu(I) complexes with H_2O_2. Indeed, an extensive fragmentation of intact full-length APP became apparent in the presence of copper and hydrogen peroxide [72]. A selective and time-dependent degradation of

H_2O_2-modified APP-Cu(I) complexes gave distinct peptide molecules with apparent molecular weights of 25 and 18 kD [72]. This suggests that H_2O_2 may cause a partial denaturation or unfolding in the copper binding site of APP that was observed to be antagonized in the presence of Zn(II). A denaturation of APP-Cu(I) by H_2O_2 may reveal previously shielded hydrophobic amino acid residues as newly preferred proteolytic substrates. Such fragments as the 18-kD fragment with 170 residues that was liberated by the reaction of APP-Cu(I) and H_2O_2 are potentially amyloidogenic and may be further degraded by intracellular peptidases to the amyloidogenic 100-amino-acid C-terminal fragment of APP [87]. Recently it was also reported that the intact Aβ domain can be generated by nonspecific proteases from such molecules [88].

Copper binding to APP and reduction leads to increased protein degradation as also had been observed for Cu/Zn-SOD [47,89]. The fragmentation process for Cu-modified APP was strongly inhibited by BC when added prior to H_2O_2 and the chelation of Cu(I) by BC blocked its initiation [72]. Here BC chelated and removed Cu(I) from the copper binding site and interfered with the ability of the protein to catalyze the peroxidation reaction of H_2O_2, implying that Cu(I) in APP participates in the mechanism. This indicates the importance of Cu(I) in the critical step, the reduction of H_2O_2, where H_2O_2 accepts an electron from Cu(I). A chelator for divalent metal ions, diethylenetriaminepentaacetic acid (DTPA), most likely interrupted the second step of the Fenton-type reaction with the Cu(II) ion (Cu(II) + $H_2O_2 \rightarrow$ Cu(I) + O_2^- + 2H$^+$). Thus, the oxidation of APP-Cu(I) complexes by H_2O_2 catalyzed by bound Cu(I) suggests that the reaction strictly occurs at the copper binding site of APP. This also implies a three-dimensional structural proximity between the site of APP-Cu(I)/H_2O_2-derived radical formation and their site of reaction (the cleavage site at the N terminus of the 18 kD fragment that is encoded by exon 14 of the APP gene) since radicals will react with most molecules present at their site of production. The postulated reaction occurs at physiological concentrations of H_2O_2 and perhaps may also occur at as low as nanomolar amounts. Aβ has been proposed to cause an overproduction of H_2O_2 or related peroxides [90] and thus may act on APP in a feedback reaction, thereby increasing oxidative stress. Most importantly, a copper-induced toxicity for neuronal cells is known to result from an increased production of hydrogen peroxide, radical production, and concomitant lipid oxidation that leads to neuronal cell death. Also, oxidative reactions of H_2O_2 may be highly vulnerable for forebrain neurons because there is a variant level of catalase activity reported in those cells [91]. Although the link between APP and Aβ toxicity remains obscure, our recent finding that the Aβ domain may serve as a functional axonal sorting signal for APP could provide the missing link in explaining the early accumulation of APP [92]. This accumulation could simply be due to a competition between Aβ and APP for the sorting receptor. Thus, the hypothesis of radical-based APP fragmentation and neurotoxicity of degradation products [85] opens novel therapeutic strategies for all forms of AD.

6. OXIDATIVE STRESS AND ANTIOXIDANTS

Free radical reactions are implicated in the pathogenesis of a growing number of disorders such as cancer, atherosclerosis, cardiovascular disease, and in a number of brain pathologies including brain ischemia, Parkinson's disease, FALS, and AD. Infections may also cause oxidant stress leading to antioxidant depletion by activated phagocytic cells.

Aerobic organisms are constantly exposed to oxygen radicals and related oxidants. Thus, damage to cellular proteins, lipids, nucleic acids, and carbohydrates can be observed under normal physiological conditions leading to degeneration of cellular components [93]. The damage of mitochondrial DNA by reactive oxygen species is believed to be a key step when a leakage of reactive oxygen species from the mito-chondrial respiratory chain attacks the genome and accelerates the formation of deletions and point mutations [94]. In addition to inevit-able byproducts of cellular processes, the membrane-associated NADPH oxidase releases O_2^- in the extracellular space [94]. Leukocytes and other phagocytic cells produce NO, O_2^-, H_2O_2, and OCl^- to destroy virus-infected cells. For instance, intracellular superoxide will find its targets such as NO that is produced by the NO synthase and serves diverse physiological functions throughout the body [95]. When NO reacts in a diffusion-limited manner with the superoxide anion peroxynitrite is generated followed by nitrosylation of proteins and lipid peroxida-tion [96].

The concentrations of these oxygen-centered species are normally kept in a harmless range by potent protective mechanisms. The adverse effects of free radical reactions are countered in part by a variety of enzymatic and nonenzymatic means. First, a sequestration of transition metals to prevent any free iron and copper ions in transferrin and ferritin or ceruloplasmin from participating in free radical reactions is one of the defense mechanisms. Second, an intracellular enzymatic antioxidant defense is provided by SOD, catalase, glutathione peroxidase, glutathione reductases, and S-transferase (reviewed in [97]). Third, nonenzymatic physiological antioxidants, such as ubiquinol, urate, bilirubin, and carnosine, help to protect against radical-induced damage. Among the dietary antioxidants are components such as vitamin E (tocopherol) that protects membranes against oxidative injury by donating a hydrogen atom to lipid peroxyl radicals [98], to vitamin C (ascorbate), as an essential cofactor of hydroxylation reactions, and to carotinoids.

Recently, it has been discovered that the Bcl-2 protooncogene product that is involved in the control of apoptotic response also has an antioxidant activity [99]. With the prerequisite that reactive oxygen species are required for apoptosis, Bcl-2 was suggested to prevent a decrease in antioxidant enzyme activity [100] and expression of Bcl-2 blocked the activation of caspases and generation of reactive oxygen species [101].

7. CONCLUSIONS

Rapid progress has been made in understanding the molecular pathogenesis of neurodegenerative diseases in recent years. As mentioned above, substantial evidence suggests that cellular dysfunctions and oxidative damage are major factors in the progression of these diseases as well as in normal aging.

A variety of therapeutic approaches are possible to improve mitochondrial function involving agents that act as free radical scavengers and neurotrophic factors in neurodegenerative disorders. In Alzheimer's patients with moderately severe impairment a treatment with Selegiline or α-tocopherol was found to slow the progression of the disease [102] as an oral antioxidant supplementation of vitamin E was suggested to protect against lipid peroxidation [103]. In Parkinson's disease alterations such as brain iron content, impaired mitochondrial function, and alterations in antioxidant protective systems are discussed as being reversible by drugs such as Selegiline [104]. An oral administration of coenzyme Q10 that acts as a cofactor of the electron transport chain exhibited antioxidant activities in Huntington's disease [105].

Although these findings suggest that antioxidants might be useful in treating neurodegenerative diseases, many questions remain. Nevertheless, the current concepts of neurodegenerative disorders as outlined here strongly suggest that antioxidants or metal chelators may play a key role in preventing or slowing these diseases.

ACKNOWLEDGMENTS

We acknowledge the support of the Deutsche Forschungsgemeinschaft (DFG) through SFB317, the BMBF through grant 030666A, and the National Health and Medical Research Council of Australia.

ABBREVIATIONS

Aβ	amyloid Aβ peptide
AD	Alzheimer's disease
ALS	amyotrophic lateral sclerosis
APP	amyloid precursor protein
BC	bathocuproine disulfonate
BSE	bovine spongiform encephalopathy
DTPA	diethylenetriaminepentaacetic acid
EPR	electron paramagnetic resonance
ESI-MS	electrospray ionization mass spectrometry
FAD	familial Alzheimer's disease
FALS	familial amyotrophic lateral sclerosis
HPLC	high performance liquid chromatography
KPI	Kunitz-type inhibitory domain
NADPH	nicotinamide adenine dinucleotide phosphate (reduced)
SOD	superoxide dismutase

REFERENCES

1. B. Halliwell and J. M. Gutteridge, *Arch. Biochem. Biophys.*, *246*, 501–514 (1986).
2. E. D. Harris, *Proc. Soc. Exp. Biol. Med.*, *196*, 130–140 (1991).
3. B. Halliwell and J. M. Gutteridge, *Mol. Aspects Med.*, *8*, 89–193 (1985).
4. J. M. Gutteridge, S. K. Paterson, A. W. Segal, and B. Halliwell, *Biochem. J.*, *199*, 259–261 (1981).
5. H. Gunshin, B. Mackenzie, U. V. Berger, Y. Gunshin, M. F. Romero, W. F. Boron, S. Nussberger, J. L. Gollan, and M. A. Hediger, *Nature*, *388*, 482–488 (1997).
6. C. K. Mukhopadhyay, Z. K. Attieh, and P. L. Fox, *Science*, *279*, 714–717 (1998).
7. X. Wang, F. Manganaro, and H. M. Schipper, *J. Neurochem.*, *64*, 1868–1877 (1995).
8. D. S. Yuan, R. Stearman, A. Dancis, T. Dunn, T. Beeler, and R. D. Klausner, *Proc. Natl. Acad. Sci. USA*, *92*, 2632–2636 (1995).
9. D. M. De Silva, C. C. Askwith, D. Eide, and J. Kaplan, *J. Biol. Chem.*, *270*, 1098–1101 (1995).
10. C. Askwith, D. Eide, A. Van Ho, P. S. Bernard, L. Li, S. Davis-Kaplan, D. M. Sipe, and J. Kaplan, *Cell*, *76*, 403–410 (1994).
11. R. Stearman, D. S. Yuan, Y. Yamaguchi-Iwai, R. D. Klausner, and A. Dancis, *Science*, *271*, 1552–1557 (1996).
12. D. Fu, T. J. Beller, and T. M. Dunn, *Yeast*, *11*, 283–292 (1995).
13. J. Chelly, Z. Tümer, T. Tonnesen, A. Petterson, Y. Ishikawa Brush, N. Tommerup, N. Horn, and A. P. Monaco, *Nature Genet.*, *3*, 14–19 (1993).
14. J. F. Mercer, J. Livingston, B. Hall, J. A. Paynter, C. Begy, S. Chandrasekharappa, P. Lockhart, A. Grimes, M. Bhave, D. Siemieniak, and T. W. Glover, *Nature Genet.*, *3*, 20–25 (1993).
15. C. Vulpe, B. Levinson, S. Whitney, S. Packman, and J. Gitschier, *Nature Genet.*, *3*, 7–13 (1993).
16. P. C. Bull, G. R. Thomas, J. M. Rommens, J. R. Forbes, and D. W. Cox, *Nature Genet.*, *5*, 327–337 (1993).
17. R. E. Tanzi, K. Petrukhin, I. Chernov, J. L. Pellequer, W. Wasco,

B. Ross, D. M. Romano, E. Parano, L. Pavone, L. M. Brzustowicz, M. Devoto, J. Peppercorn, A. I. Bush, I. Sternlieb, M. Pirastu, J. F. Gusella, O. Evgrafov, G. K. Penchaszadeh, B. Honig, I. S. Edelman, M. B. Soares, I. H. Scheinberg, and T. C. Gilliam, *Nature Genet.*, 5, 344–350 (1993).

18. Y. Yamaguchi, M. E. Heiny, and J. D. Gitlin, *Biochem. Biophys. Res. Commun.*, 197, 271–277 (1993).

19. M. Solioz and C. Vulpe, *Trends Biochem. Sci.*, 21, 237–241 (1996).

20. J. Camakaris, M. J. Petris, L. Bailey, P. Shen, P. Lockhart, T. W. Glover, C. Barcroft, J. Patton, and J. F. Mercer, *Hum. Mol. Genet.*, 4, 2117–2123 (1995).

21. M. J. Petris, J. F. Mercer, J. G. Culvenor, P. Lockhart, P. A. Gleeson, and J. Camakaris, *EMBO J.*, 15, 6084–6095 (1996).

22. B. Zhou and J. Gitschier, *Proc. Natl. Acad. Sci. USA*, 94, 7481–7486 (1997).

23. G. Multhaup, A. Schlicksupp, L. Hesse, D. Beher, T. Ruppert, C. L. Masters, and K. Beyreuther, *Science*, 271, 1406–1409 (1996).

24. G. Multhaup, A. Schlicksupp, L. Hesse, D. Beher, T. Ruppert, C. L. Masters, and K. Beyreuther, *Alzheimer's Dis. Biol.*, 529–535 (1997).

25. J. S. Valentine and E. B. Gralla, *Science*, 278, 817–818 (1997).

26. D. R. Brown, K. Qin, J. W. Herms, A. Madlung, J. Manson, R. Strome, P. E. Fraser, T. Kruck, A. von Bohlen, W. Schulz-Schaeffer, A. Giese, D. Westaway, and H. Kretzschmar, *Nature*, 390, 684–687 (1997).

27. S. Labbe, Z. Zhu, and D. J. Thiele, *J. Biol. Chem.*, 272, 15951–15958 (1997).

28. C. E. Ooi, E. Rabinovich, A. Dancis, J. S. Bonifacino, and R. D. Klausner, *EMBO J.*, 15, 3515–3523 (1996).

29. R. D. Klausner, J. Harford, and J. van Renswoude, *Proc. Natl. Acad. Sci. USA*, 81, 3005–3009 (1984).

30. S. A. Lipton and P. A. Rosenberg, *N. Engl. J. Med.*, 330, 613–622 (1994).

31. V. L. Dawson, T. M. Dawson, E. D. London, D. S. Bredt, and S. H. Snyder, *Proc. Natl. Acad. Sci. USA*, 88, 6368–6371 (1991).

32. J. S. Beckman, T. W. Beckman, J. Chen, P. A. Marshall, and B. A. Freeman, *Proc. Natl. Acad. Sci. USA*, 87, 1620–1624 (1990).

33. J. S. Beckman, J. Chen, H. Ischiropoulos, and J. P. Crow, *Meth. Enzymol.*, *233*, 229–240 (1994).

34. B. A. Freeman and J. D. Crapo, *Lab. Invest.*, *47*, 412–426 (1982).

35. H. Sies, *Eur. J. Biochem.*, *215*, 213–219 (1993).

36. I. Fridovich, *Annu. Rev. Biochem.*, *44*, 147–159 (1975).

37. B. Halliwell and J. M. Gutteridge, *Meth. Enzymol.*, *186*, 1–85 (1990).

38. M. A. Takahashi and K. Asada, *Arch. Biochem. Biophys.*, *226*, 558–566 (1983).

39. R. E. Lynch and I. Fridovich, *J. Biol. Chem.*, *253*, 4697–4699 (1978).

40. B. Halliwell and J. M. Gutteridge, *FEBS Lett.*, *307*, 108–112 (1992).

41. W. A. Horton, R. Eldridge, and J. A. Brody, *Neurology*, *26*, 460–465 (1976).

42. H. X. Deng, A. Hentati, J. A. Tainer, Z. Iqbal, A. Cayabyab, W. Y. Hung, E. D. Getzoff, P. Hu, B. Herzfeldt, R. P. Roos, C. Warner, G. Deng, E. Soriano, C. Smyth, H. E. Parge, A. Ahmed, A. D. Roses, R. A. Hallewell, M. A. Pericak-Vance, and T. Siddique, *Science*, *261*, 1047–1051 (1993).

43. R. H. Brown, Jr., *Cell*, *80*, 687–692 (1995).

44. P. C. Wong, C. A. Pardo, D. R. Borchelt, M. K. Lee, N. G. Copeland, N. A. Jenkins, S. S. Sisodia, D. W. Cleveland, and D. L. Price, *Neuron*, *14*, 1105–1116 (1995).

45. M. E. Gurney, H. Pu, A. Y. Chiu, M. C. Dal Canto, C. Y. Polchow, D. D. Alexander, J. Caliendo, A. Hentati, Y. W. Kwon, H. X. Deng, W. Chen, P. Zhai, R. L. Sufit, and T. Siddique, *Science*, *264*, 1772–1775 (1994).

46. E. K. Hodgson and I. Fridovich, *Biochemistry*, *14*, 5294–5299 (1975).

47. K. Sato, T. Akaike, M. Kohno, M. Ando, and H. Maeda, *J. Biol. Chem.*, *267*, 25371–25377 (1992).

48. M. B. Yim, J. H. Kang, H. S. Yim, H. S. Kwak, P. B. Chock, and E. R. Stadtman, *Proc. Natl. Acad. Sci. USA*, *93*, 5709–5714 (1996).

49. M. Wiedau-Pazos, J. J. Goto, S. Rabizadeh, E. B. Gralla, J. A. Roe, M. K. Lee, J. S. Valentine, and D. E. Bredesen, *Science*, *271*, 515–518 (1996).

50. L. Hesse, D. Beher, C. L. Masters, and G. Multhaup, *FEBS Lett.*, *349*, 109–116 (1994).

51. D. R. Rosen, T. Siddique, D. Patterson, D. A. Figlewicz, P. Sapp, A. Hentati, D. Donaldson, J. Goto, J. P. O'Regan, H. X. Deng, Z. Rahmani, A. Krizus, D. McKenna-Yasek, A. Cayabyab, S. M. Gaston, R. Berger, R. E. Tanzi, J. J. Halperin, B. Herzfeldt, R. Van den Bergh, W. Y. Hung, T. Bird, G. Deng, D. W. Mulder, C. Smyth, N. G. Laing, E. Soriano, M. A. Pericak-Vance, J. Haines, G. A. Rouleau, J. S. Gusella, H. R. Horvitz, and R. H. Brown, Jr., *Nature*, *362*, 59–62 (1993).

52. J. Busciglio and B. A. Yankner, *Nature*, *378*, 776–779 (1995).

53. B. Rumble, R. Retallack, C. Hilbich, G. Simms, G. Multhaup, R. Martins, A. Hockey, P. Montgomery, K. Beyreuther, and C. L. Masters, *N. Engl. J. Med.*, *320*, 1446–1452 (1989).

54. C. J. Epstein, K. B. Avraham, M. Lovett, S. Smith, O. Elroy Stein, G. Rotman, C. Bry, and Y. Groner, *Proc. Natl. Acad. Sci. USA*, *84*, 8044–8048 (1987).

55. A. I. Bush, G. Multhaup, R. D. Moir, T. G. Williamson, D. H. Small, B. Rumble, P. Pollwein, K. Beyreuther, and C. L. Masters, *J. Biol. Chem.*, *268*, 16109–16112 (1993).

56. K. Paliga, G. Peraus, S. Kreger, U. Durrwang, L. Hesse, G. Multhaup, C. L. Masters, K. Beyreuther, and A. Weidemann, *Eur. J. Biochem.*, *250*, 354–363 (1997).

57. T. A. Bayer, K. Paliga, S. Weggen, O. D. Wiestler, K. Beyreuther, and G. Multhaup, *Acta Neuropathol. (Berl)*, *94*, 519–524 (1997).

58. W. Wasco, S. Gurubhagavatula, M. D. Paradis, D. M. Romano, S. S. Sisodia, B. T. Hyman, R. L. Neve, and R. E. Tanzi, *Nature Genet.*, *5*, 95–100 (1993).

59. H. von der Kammer, C. Loffler, J. Hanes, J. Klaudiny, K. H. Scheit, and I. Hansmann, *Genomics*, *20*, 308–311 (1994).

60. L. C. Cork, C. Masters, K. Beyreuther, and D. L. Price, *Am. J. Pathol.*, *137*, 1383–1392 (1990).

61. G. Giaccone, F. Tagliavini, G. Linoli, C. Bouras, L. Frigerio, B. Frangione, and O. Bugiani, *Neurosci. Lett.*, *97*, 232–238 (1989).

62. D. Wang and D. G. Munoz, *J. Neuropathol. Exp. Neurol.*, *54*, 548–556 (1995).

63. G. Multhaup, T. Ruppert, A. Schlicksupp, L. Hesse, D. Beher, C.

L. Masters, and K. Beyreuther, *Biochem. Pharmacol.*, *54*, 533–539 (1997).

64. A. I. Bush, W. H. Pettingell, G. Multhaup, M. d. Paradis, J. P. Vonsattel, J. F. Gusella, K. Beyreuther, C. L. Masters, and R. E. Tanzi, *Science*, *265*, 1464–1467 (1994).

65. G. Multhaup, A. I. Bush, P. Pollwein, and C. L. Masters, *FEBS Lett.*, *355*, 151–154 (1994).

66. G. Multhaup, *Biochimie*, *76*, 304–311 (1994).

67. Y. Komiyama, T. Murakami, H. Egawa, S. Okubo, K. Yasunaga, and K. Murata, *Thromb. Res.*, *66*, 397–408 (1992).

68. W. E. Van Nostrand, *Thromb. Res.*, *78*, 43–53 (1995).

69. Q. X. Li, G. Evin, D. H. Small, G. Multhaup, K. Beyreuther, and C. L. Masters, *J. Biol. Chem.*, *270*, 14140–14147 (1995).

70. S. Whyte, L. Jones, E. J. Coulson, R. D. Moir, A. I. Bush, K. Beyreuther, and C. L. Masters, *Alzheimer's Dis. Biol.*, 417–422 (1997).

71. M. Simons, E. Ikonen, P. J. Tienari, A. Cidarregui, U. Monning, K. Beyreuther, and C. G. Dotti, *J. Neurosci. Res.*, *41*, 121–128 (1995).

72. G. Multhaup, T. Ruppert, A. Schlicksupp, L. Hesse, E. Bill, R. Pipkorn, C. L. Masters, and K. Beyreuther, *Biochemistry* (in press).

73. M. C. Linder and M. Hazegh-Azam, *Am. J. Clin. Nutr.*, *63*, 797S–811S (1996).

74. L. C. Stewart and J. P. Klinman, *Annu. Rev. Biochem.*, *57*, 551–592 (1988).

75. K. M. Beem, W. E. Rich, and K. V. Rajagopalan, *J. Biol. Chem.*, *249*, 7298–7305 (1974).

76. I. H. Hung, M. Suzuki, Y. Yamaguchi, D. S. Yuan, R. D. Klausner, and J. D. Gitlin, *J. Biol. Chem.*, *272*, 21461–21466 (1997).

77. T. Yamazaki, D. J. Selkoe, and E. H. Koo, *J. Cell Biol.*, *129*, 431–442 (1995).

78. C. Haass, M. G. Schlossmacher, A. Y. Hung, C. Vigo Pelfrey, A. Mellon, B. L. Ostaszewski, I. Lieberburg, E. H. Koo, D. Schenk, D. B. Teplow, and D. J. Selkoe, *Nature*, *359*, 322–325 (1992).

79. E. H. Koo and S. L. Squazzo, *J. Biol. Chem.*, *269*, 17386–17389 (1994).

80. V. C. Culotta, L. W. Klomp, J. Strain, R. L. Casareno, B. Krems, and J. D. Gitlin, *J. Biol. Chem.*, *272*, 23469–23472 (1997).

81. R. A. Pufahl, C. P. Singer, K. L. Peariso, S. Lin, P. J. Schmidt, C. J. Fahrni, V. C. Culotta, and J. E. Penner-Hahn, *Science*, *278*, 853–856 (1997).

82. S. S. Percival and E. D. Harris, *J. Nutr.*, *119*, 779–784 (1989).

83. S. S. Sisodia, E. H. Koo, K. Beyreuther, A. Unterbeck, and D. L. Price, *Science*, *248*, 492–495 (1990).

84. M. Miyata and J. D. Smith, *Nature Genet.*, *14*, 55–61 (1996).

85. M. R. Kozlowski, A. Spanoyannis, S. P. Manly, S. A. Fidel, and R. L. Neve, *J. Neurosci.*, *12*, 1679–1687 (1992).

86. B. S. Berlett and E. R. Stadtman, *J. Biol. Chem.*, *272*, 20313–20316 (1997).

87. T. Dyrks, E. Dyrks, T. Hartmann, C. Masters, and K. Beyreuther, *J. Biol. Chem.*, *267*, 18210–18217 (1992).

88. L. O. Tjernberg, J. Naslund, J. Thyberg, S. E. Gandy, L. Terenius, and C. Nordstedt, *J. Biol. Chem.*, *272*, 1870–1875 (1997).

89. T. Ookawara, N. Kawamura, Y. Kitagawa, and N. Taniguchi, *J. Biol. Chem.*, *267*, 18505–18510 (1992).

90. C. Behl, J. B. Davis, R. Lesley, and D. Schubert, *Cell*, *77*, 817–827 (1994).

91. S. Moreno, E. Mugnaini, and M. P. Ceru, *J. Histochem. Cytochem.*, *43*, 1253–1267 (1995).

92. P. J. Tienari, N. Ida, E. Ikonen, M. Simons, A. Weidemann, G. Multhaup, C. L. Masters, C. G. Dotti, and K. Beyreuther, *Proc. Natl. Acad. Sci. USA*, *94*, 4125–4130 (1997).

93. R. E. Pacifici and K. J. Davies, *Gerontology*, *37*, 166–180 (1991).

94. I. Fridovich, *Annu. Rev. Biochem.*, *64*, 97–112 (1995).

95. H. H. Schmidt and U. Walter, *Cell*, *78*, 919–925 (1994).

96. H. Y. Yun, V. L. Dawson, and T. M. Dawson, *Crit. Rev. Neurobiol.*, *10*, 291–316 (1996).

97. B. N. Ames, M. K. Shigenaga, and T. M. Hagen, *Proc. Natl. Acad. Sci. USA*, *90*, 7915–7922 (1993).

98. P. B. McCay, *Annu. Rev. Nutr.*, *5*, 323–340 (1985).

99. F. Virgili, M. P. Santini, R. Canali, R. R. Polakowska, A. Haake, and G. Perozzi, *Free Rad. Biol. Med.*, *24*, 93–101 (1998).

100. M. M. Briehl, A. F. Baker, L. M. Siemankowski, and J. Morreale, *Oncol. Res.*, *9*, 281–285 (1997).

101. J. B. Schulz, P. L. Huang, R. T. Matthews, D. Passov, M. C. Fishman, and M. F. Beal, *J. Neurochem.*, *67*, 430–433 (1996).

102. M. Sano, C. Ernesto, R. G. Thomas, M. R. Klauber, K. Schafer, M. Grundman, P. Woodbury, J. Growdon, D. W. Cotman, E. Pfeiffer, L. S. Schneider, and L. J. Thal, *N. Engl. J. Med.*, *336*, 1216–1222 (1997).

103. L. J. McIntosh, M. A. Trush, and J. C. Troncoso, *Free Rad. Biol. Med.*, *23*, 183–190 (1997).

104. P. Jenner and C. W. Olanow, *Neurology*, *47*, S161–S170 (1996).

105. M. F. Beal and R. T. Matthews, *Mol. Aspects Med.*, *18*(Suppl.), S169–S179 (1997).

12

Thiyl Radicals in Biochemically Important Thiols in the Presence of Metal Ions

Hans-Jürgen Hartmann, Christian Sievers, and Ulrich Weser

Anorganische Biochemie, Physiologisch-chemisches Institut der Universität Tübingen, D-72076 Tübingen, Germany

1. INTRODUCTION

Among the biochemically relevant free radicals sulfur-centered species play an important role in many physiological and pathophysiological reactions [1,2]. Endogenous thiols are prominent components that control cellular damage induced by transiently formed reduced oxygen species, including $^{\bullet}OH$, $O_2^{\bullet-}$, and H_2O_2. Thiols are also present in the active site of many enzymes and are frequently involved in electron transfer reactions. The most prominent sulfur-centered radical is the thiyl radical ($RS^{\bullet}$) generated by hydrogen atom abstraction from the corresponding thiol by hydroxyl radicals or carbon-centered radicals [1,3]. The latter process is known as the so-called repair reaction [1]. Thiyl radicals may also be produced by a one-electron reduction involving transition metal ions (Cu^{2+}, Fe^{3+}, etc.) during thiol oxidation to disulfide or its oxygenated forms [1,4]. In addition, thiyl radicals are able to act as oxidizing agents, e.g., on transition metal complexes [5]. Besides thiyl radicals a variety of other sulfur-centered radical species including perthiyl radicals ($RSS^{\bullet}$), the radical anion ($RSSR$)$^{\bullet-}$, the radical cations $R_2S^{\bullet+}$, $(R_2S)_2^{\bullet+}$ and ($RSSR$)$^{\bullet+}$ are described [1]. Sulfoxyl species, including $RSO^{\bullet}$, $RSOO^{\bullet}$, $RSO_2^{\bullet}$, and $RSO_2OO^{\bullet}$ are also known [6,7]. Thiyl radicals have been comprehensively dealt with elsewhere [1,2,8,9].

This contribution is focused on thiyl radicals associated with transition metal ions some of which are able to stabilize these radicals in biological systems. Of special interest was the reactivity of iron, cobalt, and manganese. Chromium and vanadium toxicity has been found to be related with the generation of thiyl radical intermediates. A major section is devoted to copper-thiolate reactions which have been examined in more detail in the authors' laboratory.

2. THIYL RADICALS

A common feature of ionic sulfur-centered radicals is their intense electronic absorption in the visible wavelength region. Due to this phenomenon pulse radiolysis was extensively used [10]. This method relies mainly on the direct observation of optical and other physical properties of the radicals. Unfortunately, thiyl radicals are extremely short-lived species and exhibit only a very low absorption between 300 and 400 nm with some exceptions. For instance, the penicillamine radical (PenS•) and the perthiyl radical (RSS•) absorb with reasonably high intensities at 330 and 380 nm, respectively [1,6]. Due to the lack of nuclear spin in the main sulfur isotope ^{32}S the application of electron paramagnetic resonance (EPR) spectroscopy is limited for sulfur-centered radicals including the thiyl species. Nevertheless, indirect identification of thiyl radicals is commonly achieved by EPR spin trapping. Direct EPR measurements are relatively rare although they would be highly desirable.

2.1. Indirect EPR Detection by Spin Trapping

In general, the measurement of thiyl radicals is monitored by means of EPR spin trapping. 5,5-Dimethyl-1-pyrroline-N-oxide (DMPO), the tetramethyl derivative tetramethylpyrroline oxide (TMPO), and N-t-butyl-α-phenylnitrone (PBN) are frequently used as spin traps [6,11,12]. Proof of the existence of thiyl radicals in biological systems has been provided by these spin-trap experiments [13–15]. The radical adducts which are stable for minutes exhibit specific coupling constants in their EPR spectra. However, the identification of different species in a mixture of radicals cannot be easily achieved. Thus, for example, if RS• and $O_2^{•-}$ are simultaneously present, overlapping of the signals prevents an exact assignment to a distinct radical species.

2.2. Direct EPR Measurements

Direct identification of the various sulfur-centered radicals by EPR is difficult. In the case of the thiyl species the direct detection is possible under special conditions. Oxidation of thiols with both Ce(IV)-H_2O_2

and Ti(III)-H_2O_2 in aqueous solution and immediate freezing to 77 K generates EPR spectra that are assigned to the formation of thiyl radicals [6,16]. Gamma irradiation of thiols at 77 K also resulted in EPR-detectable thiyl radicals [17,18]. Similarly, thiyl radicals were generated from the thiolate sulfur via oxidation by photochemically formed superoxide at 77 K. The radical species are detectable during controlled heating to 150–200 K at g = 2.036 [19].

3. THIYL RADICALS IN THE PRESENCE OF IRON, COBALT, NICKEL, MANGANESE, CHROMIUM, AND VANADIUM

There are several reports on the reactivity of heme iron with thiol compounds where thiyl radicals are suggested to be involved. In this context, the oxidation of cysteine by horseradish peroxidase in the presence of oxygen forms a thiyl free radical as demonstrated with the spin-trapping EPR technique [15]. Thiyl and ascorbate radicals from glutathione (GSH) and ascorbate, respectively, were also detected during the peroxidase-catalyzed formation of a crystal violet radical [20]. Furthermore, horseradish peroxidase-catalyzed metabolism of the analgesic acetaminophen occurs via a one-electron mechanism [21]. When either GSH, cysteine, or N-acetylcysteine were present the thiols reduced the acetaminophen-derived radicals to generate the thiyl species that where detected by EPR spin trapping. A similar reactivity was observed during prostaglandin synthase-catalyzed oxidation of the drug in the presence of the thiols. Clozapine-induced agranulocytosis is suggested from the oxidation of the drug by human myeloperoxidase and horseradish peroxidase [22]. Evidence for one-electron transfer reactions or the intermediate formation of a clozapine radical in the course of the peroxidase-mediated metabolism of the antipsychotic drug was derived from the observation of thiyl and ascorbyl radicals in the presence of GSH and ascorbate, respectively. In addition, the phenolic metabolites of polycyclic aromatic hydrocarbons enhanced the peroxidase-mediated formation of thiyl radicals [23]. In this study, GSH was shown to be oxidized by the enzyme to its thiyl species that can react with a number of chemicals, including the carcinogenic metabolite benzopyrene-7,8-dihydrodiol to give GSH conjugates. The latter were suggested to contribute to the detoxification of these chemicals. A thiyl radical has also been postulated during the GSH-dependent reduction of peroxides during ferryl myoglobin and metmyoglobin interconversion [24]. In a study from the same group the reactivity of several thiols and disulfides with different redox states of myoglobin was investigated [25]. The thiyl radicals of GSH, cysteine, and N-acetylcysteine were formed in the course of the interaction with ferryl myoglobin and detected by EPR in conjugation with the spin-trap DMPO.

It was of interest that methemoglobin may function as a biological Fenton reagent in the presence of H_2O_2 [26]. A peroxidase activity of the methemoglobin/H_2O_2 system was suggested by the oxidation of DMPO. When either GSH or ascorbate served as an electron donor, the corresponding thiyl and ascorbate radicals were formed when competing with the DMPO substrate. A hemoglobin thiyl free radical adduct in the blood of rats dosed with DMPO and hydroperoxides or H_2O_2 was detected in vivo [27]. Pretreatment of the rats with buthionine sulfoximine or diethylmaleate as well as a GSH reductase inhibitor significantly enhanced the radical adduct concentration. The same hemoglobin thiyl radical adduct was measured in the blood of rats following administration of either aniline, phenylhydroxyaniline, nitrosobenzene, or nitrobenzene. By using red blood cells or hemoglobin preparations similar results were obtained in vitro [28].

Evidence for a one-electron oxidation of estradiol into its reactive phenoxyl radical intermediate by heme iron-containing lactoperoxidase was deduced [29]. The metabolite abstracts hydrogen from reduced GSH generating GS•, which is spin-trapped by DMPO. In the absence of DMPO molecular oxygen is consumed by a sequence of reactions initiated by GS•. Similarly, the phenoxyl radical abstracts hydrogen from NADH to generate the NAD radical, which reduces molecular oxygen to superoxide. Subsequent intracellular accumulation of H_2O_2 is suggested to explain the hydroxyl radical-induced DNA base lesions reported for female breast cancer.

Irreversible inactivation of lactoperoxidase by mercaptomethylimidazole was postulated through the generation of a thiyl radical. The one-electron oxidation produced by the inhibitor was identified as the EPR-detectable DMPO adduct of the sulfur-centered radicals [30]. The GSH-dependent reduction of ferryl leghemoglobin during the reaction of ferric or ferrous leghemoglobin with H_2O_2 is associated with the formation of GS* as demonstrated by EPR measurements [31]. A variety of thiols reacted in a similar way [32]. The formation of a novel sulfur species formed by nucleophilic attack of the thiol group on the tetrapyrrole ring was shown concomitant with the formation of thiyl radicals that have been detected by EPR spin trapping.

During the cytochrome P450-catalyzed conversion of aldehydes to hydrocarbons a mechanism was proposed in which the perferryl iron oxene, resulting from heterolytic cleavage of the O−O bond of the iron peroxy intermediate, abstracts one electron from the C=O double bond of the carbonyl group of the aldehyde [33]. The reduced perferryl attacks the 1-carbon of the aldehyde to form a thiyl-iron-hemiacetal. The latter intermediate can fragment to form an alkyl radical and a thiyl-iron-formyl radical. The alkyl radical then abstracts the formyl hydrogen to produce the hydrocarbon and carbon dioxide. A further heme-thiol interaction has been described using the N-acetyl heme octapeptide from cytochrome c N-acetyl microperoxidase 8 and cysteine [34]. From the reaction between these compounds it is suggested that reduction of Fe(III) to Fe(II) by bound thiol produces thiyl radicals.

Investigation of the function and properties of the iron-sulfur center in spinach ferredoxin:thioredoxin reductase indicated the occurrence of a thiyl radical intermediate also in a nonheme iron protein with a $[4Fe-4S]^{2+}$ cluster [35]. The results from spectroscopic experiments argue against a role for the cluster in mediating electron transport from ferredoxin to the active site disulfide. An alternative role for the cluster in stabilizing the one-electron-reduced intermediate is suggested by spectroscopic studies of a modified form of the enzyme in which one of the cysteines of the active site dithiol has been alkylated with N-ethyl-maleimide. Spectroscopic and redox properties of the modified enzyme are interpreted in terms of a $[4Fe-4S]^{2+}$ cluster covalently attached through a cluster sulfide to a cysteine-based thiyl radical formed on one of the active site thiols. A mechanistic scheme for the enzyme is proposed with similarities to that established for the well-characterized NAD(P)H-dependent flavin-containing disulfide oxidoreductases, but involving sequential one-electron redox processes with the role of the $[4Fe-4S]^{2+}$ cluster being to stabilize the thiyl radical formed by the initial one-electron reduction of the active site disulfide. A new biological role for Fe-S clusters involving both the stabilization of a thiyl radical intermediate and cluster site-specific chemistry involving a bridging sulfide is suggested.

In contrast to the above mentioned protein-bound iron where thiyl radicals are observed during specific reactions, unspecific reactivities of thiol compounds with iron in a "Fenton-like" manner were reported. Thus, the effects of GSH on the generation of radicals and the oxidation of DNA in the presence of ferrous iron were studied [36]. Hydroxyl and thiyl radicals were detected by EPR spin trapping when GSH-Fe(II) reacted with H_2O_2. DMPO spin trapping was also used to detect thiyl radicals from thiol-containing angiotensin converting enzyme inhibitor, captopril, and from its stereoisomer epicaptopril in the process of $\cdot$OH radical scavenging [37]. These thiol compounds reduce Fe(III) with the concomitant formation of their respective thiyl radicals. At the same time, $\cdot$OH radicals were produced by a thiol-driven Fenton mechanism. The reaction of Fe(II)/H_2O_2 generated hydroxyl radicals or Ce(IV) with bovine serum albumin was of further interest [38]. During this radical-induced damage a protein-derived thiyl radical was selectively formed and characterized by DMPO spin trapping. According to the observation of an additional carbon-centered radical, it is suggested that damage can be transferred from the thiol group to carbon sites in the protein.

In the course of the reduction of Fe(III) bleomycin in the presence of cysteine or GSH, thiyl radical species derived from the corresponding thiols are detected by EPR spin trapping [39]. The iron chelator bleomycin is used as a DNA damaging antitumor drug in the form of the Fe(II) complex.

Cobalt-thiyl radical interactions are related essentially to class II ribonucleotide reductases. These enzymes operate with adenosyl cobalamin as the precursor of a putative transient thiyl protein radical [40]. In the class I enzymes that use O_2 and a diferric μ-oxo center to generate a stable tyrosyl radical the additionally occurring thiyl radical does not directly interact with the dinuclear iron center. Surprisingly, the critical cysteine residues in either catalytic domain of class I and II enzymes have been conserved even though the cofactor that generates the radical remains unaffected [41]. Evidence for a thiyl radical cob(II)-alamin interaction in *Lactobacillus leichmanii* class III enzyme was confirmed by rapid freeze-quench EPR [42,43] and by simulations of X- and Q-band EPR spectra of the radical intermediate [44a].

The methyl coenzyme M reductase from *Methanobacterium thermoautotrophicum* catalyzes the reduction of methyl coenzyme M ((2-methylthio)ethanesulfonate) with coenzyme B (7-thioheptanoylthreoninephosphate) to methane and the heterodisulfide of CoM (2-thioethanesulfonate) and CoB [44b]. In the course of this reaction in which the nickel porphinoid coenzyme F_{430} is involved, a mechanism is proposed that uses a CoM-derived thiyl radical intermediate.

During the thiol-mediated oxidation of nonphenolic lignin model compounds by manganese peroxidase it was found that in the presence of Mn(II), H_2O_2, and thiols the enzyme converts veratryl alcohol (3,4-dimethoxybenzyl alcohol), anisyl alcohol, and benzyl alcohol to their corresponding aldehydes [45]. It is suggested that the thiol is oxidized by enzyme-generated Mn(III) to a thiyl radical that abstracts a hydrogen from the substrate, forming a benzylic radical. The latter reacts with another thiyl radical to yield an intermediate that decays into the benzaldehyde product. A series of other compounds is likewise oxidized using the same system. Horseradish peroxidase has been shown to catalyze the oxidation of veratryl alcohol and benzyl alcohol to the respective aldehydes in the presence of GSH, $MnCl_2$, and an organic acid metal chelator such as lactate [46]. In addition to GSH, dithiothreitol, cysteine, and mercaptoethanol are capable of promoting veratryl alcohol oxidation. Spectral evidence indicates that horseradish peroxidase compound II is formed during the oxidation reaction. Furthermore, EPR studies indicate that GSH is oxidized to the thiyl radical. It is assumed that the ultimate oxidant of veratryl alcohol is an Mn(III)-GSH or Mn(II)-GS$^\bullet$ complex.

Nucleotide oxidation in the presence of a thiol and Mn(II) leads to thiyl radicals [47]. A purely chemical system for NAD(P)H oxidation to biologically active NAD(P)$^+$ has been developed and characterized. EDTA, Mn(II), and mercaptoethanol, combined at physiological pH, induce nucleotide oxidation through a chain length also involving molecular oxygen, which eventually undergoes quantitative reduction to H_2O_2. Mn(II) is specifically required for activity, while both EDTA and mercaptoethanol can be replaced by analogs. Optimal molar ratios of chelator/metal ion (2:1) yield an active coordination compound that catalyzes thiol autoxidation to thiyl radical. The transient formation of a nucleotide radical from the reaction between NAD(P)H and thiyl radicals is supposed.

Chromium and vanadium toxicity appears to be related to thiyl radical interconversions. In this context, Cr(VI) carcinogenesis was assumed to depend on the presence of cellular redox components including thiols that reduce the hexavalent metal ion into reactive species capable of interacting with DNA [48,49]. In vitro Cr(VI)-induced DNA damage was investigated in the presence of GSH, H_2O_2, or a mixture of both agents [50,51]. The reaction of Cr(VI) with GSH led to the formation of GS$^\bullet$ and two Cr(V)-GSH complexes that were suggested to react with H_2O_2 in a Fenton-type manner to produce hydroxyl radicals as the DNA-damaging agent. When Cr(VI) and GSH alone reacted with DNA Cr-DNA adducts were obtained with no detectable DNA strand breakage.

The in vivo toxicity of V(V) has been found to correlate with the depletion of cellular GSH and related nonprotein thiols [52]. The oxidation of GSH, cysteine, N-acetylcysteine, and penicillamine by V(V) was investigated. In the course of this process the corresponding thiyl radicals and V(IV) complexes were generated. The authors suggest that free radical reactions play a significant role in the depletion of cellular thiols by V(V) and hence in its toxicity.

4. COPPER(I)-STABILIZED THIYL RADICALS

4.1. Reactions of Superoxide Radicals with Thiol Compounds

Besides their ability to provide reduction equivalents thiols are known to react with radicals including $^\bullet$OH [53] and $O_2^{\bullet-}$ [54]. The imbalance of activated oxygen species is collectively known as oxidative stress. GSH is probably the most important intracellular antioxidant, present in millimolar concentrations [55]. Its protective role is dual. The indirect action is based on its supply of reducing equivalents to the enzymatically catalyzed redox cycling with peroxidases. The direct action of GSH is based on its reactivity as thiol with radicals. Pulse radiolytic measurements revealed a rate constant of 1.3×10^{10} M^{-1} s^{-1} for reaction of the hydroxyl radical with GSH [53]. The second-order rate constant (k_2) for the oxidation of GSH by $O_2^{\bullet-}$ generated by the xanthine/xanthine oxidase system was estimated to be 6.7×10^6 M^{-1} s^{-1} [54]. A comprehensive report on the reaction of superoxide with GSH is also given by Winterbourne and Metodiewa [56].

A second class of cysteinyl thiolate-rich cellular molecules are the metallothioneins (MTs). The fact that they bind Zn and Cu under physiological conditions suggests that the proteins are involved in the metabolic control of either metal. Further proposed functions of the MT include metal detoxification, sulfur metabolism, control of the intracellular redox potential, and scavenging of activated oxygen species [57–60]. The molar ratio of GSH to MT in a cell is 200:1, but under stress conditions MT synthesis can be enhanced and GSH levels reduced; thus the antioxidative role of MT would, in a way, be similar to that of GSH. The in vitro oxygen radical scavenging activities of MT were proven several times. Nevertheless, the significance of their contribution to cellular antioxidant systems is still a matter of debate [59].

It was shown that MT provides cells with radioresistance and supposed that the cysteine residues were responsible for this protection [61]. The radioprotective properties of this protein were investigated in more detail by Thornalley and Vasak [62]. The kinetics and reaction mechanism of rabbit MT-I with $\cdot$OH and $O_2^{\cdot-}$ at second-order rate constants of 10^{12} M^{-1} s^{-1} and 5×10^5 M^{-1} s^{-1}, respectively, was measured. It was suggested that the thiolate groups of the cysteine residues were the primary attack targets of the radical species.

Yeast Cu(I)-thionein was examined for its role as a scavenger of these activated oxygen species employing the well-defined pulse radiolysis method. The 53-amino-acid polypeptide contains 12 cysteine residues. Unlike the vertebrate MTs in which Zn, Cu, and Cd are simultaneously bound, the yeast protein exclusively coordinates 6 tenaciously and 2 loosely bound Cu(I) per molecule [63,64]. Investigations of this Cu protein demonstrated its antiinflammatory properties in biological systems and suggested a possible protective role against oxidative stress [65]. Pulse radiolytic investigations reveal that yeast Cu(I)-thionein is an excellent scavenger for reduced oxygen species including $O_2^{\cdot-}$ and $\cdot$OH radicals [66]. A considerable superoxide dismutase activity was observed (Table 1) that was about one order of magnitude higher than that obtained with rabbit liver MT-1 [62]. The $\cdot$OH quenching properties render the protein a further function as a protector for radical-induced damage. The rate constant for the reaction of $\cdot$OH with yeast Cu-thionein was determined as $k_2 = 2.2 \times 10^{11}$ M^{-1} s^{-1}.

TABLE 1

Rate Constants for the Reaction of Superoxide
with Yeast Metallothionein

	Cu (μM)	EDTA (μM)	No. of pulses	k_2 ($M^{-1}s^{-1}$)
Control				5.05×10^5
Cu(II)-formate	18		1	1.09×10^9
		20	1	1.72×10^5
Cu(I)$_6$-thionein	18		1	5.35×10^7
	18	20	1	7.56×10^6
	18	20	3	7.00×10^6
	18	20	5	4.86×10^6
	18	20	10	1.58×10^6
Apothionein			1	8.40×10^5
Zn(II)$_4$-thionein			1	4.79×10^5

Rate constants obtained form the decay kinetics of pulse radiolytically generated superoxide (50 Gy/pulse = 31 μM superoxide)
in the presence of different thioneins at 3 μM concentration. The
decay kinetics were measured in oxygen-saturated 0.1 M formate
(pH 7.2). $T = 17°C$ at 260 nm. Values of the rate constants k_2 are
given as an average of four independent measurements each.
(Reproduced by permission from [66].)

4.2. Copper(I)-Stabilized Thiyl Radicals in Mononuclear and Oligonuclear Copper-Thiolate Species

In the above-mentioned pulse radiolytic study using yeast Cu-thionein
it was shown that superoxide reacts efficiently with this Cu(I)- and
cysteine-rich protein [66]. During this process no marked Cu(I)-thiolate
oxidation occurred leading to the assumption that the thiolate-sulfur
was possibly involved in the oxidation reduction cycle rather than the
metal component according to:

$$Cu(I)^-SR + O_2^{\bullet-} \longrightarrow Cu(I)^{\bullet}SR + O_2^{2-}$$
$$Cu(I)^{\bullet}SR + O_2^{\bullet-} \longrightarrow Cu(I)^-SR + O_2$$

This oxidation-reduction cycle would imply the transient occurrence of a thiyl radical. On the basis of this hypothesis it was of interest as to whether or not the formation of thiyl radicals as intermediate species in Cu-thionein could be convincingly demonstrated employing EPR measurements. Quite frequently, spin-trap reagents are used for this kind of measurements. Unfortunately, the spin-trap adducts of thiyl and/or superoxide radicals cannot be unequivocally assigned, respectively. Thus, the direct proof of these radicals seemed to be highly desirable. Thiyl radicals are produced from thiols in vitro by γ-irradiation of aqueous samples at 77 K and subsequent annealing of the temperature to higher values [17,18].

A similar EPR signal was observed when GSH was UV-irradiated for 30 min at 77 K and subsequently heated to 150–200 K [19]. The thiyl radical spectrum sometimes is overlapped in part by the superoxide signal at $g = 2.013$. Due to the photolysis of water superoxide is directly formed during the irradiation of plain buffer under both aerobic and anaerobic conditions.

It should be pointed out that no direct formation of thiyl radicals could be observed under the present irradiation conditions in the frozen aqueous matrix at 77 K. However, there are reports on the identification of sulfur radicals during the photolysis of mercaptanes in nonaqueous matrices [67,68].

Oxidation of GSH in the presence of Ce(IV) at 77 K resulted in an unaffected EPR signal that was assigned to the thiyl radical at $g = 2.036$. It should be emphasized that in earlier investigations on Ce(IV) oxidation of thiols different EPR spectra were obtained. However, completely different experimental conditions have been used in these studies [4,16].

Upon heating the aqueous frozen UV-irradiated GSH to 200 K the intensity of the superoxide signals leveled off concomitantly with the formation of thiyl radicals. This is demonstrated in more detail using the copper complex of GSH, which may be considered to be a model compound of Cu-thionein. The temperature-dependent transient conversion of $O_2^{\bullet-}$ to $RS^\bullet$ is seen (Fig. 1). An initial thiyl radical formation can be excluded as no $RS^\bullet$ signal is observed at 77 K.

Essentially the same characteristic EPR signals were obtained employing Cu-thionein (Fig. 2d). No detectable Cu(II) signals were noticed in both the protein (Fig. 2d) and Cu-GSH (Fig. 2b) during the

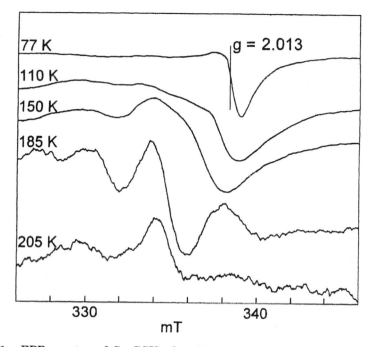

FIG. 1.   EPR spectra of Cu-GSH after 30 min. UV irradiation at 77 K and annealing to various temperatures. (Reproduced by permission from [19].)

treatment, indicating that Cu(I) remained in its original oxidation state. Nevertheless, the transient occurrence of oxidized copper species cannot be fully excluded. Indeed, a considerable portion of cupric copper is measured in Ce(IV)-oxidized Cu-GSH (Fig. 2a). However, no such effect is observed with Cu-thionein (Fig. 2c).

Additional proof for the intactness of the Cu(I)-thiolate chromophores of Cu-thionein was provided by circular dichroism and luminescence emission measurements before and after 30 min UV irradiation. As already seen in the former pulse radiolysis experiment [66], the characteristic chiroptical and the luminescence properties of the protein remained essentially unchanged in the thawed samples following irradiation at 77 K. Either specific Cotton extremum remained at exactly the same position and with the same amplitude. These Cotton bands are assigned to the hexanuclear Cu-thiolate chromophore [69,70] and allow the conclusion of the intactness of this cluster. Examination

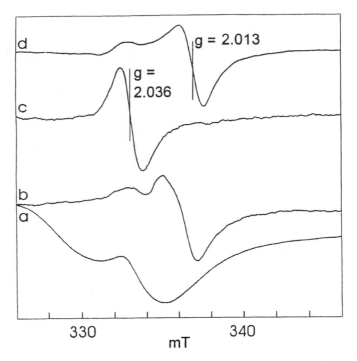

FIG. 2.   EPR spectra of Cu-GSH (a, b) and yeast Cu-thionein (c, d) at 150 K. (a) and (c) represent the Ce(IV)-oxidized thiolate compounds: (b) and (d) the UV-irradiated Cu-GSH and Cu-thionein, respectively. (Reproduced by permission from [19].)

of the luminescence near 600 nm resulted in the same characteristic properties, giving additional proof of the intactness of the Cu-thiolate chromophores. These observations demonstrate unequivocally that indeed thiyl radicals are generated from Cu-thionein in vitro.

Some further studies deal with the occurrence of thiyl radicals concomitant with the reactivity of thiol-copper complexes. Thus, the ability of $(Cu_{14}(\text{D-penicillamine})_{12} Cl)^{5-}$ to act as a superoxide dismutating agent was examined [71]. The red-violet complex proved capable of inhibiting various reactions mediated by superoxide. X-ray photoelectron spectrometric studies revealed that the actual oxidation state of copper before and after the reaction with $O_2^{\bullet-}$ was +1 while the binding energy of the 2p core electrons of sulfur remained around 163 eV. The possibility that a stabilized sulfur radical, present as Cu(I)•SR, is the species undergoing the redox cycle rather than the copper is suggested. A similar reactivity was observed by Reguli and Misik [72]. Superoxide scavenging by the thiol-copper complex of captopril, an angiotensin converting enzyme inhibitor, was accompanied by the formation of thiyl radicals that were measured by EPR spectroscopy in the presense of the DMPO spin trap.

Cu-stabilized thiyl radicals are the subject of reports concerning the modelling of the copper binding centers of type 1 copper proteins. Evidence for such a stabilization was obtained from stopped flow kinetic measurements on the reaction of mercapto amino acids with the Cu(II)-complexes of 2,2′,2″-tris(dimethylamino)triethylamine and tris-(2-pyridylmethylamine) [73]. A mechanism involving reductive elimination of sulfur to give Cu(I) and a thiyl radical is supported by the authors. Also discussed is the possibility of a stabilized Cu(I)-thiyl radical system that may be formed during the GSH-mediated reduction of Cu(II)-tetrathia macrocyclic complexes [74]. The metal-stabilized thiyl radicals were postulated due to the changes of the electronic absorption in the 600-nm wavelength region of the adduct consisting of the Cu complex and GSH.

The fast copper-catalyzed reduction of ferricytochrome c by GSH is assumed to involve the so-called GSH free radical equilibrium (GS• + GS⁻ → GSS⁻•G) [75]. The authors conclude that Cu(II), under the conditions applied, does not interact with GSH to form the unbound GS• radical. They suggest that the Cu(I)-thiolate complex is the reducing entity and that reduction in this case is not mediated by the above equilibrium.

4.3. Reaction of Copper(I)-Thionein with Ferricytochrome c

From the results described in Sec. 4.1 it was assumed that the dismutation of $O_2^{\bullet-}$ by yeast Cu-thionein may be attributed to the redox changes of the thiolate-coordinated copper. Although this reaction mechanism cannot be fully excluded, the observation of thiyl radicals from the UV-irradiated protein favors the attractive conclusion to allocate the site of the oxidation reduction process to the thiolate sulfur which is transiently oxidized to RS•. The role of the coordinated cuprous copper may be considered to act as a suitable thiolate-stabilizing component. This suggestion is consistent with earlier results obtained with yeast Cu-thionein, which successfully reduced ferricytochrome c to the ferro species without any detectable physicochemical changes of both the protein moiety and the Cu-thiolate cluster [19,76] (Fig. 3). Circular dichroism and luminescence measurements did not show any detectable deterioration of the oligonuclear Cu-thiolate-binding centers.

It was concluded that electrons were clearly transferred from Cu(I)-thionein to cytochrome c attributable to the developing α-absorption band of ferrocytochrome c. In the course of this oxidation-reduction reaction ferricytochrome c accepted one electron per 10 copper calculated on a molar basis. It would not be surprising to find that this oxidation-reduction capability of Cu-thionein may also occur in metabolic processes other that those reported here.

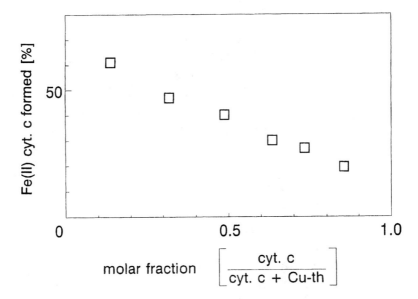

FIG. 3.   Reduction of ferricytochrome c in the presence of 10 μM yeast Cu(I)-thionein and variable cytochrome c concentrations. (Reproduced by permission from [19].)

5. IMPROVED STABILIZATION OF COPPER (I)-THIYL RADICALS EMBEDDED IN INORGANIC AND BIOLOGICAL MATRICES

Unlike in lapislazuli and ultramarine, where sulfur radicals are stabilized by the polyaluminumsilicate matrix [77], no such phenomenon is presently known in biological systems. Thus, it was attempted to mimic this phenomenon using solid thiols including cysteine, penicillamine, GSH, N-acetylcysteine, mercaptosuccinic acid, and N-2-mercaptopropionylglycine. Special emphasis was placed on the ubiquitously occurring hexanuclear Cu(I)-thiolate center of Cu(I)-thionein [69,70] and the prolongation of the respective lifetimes of the observed thiyl radicals.

The photochemical generation of transiently formed thiyl radicals from the thiolate residues of yeast Cu(I)-thionein, GSH, and Cu(I)-GSH at 77 K was shown employing frozen aqueous solutions [19]. During this process transiently formed superoxide radicals gave rise to the formation of thiyl radicals, which were detected at g = 2.036. It was of interest to improve the stability of these thiyl radicals using a number of different derivatives of cysteine, the disulfide-bridged species, and Cu(I)-thiolates in the solid state. UV irradiation using a 450-W xenon lamp seemed convenient to generate these radicals from the solid compounds.

UV irradiation of penicillamine, GSH, and cysteine resulted in the formation of a blue color, which remained unchanged for more than 6 months [78]. The optical maximum was at 601 ± 3 nm as monitored by reflection spectrometry. This phenomenon was noted with all examined solid thiols. The stabilized radical in the blue species of ultramarine exhibited a reflection maximum in the same range [77].

In the EPR spectra the thiyl radicals formed from the UV-irradiated solid thiols are clearly detectable and remained unchanged for months. Additionally, lapislazuli and ultramarine pigments were measured (Table 2). The g values are nearly identical and are consistent with those from cysteine and other mercaptanes in frozen solution that were obtained in earlier studies by UV irradiation at 77 K [69,79,80]. UV irradiation-induced thiyl radical generation from solid thiols at 293 K was not successful, probably due to the low-energy UV source of 15 W only [79]. Surprisingly, the g values of both lapislazuli and the ultra-

TABLE 2

g Values of Thiyl Radicals
from Different Sulfur-Containing
Compounds and Yeast Cu(I)-Thionein

Compound	g
Lapislazuli	2.025
Ultramarine, blue	2.027
Ultramarine, violet	2.027
Ultramarine, pink	2.025
Cysteine	2.026
Penicillamine	2.028
Glutathione	2.025
N-Acetylcysteine	2.027
Mercaptosuccinic acid	2.026
N-2-Mercaptopropinylglycine	2.027
Cu(I)-thionein	2.026

In order to generate the thiyl radicals all thiol
compounds and Cu-thionein were UV-irradiated
for 1 h at 273 k. The inorganic samples containing
stable thiyl radicals were measured directly. (Reproduced by permission from [19].)

marines were in the same range as the thiols and agreed in the case of
the ultramarines with those of earlier data [81].

Unlike the free thiols the oxidized disulfides including cystine and
GSH disulfide showed insignificant portions of thiyl radicals that were
roughly near 1% of the former compounds. Unfortunately, UV irradiation of the respective Cu(I)-thiolate compounds did not result in any
detectable EPR signal or blue coloring.

After UV treatment of Cu(I)-thionein, a sharp EPR signal at $g =$
2.026, which was identical to the signal observed upon UV irradiation
of GSH, was seen (Fig. 4). In contrast to the thiols, the thiyl radical from
Cu(I)-thionein was undetectable after 48 h. Nevertheless, the EPR
signal was clearly noticed within 24 h.

High-energy irradiation of Cu(I)-thionein using γ rays increased

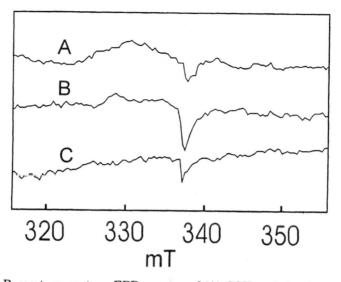

FIG. 4. Room temperature EPR spectra of (A) GSH and (B, C) yeast Cu(I)-thionein after UV irradiation for 1 h at 293 K. Spectra A and B were recorded immediately after irradiation. C represents Cu(I)-thionein (B) after 24 h storage at 293 K. The diminished signal magnitude is due to the limited sample concentration of Cu(I)-thionein in the EPR tube. For comparative reasons the same cystine concentration in Cu(I)-thionein and glutathione was chosen. (Reproduced by permission from [78].)

by twofold the yield of the observed thiyl radicals at $g \sim 2.026$. It is of utmost importance to realize that after either irradiation technique no Cu(II) was detected by EPR, indicating that all electron dislocation must be assigned to the sulfur radical species. In comparison with the EPR data obtained from lapislazuli and the ultramarines, it should be emphasized that the g values were identical at 2.026 (Table 2). However, due to the mobility of the sulfur species in the inorganic matrices the shape of the EPR spectra of these sulfur radicals was different.

The intensities of UV-generated thiyl radicals from cysteine, penicillamine, GSH, and Cu(I)-thionein were compared. The approximate quantification of the specific thiyl radical concentrations showed the highest yield when Cu(I)-thionein was used. Calculation of the intensities revealed a radical concentration ratio of roughly 1:1.5:2:5 (Cys/Pen/GSH/Cu-th). Due to the limited Cu(I)-thionein concentration in the EPR tube a high noise level was inevitable. Nevertheless, the tendency of increasing radical formation in the protein is clearly noticed. This remarkable stabilization in the case of Cu(I)-thionein could be a consequence of the smaller mobility of the larger protein molecule. It is assumed that the EPR signal of Cu(I)-thionein was not caused from free thiols. This could be excluded by comparing the characteristic spectrometric data obtained from circular dichroism and luminescence emission before and after the experimental procedure.

Besides Cu(I)-thionein the contribution of differently arranged thiolate containing Cu(I) clusters was examined. UV irradiation of the Cu-penicillamine complex $Na_5(Cu_{14}(Pen)_{12}Cl)$ did not yield any sulfur radical species. Furthermore, irradiation of polymeric chains of Cu(I)-thiourea tetrahedra and the established di(tetrabutylammonium)hexathiophenolatotetracuprate(I) arranged as $(Cu_4(SPh)_6)^{2-}$ adamantane-type clusters did not contribute to any detectable sulfur radical species. Thus, the intriguing phenomenon remains that substantial portions of thiyl radicals can be detected using the hexacopper thiolate binding centers in Cu(I)-thionein.

Whether or not an increased strain in the respective oligonuclear Cu(I)-thiolate clusters may be attributed to the observed facilitated thiyl radical generation in Cu(I)-thionein needs to be examined. Contrary to the observation of thiyl radicals in frozen aqueous Cu(I)-GSH samples, no such radical species were seen when the solid Cu(I) compound was used. Likewise, polymeric forms of Cu(I)-thiophenolate remained inactive after high-energy irradiation.

ACKNOWLEDGMENTS

Parts of these studies were supported by grants from the Deutsche Forschungsgemeinschaft and the Fonds der chemischen Industrie, Frankfurt, Germany.

ABBREVIATIONS

CoB   cobalamin
CoM   coenzyme M (= 2-thioethane sulfonate)

DMPO   5,5-dimethyl-1-pyrroline-N-oxide
EDTA   ethylenediamine-N,N,N',N'-tetraacetate
EPR   electron paramagnetic resonance
GSH   glutathione
MT   metallothionein
NAD   nicotinamide adenine dinucleotide
NADH   nicotinamide adenine dinucleotide (reduced)
NADP   nicotinamide adenine dinucleotide phosphate
PBN N-t-butyl-α-phenylnitrone
Pcn   penicillamine
PenS•   penicillamine radical
RS•   thiyl radical
RSS•   perthiyl radical
TMPO   tetramethylpyrroline oxide
UV   ultra violet

REFERENCES

1. K. D. Asmus, *Meth. Enzymol.*, *186*, 168–180 (1990).

2. R. P. Mason and D. N. R. Rao, *Meth. Enzymol.*, *186*, 318–329 (1990).

3. K. D. Asmus, in *Radioprotectors and Anticarcinogens* (O. F. Nygaard and M. G. Simic, eds.), Academic Press, New York, 1983, pp. 23–42.

4. B. C. Gilbert, H. A. H. Lauc, R. O. C. Norman, and R. C. Sealy, *J. Chem. Soc. Perkin II*, *46*, 892–900 (1975).

5. P. Huston, J. H. Espenson, and A. Bakac, *J. Am. Chem. Soc.*, *114*, 9510–9516 (1992).

6. M. Tamba, G. Simone, and M. Quintiliani, *Int. J. Radiat. Biol.*, *50*, 595–600 (1986).

7. M. D. Sevilla, D. Becker, and M. Yan, *Int. J. Radiat. Biol.*, *57*, 65–81 (1990).

8. P. Wardman, in *Glutathione Conjugation* (H. Sies and B. Ketterer, eds.), Academic Press, London 1988, pp. 43–72.

9. P. Wardman and C. von Sonntag, *Meth. Enzymol.*, *251*, 31–45 (1995).

10. K. D. Asmus, *Meth. Enzymol.*, *105*, 167–178 (1984).

11. G. R. Buettner, *FEBS Lett.*, *177*, 295–299 (1984).

12. P. Graceffa, *Biochim. Biophys. Acta*, *954*, 227–230 (1988).

13. P. Graceffa, *Arch. Biochem. Biophys.*, *225*, 802–808 (1983).

14. G. Saez, P. J. Thornalley, H. A. O. Hill, R. Hems, and J. V. Bannister, *Biochim. Biophys. Acta*, *719*, 24–31 (1982).

15. L. S. Harman, C. Mottley, and R. P. Mason, *J. Biol. Chem.*, *259*, 5606–5611 (1983).

16. W. Wolf, J. C. Kertesz, and W. C. Landgraf, *J. Magn. Reson.*, *1*, 618–632 (1969).

17. M. D. Sevilla, D. Becker, S. Swarts, and J. Herrington, *Biochem. Biophys. Res. Commun.*, *144*, 1037–1042 (1987).

18. D. Becker, S. Swarts, M. Champagne, and M. D. Sevilla, *Int. J. Radiat. Biol.*, *53*, 767–786 (1988).

19. D. Deters, H. J. Hartmann, and U. Weser, *Biochim. Biophys. Acta*, *1208*, 344–347 (1994).

20. F. R. Gadelha, P. M. Hanna, R. P. Mason, and R. Docampo, *Chem. Biol. Interact.*, *85*, 35–48 (1992).

21. D. Ross, E. Albano, U. Nilsson, and P. Moldéus, *Biochem. Biophys. Res. Commun.*, *125*, 109–115 (1984).

22. V. Fischer, J. A. Haar, L. Greiner, R. V. Lloyd, and R. P. Mason, *Mol. Pharmacol.*, *40*, 846–853 (1991).

23. G. L. Foureman, H. T. Knecht, and T. E. Eling, *Carcinogenesis*, *13*, 515–518 (1992).

24. D. Galaris, E. Cadenas, and P. Hochstein, *Free Rad. Biol. Med.*, *6*, 473–478 (1989).

25. F. J. Romero, I. Ordonez, A. Arduini, and E. Cadenas, *J. Biol. Chem.*, *267*, 1680–1688 (1992).

26. G. D. Mao, P. D. Thomas, and M. J. Poznansky, *Free Radic. Biol. Med.*, *16*, 493–500 (1994).

27. K. R. Maples, C. H. Kennedy, S. J. Jordan, and R. P. Mason, *Arch. Biochem. Biophys.*, *277*, 402–409 (1990).

28. K. R. Maples, P. Eyer, and R. P. Mason, *Mol. Pharmacol.*, *37*, 311–318 (1990).

29. H. J. Sipe, S. J. Jordan, P. M. Hanna, and R. P. Mason, *Carcinogenesis*, *15*, 2637–2643 (1994).

30. U. Bandyopadhyay, D. K. Bhattacharyya, R. Chatterjee, and R. K. Banerjee, *Biochem. J.*, *306*, 751–757 (1995).

31. A. Puppo, C. Monny, and M. J. Davies, *Biochem. J.*, *289*, 435–438 (1993).

32. A. Puppo and M. J. Davies, *Biochim. Biophys. Acta*, *1246*, 74–81 (1995).

33. J. R. Reed, D. R. Quilici, G. J. Blomquist, and R. C. Reitz, *Biochemistry*, *34*, 16221–16227 (1995).

34. H. M. Marques and A. Rousseau, *Inorg. Chim. Acta*, *248*, 115–119 (1996).

35. C. R. Staples, E. Ameyibor, W. Fu, L. Gardet-Salvi, A. L. Stritt-Etter, P. Schurmann, D. B. Knaff, and M. K. Johnson, *Biochemistry*, *35*, 11425–11435 (1996).

36. N. Spear and S. D. Aust, *Arch. Biochem. Biophys.*, *324*, 111–116 (1995).

37. V. Misik, I. T. Mak, R. E. Stafford, and W. B. Weglicki, *Free Radic. Biol. Med.*, *15*, 611–619 (1993).

38. M. J. Davies, B. C. Gilbert, and R. M. Haywood, *Free Radic. Res. Commun.*, *18*, 353–367 (1993).

39. W. E. Antholine, B. Kalyanaraman, J. A. Templin, R. W. Byrnes, and D. H. Petering, *Free Radic. Biol. Med.*, *10*, 119–123 (1991).

40. E. Mulliez and M. Fontecave, *Chem. Berichte Recueil*, *130*, 317 (1997).

41. A. Tauer and S. A. Benner, *Proc. Natl. Acad. Sci. USA*, *94*, 53–58 (1997).

42. W. F. Orme-Johnson, H. Beinert, and R. L. Blakeley, *J. Biol. Chem.*, *249*, 2338–2343 (1974).

43. S. Licht, G. J. Gerfen, and J. Stubbe, *Science*, *271*, 477–481 (1996).

44. (a) G. J. Gerfen, S. Licht, J. P. Willems, B. M. Hoffman, and J. Stubbe, *J. Am. Chem. Soc.*, *118*, 8192–8197 (1996). (b) U. Ermler, W. Grabarse, S. Shima, M. Goubeaud, and R. K. Thauer, *Science*, *278*, 1457–1462 (1997).

45. H. Wariishi, K. Valli, V. Renganathan, and M. H. Gold, *J. Biol. Chem.*, *264*, 14185–14191 (1989).

46. J. P. McEldon and J. S. Dordick, *J. Biol. Chem.*, *266*, 14288–14293 (1991).

47. F. Paoletti, A. Mocali, and D. Aldinucci, *Chem. Biol. Interact.*, *76*, 3–18 (1990).

48. J. Aiyar, H. J. Berkovits, R. A. Floyd, and K. E. Wetterhahn, *Environ. Health Perspect.*, *92*, 53–62 (1991).

49. J. Aiyar, H. J. Berkovits, R. A. Floyd, and K. E. Wetterhahn, *Chem. Res. Toxicol.*, *3*, 595–603 (1990).

50. K. E. Wetterhahn, J. W. Hamilton, J. Aiyar, K. M. Borges, and R. Floyd, *Biol. Trace Elem. Res.*, *21*, 405–411 (1989).

51. K. E. Wetterhahn and J. W. Hamilton, *Sci. Total. Environ.*, *86*, 113–129 (1989).

52. X. Shi, X. Sun, and N. S. Dalal, *FEBS Lett.*, *271*, 185–188 (1990).

53. M. Quintiliani, R. Bardiello, M. Tamba, A. Esfandi, and G. Gorin, *Int. J. Radiat. Biol.*, *32*, 195–202 (1977).

54. K. Asada and S. Kanematsu, *Agr. Biol. Chem.*, *40*, 1891–1892 (1976).

55. H. Sies, *Naturwissenschaften*, *76*, 57–64 (1989).

56. C. C. Winterbourne and D. Metodiewa, *Arch. Biochem. Biophys.*, *314*, 284–290 (1994).

57. H.-J. Hartmann, A. Gärtner, and U. Weser, *Z. Physiol. Chem.*, *365*, 1355–1359 (1984).

58. D. H. Hamer, *Annu. Rev. Biochem.*, *55*, 913–951 (1986).

59. M. Sato and I. Bremner, *Free Radic. Biol. Med.*, *14*, 471–483 (1993).

60. K. T. Tamai, E. B. Gralla, L. M. Ellerby, J. S. Valentine, and D. J. Thiele, *Proc. Natl. Acad. Sci. USA*, *90*, 8013–8017 (1993).

61. A. Bakka, A. S. Johnsen, L. Endresen, and H. E. Rugstad, *Experientia*, *38*, 381–383 (1982).

62. P. J. Thornalley and M. Vasak, *Biochim. Biophys. Acta*, *827*, 36–44 (1995).

63. D. R. Winge, K. B. Nielson, W. R. Gray, and D. H. Hamer, *J. Biol. Chem.*, *260*, 14464–14470 (1985).

64. U. Weser and H.-J. Hartmann, *Biochim. Biophys. Acta*, *953*, 1–5 (1988).

65. R. Miesel, H.-J. Hartmann, and U. Weser, *Inflammation*, *14*, 471–483 (1990).

66. K. Felix, E. Lengfelder, H.-J. Hartmann, and U. Weser, *Biochim. Biophys. Acta*, *1203*, 104–108 (1993).

67. P. S. H. Bolman, I. Safarik, D. A. Stiles, W. J. R. Tyerman, and O. P. Strausz, *Can. J. Chem.*, *48*, 3872–3876 (1970).

68. A. J. Elliot and F. C. Adam, *Can. J. Chem.*, *52*, 102–110 (1973).

69. Y.-J. Li and U. Weser, *Inorg. Chem.*, *31*, 5526–5533 (1992).

70. H.-J. Hartmann, Y.-J. Li, and U. Weser, *Biometals*, *5*, 187–191 (1992).

71. M. Younes and U. Weser, *Biochem. Biophys. Res. Commun.*, *78*, 1247–1253 (1977).

72. J. Reguli and V. Misik, *Free Rad. Res.*, *22*, 123–130 (1995).

73. H. K. Baek, R. L. Cooper, and R. A. Holwerda, *Inorg. Chem.*, *24*, 1077–1081 (1985).

74. V. V. Palischuk, P. E. Strizhak, and K. B. Yatsimirskii, *Inorg. Chim. Acta*, *167*, 47–49 (1990).

75. W. A. Prütz, J. Butler, and E. J. Land, *Biophys. Chem.*, *49*, 101–111 (1994).

76. H.-J. Hartmann, H. Rupp, and U. Weser, in *Metalloproteins* (U. Weser, ed.), Thieme, Stuttgart, 1979, pp 207–213.

77. F. Seel, G. Schäfer, H.-J. Güttler, and G. Simon, *Chem. Zeit.*, *8*, 64–70 (1974).

78. C. Sievers, D. Deters, H.-J. Hartmann, and U. Weser, *J. Inorg. Biochem.*, *62*, 199–205 (1996).

79. J. J. Windle, A. K. Wiersema, and A. L. Tappel, *J. Chem. Phys.*, *41*, 1996–2002 (1964).

80. J. Skelton and F. C. Adam, *Can. J. Chem.*, *49*, 3536–3543 (1971).

81. Y. Matsunaga, *Can. J. Chem.*, *37*, 994–995 (1959).

13

Methylmercury-Induced Generation of Free Radicals: Biological Implications

Theodore A. Sarafian

Department of Pathology and Laboratory Medicine,
University of California-Los Angeles, School of
Medicine, Los Angeles, CA 90095, USA

1. INTRODUCTION: THE TOXICITY AND PATHOLOGY OF MERCURY

Mercury and its organocompound derivatives are extremely hazardous environmental and workplace toxicants [1–3]. Most alarming about the documented sporadic cases of mercury poisoning is the insidious nature of the exposure and accumulation of mercurials, whereby protracted periods of asymptomatic states disguise the extent and irreversibility of injury [4]. By the time symptoms appear, treatment is often ineffective and pathology continues to progress.

A poignantly tragic case of mercury poisoning occurred in the summer of 1996, involving a scientist studying metal toxicity [5]. A single, limited, and unremarkable exposure to dimethylmercury had no appreciable effect for several months. Gradually, sensory functions became disturbed, characterized by numbness and tingling sensations. Common motor tasks such as walking and vocalization became increasingly difficult. Hospitalization became necessary 5 months after exposure and diagnosis of mercury poisoning required an additional week. All efforts at treatment were unsuccessful and the patient lapsed into a coma 10 days after diagnosis and died 4 months later.

The paucity of information on dimethylmercury and the ineffectiveness of current treatment strategies made this a very frustrating and painful tragedy for family, friends, clinicians, and researchers. The extremely toxic nature of the colorless and odorless dimethylmercury is cause for a great deal of concern over the potential hazards this compound presents. It is now known that dimethylmercury is exceptionally permeable to gloves, skin, and tissues. The vast majority of absorbed dimethylmercury is rapidly demethylated to yield methylmercury (MeHg) [6]. Yet, once localized to the nervous system, there is no known effective means for accelerating elimination or reversing neuropathological injury. No suitable explanation has been established for the latent period between organomercurial exposure and neurotoxic symptoms.

A wide variety of mechanisms and pathways have been investigated in an effort to understand the neuropathology caused by mercurials [7,8]. To date, no single mechanism explains the full spectrum of neuropathological consequences. This chapter explores the evidence and rationale forming the hypothesis that oxidative stress underlies the neuropathogenesis of organomercurials. Since MeHg is one of the most common and most toxic environmental forms of mercury, the majority of experimental studies have focused on this agent. However, since many organomercurials share properties with MeHg, other compounds will be occasionally addressed in this chapter. Notably, a small percentage (3–6%) of organic mercury accumulated in brain tissue is converted to inorganic mercury, Hg(II) [9–11]. Therefore, studies on oxidative stress caused by inorganic mercury have been included in some sections.

2. MERCURY CHEMISTRY

2.1. Inherent Instability

Although the mercury atom is classified with zinc and cadmium in group 12 (formerly IIB) of the periodic table, it possesses distinctly different chemical properties [12]. Elemental mercury is the only metal with a melting point below room temperature. Because of its atomic size and number of electrons, its outer electron valences are not rigidly oriented and consequently display high polarizability. This feature also contributes to its high reactivity with thiol groups [13,14] as well as the tendency of carbon-mercury bonds to disproportionate, particularly for aryl hydrocarbon alkyl groups [15]. Since the carbon-mercury bond is rather weak (energy of formation = 15–19 kcal/mol) [16], organomercury compounds are unstable in natural waters. Phenylmercury and sulfur-mercury compounds are readily converted to inorganic Hg by sunlight [17]. MeHg and ethylmercury are degraded by UV light to inorganic mercury by a process involving singlet oxygen ($^1O_2$) [18–20]. In solutions containing ascorbate, MeHg was degraded to Hg^{2+} over a period of 10–20 min [21]. Phenylmercury generates both Hg^{2+} and Hg^0. These reactions can be accelerated by copper ions and by proteins such as γ-globulin and inhibited by dimethylsulfoxide (DMSO) and mannitol, suggesting the involvement of an •OH radical. While the vast majority of mercury entering aquatic environments is inorganic, much of the mercury found in marine organisms is organic, usually in the form of methylmercury [22].

2.2. Aquatic Biotransformations

Microorganisms promote a wide range of chemical reactions involving mercury that generally reflect the particular sensitivies of the organism. Some aquatic microorganisms, including methanogenic bacteria, convert inorganic to methylmercury [23]. Several different genera of bacteria can convert $HgCl_2$ to MeHg, whereas other microorganisms have the ability to degrade MeHg in sediments to inorganic Hg^0 and methane. MeHg can also be generated chemically via transalkylation with alkyltin and alkyllead [24]. The rate of Hg methylation in contaminated water systems has been estimated to be between 15 and 150 ng/g sediment/day [25,26]. Revelation of the magnitude and ubiquity of the biomethylation process in natural waters and sediment systems has led to great concern over the ecological threats posed by production and release into the environment of inorganic mercury (Fig. 1).

2.3. Chemical Reactions in Biological Systems

MeHg can be demethylated in animals producing variable amounts of inorganic Hg in different tissues [27]. In the mammalian brain, the degree of demethylation varies as a function of species, brain region, and cell type. Individual variation may also be high. In rabbits, very

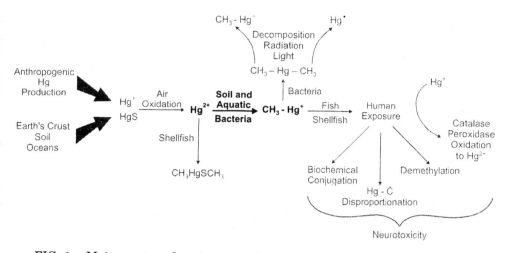

FIG. 1.   Major routes of environmental accumulation and transformation of mercurial, leading ultimately to human exposure and neurotoxicity.

little inorganic mercury is generated whereas in monkeys exposed to high daily doses, 10–33% of total mercury was in inorganic form.

Although victims of Minimata disease were exposed primarily to MeHg, the majority of Hg found in brain following autopsy was inorganic, suggesting that biotransformation-mediated demethylation had taken place [28]. Rat liver microsomes, phagocytic cells, and intestinal microflora accelerate conversion of organic to inorganic mercury by mechanisms probably mediated by electron transport systems and reactive oxygen species (ROS) [19]. Thus, organic mercury compounds are unstable in aqueous and biological systems and much of this instability is linked to oxidative/free radical reactions.

3.   MERCURIAL-INDUCED OXIDATIVE STRESS IN BIOLOGICAL SYSTEMS

3.1.   Mitochondria

In eukaryotic organisms mitochondria are the principal cellular site for production of O_2^- and H_2O_2, converting 1–5% of cellular O_2 to O_2^- during normal metabolism [29]. Mitochondria are dependent on several

sulfhydryl-dependent proteins for normal function [30] and are implicated in the regulation of programmed cell death [31,32]. Agents such
as amytal and antimycin A, which block electron transport, stimulate
superoxide anion production [33]. This process is thought to occur by
the diversion of electrons at the site of transport chain interruption.
There is strong evidence for mercury-induced generation of ROS from
studies on kidney mitochondria.

The kidneys are a primary target of inorganic mercury injury, in
part due to high levels of accumulation [34]. Nephrotoxicity, particularly in the renal proximal tubule, is characterized by ATP depletion,
mitochondrial swelling, lipid peroxidation, and glutathione depletion
[35]. By impairing the efficiency of oxidative phosphorylation and electron transport at the ubiquinone-cytochrome b5 step, Hg(II) causes an
increase in H_2O_2 formation. Several studies have shown that Hg(II)
structurally disrupts the inner mitochondrial membrane and increases
permeability to K^+ and Mg^{2+}, thus diminishing oxidative phosphorylation [36–39].

While these studies focused on the effects of inorganic Hg in vitro
and in vivo, similar effects would be expected from organic Hg exposure.
A variable percentage of organic Hg is converted to Hg^{2+} in most tissues
[19,20,27,40,41] and the MeHg targets many of the same protein sites as
Hg(II). Both organic and inorganic mercury have been shown to decrease cellular ATP levels via disruption of mitochondrial activity [42–
44]. In purified cell types from rat CNS, MeHg inhibited respiration and
increased ROS production in brain mitochondria [45].

Studies by Verity and coworkers indicate that MeHg acts as an
uncoupling agent, stimulating state 4 respiration [44]. Uncouplers are
known to deregulate and accelerate electron transfer rates in the electron transport chain [33,46]. This, in turn, results in increased premature shedding of electrons to molecular oxygen to form superoxide
anion. Thus, MeHg could cause increased O_2^- and H_2O_2 production via
uncoupling activity [47] (Fig. 2).

3.2. Plasma Membrane

Studies by Welsh and others have shown that vitamin E can protect
against MeHg toxicity in vivo [48–51]. Since vitamin E serves an antioxidant function that primarily protects cell membranes, the protective

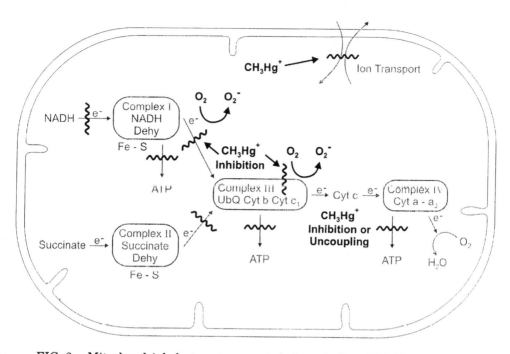

FIG. 2.   Mitochondrial electron transport chain and sites of MeHg interaction (~~~~~) implicated in superoxide anion generation and inhibition of ATP production.

effects of vitamin E suggest that oxidative reactions in the plasma membrane may play an important role in MeHg toxicity. The hydrophobicity of MeHg enables interaction with both the protein and lipid bilayer components of the membrane. Free radicals generated by MeHg disproportionation or other Hg-mediated events activate chain reaction free radical generation and peroxidation of unsaturated lipids prevalent in the nervous system [52]. Vitamin E, serving as a radical scavenger, could suppress this oxidative damage either by terminating the peroxidative lipid chain reactions or by complexing with and inactivating the initiating Hg or methyl radical. A schematic representation of possible mechanisms of MeHg-mediated membrane disruption is provided in Fig. 3.

Lipid peroxidation has been observed in rats treated with $HgCl_2$ [53]. Parenteral administration of $HgCl_2$ stimulated lipid peroxidation in liver, kidney, lung, testis, and serum within 9 h. Vijayalakshmi and

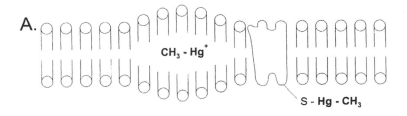

FIG. 3.   (A) Cell plasma membrane showing physical disruption of lipid bilayer and covalent mercaptide formation with membrane proteins. (B) Hypothetical propagation of membrane free radical chain reactions initiated by MeHg, leading to gradual antioxidant depletion and neuronal destruction.

Sood report that vitamins and monothiols have ameliorative capacities in the restoration of brain glutathione metabolism following MeHg exposure [54].

Both inorganic and organic mercury have been reported to cause lipid peroxidation in liver, kidney, spleen, and brain [55,56]. Studies by Sarafian and Verity have demonstrated MeHg-induced lipid peroxidation by measurement of thiobarbituric acid-reactive substances (TBARS) in suspensions of cerebellar granule neurons [57] and in differentiated cultures of neurons (unpublished observation). While the measured lipid peroxidations in these acute in vitro models were not major causal factors in cell death, the findings suggest a mechanism of action of mercury compounds in contributing to neurotoxicity in chronic in vivo studies and in incidences of human exposure.

3.3. Glutathione

Glutathione (GSH) serves as a primary line of cellular defense against mercury compounds. Several accounts of Hg-induced alteration in GSH levels and GSH metabolism have been reported in the literature [35, 57–60]. Unlike rat blood cells, in which MeHg binds primarily to hemoglobin, in human blood cells MeHg binds to GSH [61,62]. Unliganded mercury (Hg(I) or Hg(II)) complexes avidly with nucleophiles and sulfhydryl compounds bind with exceptionally high affinity ($K_a = 10^{15}$–10^{18}) [63]. MeHg readily permeates the cell membrane, either directly or as a cysteine conjugate [64–66], and complexes with intracellular reduced GSH, effectively lowering GSH concentration and thus impairing GSH function. Among the functions of GSH are (1) to maintain protein thiols in a reduced state, mandatory for normal protein function and enzyme activity; (2) to combine with and neutralize H_2O_2 via GSH peroxidase; and (3) to combine with and eliminate xenobiotic compounds that can damage cells. Quinones such as the vitamin K analog menadione, which generate ROS by enzymatic cycling events in the cell, drastically lower the GSH concentration [67,68]. Like MeHg, these quinones also form complexes with GSH directly and further compromise this defense system. Thus the interaction of MeHg and GSH produces oxidative stress via several mechanisms (Fig. 4). Although exceptions have been reported [69], GSH levels are generally much lower in neurons than in glial cells [70,71]. Consequently, MeHg-induced depletion of GSH would be more likely to induce oxidative stress and have greater impact in neurons than in glial cells.

3.4. Phospholipase A$_2$

MeHg has been demonstrated to increase phospholipase activity in cultures of cerebellar granule neurons and in liposomes [72]. Phospholipase enzymes variously function to repair damaged membrane and to generate cellular second messengers [73,74]. In the case of PLA$_2$, which

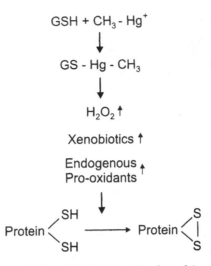

FIG. 4.   Interaction of MeHg (CH_3-Hg^+) with glutathione (GSH) and accompanying oxidative consequences leading to protein oxidation and dysfunction.

is activated by Ca^{2+} and a number of toxic stimuli, reaction with membrane phospholipid generates arachidonic acid and lysophosphatidic acid. Arachidonic acid, in turn, is rapidly acted upon by lipoxygenase and cyclooxygenase [75]. Both enzymes produce O_2^- as obligatory products of reaction. Thus, activation of the PLA_2 signaling pathway is associated with elevation of ROS, often accompanied by cytotoxicity [76–80]. While in vitro studies with cerebellar granule neurons revealed PLA_2 activation, a causal link between PLA_2 and MeHg-induced neurotoxicity was not demonstrated [81]. The PLA_2 inhibitor mepacrine failed to attenuate MeHg-induced cytotoxicity despite substantial decreases in arachidonic acid release. This observation, however, does not preclude a cytotoxic mechanism involving PLA_2-activated oxidative stress since some types of oxidative cell damage might not lead to rapid cell death.

3.5.   Calcium

Numerous studies indicate that MeHg causes disturbances in neuronal calcium homeostasis and several mechanisms have been proposed [82].

Elevated levels of cytoplasmic calcium are known to exacerbate oxidative/free radical injury [83]. Among the consequences of elevated neuronal calcium are (1) activation of the phospholipase A_2 pathway leading to arachidonic acid production [84], (2) conversion of xanthine dehydrogenase to xanthine oxidase, which then acts to produce O_2^- and H_2O_2 [85], and (3) activation of nitric oxide synthase (NOS) enzymes, including neuronal NOS [86]. The latter pathway is of particular interest with respect to MeHg [87]. Excess production of NO appears to be a central factor in the neuropathology associated with glutamate and N-methyl-D-aspartate (NMDA) excitotoxicity. Recent studies in rats reveal that chronic exposure to MeHg (5–10 mg/kg, s.c.) stimulates Purkinje cell NOS to 160% of control [88]. Furthermore, degeneration of cerebellar granule neurons was preceded by increases in cellular NOS activity and cerebrospinal fluid levels of nitrite and nitrate [89].

3.6. Iron

A number of reports suggest that transition metals, especially iron, may contribute to the neuropathogenesis of MeHg [90]. Fenton-mediated free radical formation and oxidation is suppressed by avid conjugation of metal by proteins [91]. In serum and cerebrospinal fluid, iron is bound primarily to transferrin [92] and copper to ceruloplasmin [93]. Among intracellular proteins, ferritin is the major storage site for iron, sequestering 4500 Fe^{3+} atoms per molecule of apoferritin [94]. Other ferrometalloproteins include cytochromes, catalase, hemoglobin, and myoglobin. Xenobiotics can cause release of iron from protein [95]. MeHg has demonstrated ability to bind avidly to numerous proteins and distort tertiary structure [96,97]. Such denaturation can lead to decreased affinity for metal cofactors or heme groups, resulting in dissociation of active iron or copper (Fig. 5). Metal displacement can also result from competitive binding of MeHg to sulfhydryl or other nucleophilic binding sites. For example, mercaptide formation between a mercurial compound, p-hydroxymercuriphenylsulfonate, and aspartate transcarbamylase results in the release of Zn^{2+} [98].

LeBel et al. have provided evidence that iron plays a role in the neurotoxicity of MeHg [99]. Seven days after i.p. injection of 5 mg/kg MeHg in rats, elevated rates of dichlorofluorescein oxidation were observed in fractions from specific brain regions known to be preferen-

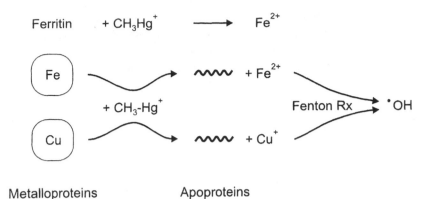

FIG. 5.   MeHg-induced denaturation of metalloproteins promoting liberation of transition metal ions and Fenton-catalyzed production of hydroxyl radical.

tially impaired by MeHg. Pretreatment of animals with the specific iron-chelating agent desferrioxamine completely suppressed the MeHg-induced DCF oxidation in these fractions. Additional evidence for involvement of iron in MeHg neurotoxicity has come from in vitro studies with cerebellar granule neuron suspensions [57]. In these studies, 3-h incubation with 20 μM MeHg reduced cell viability from 94% to 14%, accompanied by an 80% decrease in reduced GSH content and 76% increase in TBARS. Simultaneous inclusion of 2.5 mM desferrioxamine, however, spared over 50% of the cells while completely eliminating MeHg-induced lipid peroxidation. GSH levels were only slightly protected by desferrioxamine. These results suggest that iron-mediated oxidations are responsible for lipid peroxidation and cell killing, whereas GSH depletion was likely a consequence of direct interaction with MeHg.

3.7. Selective Neuronal Vulnerability

Compared to other cell types, neurons have several properties that make them particularly vulnerable to oxidative stress [52].

  1. Their high-energy demands for neurotransmitter and ion transport require high metabolic respiration rates and oxygen

consumption. Production of reactive oxygen species such as superoxide anion is an inescapable side reaction of this activity [93,100].

2. Membrane polyunsaturated lipid content is high, making the cell highly susceptible to lipid peroxidation.

3. Some neurons produce high amounts of oxidizable substrates associated with neurotransmitter pathways [101].

4. Neurons synthesize significant amounts of NO, which serves a neurotransmitter function [102]. This free radical can add to the oxidant stress burden of exposed cells.

Compounding this inherent sensitivity to oxidative stress, and for reasons as yet unclear, neurons frequently possess relatively low levels of antioxidant defense systems [103]. Included among these defenses are the levels of reduced GSH, metallothionein III, hemoxygenase I, catalase, and GSH peroxidase [104,105]. For this reason, agents such as MeHg, which can penetrate the blood-brain barrier and promote oxidative stress, are generally highly neurotoxic. Heavy metals and chemical solvents are among such xenobiotics [106]. Neurons appear to be dependent on antioxidant defense systems provided by astrocytes [107–109]. The fact that neuronal injury and death is characteristic of MeHg neurotoxicity is consistent with a role for oxidative stress in such neurotoxicity.

4. PROTECTIVE MECHANISMS AGAINST MERCURY TOXICITY

4.1. Antioxidant Compounds

Perhaps the strongest evidence linking oxidative/free radical pathways to MeHg neurotoxicity comes from studies demonstrating protective effects of antioxidant vitamins. Early studies revealed that low levels (0.5 ppm) of selenium could suppress or delay MeHg neurotoxicity in rats, cats, and quail [110]. Two mechanisms were proposed for this effect. (1) The first mechanism is direct binding to mercury, based on the known high affinity of selenium for heavy metals [111–113]. Such interaction would sequester the mercurial and result in lower effective concentrations. (2) Selenium serves as an obligate cofactor for GSH peroxidase, an important neuronal enzyme for removal of H_2O_2 [114]. Since selenium could protect animals against much higher concentrations of MeHg, the first mechanism seems unlikely. If the second mechanism is valid, it implies that rapid removal of H_2O_2 is important to prevent accumulation of MeHg-mediated neurological injury.

Vitamin E prevents lipid peroxidation by scavenging free radicals generated in the hydrophobic cell membrane environment [115]. Vitamin E has been shown to reduce toxicity of MeHg in Japanese quail and in rat [48,49] and to prevent neuropathology in golden hamsters [116]. Vitamin E also protects against MeHg-induced lipid peroxidation in mouse liver [117]. Some protection by vitamin E against MeHg toxicity has been observed using in vitro culture conditions [57,118]. Protection against MeHg neurotoxicity has also been reported for desferrioxamine (as discussed earlier), N,N'-diphenyl-p-phenylenediamine [48], cysteine [119,120], and GSH glycoside [121]. The latter compound offered a novel means to upregulate intracellular GSH levels.

Protective effects of exogenously applied antioxidants imply a role for reactive oxygen species in overall injury. However, these findings do not distinguish between primary effects involving intracellular ROS generation and secondary oxidative effects resulting from damaged or disrupted cells.

4.2. Protection by Gene Induction

The induction by MeHg of antioxidant cellular defense systems provides strong evidence for a primary oxidative mechanism of action. There are several reports of MeHg-induced upregulation of defense systems.

γ-Glutamylcysteine synthetase (γ-GCS), the rate-limiting enzyme in GSH synthesis, is upregulated by a wide variety of oxidant agents [122]. γ-GCS mRNA is increased in kidney and brain by MeHg [60,123]. The increase in brain was localized to the hippocampus and cerebellum selectively within regions sensitive to MeHg. The relative inability to upregulate GSH synthesis in specific neuronal cell types may provide the basis for selective vulnerability. MeHg-resistant PC12 cells were selected by sequential MeHg exposures [124]. These cells displayed decreased accumulation and increased efflux of MeHg. Resistant cells also manifested a fourfold higher level of GSH.

mRNA levels for metallothionein (MT) I and II are increased in most cells by the classical inducers zinc and cadmium, but can also be induced by dexamethasone and H_2O_2 [125,126]. In astrocytes MT I can be induced by both $HgCl_2$ and MeHg [127]. Six-hour exposure to 2 μM MeHg in astrocyte culture results in a twofold increase in MT I mRNA levels. Other studies [128], however, report no increase in MT I, II, or III in astrocyte cultures treated with MeHg. This discrepancy may reflect differences in culture conditions or mercurial chemical purity. $HgCl_2$ has been shown to increase MT I and MT II mRNA but decrease MT III mRNA in embryonic mouse neuronal cultures [129]. Rat cerebellar granule neurons were reported to be devoid of metallothionein expression, suggesting a basis for selective vulnerability of these cells to MeHg [130].

Cadmium chloride-induced expression of MT I and MT II has been shown to correlate with protection against MeHg-induced disruption of astrocyte volume regulation properties [131]. The heme-oxygenase enzyme is a 32-kD stress response protein that cleaves and detoxifies the porphyrin ring system of heme moieties liberated from denatured heme proteins [132,133]. Radiation, H_2O_2, and heavy metals including $HgCl_2$ are known to induce synthesis of this enzyme [39,134]. A metal-dependent transcription-enhancer sequence has been reported approximately 10 kilobases upstream of the transcription start site for heme oxygenase I [135]. In brain, only subclasses of Bergmann glia, ependymal, and leptomeningeal cells are reported to induce heme-oxygenase I expression [136,137].

4.3. Protection by Genetic Engineering

The previous section addressed cellular/biological mechanisms for defending against MeHg and accompanying oxidative stress. This section considers strategies for the rationally directed engineering of cells to confer protection against MeHg. One such attempt involved in vitro transfection studies with the antiapoptotic gene, bcl-2 [138,139]. In studies aimed at understanding the mechanism of action of this gene product, the bcl-2 gene was transfected into several neuronal cell lines. The consequent overexpression of the bcl-2 protein was associated with several changes in cellular antioxidant status, including decreased basal and stimulated ROS generation and increased GSH level detected by monochlorobimane fluorescence. As anticipated, these cells displayed significant resistance to necrotic cell death-inducing toxicants, including MeHg [140] (Fig. 6). Bcl-2-transfected cells displayed 22% less cell death than control-transfected cells 3 h after exposure to 10 μM MeHg and 56% less cell death after 24 hr exposure to 5 μM MeHg. In these studies, suppression of cell death correlated best with inhibition of ROS accumulation. Since bcl-2 functions normally in a pathway preventing apoptosis and overexpression results in tumor formation, the bcl-2 protein by itself is unlikely to be a viable genetic engineering tool to bolster endogenous defense systems.

An alternative model is the manipulation of metallothionein expression. Metallothionein has multiple subtypes differentially expressed in brain cells. MT I and MT II gene expression is regulated by Zn^{2+} via the transcription factor MTF-1 [141]. Zn^{2+} and other metals promote binding of MTF-1 to metal response elements in upstream promoter regions. Metallothioneins serve a dual function in metal detoxification in that they not only bind to and sequester heavy metals but also act as antioxidants in scavenging free radicals and in maintaining reduced forms of cellular protein. Individual cells or knock-out mice that do not synthesize MT I or II due to specific gene targeting manipulations are more sensitive to metal toxicity than control cells or animals. Transgenic mice, engineered to overexpress MT-1, appear to display resistance to cadmium [142–144].

Recent studies using another antioxidant gene, thioredoxin peroxidase (also known as natural killer enhancing factor, NKEF, and thiol-specific antioxidant, TSA), overexpressed in an endothelial cell line, revealed protection against some parameters of MeHg-mediated

FIG. 6.   Effect of 3 h exposure to 5 or 10 μM MeHg with or without 1 mM diethylmaleate (DEM) on production of reactive oxygen species (A), reduced glutathione (B), and cell viability (C) in control and Bcl-2-transfected GT1-7 cells. Values represent means of 6–12 determinations $\pm$ SEM. $*p < 0.01$, $**p < 0.001$ comparing Bcl-2 to control cells. (Reproduced with permission from [140].) DCF = 2′,7′-dichlorofluorescein. MCB = monochlorobimane.

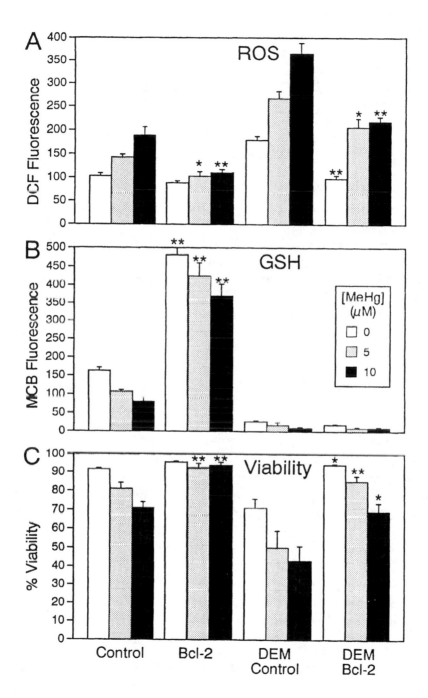

injury [145]. In these cells, MeHg inhibition of thymidine incorporation
was attenuated by thioredoxin peroxidase.

Countless other systems have been genetically manipulated by
either cell transfection or transgenic mouse methodologies. Strategies
using antioxidant, heat shock, or other defense-related genes could
prove useful in abrogating MeHg toxicity and delineating the mechanism of neurotoxic action. The possibilities are virtually limitless, yet
few reports are available on the application of such systems to MeHg
research. In order to be therapeutically effective against MeHg neurotoxicity, genetic manipulation strategies must be directed to brain cells
in vivo, which represents a difficult challenge. Viral systems such as
replication-defective adenovirus vectors, or herpes simplex virus [146–
149], have generally proven effective for stable modification of gene
expression in brain but suffer from complications of inflammatory responses involving activated microglial cells and infiltrating lymphocytes and monocytes [150]. The use of UV-irradiated adenoviral vectors,
which express no adenoviral proteins, may provide a solution to this
problem [151,152].

5. BIOLOGICAL IMPLICATIONS

MeHg exerts its greatest damage in the nervous system, often after a
considerable (weeks to months) latent period following onset of exposure. The initial symptom is generally numbness in extremities followed by constriction of visual field, loss of hearing, and impaired coordination and motor function [153]. These deficits correspond to locations
of major neuropathological injury—the cerebellar granule layer and
the motor and calcarine regions of the cerebral cortex [154,155]. Cellular
changes include neuronophagia and neuronal death, astrogliosis, microglial activation, and demyelination. Dilatation of perivascular spaces
may reflect disruption of the blood-brain barrier [156]. In both the
cerebellum and cortex, damage is greatest in the deeper foliar regions.
In general, small neurons such as cerebellar granule neurons display
the greatest sensitivity to MeHg.

MeHg toxicity in neuronal culture can be prevented by NMDA receptor antagonists [119]. MeHg disturbs glutamate metabolism, which

may produce excitotoxic injury [127,157,158]. Since neurotoxicity mediated by the NMDA receptor appears to be primarily caused by NO production [87], suppression of MeHg neurotoxicity by MK-801 and APV suggests that the free radical NO may play an important role in injury.

One explanation for the selective sensitivity of cerebellar granule neurons to MeHg in vivo that is consistent with an oxidative stress mechanism of injury is based on the low cytoplasmic volume of this morphologically unique cell type. The high surface-to-volume ratio of the cerebellar granule cells mandates low ratios of GSH and antioxidant enzymes relative to protein and DNA. Consequently, the primary lines of cellular defense are easily overpowered by the tenacious MeHg molecule resulting in oxidative damage to sensitive cellular components. Thus a proposed rational approach to therapy, in addition to the conventional approach of expediting elimination of the mercurial, may be to fortify endogenous antioxidant defenses. Positive results have been reported in an experimental animal model using this concept [54,120,121].

The latter strategy would be most effective if directed to the specific prooxidant species generated by MeHg or the cellular defense pathway designed to ameliorate this type of oxidative stress. Further benefit would be achieved by directing the antioxidant fortification to subcellular locations where damaging molecules are generated or where they inflict the greatest damage. The mitochondria are one candidate for such a location [121]. Clearly, further research is required to improve our understanding and control of these toxicological issues.

6. CONCLUSIONS

Oxidative stress and free radical formation by a number of mechanisms are likely to contribute to the neurotoxicity of MeHg. While the evidence presented here suggest a cumulative oxidative stress induced by several mechanisms, there is insufficient evidence that oxidative stress is the primary or predominant cause for the neuropathology and symptomatology of MeHg poisoning. Antioxidants offer partial protection against neurotoxicity in vitro and in vivo. Unique elements required for nervous system function are highly sensitive to xenobiotic

chemicals. Hence, the blood-brain barrier is crucial to cellular communication systems which are the fundamental basis for nervous system
function. Among the unique characteristics of this organ are (1) exceptionally high energy requirements, mandating high mitochondrial activity; (2) high iron content; (3) prevalent and varied unsaturated lipids;
and (4) free radical-mediated neurotransmission. Agents that promote
free radical injury and that can penetrate the blood-brain barrier defenses are exceptionally damaging to the nervous system.

Despite intensive research efforts, the neurodegenerative diseases such as Alzheimer's, Parkinson's, and amyotrophic lateral sclerosis have as yet no well-delineated etiology. However, there is mount-
ing evidence to indicate that oxidative stress plays a role in these and
other disorders [159,160]. Currently, there is no strong evidence to suggest a link between mercury and any of these diseases. Reports indicate
that mercury levels in the brain and blood of patients with these disorders are the same as levels from control tissues [161,162]. Nevertheless, it seems reasonable to hypothesize that toxicants such as MeHg,
which penetrate the blood-brain barrier and promote oxidative stress,
may exacerbate the neuropathological damage caused by the undefined
etiological agents associated with these disorders. Development and
refinement of antioxidant strategies for treating mercury-induced neurotoxicity could have beneficial impact on a broad range of neurological
conditions.

ACKNOWLEDGMENTS

I would like to express my sincere gratitude to Tom Clarkson, Ken
Reuhl, Mickey Aschner, Elaine Faustman, and Bill Atchison for informative discussions in the conceptual formulation of this chapter. I
am particularly grateful to Tony Verity for manuscript reading as well
as his supportive mentorship, collaboration, and friendship spanning
many years. I also appreciate the always expeditious manuscript preparation by Fernando Casimiro and the excellent artwork of Carol Gray
and Donna Crandall of the UCLA Mental Retardation Research Center.
I wish to thank Nadia Rajper, Ronnie Karayan, Amy Ling, Bianca

Grigorian, Angela Chang, Priya Pilutla, and Nancy Pashayan for helping with literature searches.

ABBREVIATIONS

APV	2-amino-5-phosphonovaleric acid
ATP	adenosine 5'-triphosphate
CNS	central nervous system
Cyt	cytochrome
DCF	dichlorofluorescein
DEM	diethylmaleate
DMSO	dimethylsulfoxide
EGTA	ethylene glycol-bis(β-aminoethylether)-N,N,N',N'-tetraacetate
γ-GCS	γ-glutamylcysteine synthetase
GSH	glutathione, reduced
i.p.	intraperitoneal
MeHg	methylmercury (CH_3Hg^+)
MK-801	(+)-5-methyl-10,11-dihydro-5H-dibenzocyclohepten-5,10-imine maleate
MCB	monochlorobimane
mRNA	messenger RNA
MT	metallothionein
MTF	metallothionein transcription factor
NADH	nicotinamide adenine dinucleotide, reduced
NKEF	natural killer enhancing factor
NMDA	N-methyl-D-aspartate
NO	nitric oxide
NOS	nitric oxide synthetase
O_2^-	superoxide anion
PC12	pheochromocytoma cell line
PLA_2	phospholipase A_2
ROS	reactive oxygen species
s.c.	subcutaneous
TBARS	thiobarbituric acid-reactive substance

TSA   thiol-specific antioxidant
UbQ   ubiquinone
UV   ultraviolet

REFERENCES

1. T. W. Clarkson, *Crit. Rev. Clin. Lab. Sci.*, *34*, 369–403 (1997).
2. C. Watanabe and H. Satoh, *Environ. Health Perspect.*, *104S2*, 367–379 (1996).
3. M. Harada, *Crit. Rev. Toxicol.*, *25*, 1–24 (1995).
4. D. C. Rice, *Neurotoxicology*, *17*, 583–596 (1996).
5. C. Holden, *Science*, *276*, 1797 (1997).
6. T. Clarkson, personal communication.
7. W. D. Atchison and M. F. Hare, *FASEB J.*, *8*, 622–629 (1994).
8. M. A. Verity, Pathogenesis of Methyl Mercury Neurotoxicity, in *Mineral and Metal Neurotoxicology* (M. Yasui, M. J. Strong, K. Ota, and M. A. Verity, eds.), CRC Press, Boca Raton, 1997, pp. 159–167.
9. T. L. Syversen, *Acta Pharmacol. Toxicol.*, *35*, 277–283 (1974).
10. M. Berlin, C.A. Grant, J. Hellberg, J. Hellstrom, and A. Schultz, *Arch. Environ. Health*, *30*, 340–348 (1975).
11. S. B. Lind, L. T. Friberg, and M. J. Nylander, *J. Trace Elem. Exp. Med.*, *1*, 49–56 (1988).
12. F. A. Cotton and R. G. Wilkinson, Zinc, Cadmium, and Mercury, in *Advanced Inorganic Chemistry: A Comprehensive Text*, 2nd ed., Interscience, New York, 1966, pp. 600–622.
13. R. B. Simpson, *J. Am. Chem. Soc.*, *83*, 4711 (1961).
14. A. Yasutake, K. Hirayama, and M. Inoue, *Arch. Toxicol.*, *64*, 639–622 (1990).
15. A. J. Carty and S. F. Malone, in *The Biogeochemistry of Mercury in the Environment* (J. O. Nriagu, ed.), Elsevier, New York, 1979, pp. 433–479.
16. H. A. Skinner, *Adv. Organomet. Chem.*, *2*, 46 (1964).
17. P. Benes and B. Havlick, in *The Biogeochemistry of Mercury in the Environment* (J. O. Nriagu, ed.) Elsevier, New York, 1979, pp. 175–202.

18. I. Suda, S. Totoki, and H. Takahashi, *Arch. Toxicol.*, *65*, 129–134 (1991).

19. I. Suda and K. Hirayama, *Arch. Toxicol.*, *66*, 398–402 (1992).

20. I. Suda, M. Suda, and K. Hirayama, *Arch. Toxicol.*, *67*, 365–378 (1993).

21. J. C. Gage, *Toxicol. Appl. Pharmacol.*, *32*, 225–238 (1975).

22. D. Gardner, *Nature*, *272*, 49–51 (1978).

23. K. Bijer and A. Jernelov, in *The Biogeochemistry of Mercury in the Environment* (J. O. Nriagu, ed.), Elsevier, New York, 1979, pp. 203–210.

24. J. J. Bisogni, in *The Biogeochemistry of Mercury in the Environment* (J. O. Nriagu, ed.), Elsevier, New York, 1979, pp. 211–230.

25. W. J. Spangler, J. L. Spigarelli, J. M. Rose, R. S. Flippin, and H. H. Miller, *Appl. Microbiol.*, *25*, 488–493 (1973).

26. H. Windom, W. Gardner, J. Stephens, and F. Taylor, *Estuar. Coast. Mar. Sci.*, *4*, 579–583 (1976).

27. N. K. Mottet, M. E. Vahter, J. S. Charleston, and L. T. Friberg, Metabolism of Methylmercury in the Brain and its Toxicological Significance, in *Metal Ions in Biological Systems*, Vol. 34 (A. Sigel and H. Sigel, eds.), Marcel Dekker, New York, 1997, pp. 371–403.

28. T. Takeuchi, K. Eto, and H. Tokunaga, *Neurotoxicology*, *10*, 651–657 (1989).

29. A. Boveris, N. Oshino, and B. Chance, *Biochem. J.*, *128*, 617–630 (1972).

30. K. F. LaNoue and A. C. Schoolwerth, *Annu. Rev. Biochem.*, *48*, 871–922 (1979).

31. P. Nicotera, M. Ankarcrona, E. Bonfoco, S. Orrenius, and S. A. Lipton, *Adv. Neurol.*, *72*, 95–101 (1997).

32. C. Richter, *Biosci. Rep.*, *17*, 53–66 (1997).

33. K. Schulze-Osthoff, A. C. Bakker, B. Vanhaesebroeck, R. Beyaert, W. A. Jacob, and W. Fiers, *J. Biol. Chem.*, *267*, 5317–5322 (1992).

34. A. Rothstein and A. D. Hayes, *J. Pharmacol. Exp. Ther.*, *130*, 166–176 (1960).

35. B. O. Lund, D. M. Miller, and J. S. Woods, *Biochem. Pharmacol.*, *45*, 2017–2024 (1993).

36. E. Chavez and J. A. Holguin, *J. Biol. Chem.*, *263*, 3582–3587 (1988).

37. S. J. Stohs and D. Bagchi, *Free Rad. Biol. Med.*, *18*, 321–336 (1995)

38. M. F. Hare and W. D. Atchison, *J. Pharmacol. Exp. Ther.*, *261*, 166–172 (1992).

39. K. A. Nath, A. J. Croatt, S. Likely, T. W. Behrens, and D. Warden, *Kidney Int.*, *50*, 1032–1043 (1996).

40. R. Yamamoto, T. Suzuki, H. Satoh, and K. Kawai, *Environ. Res.*, *41*, 309–318 (1986).

41. I. Suda and H. Takahashi, *Toxicol. Appl. Pharmacol.*, *82*, 45–52 (1986).

42. A. L. Nieminen, T. L. Dawson, G. J. Gores, T. Kawanishi, B. Herman, and J. J. Lemasters, *Biochem. Biophys. Res. Commun.*, *167*, 600–606 (1990).

43. T. Sarafian and M. A. Verity, *Neurochem. Pathol.*, *3*, 27–39 (1985).

44. M. A. Verity, W. J. Brown, and M. Cheung, *J. Neurochem.*, *25*, 759–766 (1975).

45. S. Yee and B. H. Choi, *Neurotoxicology*, *17*, 17–26 (1996).

46. T. A. Paget, M. Fry, and D. Lloyd, *Biochem. J.*, *243*, 589–595 (1987).

47. N. Sone, M. K. Larsstuvold, and Y. Kagawa, *J. Biochem.*, *82*, 859–868 (1977).

48. S. O. Welsh, *J. Nutr.*, *109*, 1673–1681 (1979).

49. S. O. Welsh and J. H. Soares, Jr., *Nutr. Rep. Int.*, *13*, 43–51, (1976).

50. R. S. Parker, *Adv. Food Nutr. Res.*, *33*, 157–232 (1989).

51. E. Niki, Y. Yamamoto, M. Takahashi, E. Komuro, and Y. Miyama, *Ann. NY Acad. Sci.*, *570*, 23–31 (1989).

52. B. Halliwell and J. M. C. Gutteridge, *Trends. Neurosci.*, *8*, 22–26 (1985).

53. Y. L. Huang, S. L. Cheng, and T. H. Lin, *Biol. Trace Element Res.*, *52*, 193–206 (1996).

54. K. Vijayalakshmi and P. P. Sood, *Cell. Mol. Biol.*, *40*, 211–224 (1994).

55. M. Yonaha, M. Saito, and M. Sagai, *Life Sci.*, *32*, 1507–1514 (1983).

56. T. J. Taylor, F. Riedess, and J. J. Kocsis, *Fed. Proc.*, *32*, 261 (1973).

57. T. Sarafian and M. A. Verity, *Int. J. Dev. Neurosci.*, *9*, 147–153 (1991).

58. A. Yasutake and K. Hirayama, *Arch. Toxicol.*, *68*, 512–516 (1994).

59. J. S. Woods and M. E. Ellis, *Biochem. Pharmacol.*, *50*, 1719–1724 (1995).

60. J. S. Woods, H. A. Davis, and R. P. Baer, *Arch. Biochem. Biophys.*, *296*, 350–353 (1992).

61. D. L. Rabenstein and C. A. Evans, *Bioinorg. Chem.*, *8*, 107–114 (1978).

62. W. J. Racz and L. J. Vandewater, *Can. J. Physiol. Pharmacol.*, *60*, 1037–1045 (1982).

63. D. L. Rabenstein and M. T. Fairhurst, *J. Am. Chem. Soc.*, *97*, 2086–2092 (1975).

64. N. Ballatori and A. T. Truong, *J. Toxicol. Env. Health*, *46*, 343–353 (1995).

65. E. M. Mokrzan, L. E. Kerper, N. Ballatori, and T. W. Clarkson, *J. Pharmacol. Exp. Ther.*, *272*, 1277–1284 (1995).

66. G. Wu, *J. Appl. Toxicol.*, *16*, 77–83 (1996).

67. T. J. Monks, R. P. Hanzlik, G. M. Cohen, D. Ross, and D. G. Graham, *Toxicol. Appl. Pharmacol.*, *112*, 2–16 (1992).

68. T. W. Grant, M. D. Doherty, D. Odowole, K. D. Sales, and G. M. Cohen, *FEBS Lett.*, *201*, 296–300 (1986).

69. S. Desagher, J. Glowinski, and J. Premont, *J. Neurosci.*, *16*, 2553–2563 (1996).

70. S. P. Raps, J. C. Lai, L. Hertz, and A. J. Cooper, *Brain Res.*, *493*, 398–401 (1989).

71. T. K. Makar, M. Nedergaard, A. Preuss, A. S. Gelbard, A. S. Perumal, and A. J. Cooper, *J. Neurochem.*, *62*, 45–53 (1994).

72. M. A. Verity, *Neurotoxicology*, *15*, 81–91 (1994).

73. J. Chang, J. H. Musser, and H. McGregor, *Biochem. Pharmacol.*, *36*, 2429–2436, (1987).

74. J. Rashba-Step, A. Tatoyan, R. Duncan, D. Ann, T. R. Pushpa-Rehka, and A. Sevanian, *Arch. Biochem. Biophys.*, *343*, 44–54 (1997).

75. P. H. Chan, S. F. Chen, and A. C. Yu, *J. Neurochem.*, *50*, 1185–1193 (1988).

76. N. G. Bazan, *Ann. NY Acad. Sci.*, *559*, 1–16 (1989).

77. P. H. Chan, S. Longar, S. Chen, A. C. Yu, L. Hillered, L. Chu, S. Imaizumi, B. Pereira, K. Moore, V. Woolworth, and R. A. Fishman, *Ann. NY Acad. Sci.*, *559*, 237–247 (1989).

78. B. K. Siesjo, F. Bengtsson, W. Grampp, and S. Theander, *Ann. NY Acad. Sci.*, *568*, 234–251 (1989).

79. D. O. Keyser and B. E. Alger, *Neuron*, *5*, 545–553 (1990).

80. W. T. Shier and D. J. DuBourdieu, *Biochem. Biophys. Res. Commun.*, *110*, 758–765 (1983).

81. M. A. Verity, T. Sarafian, E. H. Pacifici and A. Sevanian, *J. Neurochem.*, *62*, 705–714 (1994).

82. W. D. Atchison and M. F. Hare, *FASEB J.*, *8*, 622–629 (1994).

83. J. M. Braughler, L. A. Duncan, and T. Goodman, *J. Neurochem.*, *45*, 1288–1293 (1985).

84. M. M. Billah, E. G. Lapetina, and P. Cuatrecasas, *J. Biol. Chem.*, *255*, 10227–10231 (1980).

85. T. Nishino, *J. Biochem.*, *116*, 1–6 (1994).

86. V. L. Dawson and T. M. Dawson, *Adv. Pharmacol.*, *34*, 323–342 (1995).

87. V. L. Dawson, V. M. Kizushi, P. L. Huang, S. H. Snyder, and T. M. Dawson, *J. Neurosci.*, *16*, 2479–2487 (1996).

88. T. Himi, M. Ikeda, I. Sato, T. Yuasa, and S. Murota, *Brain Res.*, *718*, 189–192 (1996).

89. T. Yamashita, Y. Ando, N. Sakashita, K. Hirayama, Y. Tanaka, K. Tashima, M. Uchino, and M. Ando, *Biochim. Biophys. Acta*, *1334*, 303–311 (1997).

90. R. A. Goyer, *Annu. Rev. Nutr.*, *17*, 37–50 (1997).

91. J. M. Gutteridge, G. J. Quinlan, S. Mumby, A. Heath, and T. W. Evans, *J. Lab. Clin. Med.*, *124*, 263–273 (1994).

92. M. W. Bradbury, *J. Neurochem.*, *69*, 443–454 (1997).

93. K. A. Koch, M. M. Pena, and D. J. Thiele, *Chem. Biol.*, *4*, 549–560 (1997).

94. T. Rouault and R. Klausner, *Curr. Topics Cell. Reg.*, *35*, 1–19 (1997).

95. T. P. Ryan and S. D. Aust, *Crit. Rev. Toxicol.*, *22*, 119–141 (1992).

96. A. Muller and W. Saenger, *Adv. Exp. Med. Biol.*, *379*, 183–189 (1996).

97. X. Wang and J. D. Horisberger, *Mol. Pharmacol.*, *50*, 687–691 (1996).

98. J. B. Hunt, S. H. Neece, H. K. Schachman, and A. Ginsburg, *J. Biol. Chem.*, *259*, 14793–14803 (1984).

99. C. P. LeBel, S. F. Ali, and S. C. Bondy, *Toxicol. Appl. Pharmacol.*, *112*, 161–165 (1992).

100. B. N. Ames, M. K. Shigenaga, and T. M. Hagen, *Biochim. Biophys. Acta*, *1271*, 165–170 (1995).

101. P. Jenner and C. W. Olanow, *Neurology*, *47S3*, s161–s170 (1996).

102. H. Y. Yun, V. L. Dawson, and T. M. Dawson, *Mol. Psychiatry*, *2*, 300–310 (1997).

103. T. A. Sarafian, D. E. Bredesen, and M. A. Verity, *Neurotoxicology*, *17*, 27–36 (1996).

104. E. Geremia, D. Baratta, S. Zafarana, R. Giordano, M. R. Pinizzotto, M. G. La Rosa, and A. Garozzo, *Neurochem. Res.*, *15*, 719–723, (1990).

105. T. K. Makar, M. Nedergaard, A. Preuss, A. S. Gelbard, A. S. Perumal, and A. J. Cooper, *J. Neurochem.*, *62*, 45–53 (1994).

106. C. P. LeBel and S. C. Bondy, *Neurotoxicol. Teratol.*, *13*, 341–346 (1991).

107. B. Drukarch, E. Schepens, C. A. Jongenelen, J. C. Stoof, and C. H. Langeveld, *Brain Res.*, *770*, 123–130 (1997).

108. M. Aschner, K. J. Mullaney, M. N. Fehm, D. E. Wagoner, Jr., and D. Vitarella, *Cell. Mol. Neurobiol.*, *14*, 637–652 (1994).

109. M. Aschner, *Ann. NY Acad. Sci.*, *825*, 334–347 (1997).

110. H. E. Ganther, C. Goudie, M. L. Sunde, M. J. Kopecky, and P. Wagner, *Science*, *175*, 1122–1124 (1972).

111. H. E. Ganther, P. A. Wagner, M. L. Sunde, and W. G. Hoekstra, in *Trace Substances in Environmental Health* (D. D. Hemphill, ed.), Univ. of Missouri, Columbia, 1973, pp. 247–252.

112. S. Skerfving, *Environ. Health Perspect.*, *25*, 57–65 (1978).

113. S. Potter and G. Matrone, *J. Nutr.*, *104*, 638–647 (1974).

114. J. T. Rotruck, A. L. Pope, H. E. Ganther, A. B. Swanson, D. G. Hafeman, and W. G. Hoekstra, *Science*, *179*, 558–590 (1973).

115. A. G. Olsson and X. M. Yuan, *Curr. Opin. Lipidol.*, *7*, 374–380 (1996).

116. L. W. Chang, M. Gilbert, and M. J. Sprecher, *Environ. Res.*, *17*, 356–366 (1978).

117. H. R. Andersen and O. Andersen, *Pharmacol. Toxicol.*, *73*, 192–201 (1993).

118. M. Kasuya, *Toxicol. Appl. Pharmacol.*, *32*, 347–354 (1975).

119. S. T. Park, K. T. Lim, Y. T. Chung, and S. U. Kim, *Neurotoxicology*, *17*, 37–45 (1996).

120. F. Ornaghi, S. Ferrini, M. Prati, and E. Giavini, *Fund. Appl. Toxicol.*, *20*, 437–445 (1993).

121. B. H. Choi, S. Yee, and M. Robles, *Toxicol. Appl. Pharmacol.*, *141*, 357–364 (1996).

122. A. Meister, *J. Biol. Chem.*, *263*, 17205–17208 (1988).

123. S. Li, S. A. Thompson, T. J. Kavanagh, and J. S. Woods, *Toxicol. Appl. Pharmacol.*, *141*, 59–67 (1996).

124. K. Miura and T. W. Clarkson, *Toxicol. Appl. Pharmacol.*, *118*, 39–45 (1993)

125. J. H. Kagi, *Meth. Enzymol.*, *205*, 613–626 (1991).

126. D. H. Hamer, *Annu. Rev. Biochem.*, *55*, 913–951 (1986).

127. M. Aschner, *Neurotoxicology*, *17*, 93–106 (1996).

128. K. K. Kramer, J. Liu, S. Choudhuri, and C. D. Klaassen, *Toxicol. Appl. Pharmacol.*, *136*, 94–100 (1996).

129. K. K. Kramer, J. T. Zoelle, and C. D. Klaassen, *Toxicol. Appl. Pharmacol.*, *141*, 1–7 (1996).

130. K. Leyshon-Sorland, B. Jasani, and A. J. Morgan, *Histochem. J.*, *26*, 161–169 (1994).

131. D. Vitarella, D. R. Conklin, H. K. Kimelberg, and M. Aschner, *Brain Res.*, *738*, 213–221 (1996).

132. N. G. Abraham, J. H. Lin, M. L. Schwartzman, R. D. Levere, and S. Shibahara, *Int. J. Biochem.*, *20*, 543–558 (1988).

133. M. D. Maines, *FASEB J.*, *2*, 2557–2568 (1988).

134. S. M. Keyse and R. M. Tyrrell, *Proc. Natl. Acad. Sci. USA*, *86*, 99–103 (1989).

135. J. Alam, S. Camhi, and A. M. Choi, *J. Biol. Chem.*, *270*, 11977–11984 (1995).

136. J. F. Ewing, S. N. Haber, and M. D. Maines, *J. Neurochem.*, *58*, 1140–1149 (1992).

137. J. F. Ewing, C. M. Weber, and M. D. Maines, *J. Neurochem.*, *61*, 1015–1023 (1993).

138. L. T. Zhong, T. Sarafian, D. J. Kane, A. C. Charles, S. P. Mah, R. H. Edwards, and D. E. Bredesen, *Proc. Natl. Acad. Sci. USA*, *90*, 4533–4537 (1993).

139. D. J. Kane, T. A. Sarafian, R. Anton, H. Hahn, E. B. Gralla, J. S. Valentine, T. Ord, and D. E. Bredesen, *Science*, *262*, 1274–1277 (1993).

140. T. A. Sarafian, L. Vartavarian, D. J. Kane, D. E. Bredesen, and M. A. Verity, *Toxicol. Lett.*, *74*, 149–155 (1994).

141. S. L. Samson and L. Gedamu, *Prog. Nucl. Acid Res. Mol. Biol.*, *59*, 257–288 (1998).

142. Y. Liu, J. Liu, M. B. Iszard, G. K. Andrews, R. D. Palmiter, and C. D. Klaassen, *Toxicol. Appl. Pharmacol.*, *135*, 222–228 (1995).

143. D. K. Lee, K. Fu, L. Liang, T. Dalton, R. D. Palmiter, and G. K. Andrews, *Mol. Reprod. Dev.*, *43*, 158–166 (1996).

144. Y. P. Liu, J. Liu, R. D. Palmiter, and C. D. Klaassen, *Toxicol. Appl. Pharmacol.*, *137*, 307–315 (1996).

145. A. T. Kim, T. A. Sarafian, and H. Shau, *Toxicol. Appl. Pharmacol.*, *146*, 135–142 (1997).

146. B. L. Davidson, E. D. Allen, K. F. Kozarsky, J. M. Wilson, and B. J. Roessler, *Nature Genet.*, *3*, 219–223 (1993).

147. C. Landau, M. J. Pirwitz, M. A. Willard, R. D. Gerard, R. S. Meidell, and S. E. Willard, *Am. Heart J.*, *129*, 1051–1057 (1995).

148. J. C. Glorioso, W. F. Goins, C. A. Meaney, D. J. Fink, and N. A. DeLuca, *Ann. Neurol.*, *35S*, S28–S34 (1994).

149. B. V. Zlokovic and M. L. Apuzzo, *Neurosurgery*, *40*, 805–812 (1997).

150. E. Peltekian, E. Parrish, C. Bouchard, M. Peschanski, and F. Lisovoski, *J. Neurosci. Metab.*, *71*, 77–84 (1997).

151. A. P. Byrnes, R. E. MacLaren, and H. M. Charlton, *J. Neurosci.*, *16*, 3045–3055 (1996).

152. J. F. Engelhardt, X. Ye, B. Doranz, and J. M. Wilson, *Proc. Natl. Acad. Sci. USA*, *91*, 6196–6200 (1994).

153. T. Takeuchi, Pathology of Minimata disease, in *Minimata Disease* (M. Kutsuma, ed.), Kumamoto U., Japan (1968), pp. 141–228.

154. K. R. Reuhl, L. W. Chang, and J. W. Townsend, *Environ. Res.*, *26*, 281–306 (1981).

155. K. R. Reuhl, L. W. Chang, and J. W. Townsend, *Environ. Res.*, *26*, 307–327 (1981).

156. J. M. Jacobs, *Environ. Health Persp.*, *26*, 107–116 (1978).

157. W. D. Atchison and M. F. Hare, *FASEB J.*, *8*, 622–629 (1994).

158. J. N. Reynolds and W. J. Racz, *Can. J. Physiol. Pharmacol.*, *65*, 791–798 (1987).

159. N. A. Simonian and J. T. Coyle, *Annu. Rev. Pharmacol. Toxicol.*, *36*, 83–106 (1996).

160. M. F. Beal, *Curr. Opin. Neurobiol.*, *6*, 661–666 (1996).

161. Y. K. Fung, A. G. Meade, E. P. Rack, and A. J. Blotcky, *J. Toxicol. Clin. Toxicol.*, *35*, 49–54 (1997).

162. Y. K. Fung, A. G. Meade, E. P. Rack, A. J. Blotcky, J. P. Claassen, M. W. Beatty, and T. Durham, *J. Toxicol. Clin. Toxicol.*, *33*, 243–247 (1995).

14

Role of Free Radicals
in Metal-Induced Carcinogenesis

Joseph R. Landolph

Departments of Molecular Microbiology and
Immunology, Pathology, and Molecular Pharmacology
and Toxicology, USC/Norris Comprehensive Cancer
Center, University of Southern California Schools
of Medicine and Pharmacy, Los Angeles,
CA 90033-0800, USA

445

1. INTRODUCTION

The purpose of this chapter is to summarize the knowledge on the involvement of free radicals in the process of metal-induced carcinogenesis, to critically evaluate this knowledge, and to propose future directions for research in this area. No attempt is made to be exhaustive in this very large body of literature. Rather, the author focuses particularly on specific chromium, nickel, arsenic, beryllium, iron, and copper compounds. These specific compounds were chosen because certain compounds of the elements chromium and nickel and the metalloid arsenic are known human occupational carcinogens. In addition, specific beryllium compounds are carcinogenic in animals and are suspected but not proven human carcinogens, making them interesting to study. Also, there are developing data that specific compounds of the redox-active metals iron and copper may accumulate during specific genetic diseases of storage and transport of these metals, leading to an abnormal overload of these metals in animals or humans. High steady-state levels of iron or copper may generate oxygen radicals due to their redox chemistry, and may lead to liver carcinogenesis in animals or humans in specific circumstances, although the evidence for their human carcinogenicity is less than for specific chromium, nickel, and arsenic compounds.

It is widely believed that free radicals, particularly oxygen radicals such as superoxide and hydroxyl radical, and activated oxygen metabolites such as hydrogen peroxide, may be potentially involved in cancer and also may be involved in many other diseases, such as neurological diseases, and during aging (reviewed in [1]). These radicals may also be generated endogenously by iron and copper compounds, although such generation of radicals by iron and copper in animals and humans requires further substantiation, and the role of such radicals, if generated during carcinogenesis, requires further experimentation to prove.

It is also well recognized that the area of free radical involvement in disease processes is often difficult to substantiate conclusively for many reasons. First, the lifetime of most radicals in biological systems is short because there are many radical-quenching systems with which they can react. Second, there are many artifacts to the measurement of free radicals, particularly in biological systems. Third, radicals often do not leave concrete and easily measurable types of damage in their wake, such as the carcinogen-DNA covalent adducts that many chemical carcinogens make. Fourth, often the data concerning the existence of oxygen or other radicals are circumstantial or inferential, as often occurs when various inhibitors are used to demonstrate the existence of free radicals in biological systems. Fifth, many radicals that are detectable are very reactive, and their various reactions often lead to other radicals or other reactive species that have not yet been identified or are not easily identified (reviewed in [2]). For all of these reasons, the literature on the roles of free radicals in biological systems must be interpreted critically and cautiously.

A further general consideration is that the molecular mechanisms of metal compound carcinogenesis are still somewhat obscure, although insights into these processes are accumulating at a rapid rate, compared to the molecular mechanisms of carcinogenesis by polycyclic aromatic hydrocarbons (PAHs), aromatic amines, aflatoxin B_1, and other organic carcinogens. For instance, we now know that carcinogenesis by PAHs in whole animals (reviewed in [3,4]) and in cultured murine and human cells (reviewed in [5,6]) is a multistep process. Carcinogenic PAHs, such as benzo(a)pyrene, 3-methylcholanthrene, and related members of this family, are metabolized by specific cytochrome P450 enzymes (reviewed in [7–9]) to reactive intermediates, including bay-region diol epoxides and hydroxy epoxides, that bind covalently to DNA [8,9], leading to mutations in bacteria [10] and mammalian cells [5,11,12] and morphological and neoplastic transformation of mammalian cells [5,11,12]. When these mutations occur in protooncogenes, they may activate these protooncogenes to oncogenes [13–16]. When these mutations occur in tumor suppressor genes, they may mutationally inactivate these tumor suppressor genes or cause chromosome breakage, leading to deletion of these tumor suppressor genes [17,18] (reviewed in [19–21]). Combinations of activation of protooncogenes to oncogenes and inactivations of tumor supressor genes, such that on the order of five to seven events of these types occur, leads to carcinogenesis and generation of the first tumor cell (reviewed in [19–23]). These events likely occur one at a time, such that a mutation in one cell may lead to the first activation of the first protooncogene to an oncogene, and then the cell bearing this oncogene may divide so that a field of cells now bears this activated oncogene. At some later point in time, in a random manner, a second mutation in either a tumor supressor gene, inactivating it, or in a second protooncogene, activating it to a second oncogene, may occur in one of the group of cells bearing the first mutated oncogene. This process of mutation, expansion of the cells bearing the mutation, and a later mutation in one of the progeny of the prior mutated cells continues until the first tumor cell arises. This first or primordial tumor cell now bears five to seven genes, each with a mutation, which causes the generation of activated oncogenes and the inactivation of tumor suppressor genes (reviewed in [6,7,24–29]). This primordial tumor cell is actually a degraded cell that lacks the mechanisms regulating growth and mechanisms coupling the growth of this cell to that of its neighboring cells. This primordial tumor cell then can further divide and then undergo further genetic change such that metastatic cells arise in this population.

We do not yet have such specific information on the molecular mechanisms of carcinogenesis induced by metal compounds. We do know that for compounds of metallic elements such as nickel and chromium, there is abundant evidence that these metals bind to DNA; nickel binds more strongly to protein, which then may bind to DNA (reviewed in [30]). Nickel may generate oxygen radicals and cause lipid peroxidation and chromosome breakage (reviewed in [39,40]). Specific insoluble chromium compounds also cause chromosome breakage [32–34] (reviewed in [30]). While both specific insoluble nickel and insoluble chromium compounds cause carcinogenesis in animals and in humans, the exact molecular mechanisms by which they cause carcinogenesis are not known (reviewed in [25–30]). For copper and iron compounds, there are specific situations in which an overload of these compounds causes carcinogenesis in animals and is believed to be linked to human liver cancer [1]. However, the molecular mechanisms by which these compounds cause cancer is still not clear and requires further investigation. A discussion of our knowledge in these areas is described in detail below.

2. THE ROLE OF FREE RADICALS IN CHROMIUM CARCINOGENESIS

Specific insoluble hexavalent chromium chompounds are the strongest carcinogens among chromium compounds. The exact ultimate intracellular carcinogen(s) in the process of chromium carcinogenesis is still not known and has been variously thought to be chromium(VI), chromium(III), oxygen radicals generated during reduction of chromium(VI) to chromium(III) when chromium(VI) functions as a redox-active metal, and, more recently, chromium(V), in the order of the history of progress of research in this field. Combinations of these species might conceivably also act together to induce carcinogenesis. Further research is required to clarify the mechanisms of chromium carcinogenesis, to define whether specific trivalent chromium compounds are carcinogenic, and to define the relative carcinogenicities of insoluble versus soluble hexavalent chromium compounds. Further research is also needed to clarify the mechanisms of chromium carcinogenesis, and to determine whether the chromium(V) radical and/or oxygen radicals are generated in significant quantities during the reduction of chromium(VI) and, if so, whether these radicals are involved in the molecular mechanisms of chromium carcinogenesis. The specific details of the research in this area of chromium carcinogenesis are given below.

2.1. Carcinogenicity of Specific Chromium Compounds

Epidemological studies have indicated that workers who mine chromium ores or work in the pigment industries where specific chromium compounds are components of the pigments, such as lead chromate, have increased frequencies of lung cancer (reviewed in [25–29,35]). These epidemiological studies do not of themselves identify the specific chromium compounds that are carcinogenic to humans. To do this, it is necessary to study the carcinogenicity of chromium in animal carcinogenesis assays, in cell transformation assays, in clastogenicity assays, and in mutagenesis assays.

2.1.1. Insoluble Chromium Compounds

A number of chromium compounds have been shown to induce morphological and neoplastic transformation of C3H/10T1/2 Cl 8 (10T1/2) mouse embryo cells [36] and in Syrian hamster embryo cells (reviewed in [25–29,35]). Our current belief is that the insoluble chromium compounds, such as lead chromate, are the most carcinogenic. The insoluble hexavalent chromium compounds are taken into cells by phagocytosis, depositing a bolus of insoluble hexavalent chromium compound into the cells, as we found in our laboratory [36] (summarized in Table 1).

2.1.2. Soluble Chromium Compounds

In the case of soluble hexavalent chromium compounds, the chromate ion containing the hexavalent chromium atom is taken up into cells by the somewhat nonspecific sulfate anion transport system. There are many biochemical systems in serum that can reduce soluble hexavalent chromium to trivalent soluble chromium, thereby inactiviting it, since soluble trivalent chromium ion is not believed to be able to enter the cell. Soluble hexavalent chromium compounds are believed at this point in time to be less carcinogenic than insoluble hexavalent chromium compounds, but more work needs to be done in this area.

Soluble trivalent chromium compounds are not taken up into cells to any significant extent and in serum are believed to be detoxification products resulting from the reduction of hexavalent chromium compounds (reviewed in [25–29,35], Table 1).

2.1.3. Are Trivalent Chromium Compounds Carcinogenic?

To our knowledge, the soluble trivalent chromium compounds are regarded as detoxification products in the extracellular milieu, since they are not taken up into mammalian cells to any appreciable extent. There is a question as to whether insoluble, trivalent chromium compounds are carcinogenic. Theoretically, since intracellular trivalent chromium ion is thought by many workers to be carcinogenic, it is possible that

TABLE 1

Important Features of Chromium Carcinogenesis

Insoluble hexavalent chromium compounds

Are taken into cells by phagocytosis, are the most carcinogenic chromium compounds. This is likely due to deposition of a bolus of chromium compound, hence chromium ion, inside the cell, and in the case of lead chromate, also lead ions.

Soluble hexavalent chromium compounds

Are taken into the cell on the nonspecific sulfate anion transport carrier. Many of these compounds are mutagenic and carcinogenic. All of these compounds have the potential to be carcinogenic because they enter cells and deposit hexavalent chromium ion, which is believed to be a proximate carcinogenic form of chromium, inside cells.

Insoluble trivalent chromium compounds

May be taken into cells by phagocytosis. May potentially pose a carcinogenic risk. However, insufficient evidence exists for their carcinogenicity because these compounds have not been tested well for carcinogenicity. If these compounds are phagocytosed by cells, they have the potential to be genotoxic because intracellular trivalent chromium is believed to be one ultimate carcinogenic form of chromium. However, trivalent chromium compounds are approximately 1000-fold less cytotoxic and mutagenic than hexavalent chromium compounds.

Soluble trivalent chromium compounds

These compounds are not taken up to any appreciable extent into mammalian cells. When present in extracellular fluids, such as after reduction of hexavalent chromium, the trivalent chromium ion is believed to be a detoxification product.

---

insoluble, trivalent chromium compounds could be phagocytosed, leading to the uptake of a bolus of these compounds, and to the intracellular release of intracellular chromium(III) ion, which attacks DNA due to its chemical reactivity. Hence, there is a possibility that insoluble trivalent chromium compounds might be carcinogenic. However, much more sound experimental work needs to be done to address this question.

These important considerations on the carcinogenicity of hexavalent chromium compounds and the state of knowledge on the lack of carcinogenicity of soluble trivalent chromium compounds and the potential for carcinogencity of insoluble trivalent chromium compounds are summarized in Table 1.

2.2. Reduction of Chromium(VI) to Chromium(V) and Chromium(III)

It is believed that the reduction of hexavalent chromium to lower valence states is responsible for its toxicity, mutagenicity for some chromium compounds, clastogenicity, and likely also its carcinogenicity (reviewed in [25–29,35]). The exact identity of the chromium ion (hexavalent, pentavalent, or trivalent chromium ion) or other chromium-generated species that attacks DNA during the process of carcinogenesis and that is responsible for chromium carcinogenesis is still not clear. Earlier workers held that the intracellular reduction of chromium(VI) led to intracellular chromium(III), which was the important chromium ion resposible for carcinogenesis. This view is still held by numerous scientists, but a current summary of this field is that production of intracellular chromium(III) leads to one important species involved in chromium carcinogenesis. Work from the laboratory of the late Professor Karen Wetterhahn stressed the importance of the formation of a transient chromium(V) ion, which is a radical, in chromium carcinogenesis (reviewed in [25–29,35]). Evidence has been presented for the production of chromium(V) radical during treatment of cells with chromium compounds. However, further work is needed in this area to establish whether chromium(V) radical is an important ultimate intracellular carcinogen and/or one of a number of ultimate carcinogenic species in chromium carcinogenesis (see Tables 1 and 5).

2.3. Evidence for the Role of Free Radicals in Chromium Carcinogenesis

As mentioned above, there is evidence that the chromium(V) ion is generated during the reduction of intracellular hexavalent chromium by mammalian cells. Chromium(V) is a radical because it has an unpaired electron. The role, if any, of chromium(V) radical in the molecular mechanism of chromium carcinogenesis is speculative and intriguing, but at this point in time is still not clear. In addition, workers in the past had presented evidence and previously held that chromium(VI) could participate in redox reactions and that oxygen radicals could be generated from this pathway in Fenton-like reactions (reviewed in [25–29]). However, in recent years little new evidence has accumulated to substantiate this pathway as having a role in chromium carcinogenesis. This hypothesis is attractive given the transition metal properties of chromium and the fact that it is a redox-active metal. Chromium can easily be postulated to participate in reactions similar to the Fenton reactions in which iron participates. However, further research needs to be conducted in the area to determine whether chromium(V) radicals or chromium-induced oxygen radicals contribute to the molecular mechanisms of chromium carcinogenesis (see Table 5 in Sect. 7).

3. THE ROLE OF FREE RADICALS IN NICKEL CARCINOGENESIS

A number of insoluble nickel compounds are carcinogenic. Current thinking and evidence is that the nickel ion, complexed likely to proteins or amino acids, is the intracellular carcinogenic form of insoluble nickel compounds. It is likely that the phagocytosis of a large quantity of insoluble nickel compounds leads to a very high concentration of intracellular nickel ions and contributes to the carcinogenicity of insoluble nickel compounds. Soluble nickel compounds are not thought to be carcinogenic in animals due to excretion of the soluble compounds and to only small uptake of these compounds. There is the possibility that under conditions of exposure to very high concentrations there is a potential that forced uptake might render soluble nickel compounds carcinogenic, but there are no data to suggest this at present. There is preliminary evidence that specific nickel compounds may stimulate arachidonic acid metabolism, but more work is needed to define whether significant oxygen radical generation occurs off this pathway in response to stimulation of this pathway by specific nickel compounds. In addition, critical experimentation needs to be conducted to determine whether generation of oxygen radicals, if it occurs, is definitively involved in the molecular mechanisms of carcinogenesis induced by specific nickel compounds. There is accumulating and convincing evidence that nickel ions complex strongly with proteins and amino acids. Under specific circumstances, these nickel complexes may generate hydroxyl radicals at the site of DNA, leading to site-directed mutagenesis. Nickel compounds are well known to generate chromosome breakage. Nickel compound-induced oxygen radical generation, by nickel-protein complexes bound to DNA, leading to consequent site-directed mutagenesis, may be one of the molecular mechanisms of nickel carcinogenesis. There is also a developing literature that nickel compounds induce lipid peroxidation, further strengthening the idea that specific nickel compounds can generate radical species. Specific details of the progress of research in this area are outlined and discussed below.

3.1. Carcinogenicity of Specific Nickel Compounds

Epidemiological studies have indicated that there are higher frequencies of nasal and respiratory cancer in workers who mine and refine ores containing nickel compounds (reviewed in [25–31]). There are often confounding aspects to this situation, since often many of the nickel workers have been shown to be cigarette smokers. It is likely that in this situation, both cigarette smoke carcinogens, such as benzo(a)-pyrene (BaP), dibenz(c,g)carbazole, the tobacco-specific nitrosamines NNN and NNK, and the cocarcinogens and tumor promoters in cigarette smoke interact in a synergistic or additive manner with the nickel compounds to increase the frequency of respiratory cancer in the nickel workers.

3.1.1. Insoluble Nickel Compounds

Injection of specific nickel compounds into animals induces fibrosarcomas at the site of injection, so that a number of specific nickel compounds are animal carcinogens (reviewed in [24–31]). In addition, many nickel compounds, particularly the insoluble nickel compounds, are able to induce morphological and neoplastic transformation of cultured Syrian hamster embryo cells (reviewed in [30]) and of cultured C3H/-

10T1/2 Cl 8 (10T1/2) mouse embryo cells [37]. There is experimental evidence in animals and in cultured cells that the insoluble nickel compounds are the most carcinogenic among the nickel compounds. Insoluble nickel compounds that are established to be carcinogenic in animals include nickel subsulfide, crystalline nickel monosulfide, and various forms of nickel oxide, including the black nickel oxide and the green preparation of nickel oxide (there are more than 30 different types of nickel oxides) (reviewed in [25–30]). In our laboratory, we have found that nickel subsulfide, crystalline nickel monosulfide, and black and green nickel oxides all are phagocytosed by the cells, all induce strong cytotoxicity, and all induce morphological transformation in 10T1/2 mouse embryo fibroblasts [37,64]. A unique feature of insoluble nickel compounds that are prepared in particles of a size of 1 μm or less is their ability to be phagocytosed into cells. This leads to a very large amount of nickel ion in the cells, approaching concentrations of 0.25–4.75 M, when Chinese hamster ovary (CHO) cells were treated with crystalline nickel monosulfide particles, calculated as if all of the nickel were to be dissolved (reviewed in [30]). We believe that the phagocytosis of this large amount of insoluble nickel compound, leading to these extremely high intracellular concentrations of insoluble nickel [30], is responsible for the strong cytotoxicity, clastogenicity, cell transformation ability, and carcinogenicity of insoluble nickel compounds [37]. It is currently believed that intracellular Ni(II) ion is the ultimate carcinogenic form of nickel, when complexed with specific ligands, such as proteins (reviewed in [25–31]; see Table 2).

3.1.2. Soluble Nickel Compounds

All soluble nickel compounds studied have been found to be noncarcinogenic in standard whole-animal carcinogenesis bioassays, including nickel sulfate and nickel chloride. It is believed that the soluble nickel compounds are not appreciably carcinogenic because they are rapidly excreted in the urine and because they are not taken up into mammalian cells in large quantities like the insoluble nickel compounds. However, soluble nickel compounds are of course cytotoxic and can also induce chromosome breakage, and hence exert genotoxic activity. It is possible that at conditions of very high exposure, soluble nickel compounds could exert cytotoxic and genotoxic effects in humans. However, there are no data at present to confirm this specula-

TABLE 2

Important Features of Nickel Carcinogenesis

Insoluble nickel compounds

1. Are phagocytosed by the cell, leading to sequestration of large concentrations of Ni(II) inside the cell.
2. Many are carcinogenic in animals and can induce morphological and neoplastic cell transformation if the particle size is less than 1 μm.
3. Many can induce chromosome breakage in cells.
4. Examples: nickel subsulfide, crystalline monosulfide, and black and green nickel oxides.

Soluble nickel compounds

1. Are taken up only sparingly by the cell, if at all.
2. Most are not carcinogenic to animals in results to date but do not induce morphological or neoplastic transformation in mouse embryo fibroblastic cell lines (e.g., C3H/10T1/2) at high concentrations in cell culture.
3. Possess either no carcinogenic potential (current data) or at best a weak carcinogenic potential at very high concentrations (theoretical projection). This theoretical projection is based on the assumption that complexes of Ni(II) with proteins or amino acids near the DNA are the ultimate carcinogenic species for nickel compounds.
4. Examples: nickel chloride, nickel sulfate.

Mechanisms of carcinogenesis by nickel compounds

May involve complexes between nickel and specific amino acids, peptides, or proteins binding to DNA generating oxygen radicals at the site of DNA, such as hydroxyl radicals, and these radicals attacking DNA, leading to mutation not easily detected in classical assays of mammalian cells, and to chromosome breakage.

tion. It certainly is clear that the effects of the insoluble nickel compounds are far stronger than those of the soluble nickel compounds because the insoluble nickel compounds are efficiently phagocytosed by mammalian cells and hence efficiently sequestered intracellularly. This leads to very high effective intracellular concentrations of nickel ion when cells are treated with small particles of insoluble nickel compounds that can be phagocytosed [30] (reviewed in [25–30]; see Table 2).

The soluble nickel compounds, including nickel sulfate, are cytotoxic too and have been found to induce morphological transformation in 10T1/2 mouse embryo cells at high concentrations under standard exposure regimens of 48 hours in work from our laboratory [64]. The soluble nickel compounds such as nickel sulfate and nickel chloride to date have been found to be noncarcinogenic in standard whole-animal carcinogenesis bioassays. It is not clear as to whether soluble nickel compounds are carcinogenic to humans or whether they potentiate the carcinogenicity of insoluble nickel compounds to humans, since human exposure is to a complex array of nickel compounds, including both soluble and insoluble nickel compounds, and the epidemiology indicating an increased risk of nasal and respiratory cancer in workers involved in the mining and refining of nickel-containing ores do not distinguish among the nickel compounds. Further epidemiological and experimental work is needed to answer this question. However, at present, there is evidence that specific insoluble nickel compounds are carcinogenic in animals and induce morphological and neoplastic transformation of cultured murine and hamster cells. Hence, specific insoluble nickel compounds are carcinogenic, including nickel subsulfide, crystalline nickel monosulfide, and black and green nickel oxides, and they likely contribute to the carcinogenicity of the process of mining and refining nickel compounds (reviewed in [25–30]). Whether soluble nickel compounds contribute to the carcinogenicity of these processes is not known at present. If they do indeed, then they likely contribute substantially less than do the insoluble compounds (summarized in Table 2).

3.2. Binding of Nickel Compounds to Protein and Then to DNA

A number of interesting papers from various laboratories, including those of Kasprzak and Costa, have indicated that nickel complexes with certain amino acids and peptides (reviewed in [30,31]). The binding of Ni(II) ion to peptides and amino acids is reported to lower the oxidation potential of nickel ions. This is reported to allow more facile oxidation of Ni(II) to Ni(III), causing oxidation of proteins and DNA. Ni(II)-

protein complex-generated oxidants can cause lipid peroxidation and crosslinking of protein to DNA (reviewed in [30]).

In 1991, Nackerdien et al. [38] showed that nickel(II) complexes in the presence of hydrogen peroxide cause damage to DNA bases in isolated human chromatin but do not damage naked DNA appreciably. Hence, these authors began to pursue the hypothesis that nickel bound to protein in such a complex was able to oxidize DNA. The laboratory of Kasprzak synthesized a small peptide based on an evolutionarily con- served site in histone H3 and showed that this peptide forms two complexes with Ni(II), NiL and NiLL, in which Ni(II) is bound through cysteine and histidine side chains in a square complex. This square planar complex is expected to catalyze oxidation reactions. Hence, these authors postulate that the binding of Ni(II) ion to histone H3 in the nuclei of cells may lead to a redox-active complex that could generate oxidants and induce oxidative damage in DNA [39]. They then found, surprisingly, that a dimeric complex of the peptide they synthesized and its weak octahedral Ni(II) complex, in the presence of hydrogen peroxide, catalyzed oxidation of guanosine to 8-oxo-2'-deoxyguanosine. These authors therefore suggested that molecular mechanisms of nickel carcinogenesis may involve the generation of oxidants by Ni(II) complexes and the resultant oxidation of cellular components [40]. Some of these nickel complexes are redox active and can generate hydroxyl radicals. Current hypotheses are that nickel can bind avidly to proteins, and that if these protein- or amino acid-nickel complexes bind to DNA, they may be able to react with molecular oxygen and thereby generate hydroxyl radicals at the site to which the nickel-protein complexes bind to DNA (reviewed in [30,31]). This is important because the generation of hydroxyl radical must be close to DNA in order for it to react with DNA. Hydroxyl radical is so reactive that it reacts with other molecules at diffusion-controlled rates [1,2,31].

3.3. The Apparent Paradox of the Nonmutagenicity of Carcinogenic Nickel Compounds

Since Ni(II) ions do indeed bind avidly to proteins such as histones, and since histones bind avidly to DNA, and since Ni(II) ions bound to his- tones can generate superoxide and eventually hydroxyl radicals at the

site of DNA [30,31], this might be expected to result in mutations in
the DNA of mammalian cells treated with nickel compounds. These
have not yet been shown to occur despite work in many mammalian
mutagenesis assays. In our laboratory, we found that nickel subsulfide
and green nickel oxide, which facilely induce morphological transformation in 10T1/2 cells, do not induce a very specific type of base substitution mutation to ouabain resistance [25–29,37]. Furthermore, we
found that nickel subsulfide, which facilely induced the formation of
anchorage-independent colonies in cultured primary human fibroblastic cells (HFCs), did not induce mutations to either ouabain resistance
or to 6-thioguanine resistance at these same concentrations that induced anchorage independence [41]. Either nickel complexes are not
mutagenic in mammalian cells, or these complexes generate specific
types of mutations that are not easily detectable in classical assays to
detect induction of mutation to ouabain resistance (specific types of
base substitution mutations) or to 6-thioguanine resistance (base substitution mutations, frameshift mutations, small deletions). In addition, nickel complex generation of superoxide and hydroxyl radicals
may be an interesting mechanism for nickel-induced chromosome
breakage, since nickel compounds have been shown to efficiently cause
chromosome breakage in mammalian cells [32–34] (reviewed in [25–
30] (see Table 2).

3.4. Inhibition of Nickel Compound-Induced Cell Transformation by Inhibitors of Arachidonic Acid Release and Metabolism. Further Evidence for the Involvement of Free Radicals in Nickel-Induced Cell Transformation

We have also shown in our laboratory that nickel compounds are able to
induce anchorage independence in diploid human fibroblasts [41]. We
have further shown that acetylsalicylic acid, a known inhibitor of cyclooxygenase, is able to inhibit nickel-induced anchorage independence in
diploid human fibroblasts [65]. This suggests that nickel compounds
may be capable of inducing arachidonic acid release and its metabolism
by cyclooxygenase, with consequent generation of hydroxyl radicals,
and that these may also be able to induce genetic change in mammalian
cells, leading to induction of anchorage independence. Further work is
needed at the level of studies of arachidonic acid metabolism under

these conditions and to trap and identify hydroxyl radicals before concrete conclusions can be drawn from these suggestive inhibitor experiments. However, the intriguing possibility exists that nickel compounds may be able to stimulate arachidonic acid release and its metabolism via cyclooxygenase, leading to generation of hydroxyl radicals and their attack on DNA to cause chromosome breakage and possibly mutations, which may inactivate tumor suppressor genes. Experimental testing of these hypotheses is currently ongoing in our laboratory.

3.5. Molecular Biology of Nickel Compound-Induced Carcinogenesis

To date, our laboratory and other laboratories have shown that insoluble nickel compounds can induce morphological transformation of cultured C3H/10T1/2 mouse embryo cells [37] and Syrian hamster cells (reviewed in [30]). We have also shown that nickel compounds induce anchorage independence in diploid human fibroblasts [41]. Our laboratory has shown that nickel subsulfide and green nickel oxide do not induce mutation to ouabain resistance in 10T1/2 cells [37] and nickel subsulfide does not induce mutation to either ouabain resistance or to 6-thioguanine resistance in human diploid fibroblasts at concentrations that induce morphological transformation in 10T1/2 cells or anchorage independence in human diploid fibroblasts [41]. Hence, we have postulated either that the nickel compounds generate oxygen radicals that are not easily shown to be mutagenic in classical mutagenesis systems in mammalian cells, or that the nickel compounds cause chromosome breakage, leading to inactivation of tumor suppressor genes, and perhaps to breakage or small deletions in elements of negative regulatory controlling elements from protooncogenes, activating them to oncogenes (reviewed in [25–29]).

To address these possibilities, we have begun studies of whether any protooncogenes are amplified or rearranged in nickel-induced transformed 10T1/2 cell lines. We have found that this is not the case, and we have tentatively ruled out this hypothesis for a limited number of oncogenes that we have studied. In addition, we have asked whether there is overexpression of known protooncogenes, and we have also ruled out substantial elevations of steady-state levels of mRNAs transcribed from known protooncogenes that we have studied [66].

Therefore, utilizing a more general technique to look at a large

number of target genes, i.e., mRNA differential display, we have
begun to determine whether specific genes are overexpressed in nickel-
induced transformed 10T1/2 cell lines and whether specific genes are
underexpressed in the transformed cell lines. In preliminary experiments, we found that numerous genes are overexpressed and numerous
genes are underexpressed in these cell lines. Our initial experimental
data, extrapolated from a small number of gels to 100% coverage of the
mRNA pool, indicate that this applies to a total of approximately 300
genes. We are now in the process of isolating, sequencing, identifying,
and characterizing these genes [67]. Further insight into the molecular
mechanisms of nickel-induced cell transformation and hence nickel
carcinogenesis will be gained shortly through the results of these experiments. First, we expect to identify secondary genes whose expression is
influenced by the approximately 5–10 primary genes that are targets
for nickel-induced genetic damage. Then we will determine into which
signal transduction or biochemical pathways these secondary genes fit
and work our way backward to identify the primary genes that nickel
has damaged. Then we propose to identify the molecular damage, muta-
tional or otherwise to these primary genes caused by nickel. It is our
current hypothesis that each of the approximately 5–10 primary genes
that are altered, whether this is an activated oncogene or an inactivated
tumor suppressor gene, influences the activity of, on the average, 60
genes by affecting major biochemical pathways involved in the control
of cellular growth and differentiation in the cell. For instance, we propose that if nickel inactivates a specific tumor suppressor gene, this
might result in the ablation of expression of 60 genes, on average.
Similarly, if nickel activates expression of an otherwise quiescent protooncogene, converting it to an oncogene, this may activate or overstimu-
late an entire signal transduction pathway, causing overexpression of
approximately 60 genes.

4. THE POSSIBLE ROLE OF FREE RADICALS IN ARSENIC CARCINOGENESIS

Specific arsenic compounds are also believed to be carcinogens, cocarcinogens, or tumor progressors in humans due to positive associations
between exposure to arsenic compounds and increased frequencies of

specific cancers in humans. However, to date, arsenic compounds have not been shown to be complete carcinogens in animal carcinogenesis bioassays when administered alone. The molecular mechanism of carcinogenesis, cocarcinogenesis, or tumor progression by specific arsenic compounds in humans remains obscure and is severely hampered by the lack of appropriate animal models. Workers have shown that dimethylarsenic can generate arsenic peroxy radicals. However, radical generation from dimethylarsenic acid occurs at very high concentrations of dimethylarsenic. Therefore, it is still a possibility that specific arsenic compounds may generate radical species during the process of arsenic carcinogenesis. However, it is not clear as to whether any of these radical species derived from arsenic compounds are generated at moderate concentrations of arsenic to which animals or humans could be exposed environmentally. It is therefore not yet clear whether any such arsenic-derived radicals could play an important role in arsenic-induced carcinogenesis, cocarcinogenesis, or tumor progression in humans (see Tables 3 and 5). Further research is needed in this area.

4.1. Carcinogenicity of Arsenic Compounds: The Paradox of Human Carcinogenesis Without Animal Carcinogenesis

Today there is a significant amount of epidemiological evidence that humans in the copper smelting industry, where arsenic compounds are contaminants of the copper ores, exhibit an increased incidence of respiratory cancer. Smelting of copper ores leads to evolution of arsenic trioxide, which is volatile. Exposure of copper smelter workers to arsenic trioxide is likely responsible for the correlation between copper smelting and an increased frequency of respiratory cancer (reviewed in [25–29,42–44]). In addition, exposure of vintners to arsenic compounds during the spraying of these compounds as pesticides led to an increased frequency of skin cancers and leukemia. Further, in Taiwan, arsenic compounds often naturally contaminate artesian wells that are used as sources of drinking water and a source of water to irrigate crops. Drinking of arsenic-contamined water correlates with an increased incidence of cancer of the lung and other internal organs. In the past, treatment of patients with psoriasis with Fowler's solution, containing arsenic compounds, correlated with an increased incidence of skin cancer. All of these cases of increased incidence of cancers in humans exposed to arsenic compounds have been discussed at length many times. In the United States, risk assessment calculations to define the risk of contracting cancer from drinking water are often dominated by the risk attributed to arsenic [42–44]. In summary, these epidemiological studies indicate that exposure of humans to arsenic by inhalation, by dermal contact, or by oral ingestion often correlates with an increased frequency of lung cancer, skin cancer, or lung and other internal cancers, respectively (summarized in Table 3).

4.2. Recent Data on the Ability of Arsenic Compounds to Induce Cell Transformation and Tumor Promotion

Despite epidemiological studies indicating that exposure of humans to arsenic compounds correlates with increased cancer rates, past whole animal carcinogenesis bioassays of various arsenic compounds have been largely negative. Therefore, in animals, arsenic compounds do not appear to be strong complete carcinogens. A resolution of this apparent paradox may be that arsenic compounds are incomplete carcinogens, such as tumor promoters, tumor initiators, or tumor progressors. Recent studies being reported in preliminary form indicate that it may require such methodologies as intratracheal instillation, which adds a wounding component, to cause arsenic administration to animals to induce tumors. This could indicate that arsenic is a very weak carcinogen or a cocarcinogen. A second possibility is that humans are more sensitive to the carcinogenicity of arsenic compounds than mice and rats. There are likely also a number of cofactors affecting the carcinogenicity of arsenic compounds in humans, such as diet, exposure to other carcinogens such as tobacco smoke, and other unknown factors (reviewed in [25–29,44–46] and summarized in Table 3).

4.3. Evidence That Arsenic Compounds May Generate Arsinic Acid Peroxy Radicals, Superoxide, or Hydroxy Radicals

The molecular mechanisms of arsenic carcinogenicity are not yet known. There have been a few previous reports that, following metabo-

TABLE 3

Important Features of Arsenic
Genotoxicity/Carcinogenicity/Cocarcinogenicity

1. Arsenic compounds do not appear to be complete carcinogens in animals.
2. Epidemiological studies indicate that exposure of humans to specific arsenic compounds correlates with an increased frequency of lung cancer, skin cancer, and liver cancer.
3. Arsenic compounds may be incomplete carcinogens or tumor progressors in animals, and may work together with other compounds in humans to induce cancer. This is considered most likely at present. Alternatively, arsenic compounds may have a species specificity and be human-specific carcinogens. There are many factors (cigarette smoking, diet, and unknown cofactors) that may interact with arsenic to induce carcinogenesis in humans.
4. Arsenic compounds are clastogens and comutagens, but the molecular mechanisms by which they exert clastogenicity and comutagenesis are not completely understood.

lism of arsenic to dimethylarsenic, this compound may be able to react with molecular oxygen to generate dimethylarsenic peroxyl radicals [47]. These novel findings have, however, been criticized because these reactions occur at very high concentrations of arsenic. Therefore, to date, there is no strong evidence that arsenic can generate arsenic radicals or other radicals at concentrations of arsenic to which humans are likely to be exposed, such as through environmental contamination. Hence, at the present time, there is not yet strong enough evidence to indicate that dimethylarsenic peroxyl radicals or other radicals are generated and are involved in the molecular mechanisms of arsenic-induced carcinogenesis or co-carcinogenesis. Nevertheless, this is still an open question because the mechanisms of arsenic carcinogenesis and cocarcinogenesis are still unknown, and the findings of Okada et al. [47] are novel and do warrant further study and consideration (Table 5).

5. BERYLLIUM PULMONARY FIBROSIS AND
    CARCINOGENESIS: IS THERE A ROLE FOR FREE
    RADICALS?

Occupational exposure to specific beryllium compounds induces acute
beryllium disease characterized by chemical pneumonitis. Low concentrations of beryllium have caused humans to develop chronic beryllium
disease, a granulomatous lung disease characterized by infiltration of
lymphocytes, pulmonary fibrosis, and reduced pulmonary function.
This may be mediated by the immune system in part. There is the
possibility that it may also be due to the generation of oxygen radicals
by immune cells in the lung, although this interesting hypothesis has
not yet been tested and clearly requires a substantial amount of study
to test it critically. Inhalation of beryllium compounds causes pulmonary tumors in rats, and beryllium is suspected to be a human carcinogen
but has not yet been proven to be so. Again, there is the interesting
possibility that beryllium-induced lung cancer in animals, and perhaps
humans, may be due in part to the generation of oxygen radicals by
immune cells. Clearly, this interesting hypothesis also requires that
sufficient experimentation be done to test it critically (summarized in
Tables 4 and 5).

  Beryllium compounds are used in the brakes of airplanes, in neu-
tron monochromators, as a window material for x-ray tubes, and as a
radiation detector in nuclear technology. Beryllium oxide is used as an
electrical insulator, as a crucible material, and in the transistor and
plastics industries (reviewed in [48]).

  Beryllium is carcinogenic in many animal models. Intravenous
injection of beryllium sulfate into rabbits induces osteosarcomas, and
beryllium oxide, beryllium silicate, and beryllium phosphate are all
animal carcinogens. Beryllium compounds have been shown to be carcinogens when administered by inhalation in rodents and primates.
They are now listed as established animal carcinogens and probable
human carcinogens by the International Agency for Research on Cancer. The epidemiological data are generally equivocal as to whether
beryllium compounds are human carcinogens. The data on beryllium
compounds as clastogens in mammalian cells are equivocal to date, but
there are data suggesting that beryllium compounds are mutagenic in
mammalian cells. Therefore, while beryllium compounds are consid-

ered animal carcinogens and possible/probable human carcinogens, there are few or no data on the molecular mechanisms of their carcinogenicity (reviewed in [49]).

In addition, beryllium compounds have toxic effects on the pulmonary system of humans when administered by the inhalation route. When humans are exposed to very high concentrations of beryllium compounds in an occupational setting, they develop acute beryllium disease, which is a chemical pneumonitis that can be fatal if the exposure is sufficiently high (reviewed in [49,50]). Chronic inhalation of lower concentrations of beryllium compounds induces a syndrome referred to as chronic beryllium disease, which is a granulomatous lung disease with lymphocytic infiltration and pulmonary fibrosis among its symptoms. Acute beryllium disease can be reproduced in a monkey model by treating monkeys with high concentrations of beryllium fluoride, beryllium sulfate, or beryllium phosphate. Exposure of monkeys to lower concentrations of beryllium compounds has led to a model for chronic beryllium disease [50]. Administration of beryllium compounds by inhalation to mice has also led to development of a model for chronic beryllium disease (reviewed extensively in [48,50] and summarized in Table 4).

TABLE 4

Pulmonary Toxicity, Genotoxicity, and Carcinogenicity of Beryllium Compounds and Possible Role of Free Radicals in These Processes

1. Beryllium compounds induce acute beryllium disease and chronic beryllium disease in humans. Inflammation is involved in these processes. A speculative hypothesis can be made that immune cells may generate oxygen radicals during these disease processes. This hypothesis needs to be tested experimentally.

2. Beryllium compounds induce pulmonary carcinogenicity in animals. The involvement of inflammation makes the hypothesis plausible that oxygen radicals may be involved in this process, but this hypothesis needs to be tested experimentally. The evidence for genotoxicity of beryllium compounds is limited. Evidence for beryllium compound-induced pulmonary carcinogenicity in humans is equivocal.

For both cancer and acute and chronic beryllium diseases, there appears to be an immune-mediated component to these diseases. One can easily make the hypothesis that the inflammatory component of these diseases results in oxygen radical generation, which plays a role in causing DNA damage, such as 8-hydroxydeoxyguanosine, that causes mutation and plays a role in the mechanism of these three diseases. However, it will obviously require experimental investigation to test this hypothesis. To date, there has not been much experimental work in this area (reviewed in [49, 50]), so there are no experimental data at the present time in favor of this interesting hypothesis (summarized in Tables 4 and 5).

6. IRON AND FREE RADICALS: IS THERE A ROLE IN CARCINOGENESIS OR TUMOR PROMOTION?

During a specific situation in which there is an overload of iron in mammalian organs, particularly in the liver, there is evidence that carcinogenesis may occur. There is also evidence that iron compounds may be involved in this process and that they may participate in redox reactions leading to the generation of superoxide and hydroxyl radicals. Work in progress in a number of laboratories suggests that iron-induced oxygen radical generation may in some cases be responsible for liver cancer in humans during situations of iron overload. A certain fraction of "endogenous carcinogenesis" may be postulated to be due to the participation of iron in the generation of oxgyen radicals and their subsequent attack on DNA, leading to mutations. This hypothesis requires rigorous experimental testing.

6.1. Iron Compounds and Carcinogenesis

Iron has long been known to be an essential element in biological systems. At the same time, due to its activity as a redox-active metal, it has also been postulated to be able to generate oxygen radicals in Fenton reactions [51]. Hence, iron compounds have also been postulated to be potential human and animal carcinogens. In the human disease hereditary hemochromatosis, iron accumulation is correlated with cirrhosis of the liver and carcinomas of the liver [52]. Until recently, there has been little experimental evidence in animals to support the hypothesis of the carcinogenicity of iron compounds. Iron has been shown to enhance the carcinogenicity of dimethylnitrosamine and diethylnitrosamine in inducing lung and nasal tumors in hamsters when administered topically as iron oxide, and iron oxide delivered by intratracheal instillation along with BaP has enhanced the carcinogenicity of BaP in hamsters. To date, the data on the carcinogenicity of iron compounds alone are weak to equivocal, and the data on carcinogenicity of iron compounds to humans in occupational settings are equivocal or negative (reviewed in [49,52] and summarized in Table 5).

6.2. Ability of Iron Compounds to Generate Free Radicals

Helmut Bartsch and his coworkers have investigated this hypothesis of the potential carcinogenicity of iron compounds very carefully. They also studied the linkage of this hypothesis to the hypothesis that in situations of iron overload, iron-mediated oxidative stress may be responsible for lipid peroxidation and the consequent generation of aldehydes, such as trans-4-hydroxy-2-nonenal, that react with DNA to form exocyclic adducts [53,54]. Some of the background for their studies were the previous studies that these exocyclic adducts, termed etheno adducts, were shown to be formed in DNA treated with the epoxide metabolites of the human carcinogens vinyl chloride and urethane [53]. Their laboratory has developed very sensitive methods utilizing immunoaffinity methods to detect exocyclic DNA adducts [54–56]. These authors have also noted that nitric oxide is overproduced when conditions of chronic infection and inflammation occur, and that several human cancers arise against a background of chronic viral infection. They have utilized a mouse model system, the SGL mouse, which produces high levels of NO after these mice are injected with lymphoma cells. In this model mouse system, they found that lipid peroxide-induced etheno-DNA adducts occur in tissue in which NO is overproduced. This provides an interesting model that inflammation leads to lipid peroxidation and, hence, production of etheno-DNA adducts in animals [57].

Next, these workers have chosen to study the human genetic disease primary hemochromatosis, in which there are iron accumulations in the patients affected, resulting in increased frequencies of liver damage and liver cancer. They have measured increased levels of 1,N-6-ethenodeoxyadenosine (EdA) adducts and 3,N-4-ethenodeoxy-cytidine (EdC) DNA adducts in the livers of patients with this disease [58]. Therefore, these authors hold that elevated iron concentrations elevate the products of lipid peroxidation, which elevate the levels of etheno-DNA adducts. They also suggest that these processes are likely part of the process of liver disease and liver cancer in these patients [58] (summarized in Table 5).

Bartsch et al. also studied the role of dietary ω-6 polyunsaturated fatty acids (PUFAs) in influencing the frequency of human colon and prostate cancer [59]. They have shown that dietary ω-6 PUFAs drastically increase the formation of etheno-DNA base adducts in white blood cells from humans. They found a positive association between dietary fat and the rise of colon and prostate cancer. They also found that high dietary ω-6 PUFA (linoleic acid, C18:12) and low amounts of ω-3 PUFA may be linked to colon cancer risk and breast cancer risk[59]. Finally, they also noted that ω-6 PUFAs enhance experimental mammary tumorigenesis and that ω-3 PUFAs are in high levels in humans. The work of Bartsch, Nair, and their collaborators has been summarized at a recent scientific meeting [60].

These authors conclude that exocyclic DNA adducts derived from lipid peroxidation are good biomarkers to determine the effects of dietary fat intake and oxidative stress on endogenous DNA damage, and may be mechanistically linked to elevated risk for specific human cancers. These authors also stress that the etheno adducts that they have identified and measured should be useful for studies on the biological significance of background levels of such adducts, the repair of such adducts in human tissues, their use in molecular epidemiology studies, ways in which to prevent formation of such adducts, and whether such intervention decreases the risk of specific human cancers [60].

7. COPPER AND FREE RADICALS: IS THERE A ROLE IN CARCINOGENESIS?

There are few or no data to suggest that, under ordinary circumstances, copper compounds are carcinogenic. To date, the data suggest that copper compounds are not carcinogenic. However, again, under specific conditions of genetic abnormalities of copper uptake, transport, and metabolism, such that copper accumulates in humans, there is increasing evidence that high overloads of copper may lead to the generation of oxygen radicals, lipid peroxidation, and formation of etheno-DNA adducts. There is concomitant evidence that under such circumstances the risk of human cancer may be increased. It is reasonable to speculate that under conditions of copper overload, generation of oxygen radicals, such as the hydroxyl radical, may be involved in increasing the incidence of specific human cancers. This is some interesting evidence in support of this hypothesis, and this hypothesis needs to be further tested rigorously and critically by further experimentation (summarized in Table 5).

7.1. Copper Compounds, Toxicity, and Carcinogenesis

To date, there have been few studies testing whether copper compounds are carcinogenic. Further work in this area could be useful, particularly in testing high concentrations of specific copper compounds, to mimic copper overloads, which occur in certain copper storage diseases, such as Wilson's disease (reviewed in [49,52]).

Bartsch and colleagues have utilized an animal model in which Long Evans-Cinnamon (LEC) rats develop liver tumors due to a copper-dependent formation of miscoding etheno-DNA adducts [54] (reviewed in [60]). In this model animal system, these LEC rats spontaneously develop hepatitis, and 100% of the animals develop hepatocellular carcinoma. The rats originally have no symptoms, but they progress such that 40% of them develop a fulminant hepatitis and 60% a prolonged hepatitis. Those rats developing a prolonged hepatitis progress to develop liver cancer in 1 year. These workers have shown that the copper content of the liver correlates with the etheno-DNA adduct levels [54] (summarized in Table 5).

Hence, it is likely that copper is participating in redox reactions, generating oxygen radicals via Fenton chemistry, and that these radicals cause lipid peroxidation. As a result of this lipid peroxidation, aldehydes are generated, and these react with cellular DNA, leading to the formation of etheno adducts in the DNA [54]. It is further likely that the etheno adducts are formed in critical genes, such as oncogenes and tumor suppressor genes, and that this leads to cancer in the livers of these animals. Further molecular biology studies in this elegant system should lead to insight as to the exact genes that are damaged, the mutations that are formed in these genes, and the biological effectiveness of these mutations in inducing tumorigenicity. Such studies are proposed and will be ongoing in the laboratory of H. Bartsch and his collaborators.

These workers have also studied the human genetic disease, Wilson's disease, in which there is an aberrantly high accumulation of copper in patients, resulting in increased frequencies of liver damage and liver cancer. They have measured increased levels of $1,N$-6-ethenodeoxyadenosine adducts and $3,N$-4-ethenodeoxycytidine DNA adducts in the livers of patients with these diseases. In Wilson's disease patients, there is a roughly linear correlation between the amounts of EdA and EdC adducts in DNA versus levels of copper in liver [58]. Therefore, these authors hold that elevated levels of etheno adducts in LEC rats and Wilson's disease patients suggests that there is a role of oxidative stress, causing lipid peroxidation, and that products of lipid peroxidation elevate the levels of etheno-DNA adducts. They also suggest that these processes are likely part of the process of liver disease and liver cancer in these patients [58] (reviewed in [60], see also Table 5).

7.2. Ability of Copper Compounds to Generate Free Radicals

There is of course substantial evidence that specific copper compounds, which dissociate to copper ions and counterions, can increase the generation of oxygen radicals in aqueous solution. Copper is an essential element that is found in many enzymes and transport proteins [61]. Iron is effective through Fenton/Haber-Weiss chemistry to generate hydroxyl radicals, and copper does this also, but not quite as effectively as iron (reviewed in [31,61]). There is likely a role for copper-generated reactive oxygen species in the evolution of Wilson's disease in humans. In this disease, there is a lack of ceruloplasmin, the major copper-chelating protein in plasma, and this leads to higher steady-state levels of low molecular weight copper. This low molecular weight copper may generate reactive oxygen species, and this may be responsible for the neurotoxic symptoms in this disease [61]. Copper has also been shown to generate hydroxyl radicals through the above-mentioned Fenton/Haber-Weiss chemistry, and these hydroxyl radicals initiate lipid peroxidation (reviewed in [62]). Hence, it is clear that copper can generate reactive oxygen species and hydroxyl radicals through Fenton/Haber-Weiss chemistry, and also can cause oxidation of DNA bases [31]. It remains for critical experiments to be done to determine whether such copper-mediated generation of reactive oxygen species occurs during conditions of experimental carcinogenesis and in humans at risk for liver cancer, such as Wilson's disease patients. Finally, experiments need to be done to test critically whether copper-induced generation of reactive oxygen species is indeed causative in experimental carcinogenesis in animals or during liver carcinogenesis in humans. The experiments of Bartsch and his collaborators mentioned above come closest to fulfilling these criteria. Further experiments to critically test correlations between carcinogenesis in animals and humans and copper-induced reactive oxygen species and the products of lipid peroxidation along the lines of the work of Bartsch and collaborators are needed (Table 5).

8. CONCLUSIONS

In chromium carcinogenesis, specific insoluble hexavalent chromium compounds cause chromosome breakage in vitro and carcinogenesis in animals and induce morphological and neoplastic cell transformation of murine cells and anchorage-independent transformation of human fibroblasts in vitro. An ultimate carcinogenic species generated from insoluble hexavalent chromium compounds is considered to be intracellular Cr(III). Whether intracellular Cr(V) generated from intracellular reduction of Cr(VI) is also an ultimate carcinogenic form of chromium is still an open question. There is some evidence that intracellular Cr(VI) functions as a transition metal in Fenton-like reactions to generate oxygen radicals, but more work needs to be done to determine whether these radicals play a significant role in chromium carcinogenesis. It is believed at present that soluble hexavalent chromium com-

TABLE 5

Summary of Knowledge on the Involvement of Radicals in Metal/Metalloid Carcinogenesis

Compounds of the elements	Evidence that these metals/metalloids can generate radicals in biological systems	Evidence that these metals/metalloids generate radicals as part of the mechanism of carcinogenesis
Chromium	Evidence for Cr(V) formation. Evidence for oxygen radical formation from Cr(VI).	Not clear; further work is needed to study both Cr(V)- and Cr(VI)-generated oxygen radical formation and their participation, if any, in carcinogenesis.
Nickel	Evidence is growing for the generation of oxygen radicals by Ni(II)-protein and Ni(II)-peptide complexes at the site of DNA and site-specific DNA damage. Evidence is growing for lipid peroxidation by Ni(II) complexes.	Critical experiments must be done to test the hypothesis that Ni(II)-induced lipid peroxidation or oxygen radical generation plays a role in Ni carcinogenesis.
Arsenic	There is some evidence for the generation of dimethylarsenic peroxy radicals at high concentrations of arsenic. Further work needs to be done in this area.	There is little or no evidence for the role of arsenic-derived or induced radicals in arsenic carcinogenesis.

Beryllium	Little or no work has been done in this area. It is an intriguing hypothesis that beryllium-mediated inflammation could generate oxygen or other radicals. Serious work should be conducted in this area.	Little or no work has been done in this area, so no evidence has been gathered for or against radical involvement in beryllium carcinogenesis.
Iron	There is evidence that in iron storage/overload diseases, lipid peroxidation occurs, leading to formation of etheno-DNA adducts.	Evidence is beginning to develop that oxygen radicals may play a role in liver carcinogenesis in iron storage diseases in animals and in humans. This area is very exciting; correlations between lipid peroxidation and carcinogenesis in animals and humans exist and are getting stronger with time. More work needs to be done to critically test the linkage between iron-generated oxygen radicals and human liver carcinogenesis.
Copper	There is growing evidence that in copper storage diseases, both in animals and in humans, radicals are generated, lipid peroxidation occurs, and etheno-DNA adducts are formed.	There is growing evidence that copper-induced oxygen radicals, lipid peroxidation, and etheno-DNA adducts are formed in the livers of rats and humans with defects in copper metabolism and copper storage diseases while liver carcinogenesis is ongoing. This evidence is strong.

pounds are somewhat less carcinogenic than insoluble hexavalent chromium compounds, but more work needs to be done in this area to test this concept rigorously. It is also believed that soluble trivalent chromium compounds are not carcinogenic, since the animal carcinogenicity data to date are negative for these compounds. At present, this is a reasonable belief, since soluble trivalent chromium compounds are not appreciably taken into mammalian cells. Work from our laboratory has shown that soluble trivalent chromium compounds are not appreciably taken up into primary human diploid fibroblastic cells in culture [63]. Insoluble trivalent chromium compounds require further study. Our laboratory has shown that they adhere tightly to the membranes of diploid human fibroblasts but that they are not internalized into the cells [63]. If the particle size is less than 1 μm, the possibility arises that these compounds could be phagocytosed by cells, allowing them to engulf significant amounts of trivalent chromium, which is believed to be one ultimate intracellular carcinogenic form of chromium. Further work needs to be conducted in this area. Interestingly, both soluble and insoluble hexavalent chromium compounds are approximately 1000-fold more cytotoxic to cultured human diploid fibroblasts, and were also approximately 1000-fold more mutagenic to diploid human fibroblasts, than soluble and insoluble trivalent chromium compounds, as shown in work in our laboratory [63]. We also showed that both soluble and insoluble hexavalent chromium compounds induced dose-dependent anchorage independence in cultured human diploid fibroblasts, over the same concentration ranges at which these compounds were cytotoxic and mutagenic to the human diploid fibroblasts. This suggests the potential of all of these hexavalent chromium compounds, both insoluble and soluble, to be carcinogenic to humans [41,63]. The essential features of this knowledge are summarized in Tables 1 and 5.

It is clear that certain insoluble nickel compounds, such as nickel subsulfide, crystalline nickel monosulfide, and green and black nickel oxides, are carcinogenic in animals and are also able to induce morphological and neoplastic transformation in cultured murine cells. The soluble nickel compounds, such as nickel sulfate, are not carcinogenic in animals in the assays employed to date, although they do induce morphological transformation in cultured 10T1/2 mouse embryo cells, but only at very high concentrations [37,64]. It is likely that the carcinogenicity of the insoluble nickel compounds of particle size less than 1 μm is due in part to the fact that they are phagocytosed, depositing a large concentration of nickel ions in the cell. Since it is believed that a complex of Ni(II) is the ultimate carcinogenic form of nickel compounds, it is likely that insoluble nickel compounds are the strongest carcinogens because they deposit the largest concentration of nickel ions inside cells. Furthermore, it is becoming clear that nickel compounds can generate oxygen radicals, lipid peroxidation, and chromosome breakage. It is likely that nickel-induced oxygen radicals and lipid peroxidation may participate in causing nickel-induced chromosome damage. Further work is needed to prove whether specific oncogenes are activated and whether specific tumor suppressor genes are inactivated in nickel-induced transformed cell lines. Work is also needed to determine whether nickel-generated oxygen radicals induce 8-hydroxyguanosine in cells and whether (and what fraction of) these oxidized base products lead to mutations in tumor suppressor genes and in protooncogenes. Our knowledge of nickel carcinogenesis is summarized in Tables 2 and 5.

Many epidemiological studies show a correlation between increased frequencies of lung, skin, liver, or other internal cancers in persons exposed to arsenic compounds through arsenic-contaminated drinking water, by working in the copper smelting industry, by spraying arsenical-containing pesticides, or by being treated with Fowler' solution for psoriasis. It is also clear that there are inconsistencies in the carcinogenic risk of exposure to arsenic compounds among populations in different geographic areas. Hence, there are modifiers of arsenic-induced carcinogenesis in humans that need to be identified. It is still not clear as to whether arsenic compounds are carcinogens, cocarcinogens, tumor promoters, or modifying factors of carcinogenesis in humans. This same statement can be made of the role arsenic compounds play in lower animals. More work must be done to understand the mechanisms of arsenic carcinogenesis or cocarcinogenesis in humans and in animals. While the possibility of dimethylarsenic-peroxy radicals being generated in biological systems treated with arsenic compounds is intriguing, attempts to demonstrate these radicals at lower, biologically plausible levels of arsenic should be conducted. Then, further investigations to determine whether these radicals play a role in arsenic carcinogenesis or cocarcinogenesis need to be conducted in a critical manner (Tables 3 and 5).

In iron and copper storage diseases and in experimental animal models utilizing iron or copper overloads, there is evidence that there is generation of the products of lipid peroxidation, likely due to generation of oxygen radicals and their attack on cellular lipids. The products of lipid peroxidation lead to the formation of etheno adducts, which are mutagenic due to their formation in animals treated with carcinogenic vinyl chloride. Further measurements of the generation of and the identity of these radicals would strengthen the concept that iron- or copper-generated oxygen radicals lead to lipid peroxidation, etheno-DNA adduct formation, and carcinogenesis. Further work is also needed to substantiate whether protooncogenes are activated to oncogenes and whether tumor suppressor genes are mutationally inactivated or deleted by these processes in model animal systems and in tissues from humans who contract liver cancer in situations of iron overload (hemochromatosis) or of copper overload (Wilson's disease). This is a very exciting area that merges the concepts of iron or copper overload with lipid peroxidation, etheno-DNA adducts, and liver carcinogenesis in humans (summarized in Table 5).

ACKNOWLEDGMENTS

The author thanks three senior scientists who had a significant impact on his scientific career via their mentorship, friendship, and colleagueship. First, the author thanks the late Dr. Melvin Calvin, University Professor in the Department of Chemistry at the University of California at Berkeley, Member of the National Academy of Sciences of the United States, and Nobel Laureate. Professor Calvin and his collaborators over many decades in his chemical carcinogenesis group made many seminal contributions to the field of chemical carcinogenesis. As my doctoral thesis advisor, Dr. Calvin taught me the basics of chemical carcinogenesis, the necessity of critical thinking and rigor in science, as well as the necessity of keeping an open mind for new advances and directions in science and for giving collaborators and competitors credit for their accomplishments. Second, I thank the late Professor Charles Heidelberger, member of the National Academy of Sciences of the United States, with whom I did postdoctoral study at the USC/Norris Comprehensive Cancer Center at the University of Southern California (USC). Professor Heidelberger taught me the basic considerations of the use of cell transformation systems to study the molecular and cellular mechanisms of chemical carcinogenesis. Third, I thank Dr. Paul Hochstein, Professor Emeritus, Department of Molecular Pharmacology and Toxicology, USC School of Pharmacy. Professor Hochstein was my collaborator and still is my friend, and he graciously taught me some of the rudiments of free radical chemistry and biochemistry and attracted me into the field of free radicals by virtue of his vast knowledge and seminal contributions to this area, his willingness and delight to share that vast knowledge with others, his mentorship of myself and other junior colleagues, and his large and generous spirit, and continual optimism. Finally, I also thank my past and present graduate students and postdoctoral fellows, with whom it has been my privilege to work. Lastly, I thank my colleagues and friends in the field of metal carcinogenesis, who have unfailingly shared their interesting science with me and showed me collegiality and courtesy over many years. These include, but are not limited to, Drs. Max Costa, F. William Sunderman, Toby Rossman, Kas Kasprzak, Michael Waalkes, and Ernest Foulkes.

ABBREVIATIONS

BaP	benzo(a)pyrene
CHO	Chinese hamster ovary
EdA	1,N-6-ethenodeoxyadenosine
EdC	3,N-4-ethenodeoxycytidine
HFC	human fibroblastic cells
LEC rats	Long Evans-Cinnamon rats
NNK	N-nitrosonornicotine
NNN	4-(methylnitrosamino)-1-(3-pyridyl)-1-butanone
PAH	polycyclic aromatic hydrocarbon
PUFA	polyunsaturated fatty acid

REFERENCES

1. B. N. Ames, M. K. Shigenaga, and T. M. Hagen, in *Biological Oxidants and Antioxidants: New Strategies in Prevention and Therapy* (L. Packer and E. Cadenas, eds.), Hippokrates Verlag, Stuttgart, 1994, pp. 193–203.

2. T. A. Dix, K. M. Hess, M. A. Medina, S. L. Tilly, and R. W. Sullivan, in *Biological Oxidants and Antioxidants: New Strategies in Prevention and Therapy* (L. Packer and E. Cadenas, eds.), Hippokrates Verlag, Stuttgart, 1994, pp. 13–23.

3. R. K. Boutwell, *Prog. Exp. Tumor Res.*, *4*, 207–250 (1964).

4. H. C. Pitot and Y. D. Dragan, *FASEB J.*, *5*, 2280–2286 (1994).

5. J. C. Barrettt and P. O. P. Ts'o, *Proc. Natl. Acad. Sci. USA*, *75*, 3297–3301 (1978).

6. J. R. Landolph, *IARC Publication No. 67, Transformation Assay with Established Cell Lines* (T. Kakunaga and H. Yamasaki, eds.), 1985, pp. 185–198.

7. J. R. Landolph, in *Role of Chemicals and Radiation in the Etiology of Neoplasia* (E. Huberman and S. H. Barr, eds.), Argonne National Laboratory Symposium, Raven Press, 1985, pp. 195–210.

8. J. A. Miller and E. C. Miller, in *Mechanisms of Carcinogenesis*, Book B (H. H. Hiatt, J. D. Watson, and J. A. Winsten, eds.), Cold Spring Harbor Laboratory, 1977, pp. 605–627.

9. T. Sugimura, *Science*, *258*, 603–607 (1992).

10. J. McCann, E. Choi, E. Yamasaki, and B. N. Ames, *Proc. Natl. Acad. Sci. USA*, *72*, 5135–5139 (1975).

11. E. Huberman, R. Mager, and L. Sachs, *Nature*, *264*, 360–336 (1976).

12. J. R. Landolph and C. Heidelberger, *Proc. Natl. Acad. Sci. USA*, *76*, 930–934 (1979).

13. C. Shih and R. Weinberg, *Cell*, *29*, 161–169 (1982).

14. A. Eva and S. Aaronson, *Science*, *220*, 955–956 (1983).

15. K. Sukumar, V. Notario, D. Martin-Zanca, and M. Barcacid, *Nature*, *306*, 658–611 (1983).

16. A. Balmain, M. Ramsden, G. T. Bowden, and J. Smith, *Nature*, *307*, 658–660 (1984).

17. B. E. Weissman, P. J. Saxon, S. R. Pasquale, G. R. Jones, A. G. Geiser, and E. J. Stanbridge, *Science*, *236*, 175–180.

18. C. A. Finlay, P. W. Hinds, and A. J. Levine, *Cell*, *57*, 1083–1086 (1989).

19. G. Klein, *Science*, *238*, 1539–1545 (1987).

20. R. Sager, *Science*, *246*, 1406–1412 (1989).

21. E. J. Stanbridge, *Science*, *247*, 12–13 (1990).

22. M. J. Bishop, *Science*, *235*, 305–311 (1987).

23. R. A. Weinberg, *Science*, *230*, 770–776 (1985).

24. T. Shuin, P. C. Billings, J. R. Lillehaug, S. R. Patierno, P. Roy-Burman, and J. R. Landolph, *Cancer Res.*, *46*, 5302–5311 (1986).

25. J. R. Landolph, *Biol. Trace Elem. Res.*, *21*, 459–467 (1989).

26. J. R. Landolph, in *CRC Critical Reviews in Toxicology: Biological Effects of Heavy Metals*, Vol. 2 (E. Foulkes, ed.), CRC Press, Boca Raton, Chapter 5, 1990, pp. 1–18.

27. J. R. Landolph, *Environ. Health Persp.*, *102*(S3), 119–123 (1994).

28. J. R. Landolph, "Arachidonic acid metabolism, oxygen radicals, and chemical carcinogenesis," in *Biological Oxidants and Antioxidants* (L. Packer and E. Cadenas, eds.), Hippokrates Verlag, Stuttgart, 1994, pp. 133–143.

29. J. R. Landolph, M. Dews, L. Ozbun, and D. P. Evans, in *Toxicology of Metals* (L. Chang, L. Magos, and T. Suzuki, eds.), CRC Press, Boca Raton, 1996, pp. 321–329.

30. M. Costa, in *Toxicology of Metals* (L. Chang, L. Magos, and T. Suzuki, eds.), CRC Press, Boca Raton, 1996, pp. 245–252.

31. K. Kasprzak, in *Toxicology of Metals* (L. Chang, L. Magos, and T. Suzuki, eds.), CRC Press, Boca Raton, 1996, pp. 299–320.

32. K. Gasiorek and M. Bauchinger, *Environ. Mutagen.*, *3*, 513–518 (1981).

33. M. L. Larramendy, N. C. Popescu, and J. A. DiPaolo, *Environ. Mutagen.*, *3*, 597–606 (1981).

34. M. Nishimura and M. Umeda, *Mutat. Res.*, *68*, 337–349 (1979).

35. C. B. Klein, in *Toxicology of Metals* (L. Chang, L. Magos, and T. Suzuki, eds.), CRC Press, Boca Raton, 1996, pp. 205–220.

36. S. R. Patierno, D. Banh, and J. R. Landolph, *Cancer Res.*, *48*, 5280–5288 (1988).

37. T. Miura, S. R. Patierno, T. Sakuramoto, and J. R. Landolph, *Environ. Mol. Mutagen.*, *14*, 65–78 (1989).

38. Z. Nackerdien, K. S. Kasprzak, G. Rao, B. Halliwell, and M. Dizdaroglu, *Cancer Res.*, *51*, 5837–5842 (1991).

39. W. Bal, J. Lukszo, M. Jezowska-Bojczuk, and K. S. Kasprzak, *Chem. Res. Toxicol.*, *8*(5), 683–692 (1995).

40. W. Bal, J. Lukszo, and K. S. Kasprzak, *Chem. Res. Toxicol.*, *9*(2), 535–540 (1996).

41. K. A. Biedermann and J. R. Landolph, *Cancer Res.*, *47*, 3815–3823 (1987).

42. A. H. Smith, C. Hopenhayn-Rich, M. N. Betes, H. M. Goeden, I. Hertz-Picciotto, H. M. Duggan, R. Wood, M. J. Kosnett, and M. T. Smith, *Environ. Health Persp.*, *97*, 259–267 (1992).

43. Z. Wang and T. Rossman, in *Toxicology of Metals* (L. Chang, L. Magos, and T. Suzuki, eds.), CRC Press, Boca Raton, 1996, pp. 221–229.

44. A. M. Fan, in *Toxicology of Metals* (L. Chang, L. Magos, and T. Suzuki, eds.), CRC Press, Boca Raton, 1996, pp. 39–54.

45. C. Ferreccio, C. Gonzalez, V. Milosavjlevic, G. Marshall, and A. M. Sancha, Abstract #S-14, in *Arsenic: Health Effects, Mechanisms of Actions, and Research Issues, Meeting Abstract Booklet*. Meeting sponsored by the U.S. National Cancer Institute, National Institute of Environmental Health Sciences, and the Environmental Protection Agency, Sept. 22–24, 1997, Marriott's Hunt Valley Inn, Hunt Valley, Maryland.

46. N. Hotta, in *Arsenic: Health Effects, Mechanisms of Actions, and Research Issues*, Abstract #S-16, Meeting sponsored by the U.S. National Cancer Institute, the National Institute of Environmental Health Sciences, and the Environmental Protection Agency, Sept. 22–24, 1997, Marriott's Hunt Valley Inn, Hunt Valley, Maryland.

47. S. Okada, in *Arsenic: Health Effects, Mechanisms of Actions, and Reseach Issues*, Abstract #S31, Meeting sponsored by the U.S. National Cancer Institute, the National Institute of Environmental Health Sciences, and the Environmental Protection Agency, Sept. 22–24, 1997, Marriott's Hunt Valley Inn, Hunt Valley, Maryland.

48. H. R. Zorn, T. W. Stiefel, J. Beuers, and R. Schlegelmilch, in *Handbook on Toxicity of Inorganic Compounds* (H. G. Seiler, H. Sigel, and A. Sigel, eds.), Marcel Dekker, New York, 1988, pp. 105–114.

49. M. D. Cohen, D. H. Bowser, and M. Costa, in *Toxicology of Metals* (L. W. Chang, L. Magos, and T. Suzuki, eds.), CRC Press, Boca Raton, 1996, pp. 254–255.

50. J. M. Benson and J. T. Zelikoff, in *Toxicology of Metals* (L. W.

Chang, L. Magos, and T. Suzuki, eds.), CRC Press, Boca Raton, 1996, pp. 254–255.

51. B. Halliwell, *FASEB J.*, *1*, 358–364, (1987).

52. L. A. Poirier and N. A. Littlefield, in *Toxicology of Metals* (L. Chang, L. Magos, and T. Suzuki, eds.), CRC Press, Boca Raton, 1996, pp. 289–298.

53. H. Bartsch, A. Barbin, M. Marion, J. Nair, and Y. Guichard, *Drug Metab. Rev.*, *26*, 349–371 (1994).

54. J. Nair, H. Sone, M. Hagao, A. Barbin, and H. Bartsch, *Cancer Res.*, *56*, 1267–1271 (1996).

55. G.-L. Chung, J.-J. C. Chen, and R. G. Nath, *Carcinogenesis*, *17*, 2105–2111 (1996).

56. G. El-Ghissassi, A. Barbin, J. Nair, and H. Bartsch, *Chem. Res. Toxicol.*, *8*, 273–283 (1995).

57. J. Nair, A. Gal, S. Tamir, S. Tannenbaum, J. G. Wogan, and H. Bartsch, *Proc. Am. Assoc. Cancer Res.*, Abstract #522 (1997).

58. J. Nair, P. L. Carmichael, R. C. Fernando, D. H. Philips, and H. Bartsch, *Proc. Am. Assoc. Cancer Res.*, Abstract, #813 (1996).

59. J. Nair, C. E. Vaca, I. Velic, M. Mutanen, L. M. Valsta, and H. Bartsch, *Cancer Epidemiology Biomarkers Prevention*, in Press (1998).

60. Bartsch, H. and Nair, J., in *Cancer: Genesis, Detection, Therapy — 7th Charles Heidelberger Meeting, Abstract Booklet*, Reisensburg Castle, Gunzburg, Germany, July 2–5, 1997, pp. 17–18.

61. S. C. Bondy, in *Toxicology of Metals* (L. W. Chang, L. Magos, and T. Suzuki, eds.), CRC Press, Boca Raton, 1996, pp. 699–706.

62. A.-L. Nieminen and J. J. Lemasters, in *Toxicology of Metals* (L. W. Chang, L. Magos, and T. Suzuki, eds.), CRC Press, Boca Raton, 1996, pp. 887–899.

63. K. A. Biedermann and J. R. Landolph, *Cancer Res.*, *50*, 7835–7842, (1990).

64. S. Ohshima, A. Verma, and J. R. Landolph (manuscript in preparation).

65. K. A. Biedermann and J. R. Landolph (manuscript in preparation).

66. A. Verma and J. R. Landolph (manuscript in preparation).

67. A. Verma, J. Ramnath, S. Ohshima, and J. R. Landolph (manuscript in preparation).

15

pH-Dependent Organocobalt Sources for Active Radical Species: A New Type of Anticancer Agents

Mark E. Vol'pin, *Iliu Yu. Levitin,[1]*
and Sergei P. Osinsky[2]

[1]Institute of Organoelement Compounds, Russian Academy of Sciences, Moscow, 117813 GSP-1, Russia

[2]Institute of Experimental Pathology, Oncology, and Radiobiology, Kiev, 252022, Ukraine

---

*Deceased on Sept. 28, 1996.

1. INTRODUCTION: PREREQUISITES FOR THE SEARCH FOR POTENTIAL ANTICANCER DRUGS AMONG ORGANOCOBALT COMPLEXES

It is widely known that the main feature of malignant neoplastic cells distinguishing them from normal ones is their capability for uncontrollable proliferation. Hence, any anticancer agent must possess cytostatic and, to some extent, cytotoxic properties. The problem is to make its action selective, directing it against tumorous rather than normal tissues. Evidently, differences between the two kinds of tissue should be exploited for this purpose. Principally, such differences exist at any level, from the molecular to the physiological.

In our opinion, the microphysiological approach, i.e., exploitation of distinctions between a tumor and an adjacent normal tissue, is yet underdeveloped in cancer treatment even though most of the differences are known since Warburg's works of the 1920s and early 1930s [1]. Tumorous tissue is characterized by slow blood flow, reduced partial pressure of dioxygen, and low extracellular pH [2,3].

It looks especially alluring to exploit the enhanced acidity of malignant neoplasm. This is particularly so, because the difference in pH between tumorous and normal tissues, which typically is 0.2–0.5 pH unit, can be greatly increased, up to 1.5–2, by artificial intensification of the glycolysis, first of all by means of glucose infusion [3]. In this way the pH in some tumors can be decreased to a level of nearly 5.5 [4]. Thus, the primary task in this approach consists in designing appropriate pH-dependent agents, i.e., relevant cytostatics or modulators of known antitumor drugs and therapeutic means. Several substances generating conventional cytotoxic compounds such as nitrogen mustards and their analogs under slight acidification have recently been reported to fit this purpose (see, e.g. [5,6]). Nevertheless, to our knowledge, none of them has passed through even in vivo examinations.

Within the above approach, the line that we have been developing for about 15 years consists of the use of acid-sensitive sources for active radical species. This is primarily based on vast evidence in the literature supporting the ability of transient free radicals, including alkyl ones, to cleave or damage biomolecules, e.g., nucleic acids, as well as intracellular structures and membranes (see, e.g. [7]).

To develop this new line in cancer treatment, we needed compounds or systems capable of generating active radical species under physiological conditions in the slightly acidic media characteristic of malignant neoplasm, i.e., only at pH values less than about 7. To our knowledge, none of the various free radical sources known satisfied this requirement; hence, such compounds or systems had yet to be found.

We decided to look for such sources among organocobalt complexes, i.e., in the range of Co(III) compounds containing a metal-carbon σ-bond. This choice was made because of their susceptibility to homolytic cleavage and the dependence of the reaction rate on the coordination environment of the metal. The latter point was theoretically established and experimentally proven, in the case of a varying trans-R ligand in a benzyl(pyridine)cobaloxime series by Halpern et al. in the 1980s [8]: homolysis is facilitated by a decreasing donor strength of the para-substituted pyridine ligand. For our purpose, it is essential that raising the acidity of the medium allows, in principle, a change of the coordination environment of the metal in the needed direction, e.g., to substitute a nitrogen base ligand by a water molecule (i.e., a weaker donor), or even to cause protonation of a donor atom (see, e.g., [9]). However, with the known types of organocobalt complexes, the effect of such substitutions is not great enough and protonation occurs only under strong acidification, i.e., beyond the physiological pH range.

2. ALKYLCOBALT(III) CHELATES WITH TRIDENTATE SCHIFF BASES

2.1. Preparation

In this connection, we assumed that steric hindrance arising in certain known alkylcobalt chelates due to bulky groups in the chelating ligands may favor the formation of some novel organometallic species, which would fit our purpose, under the solvolytic conditions of template synthesis — a technique first applied to the preparation of organocobalt complexes by Schrauzer et al. [10]. Indeed, this approach allowed us to obtain organocobalt complexes of a new type (Scheme 1, formulas III

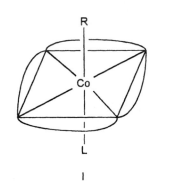

I

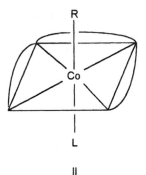

II

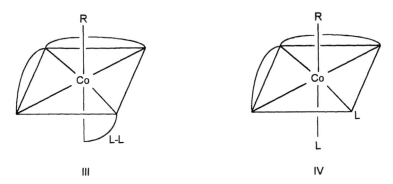

III   IV

SCHEME 1. Typical coordination geometries of monoorganocobalt(III) complexes. (Reproduced by permission from [24].)

and IV) [11]. They are octahedral, like most compounds of trivalent cobalt, and include a primary or secondary alkyl group bound to the metal. Unlike well-known types of organocobalt complexes, i.e., those with macrocyclic or chelating ligands (Scheme 1, formulas I and II), they contain a tridentate ligand rather than a tetradentate or bis-bidentate one such as *trans*-bis-dioxamates. It is the anionic residue of a Schiff base composed of an aliphatic diamine, which is capable of chelation and has at least one primary amino group, and a ketoenol, i.e., a β-diketone or an aromatic *o*-hydroxyketone, in a 1:1 ratio. The two remaining coordination sites are occupied by either a chelating biden-tate ligand (in III) or two Lewis bases (in IV). To our knowledge, no other research groups have dealt with such complexes yet.

Scheme 2 summarizes the results of our template syntheses [11–15]. It shows that the structures of the organometallic products depend on the nature of the complexing agents, i.e., on their substituents. Steric factors seem to operate decisively. Thus, chelates of the new type (formulas VI and VII) can be obtained from ketoenols, but not from their aldehyde analogs. Further, the fifth and sixth coordination sites in the product are usually occupied by the chelating diamine (in VI). In

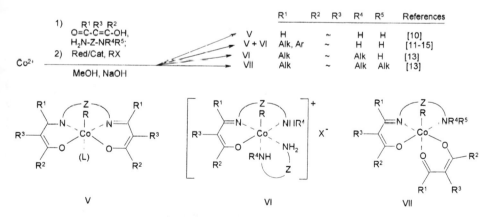

	R^1	R^2	R^3	R^4	R^5	References
V	H	~		H	H	[10]
V + VI	Alk, Ar	~		H	H	[11-15]
VI	Alk	~		Alk	H	[13]
VII	Alk	~		Alk	Alk	[13]

SCHEME 2.   Template synthesis of alkylcobalt(III) chelates with Schiff bases. Z is a 2- or 3-link saturated hydrocarbon bridge. Red/Cat = NaBH$_4$/Pd or H$_2$/Ni. RX is a *prim*- or *sec*-alkyl bromide or iodide. The ~ symbol means either that R^2 = alkyl and R^3 = H or alkyl, or that R^2, R^3 represent the fragment of the annelated *o*-arylene. (Reproduced by permission from [15].)

this case, the complexes are cationic. However, if the complex contains a tertiary amino group, such a coordination proved to be sterically hindered, so that the product (VII) includes a ketoenolate as the bidentate ligand, and hence is neutral.

It is noteworthy that the tridentate ligands in complexes derived from diamines with both primary amino groups correspond to Schiff bases that are quite unstable toward disproportionation, particularly in protic media. Concurrent formation of two organometallic products was observed in this case, including the well-known neutral complexes containing tetradentate Schiff bases (V). Despite this and other complications, the alkylcobalt chelates with tridentate ligands can usually be prepared in satisfactory yields.

A partially successful attempt to elucidate the route of the template synthesis was made by using ethylenediamine and o-hydroxyacetophenone as complexing agents [13].

To diversify further a variety of the complexes prepared by template synthesis, we attempted to carry out ligand substitutions. The direct replacement proved unattainable because of the relative kinetic inertness of the starting complexes, on the one hand, and their rather poor thermostability, on the other. Nevertheless, we found that substitution of the chelating diamine can be performed as a proton-assisted process under mild, strictly controlled conditions [16]: the diamine ligand was replaced by weaker bases, both mono- and bidentate ones, under the action of buffers composed of such a base and its conjugated acid.

Monodentate ligands in complexes obtained in this way are bound more loosely, so that they can be readily replaced. Such direct substitutions are particularly suitable for introducing ligands that effectively bind Co(III), namely, typical chelating molecules and anions, or strong uniacidic bases, such as the hydroxide anion.

Both the proton-assisted and direct ligand substitutions were used to prepare a wide variety of alkylcobalt chelates with tridentate Schiff bases ranging in charge from cationic through neutral to anionic. Furthermore, counterion exchange procedures can be applied to the cationic complexes [16]. Synthetic routes to and types of complexes obtained (formulas IX–XIV) as well as examples studied are shown in Scheme 3. The whole range of alkylcobalt chelates with tridentate Schiff bases that have been prepared by now is presented in Scheme 4.

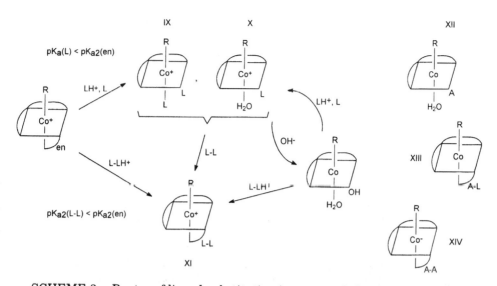

SCHEME 3. Routes of ligand substitution in organocobalt(III) chelates with tridentate Schiff bases and types of complexes thus synthesized. L, neutral monodentate Lewis bases as exemplified by pyridine, NH_3, R^1NH_2, R^1R^2NH, NH_2OH, and PPh_3. A, anionic monodentate ligand such as OH^-, SCN^-. L-L, neutral chelating bidentate ligand such as 2,2'-bipyridyl, 1,10-phenanthroline, adenine, guanine, o-$C_6H_4(NH_2)_2$, en, N-Me-en, N,N-Me_2-en, 1,2-bis(diphenyl-phosphino)ethane. A-L, (mono)anionic chelating bidentate ligand such as quin-olinate-8, o-OC_6H_4COMe, glycinate, alaninate, phenylalaninate, prolinate, ly-sinate, 1/2 cystinate. A-A, dianionic chelating bidentate ligand exemplified by catecholate, 3,5- and 3,6-$(t$-$Bu)_2$ catecholate.

2.2. Structure and Characterization

We studied the structure and reactivity of alkylcobalt chelates with tridentate Schiff bases. They were mostly represented by the homolo-gous series [RCo(acacen)(en)]Y (XV with R^1, R^3 = H, R^2, R^4 = Me, Z = $(CH_2)_2$, and L-L' = en, or XVA) and [RCo(7-Me-salen)(en)]Y (XVII with R^1, X = H, R^2 = Me, Z = $(CH_2)_2$, and L-L' = en, or XVIIA) which were selected due to their ready availability and fairly high stability. Please note that the symbolic terms of the Schiff base ligands here inevitably deviate from the conventional ones because the latter do not take into account the number of carbonylydenate residues.

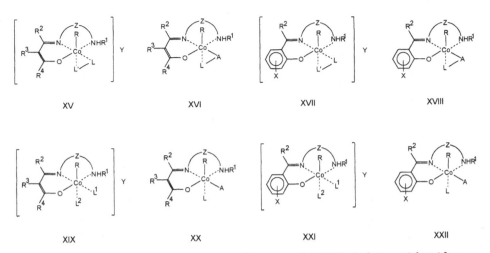

SCHEME 4. Main structural types of organocobalt(III) chelates with triden-
tate Schiff bases. R, *prim*- or *sec*-alkyl; R^1, R^3, H or alkyl; R^2, R^4, alkyls, or alkyl
and aryl; X, functional group such as $-NH_2$, $-SO_3Na$, $-COONa$; Y, counter
(an)ion such as Cl^-, Br^-, I^-, NO_3^-, ClO_4^-, acetate, benzoate, n-$C_{17}H_{34}COO^-$, 1/2
o-$C_6H_4(COO)_2^{2-}$; Z, 2- or 3-link saturated hydrocarbon bridging group such as
$(CH_2)_n$, $n = 2$, 3, or CH_2CHMe; L_1, L_2, neutral monodentate ligands exemplified
by pyridine(s), or by NH_3 and H_2O in cis and trans-R positions, respectively;
L, A, neutral and anionic monodentate ligand, respectively, such as H_2O, and
OH^- or SCN^-; L-L', neutral chelating bidentate ligand such as 2,2'-bipyridyl,
1,10-phenanthroline, adenine, guanine, o-$C_6H_4(NH_2)_2$, en, *N*-Me-en, *N,N*-Me$_2$en;
L-A, (mono)anionic chelating bidentate ligand such as quinolinate-8,
o-OC_6H_4COMe, glycinate, alaninate, phenylalaninate, prolinate, lysinate, 1/2
cystinate.

    The main structural features of our complexes established by
x-ray analysis [11,13,17] can be described as follows and are exemplified
in Fig. 1. The tridentate ligand is coordinated in the meridional plane
with respect to the Co-C bond. Next, the bonds of the metal ion with
Lewis bases in the trans and cis-R positions are sharply nonequivalent,
i.e., the trans-R bond is much weaker than the cis one. In this case,
such a difference is displayed as an asymmetrical coordination of the
chelating diamine: the trans-R Co-N bond is much longer. Its loosening
is obviously due to a strong trans-labilizing influence of the carbanion
ligand, i.e., the alkyl group bound to the metal. In the complexes with
two monodentate Lewis bases, the weakness of trans-R coordination

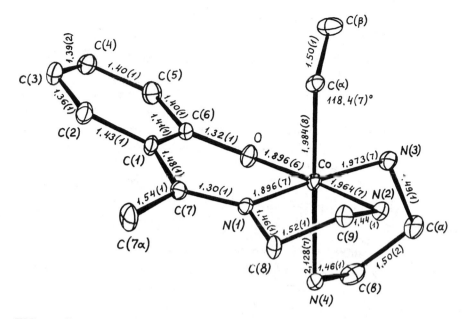

FIG. 1.   Structure of the cation in [EtCo(7-Me-salen)(en)]Br (XVIIA with R = Et and Y = Br). (Reproduced by permission from [17].)

was revealed by ligand exchange kinetics [16]. Namely, these replacements in the trans- and cis-R positions are fast and slow in the nuclear magnetic resonance (NMR) time scale, respectively.

The crystal lattices of such cationic complexes are effectively stabilized by numerous hydrogen bonds formed by the coordinated amino groups with the halide anions. As a consequence, such complexes, while being rather unstable in solution, can be stored for a long time in the solid state [18–20].

Alkylcobalt chelates with tridentate Schiff bases were thoroughly characterized using various techniques, in particular ¹H NMR and infrared (IR) spectroscopy [13,15,16], thin-layer chromatography (TLC) and high-performance liquid chromatography (HPLC) [19], and, in the case of cationic complexes, capillary electrophoresis [20]. Relevant procedures were developed to assay the complexes in question as well as via their degradation with a strong acid [16] and iodine [15,21].

2.3. Acid-Induced Decomposition

The main reactivity feature of our complexes is their ready decomposition under the action of acids and, in particular, the strictly homolytic pathway this follows [11,22–27]; other aspects of their reactivity have been considered elsewhere [28,29]. Neither protolysis of the metal-carbon bond nor β-elimination contributes to this process. These points were proven by using the isotope labeling technique and by the lack of dihydrogen in the products, respectively (Scheme 5).

The generation of alkyl free radicals in the course of these decompositions was directly proven by spin-trapping techniques [22]. Carrying out these reactions in buffer solutions in the presence of *tert*-nitrosobutane or C-phenyl-*tert*-butylnitrone (PBN), we detected electron spin resonance (ESR) signals characteristic of spin adducts of relevant alkyl radicals. Kinetic measurements using PBN have shown that the intensity of the signal is pH-dependent. Their quantitative evaluation allows the conclusion that stationary concentrations of alkyl free radicals produced at low conversions increase with rising acidity of the medium (Scheme 6). Hence, protons are definitely involved in the process leading to the formation of free radicals.

Obviously, such a ready homolytic decomposition under the action of protons is unusual for organometallic complexes. We thoroughly studied the kinetics and mechanism of this process using a number of

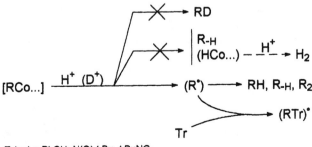

SCHEME 5. Evidence for homolysis of the M–C bond in alkylcobalt(III) chelates with tridentate Schiff bases as the only pathway for their decomposition in acidic media.

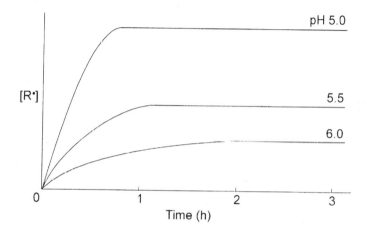

$$RCo^{III}... \xrightarrow{H^+} R^\bullet + Co^{II}...$$

$$R^\bullet + PhCH=N(O)\text{-}t\text{-}Bu \xrightarrow{k} PhCHRN(\overset{\bullet}{O})\text{-}t\text{-}Bu$$
$$\qquad\qquad PBN \qquad\qquad\qquad (PBN\text{-}R)^\bullet$$

$$d[(PBN\text{-}R)^\bullet]/dt = k[PBN][R^\bullet]; \quad [PBN]_0 \gg [RCo]_0$$
$$d[(PBN\text{-}R)^\bullet]/dt \approx k[PBN]_0[R^\bullet] = k'[R^\bullet] \quad (k' = k[PBN]_0)$$

SCHEME 6. Kinetic evidence for the involvement of protons in the generation of alkyl free radicals. The concentration of the ethyl free radical versus time was obtained in accordance with the reaction and rate equations presented in this scheme, namely, by graphic derivation from the relevant experimental data on the time dependence of the ESR signal intensity of the spin adduct in the case of [EtCo(7-Me-salen)(en)]Br (XVIIA with R = ethyl, Y = Br) in phosphate buffer solutions at 20°C.

techniques [22–27]. The overall pattern is very complicated because of intervening radical chain reactions and acid–base equilibria. They arise in the absence of radical scavengers, and in strongly acidic and alkaline media, respectively. However, neither of these complications needs to be taken into account for our purpose because they do not seem to operate under true physiological conditions. Hence, we can consider here only the mainstream reaction occurring at neutral to mildly acidic pH (Scheme 7). Its first step consists of the reversible protonation-decoordination of the amine ligands, in this case ethylene-

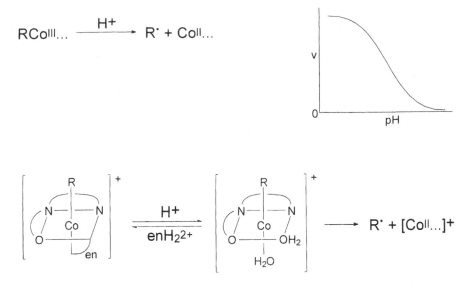

SCHEME 7.   Typical kinetic pattern, mechanism, and driving force of proton-induced homolytic decomposition of alkylcobalt(III) chelates with tridentate Schiff bases in slightly acidic media.

diamine. The resulting diaquo complex is much more susceptible to homolytic cleavage.

Thus, the main intrinsic cause of the homolytic decomposition of our complexes under the action of acids is indeed the reduction of the overall donor strength of the ligands other than alkyl due to proton-induced changes in the coordination sphere. Evidently, in this case the effect should be large, since the change here encompasses two coordination sites rather than one.

Figure 2 exemplifies pH profiles of this reaction. One can see that it does proceed under quasi-physiological conditions and that the pH dependence follows an S-shaped manner. Naturally, the rate constant and the exact pH range can be controlled by varying the ligands.

2.4.   Modeling Biological Action

Eventually, the above kinetic data mean that the isopropyl complex totally decomposes in solution under standard physiological values of

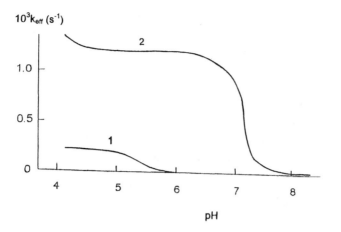

FIG. 2. pH profile of first-order rate constants for the decomposition of
[RCo(7-Me-salen)(en)]Br complexes (XVIIA with Y = Br) under quasi-physiological conditions (H_2O-ethanol 9:1 v/v solutions, I = 1 M, 25°C; in air): R = ethyl
(curve 1) and isopropyl (curve 2, which is reproduced by permission from [24]).

pH (close to 7.5) and temperature (37°C) within a few minutes. On the
other hand, pharmacokinetic studies using rats treated with this preparation (see below) have shown that animal tissues, including tumors,
retain the original complex at least partially intact for several hours
after treatment. In this connection, we have undertaken model kinetic
experiments using blood plasma and milk emulsion as a reaction medium and found that they greatly increase the stability of the complex
(Fig. 3) [27]. It should be noted that solutions of the most conventional
surfactants do not display such an effect. There is indirect evidence
suggesting that such a stabilization of the complex occurs as soon as it
enters the hydrophobic nucleus of a lipoprotein particle. It is noteworthy that the kinetics of the decomposition in blood plasma is complicated: its first-order transform includes an inflection point (Fig. 3).
Hence, one may assume that the complex can be arranged either in two
different positions in the lipophilic nuclei or in two different types of the
latter. Evidently, these findings may be useful in searching for ways to
stabilize the therapeutically useful forms of these somewhat labile
complexes.
    To obtain a simple chemical model of the suggested biological
activity of our complexes, we examined the behavior of a nucleic acid

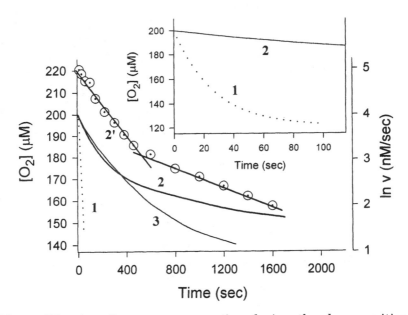

FIG. 3. Kinetics of oxygen consumption during the decomposition of [i-PrCo(acacen)(en)]Br (XVA1, $c = 100$ μM) in 50 mM phosphate buffer, pH 7.40 (1), blood plasma (2), and milk emulsion (3) at 37°C. Line 2' represents the transformation of curve 2 linearizing the first-order kinetics. The process was monitored with the oximetry (Clark electrode) technique, the applicability of which is due to the very fast reaction of dioxygen with the free radicals R• generated during the course of the homolytic decomposition. These results were further supported by independent spectrophotometric measurements.

under conditions of their acid-induced decomposition [30]. First, the nucleic acid is cleaved to a certain extent even under anaerobic conditions, i.e., when alkyl free radicals are the only active species. Second, the degradation of the nucleic acid greatly increases if both ascorbic acid and dioxygen are available. Evidently, a cobalt(II) complex resulting from the homolytic decomposition of alkylcobalt(III) chelates with tridentate Schiff bases catalyzes the autoxidation of the biogenic reductant and the generation of active oxygen species capable of cleaving nucleic acids. Hence, under quasi-physiological conditions, our complexes act in two steps, i.e., as both stoichiometric and catalytic sources of active radical species, so that the overall process of cleaving a nucleic acid can be presented as shown in Scheme 8.

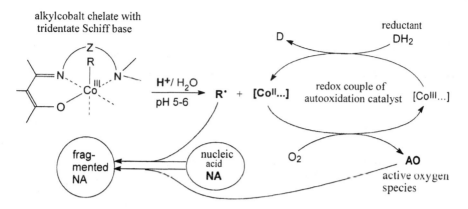

SCHEME 8.   Mechanism of the nuclease action of the pH-dependent organo-
cobalt(III) source for free radical species. (Reproduced by permission from [30].)

3.   BIOMEDICAL EXAMINATIONS

One could a priori expect that the pH-dependent sources of active radical species might discriminate acidic tumors from normal tissues and
selectively attack the former. Complexes of the [RCo(acacen)(en)Br]
series (XVA with Y = Br, i.e., cationic complexes XV with R^1, R^3 = H, R^2,
R^4 = Me, Z = $(CH_2)_2$, L-L' = en, and Y = Br), where R is a C2–C4 primary
or secondary alkyl, were selected for these early tests because of their
appropriate decomposition rates, ready availability, relatively easy
handling, and fair solubility in water. Here they are exemplified by the
complex with R = *i*-Pr (XVA1).

Complex XVA1 was prepared and characterized as mentioned
above. Its samples contained 95 ± 1% of the complex according to [1]H
NMR spectroscopy. The main admixture was NaBr (2.5 ± 0.5%, as
determined by ionometry), which had been intentionally introduced to
favor crystallization of the target complex. The nature of the other
impurities (~2% in total) has not yet been established in detail. Apparently, they are "inorganic" cobalt complexes, i.e., they contain no metal-
carbon bond. Complex XVA1 was dissolved in *aqua pro injectionibus*
immediately before use. The solubility of complex XVA1 in water at
ambient temperature is 8.5 mg/mL.

3.1. Acute Toxicity

To study the acute toxicity of complex XVA1, graded doses dissolved in *aqua pro injectionibus* were administered i.v. or i.p. into rats (IEPOR strain bred) as well as F1 or BALB/c mice, and toxic reactions were observed for 14 days. There were 8–10 animals/dose point. The $LD_{50/14}$ (drug dose to kill 50% of animals within 14 days) was determined from the probit analysis.

Acute toxicity was found to be as follows: rats, (a) i.p. — $LD_{50/14}$ = 73.8 mg/kg, median $LD_{50/14}$ = 65 mg/kg, maximal tolerable dose = 60 mg/kg; (b) i.v. — $LD_{50/14}$ = 30 mg/kg, median $LD_{50/14}$ = 35 mg/kg, maximal tolerable dose = 20 mg/kg; mice, i.p. — $LD_{50/14}$ = 175 mg/kg, median $LD_{50/14}$ = 190 mg/kg, maximal tolerable dose = 165 mg/kg.

The body weights of the rats were reduced by 3–5% as compared with controls after the XVA1 regimen at the dose of 20–40 mg/kg. The combined administration of XVA1 with conventional antitumor drugs and means has not increased the weight loss. It must be noted that flabbiness, failing appetite, and strong thirst were observed within the first hours after XVA1 injection. The flabbiness and the thirst fully disappeared within the next 2–5 h, whereas the failing appetite was observed for 2–3 days after XVA1 injection. A nonsignificant cyanosis of digestive organs and congestive liver were observed in all dead animals.

3.2. Pharmacokinetics

HPLC was used for the assay of XVA1 concentrations in tumor and normal tissues: column — Diasorb 130 T C16, 150 × 4 mm (i.d.); eluent — a 80:20 (v/v) mixture of methanol with 0.025 M aqueous acetic acid adjusted to pH 7.5 with ethylenediamine; flow rate — 0.7 mL/min; photometric detection by absorbance at 313 nm [19]. Rats with transplanted Guerin carcinoma were sacrificed at various times after the i.p. injection of XVA1 at a dose of 40 mg/kg, and samples of tumor, blood, liver, kidney, lung, spleen, and muscle were immediately stored in liquid nitrogen. After special preparation of whole blood and homogenized tissues, the concentrations of XVA1 were determined using an inner standard.

It was observed that in all studied tissues (except muscle) the maximum concentration of XVA1 occurred 15–30 min after injection. A considerable decrease in concentration of XVA1 was found in the second hour in blood, kidney, liver, and muscle, and in the third hour in the tumor. The highest concentrations have been observed in blood, kidney, and liver. Six hours after the injection XVA1 was no longer detected in kidney and lung, and after 24 and 48 h in any of the other tissues studied. It is worth mentioning that the total concentration of XVA1 in all studied tissues of each rat was substantially lower than the dose injected. The elimination half-life of XVA1 in serum was determined as 34.2 min. The total area under the curve representing XVA1 concentrations in serum was calculated as 249 μg/mL·h.

3.3. Direct Anticancer Activity and Modifying Effects on Other Antitumor Therapies

3.3.1. Experimental Procedures

(a) Animals and Tumors. Female rats (strain IEPOR bred, with a body weight of 150–200 g) bearing a subcutaneous Guerin carcinoma, Walker 256 carcinoma, or Sarcoma 45 were used. Tumors were transplanted into the right flank except Sarcoma 45 which was transplanted into the limb for thermoradio treatment. In addition, female mice (strain IEPOR bred, with a body weight of 20–25 g) bearing a subcutaneously transplanted Sarcoma 180 were used and female C57Bl/6 mice (20–25 g) were employed to study the antimetastatic activity of the XVA1 complex. Lewis lung carcinoma (3LL) was maintained as a solid intramuscular tumor by serial transplantation at 2-week intervals. The single-cell suspension was obtained by a standard mechanical procedure and injected intramuscularly into the hind leg of the mice (0.5×10^6 cells/mouse). In separate experiments the single-cell suspension of 3LL carcinoma was injected into the pad of a hind leg (0.5×10^6 cells/mouse). This leg with palpable tumor was amputated on the 13th day after implantation. All animals were kept in Macrolon cages bedded with dust-free wood granulate, and had free access to a standard diet and tap water. Rats undergoing glucose infusion as well as being treated with hyperthermia and radiation were anaesthetized with sodium pentobarbital. All experiments had been approved by the regional animal ethics committee.

(b)   Tumor Heating and Infusions.   Local hyperthermia was performed by microwave unit Luch-2 (Medradioelectronics, Ukraine) operating at 2450 MHz with a 20-W power output. Sarcoma 45 transplanted into the limb was heated by the water bath technique. The details of microwave heating and temperature measurement were described earlier [31]. A 20% aqueous solution of glucose was infused into the tail vein via an infusion pump at a rate of 80 mg/kg/min within 60 min. The glucose concentration in blood reached a value of ~15 mM on the 60th min of infusion, and returned to the pretreatment value over the following 90 min after cessation of infusion.

(c)   Administration Schedule.   A treatment was initiated when tumors were approximately 1.0 cm^3 (Guerin carcinoma and Sarcoma 45), 0.5 cm^3 (Walker 256 carcinoma), or 100 mm^3 (Sarcoma 180) in volume. XVA1 was given 90 min before starting the glucose infusion, 60 min before the onset of hyperthermia or irradiation, and 90 min before platidiam (*cis*-DDP-based preparation of Lachema, Czechia) injection. Each treatment, either single or combined, was performed twice with a 2-day break (Guerin carcinoma, Sarcoma 180) or daily (Walker 256 carcinoma). Irradiation, alone and supplemented with local heating or XVA1, was used in a single fraction (Sarcoma 45). XVA1 and platidiam were given i.p. (the doses are indicated in the tables). Irradiation was carried out using a 190-kV x-ray machine (RUM-11, Russia) at a dose rate of 1.72 Gy/min.

Two variations of experiments with 3LL carcinoma were tested. (1) The treatment was started when a tumor reached a volume of 110 mm^3 (e.g., on day 7 after tumor implantation). Two schedules of treatment were performed: (a) only one injection was given; (b) all preparations were given twice: on the 7th and 13th day after tumor implantation. (2) The treatment was started on the day after implantation. All preparations were given 10 times during days 1–21. (3) Regarding the palpable tumor, treatment was initiated on the day following the amputation of the leg with the palpable tumor (on the 13th day after implantation). All preparations were given 10 times during days 14–28.

(d)   Assay of Tumor Response to Treatment.   The progress of each tumor was measured 5 times weekly until the tumor reached a volume of 25 cm^3 (rats tumors) or 500 mm^3 (Sarcoma 180). Tumor volumes were calculated by an ellipsoid approximation using the three orthogonal

diameters. The tumor growth delay (TGD) was calculated as the days taken by each individual tumor to reach 25 cm^3 or 500 mm^3 compared with the untreated control. Other criteria of the response were percentages of animals with complete regression of tumor on the 30th day after treatment cessation and surviving 120 (rats) and 90 (mice) days after treatment cessation (cured animals). Estimation of the antimetastatic effect was estimated based on the number of days the mice survived (experiment 1), or in accordance with the following schedule: the mice were killed on the 24th (experiment 2) or 28th day (experiment 3) after tumor implantation. In all cases the primary tumors were excised and weighted, the lungs removed and fixed in Bouin's solution, and metastatic nodules counted and weighted. Each treatment group consisted of 6–10 animals and the experiments were repeated 3–5 times.

(e) Statistical Analysis. Values in this study are means ± standard error of the mean (s.e.m.). The significance of the differences between the various groups was calculated by Student's *t* test. Differences were considered as significant if $p < 0.05$.

3.3.2. *Antitumor Activity of the [i-PrCo(acacen)(en)]Br Complex (XVA1) and Its Modifying Effects on Other Antitumor Therapies*

It was observed (Table 1) that XVA1 administered i.p. at 20 mg/kg 90 min before a 60-min-long glucose infusion produced a tumor growth delay of Guerin carcinoma of ~10.5 days. Similarly, XVA1 given 60 min before hyperthermia (41°C, 60 min) produced a TGD of ~10.0 days. XVA1 administered just before heating at either 43°C or 41°C was less effective producing a TGD of ~7.5 or 5.0 days, respectively. Also, the local tumor control rate was obtained as 40% when XVA1 was supplemented with glucose infusion. It is noteworthy that the percentage of cured rats in this case was found to be ~35%.

Table 2 shows the Guerin carcinoma response upon treatment with combinations of XVA1, glucose infusion, platidiam, and hyperthermia. Overall, the data of Table 2 on local tumor control and cured animals distinctly indicate a modifying effect of XVA1 on the Guerin carcinoma response to platidiam and heating. A maximal tumor response was obtained when XVA1 was given with platidiam and glucose

TABLE 1

Delay in Tumor Growth of Guerin Carcinoma
Following Various Treatments

Treatment	No. of rats	Days to 25 cm³	Growth delay (days)
Control	39	13.5 ± 0.9	—
XVA1, 40 mg/kg	15	19.0 ± 1.0	5.5
XVA1, 20 mg/kg + glucose infusion	22	24.0 ± 1.5	10.5
Local heating, 43°C	35	19.0 ± 1.5	5.5
Local heating, 41°C	25	16.0 ± 1.0	2.5
XVA1, 20 mg/kg just before local heating, 43°C	17	21.0 ± 2.5	7.5
XVA1, 20 mg/kg just before local heating, 41°C	18	18.5 ± 2.0	5.0
XVA1, 20 mg/kg 60 min before local heating, 41°C	25	23.5 ± 2.5	10.0

Source: Taken with permission from [32].

infusion (or local heating): 47.5% and 50% of cured rats were observed, respectively.

Table 3 demonstrates the Walker 256 carcinoma response on the treatment with combinations of XVA1, glucose infusion, and platidiam. One can see that we did not obtain the antitumor effect of XVA1 supplemented with glucose infusion in the case of Walker 256 carcinoma. This fact may be explained by the following observation. Acidification of this tumor due to glucose infusion is not as significant as acidification of the Guerin carcinoma, i.e., the ΔpH values are 0.50 ± 0.10 and 0.95 ± 0.15, respectively. At the same time, XVA1 manifested a significant synergism with platidiam in the treatment of Walker 256 carcinoma. Platidiam at a dose of 1.5 mg/kg given 90 min after administration of XVA1 produced a rate of cured rats of 92%. The application of platidiam alone at a doses of 0.75 and 0.30 mg/kg resulted in weaker antitumor effects, i.e., 25% and 10% of cured rats, respectively. The combination of XVA1

TABLE 2

Chemo- and Thermochemosensitizing Activity of the Complex XVA1 in Rats Bearing Guerin Carcinoma

Treatment	No. of rats	Days to 25 cm^3	Growth delay (days)	CR (%)	Cured rats (%)
Untreated control	39	13.5 ± 0.9	—	0	0
Platidiam, 3 mg/kg	35	36.5 ± 2.5	23.0	51	45
Platidiam, 1.5 mg/kg	20	27.5 ± 2.0	14.0	32	25
XVA1, 20 mg/kg + platidiam, 1.5 mg/kg	28	35.0 ± 2.0	21.5	50	43
XVA1, 20 mg/kg + platidiam, 1.5 mg/kg + glucose infusion	23	39.0 ± 3.2	25.5	60	47.5
Platidiam, 1.5 mg/kg + glucose infusion	19	31.0 ± 1.9	17.5	45	35
XVA1, 20 mg/kg + platidiam, 1.5 mg/kg + heating 41°C	20	33.5 ± 2.3	20.0	55	50
Platidiam, 1.5 mg/kg + heating, 41°C	19	31.0 ± 2.0	17.5	40	32
Platidiam, 1.5 mg/kg + heating, 43°C	17	33.0 ± 1.7	19.5	52	40

Source: Taken with permission from [32].

TABLE 3

Chemosensitizing Activity of the Complex XVA1 in Rats Bearing Walker 256 Carcinoma

Treatment	No. of rats	Days to 25 cm^3	Growth delay (days)	CR (%)	Cured rats (%)
Untreated control	25	9.0 ± 0.6	—	0	0
XVA1, 40 mg/kg	15	9.5 ± 0.6	0.5	0	0
XVA1, 20 mg/kg + glucose infusion	17	9.7 ± 0.5	0.7	0	0
Platidiam, 3 mg/kg	35	22.5 ± 1.2	14.0	92	85
Platidiam, 1.5 mg/kg	20	20.0 ± 1.0	12.5	85	80
Platidiam, 0.75 mg/kg	28	17.0 ± 0.9	8.75	75	62.5
Platidiam, 0.3 mg/kg	23	14.5 ± 1.6	4.75	25	20
XVA1, 40 mg/kg + platidiam, 1.5 mg/kg	20	21.0 ± 1.3	13.5	94	87
XVA1, 40 mg/kg + platidiam, 0.75 mg/kg	27	18.0 ± 1.4	11.5	80	78.5
XVA1, 40 mg/kg + platidiam, 0.3 mg/kg	19	17.5 ± 1.8	7.5	50	40

Source: Taken with permission from [32].

with platidiam at doses of 0.75 and 0.30 mg/kg produced 77% and 40% of cured rats, respectively. The enhancement ratio was found to be ~3.5–4.0 in these cases. The fact that platidiam at a dose of 0.3 mg/kg given alone produced a minimal effect on the tumor with only 10% of cured rats, and when it was supplemented with XVA1 40% of cured rats resulted, is very important and encouraging. Thus, it seems possible to apply reduced doses of platinum agents for tumor treatment, which may be essential because of their well-known nephrotoxicity.

Table 4 demonstrates the Sarcoma 180 response on the treatment with combinations of XVA1 and platidiam. One can see that XVA1 manifested a significant enhancement of the platidiam antitumor effect in the treatment of Sarcoma 180.

Table 5, presenting the Sarcoma 45 response on the radiation alone and supplemented with XVA1, shows the significant potentiation of the radiation effect on the tumor by means of XVA1. The enhancement ratio was found to be more than 2.0. It is important that the modifying effect of XVA1 was most significant when it was given before radiation. Moreover, it was observed (Table 6) that XVA1 significantly enhanced the antitumor effect of single thermoradio treatment; the TGD was found to be 8.5 and 17.5 days and the life span 46 and 64 days for groups "10 Gy + 41°C" and "XVA1 + 10 Gy + 41°C," respectively.

TABLE 4

Chemosensitizing Activity of the Complex XVA1
in Mice Bearing Sarcoma 180

Treatment	No. of mice	Days to 500 cm³	Growth delay (days)	CR (%)	Cured mice (%)
Untreated control	25	7.75 ± 0.6	—	17	12.5
Platidiam, 5 mg/kg	35	9.5 ± 0.6	2.0	47	27
Platidiam, 2.5 mg/kg	20	8.4 ± 1.25	1.0	21	15
XVA1, 60 mg/kg + platidiam, 2.5 mg/kg	29	12.0 ± 1.2	4.5	47.5	42.1

Source: Taken with permission from [32].

TABLE 5

Radiosensitizing Activity of the Complex XVA1
in Rats Bearing Sarcoma 45

Treatment	No. of rats	Days to 25 cm^3	Growth delay (days)	Life span (days)	Cured rats (%)
Untreated control	20	18.3 ± 0.8	—	36.1 ± 2.2	0
XVA1, 40 mg/kg	10	23.8 ± 1.2	5.5	33.0 ± 2.0	0
10 Gy	20	28.1 ± 1.6	9.8	54.7 ± 5.1	10
15 Gy	10	33.0 ± 2.0	14.7	60.2 ± 6.4	20
20 Gy	10	41.4 ± 1.6	23.1	74.0 ± 5.5	30
10 Gy + XVA1, 20 mg/kg	20	23.0 ± 1.2	4.7	48.3 ± 4.9	10
XVA1, 20 mg/kg + 10 Gy	20	45.0 ± 3.7	26.7	64.4 ± 8.3	30

Source: Taken with permission from [33].

TABLE 6

Thermoradiosensitizing Activity of the Complex XVA1
in Rats Bearing Sarcoma 45

Treatment	No. of rats	Days to 25 cm^3	Growth delay (days)	Life span (days)	Cured rats (%)
Untreated control	10	18.3 ± 0.8	—	36.1 ± 2.2	0
10 Gy	10	28.1 ± 1.6	9.8	54.7 ± 5.1	10
10 Gy + local heating, 41°C	10	23.0 ± 1.2	4.7	48.3 ± 4.9	10
XVA1, 20 mg/kg + 10 Gy + local heating, 41°C	10	45.0 ± 3.7	26.7	64.4 ± 8.3	30

Complex XVA1 was injected 60 min before radiation, which was performed 60 min
before heating.

TABLE 7

General Antitumor and Antimetastatic Effects of the Complex XVA1
and Platidiam in Mice Bearing Lewis Lung Carcinoma

Treatment	No. of mice	Inhibition of primary tumor growth (%)	Inhibition of metastasis formation	
			In number of metastases per mouse (%)	In volume of metastases per mouse (%)
Untreated control	25	—	—	—
XVA1, 60 mg/kg × 1	16	11	52.5	74
XVA1, 60 mg/kg × 2	16	30	88	86
XVA1, 12 mg/kg × 10	16	47	66	80
Platidiam, 5 mg/kg × 1	16	34	75	69
Platidiam, 5 mg/kg × 2	16	60	93	67

The results summarized in Table 7 show that treatment with
XVA1 did not significantly reduce the growth of the primary 3LL carcinoma but did inhibit the development of spontaneous metastases,
which was most distinctly observed by the estimation of the volume of
all metastases appearing in the lung. Comparison of the XVA1 effects
with those caused by platidiam allowed the assumption that XVA1 has
a significant antimetastatic effect that can be further enhanced by
modulation of the therapeutic protocol and doses of the complex.

Table 8 (experiments with amputation of the leg with a primary
tumor) shows that XVA1 significantly inhibited the development of spontaneous metastases as estimated by the number and volume of metastases. It is important to note that the antimetastatic effect of XVA1 was
more pronounced than that due to platidiam administration.

TABLE 8

Specific Antimetastatic Effect of the Complex XVA1
and Platidiam in Mice Bearing Lewis Lung Carcinoma

Treatment	No. of mice	Inhibition of metastasis formation	
		In number of metastases per mouse (%)	In volume of metastases per mouse (%)
Untreated control	16	—	—
XVA1, 12 mg/kg × 10	16	66	77.5
Platidiam, 1 mg/kg × 10	16	42	75

3.4. Some Biological Effects of the [*i*-PrCo(acacen)(en)]Br Complex (XVA1) Related to Its Antitumor Activity and Its Modifying Influence on Other Antitumor Therapies

The reason for the chemo-, radio-, chemothermo-, and radiothermo-sensitizing activity of XVA1 has yet to be elucidated. Evidently, the efficacy of combined treatments strongly depends on the time schedule of combined procedures. In effect, the best results were obtained when XVA1 and platidiam or hyperthermia (as well as glucose infusion) were separated in time, in particular by 60–90 min. The reasons for this phenomenon are not clear at present. In this connection, it is worthwhile to cite some interesting published data on the combined application of tirapazamine with *cis*-DDP. It was recently shown [34] that tirapazamine greatly enhanced the anticancer effect of *cis*-DDP when given 2.5 h before *cis*-DDP. Although a casual coincidence cannot be excluded, this analogy may indicate a key for an explanation of the mechanism of the modifying action, both of tirapazamine and XVA1. It is known that the mechanism of tirapazamine action involves the reduction of the drug to a radical which, in turn, abstracts hydrogen from DNA, forming double-strand breaks and chromosome aberrations [34].

In the meantime, it can be tentatively assumed from the ability of transient free radicals, including alkyls, to cleave or damage bio-molecules, e.g., nucleic acids [7], that the inhibiting action of R-Co(III) complexes on DNA in tumor cells is responsible for this phenomenon. DNA is considered to be an important target for both radiation and *cis*-DDP. For ionizing radiation, increases in DNA single-strand breaks (SSBs) have been associated with increased cell killing; for *cis*-DDP, the formation of platinum-DNA adducts is implicated as being respon-sible for cytotoxicity.

In order to clarify the possible mechanisms of the chemo-poten-tiating effect of XVA1, we have studied the production of SSBs in Guerin carcinoma cells under combined application of platidiam with and un-der administration of XVA1 alone. SSBs were assayed by alkaline elu-tion 1 and 4 h after single platidiam and/or XVA1 doses. The method of alkaline elution for the determination of SSBs is our modification of the original procedure of Kohn et al. [35].

Table 9 reveals that supplementation of XVA1 produces an in-

TABLE 9

DNA Single-Strand Breaks (SSBs) in Guerin Carcinoma[a] Cells
Upon Platidiam and XVA1 Administration

Treatment	No. of tumors	Velocity of DNA alkaline elution (K), $K \times 10^2$ mL^{-1}
Untreated control	10	3.4 ± 0.35
1 h after platidiam, 1.5 mg/kg	10	10.2 ± 0.8
4 h after platidiam, 1.5 mg/kg	9	7.1 ± 0.6
1 h after XVA1, 40 mg/kg + platidiam, 1.5 mg/kg	10	14.4 ± 0.8
4 h after XVA1, 40 mg/kg + platidiam, 1.5 mg/kg	11	9.5 ± 0.8
1.5 h after XVA1, 40 mg/kg	9	10.0 ± 0.7

[a]Tumor volume 1–1.5 cm^3.
Source: Taken with permission from [32].

crease in SSBs as compared with platidiam alone at 1 h as well as 4 h after platidiam injection by a factor of 1.4 and 1.3, respectively. XVA1 alone produces a significant number of SSBs similar to that observed after injection of platidiam alone. These data suggest that an increased production of SSBs resulting from the combined action of XVA1 with platidiam may be one of the main steps in the mechanism responsible for the chemo-potentiating effect of XVA1 in our tumor systems. The radio-potentiating activity of XVA1 observed in our experiments with Sarcoma 45 can be explained in a similar manner. It is well known that the increase in SSBs has been associated with increased cell killing due to ionizing radiation. The production of SSBs due to administration of XVA1 alone may indicate the interaction of free radicals produced by XVA1 homolysis in an "acidic" tumor with DNA of tumor cells.

It is pertinent to mention here that some "inorganic," i.e., containing no metal-carbon bond, cobalt(III) complexes are known to display cytotoxicity and a radiosensitizing effect in cell cultures as well as an antitumor action in vivo. It was suggested that such activities were stipulated by the ability of the metal ion either to form a coordinate covalent linkage to DNA nucleobases or to release, via a reduction-to-Co(II) step occurring under hypoxic conditions, cytotoxic "radiomimetic" chelating ligands, e.g., nitrogen mustards, attacking DNA [36–39]. However, none of such "inorganic" cobalt complexes has been proven, even suggested, to be a pH-dependent agent. Thus, the pH-dependent antitumor activity is a specific immanent characteristic of our organometallic complexes. Nevertheless, similar mechanisms may well also contribute to their antitumor effect.

Furthermore, there are data concerning the ability of some cobalt compounds to affect tissue perfusion, which can correct the antitumor activity and activity-modifying effects of these substances in vivo. It was shown that hypodermic injection of cobalt chloride, citrate, or salicylate in humans dilates the vessels of the face and ears, beginning within 3 min and persisting for 50 min [40]. The blood pressure falls by about 10 mm Hg. These data have stipulated tests of the effect of XVA1 on the tumor microenvironment. In our experiments, done in collaboration with Dr. D. B. Kelleher from the University of Mainz, Germany, it was shown that XVA1 insignificantly reduced the mean arterial blood pressure in rats bearing transplanted DS-sarcoma to a minimum value of 130 ± 15 mm Hg 10 min after the injection (the pretreatment value was 140 ± 10 mm Hg), with recovery thereafter to 145 ± 8 mm Hg at 90 min after the injection. The introduction of XVA1 resulted in an initial rapid decrease in tumor red blood cell flux (which was measured by laser Doppler fluxometry) through tumor tissue that reached a minimum of 62% of the pretreatment value during the first 10 min following the injection. Thereafter, tumor red blood cell flux rose until it reached a level of 80% of the pretreatment value 90 min after the injection. It must be noted that the red blood cell flux through DS-sarcoma remains relatively stable under control conditions over the measurement period, e.g., without XVA1 administration. Our data correspond to the above-mentioned results [40] on the inhibition of tissue perfusion after administration of simple cobalt compounds. Hence, it can be assumed that administration of all cobalt-containing substances may result in the inhibition of tissue perfusion due to the influence of cobalt ions on the blood pressure by dilatation of blood vessels. This circumstance must be taken into account in the investigations of compounds containing cobalt(III) as potential radiosensitizers or bioreductive drugs because a microphysiological effect of cobalt may contribute to the biological activity of the agents in question.

4. CONCLUSION AND OUTLOOK

The described medicobiological examinations have clearly verified the fruitfulness of this new line in cancer therapy and established its fundamental and potentially practical value. Relevant bactericide and fungicide applications of the complexes can also be anticipated.

Finally, it is pertinent to compare antitumor techniques based on the use of various sources of highly reactive species (Table 10). The fundamental acting principles of the two other techniques are physical means, namely, radiations that are potentially harmful to living bodies and have only a limited penetration capacity. For this reason, their application is restricted to undissipated tumors that can be exposed to radiation. There should be no such limitation for our method described here.

TABLE 10

Comparison of Cancer Therapy Methods Based on the Cytotoxic Action of Radical Species

	Therapeutic method		
	Radiotherapy	Photodynamic therapy	Use of pH-dependent sources of free radicals
Active species	Hydroxyl radical	Singlet dioxygen	Carbon-centered free radicals and active oxygen species
Method of generation	Radiolysis of water	Photocatalysis	pH-dependent chemical reaction, i.e., homolysis of a transition metal-carbon bond, and succeeding catalytic autoxidation processes
(Chemical nature of the) drugs	—	Tetrapyrrole macrocyles	Organocobalt chelates and tridentate Schiff bases (RCo's)
Prerequisites for medicinal action	Spatial distribution of radiation	Selective concentration in tumor	Lowering pH in tumors as compared with normal tissues
Complementary factor(s)	Electron affinity compounds, tumor oxygenation	Light (from a laser source)	Hyperglycemia (glucose infusion), hyperthermia (local microwave heating). Furthermore, RCo's have been found effective in modifying chemo- and radiotherapy
Suggested scope of application	Radiosensitive tumors without dissemination	Solid tumors accessible to a laser probe	Tumors of any location

ACKNOWLEDGMENTS

We are greatly indebted to our long-time coworkers, Drs. Larisa Bub-novskaya, Irina Ganusevich and Andrey Sigan, and Mrs. Marina Tsikalova for their permanent hard efforts. We are also thankful to Profs. Vladimir Bakhmutov, Vitaly Roginsky, and Anatoly Yatsimirsky, and Drs. Evgeny Mironov and Alexandr Yanovsky for their contributions. Partial financial support of the International Science Foundation, INTAS, the Cancer Research Institute (New York), and the Moscow City Administration is gratefully acknowledged.

ABBREVIATIONS

acacen	acetylacetonate and ethylenediamine in a 1:1 ratio
Alk	alkyl
Ar	aryl
cis-DDP	cis-diamminedichloroplatinum
CR	share of cases of complete regression of tumor
DS-sarcoma	the kind of sarcoma first described by D. Schmaehl et al. (cf. [41])
en	ethylenediamine (1,2-diaminoethane)
ESR	electron spin resonance
Et	ethyl group
HPLC	high-performance liquid chromatography
IEPOR	Institute of Experimental Pathology, Oncology, and Radiobiology (Kiev, Ukraine)
i.d.	internal diameter
i.p.	intraperitoneally
i-Pr	isopropyl
IR	infrared
i.v.	intravenously
$LD_{50/14}$	lethal dose for 50% of the treated animals within 14 days
3LL	Lewis lung carcinoma
Me	methyl group
MR	magnetic resonance

NA	nucleic acid
NMR	nuclear magnetic resonance
PBN	C-phenyl-*tert*-butylnitrone
Ph	phenyl group
platidiam	*cis*-DDP-based preparation of "Lachema", Czechia
salen	Schiff base ligand (anionic residue) composed of salicylidenate and ethylenediamine in a 1:1 ratio
s.e.m.	standard error of the mean value
SSB	single-strand break of DNA
t-Bu	*tert*-butyl
TGD	tumor growth delay
TLC	thin-layer chromatography

REFERENCES

1. O. Warburg, *The Metabolism of Tumors*, Arnold Constable, London, 1930.

2. P. Vaupel, *Blood Flow, Oxygenation, Tissue pH Distribution, and Bioenergetic Status of Tumors*, Ernst Schering Research Foundation, Berlin, 1994.

3. M. von Ardenne, *Systematische Krebs-Mehrschritt-Therapie*, Hippokrates Verlag, Stuttgart, 1997.

4. S. Osinsky and L. Bubnovskaya, *Arch. Geschwulstforsch.*, *54*, 463–469 (1984).

5. I. Tietze, M. Neumann, S. Moller, R. Fischer, K.-H. Glusenkamp, and E. Jahde, *Cancer Res.*, *49*, 4179–4184 (1989).

6. M. von Ardenne, P. G. Reitnauer, and W. Schulze, *Pharmazie*, *47*, 28–30 (1992).

7. G. Pratviel, J. Bernadon, and B. Meunier, *Angew. Chem. Int. Ed. Engl.*, *34*, 746–769 (1995).

8. J. Halpern, *Pure Appl. Chem.*, *55*, 1059–1068 (1983).

9. F. D. W. Kemmit and D. R. Russel, in *Comprehensive Organometallic Chemistry*, Vol. 5 (G. Wilkinson, ed.), Pergamon Press, Oxford, 1982, pp. 2–276.

10. G. N. Schrauzer, J. W. Sibert, and R.J. Windgassen, *J. Am. Chem. Soc.*, *90*, 6681–6688 (1968).

11. I. Levitin, A. Sigan, E. Kazarina, G. Alexandrov, Yu. Struchkov, and M. Vol'pin, *J. Chem. Soc. Chem. Commun.*, 441–442 (1981).

12. I. Ya. Levitin, R. M. Bodnar, and M. E. Vol'pin. In: *Inorganic Synthesis*, Vol. 23 (S. Kirschner, ed.), John Wiley and Sons, New York, 1985, pp. 163–171.

13. I. Ya. Levitin, M. V. Tsikalova, V. I. Bakhmutov, A. I. Yanovsky, Yu. T. Struchkov, and M. E. Vol'pin, *J. Organomet. Chem.*, *330*, 161–178 (1987).

14. I. Ya. Levitin, A. L. Sigan, M. V. Tsikalova, M. E. Vol'pin, M. S. Tsar'kova, A. A. Kuznetsov, and I. A. Gritskova (Institute of Organoelement Compounds). Russ. Pat. 2,070,202 (Appl. 1991/4,939,599).

15. I. Levitin, M. Tsikalova, N. Sazikova, V. Bakhmutov, Z. Starikova, A. Yanovsky, and Yu. Struchkov, *Inorg. Chim. Acta* (in press).

16. I. Ya. Levitin, A. N. Kitaigorodskii, A. T. Nikitaev, V. I. Bakhmutov, A. L. Sigan, and M. E. Vol'pin, *Inorg. Chim. Acta*, *100*, 65–77 (1985).

17. A. I. Yanovsky, G. G. Alexandrov, Yu. T. Struchkov, I. Ya. Levitin, R. M. Bodnar, and M. E. Vol'pin, *Koord. Khim.*, *9*, 825–834 (1983).

18. K. B. Yatsimirsky, V. A. Pokrovsky, Yu. E. Kovalenko, Ya. D. Lampeka, A. L. Sigan, and M. V. Tsikalova, *Metalloorg. Khim.*, *2*, 611–614 (1989) (in Russian); *Organomet. Chem. USSR*, *2*, 311–313 (1989).

19. P. N. Nesterenko, I. Ya. Levitin, M. V. Tsikalova, and M. E. Vol'pin, *Inorg. Chim. Acta*, *240*, 665–668, (1995).

20. P. Nesterenko, I. Levitin, N. Chernoglazova, E. Paskonova, N. Penner, and M. Tsikalova, *Inorg. Chim. Acta* (submitted).

21. I. Ya. Levitin, M. V. Tsikalova, and E. A. Paskonova, unpublished data.

22. I. Ya. Levitin, A. L. Sigan, R. M. Bodnar, R. G. Gasanov, and M. E. Vol'pin, *Inorg. Chim. Acta*, *76*, L169–L171 (1983).

23. A. D. Ryabov, I. Ya. Levitin, A. T. Nikitaev, A. N. Kitaigorodskii, V. I. Bakhmutov, I. Yu. Gromov, A. K. Yatsimirsky, and M. E. Vol'pin, *J. Organomet. Chem.*, *292*, C4–C8 (1985).

24. I. Ya. Levitin, A. K. Yatsimirsky, and M. E. Vol'pin, *Metalloorg. Khim.*, *3*, 865 (1990) (in Russian); *Organomet. Chem. USSR*, *3*, 442 (1990).

25. A. K. Yatsimirsky, I. Ya. Levitin, O. I. Kavetskaya, M. V. Tsikalova, V. I. Bakhmutov, S. V. Vitt, and M. E. Vol'pin, *Teor. i eksp. khim.*, *27*, 293–300 (1991).

26. I. Ya. Levitin, A. K. Yatsimirsky, M. V. Tsikalova, O. I. Kavetskaya, I. O. Nenasheva, and M. E. Vol'pin, Novel mechanisms of homolytic decomposition of organocobalt complexes, *Abstracts of the Xth FECHEM Conference on Organometallic Chemistry*, Agia Pelagia, Crete, Greece, Sept. 5–10, 1993, p. 29.

27. I. Ya. Levitin, E. A. Mironov, V. A. Roginsky, and T. K. Barsukova (submitted).

28. M. E. Vol'pin, I. Ya. Levitin, A. L. Sigan, and A. T. Nikitaev, *J. Organomet. Chem.*, *279*, 263–280 (1985).

29. I. Ya. Levitin, Dissertation (Dr. Chem. Sc.), Inst. of Organoelement Compounds, Moscow, 1987.

30. I. Ya. Levitin, V. M. Belkov, G. N. Novodarova, Z. A. Shabarova, and M. E. Vol'pin, *Mendeleev Commun.*, 153–155 (1996).

31. S. Osinsky, A. Rikberg, L. Bubnovskaya, and V. Trushina, *Int. J. Hyperthermia*, *9*, 297–301 (1993).

32. S. Osinsky, I. Levitin, L. Bubnovskaya, N. Kornuta, I. Ganusevich, M. Tsikalova, and M. Vol'pin, *Med. Biol. Environ.*, *25*, 75–79 (1997).

33. S. Osinsky, I. Levitin, L. Bubnovskaya, I. Ganusevich, M. Tsikalova, E. Zhavrid, Yu. Istomin, and M. Vol'pin, *Anticancer Res.*, *17*, 3457–3462 (1997).

34. M. J. Dorie and J. M. Brown, in *Tumor Oxygenation* (P. Vaupel, D. K. Kelleher and M. Gunderoth, eds.), Gustav Fischer Verlag, Stuttgart, 1995, pp. 125–135.

35. K. W. Kohn, R. A. Ewig, L. C. Erickson, and L. A. Zwelling, in *DNA Repair: A Laboratory Manual of Research Techniques* (E. C. Friedberg and P. C. Hanawalt, eds.), Marcel Dekker, New York, 1981, pp. 379–401.

36. B. Teicher, J. Jacobs, K. Cathcart, M. Abrams, J. Vollano, and D. Picker, *Radiat. Res.*, *109*, 36–46 (1987).

37. J. O'Hara, E. Douple, M. Abrams, D. Picker, C. Giandomenico, and J. Vollano, *Int. J. Radiat. Oncol. Biol. Phys.*, *16*, 1049–1052 (1989).

38. B. Teicher, M. Abrams, K. Rosbe, and T. Herman, *Cancer Res.*, *50*, 6971–6975 (1990).

39. W. Wilson, J. Moselen, S. Cliffe, W. Denny and D. Ware, *Int. J. Radiat. Oncol. Biol. Phys.*, *29*, 323–327 (1994).

40. T. Sollman, in *A Manual of Pharmacology and Its Application to Therapeutics and Toxicology*, W. B. Saunders, Philadelphia, 1957, pp. 1306–1309.

41. D. Schmähl and T. Rieseberg, *Naturwissenschaften*, *43*, 475 (1956).

16

Detection of Chromatin-Associated Hydroxyl Radicals Generated by DNA-Bound Metal Compounds and Antitumor Antibiotics

G. Mike Makrigiorgos

Joint Center for Radiation Therapy and Dana Farber
Cancer Institute, Department of Radiation Oncology,
Harvard Medical School, 330 Brookline Avenue,
Boston, MA 02215, USA

1. INTRODUCTION

Exposure of populations to inorganic carcinogens as a result of pathological states and industrial or environmental conditions represents an
acknowledged cancer hazard [1–6] and the mechanism of carcinogenicity continues to be scrutinized. One important mechanism of action of
the metals copper, iron, and possibly chromium is the production of
chromatin lesions via oxidizing free radicals, hydroxyl radicals (HO·) in
particular [1–18]. Although the predominance of HO· over chromatin-
damaging radicals other than HO· in some cases is uncertain [19–21],

the importance of HO· and the mutagenic and carcinogenic effects of HO· attack in chromatin are well documented [22–31]. As such, knowledge concerning the fate of the generated HO· is imperative for an adequate understanding of the mutagenicity and carcinogenicity of these metals.

The generation of unrepaired chromatin damage that propagates to the chromosomal and cellular levels has been postulated to be related to the *distribution* of HO·-induced lesions in chromatin [32]. The major factors that determine HO· distribution in the presence of agents activating metal catalysis of HO· include (1) conformation of chromatin [33,34], (2) the location of the metal binding site(s) in chromatin, (3) the metal affinity to chromatin, and (4) the metal chelators present that can modulate the free radical reactions and deliver or abstract the metal from chromatin [17]. These factors are interrelated, making it a formidable problem to study metal-catalyzed HO· within chromatin both in cell-free systems, where most of the studies are currently being conducted, and in whole cells. Summarizing the complexity involved, a workshop focusing on metal carcinogenesis [5] concluded that: "research in the area of metal carcinogens has progressed slowly due to the inherent difficulties of studying metal interactions in cells. New methodologies need to be developed in order to document the nature of the interactions of metals with cellular components. Since the field is not well developed, many basic studies still need to be conducted."

2. CHROMATIN-ASSOCIATED HYDROXYL RADICALS

2.1. Diffusion Distance of Hydroxyl Radicals in Tissues

Hydroxyl radical is a species with high electrophilicity and high thermochemical reactivity and, as such, has a high reactivity toward most biomolecules including DNA [35]. For example, the hydroxyl rate constant for reaction with isolated DNA bases is of the order of 10^9–10^{10} $M^{-1} s^{-1}$ [35]. As a result, while in dilute aqueous solutions the diffusion distance of HO· can be of the order of micrometers, the intracellular diffusion distance of HO· is expected to be very short due to the high concentrations of HO· scavengers. This distance has been estimated to be of the order of 8–60 Å [36].

2.2.   A Subpopulation of Hydroxyl Radicals with Major Biological Significance (Chromatin-Associated HO•)

From the above it follows that the only HO• molecules that can diffuse enough to attack DNA in vivo are those generated in the immediate proximity of the nucleic acid, i.e., the chromatin-associated HO•. When the generated HO• is randomly distributed across the cell (e.g., HO• generated by γ radiation), only a small proportion of HO• will be chromatin-associated and will cause DNA damage. There are agents, however, such as metals and metal compounds, that may generate HO• upon binding to DNA, in which case a high proportion of the generated HO• would be expected to be chromatin-associated.

Detection of chromatin-associated HO• has been a notoriously difficult task, as these radicals interact with the nucleic acid at almost diffusion-controlled rates; therefore, they cannot be trapped by indicator/detector molecules to an extent adequate to generate a detectable signal. As a result, current methods for studying metal-catalyzed HO• in chromatin model systems only provide limited information. For example, while spin-trap ESR/EPR methods [37] are well suited to the indiscriminate detection of all HO• radicals in model systems, they cannot identify specifically the chromatin-associated HO•, i.e., the subpopulation of HO• that is most likely to damage DNA in vivo. Since the distribution of chromatin-associated HO• is strongly dependent on the metal binding site(s) within chromatin, which can be site-specific [5,23,38–41], the frequency with which certain sites are accessed by chromatin-associated HO• can be radically different from that with which they are accessed by HO• produced in the bulk solution [33]. Therefore, measuring the characteristics of HO• production in the bulk solution can have a poor correlation with HO•-produced chromatin damage. Similarly, when used alone, most chromatin damage detection methods do not reveal which form of free radical caused the damage and whether the damage was even caused by HO• or other free radicals. In addition, the ability of metals to access specific chromatin sites (histones, DNA sequences) and the cell's capacity to repair the concomitant damage may vary significantly across chromatin [5,42]. For example, unlike lesions in the internucleosomal DNA region, nucleosomal lesions caused by a variety of carcinogenic agents are much less accessible to repair enzymes [42–48]. Consequently, although some chromatin damage detection methods are very sensitive [25,49–51], their use in studying the behavior of chromatin-associated HO• is limited.

3.   DETECTION OF CHROMATIN-ASSOCIATED HYDROXYL RADICALS

3.1.   Indirect Methods Based on the Spectrum of Hydroxyl Radical Base Damages

An attempt to overcome the outlined problems in the detection of chromatin-associated HO• is the application of gas chromatography–mass spectrometry (GC-MS) [25] to determine the spectrum of base damages generated by a DNA-damaging agent. HO• produces a unique spectrum of base damages (i.e., a "fingerprint" of damage) following attack of the nucleic acid. The ratio between two types of base damage (e.g., 8-hydroxyadenine versus 8-hydroxyguanine) has been proposed to be a verification of the presence of HO•-generated chromatin damage [10]. The GC-MS method is very sensitive and informative; however, it is laborious and requires sophisticated and expensive instrumentation. In addition, like most chromatin damage assays, it scores the average oxidative damage over the whole sample and not the distribution of lesions within specific chromatin regions. Since metal-catalyzed chromatin damage may be site-specific, it would be desirable to quantitate chromatin-associated, metal-catalyzed HO• and the damage caused within specific chromatin sites in order to make meaningful correlations of probable *cause* (HO• radicals) and *effect* (damage).

3.2.   Direct Quantitation of Chromatin-Associated HO• via a Chromatin-Bound Molecular Fluorescent Probe (SECCA). Specificity of SECCA Response to HO•

In an attempt to address the problem of chromatin-associated HO• detection, we have initiated a direct approach by which an HO• molecular probe is conjugated to desired positions in the immediate vicinity of a nucleic acid, or directly on the nucleic acid itself [52–55]. Coumarin-3-

carboxylic acid (CCA) has been used as the hydroxyl radical "trap." CCA
is a small, nonfluorescent, aromatic hydrocarbon that, upon reaction
with HO• generates hydroxylated products among which is the highly
fluorescent 7-hydroxy-CCA [52] (Fig. 1). Earlier work by Ashawa et al.
[56] utilized the conversion of free coumarin in solution to fluorescent
7-hydroxycoumarin as a chemical radiation dosimeter. We synthesized
SECCA, a succinimidyl ester of coumarin-3-carboxylic acid that covalently couples to primary aliphatic amines on biomolecules [52]. By
coupling SECCA to polylysine, proteins, or an amino-modified nucleic
acid (see Sec. 4) we demonstrated that, upon generation of HO•, SECCA-
biomolecule complexes become fluorescent, presumably due to hydroxylation of the SECCA aromatic ring by HO• (Fig. 2A–C). Since SECCA

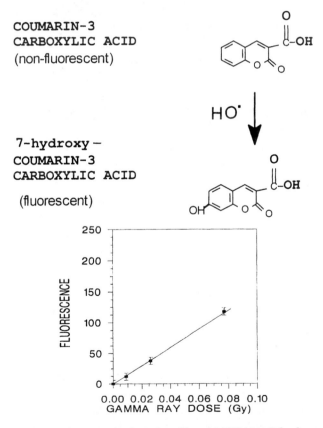

FIG. 1.   Conversion of coumarin-3-carboxylic acid (CCA) to 7-hydroxy-coumarin-
3-carboxylic acid (7-hydroxy-CCA) by hydroxyl radicals generated via γ-irradiation of an aqueous CCA solution.

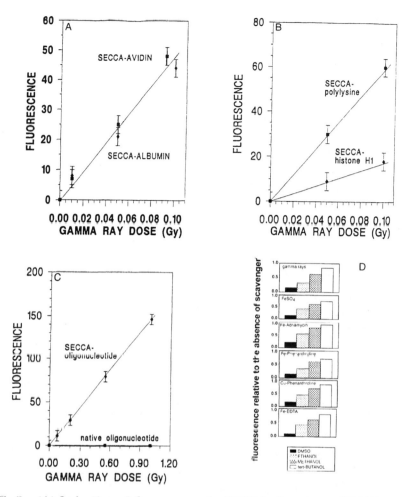

FIG. 2. (A) Induction of fluorescence in SECCA-albumin or SECCA-avidin conjugates irradiated with ^{137}Cs γ-rays at doses of 0.01–0.1 Gy. The fluorescence, in arbitrary units, is derived by subtracting the fluorescence of unirradiated samples. (B) Induction of fluorescence in solution with conjugates of SECCA to polylysine or to histone-H1 irradiated with ^{137}Cs γ-rays at doses of 0.01–0.1 Gy. The fluorescence, in arbitrary units, is derived by subtracting the fluorescence of unirradiated samples. (C) Induction of fluorescence in solution with conjugates of SECCA to amino-modified oligonucleotide irradiated with ^{137}Cs γ rays at doses of 0.01–0.1 Gy. The fluorescence, in arbitrary units, is derived by subtracting the fluorescence of unirradiated samples. The fluorescence of irradiated native oligonucleotide is also depicted. (D) Relative scavenging ability of DMSO, ethanol, methanol, and *tert*-butanol (0.1 mmol/dm^3) in solution exposed to 9.2 Gy of γ rays or containing Fe(II)-adriamycin/Fe(II)-phenanthroline/Fe(II)-EDTA /Cu(II)-phenanthroline with ascorbate and H$_2$O$_2$. Relative fluorescence is the percent ratio of fluorescence induced in presence versus absence of scavenger [64]. (Reprinted with permission from [64].)

is ~4 Å, the detected HO$^{\bullet}$ can be expected to be at a distance of 0–4 Å from the binding site of the probe, i.e., in the immediate vicinity of the biomolecule (chromatin-associated HO$^{\bullet}$).

A number of studies have been performed to verify that the presence of HO$^{\bullet}$ is a necessary and sufficient condition for generation of fluorescent 7-hydroxy-SECCA following free radical attack on SECCA, free in solution or bound to biomolecules. While hydroxylation of the phenolic ring of aromatic hydrocarbons is frequently taken as a reliable indication of HO$^{\bullet}$ [57–61], other species (such as singlet oxygen) may also, in specific situations, hydroxylate the aromatic hydrocarbon ring [57,62,63]. However, this usually occurs with a different *regioselectivity* than that of HO$^{\bullet}$, i.e., the distribution of isomeric products formed is distinct from that formed by HO$^{\bullet}$. We have shown that $^{1}O_{2}$ does not produce 7-hydroxy-SECCA; neither do peroxyl radicals [64]. In experiments with γ-ray irradiation, our data indicated that HO$^{\bullet}$ is required for generation of the fluorescent 7-hydroxy-SECCA and that direct radiation action, or primary and secondary radiation radicals other than HO$^{\bullet}$, does not induce fluorescence [52]. Another commonly used test to detect the involvement of HO$^{\bullet}$ in free radical processes is to observe the behavior of the system in the presence of HO$^{\bullet}$ scavengers. Then, if the change in the observed endpoint corresponds to the hydroxyl rate constant of the scavenger, one can presume the presence of HO$^{\bullet}$. This seemingly simple approach becomes complicated by the fact that occasionally the generated scavenger radicals may propagate or interfere with the reaction or even modify the endpoint measured [59,60]. We applied the scavengers dimethylsulfoxide (DMSO), methanol, ethanol, and *tert*-butanol because these are commonly used in free radical mechanistic studies, and their resulting radicals are relatively unreactive. A number of reactions reported to produce copper-, or iron-catalyzed HO$^{\bullet}$ were tested with free SECCA at a single scavenger concentration (0.1 mM of DMSO, methanol, ethanol, and *tert*-butanol) and compared to γ-ray-generated HO$^{\bullet}$ [64]. With minor differences, the dependence of the fluorescent signal on the scavengers was similar in all cases (Fig. 2D), and the extent to which each HO$^{\bullet}$ scavenger reduced the induced fluorescence corresponded approximately to its HO$^{\bullet}$ constant. Since for γ rays the origin of fluorescence is well characterized [52], the scavenging studies indicate that HO$^{\bullet}$ is the main species responsible for the induction of fluorescence of 7-hydroxy-SECCA by

Cu(II) and Fe(II/III) and their complexes with adriamycin, phenanthroline, and EDTA in the presence of ascorbate and hydrogen peroxide.

4. METHODS FOR ATTACHING SECCA TO IN VITRO CHROMATIN MODELS

4.1. Binding of SECCA on Polylysine, Bovine Serum Albumin, Avidin, and Histone

The succinimidyl esters of SECCA and 7-hydroxy-SECCA were synthesized as described [52]. Both of these compounds are now commercially available from Molecular Probes (Eugene, Oregon). Coupling of SECCA to aliphatic amines of polylysine, bovine serum albumin (BSA), avidin, and histone H1 [52] was performed by reacting SECCA to an appropriate concentration of each biomolecule in 0.1 mol/dm^3 sodium bicarbonate for 3 h at room temperature (pH 8.5, except for histone where pH 8.0 was used due to solubility problems). Purification was done by gel filtration (Sephadex G-50) and the SECCA-biomolecule complexes were recovered in phosphate-buffered saline (PBS), pH 7.4. The same procedures were used to bind to the same biomolecules the expected fluorescent product following irradiation of SECCA and 7-hydroxy-SECCA, and these conjugates were used as controls.

4.2. Binding of SECCA on Polylysine-DNA Complexes, and on Histones within Nucleohistone or Nucleosomes

To form SECCA-polylysine-DNA complexes, the SECCA-polylysine was added dropwise to calf thymus DNA with vigorous stirring. An irreversible complex formed that possessed a stoichiometric ratio of polylysine to DNA 20% by weight [53,54,65]. Selected samples were also associated with DNA using SECCA-polylysine containing a small amount (~1%) of 7-hydroxy-SECCA-polylysine. These samples are the expected products of incubation of SECCA-polylysine-DNA with HO$^\bullet$-generating reagents and have been used as positive fluorescence controls. To form SECCA-histone H1-DNA complexes (nucleohistone),

DNA (1 mg/mL) was dissolved in 1 M NaCl, pH 6.5, and mixed with an equal volume of histone (0.63 mg/mL), either unlabeled or labeled with SECCA/7-OH-SECCA, dissolved in the same solution. The mixture was dialyzed (3500 cutoff molecular weight membranes) in successively lower NaCl concentrations (1 mol/dm^3 for 4 h, 0.4 mol/dm^3 for 18 h, 0.3 mol/dm^3 for 4 h, 0.15 mol/dm^3 for 4 h) and finally overnight in a 10-fold dilution of PBS (0.1 × PBS), pH 7.4. At the histone-to-DNA ratio used (0.63), no aggregation of material was observed [53,54]. The preparation was then either sedimented by centrifugation at 130,000g or dialyzed in 0.1 × PBS (100,000 cutoff molecular weight membranes) in order to separate DNA-bound histone from unbound histone.

A salt dialysis procedure was also used for the reconstitution of nucleosomes [66]. Briefly, equimolar amounts of the four core histones (H2A, H2B, H3, H4) and plasmid pUC19 supercoiled DNA (histone/DNA = 0.8, w/w) were mixed in 5 mM Tris buffer (pH 7.2) containing 2 M NaCl, 0.2 mM phenylmethylsulfonylfluoride, and 0.01% mercaptoethanol and transferred to a dialysis bag (Spectrapor, molecular weight cutoff = 2000). Reconstitution was accomplished by a gradual lowering of the salt concentration (2, 1.2, 1, 0.8, 0.6, 0.4, 0.2, 0.05 M) in eight steps, each of 1.5 h duration. The final 0.05 M dialysis was allowed to continue overnight. Labeling of core histones or core particles with nonfluorescent SECCA or its fluorescent analog 7-hydroxy-SECCA was performed after formation of nucleosomes as follows. Nucleosomes were dialyzed in 10 mM phosphate buffer and 10 mM NaCl (pH 8.0) before labeling. Microliter amounts of concentrated SECCA/7-hydroxy-SECCA solutions, dissolved in acetonitrile, were added in different molar ratios to nucleosomes and allowed to react for 4 h at 4°C. The SECCA-conjugated nucleosomes were then exhaustively dialyzed against 10 mM phosphate and 10 mM NaCl (pH 7.4) in order to eliminate unreacted SECCA/7-hydroxy-SECCA. It was verified that the probe binds to the core histones within nucleosomes and that there is no nonspecific binding to DNA [66]. The samples were processed for electron microscopy as described by Slayter et al. [67]. Figure 3 shows electron micrographs (EMs) of unlabeled (A and C) and SECCA-labeled (B and D) structures reconstituted with pUC19 DNA and calf thymus core histones. Almost all of the core histones and supercoiled DNA used have associated into "beads-on-a-string-like" structures, typical of core particles [68]. Furthermore, the images for both the labeled and unlabeled nucleosomes are very similar. Therefore it appears that morphologi-

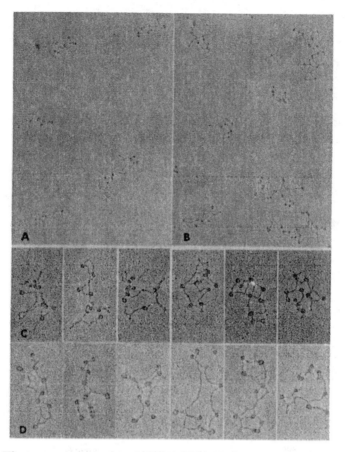

FIG. 3.   Electron micrographs of pUC19 DNA nucleosomes reconstituted with unlabeled or SECCA-labeled core histones (1 SECCA/histone). Nucleosomes were reconstituted by salt dialysis procedures [66]. Samples were then fixed with glutaraldehyde, dialyzed in ammonium acetate, and processed for electron microscopy. A and C represent unlabeled nucleosomes, whereas B and D represent SECCA-labeled nucleosomes. (Reprinted with permission from [66].)

cally there are no obvious differences caused by the binding of the probe to the histones that form the bead-like structures. To further examine whether the structures shown in Fig. 3 represent nucleosomes, their micrococcal nuclease digestion pattern was determined. Consistent with formation of nucleosomes when unlabeled nucleosomes are exposed to the enzyme for 30 or 60 min, bands of approximately 140 base pairs are evident on gel electrophoresis [66]. These bands are absent when supercoiled pUC19 DNA is digested under identical conditions.

4.3. DNA Attachment of SECCA

To couple SECCA to an amino-modified oligonucleotide, the amino-modified oligomer (NH_2-ATCGATCG, MW = 2506 Da) was suspended in Na_2CO_3/$NaHCO_3$ buffer, pH 9.0, and reacted for 36 h with 1000-fold excess of SECCA dissolved in dimethylformamide (DMF) [52]. Unreacted SECCA was removed by G-50 column chromatography, yielding approximately 60–70% labeled oligonucleotide conjugates (SECCA-OLIGO). The conjugates were separated from unlabeled oligonucleotide by column chromatography (reverse phase C-18).

5.   METAL COMPOUNDS AND ANTITUMOR ANTIBIOTICS THAT GENERATE CHROMATIN-ASSOCIATED HO•

5.1.   γ-Radiation, Copper/Ascorbate/Hydrogen Peroxide, and Fe-EDTA/Ascorbate/Hydrogen Peroxide Systems

5.1.1.   Induction of Fluorescence by Radiation

The induction of fluorescence by γ-ray irradiation of SECCA-H1 and SECCA-nucleohistone is depicted in Fig. 4A. A linear induction of fluorescence is observed in the dose range examined (0.3–30 Gy) [53]. The induction of fluorescence/Gy (the slope of the curves) is approximately 10-fold less for the nucleohistone relative to isolated histone. Three factors might be expected to contribute to this effect: (1) the reduced quantum efficiency following complexing of histone with DNA (reduction by approximately 30% [53]), (2) the scavenging of •OH by DNA, and (3) the reduced collision frequency of •OH with the histone following the association of histone with the DNA macromolecule. The fluorescence excitation spectra of SECCA-labeled nucleohistone irradiated with graded doses of γ rays are depicted in Fig. 4B, curves 1–6. The spectra have been recorded at an emission wavelength of 450 nm (slit width 10) and have been corrected for the scatter contribution of the

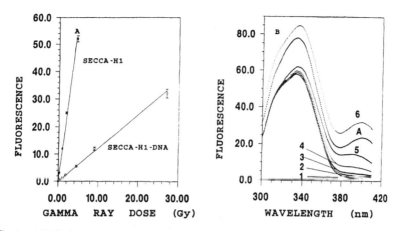

FIG. 4.   (A) Induction of fluorescence in a γ-irradiated solution of SECCA-H1 and SECCA-H1-DNA (nucleohistone) [53]. (B) Fluorescence excitation spectra of SECCA-H1 complexed with DNA (nucleohistone) following irradiation with graded γ-ray doses; curves 1–6 irradiated with 0, 0.9, 1.8, 4.5, 9.0, 27.0 Gy, respectively. Curve A represents unirradiated SECCA-containing nucleohistone sample in which 7-OH-SECCA-H1 labeled DNA was added. Emission of these spectra was recorded at 450 nm [53]. (Reprinted with permission from [53].)

solvent. The excitation spectrum of unirradiated SECCA-nucleohistone reflects the inherent fluorescence of the biomolecules and of SECCA, with a peak around 332 nm (Fig. 4B, curve 1). The data also indicate that the fluorescence of 7-OH-SECCA starts to build up when the irradiation dose is increased (Fig. 4B, curves 2–6). In addition, when a small amount $(10^{-9}\,\text{mol/dm}^3)$ of 7-OH-SECCA-nucleohistone is added to a sample of unirradiated SECCA-nucleohistone, the resulting spectrum corresponds to that of irradiated SECCA-nucleohistone (Fig. 4B, curve A), indicating indirectly that the fluorescent product formed during irradiation of SECCA-nucleohistone is indeed 7-OH-SECCA.

5.1.2. Induction of Fluorescence by Copper/Ascorbate/ Hydrogen Peroxide and Fe-EDTA/Ascorbate/ Hydrogen Peroxide-Catalyzed Free Radicals

Figure 5A shows the induction of fluorescence on SECCA-polylysine-DNA complexes following addition of various concentrations of copper(II) sulfate, ascorbate, and hydrogen peroxide [55]. The induction of

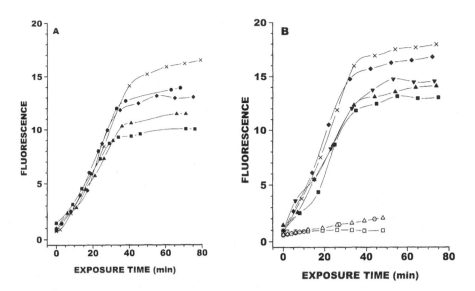

FIG. 5.   (A) Induction of fluorescence of SECCA-labeled polylysine DNA conjugates incubated with 10 μmol/dm³ copper sulfate, 100 μmol/dm³ ascorbate, and 1 mmol/dm³ hydrogen peroxide in presence of DMSO (0 mmol/dm³: ×; 1 mmol/dm³: ●; 10 mmol/dm³: ♦; 100 mmol/dm³: ▲; 1000 mmol/dm³: ■) [55]. (B) Induction of fluorescence on SECCA-labeled polylysine-DNA conjugates incubated with 10 μmol/dm³ copper sulfate, 100 μmol/dm³ ascorbate, and 1 mmol/dm³ hydrogen peroxide and in presence of no scavenger (×), *tert*-butanol (10 mmol/dm³: ♦), methanol (10 mmol/dm³: ▲), ethanol (10 mmol/dm³: ▼), DMSO (10 mmol/dm³: ■), EDTA (0.1 mmol/dm³: ○), histidine (0.1 mmol/dm³: △) or catalase (8.3 × 10⁻⁵ mmol/dm³: □) [55]. (Reprinted with permission from [55].)

fluorescence has characteristics similar to those of free SECCA and SECCA-polylysine [55]. However, upon addition of DMSO only a *moderate* (~20%) decrease in the fluorescence is observed (Fig. 5A). Similar results are obtained for the hydroxyl radical scavengers, methanol, ethanol, and *tert*-butanol (Fig. 5B, 10 mmol dm⁻³). On the other hand, upon addition of EDTA (0.1 mmol/dm³), histidine (0.1 mmol/dm³), or catalase (8.3 × 10⁻⁵ mmol/dm³), the fluorescence is nearly eliminated (Fig. 5B). As with free SECCA (Fig. 1) there is a fluorescence induction when SECCA is positioned in close proximity to DNA via SECCA-polylysine-DNA complexes. Unlike free SECCA, however, when SECCA is attached close to DNA, DMSO (Fig. 5A), as well as the other hydroxyl radical scavengers (Fig. 5B), fails to significantly reduce the fluorescence. This is consistent with the notion that hydroxyl radical is *site specifically* produced when copper is bound to the nucleic acid [38,41]. In agreement with these results, in vitro systems that score DNA damage (i.e., the *result* of copper-catalyzed hydroxyl radical production) have repeatedly demonstrated the inability of scavengers to reduce the damage [38,41,69,70]. On the other hand, agents acting on any of the components of the copper sulfate-ascorbate-hydrogen peroxide system can eliminate DNA damage [38,41,69,70]. In agreement with this, in the present experiments EDTA, histidine, and catalase all result in the substantial elimination of fluorescence (Fig. 5B). Therefore it appears that the fluorescence induction caused by copper/ascorbate/hydrogen peroxide on SECCA-polylysine-DNA complexes detects hydroxyl radical produced very close to DNA, i.e., chromatin-associated HO•. In contrast, when a similar system (SECCA-histone H1-DNA) was exposed to radicals catalyzed by Fe-EDTA/ascorbate /hydrogen peroxide, the fluorescence of SECCA was completely eliminated by DMSO [64], indicating the absence of chromatin-associated HO• in this case.

When SECCA-labeled core histones within nucleosomes were exposed to radiation or copper/ascorbate/hydrogen peroxide (Fig. 6A and B), the induction of fluorescence was proportional to SECCA content and radiation dose (A) as well as to exposure time to copper–ascorbic acid–hydrogen peroxide (B) [66]. In Fig. 6C, the relative reduction of fluorescence between samples that contain no DMSO (value = 1) and samples with 0.1 M DMSO is shown. The results indicate that copper/ascorbate/hydrogen peroxide generates a much higher proportion of chromatin-associated HO• than γ radiation.

5.2. Fe-Bleomycin and Fe/Cu-Adriamycin/Ascorbate/ Hydrogen Peroxide Systems

Figures 7 and 8 show the application of the SECCA-polylysine-DNA and SECCA-histone H1-DNA models to examine the generation of chromatin-associated HO• by bleomycin-Fe(II) plus ascorbate/hydrogen peroxide [65] or by adriamycin-Fe(III)/Cu(II) plus ascorbate/hydrogen peroxide [64]. Although the range of free radicals generated by these agents is expected to be very different, both appear to have a common aspect:

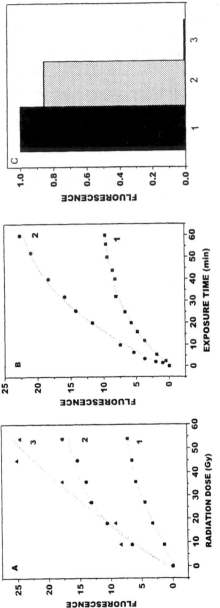

FIG. 6.   Induction of fluorescence and excitation spectra obtained under various conditions in SECCA-labeled nucleosomes [66]. (A) Induction of fluorescence following γ irradiation. Nucleosomes were formed and then labeled with SECCA at SECCA/histone ratios of 0.3 (curve 1), 0.7 (curve 2) and 1 (curve 3). (B) Induction of fluorescence following exposure to 8.9 μM Cu(II)–0.8 mM ascorbic acid–2.0 mM hydrogen peroxide; curve 1 = 0.6 SECCA/histone; curve 2 = 1 SECCA/histone. (C) Comparison of DMSO-caused reduction in fluorescence occurring in SECCA nucleosomes (1 SECCA/histone) when exposed for 30 min to either 9 μM Cu(II)–0.9 mM ascorbic acid–1 mM hydrogen peroxide or γ radiation (at 90 cGy/min). In the absence of DMSO, both the copper system and the γ-ray system have been set to be equal to 1 (bar 1). The other two bars represent the presence of 0.1 M DMSO in copper system (bar 2) and with γ radiation (bar 3). (Reprinted with permission from [66].)

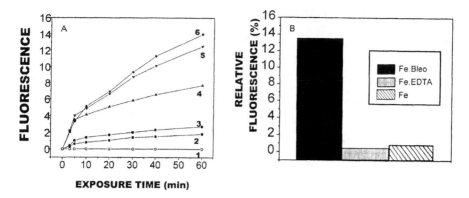

EXPOSURE TIME (min)

FIG. 7.   (A) Induction of fluorescence in SECCA-polylysine-DNA sample incubated with 10 μmol dm^{-3} Fe(II)-bleomycin (1:2) in presence of 100 μmol dm^{-3} ascorbate and varying concentrations of H_2O_2 [65]. Curve 1, bleomycin without iron plus ascorbate plus 1000 μmol dm^{-3} H_2O_2; Curve 2, Fe(II)-bleomycin, ascorbate, and no H_2O_2; Curve 3, Fe(II)-bleomycin, ascorbate, and 10 μmol dm^{-3} H_2O_2; Curve 4, Fe(II)-bleomycin, ascorbate, and 100 μmol dm^{-3} H_2O_2; Curve 5, Fe(II)-bleomycin, ascorbate, and 1000 μmol dm^{-3} H_2O_2; Curve 6, Fe(II)-bleomycin, ascorbate, and 2000 μmol dm^{-3} H_2O_2. (B) Induction of fluorescence in SECCA-polylysine-DNA sample incubated with Fe(II)-bleomycin/Fe(II)-EDTA/Fe(II) in presence of ascorbate, H_2O_2, and DMSO. Relative fluorescence is the fluorescence in the presence of DMSO divided by the fluorescence in the absence of DMSO. Reprinted with permission from [65].

they generate significant amounts of chromatin-associated HO•, which cannot be scavenged by DMSO, ethanol, methanol, or *tert*-butanol (Figs. 7B and 8C). These results seem to be in agreement with the hypothesis by Gajewski et al. [21], who predicted the generation of DNA-associated HO• by bleomycin by observing the distribution of damaged DNA bases. Such generation of HO• does not exclude the simultaneous presence of other mechanisms for DNA damage by bleomycin-Fe(II) or adriamycin-Fe(III) [71–73] whose relative contribution in the overall DNA damage by bleomycin may be larger than the one of HO•. However, it must be noted that the significance of each mechanism of DNA damage for the production of cellular endpoints such as carcinogenesis is not a sole function of the quantity of damage it causes DNA; thus, the quality (or the "fingerprint") of DNA damage produced by each mechanism is recognized as another major factor [10,32]. Accordingly, the

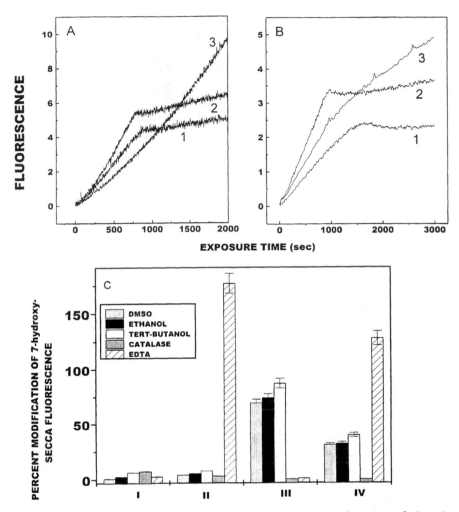

FIG. 8. Induction of 7-hydroxy-SECCA fluorescence as function of time in
incubation mixture containing SECCA-H1-DNA, 10 μmol/dm^3 SECCA, metal-
adriamycin complex, and various concentrations of ascorbate and hydrogen
peroxide [64]. (A) Cu(II)-adriamycin complex (1:1) with 1: 100 μmol/dm^3 ascorbate and 1 mmol/dm^3 hydrogen peroxide; 2: 100 μmol/dm^3 ascorbate and 2
mmol/dm^3 hydrogen peroxide; 3: 500 μmol/dm^3 ascorbate and 1 mmol/dm^3
hydrogen peroxide. (B) Fe(III)-adriamycin complex (1:3) with 1: 100 μmol/dm^3
ascorbate and 1 mmol/dm^3 hydrogen peroxide; 2: 100 μmol/dm^3 ascorbate and 5
mmol/dm^3 hydrogen peroxide; 3: 500 μmol/dm^3 ascorbate and 1 mmol/dm^3
hydrogen peroxide. (C) Modification of 7-hydroxy-SECCA fluorescence induced
following exposure to Cu(II)- or Fe(III)-adriamycin complex in presence of 500

identification of *all distinct possible mechanisms* for the production of DNA damage (such as via the chromatin-associated HO• detected by the present method) is important for the prediction of the range of effects by genotoxic agents such as bleomycin and adriamycin.

5.3. Are Certain DNA Sequences Particularly Susceptible to Hydroxyl Radical Attack?

When HO• is catalyzed randomly in the bulk solution by Fe-EDTA it causes DNA cleavage to occur almost randomly in naked DNA [74]. Folding of DNA into nucleosomes, however, severely restricts the ability of HO• to attack certain DNA sites and causes cleavage to occur with a 10-base-pair periodicity within nucleosomes [74]. Furthermore, unlike Fe-EDTA, which remains in the bulk solution [35], several metal compounds such as copper or bleomycin-Fe(II) bind to specific chromatin positions and generate sequence-specific damage [5,35,38–41,74]. Accordingly, it can be hypothesized that a portion of this sequence-dependent damage may be due to chromatin-associated HO•. If such sequence-dependent, chromatin-associated HO• is indeed generated, then one may postulate an increased susceptibility of certain DNA genes for hydroxyl radical damage. As an example, the transcription factor (TFIIIA) binding site of the naturally occurring 5S rRNA gene of *Xenopus* species has a known affinity for transition metals [75] and this could render functional DNA sequences on this gene particularly susceptible to HO• attack. The existence and extent of such hydroxyl radical "hot spots" on 5S rRNA is currently investigated by attaching the hydroxyl radical probe, SECCA, to specific sequence positions along the 430-base-pair sequence of this gene. Specific placement of SECCA is accomplished by using DNA synthesizer-based DNA templates that contain a SECCA-modified phosphoramidite of dUTP at specific sequence positions (Fig. 9A). The SECCA-labeled templates are then used

---

μmol/dm³ ascorbate and 1000 μmol/dm³ hydrogen peroxide and several modifiers. *Percent modification* is the percent ratio of the induced fluorescence in the presence versus the absence of each modifier. I. Cu(II)-adriamycin complex (1:1), free SECCA; II. Fe(III)-adriamycin (1:3), free SECCA; III. Cu(II)-adriamycin complex (1:1), SECCA-H1-DNA; IV. Fe(III)-adriamycin (1:3), SECCA-H1-DNA. (Reprinted with permission from [64].)

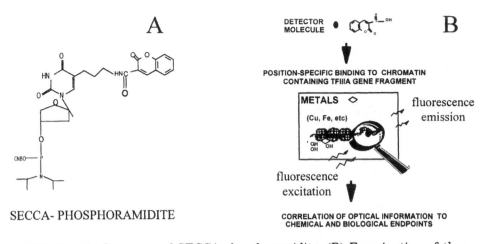

FIG. 9. (A) Structure of SECCA-phosphoramidite. (B) Examination of the presence of sequence-specific *hydroxyl radical "hot spots"* in chromatin, via site-specific incorporation of SECCA within nucleosomes.

to replace the corresponding DNA stretches on the 430-base-pair sequence using recombinant DNA techniques and nucleosomes are reconstituted in the presence of histones (Fig. 9B). In parallel studies, DNA cleavage is examined using a DNA sequencing procedure to correlate the sequence-specific fluorescent signals with the sequence-specific cleavage. Such investigations may allow us to determine the extent to which the metal affinity of the 5S rRNA gene causes an increased susceptibility to damage, whether damage is caused by HO$^\bullet$, and what the influence of nucleosomal conformation might be.

6. CONCLUSIONS

The long-term aim of our studies is to correlate *local variations in the distribution of HO*$^\bullet$ in chromatin with the *carcinogenic potential* of metal compounds. This information is essential for understanding the mechanisms of carcinogenicity of transition metals. A "direct" approach, using SECCA-binding to chromatin, has been developed for detection and study of *chromatin-associated, metal-catalyzed HO*$^\bullet$, a subpopulation of HO$^\bullet$ that may be particularly important for chromatin damage, mutagenesis, and carcinogenesis and that cannot be studied adequately by current methods. This technology allows the analytical investigation of chromatin-associated HO• on chromosomal proteins and DNA and may provide a methodology for the estimation of the susceptibility of specific DNA sequences to metal-catalyzed HO• attack.

ACKNOWLEDGMENTS

The author appreciates the contributions of Drs. A. I. Kassis, E. A. Bump, S. Chakrabarti, H. S. Slayter, J. Baranowska-Kortylewicz, and G. Svensson. Work on metals has been supported in part by NIH grants R01 63334 and K04 69296.

ABBREVIATIONS

BSA	bovine serum albumin
CCA	coumarin-3-carboxylic acid
DMF	dimethylformamide
DMSO	dimethylsulfoxide
EDTA	ethylenediamine-N,N,N',N'-tetraacetate
EPR	electron paramagnetic resonance
ESR	electron spin resonance
GC-MS	gas chromatography–mass spectrometry
HO•	hydroxyl radical
PBS	phosphate-buffered saline
PMSF	phenylmethylsulfonylfluoride
SECCA	succinimidyl ester of coumarin-3-carboxylic acid
Tris	tris(hydroxymethyl)methylamine

REFERENCES

1. E. S. Copeland, Workshop report from the Division of Research Grants, National Institutes of Health. *Cancer Res.*, *43*, 5631–5637 (1983).

2. S. D. Aust, L. A. Morehouse, and C. E. Thomas, *J. Free Rad. Biol. Med.*, *1*, 3–25 (1985).

3. L. A. Loeb, *Cancer Res.*, *49*, 5489–5496 (1989).

4. C. B. Klein, K. Frenkel, and M. Costa, *Chem. Res. Toxicol.*, *4*, 592–604 (1991).

5. K. E. Wetterhahn, B. Demple, M. Kulesz-Martin, and E. S. Copeland, eds. Workshop report from the Division of Research Grants, National Institutes of Health. *Cancer Res.*, *52*, 4058–4063 (1992).

6. K. Frenkel, *Pharmacol. Ther.*, *53*, 127–166 (1992).

7. B. Halliwell, *FEBS Lett.*, *92*, 321–326 (1978).

8. J. C. Fantone and P. A. Ward, *Am. J. Pathol.*, *107*, 395–418 (1982).

9. S. A. Lesko, J. L. Drocourt, and S. U. Yang, *Biochemistry*, *21*, 5010–5015 (1982).

10. B. Halliwell and J. M. C. Gutteridge, *Arch. Biochem. Biophys.*, *246*, 501–514 (1986).

11. S. Kawanishi, S. Inoue, and S. Sano, *J. Biol. Chem.*, *261*, 5952–5958 (1986).

12. S. Kawanishi, S. Inoue, and K. Yamamoto, *Biol. Trace Elem. Res.*, *21*, 367–372 (1989).

13. K. D. Sugden, R. B. Burris, and S. J. Rogers, *Mut. Res.*, *244*, 239–244 (1990).

14. X. G. Shi and N. S. Dalal, *Arch. Biochem. Biophys.*, *277*, 342–350 (1990).

15. J. Aiyar, H. J. Berkovits, R. A. Floyd, and K. E. Wetterhahn, *Chem. Res. Toxicol.*, *3*, 595–603 (1990).

16. A. M. Standeven and K. E. Wetterhahn, *Chem. Res. Toxicol.*, *4*, 616–625 (1991).

17. B. Halliwell and O. I. Aruoma, *FEBS Lett.*, *281*, 9–19 (1991).

18. L. K. Tkeshelashvili, T. McBride, K. Spence, and L. A. Loeb, *J. Biol. Chem.*, *266*, 6401–6406 (1991).

19. Y. Sugiura, T. Suzuki, J. Kuwahara, and H. Tanaka, *Biochem. Biophys. Res. Commun.*, *105*, 1511–1518 (1982).

20. K. Yamamoto and S. Kawanishi, *J. Biol. Chem.*, *264*, 15435–15440 (1989).

21. E. Gajewski, O. I. Aruoma, M. Dizdaroglu, and B. Halliwell, *Biochemistry*, *30*, 2444–2448 (1991).

22. J. D. Chapman, A. P. Reuvers, J. Borsa, and C. L. Greenstock, *Radiat. Res.*, *56*, 291–306 (1973).

23. J. R. F. Muindi, B. K. Sinha, L. Gianni, and C. E. Myers, *FEBS Lett.*, *172*, 226–230 (1984).

24. R. Zimmerman and P. Cerutti, *Proc. Natl. Acad. Sci. USA*, *81*, 2085–2087 (1984).

25. M. Dizdaroglu and D. S. Bergtold, *Anal. Biochem.*, *156*, 182–188 (1986).

26. R. M. Schaaper, R. M. Koplitz, L. K. Tkeshelashvili, and L. A. Loeb, *Mut. Res.*, *177*, 179 188 (1987).

27. M. Vuillaume, *Mut. Res.*, *186*, 43–72 (1987).

28. L. A. Loeb, E. A. James, A. M. Waltersdorph, and S. J. Klebanoff, *Proc. Natl. Acad. Sci. USA*, *85*, 3918–3922 (1988).

29. M. L. Wood, M. Dizdaroglu, E. Gajewski, and J. M. Essigmann, *Biochemistry*, *29*, 7024–7032 (1990).

30. T. J. McBride, B. D. Preston, and L. A. Loeb, *Biochemistry*, *30*, 207–213 (1991).

31. S. Shibutani, M. Takeshita, and L. L. Grollman, *Nature*, *349*, 431–434 (1991).

32. J. F. Ward, *Radiat. Res.*, *104*, S103–S111 (1985).

33. H. U. Enright, J. W. Miller, and R. P. Hebbel, *Nucl. Acids Res.*, *20*, 3341–3346 (1992).

34. L. E. Pope and D. S. Sigman, *Proc. Natl. Acad. Sci. USA*, *81*, 3–7 (1984).

35. G. V. Buxton, C. L. Greenstock, W. P. Hellman, and A. B. Ross, *J. Phys. Chem. Ref. Data*, *17*, 513–886 (1988).

36. R. Roots and S. Okada, *Radiat. Res.*, *64*, 306–320 (1975).

37. R. P. Mason and C. Mottley. In: M. C. R. Symonds, ed. Royal Society of Chemistry, London, 1987, pp. 10B,185.

38. A. Samuni, J. Aronovitch, D. Godinger, M. Chevion, and G. Czapski, *Eur. J. Biochem.*, *137*, 119–124 (1983).

39. O. I. Aruoma, M. Grootveld, and B. Halliwell, *J. Inorg. Biochem.*, *29*, 289–299 (1987).

40. S. A. Kazakov, T. G. Astashkina, S. V. Mamaev, and V. V. Vlassov, *Nature*, *335*, 186–188 (1988).

41. M. Chevion, *Free Rad. Biol. Med.*, *5*, 27–37 (1988).

42. V. A. Bohr, D. H. Phillips, and P. C. Hanawalt, *Cancer Res.*, *47*, 6426–6436 (1987).

43. T. Cech and M. L. Pardue, *Cell*, *11*, 631–640 (1977).

44. K. D. Tew, S. Sudhakar, P. S. Schein, and M. E. Smulson, *Cancer Res.*, *38*, 3371–3378 (1978).

45. D. T. Goodhead, R. J. Munson, J. Thacker, and R. Cox, *Int. J. Radiat. Biol.*, *37*, 135–167 (1980).

46. K. T. Wheeler and J. V. Wierowski, *Radiat. Res.*, *93*, 312–318 (1983).

47. N. L. Oleinick, S. M. Chiu, and L. R. Friedman, *Radiat. Res.*, *98*, 629–641 (1984).

48. Z. G. Wang, X. H. Wu, and E. C. Friedberg, *J. Biol. Chem.*, *266*, 22472–22478 (1991).

49. R. A. Floyd, *Carcinogenesis*, *11*, 1447–1450 (1990).

50. K. Frenkel, Z. J. Zhong, H. C. Wei, J. Karkoszka, U. Patel, K. Rashid, M. Georgescu, and J. J. Solomon, *Anal. Biochem.*, *196*, 126–136 (1991).

51. H. Wei and K. Frenkel, *Cancer Res.*, *51*, 4443–4449 (1991).

52. G. M. Makrigiorgos, J. Baranowska-Kortylewicz, E. Bump, S. K. Sahu, R. M. Berman, and A. I. Kassis, *Int. J. Radiat. Biol.*, *63*, 445–458 (1993).

53. G. M. Makrigiorgos, M. Folkard, C. Huang, E. Bump, J. Baranowska-Kortylewicz, S. K. Sahu, B. D. Michael, and A. I. Kassis, *Radiat. Res.*, *138*, 177–185 (1994).

54. G. M. Makrigiorgos, E. Bump, C. Huang, J. Baranowska-Kortylewicz, and A. I. Kassis, *Int. J. Radiat. Biol.*, *66*, 247–257 (1994).

55. G. M. Makrigiorgos, E. Bump, C. Huang, and A. I. Kassis, *Free Rad. Biol. Med.*, *18*, 669–678 (1995).

56. S. C. Ashawa, U. R. Kini, and U. Madhvanath, *Int. J. Appl. Radiat. Isot.*, *30*, 7–10 (1979).

57. R. Richmond, B. Halliwell, J. Chauhan, and A. Darbre, *Anal. Biochem.*, *118*, 328–335 (1981).

58. R. A. Floyd, J. J. Watson, and P. K. Wong, *J. Biochem. Biophys. Meth.*, *10*, 221–235 (1984).

59. Z. Maskos, J. D. Rush, and W. H. Koppenol, *Free Rad. Biol. Med.*, *8*, 153–162 (1990).

60. Z. Maskos, J. D. Rush, and W. H. Koppenol, *Arch. Biochem. Biophys.*, *296*, 521–529 (1992).

61. L. P. Candeias, K. B. Patel, M. R. L. Stratford, and P. Wardman, *FEBS Lett.*, *333*, 151–153 (1993).

62. B. Kalyanaraman, S. Ramanujam, R. J. Singh, J. Joseph, and J. B. Feix, *J. Am. Chem. Soc.*, *115*, 4007–4012 (1993).

63. K. Briviba, T. P. Devasagayam, H. Sies, and S. Steenken, *Chem. Res. Toxicol.*, *6*, 548–553 (1993).

64. S. Chakrabarti, A. Mahmood, A. I. Kassis, E. Bump, A. Jones, and G. M. Makrigiorgos, *Free Rad. Res.*, *25*, 207–220 (1996).

65. S. Chakrabarti, G. M. Makrigiorgos, K. O'Brien, E. Bump, and A. I. Kassis, *Free Rad. Biol. Med.*, *20*, 777–783 (1996).

66. S. Chakrabarti, H. Slayter, A. I. Kassis, E. Bump, S. K. Sahu, and G. M. Makrigiorgos, *Int. J. Radiat. Biol.* (in press).

67. H. S. Slayter, T. Y. Shih, A. J. Adler, and G. D. Fasman, *Biochemistry*, *11*, 3044–3054 (1972).

68. F. Thoma, T. H. Koller, and A. Klug, *J. Cell Biol.*, *83*, 403–427 (1979).

69. P. G. Parsons and L. E. Morrison, *Cancer Res.*, *42*, 3783–3788 (1982).

70. A. Samuni, M. Chevion, and G. Czapski, *Radiat. Res.*, *99*, 562–572 (1984).

71. L. E. Rabow, J. Stubbe, and J. W. Kozarich, *J. Am. Chem. Soc.*, *112*, 3196–3203 (1990).

72. L. O. Rodriguez and S. M. Hecht, *Biochem. Biophys. Res. Commun.*, *104*(4), 1470–1476 (1982).

73. S. A. Akman, J. H. Doroshow, T. G. Burke, and M. Dizdaroglu, *Biochemistry*, *31*, 3500–3506 (1992).

74. B. L. Smith, G. B. Bauer, and F. Povirk, *J. Biol. Chem.*, *269*, 30587–30594 (1994).

75. M. A. Martinez-Balbas, E. J. Jimenez-Garcia, and F. Azorin, *Nucl. Acid Res.*, *23*, 2464–2471 (1995).

17

Nitric Oxide (NO): Formation and Biological Roles in Mammalian Systems

Jon M. Fukuto[1] and David A. Wink[2]

[1]Department of Pharmacology, UCLA School of Medicine, Center for the Health Sciences, Los Angeles, CA 90095-1735

[2]Tumor Biology Section, Radiation Biology Branch, National Cancer Institute, Bethesda, MD 20892, USA

1. INTRODUCTION

Before about 1988, nitric oxide (NO) was of interest only to a fairly select group of researchers. Cardiovascular pharmacologists were aware of the biological activity of NO since it was determined that NO-donating drugs such as nitroglycerin were potent vasorelaxants and as such represent a major class of drugs for the treatment of coronary artery disease. Environmental chemists have a longstanding interest in NO because it is, along with its oxidative breakdown products, a toxic component of air pollution. Inorganic chemists have been interested in NO for a long time because it possesses novel reaction chemistry and can serve as a unique ligand in a variety of metal coordination complexes and biochemists have been able to utilize some of the unique aspects of NO coordination chemistry to explore dioxygen-binding proteins. Bioinorganic chemists investigating the mechanisms of nitrogen oxide reduction, i.e., nitrate (NO_3^-) and nitrite (NO_2^-) reductases, have been interested in NO since it is a possible intermediate in the reduction process. In spite of the above mentioned interest in NO chemistry, biochemistry, toxicology, and pharmacology, there wasn't any indication that NO might be a physiologically endogenous species until the late 1980s and early 1990s when several lines of investigation converged to find that NO was biosynthesized in a variety of cells and served in a wide array of physiological functions. Since that time, NO has been the subject of intense research and it has been established that NO is, among other things, an endogenous mediator of vascular tone, an immune response agent, and involved in neurotransmission in both the peripheral and central nervous systems. Furthermore, endogenously generated NO has been implicated to participate in the development of a variety of diseases such as amyotrophic lateral sclerosis, diabetes, arthritis, Huntington's and Parkinson's disease, just to name a few. The established diversity and ubiquity associated with NO physiology and pathophysiology is truly remarkable and it is likely that many other roles for NO will be discovered in the future.

Herein we will discuss the biosynthesis and physiology of NO and how both relate to its unique and fascinating chemistry. However, due to the sheer volume of literature regarding NO chemistry, physiology, and enzymology, a comprehensive treatment of these subjects is beyond the scope of this chapter. There are a number of excellent reviews of NO enzymology [1–4], chemistry [5–8], and general physiology [9–11] and the reader should refer to these for a more detailed treatment of these subjects. The intent of this chapter is to examine the biosynthesis and physiology/pathophysiology of NO from a chemical perspective. Thus, prior to embarking on a discussion of the enzymology and biology of NO, it is worthwhile to review some of its basic and biologically relevant chemistry.

2. BIOLOGICALLY RELEVANT CHEMISTRY OF NITRIC OXIDE

2.1. NO Bonding

Much of the physiology and specificity of action of NO is due to its unique and novel chemistry. NO is a colorless gas at room temperature

$$\cdot \ddot{N} \cdot \; + \; \cdot \ddot{O} \cdot \; \longrightarrow \; \cdot \dot{N} = \ddot{O} \colon$$

FIG. 1.   Lewis dot structure of NO.

(b.p. $-151.7°C$) with a maximum solubility in water of about 1–2 mM at room temperature. Using Lewis dot formalism to depict the bonding in NO, one can immediately see that a diatomic molecule formed between an oxygen atom and a nitrogen atom should be a radical species. Indeed, NO is a paramagnetic species possessing a single unpaired electron (Fig. 1). Considering that NO is a radical, it may be expected that it should exist as a dimer. Afterall, as depicted in the figure, the nitrogen atom possesses only seven valence electrons and dimerization would result in a product that will satisfy the octet rule (i.e., ON-NO). However, except at very low temperatures, NO exists primarily as a paramagnetic monomer. In order to reconcile this apparent paradox one must turn to molecular orbital theory. The molecular orbitals for NO are shown in Fig. 2 and the electron configuration indicates that the unpaired electron in NO resides in an antibonding orbital. Thus, the overall bond order for NO is 2.5. The bond order for an NO dimer (ONNO) would therefore be 5, or 2.5 per NO molecule. This means that there is no net gain of bond order when two NO molecules form a dimer and entropic considerations would favor the monomers.

2.2.   Reaction of NO with Dioxygen

Aqueous NO solutions are stable almost indefinitely in the absence of O_2. However, in the presence of O_2, solutions of NO decompose. Like NO, O_2 is a radical species. However, since O_2 has one more valence electron than NO, both of the degenerate π^* orbitals are occupied by a single electron (see Fig. 2) and thus the groundstate of O_2 is a triplet (two unpaired electrons with parallel spins). Since O_2 is a triplet and NO is a doublet (one unpaired electron), NO will react twice with O_2. The reaction of NO with O_2 generates, initially, nitrogen dioxide (NO_2) presumably via the sequence of reactions (1)–(3).

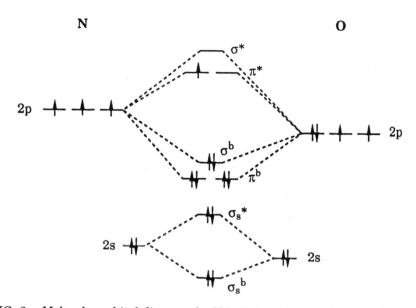

FIG. 2. Molecular orbital diagram for NO. Note: the π^b orbitals are lower in energy than the σ^b because electrons in the π^b orbitals are further removed from the filled $\sigma_s{}^*$ and $\sigma_s{}^b$ orbitals. The π^b and σ^b orbital energies for O_2 are reversed from that depicted here.

$$NO + O_2 \longrightarrow {}^\bullet OONO \tag{1}$$
$${}^\bullet OONO + NO \longrightarrow ONOONO \tag{2}$$
$$ONOONO \longrightarrow 2NO_2 \tag{3}$$

As indicated by the above reactions, the generation of NO_2 from NO and O_2 is overall third order, first order in O_2 and second order in NO. The rate expression for the NO/O_2 reaction is $-d[NO]/dt = 4k[NO]^2[O_2]$ with a third-order rate constant of 2×10^6 $M^{-2}s^{-1}$ [12–14]. The kinetics of the NO/O_2 reaction indicate that O_2-mediated NO degradation would be slow at low NO concentrations relative to other biological reactions. The rate-limiting step in the NO/O_2 reaction appears to be independent of the media. This would suggest that the rate constant is similar no matter what the media; thus, the rate of the NO/O_2 reaction would primarily be dependent only on the concentration of both NO and O_2 [12]. Since both of these diatomic molecules are 10–20 times more soluble in hydrophobic media, this would predict that the primary site of the NO/O_2 reaction in biological systems would be in the membranes and not in aqueous solution. Another important difference between hydrophobic and aqueous solutions is the intermediates formed in the reaction. Though the rate constants between hydrophobic and aqueous media are similar, different reports suggest that the formation of N_2O_3 differs (reviewed in [6]). It has been shown that in aqueous media free NO_2 is not present, whereas it is readily produced in hydrophobic media. The nature of the chemical intermediates in the aqueous reaction is unclear at present; however, the formation of N_2O_3 is the predominant reactive nitrogen species produced in aqueous solutions. In biological systems, it is likely that the predominent media for the NO/O_2 reaction will be hydrophobic compartments and NO_2 will play a significant role. Like NO, NO_2 is a radical species. However, unlike NO, NO_2 is a fairly potent oxidant with similar oxidizing properties in solution as bromine [15] ($E^0 = 1.04$ V). Nitrogen dioxide can rapidly dimerize to form dinitrogen tetraoxide, N_2O_4 (reaction (4)). Since both NO and NO_2 are radicals, they will react rapidly with each other to form dinitrogen trioxide, N_2O_3 (reaction (5)). In aqueous solution, N_2O_4 will react with water to give nitrite and nitrate in equimolar amounts (reaction (6)). N_2O_3 will react with water to give exclusively nitrite (reaction (7)). The decomposition of NO in aerobic aqueous solution results in the near quantitative generation of NO_2^- via reactions (1)–(3) and (5) and (7) [14,16,17]. That is, any NO_2 formed from reactions (1), (2) and (3) is quickly scavenged by excess NO (reaction (5)) to generate N_2O_3 which hydrolyzes to NO_2^-.

$$NO_2 + NO_2 \longrightarrow N_2O_4 \qquad (k = 4.5 \times 10^8 \text{ M}^{-1}\text{ s}^{-1} \text{ [18]}) \quad (4)$$

$$NO_2 + NO \longrightarrow N_2O_3 \qquad (k = 1.1 \times 10^9 \text{ M}^{-1}\text{ s}^{-1} \text{ [19]}) \quad (5)$$

$$N_2O_4 + H_2O \longrightarrow NO_2^- + NO_3^- + 2H^+ \quad (k = 1000 \text{ s}^{-1} \text{ [19]}) \quad (6)$$

$$N_2O_3 + H_2O \longrightarrow 2NO_2^- + 2H^+ \qquad (k = 1000 \text{ s}^{-1} \text{ [19]}) \quad (7)$$

Since NO_2 levels are kept low by reaction (5), significant dimerization of NO_2 (reaction (4)) does not occur. Both N_2O_3 and N_2O_4 are electrophilic species that can react with a variety of nucleophiles. In fact, the reaction of either species with water (which serves as the nucleophile in this reaction) is indicative of their electrophilic character. In one instance, water is electrophilically nitrosated to give NO_2^- (reaction (7)); in the other instance, water is electrophilically nitrated to give NO_3^- (reaction

(6)). The reaction of N_2O_3 with other nucleophilic species such as thiols or amines is also possible and will be discussed later in the chapter.

2.3. Reaction of NO with Superoxide and Subsequent Chemistry

One electron reduction of O_2 generates superoxide (O_2^-) which is also a free radical species. NO and O_2^- react with each other in a radical-radical combination reaction at a near diffusion-controlled rate [20,21] to generate peroxynitrite (^-OONO) (reaction (8)).

$$NO + O_2^- \longrightarrow {}^-OONO \qquad (k = 4\text{--}7 \times 10^9 \text{ M}^{-1}\text{ s}^{-1} \text{ [20,21]}) \qquad (8)$$

Peroxynitrite is stable as the anion and therefore basic solutions of ^-OONO can be stored and kept for months. However, the pK_a of ^-OONO is 6.7 [22] and the conjugate acid of ^-OONO, peroxynitrous acid (HOONO), is not stable and can either spontaneously decompose to give NO_3^- or serve as a potent oxidizing agent. Due to the possible physiological relevance of the NO/O_2^- reaction there has been considerable interest in ^-OONO/HOONO chemistry and numerous reviews have appeared on the subject [23–25]. Oxidations by ^-OONO can occur by either one- or two-electron chemistry [26]. For example, oxidation of methionine by HOONO can occur by either a one-electron or two-electron pathway. In the two-electron oxidation pathway, the methionine sulfur atom reacts as a nucleophile and attacks the electrophilic oxygen of HOONO, resulting in the generation of the corresponding sulfoxide and inorganic nitrite. One-electron oxidation of methionine occurs via an initial electron transfer reaction from sulfur to HOONO, which leads to the initial formation of a thiyl radical and eventual generation of ethylene.

Peroxynitrite is also capable of nitrating aromatic rings [27,28]. Metal-catalyzed nitration of, for example, the phenol ring of tyrosine has been demonstrated and has been proposed to occur via generation of the equivalent of nitronium ion (NO_2^+). Metal-independent nitration of tryptophan has also been observed [29] and two mechanisms have been proposed. One proposal involves a direct electrophilic aromatic substitution reaction by an activated form of HOONO. Alternatively, HOONO may abstract a hydrogen atom to form the tryptophan radical, water, and NO_2. Trapping of the tryptophan radical by NO_2 would then result in tryptophan nitration.

Along with its properties as a potent oxidant, ^-OONO is nucleophilic as well. The nucleophilic character of ^-OONO is exemplified by its rapid reaction ($k = 3 \times 10^4$ M^{-1} s^{-1}) with carbon dioxide in which an unstable nitrosoperoxycarbonate ($ONOOCO_2^-$) is proposed to be formed [30,31]. This reaction represents one of the fastest known reactions for ^-OONO. The nitrosoperoxycarbonate species will spontaneously rearrange to a nitrocarbonate ion ($O_2NOCO_2^-$), which can decompose to generate NO_3^-. Thus, CO_2 can serve to increase the rate of ^-OONO decomposition. Carbon dioxide also serves to activate ^-OONO as a nitrating agent which is evidenced by the fact that the presence of CO_2 will increase the yields of aromatic ring nitration by ^-OONO several fold [31–33].

The chemical fate of ^-OONO formed via the NO/O_2^- reaction is highly dependent on the flux of the two reactants because a secondary reaction between HOONO and either excess NO or O_2^- appears to occur (reactions (9) and (10)) [34]. As might be expected, excess NO or O_2^- inhibits the oxidation chemistry of HOONO (although oxidative chemistry can still occur via NO_2-mediated processes).

$$HOONO + O_2^- + H^+ \longrightarrow NO_2 + O_2 + H_2O \qquad (9)$$

$$HOONO + NO \longrightarrow NO_2 + NO_2^- + H^+ \qquad (10)$$

Tangentially related to the NO/O_2^- chemistry described above is the reaction between NO and dioxygen bound to hemoproteins. For example, NO reacts with oxymyoglobin ($MbFe(II)O_2$) to give metmyoglobin and NO_3^- (reaction (11)) [35].

$$[MbFe^{II}O_2 \leftrightarrow MbFe^{III}O_2^-] + NO \longrightarrow MbFe^{III} + NO_3^-$$
$$(k = 3.7 \times 10^7 \text{ } M^{-1} \text{ } s^{-1}) \qquad (11)$$

Since the ferrous-oxygen complex of myoglobin is probably more accurately depicted as a ferric O_2^- complex, this reaction can be viewed similarly to (8). Since it is easy to spectroscopically distinguish oxymyoglobin (or oxyhemoglobin) from metmyoglobin (or methemoglobin), the chemistry depicted in reaction (11) serves as the basis for a convenient in vitro assay for NO (e.g., [36]).

2.4. Reaction of NO with Metals and Metalloproteins

From a physiological perspective, one of the most important aspects of NO chemistry is its interaction with metals and/or metalloproteins. Nitric oxide is an exceedingly good ligand for reduced iron hemoproteins such as deoxyhemoglobin ($HbFe^{II}$). The reaction of NO with reduced iron hemes is fairly unique in that coordination of NO to the heme should result in liberation of the transaxial ligand [37] (Fig. 3). Unlike other simple diatomic ligands such as O_2 or carbon monoxide, NO will also bind oxidized iron heme proteins as well (although with lower affinity compared to ferrous hemes) (e.g., see [38]). The ferric-NO adduct (Fe^{III}-NO) can also be depicted as a ferrous-nitrosonium adduct (Fe^{II}-$^+$NO). In fact, this species is capable of nitrosating a variety of nucleophiles (reaction (12)) [39,40].

$$[Fe^{III}\text{-}NO \leftrightarrow Fe^{II}\text{-}^+NO] + X^- \longrightarrow X\text{-}NO + Fe^{II} \qquad (12)$$

The unique coordination chemistry between NO and hemoproteins is the basis for much of the biological activity associated with NO (discussed later). However, NO can also interact with other biologically relevant metal centers as well. For example, NO can interact with some copper proteins. Type 1 copper proteins form complexes with NO in the oxidized form, type 2 copper proteins do not interact with NO, and type 3 copper proteins complex NO in either the oxidized or the reduced state [41]. The interaction between NO and cobalamin has also been examined [42] and it was found that NO is capable of oxidizing Co^{II} to Co^{III}, with presumed formation of $^-$NO (discussed in detail later). Moreover, NO appears to form a highly reversible complex with Co^{III}.

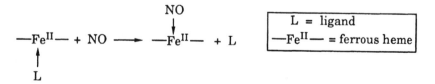

FIG. 3. Release of an axially bound ligand upon coordination of NO to a ferrous heme.

2.5.   Reaction of NO and Related Species with Thiols

The chemical interactions between nitrogen oxides and thiols has been examined in depth. Under aerobic conditions, NO will react with thiols to generate nitrosothiols [43–46]. Much of this chemistry is oxygen-dependent and likely involves oxidized nitrogen species such as NO_2 and N_2O_3. For example, oxidation of NO by oxygen results in the generation of N_2O_3 (reactions (1), (2), (3), and (5) above), which can then nitrosate thiols to form an S-nitrosothiol (reaction (13)).

$$N_2O_3 + RSH \longrightarrow RSNO + NO_2^- + H^+ \tag{13}$$

Under anaerobic conditions, NO also reacts with thiols to give as the primary products reduced nitrogen species such as nitrous oxide (N_2O) and oxidized thiols [47–50]. One mechanism for the reduction of NO by thiols has been proposed to occur via the reaction sequence indicated in Fig. 4 [49].

2.6.   Nitroxyl (HNO) Chemistry

One electron reduction of NO generates NO^-. This species and its conjugate acid, HNO, are referred to as nitroxyl (not to be confused with the stable free radical species known as nitroxides). NO^- is isoelec-

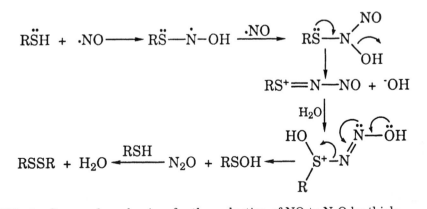

FIG. 4.   Proposed mechanism for the reduction of NO to N_2O by thiols.

tronic with O_2 and can therefore exist in either the ground state triplet or the excited singlet spin state. The reduction potential for the redox couple $NO/{}^1NO^-$ is -0.35 V and for the $NO/{}^3NO^-$ is 0.39 V [51]. Thus, singlet NO^- is a much better reducing agent than the triplet form. The pK_a of HNO generated by pulse radiolysis has been reported to be 4.7 [52] (the species generated by pulse radiolysis is thought to be the singlet form and the triplet form is expected to be more acidic [53]). Nitroxyl is a metastable species because it can react with another NO^- (or HNO) to give hyponitrous acid, which then rapidly dehydrates to generate N_2O (reactions (14) and (15)).

$$2HNO \longrightarrow HON{=}NOH \tag{14}$$

$$HON{=}NOH \longrightarrow N_2O + H_2O \tag{15}$$

$$\text{(combined (14) and (15), } k = 2\text{--}8 \times 10^9 \text{ M}^{-1} \text{ s}^{-1} \text{ [54])}$$

Thus, detection of N_2O is often used as an indication of HNO intermediacy (although this is certainly not rigorous, e.g., see Fig. 4).

Besides its propensity to dimerize and generate N_2O, HNO will also react with NO sequentially to generate $N_2O_2^-$ and $N_3O_3^-$ with the latter species decomposing to N_2O and NO_2^- (reactions (16)–(18)).

$$NO^- + NO \longrightarrow N_2O_2^- \qquad (k = 1.7 \times 10^9 \text{ M}^{-1} \text{ s}^{-1} \text{ [52]}) \quad (16)$$

$$N_2O_2 + NO \longrightarrow N_3O_3 \qquad (k = 4.9 \times 10^6 \text{ M}^{-1} \text{ s}^{-1} \text{ [52]}) \quad (17)$$

$$N_3O_3^- \longrightarrow N_2O + NO_2^- \qquad (k = 87 \text{ s}^{-1} \text{ [52]}) \quad (18)$$

The second order rate constant for reaction (16) indicates that NO^- can be a highly efficient scavenger of NO. As mentioned earlier, NO reacts extremely rapidly with O_2^- to generate ^-OONO (reaction (8)). An isoelectronic process would be the reaction of ^-NO with O_2. It may be expected that ^-OONO should be formed from the reaction of NO^- with O_2. Indeed, ^-OONO can be generated from this reaction, but it is highly dependent on the spin state of NO^-. Donald et al. [53] have shown that triplet NO^- can react with O_2 to give ^-OONO but singlet NO^- does not react with O_2 at an appreciable rate (reactions (19) and (20)).

$$^3NO^- + O_2 \longrightarrow {}^-OONO \tag{19}$$

$$^1NO^- + O_2 \longrightarrow \text{no reaction (or slow reaction)} \tag{20}$$

Thus, the generation of HNO in an aerobic system may not always lead to ^-OONO formation.

HNO has electrophilic properties as evidenced by its reaction with thiols to generate, presumably, an N-hydroxysulfenamide (reaction (21)) [55].

$$HNO + RSH \longrightarrow RS\text{-}NHOH \ (N\text{-hydroxysulfenamide}) \qquad (21)$$

This intermediate can further react with excess thiol to give hydroxylamine and the corresponding disulfide (reaction (22)).

$$RS\text{-}NHOH + RSH \longrightarrow RSSR + NH_2OH \qquad (22)$$

Alternatively, the N-hydroxysulfenamide can rearrange in a unimolecular process to generate a sulfinamide product (Fig. 5) [56].

The above thiol chemistry indicates that HNO can be electrophilic. Another example of the electrophilic nature of HNO can be seen in its reaction with olefinic species. In the singlet spin state, NO^- is capable of comparable chemistry to that observed for $^1O_2$. For example, $^1NO^-$ reacts with 1,3-dienes in an apparent $4 + 2$ cycloaddition reaction as well as with olefins in an "ene-like" reaction (Fig. 6) [57].

As indicated by its negative reduction potential ($E^0 = -0.35$ V for the singlet species), NO^- can act as a single-electron reducing agent. An example of this is represented by the ability of HNO to reduce the cupric form of the enzyme superoxide dismutase (SOD) to the cuprous form [58–60] (reaction (23)).

$$NO^- + SODCu^{II} \longrightarrow NO + SODCu^{I} \qquad (23)$$

In these studies, the source of HNO dictated that its spin state would have been singlet and therefore highly reducing. Nitroxyl can also react

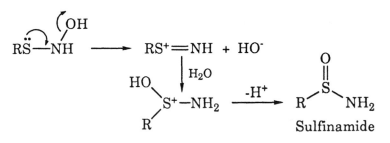

FIG. 5. Mechanism of formation of sulfinamide from the reaction of HNO with thiols.

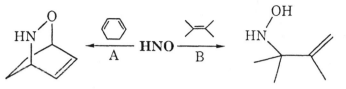

FIG. 6.   Reaction of HNO with diene in a 4 + 2 cycloaddition reaction (A) and with an olefin in an ene-type reaction (B).

with oxidized heme proteins such as methemoglobin or metmyoglobin to generate the reduced ferrous nitrosyl adducts [61] (reaction (24)).

$$NO^- + HbFe^{III} \longrightarrow HbFe^{II}\text{-}NO \tag{24}$$

2.7.   Other Nitrogen Oxides

We have briefly discussed some of the chemistry that may have some physiological importance for NO, NO_2, N_2O_3, and HNO. These species are related to each other by the redox scheme shown in Fig. 7. Among the other nitrogen oxides not previously discussed, NO_3^- would be expected to be innocuous and/or inert under physiological conditions. The conjugate acid of NO_2^-, nitrous acid (HONO, $pK_a = 3.37$), can serve as the precursor to nitrosating species (possibly via the formation of N_2O_3). Hydroxylamine (NH_2OH) is a reasonable reducing agent and is capable of reacting with, for example, ferric ion to generate ferrous ion and N_2O [15].

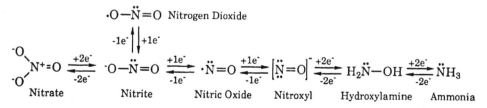

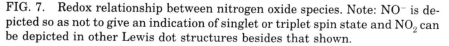

FIG. 7.   Redox relationship between nitrogen oxide species. Note: NO^- is depicted so as not to give an indication of singlet or triplet spin state and NO_2 can be depicted in other Lewis dot structures besides that shown.

3. NITRIC OXIDE BIOSYNTHESIS

3.1. Enzymology of NO Biosynthesis

The biosynthesis of NO is accomplished by a family of enzymes generally referred to as the nitric oxide synthases (NOS). These enzymes
oxidatively convert the amino acid L-arginine to NO and L-citrulline in
an O_2- and NADPH-dependent reaction (Fig. 8). The conversion of one
of the terminal guanidinium nitrogens of L-arginine to NO represents
a net 5e⁻ oxidation (because the formal oxidation state of the nitrogen
changes from −3 to +2). This process is stepwise as N-hydroxy-L-arginine
(NOHA) has been demonstrated to be a biosynthetic intermediate [62–
65]. Thus, NOS is capable of both hydroxylating a guanidinium function, to generate NOHA, and further oxidizing the N-hydroxyguanidine
to generate NO and the corresponding urea. The overall process has no
chemical or biochemical precedence and, therefore, represents a unique
transformation. The intimate chemical steps involved in this oxidation
will be discussed in detail later. There are three major classes of NOS:
endothelial cell-derived NOS (eNOS), neuronal NOS (nNOS), and immunological NOS (iNOS). The eNOS isoform is membrane-bound as a
result of posttranslational myristoylation [66–69] or palmitylation [70],
and has been found to be localized in the caveoli of the plasma membrane [71]. On the other hand, the nNOS and iNOS isoforms are both
soluble proteins. All three isoforms — nNOS, eNOS, and iNOS — have
molecular weights of approximately 125–160 kDa (depending on the

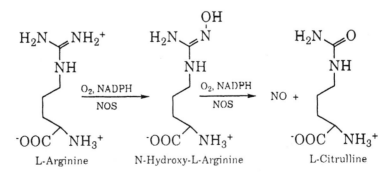

FIG. 8.   Reaction catalyzed by NOS.

animal species). However, all of these isoforms are only active as the homodimer. In general, eNOS and nNOS are considered to be constitutively expressed, although this is not entirely accurate since levels of these isoforms can change in response to a variety of physiological events (for a review, see [4]). The iNOS isoform is inducible and can be expressed at high levels in a variety of cells during an immune response. Both eNOS and nNOS are regulated by Ca^{2+} levels (e.g., see [72,73]). Thus, an influx of Ca^{2+} into a cell results in the activation of eNOS/nNOS via an interaction with the Ca^{2+} binding protein calmodulin. That is, Ca^{2+} binds calmodulin which then binds eNOS or nNOS, thus allowing enzyme turnover and NO generation. Unlike eNOS and nNOS, iNOS is not regulated by Ca^{2+} since the Ca^{2+}/calmodulin complex is essentially an enzyme subunit [74,75]. Therefore, once expressed, iNOS appears capable of generating NO continuously provided that an ample supply of substrate and cofactors are available [76].

It should also be mentioned that there are other isoforms of NOS that have been recently described and that may be distinct from eNOS, nNOS and iNOS. Evidence has been presented for a distinct NOS present in the rat kidney that is Ca^{2+}-independent, inducible during pregnancy, and not inhibitable by species known to potently inhibit the other isoforms [77]. Also, an as-yet-uncharacterized NOS has recently been found in mitochondria [78–80]. It is possible, if not likely, that other NOS isoforms will be discovered in the future.

In spite of the fact that the biosynthetic pathway for NO generation is fairly well established (Fig. 8), the intimate chemical mechanism by which NOS oxidizes L-arginine to generate NO is not established and is currently a matter of significant speculation. However, the catalytic mechanism for all the NOS isoforms is likely to be the same because they all have identical cofactor and prosthetic group requirements. All NOS isoforms thus far characterized are iron hemoproteins that contain flavin mononucleotide (FMN), flavin adenine dinucleotide (FAD), and tetrahydrobiopterin (H_4B). All isoforms also require O_2 and NADPH and utilize L-arginine as substrate. Moreover, the NOS enzymes can be considered to be monooxygenases since isotope labeling studies indicated that the oxygen atoms in NO and citrulline originate from O_2 rather than H_2O [81,82]. Thus, the initial hydroxylation of arginine to NOHA involves an activation of O_2 and the incorporation of an oxygen atom from O_2 into NOHA. Based on the cofactor-prosthetic group requirements and the fact that it is a monooxygenase, NOS is similar to the cytochrome P450 monooxygenase system and/or the aromatic amino acid hydroxylases, phenylalanine hydroxylase, tyrosine hydroxylase, and tryptophan hydroxylase. Thus it is worthwhile to briefly review these systems as a prelude to speculating on the mechanism of NOS catalysis.

The cytochrome P450 enzyme system consists of two proteins, a heme-containing cytochrome P450 (referred to, simply, as P450) and an FAD-, FMN-containing, NADPH-binding cytochrome P450 reductase (referred to as P450 reductase) (for a review of P450 enzymology, see [83]). This system is capable of oxidizing a wide variety of substrates. For example, it can hydroxylate hydrocarbons, epoxidize olefins, and oxidize heteroatoms. The reductive activation of O_2 occurs at the heme site of P450 while the P450 reductase serves as a conduit for the electrons from NADPH. Although it is counterintuitive to think that electrons are required to carry out oxidation chemistry, reactions (25) and (26) illustrate how this may be possible using hydroxylation of a substrate as an example.

$$O_2 + 2e^- + 2H^+ \longrightarrow H_2O + \text{``O''} \text{ (oxygen atom or equivalent) (25)}$$
$$\text{``O''} + \text{substrate-H} \longrightarrow \text{substrate-OH} \hspace{3.5cm} (26)$$

Thus, reductive activation of O_2 results in the simultaneous generation of H_2O and a potent two electron oxidant with similar chemistry to that expected for an oxygen atom. However, a "free" oxygen atom is not formed in this process; rather an oxygen atom bound to the heme iron is formed. The actual activated oxygen species responsible for much of the oxidative capacity of P450 is thought to be this high-valent iron-oxo species. The overall catalytic cycle for cytochrome P450 is illustrated in Fig. 9.

Not all P450-mediated oxidations occur through the generation of a high-valent iron-oxo species. A specific P450 known as aromatase is involved in one of the steps in the conversion of testosterone to estradiol and utilizes, as the oxidizing agent, the ferric peroxo intermediate in the P450 cycle (e.g., [84]). This species is especially nucleophilic and is capable of oxidizing substrates that have electrophilic centers such as aldehydes [85,86]. The proposed mechanism for P450 aromatase activity is shown in Fig. 10.

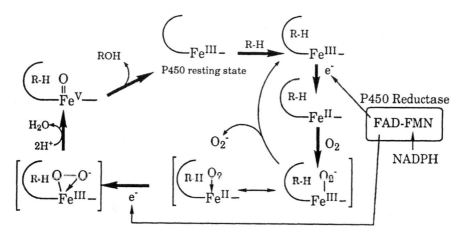

FIG. 9. Catalytic cycle for P450. R-H = substrate. The oxidizing species, FeᵛO, would be more accurately depicted as an $Fe^{IV}O$-porphyrin radical cation. However, for the sake of simplicity it is shown as an Fe^V species.

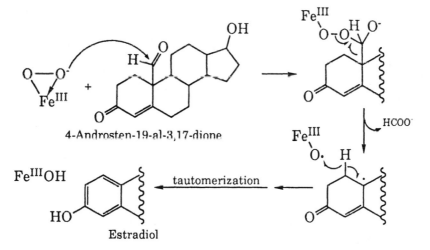

FIG. 10. Proposed mechanism for the oxidation of an aldehyde (4-androsten-19-al-3,17-dione) to an olefin by the ferric peroxo-P450 intermediate. The mechanism is shown as stepwise but is just as likely to be a concerted one.

Thus, there are two established mechanisms by which P450 is capable of oxidizing substrates; with most substrates oxidation is likely to occur through the generation of an iron-oxo species and with electrophilic substrates; oxidation can occur via the nucleophilic ferric peroxo intermediate. The oxidation of L-arginine and N-hydroxy-L-arginine by NOS may involve either or both of these oxidizing intermediate species and will be discussed later.

The aromatic amino acid hydroxylases are distinct from the P450 system in that they require H_4B for catalytic activity and contain a nonheme iron. Like P450, however, these enzymes activate O_2 reductively using NADPH (e.g., [87,88]). The activation of O_2 by the aromatic amino acid hydroxylases appears to occur via a direct interaction with the H_4B moiety. That is, O_2 reacts directly with H_4B to generate a hydroperoxy species that serves as the ultimate oxidant for the hydroxylation of the aromatic ring. The resultant oxidized pterin, dihydrobiopterin (H_2B), can then be recycled back to H_4B by one of two possible NADPH-dependent enzymes (Fig. 11). The catalytic cycle depicted below for the aromatic amino acid hydroxylases does not indicate a role for the nonheme iron contained in these enzymes. It is clearly possible, if not likely, that the nonheme iron participates in either the chemistry

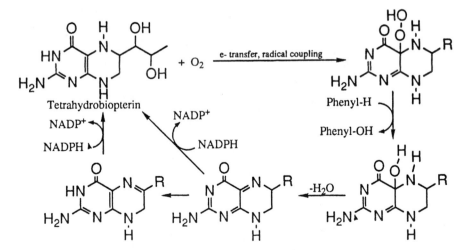

FIG. 11. Catalytic cycle for the aromatic amino acid hydroxylases. Phenyl-H represents the substrate, phenyl-OH the aromatic ring hydroxylated product.

of O_2 activation or the regulation of these enzymes (or both). However, an in-depth discussion of this topic is beyond the scope of this chapter.

Based on a variety of studies, it is evident that NOS activates O_2 in a manner similar to the P450 enzymes (e.g., [89]). That is, O_2 binds the heme and is reduced by NADPH via an electron conduit made up of FAD and FMN. Unlike mammalian P450s, however, all of the cofactors and prosthetic groups are contained in a single protein. (Of note, there are several well-characterized nonmammalian P450s in which both the P450 reductase domain and the heme domain are part of a single protein; e.g., see [90].) In the isoforms that are regulated by Ca^{2+}, the Ca^{2+}-calmodulin complex serves as a type of electron switch allowing electron flow from the flavins to the heme [91] and/or electron flow into the flavins [92].

The role of H_4B in NOS catalysis remains one of the most enigmatic issues in NOS enzymology. Since O_2 activation apparently occurs through the iron heme, as in P450, H_4B may not be acting in the same capacity as it does in the aromatic amino acid hydroxylases. That is, H_4B may not be involved in direct activation. Work by various laboratories indicates that, at the very least, H_4B may have an allosteric role in promoting enzyme dimerization [93,94]. As mentioned earlier, nNOS and iNOS are active only as the homodimers and it has been shown that dimerization of the monomers is dependent on H_4B (along with heme and L-arginine). It has also been reported that H_4B may modulate the heme environment [95,96]. Moreover, H_4B binding to NOS appears to change the conformation of NOS to a state with a high affinity for L-arginine [97].

It seems unlikely that H_4B serves only as an allosteric modulator of NOS monomer dimerization or L-arginine binding. It would be expected that the activity of NOS would be somehow linked to the redox properties of H_4B (afterall, all other known H_4B-requiring processes utilize the redox activity of H_4B). An early study indicated that H_4B is not stoichiometrically redox-active during NOS turnover since its re-reduction does not occur via established pathways involving known pterin reductases [98]. Later it was shown by the same group that H_4B/H_2B recycling can occur on the NOS enzyme in an NADPH-dependent process [99]. Thus, a redox and/or catalytic role for H_4B in NOS turnover remains a distinct possibility. Several reports have proposed that a role for H_4B may be to prevent and/or reverse the autoinhibition of

NOS by the product, NO, by serving as a reducing agent that can reductively destroy the inhibitory complex [100,101]. This would indicate a nonstoichiometric yet redox-mediated role for H_4B. Interestingly, it was reported that purified NOS was capable of converting L-arginine to NOHA in the absence of exogenously added reducing agents, thus indicating that NOS contains an endogenously present reducing agent [102]. By a process of elimination, H_4B has been suggested to be the endogenous reducing agent in these studies. Further recent work indicates that H_4B-free NOS is capable of reducing O_2 to H_2O_2 even in the presence of L-arginine [103]. However, when H_4B is added to the enzyme, it is then capable of NO/citrulline generation. Thus, it is clear that H_4B is required to couple O_2 reduction/activation to NO formation. Interestingly, H_4B appears to destabilize the ferrous-O_2 catalytic intermediate indicating a clear relationship between NOS heme chemistry and H_4B [89]. However, the chemical interactions responsible for these relationships have yet to be established, and it is clear that further work needs to be done to unequivocally determine the role of H_4B in NOS catalysis.

3.2.   Chemistry of NO Biosynthesis

Oxidation of L-arginine to N-hydroxy-L-arginine represents a two electron N-oxidation reaction. Chemically similar transformations of guanidine-containing xenobiotics have been reported for P450. For example, the guanidinium functionality of guanabenz and debrisoquine, both guanidine-containing drugs, can be N-hydroxylated to generate an N-hydroxyguanidine metabolite [104–106]. Furthermore, it is well established that the heme moiety in NOS is required for catalysis and is likely to be the site of O_2 activation just as in P450. Thus, the first step in NO biosynthesis, arginine hydroxylation, is reasonably based on known P450 chemistry and can be envisioned to occur via a mechanism similar to that proposed for other N-oxidation processes (e.g., see [107]) (Fig. 12). It has been determined that only one NADPH (two electrons) is required for the conversion of L-arginine to NOHA [62]. This is consistent with the above mechanism since the activation of O_2 to form an iron-oxene requires two electrons. However, the second step, oxidation of NOHA to NO and citrulline, has been reported to require only

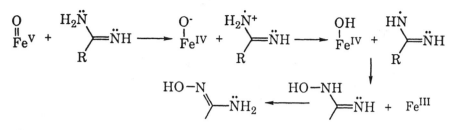

FIG. 12.   Possible mechanism for the hydroxylation of L-arginine.

one electron from NADPH [62]. Thus, the catalytic process for the second step is clearly distinct from the first step.

The conversion of NOHA to citrulline and NO represents a three-electron oxidation and therefore must contain an odd electron step. Interestingly, chemical substrates structurally analogous to NOHA have been reported to be capable of undergoing P450-catalyzed oxidation to generate nitrogen oxides and products analogous to citrulline [108–110]. These results further establish the possible mechanistic relationship between NOS and P450. Chemical model studies on N-hydroxy-guanidine oxidation indicate that it is unlikely that there are any amino acid intermediates between NOHA and citrulline [111,112]. As with the first enzymatic step, the second step appears to involve oxygen activation via the heme. The reported requirement for only one electron in this second step can be reconciled if the substrate itself serves as an electron donor. That is, oxidation of the enzyme-bound NOHA by the metal center may occur, thus precluding the requirement for a second electron from NADPH. It has been proposed that the oxidation of NOHA may occur via a P450 aromatase-like mechanism (e.g., [113,114]). Several reasonable mechanisms for the NOS-mediated conversion of NOHA to NO have been proposed [115,116] and are schematically outlined in Fig. 13.

It should be emphasized that the above mechanisms are speculative. In fact, two studies have proposed that NO may not actually be the initial enzyme product and may instead be the four-electron oxidation product, HNO [60,117]. As mentioned previously, HNO is a good reducing agent and can easily be converted to NO in the presence of a reasonable electron acceptor (such as the heme iron of NOS). Moreover,

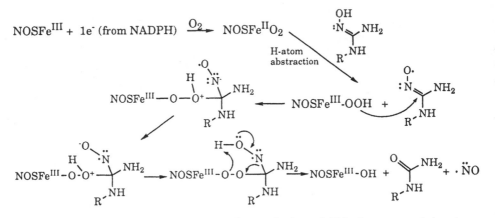

FIG. 13. Possible mechanism for the oxidation of N-hydroxy-L-arginine to citrulline and NO.

NOS has been found to be capable of generating HNO via anaerobic peroxidatic oxidation of NOHA [118].

4. PHYSIOLOGY AND PATHOPHYSIOLOGY OF NITRIC OXIDE: A CHEMICAL PERSPECTIVE

The role or effect of NO in biological systems is purely a function of the chemistry it can perform. As indicated earlier in the chapter, NO is chemically unique and it is not surprising that nature would take advantage of some of these unique chemical properties as a way of accomplishing functional specificity. One of the best ways to categorize the physiological chemistry of NO biology is to use the convention of Wink et al. [6] (known as the chemical biology of NO) whereby the chemical effects of NO can be envisioned as being either direct or indirect. Direct effects of NO are those that involve NO itself without prior conversion to other reactive species. These effects include such things as ligation to metals and direct interaction with biological radical species. Indirect effects concern the reactivity of NO derivatives. For example, NO may react with O_2 and O_2-derived species to form agents with chemical properties distinct from NO and whose reactivity is responsible for the alteration of cell function. The indirect chemistry of NO can be further categorized as either oxidative or nitrosative in nature. That is, the NO-derived species can react with biological molecules resulting in either oxidation or heteroatom nitrosation. Below, we will use these terms to address the various chemical effects associated with NO and indicate how they may be involved in its physiology and pathophysiology.

4.1. EDRF and the Activation of Soluble Guanylate Cyclase (sGC)

Probably the best understood example of a direct effect of NO is its ability to raise levels of the second messenger cGMP leading to smooth muscle relaxation. Although unknown to them at the time, Furchgott and Zawadzki were the first to describe the vascular actions of endogenously generated NO when they reported that endothelial cells exposed to acetylcholine could release a chemical entity that could diffuse into smooth muscle cells and elicit smooth muscle relaxation [119]. Since the identity of this species was not known at the time, it was referred to as endothelium-derived relaxing factor (EDRF). The identity of EDRF remained a mystery for almost a decade. Then, following original proposals by Ignarro et al. [120] and Furchgott [121], several groups identified EDRF as NO [122,123]. Although the discovery of endogenously generated NO occurred in the late 1980s, NO has actually been used therapeutically for over 100 years. For example, nitroglycerin has been used in the treatment of angina pectoris since the 1880s and the utility of nitroglycerin in this regard is due to its ability to release NO when it is bioactivated.

The vasorelaxant effect elicited by NO is due to its interaction with the enzyme soluble guanylate cyclase (sGC) (for a review on this topic, see [124–127]). Soluble guanylate cyclase is a heme protein that catalyzes the conversion of GTP to cGMP and it is the actions of cGMP that lead to smooth muscle relaxation. Soluble guanylate cyclase possesses low basal activity. However, exposure of sGC to NO results in an increase in activity by as much as 100–200 fold. Nitric oxide appears unique in its ability to elicit this degree of activation of sGC (although recent work seems to indicate that CO, in the presence of certain other activators, may be capable of significant activation of sGC as well [128]). The involvement of an NO-heme complex in the activation of sGC was discovered in the late 1970s [129]. Further work on this phenomenon resulted in a model explaining NO activation of sGC whereby the coordination of NO to the ferrous heme center causes a structural modification of the coordination of the protein to the iron porphyrin that results in enzyme activation [124]. This model, based purely on biochemical experiments, was found to be thermodynamically reasonable [37] and was later confirmed by spectroscopic studies with the purified enzyme [130–133]. Thus, in the resting state of sGC, the iron heme exists as a six-coordinate ferrous complex. As mentioned earlier, NO binding to ferrous heme results in the liberation of the proximal (trans) histidine ligand and formation of a pentacoordinate nitrosyl complex. Although the intimate details of the catalytic mechanism of sGC are currently unknown, there is little doubt that the release of the proximal histidine ligand from the porphyrin is crucial to the activation mechanism (Fig. 14).

The ability of NO to promote proximal ligand release is clearly related to its ability to activate sGC. Since NO is unique in its ability to perform this chemistry, it is not surprising that nature took advantage of this fact in developing a specific messenger for this important signaling system.

The discovery of EDRF by Furchgott and Zawadzki [119] was based on the activity of acetylcholine on isolated vessel preparations.

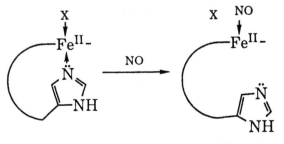

Guanylate Cyclase   Activated Guanylate Cyclase

FIG. 14.   NO-promoted release of proximal histidine ligand in the activation of guanylate cyclase. Distal ligand is shown as X and is possibly another histidine [130].

Since acetylcholine does not circulate through the vascular system, this preparation did not represent a physiologically relevant phenomenon. However, other agents and phenomena have also been found to elicit NO release from endothelial cells. For example, bradykinin or shear stress are capable of causing NO release with subsequent smooth muscle relaxation (e.g., [134,135]). The one common property of the physiological or pharmacological agents capable of eliciting NO release from endothelial cells is their ability to increase intracellular Ca^{2+} levels. This is not at all surprising in view of the fact that eNOS activity is Ca^{2+}/CaM dependent. Thus, agents or phenomena capable of eliciting Ca^{2+} influx into endothelial cells (or release from intracellular stores) results in an activation of eNOS and the production of NO. Nitric oxide then diffuses into neighboring smooth muscle cells, activating sGC within those cells. This causes an increase in cGMP levels leading to smooth muscle relaxation.

4.2. Activation of sGC and the Central Nervous System

The vascular system is not the only place where the activation of sGC by NO is important. In 1988, Garthwaite and coworkers reported that an EDRF-like species was released from cerebellar cells when N-methyl-D-aspartate (NMDA) receptors were activated by glutamate [136]. Further studies indicated that, indeed, this EDRF-like species was NO [137]. As alluded to earlier, the brain contains a Ca^{2+}-dependent, soluble NOS isoform referred to as nNOS. To be sure, the brain also contains eNOS and iNOS as well. One of the major ways to activate nNOS in the brain is through the stimulation of the NMDA receptor whose principle ligand is the excitatory amino acid neurotransmitter, glutamate. Thus, stimulation of the NMDA receptor leads to an influx of Ca^{2+}, which leads to the activation of nNOS. The NO formed in these neurons can then diffuse into other neighboring cells, resulting in the activation of sGC and a rise in intracellular cGMP. The specifics regarding the function of nNOS-derived NO and/or cGMP in the brain are unclear. Recent studies implicate nNOS-derived NO in the actions of retinal photoreceptor cells [138] and the function of olfactory cells [139]. Biochemically, NO/cGMP in the CNS may activate protein kinases, regulate phosphodiesterase activity, and regulate the activity of cyclooxygenase. A comprehensive listing of all of the possible functions of nNOS-derived NO is beyond the scope of this chapter; several excellent reviews on this topic have appeared recently and the reader is referred to these for a more detailed discussion [140–145]. An observation of particular note is that mice lacking nNOS exhibit aggressive and inappropriate sexual behavior without any other gross changes, suggesting that NO is involved in these behavioral characteristics ([146] and references therein). Also, neuronal cells from mice lacking nNOS appear to be especially resistant to NMDA toxicity, thus implicating nNOS-derived NO in certain forms of neurotoxicity [147].

Long-term potentiation (LTP) is an enduring increase in synaptic strength observed after stimulation of an excitatory pathway. It is generally thought that LTP is involved in memory and learning. One of the physiological requirements for LTP is the presence of a retrograde messenger, i.e., there must exist a messenger that can diffuse from the postsynaptic terminal back to the presynaptic terminal to increase transmitter release. In the hippocampus, LTP involves the NMDA receptor [148]; therefore, it has been proposed that NO may serve as the retrograde messenger (e.g., [149–151]). However, whether or not NO is involved in LTP and to what degree it is involved remains somewhat controversial (for a discussion of this area, see [141]). Interestingly, mice lacking nNOS exhibit normal LTP responses even though NOS inhibitors block the response [152]. This observation led to studies culminating in the somewhat surprising suggestion that LTP in the hippocampus may be a result of eNOS-derived NO rather than NO from nNOS.

4.3. Activation of sGC in the Peripheral Nervous System

Neurons that do not respond to acetylcholine or adrenaline innervate a variety of smooth muscle tissues such as the gastrointestinal tract, esophagus, and corpus cavernosum. These neurons have been designated as nonadrenergic, noncholinergic neurons, or NANC neurons. Although there is no single NANC neurotransmitter, abundant evidence supporting NO as a NANC neurotransmitter has been accumulated (for a recent review, see [153]). Neurotransmission by NO has been termed nitrergic and it is clearly and fundamentally distinct from other neurotransmitter systems (i.e., adrenergic, cholinergic, etc.). In its capacity as a NANC neurotransmitter, NO does not resemble the classic paradigm for neurotransmitters. That is, NO release is a result of de novo synthesis and not from stored pools; the "NO-receptor" is a cytoplasmic enzyme, sGC, and not a membrane protein; and NO degradation is not due to the actions of specific enzymes but rather due to simple chemical degradation. Nitrergic transmission involves an initial influx of Ca^{2+} into the nerve terminal which results in the activation of nNOS. Nitric oxide then diffuses from the nerve varicosity into the smooth muscle tissue where it activates sGC, raises cGMP levels, and elicits smooth muscle relaxation.

4.4. Immunological NO

The relationship between NO biosynthesis and cGMP-mediated events is well established and represents a major function of NO as a signal transduction or cell-signaling agent. Along with its role as a specific and diffusable activator of sGC, NO is biosynthesized for other purposes as well. In an immune response, NO is made by certain cells in response to a variety of infectious agents (e.g., see [154–159]). Thus, activation of immune response cells such as macrophages by cytokines (i.e., IFN-γ, IL-1, IL-2, TNF-α) and/or bacterial products (i.e., LPS) can result in the induction of iNOS, the Ca^{2+}-independent, "high-output" NOS isoform. The induction of iNOS is not, however, restricted to immune response cells. For example, iNOS can be induced in smooth muscle cells, endothelial cells, or hepatocytes; in severe cases of sepsis, iNOS-derived NO from some of these cells can affect the vascular system causing severe hypotension (shock). While it is clear that activated immune response cells such as macrophages produce relatively large amounts of NO, the exact function of NO generated from these cells is currently a matter of some speculation and controversy. Clearly, the biochemical actions of NO generated at physiologically high concentrations, such as in the case of an immune response, supersede simple activation of sGC and likely involves other direct chemical interactions of NO and NO-derived species with a variety of target cell systems. Interestingly, induction of iNOS in human macrophages in vitro is different from that observed in other species. In fact, there is doubt as to whether human macrophages can be induced to synthesize iNOS at all (e.g., see [160]). However, macrophages taken from patients with an inflammatory or infectious disease are capable of generating NO (for an excellent discussion of this topic, see [161] and references therein).

Presumably, one of the purposes of NO generated in an immune response is to act as a cytotoxic and/or cytostatic species toward target cells such as invading pathogens or cancer cells. Direct evidence for this has been obtained using mice lacking iNOS. These animals failed to restrain the replication of bacteria in vivo and cancer cells in vitro [162]. Immunological NO may also serve as a signaling molecule that can either promote a beneficial adaptation of nontarget cells or a deleterious change in the immune-targeted cell. The mechanism by which NO acts in these regards is a matter of considerable speculation. Regardless, the mechanism of action of NO in immune response will be a function of its distinct chemical properties and the environment in which it is generated. The effect of NO as a cytotoxic agent can be beneficial or deleterious depending on whether it is part of a normal immune function or whether it is generated pathophysiologically and acts to destroy normal cells. Although we will continue to refer to high levels of NO as "immunological" (in reference primarily to the fact that high NO levels are usually obtained through iNOS induction), we do not mean to infer that high fluxes of NO are always part of a normally beneficial immune response. That is, "immunological NO" may also have deleterious effects as well. In the continuing discussion below, we will address the possible chemical mechanisms associated with NO at physiologically high levels (such as those generated in an immune response) and relate these chemical issues to its possible role as an effector molecule in immune function and as a toxic species implicated in a variety of disorders and diseases.

4.4.1. Direct Effects of NO with Metals

As previously mentioned, one of the most important direct actions of NO under normal physiological circumstances is its ability to activate sGC via coordination to the ferrous heme prosthetic group. The direct actions of NO as a pathogen, however, are likely mediated through other chemical processes. For example, NO is capable of directly altering the function of a variety of heme proteins such as cytochrome P450

[163,164], which can lead to an alteration in drug metabolism and hormone biosynthesis. Nitric oxide is also capable of inhibiting catalase [165,166], which may lead to increased and deleterious levels of hydrogen peroxide. However, the inhibition of mitochondrial function may be a primary mechanism by which NO exerts its direct cytotoxic/cytostatic effects. Hibbs and coworkers originally proposed that NO may directly alter mitochondrial activity by inhibiting the function of iron-sulfur (Fe-S) cluster-containing proteins [158]. Thus, NO has the potential to interact with several mitochondrial components because NADH-Q reductase, succinate reductase, and cytochrome reductase all contain Fe-S moieties. In fact, all of these mitochondrial components have been reported to be inhibited by NO (however, the degree of inhibition and the susceptibility of each component to inhibition by NO is not fully understood) (e.g., see [167–172]). The interaction of NO with Fe-S clusters results in the formation of free iron in the form of iron nitrosyls [158]. Along with the possible interaction of NO with Fe-S proteins in the respiratory chain, the terminal electron acceptor protein in mitochondrial respiration, cytochrome c oxidase, has also been shown to be inhibited by NO [173,174]. The inhibition of cytochrome oxidase likely involves simple coordination of NO to the prosthetic heme moiety thus inhibiting O_2 binding. Interestingly, an NOS isoform has been found to exist in mitochondria, which presents the possibility that NO may have a normal regulatory capacity in mitochondria as well.

An Fe-S protein that appears to be particularly sensitive to the effects of NO is the citric acid cycle enzyme aconitase (e.g., [175]). Inhibition of activity by NO occurs presumably via the interaction of NO with the Fe-S cluster of aconitase. However, further investigation of this phenomenon indicates that the species responsible for aconitase inhibition is not NO but that the NO-O_2^- adduct, $^-$OONO, is the ultimate inhibitor [176,177]. Thus, the effects of NO on aconitase activity (and possibly other Fe-S proteins) may not be entirely due to direct actions of NO but may also involve indirect actions as well. Ferrochelatase, another Fe-S-containing enzyme, is also affected by NO [178]. This enzyme is responsible for iron insertion into porphyrins and, if inhibited, could dramatically alter heme protein activity within a cell. Clearly, the interaction of NO with Fe-S clusters may represent a critical event in NO pathophysiology. Moreover, based on the finding that NO could be made in mitochondria, this interaction may also be of normal physiological relevance as well (although it remains to be determined as to what mitochondrial-derived NO is doing).

Nitric oxide is capable of altering iron metabolism as evidenced by its ability to liberate iron from its cellular stores [167,179]. The ability of NO to act in this regard is thought to contribute to its overall toxicology. However, NO can also serve as a signal in controlling iron homeostasis. Nitric oxide is capable of activating the iron regulatory protein (IRP), which is an Fe-S containing protein of the aconitase family of proteins [180,181]. Activation of IRP by NO results in its binding to the iron responsive element (IRE), which is a regulator of ferritin and transferrin receptor mRNA expression.

It has become increasingly clear that proteins containing Fe-S clusters are potential targets for the actions of NO (e.g., [182]). However, the intimate chemical details of the interaction between these proteins and NO is not yet clear. Whether all of these interactions constitute direct or indirect effects of NO (or both) is a matter of some speculation. It may well be that the mechanism by which NO elicits its effects will be dependent on the nature of the Fe-S protein in question. For example, IRP activation by NO has been proposed to be due to NO (or a closely related species) and not to $^-$OONO [181], whereas aconitase inhibition has been reported to be due primarily to $^-$OONO (mentioned above).

4.4.2. Direct Effects on Free Radical Chemistry

As mentioned earlier, NO is a free radical species and as such has the potential to participate in free radical events. As a fairly unreactive and potentially long-lived radical, NO has the ability to act as a free radical chain-terminating species. For example, the lipid peroxidation free radical chain process can be terminated by NO (reactions (27)–(31)).

$$L\text{-}H + X^\bullet \longrightarrow L^\bullet + HX$$
(27)
(initiation, $X^\bullet$ = free radical chain initiator)

$$L^\bullet + O_2 \longrightarrow LOO^\bullet \text{ (propagation)}$$
(28)

$$LOO^\bullet + L\text{-}H \longrightarrow LOOH + L^\bullet \text{ (propagation)}$$
(29)

$$L^\bullet + NO \longrightarrow L\text{-}NO \text{ (termination)}$$
(30)

$$LOO^\bullet + NO \longrightarrow LOONO \text{ (termination)}$$
(31)

In this capacity, NO can be viewed as a protective species in that it limits the damage to membrane lipids. Indeed, many examples of the cytoprotective actions of NO have been reported [183–192]. Another way in which NO may act to limit the damage due to free radical chain processes is by sequestering the catalysts required for the generation of the free radical chain initiator (designated as X• in reaction (27) above). One possible method for generating a free radical chain initiating species is via the Fenton reaction (reaction (32)).

$$H_2O_2 + Fe^{II} \longrightarrow Fe^{III} + {}^{\bullet}OH + HO^- \tag{32}$$

One study indicates that NO is capable of inhibiting the Fenton reaction and is thus able to inhibit free radical chain initiation [193]. This inhibition is thought to occur through the formation of an unreactive Fe^{II}-NO adduct. In direct contrast to this, it has been reported that NO is capable of reducing Fe^{III} to Fe^{II} (reaction (12)) forming an Fe^{II}-NO adduct that was able to promote Fenton-driven oxidation [166]. This apparent inconsistency can possibly be explained by noting the differences between the experimental conditions of the two studies. The earlier study [193] examined the effect of NO on Fe-EDTA complexes whereas the later study [166] dealt exclusively with the chemistry of iron-aquo complexes (at slightly acidic pH). Significantly, NO is not able to reduce Fe^{III}-EDTA complexes even though it can reduce the Fe^{III} aquo complex (Fukuto, unpublished data). Clearly, the nature of the ligation of the iron species is extremely important in determining its interaction with NO. Thus, depending on the circumstances, NO may act to protect against free radical damage or it may act to promote it.

4.4.3. Other Possible Direct Effects

Cellular protection by NO exposure may not always be a result of its ability to act directly as a free radical scavenger or sequester metal initiators. A particularly significant example of a protective action of NO unrelated to direct chemical interactions is the observation that preexposure of cells to NO protects them from subsequent nitrosative or oxidative insult [194]. This effect is likely due to the ability of NO to act as a signaling molecule upregulating protective proteins.

The above discussion of the direct effects of NO dealt primarily with the possible interaction between NO and metal or radical species. However, recent work has alluded to the possibility that NO may also directly react with protein thiols (or, more likely, thiolates). In this process, NO is reduced and the thiols are oxidized (see Fig. 4). This chemistry has been proposed to occur in the inhibition of aldehyde dehydrogenase by NO [50] and the oxidation of the free sulfhydryl function of human serum albumin [49]. This reaction most assuredly occurs through the thiolate species, which indicates that only protein thiols with sufficiently low pK_a values will react at a reasonable rate.

4.4.4. Indirect Oxidative Effects

As mentioned earlier, NO is capable of reacting with O_2 to generate higher oxides (reactions (1)–(7)). The reaction of NO with O_2 results in the initial formation of NO_2 with subsequent generation of species such as N_2O_3 and NO_2^-. Due to the fact that generation of NO_2 from the O_2/NO reaction is second order in NO, the physiological relevance of this chemistry increases as the flux of NO increases. Thus, the most likely scenario for significant NO_2 generation would be under cytotoxic/ pathophysiological conditions where high fluxes of NO are generated near hydrophobic areas.

Like NO, NO_2 is a paramagnetic radical species. However, unlike NO, NO_2 is a fairly potent oxidizing agent (for the NO_2/NO_2^- redox couple $E^0 = 1.04$ V). For example, NO_2 is capable of nitrating the phenolic group of tyrosine via a presumed two-step radical process (Fig. 15) [195,196].

Under conditions whereby NO and O_2^- are generated simultaneously and in the vicinity of each other, $^-$OONO can be formed. Proto-

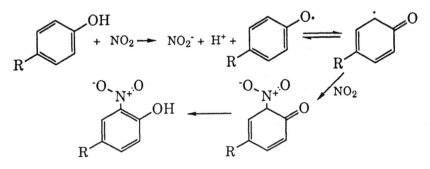

FIG. 15.   Oxidation of phenol function of tyrosine by NO_2 to give nitrotyrosine.

nation of $^-$OONO (pK_a = 6.7) generates peroxynitrous acid, which can rearrange to generate innocuous NO_3^-. However, $^-$OONO can also serve as a potent oxidizing agent capable of reacting with a variety of biological molecules. As mentioned earlier, $^-$OONO can oxidize thiols, guanosine, and nitrate aromatic rings (such as the phenol in tyrosine). Moreover, $^-$OONO can also serve as a precursor to NO_2 since excess NO or O_2^- will react with $^-$OONO to give NO_2 [34].

In the presence of a Lewis acid metal, $^-$OONO is capable of nitrating tyrosines. For example, it has been demonstrated that the catalytic cupric ion in the enzyme SOD is capable of serving as such a catalyst. The ability of $^-$OONO to perform this chemistry has been proposed to account for much of the pathophysiology of NO/O_2^- since tyrosine modification would drastically alter the function and properties of proteins (e.g., see [197–199]). More specifically, the nitration of tyrosine by SOD has been proposed to be a cause of amyotrophic lateral sclerosis [200].

Recent studies have indicated that one of the most likely physiological fates of $^-$OONO is reaction with CO_2 to form $^-O_2COONO$ [30]. The reaction of $^-$OONO with CO_2 increases the rate of its decomposition to NO_3^- and thus may be expected to be a mechanism of detoxification. However, the $^-O_2COONO$ intermediate is also a potent oxidant and therefore reaction of $^-$OONO with CO_2 could represent a mechanism of increased $^-$OONO toxicity, as several studies have suggested [31,33].

4.4.5. Indirect Nitrosative Effects

Air oxidation of NO leads to NO_2 formation, which may be a source of oxidative stress. However, further reaction of NO_2 with NO leads to the formation of N_2O_3 (reaction (5)). Dinitrogen trioxide is an electrophilic species that can nitrosate a variety of biological nucleophiles such as thiols and amines. Since thiols are inherently the most potent physiological nucleophile, it would not be surprising that the interaction between thiols and N_2O_3 is an important and relevant physiological event. The nitrosation of a thiol generates an S-nitrosothiol (reaction (33)).

$$N_2O_3 + RSH \longrightarrow RSNO + NO_2^- + H^+ \qquad (33)$$

As mentioned earlier, other possible mechanisms for thiol nitrosation exist that are independent of N_2O_3 generation and potentially physiologically relevant [33,39,40]. The modification of critical thiols in proteins can have deleterious effects. For example, zinc finger proteins contain critical thiol ligands and can be inhibited by NO [201,202], oxidation of the thiols of metallothionein by NO can lead to the release of toxic heavy metals [203], depletion of glutathione via thiol nitrosation can lead to an increased susceptibility to oxidative stress [204], and inhibition of DNA-alkyltransferases results from S-nitrosation [205]. Thiol nitrosation may not always have toxicological implications because S-nitrosation may also play a role in normal cell signaling (e.g., see [206]).

Another possible site of nitrosation is amine functions. Nitrosation of amines by, for example, N_2O_3 results in the formation of a nitrosamine (reaction (34)).

$$\text{R-NH}_2 + N_2O_3 \longrightarrow \text{R-NH-NO} + NO_2^- + H^+ \qquad (34)$$

Depending on the structure of the amine, the fate of the nitrosamine can vary. Primary amines (like the one depicted in reaction (34)) can decompose to generate carbocations (or the equivalent thereof) (reactions (35) and (36)).

$$\text{R-NH-NO} \longrightarrow \text{R-N=N-OH} \qquad (35)$$

$$\text{R-N=N-OH} + H^+ \longrightarrow [R^+ + N_2 + H_2O] \longrightarrow \text{R-OH} + H^+ \quad (36)$$

Thus, nucleic acids are capable of being deaminated presumably via the nitrosation chemistry depicted above [207,208]. Again, it is worth noting that a similar chemical transformation can be accomplished without the need for O_2 (thus without the intermediacy of N_2O_3). Ferric nitrosyl porphyrin complexes are capable of nitrosating nucleic acid bases (see reaction (11)) to form the corresponding nitrosamine. This intermediate then undergoes the reactions shown above (reactions (35) and (36)), resulting in deamination [40].

4.4.6.   Reduced NO Species

Thus far, the discussion regarding NO metabolism, fate, and biological activity has concentrated mainly on NO itself and possible oxidized NO species. However, as shown in Fig. 7, NO has reduced counterparts as well. The reduction of NO or the generation of HNO, for example, has been a subject of some interest. Stamler and Arnelle have proposed that the direct interaction of S-nitrosothiols with thiols can result in the generation of HNO and the corresponding disulfide (reaction (37)) [209].

$$RS\text{-}NO + R'\text{-}SH \longrightarrow R\text{-}SS\text{-}R' + HNO \qquad (37)$$

It has also been proposed that HNO may actually be produced by NOS [60,117]. Moreover, HNO has been found to be a primary and likely product from the oxidation of the NO biosynthetic intermediate NOHA [111,112]. Therefore, the recent results indicating that NOHA can be released from NO-producing cells both in vitro [210] and in vivo [211, 212] present the possibility that NOHA can serve as a precursor to physiological HNO generation.

The biological activity of HNO has been examined by several groups. Using HNO donor molecules, it has been reported that HNO is capable of inhibiting thiol containing proteins such as aldehyde dehydrogenase [213]. Of special significance, recent work has reported that HNO is extremely cytotoxic and capable of causing double-stranded DNA breaks [214]. Although the mechanism(s) of its toxicity is currently unknown, it appears that significant generation of HNO in a biological system can have severe deleterious effects (possibly via some of the chemistry outlined earlier). Also of potential significance is the rapid reaction between HNO and NO ($k = 1.7 \times 10^9 \, M^{-1} s^{-1}$). Thus, HNO has the capability of being one of the best NO scavengers available with comparable reactivity to O_2^- ($k = 4$–$7 \times 10^9 \, M^{-1} s^{-1}$).

5. CONCLUSIONS

The physiological utility of NO is based on its unique and fascinating chemical properties. Herein we have attempted to provide an overview of some of the chemistry of NO and related species as a primer for understanding the complexities of its biosynthesis as well as its role and function as a cardiovascular agent, neurotransmitter, signaling molecule, and cytotoxic/cytostatic agent. As indicated above, the fate and function of NO is highly dependent on the environment in which it is formed as well as its concentration. The primary function of low fluxes of NO, such as those generated for the purpose of maintaining vascular tone, is as an activator of the enzyme sGC, which is exquisitely sensitive and selective to NO activation. In these cases, the biological activity of NO is carried on by the messenger cGMP. The role and effect of higher fluxes of NO can be myriad and may be either beneficial or potentially deleterious. High fluxes of NO can be both cytotoxic and cytostatic and, therefore, can be crucial to a normal immune response. However, due to the inherent toxicity of NO at high fluxes, biological circumstances leading to unwanted and high NO levels can lead to the development of a variety of diseases and disorders such as diabetes, arthritis, neurodegenerative diseases, and many others. It is evident that an increase in the understanding of the physiological chemistry of NO will lead to an increase in an understanding of the basic biological function of NO as well as its role in the development of many diseases.

ABBREVIATIONS

cGMP	cyclic guanosine monophosphate
CNS	central nervous system
EDRF	endothelium-derived relaxing factor
EDTA	ethylenediamine-N,N,N',N'-tetraacetate
eNOS	endothelial nitric oxide synthase
FAD	flavin adenine dinucleotide
FMN	flavin mononucleotide
GTP	guanosine triphosphate
H_2B	dihydrobiopterin
H_4B	tetrahydrobiopterin
Hb	hemoglobin
HbFe^{II}	deoxyhemoglobin
IFN	interferon
IL	interleukin
iNOS	immunological nitric oxide synthase
IRE	iron responsive element
IRP	iron regulatory protein
LPS	lipopolysaccharide
LTP	long-term potentiation
MbFe^{III}	metmyoglobin
MbFe$^{II}O_2$	oxymyoglobin

NADPH	nicotinamide adenine dinucleotide phosphate, reduced
NANC	nonadrenergic, noncholinergic
NMDA	N-methyl-D-aspartate
nNOS	neuronal nitric oxide synthase
NOHA	N-hydroxy-L-arginine
NOS	nitric oxide synthase
P450	cytochrome P450
sGC	soluble guanylate cyclase
SOD	superoxide dismutase
TNF	tumor necrosis factor

REFERENCES

1. D. J. Stuehr and O. W. Griffith, in *Advances in Enzymology*, Vol. 65 (A. Meister, ed.), John Wiley and Sons, New York, 1992, p. 287–347.

2. D. J. Stuehr, *Annu. Rev. Pharmacol. Toxicol.*, *37*, 339–359 (1997).

3. O. W. Griffith and D. J. Stuehr, *Annu. Rev. Physiol.*, *57*, 707–736 (1995).

4. U. Forstermann, I. Gath, P. Schwarz, E. I. Closs, and H. Kleinert, *Biochem. Pharmacol.*, *50*, 1321–1332 (1995).

5. F. T. Bonner and M. N. Hughes, *Comm. Inorg. Chem.*, *7*, 215–234 (1988).

6. D. A. Wink, I. Hanbauer, M. B. Grisham, F. Laval, R. Nims, J. Laval, J. Cook, R. Pacelli, J. Liebmann, M. Krishna, P. C. Ford, and J. B. Mitchell, *Curr. Top. Cell. Regul.*, *34*, 159–187 (1996).

7. A. R. Butler, F. W. Flitney, and D. L. H. Williams, *Trends Pharmacol. Sci.*, *16*, 18–22 (1995).

8. J. M. Fukuto, *Adv. Pharmacol.*, *34*, 1–15 (1995).

9. C. Nathan, *FASEB J.*, *6*, 3051–3064 (1992).

10. S. Moncada, R. M. J. Palmer, and E. A. Higgs, *Pharmacol. Rev.*, *43*, 109–141 (1991).

11. J. F. Kerwin Jr., J. R. Lancaster, Jr., and P. L. Feldman, *J. Med. Chem.*, *38*, 4343–4362 (1995).

12. P. C. Ford, D. A. Wink, and D. M. Stanbury, *FEBS Lett.*, *326*, 1–3 (1993).

13. D. A. Wink, J. F. Darbyshire, J. F. Nims, R. W. Saavedra, and P. C. Ford, *Chem. Res. Toxicol.*, *6*, 23–27 (1993).

14. R. S. Lewis and W. M. Deen, *Chem. Res. Toxicol.*, *7*, 568–574 (1994).

15. F. A. Cotton and G. Wilkinson, in *Advanced Inorganic Chemistry*, 5th ed., John Wiley and Sons, New York, 1988, p. 324.

16. M. Feelisch, *J. Cardiovasc. Pharmacol.*, *17*(Suppl. 3), S25–S33 (1991).

17. L. J. Ignarro, J. M. Fukuto, J. M. Griscavage, N. E. Rogers, and R. E. Byrns, *Proc. Natl. Acad. Sci. USA*, *90*, 8103–8107 (1993).

18. R. E. Huie, *Toxicology*, *89*, 193–216 (1994).

19. S. E. Schwartz and W. H. White, *Adv. Environ. Sci. Technol.*, *12*, 1–116 (1983).

20. R. E. Huie and S. Padmaja, *Free Rad. Res. Commun.*, *18*, 195–199 (1993).

21. S. Goldstein and G. Czapski, *Free Rad. Biol. Med.*, *19*, 505–510 (1995).

22. W. H. Koppenol, J. J. Moreno, W. A. Pryor, H. Ischiropoulos, and J. S. Beckman, *Chem. Res. Toxicol.*, *5*, 834–842 (1992).

23. J. O. Edwards and R. C. Plumb, *Prog. Inorg. Chem.*, *41*, 599–635 (1994).

24. J. S. Beckman, J. Chen, H. Ischiropoulos, and J. P. Crow, *Meth. Enzymol.*, *233*, 229–240 (1994).

25. W. A. Pryor and G. L. Squadrito, *Am. J. Physiol.*, *268*, L699–L722 (1995).

26. W. A. Pryor, X. Jin, and G. L. Squadrito, *Proc. Natl. Acad. Sci. USA*, *91*, 11173–11177 (1994).

27. J. S. Beckman, H. Ischiropoulos, L. Zhu, M. van der Woerd, C. Smith, J. Chen, J. Harrison, and M. Tsai, *Arch. Biochem. Biophys.*, *298*, 438–445 (1992).

28. H. Ischiropoulos, L. Zhu, J. Chen, H. M. Tsai, J. C. Martin, C. D. Smith, and J. S. Beckman, *Arch. Biochem. Biophys.*, *298*, 431–437 (1992).

29. B. Alvarez, H. Rubbo, M. Kirk, S. Barnes, B. A. Freeman, and R. Radi, *Chem. Res. Toxicol.*, *9*, 390–396 (1996).

30. S. V. Lymar and J. K. Hurst, *J. Am. Chem. Soc.*, *117*, 8867–8868 (1995).

31. R. M. Uppu, G. L. Squadrito, and W. A. Pryor, *Arch. Biochem. Biophys.*, *327*, 335–343 (1996).

32. S. V. Lymar, Q. Jiang, and J. K. Hurst, *Biochemistry*, *35*, 7855–7861 (1996).

33. A. Gow, D. Duran, S. R. Thom, and H. Ischiropoulos, *Arch. Biochem. Biophys.*, *333*, 42–48 (1996).

34. A. M. Miles, D. S. Bohle, P. A. Glassbrenner, B. Hansert, D. A. Wink, and M. B. Grisham, *J. Biol. Chem.*, *271*, 40–47 (1996).

35. M. P. Doyle and J. W. Hoekstra, *J. Inorg. Biochem.*, *14*, 351–358 (1981).

36. M. E. Murphy and E. Noack, *Meth. Enzymol.*, *233*, 240–250 (1994).

37. T. G. Traylor and V. S. Sharma, *Biochemistry*, *31*, 2847–2849 (1992).

38. M. Hoshino, K. Ozawa, H. Seki, and P. C. Ford, *J. Am. Chem. Soc.*, *115*, 9568–9575 (1993).

39. R. S. Wade and C. E. Castro, *Chem. Res. Toxicol.*, *3*, 289–291 (1990).

40. C. E. Castro and E. W. Bartnicki, *J. Org. Chem.*, *59*, 4051–4052 (1994).

41. A. C. F. Gorren, E. de Boer, and R. Wever, *Biochim. Biophys. Acta*, *916*, 38–47 (1987).

42. L. G. Rochelle, S. J. Morana, H. Kruszyna, M. A. Russell, D. E. Wilcox, and R. P. Smith, *J. Pharmacol. Exp. Ther.*, *275*, 48–52 (1995).

43. D. A. Wink, R. W. Nims, J. F. Darbyshire, D. Christodoulou, I. Hanbauer, G. W. Cox, F. Laval, J. Laval, J. A. Cook, M. C. Krishna, W. G. DeGraff, and J. B. Mitchell, *Chem. Res. Toxicol.*, *7*, 519–525 (1994).

44. S. Goldstein and G. Czapski, *J. Am. Chem. Soc.*, *118*, 3419–3425 (1996).

45. V. G. Kharitonov, A. R. Sundquist, and V. S. Sharma, *J. Biol. Chem.*, *270*, 28158–28164 (1995).

46. M. Keshive, S. Singh, J. S. Wishnok, S. R. Tannenbaum, and W. M. Deen, *Chem. Res. Toxicol.*, *9*, 988–993 (1996).

47. W. A. Pryor, D. F. Church, C. K. Govindan, and G. Crank, *J. Org. Chem.*, *47*, 159–161 (1982).

48. N. Hogg, R. J. Singh, and B. Kalyanaraman, *FEBS Lett.*, *382*, 223–228 (1996).

49. E. G. DeMaster, B. J. Quast, B. Redfern, and H. T. Nagasawa, *Biochemistry*, *34*, 11494–11499 (1995).

50. E. G. DeMaster, B. Redfern, B. J. Quast, T. Dahlseid, and H. T. Nagasawa, *Alcohol*, *14*, 181–189 (1997).

51. D. N. Stanbury, *Adv. Inorg. Chem.*, *33*, 69–138 (1989).

52. M. Gratzel, S. Taniguchi, and A. Henglein, *Ber. Bunsenges. Phys. Chem.*, *74*, 1003 (1970).

53. C. E. Donald, M. N. Hughes, J. M. Thompson, and F. T. Bonner, *Inorg. Chem.*, *25*, 2676–2677 (1986).

54. D. A. Bazylinski and T. C. Hollocher, *Inorg. Chem.*, *24*, 4285–4288 (1985).

55. M. P. Doyle, S. N. Mahapatro, R. D. Broene, and J. K. Guy, *J. Am. Chem. Soc.*, *110*, 593–599 (1988).

56. P. S.-Y. Wong, J. Hyun, J. M. Fukuto, F, N. Shirota, E. G. DeMaster, D. W. Shoeman, and H. T. Nagasawa, *Biochemistry*, *37*, 5362–5371 (1998).

57. H. E. Ensley and S. Mahadevan, *Tetrahedron Lett.*, *30*, 3255–3258 (1989).

58. M. E. Murphy and H. Sies, *Proc. Natl. Acad. Sci. USA*, *88*, 10860–10864 (1991).

59. J. M. Fukuto, A. J. Hobbs, and L. J. Ignarro, *Biochem. Biophys. Res. Commun.*, *196*, 707–713 (1993).

60. A. J. Hobbs, J. M. Fukuto, and L. J. Ignarro, *Proc. Natl. Acad. Sci. USA*, *91*, 10992–10996 (1994).

61. D. A. Bazylinski and T. C. Hollocher, *J. Am. Chem. Soc.*, *107*, 7982–7986 (1985).

62. D. J. Stuehr, N. S. Kwon, C. F. Nathan, O. W. Griffith, P. L. Feldman, and J. Wiseman, *J. Biol. Chem.*, *266*, 6259–6263 (1991).

63. P. Klatt, K. Schmidt, G. Uray, and B. Mayer, *J. Biol. Chem.*, *268*, 14781–14787 (1993).

64. R. A. Pufahl, P. G. Nanjappan, R. W. Woodard, and M. A. Marletta, *Biochemistry*, *31*, 6822–6828 (1992).

65. G. C. Wallace, P. Gulati, and J. M. Fukuto, *Biochem. Biophys. Res. Commun.*, *176*, 528–534 (1991).

66. J. S. Pollock, V. Klinghofer, U. Forstermann, and F. Murad, *FEBS Lett.*, *309*, 402–404 (1992).

67. W. C. Sessa, C. M. Barber, and K. R. Lynch, *Circ. Res.*, *72*, 921–924 (1993).

68. L. Busconi and T. Michel, *J. Biol. Chem.*, *268*, 8410–8413 (1993).

69. J. Liu and W. C. Sessa, *J. Biol. Chem.*, *269*, 11691–11694 (1994).

70. L. J. Robinson, L. Busconi, and T. Michel, *J. Biol. Chem.*, *270*, 995–998 (1995).

71. O. Feron, L. Belhassen, L. Kobzik, T. W. Smith, R. A. Kelly, and T. Michel, *J. Biol. Chem.*, *271*, 22810–22814 (1996).

72. D. S. Bredt and S. H. Snyder, *Proc. Natl. Acad. Sci. USA*, *87*, 682–685 (1990).

73. J. S. Pollock, U. Forstermann, J. A. Mitchell, T. D. Warner, H. H. H. W. Schmidt, M. Nakane, and F. Murad, *Proc. Natl. Acad. Sci. USA*, *88*, 10480–10484 (1991).

74. H. J. Cho, Q.-W. Xie, J. Calaycay, R. A. Mumford, K. M. Swiderek, T. D. Lee, and C. Nathan, *J. Exp. Med.*, *176*, 599–604 (1992).

75. R. Stevens-Truss and M. A. Marletta, *Biochemistry*, *34*, 15638–15645 (1995).

76. S. M. Morris and T. R. Billiar, *Am. J. Physiol.*, *266* (*Endocrinol. Metab.*, *29*), E829–E839 (1994).

77. R. Singh, S. Pervin, N. E. Rogers, L. J. Ignarro, and G. Chaudhuri, *Biochem. Biophys. Res. Comm.*, *232*, 672–677 (1997).

78. T. E. Bates, A. Loesch, G. Burnstock, and J. B. Clark, *Biochem. Biophys. Res. Commun.*, *213*, 896–900 (1995).

79. T. E. Bates, A. Loesch, G. Burnstock, and J. B. Clark, *Biochem. Biophys. Res. Commun.*, *218*, 40–44 (1996).

80. L. Kobzik, B. Stringer. J.-L. Balligand, M. B. Reid, and J. S. Stamler, *Biochem. Biophys. Res. Commun.*, *211*, 375–381 (1995).

81. N. S. Kwon, C. F. Nathan, C. Gilker, O. W. Griffith, D. E. Matthews, and D. J. Stuehr, *J. Biol. Chem.*, *265*, 13442–13445 (1990).

82. A. M. Leone, R. M. J. Palmer, R. G. Knowles, P. L. Francis, D. S. Ashton, and S. Moncada, *J. Biol. Chem.*, *266*, 23790–23795 (1991).

83. Y. Watanabe and J. T. Groves, in *The Enzymes*, 3rd ed., Vol. 20 (D. S. Sigman, ed.), Academic Press, San Diego, 1992, pp. 405–452.

84. A. D. N. Vaz, E. S. Roberts, and M. J. Coon, *J. Am. Chem. Soc.*, *113*, 5886–5887 (1991).

85. M. F. Sisemore, J. N. Burstyn, and J. S. Valentine, *Angew. Chem. Int. Ed. Eng.*, *35*, 206–208 (1996).

86. M. Selke, M. F. Sisemore, and J. S. Valentine, *J. Am. Chem. Soc.*, *118*, 2008–2012, (1996).

87. S. Kaufman, *Annu. Rev. Nutr.*, *13*, 261–286 (1993).

88. T. A. Dix and S. J. Benkovic, *Acc. Chem. Res.*, *21*, 101–107 (1988).

89. H. M. Abu-Soud, R. Gachhui, F. M. Raushel, and D. J. Stuehr, *J. Biol. Chem.*, *272*, 17349–17353 (1997).

90. A. J. Fulco, *Annu. Rev. Pharmacol. Toxicol.*, *31*, 177–203 (1991).

91. H. M. Abu-Soud and D. J. Stuehr, *Proc. Natl. Acad. Sci. USA*, *90*, 10769–10772 (1993).

92. H. M. Abu-Soud, L. L. Yoho, and D. J. Stuehr, *J. Biol. Chem.*, *269*, 32047–32050 (1994).

93. K. J. Baek, B. A. Thiel, S. Lucas, and D. J. Stuehr, *J. Biol. Chem.*, *268*, 21120–21129 (1993).

94. E. Tzeng, T. R. Billiar, P. D. Robbins, M. Loftus, and D. J. Stuehr, *Proc. Natl. Acad. Sci. USA*, *92*, 11771–11775 (1995).

95. J. Wang, D. J. Stuehr, and D. L. Rousseau, *Biochemistry*, *34*, 7080–7087 (1995).

96. I. Rodriguez-Crespo, N. C. Garber, and P. R. Ortiz de Montellano, *J. Biol. Chem.*, *271*, 11462–11467 (1996).

97. B. Mayer and E. R. Werner, *Nauyn-Schmied. Arch. Pharmacol.*, *351*, 453–463 (1995).

98. J. Giovanelli, K. L. Campos, and S. Kaufman, *Proc. Natl. Acad. Sci. USA*, *88*, 7091–7095 (1991).

99. C. F. Witteveen, J. Giovanelli, and S. Kaufman, *J. Biol. Chem.*, *271*, 4143–4147 (1996).

100. J. M. Griscavage, J. M. Fukuto, Y. Komori, and L. J. Ignarro, *J. Biol. Chem.*, *269*, 21644–21649 (1994).

101. J. Hyun, Y. Komori, G. Chaudhuri, L. J. Ignarro, and J. M. Fukuto, *Biochem. Biophys. Res. Commun.*, *206*, 380–386 (1995).

102. K. L. Campos, J. Giovanelli, and S. Kaufman, *J. Biol. Chem.*, *270*, 1721–1728 (1995).

103. A. C. F. Gorren, B. M. List, A. Schrammel, E. Pitters, B. Hem-

mens, E. R. Werner, K. Schmidt, and B. Mayer, *Biochemistry*, *35*, 16735–16745 (1996).

104. B. Clement, M.-H. Schultze-Mosgau, and H. Wohlers, *Biochem. Pharmacol.*, *46*, 2249–2267 (1993).

105. B. Clement, M. H. Schultze-Mosgau, P. H. Richter, and A. Besch, *Xenobiotica*, *24*, 671–688 (1994).

106. B. Clement, M. Demedmaeker, and S. Linne, *Chem. Res. Toxicol.*, *9*, 682–688 (1996).

107. B. Clement, M. Immel, H. Pfunder, S. Schmitt, and M. Zimmerman, in *N-Oxidation of Drugs* (P. Hlavica and L. A. Damani, eds.), Chapman Hall, London, 1991, pp. 185–205.

108. V. Andronik-Lion, J. L. Boucher, M. Delaforge, Y. Henry, and D. Mansuy, *Biochem. Biophys. Res. Commun.*, *185*, 452–458 (1992).

109. B. Clement and F. Jung, *Drug Met. Disp.*, *22*, 486–497 (1994).

110. A. Jousserandot, J.-L. Boucher, C. Desseaux, M. Delaforge, and D. Mansuy, *Bioorg. Med. Chem. Lett.*, *5*, 423–426 (1995).

111. J. M. Fukuto, G. C. Wallace, R. Hszieh, and G. Chaudhuri, *Biochem. Pharmacol.*, *43*, 607–613 (1992).

112. J. M. Fukuto, D. J. Stuehr, P. L. Feldman, M. P. Bova, and P. Wong, *J. Med. Chem.*, *36*, 2666–2670 (1993).

113. M. A. Marletta, *J. Biol. Chem.*, *268*, 12231–12234 (1993).

114. P. L. Feldman, O. W. Griffith, and D. J. Stuehr, *Chem. Eng. News*, *71*, 26–38 (1993).

115. H.-G. Korth, R. Sustman, C. Thater, A. R. Butler, and K. U. Ingold, *J. Biol. Chem.*, *269*, 17776–17779 (1994).

116. D. Mansuy, J. L. Boucher, and B. Clement, *Biochimie*, *77*, 661–667 (1995).

117. H. H. H. W. Schmidt, H. Hofman, U. Schindler, Z. S. Shutenko, D. D. Cunningham, and M. Feelisch, *Proc. Natl. Acad. Sci. USA*, *93*, 14492–14497 (1996).

118. R. A. Pufahl, J. S. Wishnok, and M. A. Marletta, *Biochemistry*, *34*, 1930–1941 (1995).

119. R. F. Furchgott and J. V. Zawadzki, *Nature*, *288*, 373–376 (1980).

120. L. J. Ignarro, R. E. Byrns, and K. S. Woods, in *Vasodilation: Vascular Smooth Muscle, Peptides, Autonomic Nerves and Endothelium* (P. M. Vanhoutte, ed.), Raven Press, New York, 1988, p. 427.

121. R. F. Furchgott, in *Vasodilation: Vascular Smooth Muscle, Peptides, Autonomic Nerves and Endothelium* (P. M. Vanhoutte, ed.), Raven Press, New York, 1988, p. 401.

122. L. J. Ignarro, G. M. Buga, K. S. Wood, R. E. Byrns, and G. Chaudhuri, *Proc. Natl. Acad. Sci. USA*, *84*, 9265–9269 (1987).

123. R. M. J. Palmer, A. G. Ferrige, and S. Moncada, *Nature*, *327*, 524–526 (1987).

124. L. J. Ignarro, *Semin. Hematol.*, *26*, 63–76 (1989).

125. H. H. H. W. Schmidt, S. M. Lohmann, and U. Walter, *Biochim. Biophys. Acta*, *1178*, 153–175 (1993).

126. F. Murad, *Adv. Pharmacol.*, *26*, 19–33 (1994).

127. A. J. Hobbs, *Trends Pharmacol. Sci.*, *18*, 484–491 (1997).

128. A. Friebe, G. Schultz, and D. Koesling, *EMBO J.*, *15*, 6863–6868 (1996).

129. P. A. Craven and F. R. DeRubertis, *J. Biol. Chem.*, *253*, 8433–8443 (1978).

130. A. E. Yu, S. Hu, T. G. Spiro, and J. N. Burstyn, *J. Am. Chem. Soc.*, *116*, 4117–4118 (1994).

131. J. N. Burstyn, A. E. Yu, E. A. Dierks, B. K. Hawkins, and J. H. Dawson, *Biochemistry*, *34*, 5896–5903 (1995).

132. J. R. Stone and M. A. Marletta, *Biochemistry*, *33*, 5636–5640 (1994).

133. G. Deinum, J. R. Stone, G. T. Babcock, and M. A. Marletta, *Biochemistry*, *35*, 1540–1547 (1996).

134. R. F. Furchgott, P. D. Cherry, J. V. Zawadzki, and D. Jothianandan, *J. Cardiovasc. Pharmacol.*, *6*, S336–S343 (1984).

135. J. P. Cooke, J. Stamler, N. Andon, P. F. Davies, G. McKinley, and J. Loscalzo, *Am. J. Physiol.*, *259*, (*Heart Circ. Physiol.*, *28*), H804–H812 (1990).

136. J. Garthwaite, S. L. Charles, and R. Chess-Williams, *Nature*, *336*, 385–388 (1988).

137. R. G. Knowles, M. Palacios, R. M. J. Palmer, and S. Moncada, *Proc. Natl. Acad. Sci. USA*, *89*, 5159–5162 (1989).

138. A. Savchenko, S. Barnes, and R. H. Kramer, *Nature*, *390*, 694–698 (1997).

139. K. M. Kendrick, R. Guevera-Guzman, J. Zorrila, M. R. Hinton, K. D. Broad, M. Mimmack, and S. Okura, *Nature, 388,* 670–674 (1997).

140. J. Garthwaite, *Trends Neurosci., 14,* 60–67 (1991).

141. J. Garthwaite and C. L. Boulton, *Annu. Rev. Physiol., 57,* 683–706 (1995).

142. V. L. Dawson and T. M. Dawson, *Neurochem. Int., 29,* 97–110 (1996).

143. S. R. Vincent, *Prog. Neurobiol., 42,* 129–160 (1994).

144. I. Paakkari and P. Lindsberg, *Ann. Med., 27,* 369–377 (1995).

145. C. Szabo, *Brain Res. Bull., 41,* 131–141 (1996).

146. R. J. Nelson, G. E. Demas, P. L. Huang, M. C. Fishman, V. L. Dawson, T. M. Dawson, and S. H. Snyder, *Nature, 378,* 383–386 (1995).

147. V. L. Dawson and T. M. Dawson, *J. Chem. Neuroanat., 10,* 179–190 (1996).

148. G. L. Collingridge, S. L. Kehl, and H. McLennan, *J. Physiol. (London), 334,* 33–46 (1983).

149. G. A. Bohme, C. Bon, J.-M. Stutzmann, A. Doble, and J.-C. Blanchard, *Eur. J. Pharmacol., 199,* 379–381 (1991).

150. C. Bon, G. A. Bohme, A. Doble, J.-M. Stutzman, and J.-C. Blanchard, *Eur. J. Neurosci., 4,* 420–424 (1992).

151. J. E. Haley, G. L. Wilcox, and P. F. Chapman, *Neuron, 8,* 211–216 (1992).

152. T. J. O'Dell, P. L. Huang, T. M. Dawson, J. L. Dinerman, S. H. Snyder, E. R. Kandel, and M. C. Fishman, *Science, 265,* 542–546 (1994).

153. R. A. Lefebvre, *Ann. Med., 27,* 379–388 (1995).

154. D. J. Stuehr and M. A. Marletta, *Proc. Natl. Acad. Sci. USA, 82,* 7738–7742 (1985).

155. D. J. Stuehr, S. S. Gross, I. Sakuma, R. Levi, and C. F. Nathan, *J. Exp. Med., 169,* 1011–1020 (1989).

156. R. Keller and R. Keist, *Biochem. Biophys. Res. Commun., 164,* 968–973 (1989).

157. J. B. Hibbs, Jr., Z. Vavrin, and R. R. Taintor, *J. Immunol., 138,* 550–565 (1987).

158. J. B. Hibbs, Jr., R. R. Taintor, Z. Vavrin, and E. M. Rachlin, *Biochem. Biophys. Res. Commun.*, *157*, 87–94 (1988).

159. J. B. Hibbs, Jr., R. R. Taintor, and Z. Vavrin, *Science*, *235*, 473–476 (1987).

160. M. Denis, *J. Leuk. Biol.*, *55*, 682–684 (1994).

161. J. MacMicking, Q. Xie, and C. Nathan, *Annu. Rev. Immunol.*, *15*, 323–350 (1997).

162. J. D. MacMicking, C. Nathan, G. Hom, N. Chartrain, D. S. Fletcher, M. Trumbauer, K. Stevens, Q. Xie, K. Sokol, N. Hutchinson, H. Chen, and J. S. Mudgett, *Cell*, *81*, 641–650 (1995).

163. D. A. Wink, Y. Osawa, J. F. Darbyshire, C. R. Jones, S. C. Eshenaur, and R. W. Nims, *Arch. Biochem. Biophys.*, *300*, 115–123 (1993).

164. J. Stadler, J. Trckfeld, W. A. Schmalix, T. Brill, J. R. Siewert, H. Greim, and J. Doehmer, *Proc. Natl. Acad. Sci. USA*, *91*, 3559–3563 (1994).

165. G. C. Brown, *Eur. J. Biochem.*, *232*, 188–191 (1995).

166. R. Farias-Eisner, G. Chaudhuri, E. Aeberhard, and J. M. Fukuto, *J. Biol. Chem.*, *271*, 6144–6151 (1996).

167. J.-C. Drapier and J. B. Hibbs, Jr., *J. Immunol.*, *140*, 2829–2838 (1988).

168. J. Stadler, T. R. Billiar, R. D. Curran, D. J. Stuehr, J. B. Ochoa, and R. L. Simmons, *Am. J. Physiol.*, *260*, 910–916 (1991).

169. Y. Geng, G. K. Hansson, and E. Holme, *Circ. Res.*, *71*, 1268–1276 (1992).

170. J. P. Bolanos, S. Peuchen, S. J. R. Heales, J. M. Land, and J. B. Clark, *J. Neurochem.*, *63*, 910–916 (1994).

171. J. Torres, V. Darley-Usmar, and M. T. Wilson, *Biochem. J.*, *312*, 169–173 (1995).

172. R. Welter, L. Yu, and C.-A. Yu, *Arch. Biochem. Biophys.*, *331*, 9–14 (1996).

173. G. C. Brown and C. E. Cooper, *FEBS Lett.*, *356*, 295–298 (1994).

174. M. W. J. Cleeter, J. M. Cooper, V. M. Darley-Usmar, S. Moncada, and A. H. V. Schapira, *FEBS Lett.*, *345*, 50–54 (1994).

175. J.-C. Drapier and J. B. Hibbs, Jr., *J. Clin. Invest.*, *78*, 790–797 (1986).

176. L. Castro, M. Rodriguez, and R. Radi, *J. Biol. Chem.*, *269*, 29409–29415 (1994).

177. A. Hausladen and I. Fridovich, *J. Biol. Chem.*, *269*, 29405–29408 (1994).

178. Y.-M. Kim, H. A. Bergonia, C. Muller, B. R. Pitt, W. D. Watkins, and J. R. Lancaster, Jr., *J. Biol. Chem.*, *270*, 5710–5713 (1995).

179. J. B. Hibbs, Jr., R. R. Taintor, and Z. Vavrin, *Biochem. Biophys. Res. Commun.*, *123*, 716–723 (1984).

180. J.-C. Drapier, H. Hirling, J. Weitzerbin, P. Kaildy, and L. C. Kuhn, *EMBO J*, *12*, 3643–3649 (1993)

181. C. Bouton, M. Raveau, and J.-C. Drapier, *J. Biol. Chem.*, *271*, 2300–2306 (1996).

182. J.-C. Drapier, *Methods: A Companion to Meth. Enzymol.*, *11*, 319–329 (1997).

183. G. M. Rubanyi, E. H. Ho, E. H. Cantor, W. C. Lumma, and L. H. P. Botelho, *Biochem. Biophys. Res. Commun.*, *181*, 1392–1397 (1991).

184. J. Kanner, S. Harel, and R. Granit, *Lipids*, *27*, 46–49 (1992).

185. D. A. Wink, I. Ingeborg, M. C. Krishna, W. DeGraff, J. Gamson, and J. B. Mitchell, *Proc. Natl. Acad. Sci. USA*, *90*, 9813–9817 (1993).

186. D. A. Wink, I. Hanbauer, F. Laval, J. A. Cook, M. C. Krishna, and J. B. Mitchell, *Ann. N.Y. Acad. Sci.*, *238*, 265–278 (1993).

187. P. C. Kuo and A. Slivka, *J. Surg. Res.*, *56*, 594–600 (1994).

188. A. T. Struck, N. Hogg, J. P. Thomas, and B. Kalyanaraman, *FEBS Lett.*, *361*, 291–294 (1995).

189. D. A. Wink, J. A. Cook, M. C. Krishna, I. Hanbauer, W. DeGraff, J. Gamson, and J. B. Mitchell, *Arch. Biochem. Biophys.*, *319*, 402–407 (1995).

190. H. Rubbo, S. Parthasarathy, S. Barnes, M. Kirk, B. Kalyanaraman, and B. A. Freeman, *Arch. Biochem. Biophys.*, *324*, 15–25 (1995).

191. R. E. Laskey and W. R. Mathews, *Arch. Biochem. Biophys.*, *330*, 193–198 (1996).

192. H. H. Guitierrez, B. Nieves, P. Chumley, A. Rivera, and B. A. Freeman, *Free. Rad. Biol. Med.*, *21*, 43–52 (1996).

193. J. Kanner, S. Harel, and R. Granit, *Arch. Biochem. Biophys.*, *289*, 130–136 (1991).

194. Y.-M. Kim, H. Bergonia, and J. R. Lancaster, Jr., *FEBS Lett.*, *374*, 228–232 (1995).

195. W. A. Prutz, H. Monig, J. Butler, and E. J. Land, *Arch. Biochem. Biophys.*, *243*, 125–134 (1985).

196. A. Van der Vleit, J. P. Eiserich, C. A. O'Neill, B. Halliwell, and C. E. Cross, *Arch. Biochem. Biophys.*, *319*, 341–349 (1995).

197. J. P. Crow and J. S. Beckman, *Curr. Top. Microbiol. Immunol.*, *146*, 57–73 (1995).

198. J. S. Beckman and J. P. Crow, *Biochem. Soc. Trans.*, *21*, 330–334 (1993).

199. J. S. Beckman, *Chem. Res. Toxicol.*, *9*, 836–844 (1996).

200. J. S. Beckman, M. Carson, C. D. Smith, and W. H. Koppenol, *Nature*, *364*, 584 (1993).

201. D. A. Wink and J. Laval, *Carcinogenesis*, *15*, 2125–2129 (1994).

202. K. D. Kroncke, K. Fechsel, T. Schmidt, F. T. Zenke, I. Dasting, J. R. Wesener, H. Betterman, K. D. Breunig, and V. Kolb-Bachofen, *Biochem. Biophys. Res. Commun.*, *200*, 1105–1110 (1994).

203. R. R. Misra, J. F. Hochadel, G. T. Smith, J. C. Cook, M. P. Waalkes, and D. A. Wink, *Chem. Res. Toxicol.*, *9*, 326–332 (1996).

204. D. A. Wink, R. W. Nims, J. F. Darbyshire, D. Christodoulou, I. Hanbaeur, G. W. Cox, F. Laval, J. Laval, J. A. Cook, M. C. Krishna, W. DeGraff, and J. B. Mitchell, *Chem. Res. Toxicol.*, *7*, 519–525 (1994).

205. F. Laval and D. A. Wink, *Carcinogenesis*, *15*, 443–447 (1994).

206. J. S. Stamler, *Curr. Top. Microbiol. Immunol.*, *196*, 19–36 (1995).

207. D. A. Wink, K. S. Kasprzak, C. M. Maragos, R. K. Elespuru, M. Misra, T. M. Dunams, T. A. Cebula, W. H. Koch, A. W. Andrews, J. S. Allen, and L. K. Keefer, *Science*, *254*, 1001–1003 (1991).

208. T. Nguyen, D. Brunson, C. L. Crespi, B. W. Penman, J. S. Wishnok, and S. R. Tannenbaum, *Proc. Natl. Acad. Sci. USA*, *89*, 3030–3034 (1992).

209. D. R. Arnelle and J. S. Stamler, *Arch. Biochem. Biophys.*, *318*, 279–285 (1995).

210. G. M. Buga, R. Singh, S. Pervin, N. E. Rogers, D. A. Schmitz, C. P. Jenkinson, S. D. Cederbaum, and L. J. Ignarro, *Am. J. Physiol.*, *271* (*Heart Circ. Physiol.*, *40*), H1988–H1998 (1996).

211. M. Hecker, C. Schott, B. Bucher, R. Bussi, and J.-C. Stoclet, *Eur. J. Pharmacol.*, *275*, R1–R3 (1995).

212. R. Wigand, J. Meyer, R. Bussi, and M. Hecker, *Ann. Rheum. Dis.*, *56*, 330–332 (1997).

213. E. G. DeMaster, B. Redfern, and H. T. Nagasawa, *Biochem. Pharmacol.*, *55*, 2007–2015 (1998).

214. D. A. Wink, M. Feelisch, J. Fukuto, D. Christodoulou, D. Jourd'heuil, M. B. Grisham, Y. Vodovotz, J. A. Cook, M. Krishna, W. G. DeGraff, S. Kim, J. Gamson, and J. B. Mitchell, *Arch. Biochem. Biophys.*, *351*, 66–74 (1998).

18

Chemistry of Peroxynitrite and Its Relevance to Biological Systems

Willem H. Koppenol

Laboratorium für Anorganische Chemie, ETH Zürich,
Zürich, Switzerland

1. INTRODUCTION AND BRIEF HISTORICAL OVERVIEW

Peroxynitrite (ONOO$^-$, oxoperoxonitrate(1-)) is an inorganic toxin of biological importance. The chemistry of peroxynitrite has a long history, but that does not imply that the reactivity of this molecule is fully understood. This is because the conjugate acid of peroxynitrite is not stable, and until rapid mixing devices became available it was difficult to study it.

Baeyer and Villiger [1] demonstrated that a mixture of nitrous acid (hydrogen dioxonitrate) and hydrogen peroxide had unusual oxidizing abilities. Trifonow [2] used the ability of such a mixture to nitrate aromatic compounds as an analytical tool for nitrite. The formula HNO_4 for "persalpetrige Säure" was proposed by Raschig in 1907 [3] and used until Gleu and Roell [4] used the correct formula of ONOOH. The early history and the chemistry of peroxynitrite up to 1992 have been described by Edwards and Plumb [5]. In their excellent review they draw attention to the fact that workers in two different areas, peroxide chemistry and nitrate photochemistry, had been studying peroxynitrite without communicating with each other. As an important issue they mention the generation of nitrite when peroxynitrite is left to decay at alkaline pH. The chemistry of peroxynitrite as it relates to the biomedical field was reviewed in 1994 by Beckman et al. [6]. In 1997 a fairly complete review was published that describes progress in the peroxynitrite field up to the beginning of 1995 [7].

The historical perspective of the review by Edwards and Plumb ends as follows: "A comprehensive understanding of this chemistry (i.e., of peroxynitrite) has not been achieved." [5]. Although progress has been made, this is still a true statement. In this chapter newer developments will be discussed and areas in need of further study will be identified. To begin with, no structure of peroxynitrite or peroxynitrous acid has been reported. Our ideas about what peroxynitrite looks like come from quantum mechanical calculations [8–10] and infrared and ultraviolet spectroscopy of potassium peroxynitrite in solid argon [11].

Peroxynitrite can isomerize to nitrate or decompose to nitrite and dioxygen. The terms *decay*, *isomerization*, and *decomposition* are defined as follows, in agreement with IUPAC conventions [12]. The disappearance of peroxynitrite by any process will be referred to as *decay*, following usage in atmospheric chemistry. The process that yields nitrate is named *isomerization*, and that which results in nitrite and dioxygen is *decomposition*. *Cis-trans isomerization* refers strictly to the interconversion of *cis*-peroxynitrite and *trans*-peroxynitrite.

2. IS PEROXYNITRITE FORMED IN VIVO?

Evidence for peroxynitrite formation from nitrogen monoxide and superoxide in vivo is based on the observation that superoxide dismutase extends the biological lifetime of nitrogen monoxide by removing superoxide and the immunological detection of nitrated tyrosine residues in inflamed tissues [13]. It has been argued that these observations do not conclusively point to peroxynitrite: (1) if one assumes that nitrogen monoxide could be reduced in vivo to NO^- (oxonitrate(1-)), then superoxide dismutase could be oxidizing oxonitrate(1-) back to nitrogen monoxide, and (2) nitration could take place via nitrogen dioxide [14,15]. However, how oxonitrate(1-) and nitrogen dioxide are formed in vivo is not clear; in addition, nitration by nitrogen dioxide [16] is not very efficient [17], as it requires two nitrogen dioxide molecules per tyrosine.

The sources of nitrogen monoxide in vivo are endothelial, neuronal, and inducible (in macrophages) nitric oxide synthases. It was recently discovered that nitrogen monoxide is also produced inside mitochondria where it regulates respiration [18] by binding to cytochrome c oxidase ([19] and refs. therein). As mitochondria also leak superoxide [20], it is reasonable to conclude that mitochondria are a possible source of peroxynitrite.

Recently, it was shown that hypochlorite produced by activated neutrophils may react with nitrite to form, most likely, nitryl chloride, which also nitrates tyrosine residues [21,22]. The kinetics of this reaction were studied by Johnson and Margerum [23] who showed that the nitryl chloride either undergoes heterolysis to the nitryl cation and chloride, or, in the presence of millimolar concentrations of nitrite, reacts with the latter to produce dinitrogen tetraoxide, which homolyzes to nitrite and nitrate at a rate of 1×10^3 M^{-1} s^{-1} [24]. When submillimolar nitrite concentrations are present, the rate law derived by Johnson and Margerum [23] predicts a rate of nitryl cation formation of 4×10^4 M^{-1} s^{-1} near neutral pH. Experimentally, a value of 7×10^3 M^{-1} s^{-1} has been determined [25]. Although nitration of tyrosine per se is no longer evidence that peroxynitrite was formed locally [15], one should ask what the source is of the nitrite in the activated neutrophil. It is known that these cells produce a large flux of superoxide that, upon dismutation, yields the hydrogen peroxide necessary for hypochlorite formation. If indeed these cells also produce nitrogen monoxide, and given the diffusion-controlled rate constant for the reaction of superoxide with nitrogen monoxide (see below), all nitrogen monoxide should be converted to peroxynitrite. Reduction of the oxidizing peroxynitrite by two electrons yields nitrite. Therefore, the nitrite present in all likelihood originates from peroxynitrite.

Whether nitration by nitryl chloride plays an important role in the nitration of tyrosine residues must and can be determined: If nitration is due to nitryl chloride, then also chlorinated tyrosines will be found [22]. If peroxynitrite is the culprit, then hydroxylated tyrosines are also formed [26].

3. SYNTHESES

The syntheses of peroxynitrite from (1) nitrite and acidified hydrogen peroxide in all its modifications [27–29], (2) the reaction of azide with ozone, [4], (3) the autoxidation of hydroxylamine [30], and (4) the reaction of hydrogen peroxide with alkylnitrites [31] have recently been reviewed in Gmelin [7]. In this review a synthesis with a low yield from the reaction of hydrogen peroxide with nitrogen monoxide is also mentioned [32]. As hydrogen peroxide does not react with nitrogen monoxide, this synthesis was a mystery until it was shown that the reaction is dependent on the formation of dinitrogen trioxide from nitrogen monoxide and dioxygen, which subsequently attacks hydrogen peroxide [33]. More recently, two new syntheses were reported; the reaction of tetramethylammonium superoxide with nitrogen monoxide in liquid ammonia yields the first pure peroxynitrite salt [34], and the reaction of nitrogen monoxide with solid potassium superoxide can yield peroxynitrite [35] that is less contaminated with nitrite than that synthesized via the methods given in Gmelin.

Irradiation of solid nitrate salts with light below 280 nm, x rays, γ rays, neutrons, or electrons results in the formation of peroxynitrite salts [5,7]. The yield is less than 1%, but the peroxynitrite is stable for years at room temperature. Initially, the yellow color of the salts was misinterpreted to indicate the presence of nitrite, mainly because this anion was found upon dissolution [5].

Photolysis of nitrite solutions [36] and pulse radiolysis of nitrite- and formate-containing solutions [37,38] yield micromolar solutions of peroxynitrite in microseconds, but these methods have no preparative value.

4. PROPERTIES, ISOMERIZATION, AND DECOMPOSITION

4.1. Thermodynamic Properties

The standard enthalpy of formation of the peroxynitrite anion, -10.8 kcal/mol [39], was recently confirmed [40,41]. Calculation of the standard Gibbs energy of formation of the peroxynitrite anion requires an estimate for the absolute entropy, $S°$. In light of a criticism [42] of our earlier estimate of 45 eu, a new estimate of 31 eu was made [43]. Values derived from these thermodynamic parameters and from stopped-flow studies, such as the pK_a of peroxynitrous acid and activation parameters, are listed in Table 1.

4.2. Spectra

The spectrum of the yellow peroxynitrite anion is characterized by a broad absorption band in the UV with a maximum at 302 nm [44]. The extinction coefficient at that wavelength has recently been redeter-

TABLE 1

Thermodynamic Data Related to O=NOO⁻ and O=NOOH

Formation	$\Delta_f H°$	$\Delta_f G°$	$S°$
O=NOO⁻$_{aq}$	-10 ± 1 kcal/mol	$+14 \pm 3$ kcal/mol	$+31 \pm 7$ eu
O=NOOH$_{aq}$	-14 ± 2 kcal/mol	$+5 \pm 3$ kcal/mol	$+47 \pm 5$ eu
O=NOOH$_g$	$+1 \pm 3$ kcal/mol	$+13 \pm 3$ kcal/mol	$+72 \pm 2$ eu

Ionization	$\Delta_{ion} H°$	$\Delta_{ion} G°$	$\Delta_{ion} S°$
	$+4 \pm 2$ kcal/mol	$+8.8 \pm 0.2$ kcal/mol ($pk_a = 6.5$)	-16 ± 5 eu

Isomerization	$\Delta_{iso} H°$	$\Delta_{iso} G°$	$\Delta_{iso} S°$
	-39 ± 1 kcal/mol	-40 ± 3 kcal/mol	$+3 \pm 8$ eu

Activation (isomerization)	$\Delta H^{\ddagger}$	$\Delta V^{\ddagger}$
	20.6 ± 1.0 kcal/mol	1.7 ± 1.0 cm³/mol

Homolysis	$\Delta_{hom} G°$
ONOOH$_{aq}$ → NO$_2^{\bullet}$$_{aq}$ + HO$^{\bullet}_{aq}$	16 kcal/mol
ONOOH$_g$ → NO$_2^{\bullet}$$_g$ + HO$^{\bullet}_g$	7.2 kcal/mol
ONOO⁻$_{aq}$ → NO$^{\bullet}_{aq}$ + O$_2^{\bullet-}$$_{aq}$	18 kcal/mol

Reduction potentials
E^0(ONOOH, H⁺,NO$_2^{\bullet}$, H$_2$O) = 2.0 ± 0.1 V at pH 0
E^0(ONOOH, H⁺,NO$_2^{\bullet}$, H$_2$O) = 1.6 ± 0.1 V at pH 7
E^0(ONOOH, H⁺,NO$_2^-$, H$_2$O) = 1.3 ± 0.1 V at pH 7

[a]Based on the assumption that the standard Gibbs energy of solution of hydrogen oxoperoxonitrate is similar to that of hydrogen peroxide [98].
Source: Data from [43,97]. See also Notes Added in Proof.

mined to be 1700 cm^{-1} M^{-1} [40], close to the earlier value of 1670 cm^{-1} M^{-1} [45]. The spectrum of peroxynitrous acid shows fine structure in the region 300–400 nm [46], but no precise extinction coefficients have been determined. In any case, peroxynitrous acid absorbs much less than peroxynitrite in that region.

Raman spectroscopy, combined with quantum mechanical calculations, showed that peroxynitrite in solution is present in the cis form [10,47] (Fig. 1). The bond order between the nitrogen and the first peroxide oxygen is approximately 1.5, which allows in principle the existence of two conformational isomers. The trans isomer has never been observed.

4.3. Formation

Superoxide reacts very quickly with nitrogen monoxide to form the peroxynitrite anion (reaction (1)):

$$O_2^{\bullet -} + NO^{\bullet} \longrightarrow ONOO^- \tag{1}$$

In the literature one finds three values for the rate constant, determined by pulse radiolysis and flash photolysis, that cluster around 5×10^9 M^{-1} s^{-1} [36–38]. We recently redetermined this rate constant by laser flash photolysis [48]. Solutions of 0.1–1 mM oxoperoxonitrate(1-) at pH 12 were irradiated at 266, 355, and 532 nm with a pulse length of 10 ns at ambient temperature (20–22°C). Irradiation at 266 nm and 355 nm, but not at 532 nm, resulted in bleaching of oxoperoxonitrate(1-) and formation of superoxide. After the flash, superoxide decayed in a second-order process with simultaneous restoration of the oxoperoxoni-

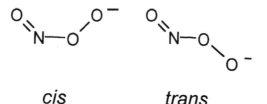

cis   trans

FIG. 1.   Conformational isomers of peroxynitrite.

trate(1-) absorption. A rate constant for the reaction of superoxide with nitrogen monoxide of $(1.9 \pm 0.2) \times 10^{10}$ M^{-1} s^{-1} was derived. As mentioned before, this rate constant is 3–4 times higher than those reported by others [36–38]. These rate constants were measured under experimental designs that involve reaction cascades to arrive at the reactants. The rapid one-step production of superoxide and nitrogen monoxide by flash photolysis has the advantage of allowing measurement of the recombination rate not complicated by other processes.

4.4. Ionization of Peroxynitrous Acid

Peroxynitrous acid is a weak acid (reaction (2)):

$$ONOOH \longrightarrow ONOO^- + H^+ \tag{2}$$

The pK_a of peroxynitrous acid has been determined by stopped-flow spectroscopy; isomerization to nitrate (reaction (3))

$$ONOOH \longrightarrow NO_3^- + H^+ \tag{3}$$

takes place too rapidly at acidic pH (1.2 s^{-1} at 25°C) [48] for the pK_a to be measured directly. The values listed in the literature [5,7,49] and references therein vary from approximately 5 to 8. The wide range is no doubt due to the instability of peroxynitrous acid. We found that the pK_a is approximately 6.5 (Table 1) at very low buffer concentrations and increases to approximately 6.8 in 50 mM phosphate buffer [48]. In other buffers containing Hepes, formate, or ammonia, the apparent pK_a increases to values near 8, possibly due to direct reactions of peroxynitrite with the buffer.

4.5. Adduct Isomerization and Decomposition

The observation that it was not possible to fit the decay of more concentrated (>0.1 mM) solutions of peroxynitrite at slightly alkaline pH to a single exponential led to the discovery of an adduct [48]. Upon dilution to total peroxynitrite and peroxynitrous acid concentrations below 0.1 mM normal behavior is observed. It was concluded that formation of an adduct between $ONOO^-$ and $ONOOH$ takes place. It may be that this adduct, which is predominant at alkaline pH, decomposes to nitrite and dioxygen. These products were observed by Pfeiffer et al. [50] under just those conditions that favor adduct formation. The structure of the adduct is unknown. Thus, below its pK_a peroxynitrous acid isomerizes to nitrate (reaction (3)), while above its pK_a and at concentrations above 0.1 mM decomposition to nitrite and dioxygen takes place (reaction (4)).

$$ONOOH \cdot ONOO^- \longrightarrow 2NO_2^- + O_2 + H^+ \qquad (4)$$

The rate constant of the isomerization to nitrate is 1.2 s^{-1} at 25°C. The value given previously, 1.3 s^{-1} at 25°C, was obtained with peroxynitrite that contained impurities (see above). The rate of decomposition is dependent on the extent of adduct formation, and therefore on the total peroxynitrite concentration and pH.

5. REACTIONS THAT ARE FIRST ORDER IN PEROXYNITROUS ACID

5.1. Isomerization to Nitrate Mechanism

As will be discussed below, some reactions of peroxynitrous acid are first order in peroxynitrous acid and zero order in the compound that is being oxidized. The rate of such a reaction is identical to the rate of isomerization. Thus, on the path to nitrate, there is an intermediate that is strongly oxidizing. The nature of this intermediate is a topic of intense interest.

In principle, there are three pathways by which peroxynitrous acid can isomerize [49]:

1. *Heterolysis, followed by reaction of the nitryl cation with water*:

$$ONOOH \longrightarrow NO_2^+ + OH^- \qquad (5)$$

$$NO_2^+ + OH^- \longrightarrow NO_3^- + H^+ \qquad (6)$$

Although the Gibbs energy change of reaction (5) should be close to zero [43,49], a high Gibbs activation energy of 45 kcal/mol has been estimated [49]. This removes heterolysis from consideration. However, at low (<2) pH the rate of decay increases [46]. It is conceivable that under those conditions peroxynitrous acid binds another proton, and that the nitryl cation and water are formed.

2. *Homolysis, followed by reattachment of the hydroxyl radical to form nitrate* [28,42,51]:

$$ONOOH \longrightarrow NO_2^\bullet + OH^\bullet \qquad\qquad (7)$$

$$NO_2^\bullet + OH^\bullet \longrightarrow NO_3^- + H^+ \qquad\qquad (8)$$

Thermodynamic and kinetic considerations [43], a small activation volume [48], and the results of a scavenger study [52] are not in agreement with homolysis; furthermore, the reaction of the hydroxyl radical with nitrogen dioxide leads, within the error of the experiment, to peroxynitrous acid, not nitric acid [53] (T. Nauser and W. H. Koppenol, unpublished), in contrast to results obtained in the gas phase [54]. Although popular in the biomedical literature, homolysis does not seem feasible.

3. *Internal rearrangement.* We proposed that *cis*-peroxynitrite undergoes a conformation change to *trans*-peroxynitrite, followed by movement of the distal peroxynitrite oxygen to the nitrogen to form nitrate [49]. This movement of the distal peroxide oxygen may already take place when, during the cis-to-trans change, the O–O bond is perpendicular to the O=N–O plane of the molecule [55]. In any case, sometime during the process of isomerization the O–O bond needs to break and a new N–O bond to be formed. It is conceivable that temporarily a diradical is formed. Quantum mechanical calculations have revealed a low-lying triplet state for peroxynitrous acid. The radical-like reactivity would result from two paired electrons in the singlet state becoming separated into distinct orbitals in the triplet state. The internal conversion between the singlet and triplet states may occur through spin-orbital coupling from torsional rotation between the singlet and triplet states in which the interconversion between those two states is rate limiting [9].

An alternative proposal is that a caged radical pair is formed [56]. If some of the hydroxyl radicals can leak from the cage and react with solutes, then the caged radical pair hypothesis is similar to homolysis, i.e., mechanism 2; if no leakage occurs then it is similar to the triplet state, i.e., mechanism 3. At present, the caged radical pair hypothesis is more a matter of semantics than of chemistry.

Anbar and Taube showed that nitrate retains the two peroxidic oxygen atoms of peroxynitrous acid [57]. It was recently shown that during isomerization 17% of the peroxynitrite oxygens exchanged with water [58]. These results were interpreted to support mechanism 3. Studies of the isomerization in deuterium oxide by the stopped-flow technique in our laboratory [56] (Padmaja, Kissner, Bounds, and Koppenol, unpublished) and elsewhere (Tsai and Beckman, unpublished) indicate that the isomerization proceeds more slowly by a factor of 1.4–1.5; this secondary isotope effect is compatible with an internal reorganization (see Notes Added in Proof).

Attempts to find a low-lying intermediate with quantum mechanical calculations have met with partial success. Houk et al. [59,60] found two similar structures, here simplified as OH··ONO, that lie approximately 15 kcal/mol above cis-peroxynitrous acid. The OH··O hydrogen bond is rather weak; in water more favorable interactions are likely. Furthermore, as discussed by Bohle and Hansert [58], oxygen exchange between OH in the intermediate and H_2O would be rapid, which appears to eliminate these structures as intermediates. A study that included water molecules showed that of all possible mechanisms, homolysis was energetically the most favorable; consequently, the hydroxyl and nitrogen dioxide radicals would have to recombine and form nitrate without exchange of oxygen [61]. Again the arguments of Bohle and Hansert [58] apply, as well as our argument given above, namely, that hydroxyl and nitrogen dioxide radicals mainly if not exclusively react to form peroxynitrous acid, not nitrate.

5.2. Reactions with Other Compounds

Stopped-flow studies have shown that many compounds are oxidized by peroxynitrous acid, but they do not affect the rate of its decay. It is assumed that peroxynitrous acid first forms an activated intermediate by a rate-limiting reaction, before it carries out the oxidation. Less than 40% of the activated peroxynitrous acid carries out an oxidation; the rest isomerizes to nitrate. Compounds that react with peroxynitrous acid in this fashion are, for instance, dimethylsulfoxide [62], hexa-cyanoferrate(II), nickel cyclam [63], 4-hydroxyphenylalanine [64], and tyrosine, phenol, and salicylate [26]. The reaction of peroxynitrous acid with phenolic compounds leads to nitrated and hydroxylated compounds, the yield of which is pH-dependent. It has been proposed that the initial event is the formation of a phenoxide radical. A nitrated product would result if the phenoxide radical reacted with the one-electron reduction product of peroxynitrous acid, i.e., nitrogen dioxide. Hydrolysis of an intermediate in this process could produce the hydroxylated product [26].

Based on a nonlinear plot of the observed rate constant versus methionine concentration, a reaction scheme was proposed that involves not only a direct reaction between peroxynitrous acid and methionine, but also that of the intermediate in the isomerization reaction [65] with methionine. However, others obtained linear plots with threonylmethionine, glycylmethionine [66], and methionine itself (Tibi, Perrin, and Koppenol, unpublished). The proposed reaction scheme [65] and its refinement [67] may apply to reactions of peroxynitrous acid with other compounds.

6. BIMOLECULAR REACTIONS

It would be advantageous if oxoperoxonitrate(1-) could be scavenged. This requires a compound that reacts with oxoperoxonitrate(1-) in a bimolecular fashion, and faster than carbon dioxide, which also forms an adduct with oxoperoxonitrate(1-) and enhances its nitrating capabilities [17,68,69]. Ascorbate reacts too slowly, and its intracellular concentration is too low [70], but it may be very good at repairing damage [71]. Three compounds have been studied that would scavenge oxoperoxonitrate(1-) at micromolar concentrations, namely, 2-phenyl-1,2-benzisoselenazol-3(2H)-one (ebselen) as well as an iron(III) and a manganese(III) porphyrin (5,10,15,20–tetrakis(N-methyl-4'-pyridyl)-porphinatoiron(III)), which all have rate constants of approximately 2.0×10^6 M^{-1} s^{-1} [72–75], respectively. The ebselen oxide formed can possibly be regenerated in vivo by glutathione reductase, whereas the iron(III) porphyrin is a true catalyst. The manganese porphyrin reacts rapidly with peroxynitrite and is also rapidly regenerated by compounds like ascorbate [76]. Slower reacting compounds such as thiols, (seleno)methionine, ascorbate, and tryptophan are listed in Table 2. Rate constants have been determined that vary from 10^2 to 10^4 M^{-1} s^{-1}. It is remarkable that there seem to be no fast reactions of peroxynitrite. With proteins peroxynitrite or peroxynitrous acid reacts in the range of 10^5–10^7 M^{-1} s^{-1} (see Table 2).

An effect of carbon dioxide on the reactivity of peroxynitrite was reported by Radi et al. [77]. Hurst and coworkers showed that the

TABLE 2

Second-Order Rate Constants
for Reactions with Peroxynitrous Acid

Compound	k (M^{-1} s^{-1})	T (°C)	Ref.
Ebselen[a]	2.0×10^6	25	[72]
Mn(III) porphyrin[b]	1.8×10^{6c}	22–24	[74]
Fe(III) porphyrin[b]	2.2×10^6	37	[73]
Cysteine	5.9×10^3	37	[99]
Methionine	9.0×10^2	25	[65]
2-Keto-4-thiomethylbutanoic acid	1.4×10^3	25	[65]
Selenomethionine	2.0×10^4		[100]
Carbon dioxide[a]	3.0×10^4	25	[68]
	5.8×10^4	25	[79]
Tryptophan	1.3×10^2	25	[101]
	1.8×10^2	37	[102]
Cytochrome c	2.3×10^5	25	[103]
Alcohol dehydrogenase	$2.6–2.5 \times 10^5$	23	[104]
Myeloperoxidase	2×10^{7d}	25	[105]
Glutathione peroxidase	8×10^{6e}	25	[107]
Aconitase	2.4×10^5	25	[118]

[a]With the peroxynitrite anion.
[b]5,10,15,20-tetrakis(N-methyl-4'-pyridyl)porphin.
[c]At pH 7.4.
[d]Extrapolated value.
[e]For the tetrameric protein.

peroxynitrite anion reacted with carbon dioxide (*not* with carbonic acid or the hydrogen carbonate anion) with a rate constant of $3 \times 10^4 \, M^{-1} \, s^{-1}$ to form, presumably, the species $ONOOCO_2^-$ [68] and estimated the lifetime of the adduct (systematic substitutive name: 1-carboxylato-2-nitrosodioxidane) to be less than 3 ms, and a reduction potential of more than 1 V for the couple $ONOOCO_2^-$, $2H^+/CO_2$, $NO_2^{\bullet}$, H_2O [78]. Carboxylatonitrosodioxidane or its dissociation products is more effective in nitrating phenolic compounds than peroxynitrous acid. On the other hand, it is less reactive toward thiols [17,79]. Carbon dioxide acts catalytically in these processes [80,81]. Merényi and Lind have argued theoretically that the adduct has a submicrosecond lifetime and that it undergoes homolysis to form $NO_2^{\bullet}$ and $CO_3^{\bullet -}$ [42]. Experimental evidence supports this proposal [81]. An alternative proposal is that the carboxylatonitrosodioxidane isomerizes to nitrocarbonate [82], but no evidence for such a rearrangement is offered. Because in plasma the concentration of carbon dioxide is 1.3 mM [79] and the rate constant for the reaction of peroxynitrite with carbon dioxide reaction is $3-6 \times 10^4$ $M^{-1} \, s^{-1}$ [68,79], this adduct is of physiological relevance.

Based simply on kinetics, and in the absence of catalysis by metal complexes or superoxide dismutase, nitration of tyrosines in cells should be a late event and take place after nearly all thiols have been exhausted. However, the carbon dioxide-enhanced nitration proceeds more quickly than thiol oxidation [78,83], which would make it a physiologically very significant process.

Some metal complexes react with peroxynitrite and catalyze the nitration of phenolic compounds. These adducts with metal complexes have not been characterized. Of all metal complexes studied so far, iron(III)EDTA is the most effective at catalyzing nitration of phenolic compounds [26,64] (EDTA = ethylenediamine-N,N,N',N'-tetraacetate).

7. SUPEROXIDE DISMUTASE AND PEROXYNITRITE

Similar to iron(III)EDTA, Cu/Zn superoxide dismutase catalyzes the nitration of tyrosine [64]. This reaction may be of importance in the disease familial amyotrophic lateral sclerosis [84], where one-quarter of all affected have one of over 50 different mutations in superoxide dismutase [85].

The formation of peroxynitrite should be avoided, except near macrophages where it is necessary. The concentration of superoxide dismutase in cells is sufficient to channel all superoxide toward dioxygen and hydrogen peroxide; the product of the rate constant of superoxide with superoxide dismutase, 2.3×10^9 M^{-1} s^{-1} [86] (Fig. 2), with an estimated intracellular superoxide dismutase concentration of 10 μM, results in a rate of superoxide disappearance of 2×10^4 s^{-1}. Nitrogen monoxide is present for signaling purposes in the nanomolar concentration range, say 10 nM. This concentration, multiplied with the rate constant of 2×10^{10} M^{-1} s^{-1} for the reaction of nitrogen monoxide with superoxide [48], yields a rate of disappearance of 2×10^2 s^{-1}, much smaller than the 2×10^4 s^{-1} calculated above. Near activated macrophages the estimated local nitrogen monoxide concentrations are much higher (5–10 μM), and the rate of oxoperoxonitrate(1-) formation is now 2–4×10^4 s^{-1}, i.e., twice the rate of superoxide disappearance through superoxide dismutase. In addition, the concentration of extracellular superoxide dismutase is likely to be lower than 5–10 μM. It is clear that superoxide dismutase cannot prevent the formation of oxoperoxonitrate(1-) near activated macrophages! Peroxynitrous acid, with its

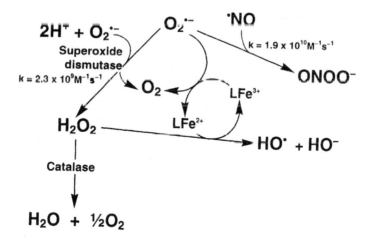

FIG. 2. Toxicity of superoxide. Under normal conditions superoxide dismutase (SOD), in cells present at concentrations of 5–10 μM, scavenges nearly all superoxide and peroxynitrite is not formed. When the local concentration of nitrogen monoxide is in the micromolar range, as is the case near activated macrophages, then SOD cannot prevent the formation of peroxynitrite. (Adapted from [106].)

pK_a near physiological pH, has been shown to cross lipid bilayers rapidly [74]. Thus, although it may be formed outside a cell, its toxicity may be felt inside.

So far peroxynitrite has been portayed as a dangerous toxin. However, there is one "useful" reaction of peroxynitrite: It can act as a substrate for prostaglandin endoperoxide synthase and thereby activate prostaglandin biosynthesis [87].

8.   PEROXYNITRITE OR HYDROXYL RADICAL?

As discussed elsewhere [106,108,109], and as indicated in Fig. 2, kinetic considerations explain why formation of the hydroxyl radical via the one-electron reduction of hydrogen peroxide by iron(II), the Fenton reaction [88,89], is kinetically not very likely. Arguments include the slow rate of rereduction of iron complexes by superoxide [90] and the unknown coordination of "chelatable" iron, which influences the rate of oxidation by peroxides [89]. In systems that produce nitrogen monoxide, *initiation* is more likely due to peroxynitrite, while iron can contribute to the *propagation* of oxyradical damage. The role of the hydroxyl radical as the initiator of tissue damage has also been questioned by Beckman [91]. Peroxynitrite reactions produce compounds that have been taken as evidence for hydroxyl radical formation, such as salicylate hydroxylation [26], initiation of lipid peroxidation [92], and spin-trap OH formation [93]. As peroxynitrite produces the same "footprints" as those expected for the hydroxyl radical, and as the formation of peroxynitrite is kinetically far more likely, it may be in order to retire the Fenton reaction. After its discovery more than 100 years ago [94,95], and its important role in biochemistry for nearly 30 years [96], a rest is well deserved.

Notes Added in Proof

In addition to peroxynitrous acid and nitryl chloride, myeloperoxidase and horseradish peroxidase in combination with nitrite and hydrogen peroxide are capable of nitrating tyrosine residues [110].

A study of the isomerization of peroxynitrous acid in deuterium oxide has demonstrated a kinetic isotope effect of 1.6 ± 0.2, an equilib-rium isotope effect of 3.3 and an activation enthalpy of 20.6 ± 1.0 kcal/mol [111]. A secondary isotope effect of 1.6 is somewhat high for homo-lysis (but does not rule it out) and supports the internal rearrangement mechanism. The activation enthalpy in deuterium oxide is within the error identical to that in water. This value is significantly higher than that published before, 18 ± 1 kcal/mol [49]. The entropy of activation is positive, but an exact value is difficult to give, due to the uncertainty caused by the extrapolation of 1/T to zero.

In a recent Forum in Chemical Research in Toxicology 6 papers highlighted aspects of peroxynitrite chemistry. Topics covered range from quantum-mechanical calculations on the structure of the reactive intermediate to homolysis and pathways of peroxynitrite decay in a biological milieu [112–118]. An activation enthalpy of 21.2 kcal/mol was reported in Ref. [113], in good agreement with that reported in Ref. [111]. An alternative scheme for nitrite and dioxygen production, based on homolysis of both the O-O *and* the N-O bonds, was proposed in Ref. [114].

ACKNOWLEDGMENTS

It is a pleasure to acknowledge many stimulating conversations regard-ing peroxynitrite with Prof. J. S. Beckman, Prof. R. Radi, Dr. R. Kissner, Dr. T. Nauser, Dr. S. Padmaja, and Dr. P. L. Bounds.

REFERENCES

1. A. Baeyer and V. Villiger, *Ber. deutschen chem. Ges., 34*, 755–762 (1901).

2. I. Trifonow, *Z. Anorg. Chemie, 124*, 136–139 (1922).

3. F. Raschig, *Ber. deutschen chem. Ges., 40*, 4580–4588 (1907).

4. K. Gleu and E. Roell, *Zeitschr. anorg. allgem. Chem., 179*, 233–266 (1929).

5. J. O. Edwards and R. C. Plumb, *Prog. Inorg. Chem., 41*, 599–635 (1993).

6. J. S. Beckman, J. Chen, H. Ischiropoulos, and J. P. Crow, *Meth. Enzymol.*, *233*, 229–240 (1994).

7. Anonymous, in *Gmelin Handbook of Inorganic and Organometallic Chemistry; Nitrogen.* Springer-Verlag, Berlin, 1996, pp. 344–358.

8. M. P. McGrath, M. M. Francl, F. S. Rowland, and W. J. Hehre, *J. Phys. Chem.*, *92*, 5352–5357 (1988).

9. H. H. Tsai, T. P. Hamilton, J. H. M. Tsai, J. G. Harrison, and J. S. Beckman, *J. Phys. Chem.*, *100*, 6942–6949 (1996).

10. H. H. Tsai, T. P. Hamilton, J. H. M. Tsai, M. van der Woerd, J. G. Harrison, M. J. Jablonsky, J. S. Beckman, and W. H. Koppenol, *J. Phys. Chem.*, *100*, 15087–15095 (1996).

11. W.-J. Lo, Y.-P. Lee, J.-H. M. Tsai, and J. S. Beckman, *Chem. Phys. Lett.*, *242*, 147–152 (1995).

12. A. D. McNaught and A. Wilkinson, *Compendium of Chemical Terminology. IUPAC Recommendations*, Blackwell, Oxford, 1997.

13. J. S. Beckman, Y. Z. Ye, P. G. Anderson, J. Chen, M. A. Accavitti, M. M. Tarpey, and C. R. White, *Biol. Chem. Hoppe-Seyler*, *375*, 81–88 (1994).

14. J. M. Fukuto and L. J. Ignarro, *Acc. Chem. Res.*, *30*, 149–152 (1997).

15. B. Halliwell, *FEBS Lett.*, *411*, 157–160 (1997).

16. W. A. Prütz, H. Monig, J. Butler, and E. J. Land, *Arch. Biochem. Biophys.*, *243*, 125–134 (1985).

17. A. Gow, D. Duran, S. R. Thom, and H. Ischiropoulos, *Arch. Biochem. Biophys.*, *333*, 42–48 (1996).

18. P. Ghafourifar and C. Richter, *FEBS Lett.*, *418*, 291–296 (1997).

19. A. C. F. Gorren, B. F. van Gelder, and R. Wever, *Ann. N.Y. Acad. Sci.*, *550*, 139–149 (1988).

20. G. Loschen, A. Azzi, C. Richter, and L. Flohé, *FEBS Lett.*, *42*, 68–72 (1974).

21. J. P. Eiserich, C. E. Cross, A. D. Jones, B. Halliwell, and A. Van der Vliet, *J. Biol. Chem.*, *271*, 19199–19208 (1996).

22. J. P. Eiserich, M. Hristova, C. E. Cross, A. D. Jones, B. A. Freeman, B. Halliwell, and A. Van der Vliet, *Nature*, *391*, 393–397 (1998).

23. D. W. Johnson and D. W. Margerum, *Inorg. Chem.*, *30*, 4845–4851 (1991).

24. M. Grätzel, A. Henglein, J. Lilie, and G. Beck, *Ber. Bunsenges. Physik. Chem.*, *73*, 646–653 (1969).

25. O. M. Panasenko, K. Briviba, L. O. Klotz, and H. Sies, *Arch. Biochem. Biophys.*, *343*, 254–259 (1997).

26. M. S. Ramezanian, S. Padmaja, and W. H. Koppenol, *Chem. Res. Toxicol.*, *9*, 232–240 (1996).

27. K. Gleu and R. Hubold, *Zeitschr. anorg. allgem. Chemie*, *223*, 305–317 (1935).

28. E. Halfpenny and P. L. Robinson, *J. Chem. Soc.*, 928–938 (1952).

29. J. W. Reed, II. II. IIo, and W. L. Jolly, *J. Am. Chem. Soc.*, *96*, 1248–1249 (1974).

30. M. N. Hughes and H. G. Nicklin, *J. Chem. Soc. (A)*, 164–168 (1971).

31. J. R. Leis, M. E. Peña, and A. Ríos, *J. Chem. Soc. Chem. Commun.*, 1298–1299 (1993).

32. G. L. Petriconi and H. M. Papée, *Can. J. Chem.*, *44*, 977–980 (1966).

33. S. Goldstein and G. Czapski, *Inorg. Chem.*, *35*, 5935–5940 (1996).

34. D. S. Bohle, P. A. Glassbrenner, and B. Hansert, *Meth. Enzymol.*, *269*, 302–311 (1996).

35. W. H. Koppenol, R. Kissner, and J. S. Beckman, *Meth. Enzymol.*, *269*, 296–302 (1996).

36. R. E. Huie and S. Padmaja, *Free Rad. Res. Commun.*, *18*, 195–199 (1993).

37. K. Kobayashi, M. Miki, and S. Tagawa, *J. Chem. Soc. Dalton Trans.*, 2885–2889 (1995).

38. S. Goldstein and G. Czapski, *Free Rad. Biol. Med.*, *19*, 505–510 (1995).

39. J. D. Ray, *J. Inorg. Nucl. Chem.*, *24*, 1159–1162 (1962).

40. D. S. Bohle, B. Hansert, S. C. Paulson, and B. D. Smith, *J. Am. Chem. Soc.*, *116*, 7423–7424 (1994).

41. M. Manuszak and W. H. Koppenol, *Thermochim. Acta*, *273*, 11–15 (1996).

42. G. Merényi and J. Lind, *Chem. Res. Toxicol.*, *10*, 1216–1220 (1997).

43. W. H. Koppenol and R. Kissner, *Chem. Res. Toxicol.*, *11*, 87–90 (1998).

44. G. Kortüm and B. Finckh, *Z. physik. Chemie*, *48B*, 32–49 (1941).

45. M. N. Hughes and H. G. Nicklin, *J. Chem. Soc. (A)*, 450–452 (1968).

46. D. J. Benton and P. Moore, *J. Chem. Soc. (A)*, 3179–3182 (1970).

47. J.-H. M. Tsai, J. G. Harrison, J. C. Martin, T. P. Hamilton, M. van der Woerd, M. J. Jablonsky, and J. S. Beckman, *J. Am. Chem. Soc.*, *116*, 4115–4116 (1994).

48. R. Kissner, T. Nauser, P. Bugnon, P. G. Lye, and W. H. Koppenol, *Chem. Res. Toxicol.*, *10*, 1285–1292 (1997).

49. W. H. Koppenol, J. J. Moreno, W. A. Pryor, H. Ischiropoulos, and J. S. Beckman, *Chem. Res. Toxicol.*, *5*, 834–842 (1992).

50. S. Pfeiffer, A. C. F. Gorren, K. Schmidt, E. R. Werner, B. Hansert, D. S. Bohle, and B. Mayer, *J. Biol. Chem.*, *272*, 3465–3470 (1997).

51. L. R. Mahoney, *J. Am. Chem. Soc.*, *92*, 5262–5263 (1970).

52. S. Goldstein and G. Czapski, *Nitric Oxide: Biol. Chem. 1*, 417–422 (1997).

53. M. Grätzel, A. Henglein, and S. Taniguchi, *Ber. Bunsenges. Physik. Chem.*, *94*, 292–298 (1970).

54. J. B. Burkholder, P. D. Hammer, and C. J. Howard, *J. Phys. Chem.*, *91*, 2136–2144 (1987).

55. W. H. Koppenol and L. Klasinc, *Int. J. Quantum Chem.*, *Quantum Biol. Symp.*, *20*, 1–6 (1993).

56. W. A. Pryor and G. L. Squadrito, *Am. J. Physiol. Lung Cell. Mol. Physiol.*, *268*, L699–L722 (1995).

57. M. Anbar and H. Taube, *J. Am. Chem. Soc.*, *76*, 6243–6247 (1954).

58. D. S. Bohle and B. Hansert, *Nitric Oxide: Biol. Chem. 1*, 502–506 (1997).

59. K. N. Houk, K. R. Condroski, and W. A. Pryor, *J. Am. Chem. Soc.*, *118*, 13002–13006 (1996).

60. K. N. Houk, K. R. Condroski, and W. A. Pryor, *J. Am. Chem. Soc.*, *119*, 2964 (1997).

61. B. S. Jursic, L. Klasinc, S. Pecur, and W. A. Pryor, *Nitric Oxide: Biol. Chem. 1*, 494–501 (1997).

62. J. P. Crow, C. Spruell, J. Chen, C. Gunn, H. Ischiropoulos, M. Tsai, C. D. Smith, R. Radi, W. H. Koppenol, and J. S. Beckman, *Free Rad. Biol. Med.*, *16*, 331–338 (1994).

63. S. Goldstein and G. Czapski, *Inorg. Chem.*, *34*, 4041–4048 (1995).

64. J. S. Beckman, H. Ischiropoulos, L. Zhu, M. van der Woerd, C. D. Smith, J. Chen, J. Harrison, J. C. Martin, and M. Tsai, *Arch. Biochem. Biophys.*, *298*, 438–445 (1992).

65. W. A. Pryor, X. Jin, and G. L. Squadrito, *Proc. Natl. Acad. Sci. USA*, *91*, 11173–11177 (1994).

66. J. L. Jensen, B. L. Miller, X. P. Zhang, G. L. Hug, and C. Schöneich, *J. Am. Chem. Soc.*, *119*, 4749–4757 (1997).

67. S. Goldstein, G. L. Squadrito, W. A. Pryor, and G. Czapski, *Free Rad. Biol. Med.*, *21*, 965–974 (1996).

68. S. V. Lymar and J. K. Hurst, *J. Am. Chem. Soc.*, *117*, 8867–8868 (1995).

69. S. V. Lymar and J. K. Hurst, *Chem. Res. Toxicol.*, *9*, 845–850 (1996).

70. D. Bartlett, D. F. Church, P. L. Bounds, and W. H. Koppenol, *Free Rad. Biol. Med.*, *18*, 85–92 (1995).

71. M. Whiteman and B. Halliwell, *Free Rad. Res.*, *25*, 275–283 (1996).

72. H. Masumoto, R. Kissner, W. H. Koppenol, and H. Sies, *FEBS Lett.*, *398*, 179–182 (1996).

73. M. K. Stern, M. P. Jensen, and K. Kramer, *J. Am. Chem. Soc.*, *118*, 8735–8736 (1996).

74. S. S. Marla, J. Lee, and J. T. Groves, *Proc. Natl. Acad. Sci. USA*, *94*, 14243–14248 (1997).

75. G. G. A. Balavoine, Y. V. Geletii, and D. Bejan, *Nitric Oxide: Biol. Chem. 1*, 507–521 (1997).

76. J. B. Lee, J. A. Hunt, and J. T. Groves, *Bioorg. Med. Chem. Lett.*, *7*, 2913–2918 (1997).

77. R. Radi, T. P. Cosgrove, J. S. Beckman, and B. A. Freeman, *Biochem. J.*, *290*, 51–57 (1993).

78. S. V. Lymar, Q. Jiang, and J. K. Hurst, *Biochemistry*, *35*, 7855–7861 (1996).

79. A. Denicola, B. A. Freeman, M. Trujillo, and R. Radi, *Arch. Biochem. Biophys.*, *333*, 49–58 (1996).

80. W. A. Pryor, J. N. Lemercier, H. W. Zhang, R. M. Uppu, and G. L. Squadrito, *Free Rad. Biol. Med.*, *23*, 331–338 (1997).

81. S. V. Lymar and J. K. Hurst, *Inorg. Chem.*, *37*, 294–301 (1998).

82. R. M. Uppu, G. L. Squadrito, and W. A. Pryor, *Arch. Biochem. Biophys.*, *327*, 335–343 (1996).

83. H. W. Zhang, G. L. Squadrito, R. M. Uppu, J. N. Lemercier, R. Cueto, and W. A. Pryor, *Arch. Biochem. Biophys.*, *339*, 183–189 (1997).

84. J. S. Beckman, M. Carson, C. D. Smith, and W. H. Koppenol, *Nature*, *364*, 584 (1993).

85. W. H. Koppenol and E. Margoliash, *Chemtracts — Biochem. Mol. Biol.*, *4*, 239–243 (1993).

86. E. M. Fielden, P. B. Roberts, R. C. Bray, D. J. Lowe, G. N. Mautner, G. Rotilio, and L. Calabrese, *Biochem. J.*, *139*, 49–60 (1974).

87. L. M. Landino, B. C. Crews, M. D. Timmons, J. D. Morrow, and L. J. Marnett, *Proc. Natl. Acad. Sci. USA*, *93*, 15069–15074 (1996).

88. S. Goldstein, D. Meyerstein, and G. Czapski, *Free Rad. Biol. Med.*, *15*, 435–445 (1993).

89. W. H. Koppenol, in *Free Radical Damage and Its Control* (C. A. Rice-Evans and R. H. Burdon, eds.), Elsevier, Amsterdam, 1994, pp. 3–24.

90. J. Butler and B. Halliwell, *Arch. Biochem. Biophys.*, *218*, 174–178 (1982).

91. J. S. Beckman, in *The Neurobiology of NO• and HO•* (C. C. Chiueh, D. L. Gilbert, and C. A. Colton, eds.), New York Academy of Sciences, New York, 1995, pp. 69–75.

92. R. Radi, J. S. Beckman, K. M. Bush, and B. A. Freeman, *Arch. Biochem. Biophys.*, *288*, 481–487 (1991).

93. S. Pou, S. Y. Nguyen, T. Gladwell, and G. M. Rosen, *Biochim. Biophys. Acta Gen. Subj.*, *1244*, 62–68 (1995).

94. H. J. H. Fenton, *J. Chem. Soc. Proc.*, *10*, 157–158 (1894).

95. W. H. Koppenol, *Free Rad. Biol. Med.*, *15*, 645–651 (1993).

96. C. Beauchamp and I. Fridovich, *J. Biol. Chem.*, *245*, 4641–4646 (1970).

97. R. Kissner, J. S. Beckman, and W. H. Koppenol, *Meth. Enzymol.* (in press).

98. W. H. Koppenol, in *Focus on Membrane Lipid Oxidation*, Vol. I (C. Vigo-Pelfrey, ed.), CRC Press, Boca Raton, 1989, pp. 1–13.

99. R. Radi, J. S. Beckman, K. M. Bush, and B. A. Freeman, *J. Biol. Chem.*, *266*, 4244–4250 (1991).

100. S. Padmaja, G. L. Squadrito, J. N. Lemercier, R. Cueto, and W. A. Pryor, *Free Rad. Biol. Med.*, *21*, 317–322 (1996).

101. S. Padmaja, M. S. Ramezanian, P. L. Bounds, and W. H. Koppenol, *Redox Rep.*, *2*, 173–177 (1996).

102. B. Alvarez, H. Rubbo, M. Kirk, S. Barnes, B. A. Freeman, and R. Radi, *Chem. Res. Toxicol.*, *9*, 390–396 (1996).

103. L. Thomson, M. Trujillo, R. Telleri, and R. Radi, *Arch. Biochem. Biophys.*, *319*, 491–497 (1995).

104. J. P. Crow, J. S. Beckman, and J. M. McCord, *Biochemistry*, *34*, 3544–3552 (1995).

105. R. Floris, S. R. Piersma, G. Yang, P. Jones, and R. Wever, *Eur. J. Biochem.*, *215*, 767–775 (1993).

106. W. H. Koppenol, *Free Rad. Biol. Med.*, *25*, 385–391 (1998).

107. K. Briviba, R. Kissner, W. H. Koppenol, and H. Sies (submitted for publication).

108. S. Liochev and I. Fridovich, *Free Rad. Biol. Med.* (1998), in press.

109. W. H. Koppenol, *Free Rad. Biol. Med.* (1998), in press.

110. J. B. Sampson, Y.-Z. Ye, H. Rosen, and J. S. Beckman, *Arch. Biochem. Biophys.*, *356*, 207–213 (1998).

111. S. Padmaja, R. Kissner, P. L. Bounds, and W. H. Koppenol, *Helv. Chim. Acta*, *81*, 1201–1206 (1998).

112. M. D. Bartberger, L. P. Olson, and K. N. Houk, *Chem. Res. Toxicol.*, *11*, 710–711 (1998).

113. G. Merényi, J. Lind, S. Goldstein, and G. Czapski, *Chem. Res. Toxicol.*, *11*, 712–713 (1998).

114. S. V. Lymar and J. K. Hurst, *Chem. Res. Toxicol.*, *11*, 714–715 (1998).

115. W. H. Koppenol, *Chem. Res. Toxicol.*, *11*, 716–717 (1998).

116. G. L. Squadrito and W. A. Pryor, *Chem. Res. Toxicol.*, *11*, 718–719 (1998).

117. R. Radi, *Chem. Res. Toxicol.*, *11*, 720–721 (1998).

118. L. Castro, M. Rodriguez, and R. Radi, *J. Biol. Chem.*, *269*, 29409–29415 (1994).

19

Novel Nitric Oxide-Liberating Heme Proteins from the Saliva of Bloodsucking Insects

F. Ann Walker,[1] José M. C. Ribeiro,[2,4]
and William R. Montfort[3]

Departments of [1]Chemistry, [2]Entomology, and
[3]Biochemistry, University of Arizona, Tucson, AZ 85721,
and the [4]Section of Medical Entomology, Laboratory
of Parasitic Diseases, NIAID, National Institutes
of Health, Bethesda, MD 20892, USA

1. INTRODUCTION

The journal *Science* named nitric oxide (NO) "molecule of the year" for 1992, and a number of reviews of the interaction of NO with living systems have appeared [1–7]. The purpose of this chapter is to summarize our recent work on nitrosylheme proteins (nitrophorins) from bloodsucking insects, to survey the known interactions of NO with heme proteins, including comparisons and contrasts between the nitrophorins and other systems, and to summarize the physical and spectroscopic methods utilized for characterizing the structure, bonding, and reactivity of heme-NO complexes in general.

Nitric oxide has been recently discovered to serve as an autacoid and neurotransmitter in vertebrates, as well as a toxic defense substance for eliminating invading organisms [1–7]. It is produced by various forms of the enzyme NO synthase (NOS), which was originally thought to include a fusion of the heme enzyme cytochrome P450 with its own reductase [1–11]. Thus, this complex enzyme contains flavin adenine dinucleotide (FAD), flavin mononucleotide (FMN), and nicotinamide adenine dinucleotide phosphate (NADP) binding sites as well as heme, and a cofactor not present in cytochromes P450, tetrahydrobiopterin. In addition, the enzyme either includes a calmodulin-like subunit that is activated by binding Ca^{2+} or, as in neuronal NOS, is activated by the association with Ca^{2+}-bound calmodulin [1–12]. The crystal structure of the domain that contains the heme and tetrahydrobiopterin centers has been reported, and although the heme is bound axially to a cysteine, as in all cytochromes P450, the protein folding pattern is quite different; P450s are largely α-helical [13], while the heme/tetrahydrobiopterin domain of NOS has a number of β sheets as well as α helices in a structure known as a curved α-β domain [14]. Thus, the heme/biopterin domain of NOS and the cytochromes P450 have achieved similar catalytic activities through convergent evolution [14].

To aid in defense against microbes, immune cells such as neutrophils and activated macrophages produce large amounts of NO [15,16], which helps to kill phagocytosed bacteria and parasites [17]. Detailed studies of inhibition of these enzymes by L-arginine analogs [18,19] and of subunit organization and conditions for dimer dissociation [20] have been reported. In some cases, activated macrophages induce production of NO synthase, which then produces NO, which can then result in metabolic dysfunction of target cells, leading to several autoimmune diseases, including insulin-dependent diabetes mellitus [21], experimental allergic encephalomyelitis [22], and possibly multiple sclerosis [22]. Inducible NOS also appears to be stimulated by proinflammatory cytokines [23,24], possibly leading to myocardial depression following cardiopulmonary bypass. The excess NO is believed to complex with the iron-sulfur centers in proteins such as aconitase, which leads to inactivation of the tricarboxylic acid pathway cycle [25] of invading bacteria and parasites. The induction of NO synthase by interferon-γ also appears to inhibit viral replication [26].

As a physiological regulatory molecule, NO is produced by a variety of tissues, such as endothelium and neural tissue; hence, NO is believed to be synonymous with the endothelium-derived relaxing factor that was described in the early 1980s [27]. It acts by activating the soluble form of the heme protein guanylyl cyclase [28]. It has been proposed that the binding of NO to the heme moiety of guanylyl cyclase triggers dissociation of the proximal histidine, causing a conformational change in the enzyme that lead to its activation [29–31]. Increased cyclic $2',3'$-guanosine monophosphate (cGMP) then activates phosphorylation events that change the state of the target tissue. In smooth muscle, relaxation occurs, with effects including lowered blood pressure [4–7] and mediation of penile erection [32]. In platelets, inhibition of aggregation occurs [28a]. A role for NO in mediating neurotransmitter release in cerebral cortex has also been found [33], and it has been suggested that NO, in conjunction with light, may be involved in resetting the human circadian clock [34].

It is interesting to note that both the synthesis and many of the physiological actions of NO involve heme proteins. Indeed, the accepted NO synthase model proposes that the arginine binding site resides in the heme pocket [6], and the action of NO in guanylyl cyclase also requires NO binding to heme [6]. Additionally, because NO is a very reactive substance, it is believed that in biological systems NO reacts with superoxide anions (Sec. 3.1), including those produced by autoxidation of hemoglobin, to produce nitrite and nitrate, or reacts with hemoglobin to form complexes that are very stable to dissociation (Sec. 3.2.1), which can decay to methemoglobin and nitrite. Cellular NO synthesis has been found to cause the loss and degradation of certain enzyme-bound hemes and to impede the action of the ferrochelatase enzyme involved in heme synthesis [35]. The interaction of nitrite with myoglobin during the curing of meat has been shown to produce a red pigment that has identical optical spectral characteristics to those of the mononitrosyl derivative of heme [36], and treatment of metmyoglobin with high concentrations of nitrite at pH values lower than 7, as in improper curing of meat, produces a green pigment known as nitrimyoglobin [37], in which the 2-vinyl group of the heme has been nitrated in the β-trans position. Thus, biochemically speaking, the synthesis, signaling, and destruction of NO involve heme proteins.

2. STRATEGIES USED BY BLOODSUCKING INSECTS TO ENSURE THAT THEY OBTAIN A SUFFICIENT BLOOD MEAL

The saliva of bloodsucking arthropods contains a vast array of substances that counteract blood coagulation, platelet function, and vasoconstriction of their hosts [38,39]. These substances allow bloodsucking arthropods to minimize the time taken for a blood meal [40,41]. Because the evolution of blood feeding is polyphyletic, convergent evolution was the scenario for these organisms' discovery of a large variety of antihemostatic compounds [42]. Since the beginning of the century, a diverse range of salivary anticoagulants in hematophagous arthropods has been recognized [43]; indeed, a few of these compounds are now characterized on a molecular level and are undergoing clinical trials (reviewed in [39]). Similarly, antiplatelet aggregation compounds are diverse, varying from large amounts of secreted apyrase (an enzyme that breaks down ATP and ADP to AMP and P_i) [39], to unique peptides that prevent collagen [44] or fibrinogen binding [45]. Evolution of salivary vasodilatory substances within hematophagous arthropods has also proven to be highly varied. Ticks secrete salivary prostaglandins, PGE_2 and PGF_2, which are potent skin vasodilators [46–50]. Within the Diptera alone, a large variety of vasodilators have been found. A peptide related to PACAP (pituitary adenyl cyclase activating peptide) [51] is the vasodilator in the phlebotomine sand fly *Lutzomyia longipalpis* [52–54], and *Simullium vittatum* black flies have a 15,000-Da novel vasodilatory peptide [55]. Within the Culicidae, the salivary vasodilator of the mosquito *Aedes aegypti* is a tachykinin (the first tachykinin described in arthropods) [56], and *Anopheles albimanus* contains a salivary catechol oxidase/peroxidase activity that destroys vasoactive amines [57].

Within the Hemiptera, the blood sucking bug *Rhodnius prolixus* has a salivary nitrovasodilator [58]. This nitrovasodilator is composed of a unique heme protein that acts as a storage and delivery system for NO [59,60]. This new class of heme proteins has accordingly been named nitrophorins [61]. We have also shown that these proteins provide an additional means of assuring the insect a sufficient blood meal by binding histamine [62,63] that is produced by the host in response to the wound, and thereby at least slowing the host defense mechanisms, as diagrammed in Fig. 1. The histamine complex dissociation constant

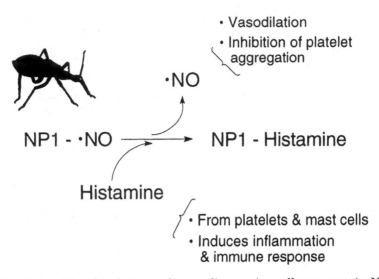

FIG. 1.   Role of the *Rhodnius prolixus* salivary nitrosylheme protein NP1 in feeding by the insect. The NP1-NO complex is stored in the insect saliva and is delivered to the host's tissue while the insect probes for a blood vessel. Dilution, binding of histamine, and an increase in pH facilitate the release of NO. Histamine is released by the host in response to tissue damage and the detection of foreign antigens. A pH change occurs on transfer of NP1-NO from the salivary gland (pH ~5) to the host's tissue (pH ~7.4). (Reproduced with permission from [63].)

is 19 nM at pH 7.0 [63], while the NO complex dissociation constant is 50 times larger at pH 7.0 [70]. Hence, histamine can displace NO from the Fe(III) form of NP1, and the tight binding of histamine provides yet another means of ensuring a successful meal for the insect. The structural basis for this tight binding is discussed below.

2.1.   General Properties of the Nitrophorins of *Rhodnius prolixus*, the "Kissing Bug"

The *Rhodnius* vasodilators have all of the typical hallmarks of nitrovasodilators. They produce a reversible vasodilatation of endothelium-less rabbit aortic rings preconstricted with norepinephrine; the vasodilatory effect is potentiated by superoxide dismutase and it is inhibited by methylene blue and hydroquinone. Furthermore, the vasodilatory activity coelutes (in a molecular sieving column) with spasmolytic activity (tested on the guinea pig ileum contracted with histamine), antiplatelet activity, and nitrogen reactive groups (as detected by the Griess reagent) [58]. We have further characterized by optical spectroscopy and electron paramagnetic resonance (EPR) methods that the cherry-red color of the glands is caused by abundant Fe(III) heme proteins that contain NO in their native states. NO can dissociate from the carrier proteins by dilution at neutral or alkaline pH, but less readily at pH 5.0. Indeed, the titration curve indicates an ionizable group with a pK_a of 6.5 [59]. Further work has led to the purification of four salivary heme proteins, the predominant being 50% of the total heme protein content and 25% of the total gland protein [61]. The amino-terminal sequence for all four heme proteins and partial sequence from proteolytic digests of the two most abundant proteins have been obtained [57]. A cDNA library was also produced from R. prolixus salivary glands and the gene for the most abundant nitrophorin, NP1, was cloned [61] and expressed [64]. The DNA and protein sequences for NP1 are shown in Fig. 2. While our work has first concentrated on NP1, the genes for NP2, NP3, and NP4 have also been cloned, sequenced, and found to be similar in sequence to that of NP1 [65]. In fact, the four nitrophorins fall into two pairs of nearly homologous protein sequences: NP1 and NP4 as one pair, and NP2 and NP3 as the other [65]. Database searches have shown little similarity between the sequences of NP1-NP4 and other proteins [61]. The presence of the noncovalently bound protoheme, the size of the protein, and the interaction of a histidine with the heme (indicated by the pH dependence of NO binding and the EPR spectral behavior of salivary homogenates [59]) first suggested a possible relationship to hemoglobins or b cytochromes. However, alignment of the sequences of NP1-NP4 with hemoglobin sequences from the insect Chironomus [66], the annelids Lumbricus and Tylorrhynchus [67,68], the mollusc Glycera [69], the parasitic nematode Ascaris [70], human β-chain hemoglobin, leghemoglobin, and vertebrate cytochromes [71] indicated an overall sequence similarity of only 38–45%, with no pronounced regions of sequence identity. No other higher overall sequence similarities could be found with other proteins, and thus we initially assumed that NP1-NP4 might be found to have globin folds. However, recent determination of the three-dimensional structure of recombinant NP1 [63] shows

```
1     GCTAAAATGAAATCA TATACAGCGTTGCTG GCCGTAGCCATTCTG  45
-23            M  K  S    Y  T  A  L  L    A  V  A  I  L   -11

46    TGCCTGTTTGCTGCA GTGGGTGTAAGTGGA AAGTGTACAAAAAAT  90
-10    C  L  F  A  A    V  G  V  S  G    K  C  T  K  N    5

91    GCACTAGCTCAAACT GGTTTTAATAAAGAC AAGTACTTCAATGGT 135
6      A  L  A  Q  T    G  F  N  K  D    K  Y  F  N  G   20

136   GATGTATGGTACGTG ACAGATTACCTAGAT TTGGAACCTGACGAC 180
21     D  V  W  Y  V    T  D  Y  L  D    L  E  P  D  D   35

181   GTTCCAAAAAGATAC TGCGCTGCTCTTGCA GCGGGTACAGCTAGT 225
36     V  P  K  R  Y    C  A  A  L  A    A  G  T  A  S   5C

226   GGTAAATTAAAAGAA GCTCTATATCACTAC GATCCAAAAACTCAA 270
51     G  K  L  K  E    A  L  Y  H  Y    D  P  K  T  Q   65

271   GACACTTTTTACGAT GTAAGTGAACTTCAA GAGGAATCTCCCGGA 315
66     D  T  F  Y  D    V  S  E  L  Q    E  E  S  P  G   80

316   AAATACACTGCAAAC TTTAAAAAAGTTGAG AAAAATGGAAATGTC 360
81     K  Y  T  A  N    F  K  K  V  E    K  N  G  N  V   95

361   AAAGTAGATGTAACG TCGGGCAACTATTAT ACCTTTACCGTTATG 405
96     K  V  D  V  T    S  G  N  Y  Y    T  F  T  V  M  110

406   TATGCTGACGATTCG TCTGCTCTTATCCAC ACTTGTTTGCATAAA 450
111    Y  A  D  D  S    S  A  L  I  H    T  C  L  H  K  125

451   GGAAACAAGGACTTG GGAGATCTCTACGCT GTATTAAATCGCAAT 495
126    G  N  K  D  L    G  D  L  Y  A    V  L  N  R  N  140

496   AAAGATACCAATGCT GGTGATAAAGTTAAA GGCGCCGTAACTGCT 540
141    K  D  T  N  A    G  D  K  V  K    G  A  V  T  A  155

541   GCTAGTTTAAAATTC AGCGACTTCATTTCC ACCAAAGATAATAAG 585
156    A  S  L  K  F    S  D  F  I  S    T  K  D  N  K  170

586   TGCGAGTACGATAAT GTTTCATTAAAATCT TTATTGACAAAATAA 630
171    C  E  Y  D  N    V  S  L  K  S    L  L  T  K  *  184

631   CTTCAGAAACTTTAT CTTAAAGATCTACCC AAAAGTTACAAACTG 675
676   GTGAAGGTTTTATAT ATTAATGTCATTCCA AATCGCTATCAGCTT 720
721   TATTACATGATGTAA TGAGTAACGAGACAT TTGATGTTCTTTTGA 765
766   AATTATTACTTCAGG AAGTATTCATGTAAC AATATTAAATAAATG 810
811   TTAGCTCGGAATTTG CC                              827
```

FIG. 2.   Nucleotide and deduced amino acid sequence of the NP1 clones. Amino acid residues confirmed by Edman sequencing are underlined. A polyadenylation consensus sequence is double underlined. (Reproduced with permission from [61].)

that its protein fold belongs to a diverse class of proteins called lipocalins [72], a family of relatively small secreted proteins that typically bind small, principally hydrophobic molecules such as retinol [73,74], prostaglandins [75], and biliverdin [76–79]. The lipocalins have little sequence homology and are almost entirely β-sheet proteins folded into β barrels that are closed at one end and bind the small hydrophobic molecule within the other end of the barrel [72]. The detailed structure of recombinant *R. prolixus* NP1 is discussed in more detail below.

2.2. The Salivary Nitrophorin of *Cimex lectularius*, the Bedbug

We have shown that the bedbug *Cimex lectularius* also has a salivary nitrophorin that shows similar, yet unique, pH-dependent reversible NO binding behavior, and optical and EPR spectral properties [80]. The size (~30 kD), amino acid sequence (Fig. 3) [81], and even the likely heme ligand of the *Cimex* nitrophorin (cys RS⁻ ?[82]) are all quite different from those of the *Rhodnius* nitrophorins. The only sequence homology found for this protein is with inositol hexaphosphate binding proteins [83]. We have as yet been unsuccessful in expressing this nitrophorin gene in a bacterial host, and thus have been unable to characterize the protein more fully, since the bedbug is only about one-fortieth the size of the kissing bug. The fact that *Rhodnius* and *Cimex* come from differ-

1	MNVPQRISKS	YLALTAEETP	DVIAVAVQGF	GFQTDKPQQG
41	PACVKNFQSL	LTSKGYTKLK	NTITETMGLT	VYCLEKHLDQ
81	NTLKNETIIV	TVDDQKKSGG	IVTSFTIYNK	RFSFTTSRMS
121	DEDVTSTNTK	YAYDTRLDYS	KKDDPSDFWI	GDLNVRVETN
161	ATHAKSLVDQ	NNIDGLMAFD	QLKKAKEQKL	FDGWTEPQVT
201	FKPTYKFKPN	TDEYDLSATP	SWTDRALYKS	GTGQTIQPLS
241	YNSLTNYKQT	EHRPVLAKFR	VTL	

FIG. 3.   Amino acid sequence of the *Cimex* nitrophorin [81].

ent families of Hemiptera suggests that the occurrence of salivary nitrovasodilators in bloodsucking insects may be fairly widespread and may serve as a common strategy for such insects in order to assure them a sufficient blood meal.

2.3. Source of NO in Insect Saliva: A Salivary Gland NO Synthase from *Rhodnius prolixus*

We have also shown that the salivary homogenates of *Rhodnius* contain NO synthase activity that is activated by tetrahydrobiopterin, Ca^{2+}, calmodulin, FAD, and requires NADPH (but not NADH) to convert arginine to citrulline and NO [57]. Furthermore, similarly to vertebrate enzymes, NO synthase activity coelutes with a diaphorase activity (a two-electron transfer from NADPH to tetrazolium dyes) in a molecular sieving column. Taken together, the data show that the activity is thus similar to that of the vertebrate soluble neural enzyme [84]. Recently, the salivary cDNA from *Rhodnius* salivary NOS was cloned and expressed, and shown to have homology to the vertebrate constitutive enzymes, including binding sites for flavins, NADPH, and calmodulin [85]. Ultrastructural localization of the associated diaphorase activity of the NO synthase indicates that it is present in cellular vacuoles, similarly to the vertebrate neural NOS enzyme [86]. Thus, we have shown the presence of NO and NOS in two families of insects, and discovered a new class of heme proteins whose function is to serve as NO carriers (nitrophorins, NP).

3. PROPERTIES OF NITRIC OXIDE

3.1. General Properties

Nitric oxide is a colorless paramagnetic gas that has a water solubility of 1.8 mM at 25°C under 1 atm pressure [5]. The solubility is unchanged within the pH range 2–13 [5]. The paramagnetism of NO is a result of its odd electron configuration, $(\sigma_1)^2(\sigma_2)^2(\pi_1)^4(\sigma_3)^2(\pi_2)^1$, but the fact that the odd electron is located in a delocalized π antibonding orbital results in less free radical reactivity than might otherwise be expected. Hence,

NO dimerizes only reluctantly [5]. Nevertheless, its gas phase chemistry is dominated by free radical reactions, mainly those involving molecular oxygen and oxides of nitrogen:

$$NO + O_2 \rightleftharpoons OONO \qquad (1)$$

$$2NO + O_2 \rightarrow N_2O_4 \rightleftharpoons 2NO_2 \qquad (2)$$

$$NO_2 + NO \rightarrow N_2O_3 \qquad (3)$$

$$N_2O_3 + H_2O \rightleftharpoons 2HNO_2 \rightleftharpoons 2H^+ + 2NO_2^- \qquad (4)$$

$$NO + O_3 \rightarrow NO_2 + O_2 \qquad (5)$$

Reaction (2) also occurs in aqueous solution, where the rate = $k[NO]^2[O_2]$ [3]. The biological half-life of NO is generally assumed to be of the order of seconds [3] but is obviously a sensitive function of concentration and the presence of stabilizing cofactors such as nitrophorins, organothiols, etc. The half-life could in principle be quite long at low concentrations of NO, in the absence of substances with which it could react. NO also reacts rapidly with superoxide ion in aqueous solution ($k \sim 3.7 \times 10^7$ M^{-1} s^{-1}), to yield peroxynitrite ion ($OONO^-$) [87,88]:

$$NO + O_2^{\bullet-} \rightarrow OONO^- \overset{H^+}{\rightleftharpoons} HOONO \qquad (6)$$

Peroxynitrous acid is surprisingly stable [89], but it can decompose to give hydroxyl radicals [88], which readily react with a variety of substrates:

$$HOONO \xrightarrow{t_{1/2} = 1.9\ sec} [OH + NO_2] \xrightarrow[2/3]{cage\ recomb.} HONO_2 \rightleftharpoons H^+ + NO_3^-$$

$$ 1/3 \Big\downarrow substrate \qquad\qquad\qquad (7)$$

$$H_2O + substrate^\bullet$$

Hydroxyl radicals are by far more reactive to biological systems than NO or superoxide ion. However, peroxynitrite itself may play a major role in cytotoxicity [90]. Peroxynitrite has also recently been shown to react rapidly with Mn(III) and Fe(III) porphyrinates to generate M(V) species that rapidly decay to oxoMn(IV) (and presumably oxoFe(IV)) intermediates that can be detected spectrophotometrically [91]. In the presence of Mn(III) porphyrinates, peroxynitrite caused extensive strand scission of plasmid DNA [91].

3.2. Interaction of NO with Heme Proteins

3.2.1. *Kinetics of NO Binding and Dissociation*

Nitric oxide is unique among diatomic molecules in being able to interact with both iron(II) and iron(III) heme proteins [29]. NO reacts rapidly according to second-order kinetics with the heme center in hemoglobins [92–94], myoglobins [95,96], catalase [96], and less rapidly with cytochrome c [96] in both the ferrous and ferric forms. The association rate constants are of the order of 10^7–10^8 M^{-1} s^{-1} for ferrous Hb [92–94] and Mb and 10^3–10^4 M^{-1} s^{-1} for most species of ferric Hb [95] and Mb [95,96] (except elephant metMb, 10^7 M^{-1} s^{-1} [95], for which the ferric form has no coordinated water). The dissociation rates are vastly different for the two oxidation states: For Fe(II), dissociation rate constants are of the order of 4×10^{-4} s^{-1} [93], leading to very large equilibrium constants for NO binding ($K_{eq} = k_f/k_r = 10^{11}$–$10^{12}$ M^{-1} [29]); for Fe(III), dissociation rate constants range from 0.65 to 40 s^{-1} [95], leading to relatively small binding constants for NO ($K_{eq} = k_f/k_r = 10^3$–10^5 M^{-1}) [29,95,96]. And unlike the binding of CO or O_2, the equilibrium constant for binding NO to *unligated* ferrous hemes is 10^3–10^4 *larger* than when the heme carries a ligand (histidine or, presumably, cysteinate) [29,97]! This has led to the hypothesis that *the loss of protein axial ligand* upon binding NO to a ferrous heme protein could begin the activation of guanylyl cyclase [29]. Guanylyl cyclase itself has also been shown to have the largest known NO dissociation rate from the Fe(II) form of the protein, $k_{obs} = 6 \times 10^{-4}$ s^{-1}, leading to a predicted half-life of the NO complex of about 2 min at 37°C [97]. This value may have implications for the mechanism of regulation of the activity of guanylyl cyclase. Recent resonance Raman reports have lent strength to this hypothesis [31]. An important related point is that if NO synthase functions by producing NO in close proximity to the Fe(II) form of the heme, then NO would bind tightly to the heme, whereas if the Fe(III) form of the enzyme is the final product, bound NO would readily dissociate at very low enzyme concentrations. For both oxidation states, dramatic shifts of the heme Soret and α and β bands occur on NO binding, which makes it simple to investigate these reactions utilizing optical detection methods [98]. The only complication, discussed further below, is that Fe(III)NO and Fe(II)NO heme centers have optical spectra with very similar wavelength maxima [98,99].

A number of picosecond laser flash photolysis studies have also been carried out on ferrous nitrosylhemoglobin and nitrosylmyoglobin [100]. These investigations have stressed the fact that geminate recombination reactions between ferrous heme and NO are much faster than those of CO or O_2 [100], suggesting that less reorganization of the heme is required in the case of NO. Nanosecond laser flash photolysis of ferric nitrosylhemoglobin (nitrosylmethemoglobin) and other ferric heme proteins have shown the geminate recombination reactions of this oxidation state are on the microsecond time scale [96].

3.2.2. Optical Absorption Spectra of Heme-NO Centers

As mentioned above, the optical absorption bands of ferrous and ferric heme proteins and model hemes in the presence and absence of bound NO are very sensitive to both axial ligand and oxidation state. For histidine-ligated heme proteins such as myoglobin, the NO complexes of Fe(II) and Fe(III) myoglobin have very similar Soret band positions [99], but the shifts are in opposite directions when NO is removed: The binding of NO to the ferrous state of whale skeletal Mb results in a shift of the Soret band from 432 to 420 nm, while the binding of NO to the ferric state results in a shift of the Soret band from 408 to 418 nm [98]. We have observed similar Soret band shifts for the nitrosylheme protein from the saliva of *Rhodnius prolixus*: 430–416 nm [101] and 403–419 nm, respectively, for the ferrous and ferric states upon binding NO [101]. Horseradish peroxidase, also a histidine-ligated heme protein, shows very similar shifts (409–419 nm) for the ferric form upon binding NO [102]. In contrast, cytochromes P450, which have a cysteine thiolate axial ligand bound to the heme, show Soret band shifts from 405 nm to a split band at 365 and 438 nm (ferrous) [103,104], and 390 to a split band at 355 and 430 nm (ferric) upon binding NO [104]; and chloroperoxidase, also a cysteinate-ligated heme protein, has the ferrous split Soret band of its NO complex at 366 and 440 nm [105].

3.2.3. Other Spectroscopic Properties of Heme-NO Centers

Other spectroscopic techniques that have been utilized to characterize nitrosylheme proteins include magnetic circular dichroism (MCD) [103,105], infrared (IR) [106–110], resonance Raman (RR) [31,111–114], Mössbauer [115–117] (Sec. 4.3), and EPR/ESR spectroscopies (Sec. 4.2).

MCD has been used to characterize the protein-provided axial ligand bound to both Fe(II) and Fe(III) forms of various heme proteins, using the NO, CO, phosphine, and other adducts to develop MCD spectra that are unique for the fifth (protein-provided) ligand [103,105]. MCD information particularly definitive for the paramagnetic forms of heme proteins is obtained at very low temperatures (4.2 K) in the near-IR region of the optical spectrum [118,119]. Infrared spectroscopy has been used to characterize the N-O stretching frequencies of ferrous (1675 and 1618–1635 cm^{-1} for five- and six-coordinate models, respectively [107], and 1615–1617 cm^{-1} for HbA [106,108]) and ferric (1925 cm^{-1} for HbA [108,110] and 1865–1910 cm^{-1} for related Fe(III)NO centers [110]), while resonance Raman spectroscopy has been used to characterize not only this stretch [111,112] but also the low-energy Fe-NO stretch for ferrous (522–527 cm^{-1} for models [113] and 551–554 cm^{-1} for HbA [111]) and ferric (601–603 cm^{-1} for models [113] and 595 cm-1 for HbA [112]). However, the axial ligand typically used in the model studies (except for that of [107]) is pyridine, which has been shown in other studies to be a poor mimic of the imidazole of histidine [120–122].

4.   SPECTROSCOPY OF THE NITROPHORINS OF *RHODNIUS PROLIXUS* IN COMPARISON TO OTHER HEME PROTEINS

4.1.   Optical Spectroscopy: pH Dependence of Absorption Bands

The Soret and α,β-band positions of recombinant NP1 in the absence and presence of NO have been investigated in detail as a function of pH [64,101]. The NO-free protein has a Soret band at 404 nm, typical of a high-spin Fe(III) heme center. Little or no change in this spectrum occurs over the pH range 5.0 (acetate buffer) to 7.0 (phosphate buffer). However, upon addition of NO, the Soret band shifts to 421 nm at pH 5.0, but 419 nm at pH 7.0; the α and β bands occur at 569 and 533 nm at both pH values. The 2-nm Soret band shift of the optical spectrum of the NO-bound protein likely reflects a functional change in the protein due to deprotonation of some amino acid side chain, probably the same side chain that is involved in the pH dependence of NO binding [59].

The small difference in the band positions of the Soret and α and β bands of the hemes of $Mb^{III}NO$ and $Mb^{II}NO$, as well as $Hb^{III}NO$ and $Hb^{II}NO$ [99], are also observed for NP1-NO and its reduced $NP1^{II}NO$ counterpart: The Soret band shifts by only about 3 nm upon reduction from Fe(III) to Fe(II), although the extinction coefficient decreases significantly [101]. Only small shifts in the α and β bands of the heme are observed as well. The small spectral changes for Fe(III)NO and Fe(II)NO centers have made for some confusion on the part of kineticists studying the mechanism of NO synthase enzymes. We find that the only reliable way to tell whether we have NP1 in the Fe(III)NO state rather than the Fe(II)NO state is by determining whether NO can be removed by blowing argon over the solution: If it is Fe(III)NO, the optical spectrum will revert to that of the high-spin Fe(III) heme center, while if it has been autoreduced to Fe(II)NO, the optical spectrum will not change when argon is blown over the sample [101,123].

The similarity in the optical spectra of Fe(III)NO- and Fe(II)NO-containing heme centers suggests that the metal center may have the same electron configuration in each case. This, as well as Mössbauer spectroscopic data discussed below and IR data reported previously for other Fe(III)NO heme centers [110], suggests that the Fe(III)NO center is perhaps better formulated as $Fe(II)(NO^+)$. Other workers have previously suggested this [110,124]. Clearly, in order for the reversible loss of NO to occur, there must be a facile valence tautomerism between the two electron configurations: $Fe(III)NO \rightleftharpoons Fe(II)(NO^+)$.

4.2. EPR Spectroscopy

By far the most common spectroscopic tool used for the investigation of nitrosylheme complexes has been EPR spectroscopy [15–25,59,125–137]. Well over 100 papers have been published to date on EPR spectroscopy of Fe(II)NO heme complexes alone, and one study of the *appearance* of high-spin metal EPR signals upon photolysis of $S = 5/2$ metal-NO complexes of heme proteins [125] has appeared, the latter of which sparked our EPR study of the *loss* of the $S = 5/2$ EPR signal of the *R. prolixus* heme protein upon binding of NO [59] (Fig. 4). The important point about the use of EPR spectroscopy for investigating ferrous and ferric heme proteins, as is evident in Fig. 4, is that Fe(III) is an odd-

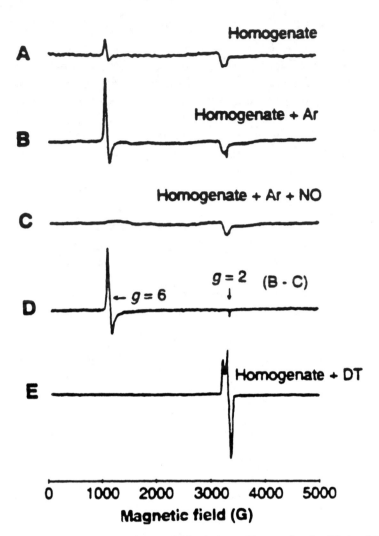

FIG. 4. EPR spectra of 100 pairs of *Rhodnius* salivary glands obtained in 125 μL of phosphate-buffered saline at pH 7.2 (A) before argon equilibration; (B) after equilibration in an argon atmosphere for 4 h, and (C) after equilibration of (B) with NO for 2 min. (D) is the difference spectrum, i.e., B–C. (E) is the homogenate as in (B) treated with dithionite (DT) to reduce Fe(III) to Fe(II), followed by equilibration with NO for 2 min. All spectra are plotted on the same scale except (E), which is shown reduced in amplitude by a factor of 3. (Reproduced with permission from [59].)

electron system, as is NO, so the binding of NO to Fe(III) produces an even-electron system and thus abolishes all EPR signals due to the metal-NO species [59,125], while Fe(II) is an even-electron system, so the binding of NO to Fe(II) produces an odd-electron system that gives rise to one of several characteristic EPR signals [126–129]. (The free diatomic NO molecule itself does not give rise to a resolved EPR spectrum under normal conditions [130], even though it is an odd electron species.) Both five- and six-coordinate heme(II)-NO complexes show very characteristic EPR spectra. In the case of five-coordinate FeNO hemes, a very characteristic three-line feature is observed on the g_2 branch, with the hyperfine coupling constant $A_{NO} = 17.2$ G for tetraphenylporphyrinatoiron nitrosyl (TPPFeNO) [131]. It is this sharp three-line spectrum, which is superimposed upon the broader spectrum of histidine-bound HbNO in the absence of inositol hexaphosphate (IHP) [132], that indicates that the proximal histidine is readily lost from the heme in the absence of the allosteric effector. Thus, as in the kinetics studies discussed above, we see that the binding of NO to ferrous heme proteins weakens the bond between the protein-provided ligand and the metal [29]. EPR-active five-coordinate heme Fe(II)-NO complexes have been reported to react with additional molecules of NO to produce the even-electron (EPR-silent) heme(II)(NO)$_2$ complex [131] that could possibly be the product of the reaction of NO with the heme of guanylyl cyclase. (However, it is likely that the binding constant for the second NO is too small at physiological temperatures to allow the formation of a bis-NO complex. Also, the smaller binding constant of aromatic nitrogen donor ligands such as histidine in the presence of bound NO [29] suggests that formation of a mono-NO complex is fully sufficient to cause the postulated loss of histidine ligand and change in conformation of guanylyl cyclase, without need for postulating formation of a bis-NO complex.) EPR spectra of a number of axial ligand complexes of (protoporphyrin IX dimethyl ester)iron(II)NO have shown a wide range of shapes [133]. The g values are characteristic of the sixth ligand (for N donors, $g_x = 2.07$–2.08, $g_y = 1.97$–1.99, $g_z = 2.004$–2.006; for S donors, $g_x = 2.09$, $g_y = 2.001$–2.003, $g_z = 2.009$–2.010; for O donors, $g_x = 2.09$–2.10, $g_y = 2.000$–2.003, $g_z = 2.04$–2.05) [133]. For N donors, superhyperfine structure from the sixth ligand is typically observed ($A_N = 5.5$–7.1 G), in addition to the typical three-line pattern observed from the NO unit ($A_{NO} = 20$–22 G when the sixth ligand is an

N or O donor, and slightly smaller, 19.6–20.2 G, when an S donor is present). The EPR spectrum of the Fe(II)NO complex of hemoglobin is sensitive to pH [127,134], allosteric effectors such as IHP [132], and the nature of the axial ligand (histidine [126–128,132–137] versus the cysteinate of cytochrome P450 and chloroperoxidase [104,129,135]).

For the Fe(II) form of NP1-NO, the EPR spectrum is very similar to that of heme proteins that have histidine as the axial ligand, although the histidine nitrogen does not give rise to resolved superhyperfine splittings, as shown in Fig. 5. Nevertheless, the spectrum was strongly suggestive of histidine being the proximal ligand of NP1, so that we expected to find this structural feature to be present in the protein, and this was found to be the case (Sec. 6.2). Use of excess dithionite to create the Fe(II) state of NP1 easily leads to loss of the His 59 ligand, as evidenced by partial replacement of the EPR spectrum of Fig. 5a with a strong three-line feature in the perpendicular region that is indicative of the five-coordinate (base-off) Fe(II)NO center that has lost its protein-provided ligand (Fig. 5b).

4.3. Mössbauer Spectroscopy

Few reports of Mössbauer spectroscopic studies of nitrosylheme complexes or proteins have appeared thus far, although this technique should be quite informative about the electronic structure of ferrous and ferric NO complexes. The Fe(II)NO complex of hemoglobin has been studied by Mössbauer spectroscopy by two research groups [115, 117]. The first of these studies resulted in the assignment of the electron configuration of the iron as $(t_{2g})^6(d_z2)^{0.5}$, presumably with the remaining unpaired electron density on the NO ligand. The Mössbauer spectrum of the siroheme(II)NO complex has also been reported [116]. No report of the Mössbauer spectrum of Fe(III)NO complexes of porphyrins or heme proteins has yet appeared, though several complexes with quadradentate chelates based on acetylacetone condensed with isothiosemicarbazides have been reported [138]. Interestingly, the isomer shifts and quadrupole splittings of these nonheme Fe(III)NO complexes are rather similar to those of the Fe(II)NO form of nitrosylhemoglobin [115], which may suggest that the electron density *at the iron center* is

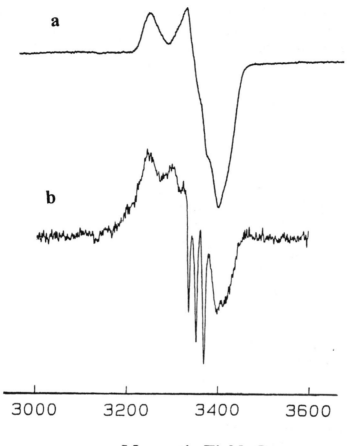

a

b

3000   3200   3400   3600

Magnetic Field, Gauss

FIG. 5.   (a) Expansion of (E) of Fig. 4. (b) Spectrum obtained after treatment of recombinant NP1 with high concentrations of dithionite, followed by exposure to NO. (Based on unpublished work.)

very similar for the two oxidation states. As mentioned above with respect to optical spectra, this could be due to the fact that in both cases the iron center is Fe(II), with the electron of the reduced form being mainly localized on the NO moiety and in the oxidized form, the NO being bound to Fe(II) as NO^+. We have carried out preliminary studies

of octaethylporphyrinatoiron(III)(N-methylimidazole)(NO) as a function of temperature and magnetic field, and find it to behave as a purely
diamagnetic Fe(II) center [139], again consistent with this complex
being formulated as Fe(II)(NO$^+$).

4.4. NMR Spectroscopy

One of the most important spectroscopic techniques for characterizing heme proteins is proton NMR spectroscopy. However, no reports of
NMR spectra of nitrosylheme proteins have yet appeared. Since the
NO-bound form of Fe(II) heme proteins is paramagnetic and undoubtedly has electron spin relaxation times that are unfavorable for observation of resolved NMR signals, it is unlikely that the heme resonances
of this form of a nitrosylheme protein can be investigated by NMR
spectroscopy. (However, amino acid residues far from the NO-heme center could certainly be observed.) On the other hand, the uncomplexed,
five-coordinate forms normally contain high-spin Fe(II) heme centers (S
= 2), which are amenable to limited NMR investigation [140]. In the
case of Fe(III) heme proteins, NO bound to Fe(III) hemes produces a
diamagnetic complex that should have heme resonances buried in the
protein proton resonance envelope, while the NO-free form of Fe(III)
heme proteins is typically a high-spin (S = 5/2) complex. Examples of
such high-spin Fe(III)-containing heme proteins include methemoglobin, metmyoglobin, and the resting state of horseradish peroxidase.
The NMR shifts of the high-spin forms of these proteins are quite large
(10–80 ppm) [141–143]. This is true of the NO-free form of recombinant
NP1 at pH 7.0 and T = 38°C, for which resolved heme and hyperfine-
shifted protein resonances extend to 70 ppm [144]. These heme resonances show a remarkably similar pattern to those of aqueous proto-
hemin at neutral pH [145], except for the broadening induced by the
much slower rotational correlation time of the protein.

   Addition of strong-field ligands such as CN$^-$ or imidazoles to these
high-spin Fe(III) centers creates the low-spin Fe(III) state (S = 1/2),
which is characterized by a smaller range of NMR shifts (20–30 ppm)
and much sharper lines than those of the high-spin forms of the same
proteins. Both one- and two-dimensional NMR techniques (1-D NOE
difference spectra, 2-D COSY, TOCSY, NOESY, and ROESY spectra)

have been extensively utilized to assign the hyperfine-shifted resonances of the heme in the cyanide-bound forms of Fe(III) heme proteins, where most, but not all, of the heme substituent resonances are found outside the diamagnetic envelope of the protein [146]. Thus, CN⁻ is being utilized in our laboratory as an even-electron, diamagnetic substitute for NO for characterizing the nitrophorins from bloodsucking insects in the low-spin Fe(III) spin state. We have also used imidazole and histamine to produce low-spin ($S = 1/2$) Fe(III) centers because of the finding that both of these ligands bind to NP1 with higher affinity than does NO [62,63]. We find that in all three of these low-spin Fe(III) complexes of NP1 there are chemical exchange processes that indicate a dynamic equilibrium between at least two different forms of the protein. This chemical exchange can be slowed on the NMR time scale by cooling the sample to 12–14°C, but the multiple species are still present. These different species do *not* involve dissociation, rotation, and reassociation of the heme, but rather some change in conformation of one or more protein side chains, or possibly cis-trans isomerization about the disulfide bonds of the protein. The two heme methyl resonances, observed at near 26 and 13 ppm at 38°C [144], are believed to be the 3- and 5-Me resonances, respectively, while the 1- and 8-Me resonances are buried in the diamagnetic region of the NMR spectrum (−1 to 11 ppm). Only one heme rotational isomer is observed [144]. We find this somewhat surprising considering the rather open-looking structure of the protein and the heme binding pocket (Sec. 6.2).

5. REDOX CHEMISTRY OF NITRIC OXIDE-HEME SYSTEMS AND THE NITROPHORINS OF *RHODNIUS PROLIXUS*

5.1. General Background

Since NO binds to both Fe(III) and Fe(II) hemes, the redox chemistry that links these two oxidation states of the nitrosylheme complexes is of interest. Ferric nitrosylhemoglobin readily undergoes rapid autoreduction to the ferrous protein. Under 1 atm NO the autoredox process is pseudo first order in Fe(III)NO, the reaction rate is enhanced by medium-intensity broad-band optical illumination, and excess NO is clearly the reductant [99]. Hemoglobin-NO autoreduces more than 10 times faster than myoglobin-NO ($k_1' = 9.67 \times 10^{-4}\,\mathrm{s}^{-1}$ and $0.64 \times 10^{-4}\,\mathrm{s}^{-1}$, respectively), and good isosbestic points are not observed [99]. However, upon "recycling" the protein, the pseudo-first-order rate constant decreases markedly ($k_1' = 0.18 \times 10^{-4}\,\mathrm{s}^{-1}$) and slightly different isosbestic points are observed, suggesting that some chemical modification of the protein may be involved. The authors point out that both lysine and histidine could be nitrosylated [99]. Presumably, free cysteine residues could also be nitrosylated, and Stamler and coworkers have shown evidence for a physiological role in lowering blood pressure for the nitrosylation of the Cysβ93 thiols of hemoglobin [147].

Electrochemical investigations of several synthetic nitrosylheme complexes have been reported [148–150]. Fe(II)NO complexes can be reversibly oxidized by one electron to Fe(III)NO in the case of octaethylporphyrin (OEP) and tetraphenylporphyrin (TPP), but for reduced hemes (chlorins and isobacteriochlorins) the optical spectra indicate that the macrocycle, rather than the metal, is oxidized [148]. The reduction potential of the OEPFe(III)NO complex is 0.60 V versus SCE (standard calomel electrode) in methylene chloride and 0.57 V in pyridine [149a], but the redox reaction in pyridine is irreversible, presumably due to displacement of NO from Fe(III) by the ligating solvent pyridine. Similar observations were reported for the TPPFe(III)NO system [149]. No other investigations of the effects of strongly binding axial ligands or the ligand concentration dependence of this redox potential have been reported. Fe(II)NO porphyrins can also be reduced in three one-electron steps [148–150]. Nitrite can also be reduced electrochemically by myoglobin in surfactant films [151]. These reduction reactions are of interest with regard to the assimilatory nitrite reductases.

Fe(III)NO heme complexes, which, as discussed above, are believed to have the electron configuration heme(II)NO$^+$, readily act as electrophilic nitrosating agents [152–158]. In aqueous solution, nitrosation can occur at -S, -N, -O, and -C centers in organic molecules [152, 153]. With particular relevance to biological systems, primary amines are readily deaminated and secondary and tertiary amines are readily N-nitrosated by NO$^+$ [3]. Nitrosation of the nitrogen functions of DNA bases can lead to carcinogenesis [154,155]. Thiols have a particularly high propensity for nitrosation (thionitrite formation) under physiological conditions [147,156–158]. Inhibition of the catalytic activity of alco-

hol dehydrogenase by NO is believed to be associated with S-nitrosyla-
tion of the zinc-bound cysteine, followed by release of zinc [159].

5.2. Redox Chemistry of the Nitrophorins of *Rhodnius prolixus*

In marked contrast to methemoglobin and metmyoglobin [99], the nitrosylheme protein of *Rhodnius prolixus* autoreduces only after prolonged treatment with gaseous NO [59,101]. Treatment with dithionite,
followed by NO, is often required to produce Fe(II)NO [59], although
high concentrations of this reducing agent cause denaturation of the
protein [82]. This observation strongly suggested that the reduction
potential of the NO complex of the nitrophorin from *R. prolixus* is quite
different from that of hemoglobin or myoglobin. We have measured the
reduction potentials of metmyoglobin and NP1 in aqueous buffers over
the pH range 5.5–7.5 by spectroelectrochemical techniques and found
that the reduction potential of NP1 is about 300 mV more negative than
that of metMb; the metMb reduction potential ranges from +28 to 0
mV, whereas that of NP1 ranges from −274 to −303 mV versus SHE
over the pH range studied [101,160]. The negative shift in potential
relative to metMb is consistent both with the sluggishness of autoreduction of NP1-NO by excess NO and with the presence of several buried
negatively charged residues in the heme pocket of NP1 (Sec. 6.2, Fig. 7).
Such negative charges have previously been shown to stabilize the
Fe(III) state, thus shifting the reduction potential in the negative direction [161].

The redox chemistry of MbNO and NP1-NO are currently under
investigation in our laboratories. Preliminary results indicate that reversible Fe(III)NO $\rightleftharpoons$ Fe(II)NO reduction is not observed for myoglobin
over the available electrochemical window (up to +900 mV versus
SHE) but that NP1 shows this reduction at roughly +800 mV relative
to SHE.

Recently, an unusual green heme protein has been isolated from
the gram-positive bacterium *Bacillus halodenitrificans* that in its native form is shown by EPR spectroscopy to be a high-spin Fe(III) heme
protein. However, it is readily photoreduced or chemically reduced by
dithionite; then it binds NO to give a five-coordinate (base-off) EPR

spectrum but reverts to the high-spin Fe(III) native form upon flushing with argon overnight [162]. The oxidized protein does not, however, bind NO and does not appear to be autoreduced by it.

6. X-RAY CRYSTALLOGRAPHY

6.1. Previous Studies of Other Heme-NO Species

The structures of a number of Fe(II)NO porphyrin complexes have been determined by x-ray crystallography [163], one of which was prepared by "reductive nitrosylation" of an Fe(III) porphyrin [163b], the same process applied to the autoreduction of metHbNO and metMbNO, discussed above. The structure of HbNO has also been reported [164]. For both Fe(II)NO model compounds, including the five-coordinate TPPFeNO [163b], and the protein [164], the Fe-N-O bond is bent; the angle ranges from 137° to 149° in model hemes [163,165] and is about 145° in HbNO [164]. To date there has been one report of the structure of Fe(III)NO heme complexes; the five-coordinate complex [Fe(OEP)(NO)]-ClO_4 and the six-coordinate complex [Fe(TPP)(NO)(H$_2$O)]ClO$_4$ have been reported [166]. In both molecules the Fe-N-O bond is linear, in line with the expected electron configuration, $(d^6)Fe(II)(NO^+)$ [124].

6.2. Structure of *Rhodnius prolixus* NP1

In order to determine the crystal and molecular structure of NP1 it was necessary to obtain more protein than was available by growing the insects, feeding them on the shaved hind quarter of an anaesthetized rabbit once a month for approximately 6 months, harvesting the salivary glands, and purifying the proteins [61]. We therefore invested time and energy to develop the protocols necessary to express the gene for NP1 in *Escherichia coli* [64]. Briefly, the gene is under control of the T7 promotor and is turned on by the effector IPTG; the protein is harvested from the cells as inclusion bodies, is renatured in a buffer containing urea and dithiothreitol, and heme is added after partial removal of the renaturing buffer by dialysis [64]. (More recently, we have also expressed NP2, NP3, and NP4 by similar procedures and have made significant progress in characterizing these proteins structurally [167].) We have determined the crystal structures of recombinant NP1 in the absence of any added ligand, and as the CN^- and the histamine complexes. The structure of the ligand-free form of NP1 was determined to 2.0 Å nominal resolution using the method of multiple isomorphous replacement [63]. The ligand-free structure was found to have what appears to be NH_3 (from the ammonium phosphate crystallization buffer) bound to the heme; this helps to explain the bright red color of the NP1 crystals in this buffer. NP1 is nearly all β-sheet, and despite little sequence similarity is a member of the lipocalin family. Lipocalins are comprised almost entirely of antiparallel β sheet that is assembled into a β barrel, with one end closed and the other open, and are used to bind lipophilic molecules [72]. Other examples of lipocalins include retinol-binding protein [73,168], insecticyanin [76,77], and bilin-binding protein [78,79]. Thus, the nitrophorins are evolutionarily distinct from the globins (the only other heme-based gas transport proteins known), as well as from all other heme proteins whose structures are known.

The NP1 fold contains the standard eight-stranded lipocalin β-barrel, three α helices, and two disulfide bonds, as shown in two views of the structure of the cyanide adduct in Fig. 6. The similarity in structure between NP1 and the bilin-binding protein [78,79], despite the lack of sequence homology between the two proteins (37% similarity, 30% identity), is striking: The α carbons of the core β barrel and the second helix in NP1 (two-thirds of the amino acids) superimpose with the analogous atoms in bilin binding protein with a root mean square deviation of only about 2 Å. However, the remaining residues, especially the loops contacting the heme and the histidine ligand, as well as the ligand itself, lie in very different positions. The heme of NP1 is ligated to His 59 and is held in place through numerous hydrophobic and electrostatic contacts. A large cavity above the heme (Fig. 6) provides ample room for ligand binding, is lined with several nonpolar amino acids and one aspartate (Asp 30), and is largely free of ordered solvent. There is no distal histidine, unlike the structures of most globins. In contrast, deep in the protein interior, behind the heme, is a buried glutamate residue (Glu 55) that is involved in an unusual series of hydrogen bonds shared among three water molecules, Tyr 17, Glu 55, Ser 72, and Tyr 105. Expulsion of these buried water molecules at lower

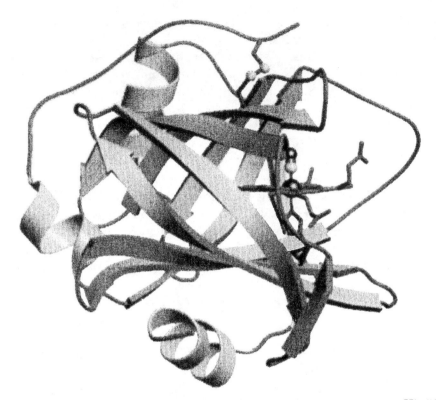

FIG. 6.   Ribbon and ball-and-stick diagrams of the NP1-CN⁻ structure. His 59 is coordinated to the heme iron. CN⁻ and iron are shown with a space-filling representation. Disulfide bonds linking Cys 2 to Cys 122 and Cys 41 to Cys 171 are indicated [63]. The view on the right is rotated approximately 90° about the vertical axis from the view on the left. Mobile loops in between β sheets are seen above and next to the heme in the left-hand view. (Reproduced with permission from [63].)

pH may provide the mechanism for increased NO binding constant at lower pH values [123].

Structural investigation of the NP1-NO complex has proved to be technically difficult, in part because of reduction of the Fe(III)NO center to Fe(II)NO in the x-ray beam, and we have thus undertaken the solution of the structure of the isosteric CN⁻ complex. We find that the Fe-C-N bond is approximately linear (173°), and is sandwiched

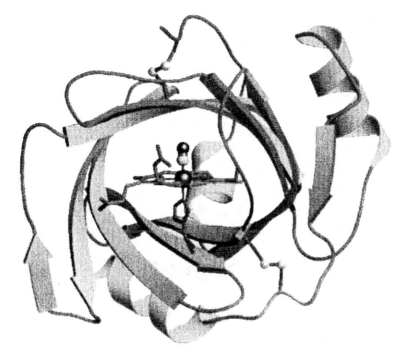

between Leu 123 and Leu 133 (3.8 Å to each). The Fe-C and C-N bonds refined to 1.9 and 1.1 Å, respectively. Based on the NP1-CN$^-$ structure, it appears likely that NO binds to NP1 simply through ligation to the heme, since no hydrogen-bonding groups are nearby that could potentially interact with the bound NO ligand. However, Asp 30 is only 5.6 Å away and may provide electrostatic stabilization for partial positive charge formation on the NO, which may occur in the Fe(II)(NO$^+$) valence tautomer discussed above.

The histamine complex of NP1 has a well-ordered structure, with histamine bound in the same distal side binding pocket. Histamine is completely buried in the protein and contacts the protein through four hydrogen bonds and extensive van der Waals contacts, in addition to being the sixth ligand to the heme iron (2.0 Å, Fig. 7). Hence, it appears that NP1 is structurally designed to not only carry and release NO but to take up histamine.

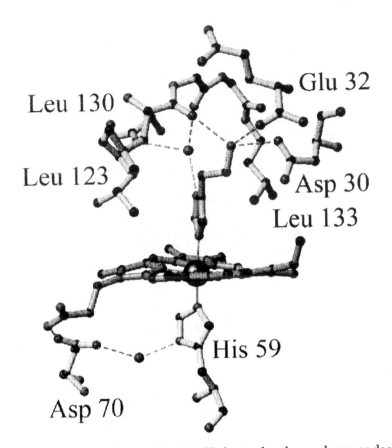

FIG. 7.   Histamine binding to NP1 [63]. Hydrogen bonds are shown as dashed
lines, nitrogen-Fe bonds as solid lines. Shown are hydrogen bonds between
histamine and Asp 30 (2.7 Å), Glu 32 (3.1 Å), Leu 130 (2.7 Å), and an ordered
water molecule (2.8 Å), which further hydrogen bonds to Thr 121 (3.2 Å, not
shown), Leu 123 (3.0 Å), and Gly 131 (2.7 Å). Van der Waals contacts are made
to Leu 123 (3.7 Å), Leu 130 (4.2 Å), and Leu 133 (3.7 Å). Also shown are hydrogen
bonds between an ordered water molecule and residues His 59 (2.7 Å), and Asp
70 (2.6 Å), and between Asp 70 and a heme propionate (2.5 Å). The other heme
propionate has been omitted for clarity. (Reproduced with permission from
[63].)

7. OTHER METAL-NITRIC OXIDE COMPLEXES OF IMPORTANCE IN INORGANIC CHEMISTRY AND BIOCHEMISTRY

Other metal-containing species, both metal centers in proteins and simple inorganic complexes, also readily react with nitric oxide: Iron-sulfur proteins such as ferredoxins [15,16] and aconitase [25] have been shown to produce EPR signals indicative of NO-bound iron, and evidence has been presented for the involvement of NO binding to the iron-sulfur cluster of the iron regulatory factor that interacts with the iron-responsive element that posttranscriptionally regulates iron metabolism [169,170]. The important citric acid cycle enzyme aconitase is inactivated by peroxynitrite [171], the product of the reaction of NO with superoxide. It appears possible to mediate iron release from ferritin by NO [172,173] (two types of Fe-NO complexes have been identified by EPR spectroscopy [174]); nitrogenase [175] also reacts with high concentrations of NO to inactivate the enzyme; reduced (Cu(I)) ceruloplasmin reacts with NO to produce a broad, structureless EPR signal around $g = 2$ [176]; NO reacts with the reduced R2 protein of ribonucleotide reductase from *E. coli* to produce apparently dimeric $S = 1$ {FeNO}^{7} centers that are EPR silent and yield N_2O [177], with the tyrosine radical of photosystem II to form an iminoxyl radical [178], and with Fe-loaded metallothionein to produce species with EPR signals at $g = 2.013$ and 2.039 [179], which are characteristic of iron complexes with the generic composition $Fe(II)(SR)_2NO$ [180] or, more likely, $Fe(SR)_2(NO)_2$ [181,182]. Synthetic iron-sulfur-NO clusters, such as $Fe_4S_4(NO)_4$ [183], Roussin's black salt ($Fe_4S_3(NO)_7^-$) [184], and $Fe_2S_2(NO)_4^{2-}$, as well as the thiolate-bridged analogs known as Roussin's esters ($Fe_2(SR)_2(NO)_4$) [185] and the dinitrosyl-dithiolate complex $Fe(SR)_2(NO)_2$, have all been shown to release NO [186]. Vanin has proposed that the endothelium-derived relaxing factor is actually a low molecular weight iron-sulfur-NO or iron-thiolate-NO complex [187]. A number of drugs that release NO have been utilized for some time to decrease blood pressure, including glyceryl trinitrate, amyl nitrite, and nitroprusside [5]; iron(II) dithiocarbamates have been tested in vitro as possible reagents for decreasing the concentration of NO in aortic tissue [188] or for trapping NO in living tissues [189].

8.   SUMMARY

The spectroscopic (UV-visible, IR, RR, MCD, Mössbauer, EPR), crystallographic, kinetic, and redox investigations that have been carried out
on model hemes, hemoglobin, myoglobin, cytochrome a_3 of cytochrome
oxidase, horseradish peroxidase, prostaglandin H synthase, cytochromes P450, chloroperoxidase, and so forth have shown us the unique
properties of heme-NO centers, as summarized above. However, in
none of these cases is the Fe(III)NO complex of any known physiological
importance. The nitrophorins of *R. prolixus* [59] (and *Cimex lectularius*
[80]) are thus far unique in this respect. It is likely that further investigations of the roles of NO in biological systems will discover additional
interesting involvements of heme proteins in these roles.

ACKNOWLEDGMENTS

The financial support of the National Institutes of Health, Grants HL
54826 (FAW, JMCR, WRM) and AI 18694 (JMCR), and the John D. and
Catherine T. McArthur Foundation (JMCR) for our research on the
interesting insect nitrophorins is gratefully acknowledged.

ABBREVIATIONS

α,β	π-π* transitions in the 500–700 nm region of the electronic spectra of hemes
ADP	adenosine 5'-diphosphate
AMP	adenosine 5'-monophosphate
ATP	adenosine 5'-triphosphate
cDNA	complementary DNA, obtained from mRNA using reverse transcriptase
cGMP	cyclic guanosine 2',3'-monophosphate
COSY	correlated spectroscopy, a two-dimensional NMR experiment that exhibits off-diagonal peaks due to protons that are scalar coupled to each other

EPR	electron paramagnetic resonance spectroscopy (same as ESR)
ESR	electron spin resonance spectroscopy (same as EPR)
FAD	flavin adenine dinucleotide
FMN	flavin mononucleotide
$g_{x,y,z}$	EPR g values
Hb	hemoglobin; may be accompanied by superscript II or III to indicate the oxidation state of the iron
HbA	human hemoglobin A
IHP	inositol hexaphosphate
IPTG	isopropyl-β-D-thiogalactopyranoside, a reagent that allows initiation of expression of a gene that is under control of the T7 promotor
IR	infrared spectroscopy
Mb	myoglobin; may be accompanied by superscript II or III to indicate the oxidation state of the iron
MCD	magnetic circular dichroism spectroscopy
metHb	met (Fe(III)-containing) hemoglobin
metMb	met (Fe(III)-containing) myoglobin
NADP	nicotinamide adenine dinucleotide phosphate
NADPH	two-electron reduced form of NADP
NMR	nuclear magnetic resonance
NOESY	nuclear Overhauser effect spectroscopy; a two-dimensional NMR experiment that exhibits off-diagonal peaks due to protons that are spatially near each other (≤ 6 Å)
NOS	nitric oxide synthase
NP1-NP4	nitrophorins 1–4, the NO-carrying heme proteins in the saliva of *R. prolixus*
OEP	octaethylporphyrin
PACAP	pituitary adenyl cyclase activating peptide
P_i	inorganic phosphate (PO_4^{3-})
protoheme	iron protoporphyrin IX, also known as heme b
ROESY	a two-dimensional rotating frame NMR experiment that provides similar information to that of the NOESY experiment
RR	resonance Raman spectroscopy
S	spin quantum number

SCE	standard calomel electrode
SHE	standard hydrogen electrode
Soret	the intense π-π^* electronic absorption band in the 400- to 450-nm region of hemes
T7 promoter	DNA site that strongly promotes the initiation of transcription by T7 polymerase
TOCSY	a two-dimensional rotating frame NMR experiment that is similar to COSY
TPP	tetraphenylporphyrin

REFERENCES

1. S. Moncada, R. M. J. Palmer, and E. A. Higgs, *Pharmacol. Rev.*, *43*, 109–142 (1991).

2. S. H. Snyder, *Science*, *257*, 494–496 (1992).

3. J. S. Stamler, D. J. Singel, and J. Loscalzo, *Science*, *258*, 1898–1902 (1992).

4. J. R. Lancaster, Jr., *Am. Scientist*, *80*, 248–259 (1992).

5. A. R. Butler and D. L. H. Williams, *Chem. Soc. Rev.*, 233–241 (1993).

6. P. L. Feldman, O. W. Griffith, and D. J. Stuehr, *Chem. Eng. News*, 26–38 (1993).

7. J. R. Lancaster, Jr., in *Encyclopedia of Inorganic Chemistry* (R. B. King, ed.), John Wiley and Sons, Chichester, 1994.

8. C. Nathan, *FASEB J.*, *6*, 3051 (1992).

9. (a) K. A. White and M. A. Marletta, *Biochemistry*, *29*, 6627 (1992); (b) S. Lamas, P. A. Marsden, G. K. Li, P. Tempst, and T. Michel, *Proc. Natl. Acad. Sci. USA*, *89*, 6348 (1992); (c) S. P. Janssens, A. Shimouchi, T. Quertermous, D. B. Bloch, and K. D. Bloch, *J. Biol. Chem.*, *267*, 14519 (1992); (d) H. H. H. W. Schmidt, R. M. Smith, M. Nakane, and F. Murad, *Biochemistry*, *31*, 3243 (1992); (e) J. M. Hevel and M. A. Marletta, *Biochemistry*, *31*, 7160 (1992); (f) Q. Xie, H. J. Cho, J. Calaycay, R. A. Mumford, K. M. Swiderek, T. D. Lee, A. Ding, T. Troso, and C. Nathan, *Science*, *256*, 225 (1992); (g) R. A. Pufahl, P. G. Nanjappan, R. W. Woodward, and M. A. Marletta, *Biochemistry*, *31*, 6822 (1992); (h) C.

J. Lowenstein, C. S. Glatt, D. S. Bredt, and S. H. Snyder, *Proc. Natl. Acad. Sci. USA*, *89*, 6711 (1992).

10. H. M. Abu-Soud, M. Loftus, and D. J. Stuehr, *Biochemistry*, *34*, 11167 (1995).

11. A. V. Hall, H. Antoniou, Y. Wang, A. H. Cheung, A. M. Arbus, S. L. Olson, W. C. Lu, C.-L. Kau, and P. A. Marsden, *J. Biol. Chem.*, *269*, 33082 (1994).

12. H. M. Abu-Soud, L. L. Yoho, and D. J. Stuehr, *J. Biol. Chem.*, *269*, 32047 (1994).

13. F. C. Bernstein, T. F. Koetzle, G. J. B. Williams, E. F. Meyer, Jr., M. D. Brice, J. R. Rodgers, O. Kennard, T. Shimanouchi, and M. Tasumi, *J. Mol. Biol.*, *112*, 535 (1977).

14. B. R. Crane, A. S. Arvai, R. Gachhui, C. Wu, D. K. Ghosh, E. D. Getzoff, D. J. Stuehr, and J. A. Tainer, *Science*, *278*, 425 (1977).

15. J.-C. Drapier, C. Pellat, and Y. Henry, *J. Biol. Chem.*, *266*, 10162 (1991).

16. J. Stadler, H. A. Bergonia, M. Di Silvio, M. A. Sweetland, T. R. Billiar, R. Simmons, and J. R. Lancaster, *Arch. Biochem. Biophys.*, *302*, 4 (1993).

17. F. Terenzi, J. M. Diaz-Guerra, M. Casado, S. Hortelano, S. Leoni, and L. Boscá, *J. Biol. Chem.*, *270*, 6017 (1995).

18. Y. Komori, G. C. Wallace, and J. M. Fukuto, *Arch. Biochem. Biophys.*, *315*, 213 (1994).

19. H. M. Abu-Soud, P. L. Feldman, P. Clark, and D. J. Stuehr, *J. Biol. Chem.*, *269*, 32318 (1994).

20. D. K. Ghosh and D. J. Stuehr, *Biochemistry*, *34*, 801 (1995); H. M. Abu-Soud, M. Loftus, and D. J. Stuehr, *Biochemistry*, *34*, 11167 (1995).

21. (a) J. A. Corbett, J. R. Lancaster, Jr., M. A. Sweetland, and M. L. McDaniel, *J. Biol. Chem.*, *266*, 21351 (1991); (b) J. A. Corbett, J. L. Wang, J. H. Hughes, B. A. Wolf, M. A. Sweetland, J. R. Lancaster, Jr., and M. L. McDaniel, *Biochem. J.*, *287*, 229 (1992). (c) J. A. Corbett, R. G. Tilton, K. Chang, K. S. Hasen, Y. Ido, J. L. Wang, M. A. Sweetland, J. R. Lancaster, Jr., J. R. Williamson, and M. L. McDaniel, *Diabetes*, *41*, 552 (1992); (d) J. A. Corbett, J. L. Wang, M. A. Sweetland, J. R. Lancaster, Jr., and M. L. McDaniel, *J. Clin. Invest.*, *90*, 2384 (1992).

22. R. F. Lin, T.-S. Lin, R. G. Tilton, and A. H. Cross, *J. Exp. Med.*, *178*, 643 (1993).

23. M. S. Finkel, C. V. Oddis, T. D. Jacob, S. C. Watkins, B. G. Hattler, and R. L. Simmons, *Science*, *257*, 387 (1992).

24. Y.-J. Geng, A.-S. Petersson, A. Wennmalm, and G. K. Hansson, *Exp. Cell Res.*, *214*, 418 (1994).

25. Y. Henry, C. Durocq, D. Servent, C. Pellat, and A. Guissani, *Eur. Biophys. J.*, *20*, 1 (1990).

26. G. Karupiah, Q. Xie, R. M. L. Buller, C. Nathan, C. Duarte, and J. D. MacMicking, *Science*, *261*, 1445 (1993).

27. R. F. Furchgott and J. V. Zawadzki, *Nature*, *288*, 373 (1980).

28. (a) V. Mollace, D. Salvemini, E. Anggard, and J. Vane, *Br. J. Pharmacol.*, *104*, 633 (1991); (b) D. L. Garbers, *Pharmac. Ther.*, *50*, 337 (1991); (c) P. S. T. Yuen and D. L. Garbers, *Annu. Rev. Neurosci.*, *15*, 193 (1992).

29. T. G. Traylor and V. S. Sharma, *Biochemistry*, *31*, 2847 (1992).

30. A. Tsai, *FEBS Lett.*, *341*, 141 (1994).

31. (a) A. E. Yu, S. Hu, T. G. Spiro, and J. N. Burstyn, *J. Am. Chem. Soc.*, *116*, 4117 (1994); (b) E. A. Dierks, S. Hu, K. M. Vogel, A. E. Yu, T. G. Spiro, and J. N. Burstyn, *J. Am. Chem. Soc.*, *119*, 7316 (1997).

32. L. J. Ignarro, *J. NIH Res.*, *4*, 59 (1992); A. L. Burnett, C. J. Lowenstein, D. S. Bredt, T. S. K. Chang, and S. H. Snyder, *Science*, *257*, 401 (1992).

33. P. R. Montague, C. D. Gancayco, M. J. Winn, R. B. Marchase, and M. J. Friedlander, *Science*, *263*, 973 (1994).

34. (a) J. M. Ding, D. Chen, E. T. Weber, L. E. Faiman, M. A. Rea, and M. U. Gillette, *Science*, *266*, 1713 (1994); (b) D. A. Oren and M. Terman, *Science*, *279*, 333 (1998); S. S. Campbell and P. J. Murphy, *Science*, *279*, 396 (1998).

35. Y.-M. Kim, H. A. Bergonia, C. Müller, B. R. Pitt, W. D. Watkins, and J. R. Lancaster, *J. Biol. Chem.*, *270*, 5710 (1995).

36. L. Jankiewicz, M. Kwaśny, K. Wasylik, and A. Graczyk, *J. Food Sci.*, *59*, 57 (1994).

37. L. L. Bondoc and R. Timkovich, *J. Biol. Chem.*, *264*, 6134 (1989).

38. J. M. C. Ribeiro, *Annu. Rev. Entomol.*, *32*, 463 (1987).

39. J. Law, J. M. C. Ribeiro, and M. Wells, *Annu. Rev. Biochem.*, *61*, 87 (1992).

40. J. M. C. Ribeiro and E. S. Garcia, *J. Exp. Biol.*, *94*, 219 (1981).

41. J. M. C. Ribeiro, P. A. Rossignol, and A. Spielman, *J. Exp. Biol.*, *108*, 1 (1984).

42. J. M. C. Ribeiro, R. H. Nussenzveig, and G. Tortorella, *J. Med. Entomol.*, *31*, 747 (1994).

43. J. W. Cornwall and W. S. Patton, *Indian J. Med. Res.*, *2*, 569 (1914).

44. C. Noeske-Jungblutt, J. Kratzschmar, B. Hacndler, A. Alugon, L. Possani, P. Verhallen, W.-D. Schleuning, *J. Biol. Chem.*, *269*, 5050 (1994).

45. S. A. Grevelink, D. E. Youssef, J. Loscalzo, and E. A. Lerner, *Proc. Natl. Acad. Sci. USA*, *90*, 9155 (1993).

46. G. A. Higgs, J. R. Vane, R. J. Hart, C. Porter, and R. G. Wilson, *Bull. Ent. Res.*, *66*, 665 (1976).

47. D. H. Kemp, J. R. Hales, A. V. Schleger, and A. A. Fawcett, *Experientia*, *39*, 725 (1983).

48. M. Shemesh, A. Hadani, A. Shklar, L. S. Shore, and F. Meleguir, *Bull. Ent. Res.*, *69*, 381 (1979).

49. J. M. C. Ribeiro, P. M. Evans, J. L. MacSwain, and J. Sauer, *Exp. Parasitol.*, *74*, 112 (1992).

50. R. G. Dickinson, J. E. O'Hagan, M. Shotz, K. C. Binnington, and M. P. Hegarty, *Aust. J. Exp. Biol. Med. Sci.*, *54*, 475 (1976).

51. O. Moro and E. A. Lerner, *J. Biol. Chem.*, *272*, 966 (1997).

52. J. M. C. Ribeiro, A. Vachereau, G. B. Modi, and R. B. Tesh, *Science*, *243*, 212 (1989)

53. E. A. Lerner, J. M. C. Ribeiro, R. J. Nelson, and M. R. Lerner, *J. Biol. Chem.*, *266*, 11234 (1991).

54. E. A. Lerner and C. B. Shoemaker, *J. Biol. Chem.*, *267*, 1062 (1992).

55. M. S. Cupp, J. M. C. Ribeiro, and E. W. Cupp, *Am. J. Trop. Med. Hyg.*, *50*, 241 (1994).

56. D. Champagne and J. M. C. Ribeiro, *Proc. Natl. Acad. Sci. USA*, *91*, 138 (1994).

57. J. M. C. Ribeiro and R. H. Nussenzveig, *J. Exp. Biol.*, *179*, 273 (1993).

58. J. M. C. Ribeiro, R. Gonzales, and O. Marinotti, *Br. J. Pharmacol.*, *101*, 932 (1990).

59. J. M. C. Ribeiro, J. M. H. Hazzard, R. H. Nussenzveig, D. E. Champagne, and F. A. Walker, *Science*, *260*, 539 (1993).

60. S. Moncada and J. F. Martin, *Lancet*, *341*, 1511 (1993).

61. D. E. Champagne, R. Nussenzveig, and J. M. C. Ribeiro, *J. Biol. Chem.*, *270*, 8691 (1995).

62. J. M. C. Ribeiro and F. A. Walker, *J. Exp. Med.*, *180*, 2251 (1994).

63. A. Weichsel, J. F. Andersen, D. E. Champagne, F. A. Walker, and W. R. Montfort, *Nature Struct. Biol.*, *5*, 304 (1998).

64. J. F. Andersen, D. E. Champagne, A. Weichsel, J. M. C. Ribeiro, C. A. Balfour, V. Dress, and W. R. Montfort, *Biochemistry*, *36*, 4423 (1997).

65. D. E. Champagne and J. M. C. Ribeiro, Manuscript in preparation.

66. M. Antoine, C. Erbil, E. Munch, S. Schnell, and J. Niessing, *Gene*, *56*, 41 (1987).

67. R. L. Garlick and A. F. Riggs, *J. Biol. Chem.*, *275*, 9005 (1982).

68. T. Suzuki and T. Gotoh, *J. Biol. Chem.*, *261*, 9257 (1986).

69. T. Imamura, T. O. Baldwin, and A. F. Riggs, *J. Biol. Chem.*, *247*, 2785 (1972).

70. D. R. Sherman, A. P. Kloek, B. R. Krishnan, B. Guinn, and D. Goldberg, *Proc. Natl. Acad. Sci. USA*, *89*, 11696 (1992).

71. B. Runnegar, *J. Mol. Evol.*, *2*, 33 (1984).

72. D. R. Flower, *Biochem. J.*, *318*, 1 (1996).

73. S. W. Cowan, M. E. Newcomer, and T. A. Jones, *Proteins: Struct. Funct. Genet.*, *8*, 44 (1990).

74. Y. Urade, A. Nagata, Y. Suzuki, Y. Fuji, and O. Hayashi, *J. Biol. Chem.*, *264*, 1041 (1989).

75. A. Nagata, Y. Suzuki, M. Igarashi, N. Eguchi, H. Ton, Y. Urade, and O. Hayashi, *Proc. Natl. Acad. Sci. USA*, *88*, 4020 (1991).

76. C. T. Riley, B. K. Barbeau, P. S. Keim, F. J. Kezdy, R. L. Heinrikson, and J. H. Law, *J. Biol. Chem.*, *259*, 13159 (1984).

77. H. M. Holden, W. R. Rypniewski, J. H. Law, and I. Rayment, *EMBO J.*, *6*, 1565 (1987).

78. R. Huber, M. Schneider, O. Epp, I. Mayr, A. Messerschmidt, X. Glugrath, and H. Kayser, *J. Mol. Biol.*, *195*, 423 (1987).

79. R. Huber, M. Schneider, I. Mayr, R. Muller, R. Deutzmann, F. Suter, H. Zuber, H. Falk, and H. Kayser, *J. Mol. Biol.*, *198*, 499 (1987).

80. J. G. Valenzuela, F. A. Walker, and J. M. C. Ribeiro, *J. Exp. Med.*, *198*, 1519 (1995).

81. J. G. Valenzuela and J. M. C. Ribeiro, *J. Exp. Biol.*, *201*, 2659 (1998).

82. Unpublished results.

83. (a) T. S. Ross, A. B. Jefferson, C. A. Mitchell, and P. W. Majerus, *J. Biol. Chem.*, *266*, 20283 (1991); (b) K. M. Laxminarayan, B. K. Chan, T. Tetaz, P. I. Bird, and C. A. Mitchell, *J. Biol. Chem.*, *269*, 17305 (1994).

84. J. M. C. Ribeiro and R. H. Nussenzveig, *FEBS Lett.*, *330*, 165 (1993).

85. M. Yuda, M. Hirai, K. Miura, H. Matsumura, K. Ando, and Y. Chinzei, *Eur. J. Biochem.*, *242*, 807 (1996).

86. R. H. Nussenzveig, D. L. Bentley, and J. M. C. Ribeiro, *J. Exp. Biol.*, *198*, 1093 (1995).

87. N. V. Blough and O. C. Zafirou, *Inorg. Chem.*, *24*, 3502 (1985).

88. P. A. King, V. E. Anderson, J. O. Edwards, G. Gustafson, R. C. Plumb, and J. W. Suggs, *J. Am. Chem. Soc.*, *114*, 5430 (1992).

89. J.-H. M. Tsai, J. G. Harrison, J. C. Martin, T. P. Hamilton, M. van der Woerd, M. J. Jablonsky, and J. S. Beckman, *J. Am. Chem. Soc.*, *116*, 4115 (1994).

90. J. M. Fukuto and L. J. Ignarro, *Acct. Chem. Res.*, *30*, 149 (1997).

91. J. T. Groves and S. S. Marla, *J. Am. Chem. Soc.*, *117*, 9578 (1995).

92. Q. H. Gibson and F. J. W. Roughton, *J. Physiol. (London)*, *136*, 507 (1957).

93. E. Antonini, M. Brunori, J. Wyman, and R. W. Noble, *J. Biol. Chem.*, *241*, 3236 (1966).

94. R. Cassoly and Q. H. Gibson, *J. Mol. Biol.*, *91*, 301 (1975).

95. V. S. Sharma, T. G. Traylor, R. Gardiner, and H. Mizukami, *Biochemistry*, *26*, 3837 (1987).

96. M. Hoshino, K. Ozawa, H. Seki, and P. C. Ford, *J. Am. Chem. Soc.*, *115*, 9568 (1993).

97. V. G. Kharitonov, V. S. Sharma, D. Magde, and D. Koesling, *Biochemistry*, *36*, 6814 (1997).

98. R. W. Romberg and R. J. Kassner, *Biochemistry*, *18*, 5387 (1979).

99. A. W. Addison and J. J. Stephanos, *Biochemistry*, *25*, 4104 (1986).

100. (a) K. A. Jongeward, D. Magde, D. J. Taube, J. C. Marsters, T. G. Traylor, and V. S. Sharma, *J. Am. Chem. Soc.*, *110*, 380 (1988); (b) J. W. Petrich, J.-C. Lambry, K. Kuczera, M. Karplus, C. Poyart, and J.-L. Martin, *Biochemistry*, *30*, 3975 (1991); (c) T. E. Carver, J. S. Olson, S. J. Smerdon, S. Krzywda, A. J. Wilkinson, Q. H. Gibson, R. S. Blackmore, J. D. Ropp, and S. G. Sligar, *Biochemistry*, *30*, 4697 (1991); (d) T. G. Traylor, D. Magde, J. Marsters, K. Jongeward, G.-Z. Wu, and K. Walda, *J. Am. Chem. Soc.*, *115*, 4808 (1993).

101. X. D. Ding, A. Weichsel, J. F. Andersen, T. Kh. Shokhireva, C. Balfour, A. J. Pierik, B. A. Averill, W. R. Montfort, and F. A. Walker, *J. Am. Chem. Soc.*, in press.

102. T. Yonetani, H. Yamamoto, J. E. Erman, J. S. Leigh, and G. H. Reed, *J. Biol. Chem.*, *247*, 2447 (1972).

103. J. H. Dawson, L. A. Andersson, and M. Sono, *J. Biol. Chem.*, *258*, 13637 (1983).

104. D. H. O'Keeffe, R. E. Ebel, and J. A. Peterson, *J. Biol. Chem.*, *253*, 3509 (1978).

105. M. Sono, K. S. Eble, J. H. Dawson, and L. P. Hager, *J. Biol. Chem.*, *260*, 15530 (1985).

106. J. C. Maxwell and W. S. Caughey, *Biochemistry*, *15*, 388 (1976).

107. T. Yoshimura, *Bull. Chem. Soc. Jpn.*, *56*, 2527 (1983).

108. V. Sampath, X.-J. Zhao, and W. S. Caughey, *Biochem. Biophys. Res. Commun.*, *198*, 281 (1994).

109. L. M. Miller, A. J. Pedraza, and M. R. Chance, *Biochemistry*, *36*, 12199 (1997).

110. Y. Wang and B. A. Averill, *J. Am. Chem. Soc.*, *118*, 3972 (1996).

111. (a) G. Chottard and D. Mansuy, *Biochem. Biophys. Res. Commun.*, *77*, 1333 (1977); (b) M. Tsubaki and N.-T. Yu, *Biochemistry*, *21*, 1140 (1982); (c) M. Walters and T. G. Spiro, *Biochemistry*, *21*, 6989 (1982).

112. B. Benko and N.-T. Yu, *Proc. Natl. Acad. Sci. USA*, *80*, 7042 (1983).

113. L. A. Lipscomb, B.-S. Lee, and N.-T. Yu, *Inorg. Chem.*, *32*, 281 (1993).

114. S. Hu and J. R. Kincaid, *J. Am. Chem. Soc.*, *113*, 2843 (1991).

115. (a) G. Lang and W. Marshall, *Proc. Phys. Soc.*, *87*, 3 (1966); (b) W. T. Oosterhuis and G. Lang, *J. Chem. Phys.*, *50*, 4381 (1969).

116. (a) J. A. Christner, E. Münck, P. A. Nanick, and L. M. Siegel, *J. Biol. Chem.*, *258*, 11147 (1983); (b) M.-C. Liu, B.-M. Huynh, W. J. Payne, H. D. Peck, Jr., D. V. Dervartanian, and J. LeGall, *Eur. J. Biochem.*, *169*, 253 (1987).

117. H. D. Pfannes, G. Bemski, E. Wajnberg, H. Rocha, E. Bill, H. Winkler, and A. X. Trautwein, *Hyperfine Int.*, *91*, 797 (1994).

118. M. R. Cheesman, A. J. Thomson, C. Greenwood, G. R. Moore, and F. Kadir, *Nature*, *346*, 771 (1990).

119. M. J. Berry, S. J. George, A. J. Thomson, H. Santos, and D. L. Turner, *Biochem. J.*, *270*, 413 (1990).

120. F. A. Walker and U. Simonis, in *Encyclopedia of Inorganic Chemistry*, Vol. 4 (R. B. King, ed.), John Wiley and Sons, Chichester, 1994; pp. 1785–1846.

121. (a) F. A. Walker, M. W Lo, and M. T. Ree, *J. Am. Chem. Soc.*, *98*, 5552 (1976); (b) F. A. Walker, D. Reis, and V. L. Balke, *J. Am. Chem. Soc.*, *106*, 6888 (1984).

122. M. J. M. Nesset, N. V. Shokhirev, P. D. Enemark, S. E. Jacobson, and F. A. Walker, *Inorg. Chem.*, *35*, 5188 (1996).

123. J. F. Andersen, A. Weichsel, C. Balfour, D. E. Champagne, and W. R. Montfort, *Structure*, *6*, 1315 (1998).

124. R. D. Feltham and J. H. Enemark, *Coord. Chem. Rev.*, *13*, 339 (1974).

125. H. Hori, M. Ikeda-Saito, G. Lang, and T. Yonetani, *J. Biol. Chem.*, *265*, 15028 (1990).

126. H. Kon, *J. Biol. Chem.*, *243*, 4350 (1968).

127. Y. Henry and R. Banerjee, *J. Mol. Biol.*, *73*, 469 (1973).

128. R. LoBrutto, Y.-H. Wei, R. Mascarenhas, C. P. Scholes, and T. E. King, *J. Biol. Chem.*, *258*, 7437 (1983).

129. R. E. Ebel, D. H. O'Keeffe, and J. A. Peterson, *FEBS Lett.*, *55*, 198 (1975).

130. Y. Henry, M. Lepoivre, J.-C. Drapier, C. Ducrocq, J.-L. Boucher, and A. Guissani, *FASEB J.*, *7*, 1124 (1993).

131. B. B. Wayland and L. W. Olson, *J. Am. Chem. Soc.*, *96*, 6037 (1974).

132. (a) H. Rein, O. Ristau, and W. Scheler, *FEBS Lett.*, *24*, 24 (1972); (b) A. Szabo and M. F. Perutz, *Biochemistry*, *15*, 4427 (1976); (c) K. Nagai, H. Hori, H. Morimoto, A. Hayashi, and F. Taketa, *Biochemistry*, *18*, 1304 (1979); (d) R. S. Magliozzo, J. McCracken, and J. Peisach, *Biochemistry*, *26*, 7923 (1987).

133. T. Yoshimura, *J. Inorg. Biochem.*, *18*, 263 (1983).

134. M. Brunori, G. Falcioni, and G. Rotilio, *Proc. Nat. Acad. Sci. USA*, *71*, 2470 (1974).

135. R. Karthein, W. Nastainczyk, and H. H. Ruf, *Eur. J. Biochem.*, *166*, 173 (1987).

136. T. H. Stevens and S. I. Chan, *J. Biol. Chem.*, *256*, 1069 (1981).

137. Y. Henry, Y. Ishimura, and J. Peisach, *J. Biol. Chem.*, *251*, 1578 (1976).

138. N. V. Gerbeleu, V. B. Arion, Yu. A. Simonov, V. E. Zavodnik, S. S. Stavrov, K. I. Turta, D. I. Gradinaru, M. S. Birca, A. A. Pasynskii, and O. Ellert, *Inorg. Chim. Acta*, *202*, 173 (1992).

139. V. Schünemann, F. A. Walker, and A. X. Trautwein, unpublished work.

140. C. M. Bougault, Y. Dou, M. Ikeda-Saito, K. C. Langry, K. M. Smith, and G. N. La Mar, *J. Am. Chem. Soc.*, *120*, 2113 (1998).

141. (a) R. Krishnamoorthi, G. N. La Mar, H. Mizukami, and A. Romero, *J. Biol. Chem.*, *259*, 265 (1984); (b) K. Rajarathanam, G. N. La Mar, M. L. Chiu, S. G. Sligar, J. P. Singh, and K. M. Smith, *J. Am. Chem. Soc.*, *113*, 7886 (1991).

142. G. N. La Mar, J. T. Jackson, L. B. Dugad, M. A. Cusanovich, and R. G. Bartsch, *J. Biol. Chem.*, *265*, 16173 (1990).

143. J. S. deRopp, P. Mandal, S. L. Brauer, and G. N. La Mar, *J. Am. Chem. Soc.*, *119*, 4732 (1997).

144. T. Kh. Shokhireva and F. A. Walker, unpublished work.

145. A. Minniear and F. A. Walker, unpublished results.

146. G. N. La Mar and J. S. de Ropp, in *Biological Magnetic Resonance*, Vol. 12, *NMR of Paramagnetic Molecules* (L. J. Berliner and J. Reuben, eds.), Plenum Press, New York, 1993; pp. 1–78.

147. (a) L. Jia, C. Bonaventura, J. Bonaventura, and J. S. Stamler, *Nature*, *380*, 221 (1996); (b) J. S. Stamler, L. Jia, J. P. Eu, T. J. McMahon, I. T. Demchenko, J. Bonaventura, K. Gernert, and C. A. Piantadosi, *Science*, *276*, 2034 (1997).

148. E. Fujita and J. Fajer, *J. Am. Chem. Soc.*, *105*, 6743 (1983).

149. (a) L. Olson, D. Schaeper, D. Lancon, and K. M. Kadish, *J. Am. Chem. Soc.*, *104*, 2042 (1982); (b) D. Lancon and K. M. Kadish, *J. Am. Chem. Soc.*, *105*, 5610 (1983).

150. I.-K. Choi, Y. Liu, D. Feng, K.-J. Paeng, and M. D. Ryan, *Inorg. Chem.*, *30*, 1832 (1991).

151. R. Lin, M. Bayachou, J. Greaves, and P. J. Farmer, *J. Am. Chem. Soc.*, *119*, 12689 (1997).

152. (a) J. H. Ridd, *Adv. Phys. Org. Chem.*, *16*, 1–49 (1978); (b) D. C. Williams (ed.), *Nitrosation*, Cambridge Univ. Press: New York, 1988, pp. 1–214.

153. R. S. Wade and C. E. Castro, *Chem. Res. Toxicol.*, *3*, 289 (1990).

154. (a) S. S. Mirvish, *Toxicol. Appl. Pharmacol.*, *31*, 325 (1975); (b) M. Miwa, D. J. Stuehr, M. A. Marletta, J. S. Whishnok, and S. R. Tannenbaum, *Carcinogenesis*, *8*, 955 (1987); (c) B. C. Challis, J. R. Outram, and D. E. G. Shuker, *IARC Sci. Publ.*, *31*, 43 (1980); (d) D. A. Wink, K. S. Kasprzak, C. M. Maragos, R. K. Elespuru, M. Misra, T. M. Dunams, T. A. Cebula, W. H. Koch, A. W. Andrews, J. S. Allen, and L. K. Keefer, *Science*, *254*, 1001 (1991).

155. (a) B. C. Challis, M. H. R. Fernandes, B. R. Glover, and F. Latif, *IARC Sci. Publ.*, *84*, 308 (1987); (b) S. R. Tannenbaum, *Ibid.*, 292.

156. J. S. Stamler, D. I. Simon, J. A. Osborne, M. E. Mullins, O. Jaraki, T. Michel, D. J. Singel, and J. Loscalzo, *Proc. Nat. Acad. Sci. USA*, *89*, 444 (1992).

157. L. J. Ignarro, *Circ. Res.*, *65*, 1 (1989).

158. J. S. Stamler, O. Jaraki, J. Osborne, D. I. Simon, J. Keaney, J. Vita, D. Singel, C. R. Valeri, and J. Loscalzo, *Proc. Natl. Acad. Sci. USA*, *89*, 7674 (1992).

159. D. Gergel and A. I. Cederbaum, *Biochemistry*, *35*, 16186 (1996).

160. X. D. Ding, Ph.D. thesis, University of Arizona, 1997.

161. R. Varadarajan, T. E. Zewert, H. B. Gray, and S. G. Boxer, *Science*, *243*, 69 (1989).

162. G. Denariaz, P. A. Ketchum, W. J. Payne, M.-Y. Liu, J. LeGall, I. Moura, and J. J. Moura, *Arch. Microbiol.*, *162*, 316 (1994).

163. (a) P. L. Piciulo, G. Rupprecht, and W. R. Scheidt, *J. Am. Chem. Soc.*, *96*, 5293 (1974); (b) W. R. Scheidt and M. E. Frisse, *J. Am. Chem. Soc.*, *97*, 17 (1975); (c) W. R. Scheidt and P. L. Piciulo, *J.*

Am. Chem. Soc., *98*, 1913 (1976); (d) W. R. Scheidt, A. C. Brinetgar, E. B. Ferro, and J. F. Kirner, *J. Am. Chem. Soc.*, *99*, 7315 (1977); (e) H. Nasri, K. J. Haller, Y. Wang, B. H. Huynh, and W. R. Scheidt, *Inorg. Chem.*, *31*, 3459 (1992).

164. J. F. Deatherage and K. Moffat, *J. Mol. Biol.*, *134*, 401 (1979).

165. (a) H. Nasri, M. K. Ellison, S. Chen, B. H. Huynh, and W. R. Scheidt, *J. Am. Chem. Soc.*, *119*, 6274 (1997); (b) M. K. Ellison and W. R. Scheidt, *J. Am. Chem. Soc.*, *119*, 7404 (1997).

166. W. R. Scheidt, Y. J. Lee, and K. Hatano, *J. Am. Chem. Soc.*, *106*, 3191 (1984).

167. J. F. Andersen, A. Weichsel, C. Balfour, and W. R. Montfort, unpublished work.

168. M. E. Newcomer, *Structure*, *1*, 7 (1993).

169. G. Weiss, B. Goossen, W. Doppler, D. Fuchs, K. Pantopoulous, G. Werner-Felmayer, H. Wachter, and M. W. Hentze, *EMBO J.*, *12*, 3651 (1993).

170. J.-C. Drapier, H. Hirling, J. Wietzerbin, P. Kaldy, and L. C. Kühn, *EMBO J.*, *12*, 3643 (1993).

171. L. Castro, M. Rodriguez, and R. Radi, *J. Biol. Chem.*, *269*, 29409 (1994).

172. D. W. Reif and R. D. Simmons, *Arch. Biochem. Biophys.*, *283*, 537 (1990).

173. Lipiński, P. and J.-C. Drapier, *JBIC*, *2*, 559 (1997).

174. M. Lee, P. Arsio, A. Cozzi, and N. D. Chasteen, *Biochemistry*, *33*, 3679 (1994).

175. M. R. Hyman, L. C. Seefeldt, T. V. Morgan, D. J. Arp, and L. E. Mortenson, *Biochemistry*, *31*, 2947 (1992).

176. G. Musci, S. Di Marco, M. C. Bonaccorsi di Patti, and L. Calabrese, *Biochemistry*, *30*, 9866 (1991).

177. C. J. Haskin, N. Ravi, J. B. Lynch, E. Münck, and L. Que, *Biochemistry*, *34*, 11090 (1995).

178. Y. Sanakis, C. Goussias, R. P. Mason, and V. Petrouleas, *Biochemistry*, *36*, 1411 (1997).

179. M. C. Kennedy, T. Gan, W. E. Antholine, and D. H. Petering, *Biochem. Biophys. Res. Commun.*, *196*, 632 (1993).

180. (a) J. R. Lancaster and J. B. Hibbs, *Proc. Natl. Acad. Sci. USA*,

87, 1223 (1990); (b) J.-C. Drapier, C. Pellat, and Y. Henry, *J. Biol. Chem.*, *266*, 10162 (1991).

181. A. R. Butler, C. Glidewell, and M.-H. Li, *Adv. Inorg. Chem.*, *32*, 336 (1988).

182. (a) A. R. Butler, C. Glidewell, A. R. Hyde, and J. C. Walton, *Polyhedron*, *4*, 797 (1985); (b) C. C. McDonald, W. D. Phillips, and H. F. Mower, *J. Am. Chem. Soc.*, *87*, 3319 (1965); (c) M. P. Boyer, J. R. Morton, and K. F. Preston, *J. Phys. Chem.*, *84*, 2989 (1980).

183. R. S. Gall, C. T.-W. Chu, and L. F. Dahl, *J. Am. Chem. Soc.*, *96*, 4019 (1974).

184. (a) G. Johansson and W. N. Lipscomb, *Acta Crystallogr.*, *11*, 594 (1958); (b) C. T.-W. Chu and L. F. Dahl, *Inorg. Chem.*, *16*, 3245 (1977); (c) A. R. Butler, C. Glidewell, and S. M. Glidewell, *Polyhedron*, *9*, 2399 (1990).

185. (a) G. Johansson and W. N. Lipscomb, *Acta Crystallogr.*, *11*, 594 (1958); (b) C. T.-W. Chu and L. F. Dahl, *Inorg. Chem.*, *16*, 3245 (1977); (c) A. R. Butler, C. Glidewell, and S. M. Glidewell, *Polyhedron*, *9*, 2399 (1990).

186. (a) F. W. Flitney, I. L. Megson, D. E. Flitney, and A. R. Butler, *Br. J. Pharmacol.*, *107*, 842 (1992). (b) A. F. Vanin, R. A. Stukan, and E. B. Manukhina, *Biochim. Biophys. Acta*, *1295*, 5 (1996).

187. (a) A. F. Vanin, *FEBS Lett.*, *289*, 1 (1991); (b) Y. P. Vedernikov, P. I. Mordvinteev, I. V. Malenkova, and A. F. Vanin, *Eur. J. Pharmacology*, *211*, 313 (1992).

188. Y. P. Vedernikov, P. I. Mordvintcev, I. V. Malenkova, and A. F. Vanin, *Eur. J. Pharmacol.*, *212*, 125 (1992).

189. A. Komarov, D. Mattson, M. M. Jones, P. K. Singh, and C.-S. Lai, *Biochem. Biophys. Res. Commun.*, *195*, 1191 (1993).

20

Nitrogen Monoxide-Related Disease and Nitrogen Monoxide Scavengers as Potential Drugs

Simon P. Fricker

AnorMED Inc., Langley, BC, V2Y 1N5 Canada

1. INTRODUCTION

The discovery that nitrogen monoxide (nitric oxide, NO) is a ubiquitous
biological messenger molecule has radically changed our view of how
cells communicate with one another. Nitric oxide has generally been
regarded as a common environmental toxin and pollutant, being a
constituent of both cigarette smoke and car exhaust emissions. Though
this is of course true, we now know that this simple diatomic molecule is
an important component of the regulatory system for controlling blood
pressure, a neurotransmitter in both the central and peripheral nervous system, and an effector molecule in the immune system [1,2] (see
Chapter 17 for a full review).

   One of the first biological functions of NO to be identified was its
role as endothelium-derived relaxing factor (EDRF). Furchgott and
Zawadski had shown that the vasorelaxant properties of acetylcholine
were dependent on blood vessels having an intact endothelium [3]. They
postulated the existence of a second messenger, EDRF, between the
endothelial cells and the vascular smooth muscle. Following this discovery, independent studies by Ignarro et al. [4] and Moncada et al. [5]
identified NO as EDRF.

   A further function for NO was discovered when Garthwaite and
coworkers observed that rat cerebellar slices, upon stimulation of the
N-methyl-D-aspartate (NMDA) receptors by glutamate, released a substance with EDRF-like properties [6]. Subsequently, Bredt and Snyder

were able to demonstrate that glutamate and related amino acids such as NMDA stimulated nitric oxide synthase (NOS) activity in rat cerebellar slices [7]. The exact role of NO in the brain has yet to be ascertained but, unlike other neurotransmitters, NO can diffuse in three dimensions and can thus act as a retrograde neurotransmitter on surrounding cells including the presynaptic neuron. This has led to the suggestion that NO in the brain may play a role in long-term potentiation and memory [8]. Nitric oxide also acts as a neurotransmitter in the peripheral nervous system for those nerves that use neither noradrenaline nor acetylcholine as a transmitter, the so-called nonadrenergic, noncholinergic (NANC) nerves. NANC nerves are found in cardiovascular, respiratory, gastrointestinal tissue, and urogenital tissue [1,9].

Concurrent investigations into macrophage function indicated that the antimicrobial and antitumor activity of macrophages was dependent on L-arginine and that the end products of arginine metabolism were nitrite and nitrate [10,11]. These were shown to be the oxidation products of NO, the cytotoxic effector molecule [12,13]. The enzyme responsible for the conversion of arginine to NO was named NOS. This enzyme catalyzes a five-electron oxidation of L-arginine to L-citrulline and NO. Nitric oxide synthase requires heme, NADPH, tetrahydrobiopterin, and calmodulin as cofactors [14,15].

Three isoforms of NOS have been identified: nNOS (neuronal NOS, NOS I), iNOS (inducible NOS, NOS II), and eNOS (endothelial NOS, NOS III) [14–16]. The neuronal NOS, the first of the isoforms to be cloned and purified [17], is found in both central and peripheral nerves. The endothelial NOS was first identified in endothelial cells but has now been found in a variety of cell types [16,18]. These enzymes are constitutive (hence they are frequently referred to as cNOS, or ecNOS and ncNOS) and are regulated by changes in intracellular calcium concentration via calmodulin. The inducible NOS was first identified in murine macrophage cells but has also now been found in a wide variety of tissues [19,20]. This enzyme is transcriptionally regulated and is induced by stimulation with cytokines and endotoxin.

In general, when we think of messenger molecules we envisage either proteins, peptides, or organic molecules that act as ligands for receptors where they bind with a precise three-dimensional interaction. The messenger function of NO is mediated by its chemical reactivity. It has an electronic configuration of $(\sigma_1)^2(\sigma_1^*)^2(\sigma_2,\pi)^6(\pi^*)$. The unpaired electron means that nitric oxide is formally a radical and is responsible for its unique properties [21]. Nitric oxide is a colorless gas at room temperature and the gas phase chemistry of NO is well known; NO will react very rapidly with oxygen to form nitrogen dioxide (NO_2). However, NO behaves differently in an aqueous environment reacting with oxygen to form nitrite [22–24]. The coordination chemistry of NO is similar to that of carbon monoxide since both are π-acceptor ligands. The ability of NO to form metal-nitrosyl complexes is in part responsible for its biological activity. NO reacts readily with iron-heme, [25], and the EDRF function of NO is mediated by its upregulation of the heme-containing enzyme guanylate cyclase [26]. NO will also react with oxygen to form N_2O_3, and superoxide (O_2^-) to form peroxynitrite, $ONOO^-$ [27]. Both of these species will react with a variety of biological molecules, including proteins, lipids, and DNA [27]. Peroxynitrite is a powerful oxidant that can decompose to form a hydroxyl-like radical [28,29]. It has been proposed that the cytotoxic actions of NO may be due to a direct inhibitory effect of NO on respiration and DNA synthesis, involving the nitrosylation of iron-sulfur centers of key enzymes such as aconitase, the oxidoreductases complex I and complex II of the mitochondrial electron transport chain, and ribonucleotide reductase [30,31]. Evidence for these targets are inhibition of aerobic respiration and DNA synthesis, and detection of nitrosylation of nonheme iron by electron paramagnetic resonance (EPR). Although recent data have indicated that NO cannot directly nitrosylate Fe/S proteins it has been shown that peroxynitrite can, and this may be the mechanism by which nitric oxide appears to act on these proteins [32,33] (see Chapter 18). It has also been proposed that NO may inhibit respiration by inhibiting cytochrome oxidase [34], another heme-containing enzyme, and effect energy metabolism via activation of poly-ADP ribosyltransferase (PARS) [35].

It is not surprising, therefore, that because of the ubiquitous and essential nature of NO, problems with NO metabolism have been implicated in the pathophysiology of a number of diseases. Both down- and upregulation of NO responses are involved in disparate disease states. This chapter will both review the involvement of NO in disease and discuss NO as a target for therapeutic intervention, with a focus on scavenging and removal of NO as a strategy for NO-mediated disease.

2. NITRIC OXIDE AND DISEASE

2.1. Diseases Related to Underproduction of Nitric Oxide

The underlying cause of essential hypertension is unclear. Initially it was attributed to an increase in vasoconstrictor activity; however, there is now an accumulating body of evidence linking hypertension in humans to abnormalities in NO metabolism [36,37]. Nitric oxide synthase inhibitors cause vasoconstriction of aortic rings [1,38] and lead to a rise in blood pressure when administered to experimental animals [39]. Hypertension is also seen in mice in which the gene encoding eNOS has been disrupted [40]. Endothelium-dependent dilatation is reduced in animals with experimental hypertension. A similar phenomenon is seen in the forearm arterial bed of patients with untreated essential hypertension; there is also a corresponding reduction in response to L-NMMA in these patients [36].

Downregulation of the L-arginine/NO/cGMP pathway may also be a contributory factor in cases of secondary hypertension, i.e., where the hypertension is a clinical symptom of an underlying disease state [41]. A well-studied example of this is renal hypertension. Evidence for this is provided by studies on two genetically manipulated strains of Dahl/Rapp rats; a salt-sensitive strain (SS/Jr) that develops renal failure and hypertension on a high-salt diet, and a corresponding salt-resistant strain (SR/Jr). The SR/Jr rats increase NO synthesis when on a high-salt diet, and giving L-arginine to the SS/Jr strain prevents both the renal failure and associated hypertension [42].

Endothelial dysfunction is an important feature of the hypertension seen in atherosclerosis. The observed decrease in endothelial-dependent relaxation in this disease is in part due to a reduction in NO production [43,44]. Experimental data from a number of sources suggest that the mechanism of reduced endothelial–dependent dilatation is not simple [36]. In some experimental models of atherosclerosis there is no obvious reduction in NO synthesis and it has been suggested that the abnormality may lie at the level of the agonist receptors and corresponding signal transduction processes responsible for controlling intracellular calcium release. An alternative explanation for the decrease in endothelial-dependent relaxation may be a reduction in the response of guanylate cyclase to NO. It has been proposed that this may be mediated by the increased level of oxidized LDL seen in atherosclerosis. Another possible cause of the endothelial dysfunction is the increase in degradation of NO by superoxide, possibly generated by the monocytes that accumulate at the site of the atheroma, leading to the formation of peroxynitrite which may contribute to tissue damage. Endothelial dysfunction has also been implicated in the hypertension associated with diabetes [45] and congestive heart failure [37].

2.2. Diseases Related to Overproduction of Nitric Oxide

An increase in NO production has been implicated in a number of diseases [1,46,47]. One of the most well-studied yet least understood is septic shock. Sepsis is characterized as a systemic response to infection that can be triggered by a variety of microorganisms including gram-negative and gram-positive bacteria, and fungi. Sepsis can progress through several stages from the initial onset of sepsis syndrome to septic shock. These stages are not discrete disease states but a continuum of increasing severity [48]. Septic shock is associated with complex changes in multiple organ systems, change in organ perfusion, severe hypotension, vascular collapse, and severe multiple organ failure. A number of chemical mediators have been identified, including tumor necrosis factor-α (TNF-α), interleukin-1 (IL-1), PAF, and prostaglandins. Several lines of evidence point to NO as one of the key mediators of the hypotension seen in septic shock [49]. Many of the symptoms of septic shock, such as hypotension, can be reproduced in animal models by administration of lipopolysaccharide (LPS; also known as endotoxin) a component of the cell wall of gram-negative bacteria such as *Escherichia coli*. The hypotension in endotoxin-induced rodent models of sepsis can be reversed by NOS inhibitors and prevented by glucocorticoids, inhibitors of iNOS induction [50–52]. There is evidence from the measurement of nitrite and nitrate levels for increased production of NO in patients with sepsis syndrome [53], and the hypotension of patients can be reversed upon treatment with the NOS inhibitor L-NMMA [54]. Nitric oxide and its oxidation products [55] may also be mediators of the multiple organ damage associated with sepsis, though conversely NO might also exert a protective effect [56]. This contradictory behavior is probably a function of the local microenvironment and the relative levels of reactive oxygen species (ROS). Nitric oxide may also be a mediator of other forms of circulatory shock such as hemorraghic and traumatic shock [57]. Generation of excess NO could also contribute to the cardiac dysfunction associated with endotoxemia and the hyperdynamic state of cirrhosis [46].

The inducible isoform of NOS is found in rodent macrophages and human neutrophils, and NO has been hypothesized to play an important role in the immune response to a number of pathogens. It is therefore not surprising that NO has been implicated as a mediator in the pathophysiology of a number of inflammatory and autoimmune disorders. Endotoxin and IL-1 induce NO synthesis in cultures of rabbit articular chondrocytes, and TNF-α and IL-1 stimulate NO synthesis in rabbit synovial fibroblasts [58]. Studies in knockout mice in which the iNOS gene has been disrupted indicate a role for NO in murine models of arthritis, and acute inflammation [59]. Evidence for the involvement of NO in rheumatoid arthritis in humans has come from measurements of nitrite and nitrate in synovial fluid [58]. These levels are increased in patients with rheumatoid arthritis and are elevated over matched serum samples, suggesting that this is a local tissue effect.

Another autoimmune disease in which NO has been shown to be an effector molecule is diabetes [60,61]. During insulinitis monocytic cells produce IL-1β, which in turn mediates destruction of pancreatic islet cells. IL-1 can stimulate NO synthesis by isolated pancreatic β cells, and HIT-T15 and RINm5F insulinoma cell lines causing DNA damage. Early administration of L-NMMA to streptozotocin-treated rats reversed the symptoms of streptozotocin-induced diabetes [62].

Increased expression of iNOS and NO production has been demonstrated in a number of other immune-related inflammatory disorders including inflammatory bowel disease [63], the immunopathology of graft-vs.-host disease [64], psoriasis [65], and asthma [66]. Elevated production of NO in response to excessive release of excitatory amino acids has been proposed to be a pathogenic mechanism for neurological disorders such as epilepsy and cerebral ischemia (stroke) and chronic neurodegenerative disorders [46].

2.3.   Nitric Oxide and Cancer

The precise role of NO in the biology of cancer remains unclear. This lack of clarity is, in part, due to the multifaceted role of NO. NO is cytotoxic toward tumor cells in vitro [12,13] but more complex effects are seen in vivo where an increasing body of evidence suggests that NO is important in controlling tumor growth and vascularization [67,68].

Nitric oxide synthases have been shown to be expressed in human and murine cancers. NOS has been found to be associated with the tumor cells of human gynecological cancers and the stroma of human breast cancers [69,70], and rat colon tumors [71]. NO has been shown to be important for maintaining the vasodilatory tone of tumors [72], regulating tumor blood flow [72,73], and tumor oxygenation and energy status [74]. There is abundant evidence to support the hypothesis of NO as an active mediator of angiogenesis. A subclone of a human colon adenocarcinoma cell line, DLD-1, engineered to produce NO had enhanced growth compared with the wild type. There was improved tumor vascularization suggesting that NO was important for the neo-vascularization process [67]. Many vasoactive vasodilatory molecules, such as Substance P and prostaglandin E_1, are known to possess angiogenic properties and NO production is induced by both of these agents [75,76]. Conversely angiogenesis can be inhibited by NOS inhibitors [77,78]. In contrast, the angiogenic response to basic fibroblast growth factor (bFGF) is not mediated by NO [79,80], suggesting at least two regulatory pathways for angiogenesis. Platelet-activating factor induced angiogenesis, and the activity of vascular endothelial growth factor (VEGF), have also recently been shown to be mediated by NO [79, 80]. In all of these cases the evidence points to the calcium-dependent cNOS as the source of NO, whereas the LPS-stimulated angiogenic activity of human monocytes appears to require iNOS [81]. Conversely, there are data indicating that NO has a negative regulatory effect on angiogenesis [82,83]. One explanation for this apparent conflict is that the inhibitory effect of NO on angiogenesis is seen in models employing embryonic (chick embryo) tissue.

The angiogenic process is intimately linked with metastasis of solid tumors. Nitric oxide increased vascular permeability in tumor-bearing mice [73,84,85], a prerequisite for metastasis. The murine breast cancer cells, EMT-6, produced NO when stimulated with LPS and IFN-γ; when the stimulated cells were injected subcutaneously into Balb/c mice there was an increase in tumor size, and in the numbers of pulmonary metastases compared with unstimulated cells [68]. This effect was inhibited by the NOS inhibitor, L-NAME. Conversely, the NO donor isosorbide mononitrate was shown to inhibit metastasis of the Lewis lung murine carcinoma [86]. Induction of macrophage iNOS in vivo by the lipopeptide CGP 31362 encapsulated in liposomes inhibited the growth of the murine sarcoma M5076 and reduced the incidence of hepatic metastases [87].

The regulation of blood flow, and concomitant energy status, by NO can be exploited therapeutically [74]. Chronic oral administration of an NOS inhibitor has been shown to retard experimental tumor growth [78]. A decrease in tumor blood leading to a hypoxic environment may allow activation of redox-active drugs [72,74]. Alternatively, an increase in blood flow will lead to an increase in tumor oxygenation, which may increase sensitivity to radiation [74]. NO donor compounds have been shown to act as radiosensitizers toward hypoxic cells [88,89]. Sodium nitroprusside increased the radiosensitivity of radioresistant human pancreatic tumor cells, [89] similarly the NONOate (see Sec. 3.1), DEA/NO, a novel class of NO donor, sensitized V79 cells [88]. The NONOates have also been shown to act as chemosensitizers, enhancing the cytotoxic action of drugs such as melphalan [90].

Nitric oxide may also be responsible for the adverse effects, hypotension, and capillary leak syndrome, of IL-2 therapy. An increase in endogenous nitrate synthesis was seen in patients on IL-2 therapy suggesting an upregulation of iNOS [91,92]. This was associated with an increase in both IFN-γ and TNF-α, cytokines known to induce iNOS in rodent macrophages. Administration of the NOS inhibitor L-NAME also abolished capillary leak in a mouse adenocarcinoma model of IL-2 therapy [93]. The regulatory role of NO in tumor biology therefore provides a number of opportunities for therapeutic intervention.

3. THERAPEUTIC STRATEGIES

3.1. Nitric Oxide Donor Drugs

Nitric oxide donor drugs have been in use since the late nineteenth century when the organic nitrates were first introduced for the treatment of angina (for a full review, see Chapter 21). Amyl nitrate was first used by Brunton in 1867 [94]. He administered the drug by inhalation and noted a rapid relief of anginal pain. The effect was transitory and in 1879 Murrell established the use of sublingual glyceryl trinitrate for relief of acute anginal attack and as a prophylactic to be taken prior to exertion [95]. The organic nitrates are now mainstay therapy for angina, and are also used for treatment of congestive heart failure and acute myocardial infarction. Isosorbide dinitrate was developed as an organonitrate with a longer duration of action. Sodium nitroprusside is an inorganic NO donor drug that is used to treat hypertensive emergencies [143].

Tolerance to organic nitrates is increasingly being regarded as a clinical problem [96]. More recently, the sydnominine derivative molsidomine has been introduced [97]. This compound may have benefit over the organic nitrates inasmuch that there is a reduced incidence of tolerance. It may have the disadvantage that the active metabolite, SIN-1, breaks down to simultaneously produce superoxide and NO, which in turn may react to give peroxynitrite. Other new NO donors include the furoxans, nitrosothiols, and the NONOates [98]. The latter represent a class of compounds with the general structure $XN(O^-)N=O$, where X is a nucleophile. These compounds give a spontaneous, controlled release of NO, the rate of which depends on the nucleophile [99]. The inorganic compound, Roussins black salt, a tetrairon-sulfur cluster nitrosyl, has been used experimentally as a source of NO [100]. Another approach has been to combine an NO-donating moiety with a known drug. An example of this is the combination of nonsteroidal antiinflammatory drugs (NSAIDs) with NO. In this case the protective effect of NO is exploited to reduce the gastrointestinal toxicity of the NSAID [101]. Inhaled NO gas has also been used clinically to treat pulmonary hypertension in both neonates and adults with adult respiratory distress syndrome (ARDS) [102] (see Chapter 17).

3.2. Nitric Oxide Synthase Inhibition

The increase in NO production in a number of disease states has prompted the search for drugs capable of modulating NO production. One major area of investigation has been the development of inhibitors of the NOS enzyme. It is apparent that in a number of diseases where NO is implicated in the mechanism of pathogenesis the induction of iNOS is the key event. On the other hand, in the early stages of both sepsis and inflammation upregulation of cNOS may exert a beneficial, protective effect. There is therefore a need for selective NOS inhibitors aimed at iNOS [46,98,103].

One of the first NOS inhibitors to be used experimentally was the arginine analog L-N^G-monomethyl-L-arginine (L-NMMA) [15]. This is a nonselective, irreversible inhibitor. A number of other arginine analogs have been used, including N^G-nitro–L-arginine (L-NNA), N^G-nitro-L-arginine methyl ester (L-NAME), N^G-amino-L-arginine (L-NAA), and N-(iminoethyl)-L-ornithine (L-NIO) (Fig. 1) [103]. None of these inhibitors shows any great selectivity for one NOS isoform over another.

More selective amino acid-based inhibitors have been synthesized including S-methyl- and S-ethylisothiocitrulline [104]. Both of these are selective for nNOS, whereas thiocitrulline itself is a potent inhibitor

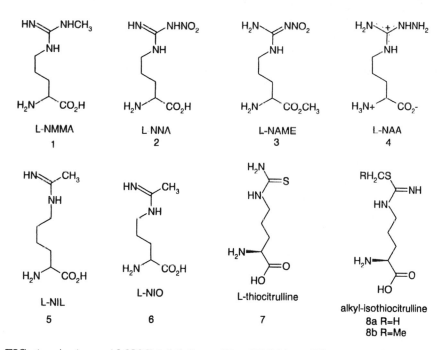

FIG. 1. Amino acid NOS inhibitors. (1) L-NMMA, L-N^G-monomethyl-L-arginine; (2) L-NNA, N^G-nitro–L-arginine; (3) L-NAME, N^G-nitro-L-arginine methyl ester; (4) L-NAA, N^G-amino-L-arginine; (5) L-NIL, L-N^G-(1-iminoethyl)lysine; (6) L-NIO, N-(iminoethyl)-L-ornithine; (7) L-thiocitrulline; (8a) S-methylisothiocitrulline; (8b) S-ethylisothiocitrulline.

of both iNOS and cNOS [105]. In order to increase selectivity of the amino acid based inhibitors a number of dipeptides of N^ω-nitroarginine and phenylalanine have been examined with selectivity for nNOS over iNOS [106]. Another selective amino acid-based inhibitor is L-N^G-(1-iminoethyl)lysine (L-NIL) [107]. L-NIL is one methylene group longer than L-NIO, which is a nonselective inhibitor. It therefore appears that subtle changes in structure can have an important effect on inhibitor selectivity.

There are three possible substrate binding sites at the NOS-active site: the guanidinium site, the heme site, and the amino acid site [98]. All of the above inhibitors are believed to compete for the arginine binding site of NOS. A number of non-amino-acid NOS inhibitors have been shown to selectively inhibit iNOS (Fig. 2). These include the guanidine derivatives aminoguanidine and methylguanidine [108], S-alkyl and S-aminoalkylisothioureas such as S-ethylisothiourea and amino-ethylisothiourea [109], and cyclic amidines such as 2-iminopiperidine [110]. The antimycotic imidazoles have been shown to inhibit NOS by binding to the heme site. Thiocitrulline, as well as potentially binding to the amino acid and guanidinium site, also seems to bind to the heme site. This mechanism appears to lower the reduction potential and inhibit electron transfer between the flavin cofactors and heme [98].

The NOS cofactors present themselves as targets for inhibitors.

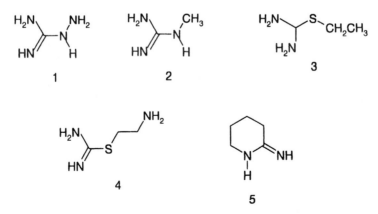

FIG. 2.   Non-amino acid NOS inhibitors. (1) Aminoguanidine; (2) methyl-guanidine; (3) S-ethylisothiourea; (4) aminoethylisothiourea; (5) iminoper-idine.

Both cNOS and nNOS are calcium dependent and can be inhibited by calmodulin antagonists such as W-7 and chlorpromazine. Tetrahydrobiopterin (BH_4) is another essential cofactor and inhibitors of the BH_4 synthetic pathway have been studied. Key enzymes in BH_4 biosynthesis that have been targeted are GTP cyclohydrolase and sepiatrin reductase. One particular inhibitor, 7-nitroindazole, binds to both the arginine and BH_4 binding sites. It has been postulated, based on this, that one method for building selectivity into an inhibitor is to target multiple domains [98].

As well as demonstrating activity in in vitro and in animal models of septic shock, the NOS inhibitor L-NMMA has been shown to raise blood pressure in the forearm capillary bed of normotensive volunteers. Furthermore, it has been shown to have possible beneficial effects on patients with septic shock [54]. This compound was the subject of a major clinical trial [49,111]. The trials were terminated (April 1998) in phase III due to lack of efficacy. This further emphasizes the need for a selective inhibitor of the inducible NOS.

3.3. Nitric Oxide Scavengers

Though inhibition of NOS is an attractive therapeutic target with many positive indications in its favor, there are a number of problems associated with the use of NOS inhibitors [98,111]. Many of these problems are associated with the dual role of NO which, as well as being a key mediator of the hypotension and tissue damage seen in septic shock, also has a protective role in several organs including liver, kidney, gut, and lung [56,112]. Nitric oxide can react with ROS such as superoxide and thus protect against ROS-induced damage. The microbiocidal activity of NO may also be an important protective factor in bacterial-induced sepsis. Nitric oxide is also important in maintaining microcirculatory homeostasis in many organs, and inhibition of NO production has been shown to exacerbate the increase in pulmonary arterial blood pressure seen in animal models of sepsis [113]. Though it is evident that L-NMMA has a positive effect on the hypotension seen in septic shock, increases in mortality have been demonstrated in numerous animal studies [111]. The use of L-NMMA has also been shown to have a negative effect on cardiac output in some animal models of sepsis. Many of these problems may be resolved by the development of specific NOS inhibitors [111] as discussed in Sec. 3.2. However, as yet, there are no inhibitors with a combination of the required specificity, necessary pharmacokinetic properties, adequate activity in in vivo disease models, or a suitable toxicological profile [98]. An alternative strategy being explored by a number of laboratories, including our own, is the use of NO scavenging molecules to remove excess NO [114,115].

The selectivity of scavengers for the NO responsible for causing pathological effects is not based on specificity for a particular enzyme, but rather on compartmental localization and rate of reaction with NO. Chemical modification of the scavenger molecule can control distribution and pharmacokinetics. A large molecule and/or a hydrophilic molecule would be unable to cross cell membranes and would therefore be restricted to extracellular compartments such as the blood and, by extravasation, interstitial fluids. The rate of NO scavenging, assuming a second-order process, would also be dependent on both the concentration of NO and the scavenger. This means that when NO concentrations are elevated, as in a number of disease states, scavenging would be promoted. This is in contrast to the NOS inhibitors that are independent of NO concentration and would therefore inhibit NO synthesis equally in regions of high and low NO synthesis.

Scavengers, in principle, provide a method of reducing excess, toxic levels of NO in compartments where NO levels have been elevated, while having a minimal effect on essential basal NO production. A variety of chemically diverse molecules are being studied as potential NO scavengers. These include organic spin-trapping agents, a range of inorganic molecules including bioinorganic ones such as vitamin B_{12} and hemoglobin, and inorganic compounds. These new opportunities and new challenges for the inorganic medicinal chemist will be discussed in the following sections.

4. ORGANIC MOLECULES AS NITRIC OXIDE SCAVENGERS

Free radicals can be detected using electron paramagnetic resonance spectroscopy (EPR), also known as electron spin resonance (ESR) [116].

Nitric oxide, though paramagnetic with an unpaired electron in a π antibonding orbital, is EPR-silent as the relaxation time of the unpaired electron is too rapid to be detected. A spin trap, a molecule that can interact with the unstable radical to produce a stable adduct, is therefore required to detect NO using this technique. Typical spin traps are hemoglobin [117], and nitroso and nitrone compounds [116]. The technique of spin trapping has been exploited in the development of NO scavengers as therapeutic drugs.

Nitronyl nitroxides will react with nitric oxide to form imino nitroxides and have been investigated as NO scavengers. The parent imidazolineoxyl N-oxide compound 2-phenyl-4,4,5,5-tetramethylimidazoline-1-oxyl 3-oxide (PTIO), and its derivatives carboxy-PTIO and carboxymethoxy-PTIO [118] are shown in Fig. 3. These compounds react with NO in a radical-radical reaction (1). The kinetics of this reaction has been studied using ESR (Fig. 4) and compared with the rates of reaction of NO with biologically occurring molecules. The resultant relative rates are as follows: hemoglobin > carboxy-PTIO > PTIO = carboxymethoxy-PTIO = L-cysteine > albumin > molecular oxygen. The reaction of NO with carboxy-PTIO is second order with a rate constant of $1.01 \times 10^4\,M^{-1}\,s^{-1}$ (Table 1). The stoichiometry for the reaction of PTIO with NO, as determined by ESR, was found to be 1:1 with the products of the reaction being nitrite and the imidazoline-N-oxyl compound 2-phenyl-4,4,5,5-tetramethylimidazoline-1-oxyl (PTI) [118], i.e.:

$$NO + PTIO \rightarrow NO_2 + PTI \tag{1}$$

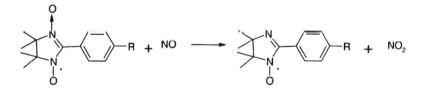

R=H: 2-phenyl-4,4,5,5-tetramethylimidazoline-1-oxyl-3-oxide (PTIO)
R=COOH: carboxy-PTIO

FIG. 3.   The reaction of 2-phenyl-4,4,5,5-tetramethylimidazoline-1-oxyl 3-oxide (PTIO) and its derivatives carboxy-PTIO and carboxymethoxy-PTIO, with nitric oxide giving the products nitrite and 2-phenyl-4,4,5,5-tetramethylimidazoline-1-oxyl (PTI).

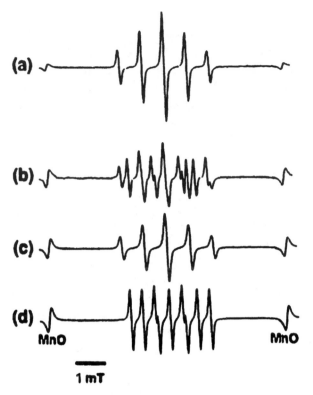

FIG. 4.   EPR spectra of PTIO: (a) PTIO in phosphate buffer, pH 7.4. (b) reaction
mixture of NO and PTIO in phosphate buffer, pH 7.4. (c) and (d) components of
(b) resolved by computer simulation identified as 2-phenyl-4,4,5,5-tetramethyl-
imidazoline-1-oxyl 3-oxide (PTIO) (c) and 2-phenyl-4,4,5,5-tetramethylimid-
azoline-1-oxyl (PTI) (d). (Reproduced with permission from [118].)

An alternative stoichiometry of 0.5:1 has been proposed [119] based on
the reaction scheme:

$$^{\bullet}NO + PTIO \rightarrow PTI + {}^{\bullet}NO_2 \qquad k = 1 \times 10^4 \ M^{-1} \ s^{-1} \qquad (2)$$

$$^{\bullet}NO + {}^{\bullet}NO_2 \rightarrow N_2O_3 \qquad k = 1.1 \times 10^9 \ M^{-1} \ s^{-1} \qquad (3)$$

$$N_2O_3 + H_2O \rightarrow 2NO_2^- + 2H^+ \qquad k = 1 \times 10^3 \ M^{-1} \ s^{-1} \qquad (4)$$

whereby one molecule of NO first reacts with PTIO (2) and a second
molecule of NO reacts with the ${}^{\bullet}NO_2$ produced in the previous reaction

TABLE 1

Comparative Reaction Rates
of Imidazolineoxyl N-Oxides with Nitric Oxide

	Rate constant ($M^{-1} s^{-1}$)
Carboxy-PTIO	1.01×10^4
PTIO	5.15×10^3
Carboxymethoxy-PTIO	5.27×10^3

PTIO, 2-phenyl-4,4,5,5,-tetramethylimidazoline-1-oxyl-3-oxide.

(3). The differences between the two sets of observations may be explained as follows. It was assumed in the first study that the reaction between NO and O_2 was (5):

$$2^\bullet NO + O_2 \to 2^\bullet NO_2 \tag{5}$$

whereas the stoichiometry of this reaction has been shown to be (6) [22–24]:

$$4^\bullet NO + O_2 + 2H_2O \to 4NO_2^- + 4H^+ \tag{6}$$

This could account for the above differences in the proposed stoichiometry of the PTIO/NO reaction.

The ability of the imidazolineoxyl N-oxides PTIO, carboxy-PTIO, and carboxymethoxy-PTIO to scavenge NO has been demonstrated in several biological systems [120]. All three compounds were able to attenuate acetylcholine-induced vasodilatation of rabbit aortic rings [118] and reverse NO-mediated hypotension in rats [121], the most potent compound being carboxy-PTIO. They have been utilized in investigations on tumor biology and therapy. For example, PTIO reduced the enhanced vascular permeability of the solid murine sarcoma S-180 [122,123]. Also, carboxy-PTIO attenuated the radiosensitizer effect of the NO donors SNP and nitrosoglutathione [89] but had no effect on blood flow in the P22 rat carcinosarcoma [72]. Carboxy-PTIO has been used to study the neurotransmitter role of NO in NANC nerves [124–126].

5.   INORGANIC MOLECULES AS SCAVENGERS

5.1.   Hemoglobin and Derivatives

The mechanism of action of NO is a good example of the importance of metal ions to the functioning of biological systems. The vasoactive properties of NO are mediated by its interaction with the heme moiety of guanylate cyclase (see Sec. 1). The chemistry of the reactions of NO with hemoglobin have been well documented [25,127,128]. Nitric oxide will react with both Fe(II) (d^6) heme and Fe(III) (d^5) heme (in hemoglobin the normal oxidation state of iron is Fe(II)). NO can react rapidly with the Fe(II) of deoxyhemoglobin to form Hb(Fe(II)NO) with a rate constant of $2-8 \times 10^7$ M^{-1} s^{-1} [25]. The resultant nitrosyl is relatively stable with a half life of the order of 3 h. In this reaction the NO acts as a two-electron donor and the remaining unpaired electron makes the nitrosyl paramagnetic, enabling its detection by EPR [25,128]. This phenomenon has been exploited to study NO production in disease states such as septic shock and graft rejection [116,117] (Fig. 5). The nitrosylhemoglobin (HbNO) can be further oxidized in the presence of oxygen to methemoglobin (Fe(III)). Nitric oxide will bind loosely to HbFe(III), reducing it slowly to Hb(Fe(II)NO). Nitric oxide will also react rapidly with oxyhemoglobin ($k = 5 \times 10^7 M^{-1} s^{-1}$) to give Fe(III) and nitrate and nitrite.

The chemistry of the reactivity of NO with hemoglobin is further complicated by the recent finding that NO also reacts with a reactive sulfhydryl group, cys93, on the hemoglobin β chain to form S-nitrosylhemoglobin (SNOHb) [129]. The rate of nitrosation was dependent on the form of the hemoglobin, with the rate of S-nitrosation being faster with oxyhemoglobin than with deoxyhemoglobin. Significant levels of the SNOHb were found in arterial blood compared with venous blood, whereas the inverse was observed for HbNO. The SNOHb was also able to trans-nitrosate thiols to produce nitrosothiols, which in turn had EDRF-like activity. Based on these data it has been suggested that hemoglobin may not only act as a scavenger for NO, but may also function as an NO donor and, in this way, be important in regulating the microcirculation.

Hemoglobin contained within red blood cells is not vasoactive [130], but numerous studies have demonstrated that cell free hemo-

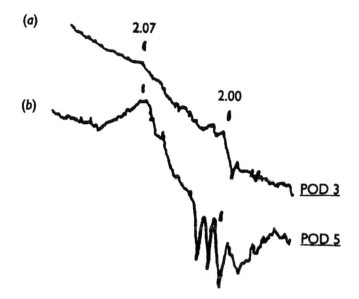

FIG. 5.   EPR spectra in blood from mice taken during allograft rejection after heart transplantation at postoperative days 3 (POD 3) and 5 (POD 5). The characteristic signal at g = 2.07 for nitrosylhemoglobin is seen at POD 5, when the first clinical signs of graft rejection appear. (Reproduced with permission from [177].)

globin has a hypertensive effect both in isolated aortic rings [131] and isolated arteries [100]. Cell free hemoglobin can also reverse the hypotension in in vivo models of septic shock [114] and reverse the endotoxin-mediated hyporesponsiveness to α-adrenergic vasoconstrictors [130]. Hemoglobin has been investigated as a potential blood substitute for transfusion [132]. Cell free hemoglobin, however, has limitations as a blood substitute as it rapidly dissociates into dimers and is cleared by the kidney, with associated nephrotoxicity. In addition, the oxygen binding capacity is very high in the absence of intracellular 2,3-glycero-phosphate found in erythrocytes, thus reducing the ability of cell free hemoglobin to deliver oxygen. In attempts to overcome these limitations of rapid clearance and high oxygen affinity, recombinant human hemoglobin and chemically modified hemoglobins have been developed as potential blood substitutes. The major chemical modification has been to crosslink the hemoglobin using diasprin [133,134] or glutaraldehyde [135] to both prevent dissociation and increase half-life. An alternative approach has been to include pyridoxal-5'-phosphate as a substitute for 2,3-glycerophosphate and to increase stability by cross-linking with a polyoxyethylene conjugate [136]. Many of these derivatives were found to have an even greater affinity for oxygen. They also had adverse vasoactive effects associated with their ability to interract with NO.

The recombinant human hemoglobin (rHb1.1) was genetically engineered with specific mutations to optimize oxygen affinity, extend intravascular residence, and prevent nephrotoxicity [137]. This was achieved by introducing a mutation in the β chain, Asn108β $\rightarrow$ Lys, to reduce oxygen affinity. The stability was increased by fusion of the two α-globin chains using an expression vector that contained two copies of the α-globin fused in tandem by a single codon encoding a glycine residue. This gave an increased in vivo half life and no apparent kidney toxicity. The recombinant hemoglobin, however, inhibited NO-mediated vasorelaxation of isolated rabbit aortic rings induced by acetylcholine and IL-1β. Concerns over drug-drug interactions led to studies on the interaction of rHb1.1 with the NO donor drugs nitroglycerin and sodium nitroprusside. It was discovered that the vasorelaxant effect of both agents was inhibited by rHb1.1 in a dose-dependent manner [131]. These data suggest limitations of recombinant human hemoglobin as a blood substitute but point to the potential use of hemoglobins as NO scavengers.

Several of the chemically modified hemoglobins have been investigated as NO scavengers. Crosslinking the α subunits of the hemoglobin tetramer with diaspirin gave a stable tetramer with a half-life of up to 30 h [133]. This modified hemoglobin has been shown to be potentially beneficial in the treatment of hemorrhagic shock and traumatic shock due to head injury [138]. It has been proposed that these effects are in part due to reaction with NO. However, use of diaspirin-crosslinked hemoglobin (DCLHb) as a blood substitute was complicated by cardiovascular effects, namely hypertension, attributed to the reaction with NO. This led to the investigation of the effect of DCLHb in sepsis. DCLHb was able to restore blood pressure in a rat model of sepsis, with no effect on cardiac output or heart rate. Regional perfusion to selected tissues was also improved [134]. Preliminary clinical studies have shown that administration of bolus doses of DCLHb to septic patients receiving vasopressor therapy led to a rapid rise in mean arterial pressure (MAP) and a reduction in catecholamine requirements. Studies with another blood substitute, Biopure 2, a glutaraldehyde crosslinked hemoglobin with an average molecular mass of 200,000, have also demonstrated a restoration of blood pressure in a rat model of septic shock. The NOS inhibitor N^G-nitro-L-arginine also restored MAP but had a deleterious effect on the already compromised renal function. Conversely, the polymerized hemoglobin improved renal function as measured by an increase in renal blood flow, glomerular filtration rate, and urinary flow [135]. In another study using a porcine model of endotoxic shock, treatment with a crosslinked hemoglobin improved MAP. However, although there was no impairment of renal blood flow, the infusion of hemoglobin significantly exacerbated the endotoxin-associated pulmonary hypertension [139].

One of the best characterized hemoglobin derivatives is the pyridoxalated hemoglobin polyoxyethylene conjugate (PHP) [136]. Whereas the hemoglobin derivatives described above have been examined in rat models of sepsis, PHP has been extensively studied in an ovine model. Rats produce significantly more NO when iNOS is stimulated by endotoxin than is found in patients with septic shock [53]. The rat model, though a good first test, is therefore not a good model of the clinical situation. A number of large animal models of septic shock, including sheep, pig, and dog, have been studied as more relevant models of human sepsis. Sepsis was induced in sheep by continuous infusion of live *Pseudomonas aeruginosa* over a 48-h period, which led to a state of hyperdynamic sepsis. PHP was administered after 24 h at doses of 50, 100, and 200 mg/kg. This was followed by an immediate increase in MAP and systemic vascular resistance (Fig. 6) [140]. PHP also reduced the amount of norepinephrine required to increase MAP, indicating a reversal of the NO-mediated hyporesponsiveness to vasoconstrictors. PHP had little effect on organ perfusion but did increase glomerular filtration rate. One concern with bacteremia-induced sepsis is that removal of NO would decrease the host response to infection. On the contrary, PHP was shown to reduce the levels of live bacteria in the ovine sepsis model [136].

One adverse effect observed with PHP was a further increase in pulmonary artery pressure over that seen in septic animals. An increase in pulmonary vascular resistance is associated with septic shock.

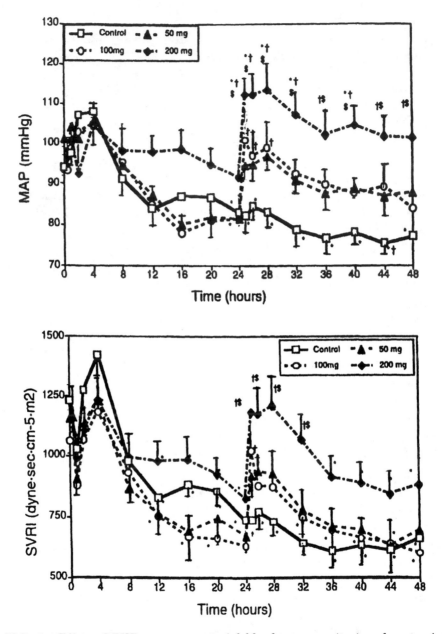

FIG. 6. Effect of PHP on mean arterial blood pressure (top) and systemic vascular resistance index (bottom) in an ovine model of sepsis. After 24 hours the sheep received either 50, 100, or 200 mg PHP as a bolus over 30 min, or vehicle. * = $p \leq 0.05$ vs. 0 h, † = $p \leq 0.05$ vs. 24 h, \$ = $p \leq 0.05$ vs. control group. (Reproduced with permission from [136].)

This is thought to be due to a loss of hypoxic pulmonary vasoconstriction, the mechanism by which the normal lung diverts blood away from poorly ventilated areas to better oxygenated parts of the lung. It has been suggested that NO may be one of the mediators of hypoxic pulmonary vasoconstriction [141]. Nitric oxide has been shown to reduce this response, and it has been hypothesised that a reduction in the level of NO may be a controlling mechanism for hypoxic pulmonary vasoconstriction. Whereas inhibitors of NOS have been shown capable of partially restoring this response, PHP was found to have no effect on the hypoxic pulmonary vasoconstriction response in the ovine sepsis model [142].

PHP has shown no adverse effects in preclinical toxicology studies and is now in clinical trials. The hypothesis on which the clinical trials are based is that normalization of hemodynamics at the systemic and microcirculatory level by NO scavenging will lead to an improvement in organ function in septic patients, and this is the proposed endpoint for the clinical trials. PHP has successfully completed Phase I clinical trials and is entering Phase II trials [114]. The preliminary data from this study indicated a restoration of systemic blood pressure, decreased vasopressor therapy, and no adverse effects on cardiac, renal, or hepatic function.

5.2. Hydroxocobalamin

Hydroxocobalamin (vitamin B_{12a}) was originally proposed as an antidote to poisoning by the cyanide released from the vasodilator sodium nitroprusside (SNP), removing the cyanide by the formation of cyanocobalamin [143]. Carbon-13 NMR studies indicated that the hydroxocobalamin was capable of reacting directly with the SNP, forming a 1:1 complex of the type -Fe-C-N-Co- [144]. Physiological studies also indicated that the hypotensive effect of SNP was attenuated by hydroxocobalamin, and it was proposed that this was due to a direct interaction with SNP [146]. More recently, it has been suggested that hydroxocobalamin can react with NO and that this may have been the cause of the inhibition of the vasorelaxant effect of SNP [145]. Hydroxocobalamin has subsequently been shown to inhibit the acetylcholine-induced relaxation of rat aortic rings and to block the neurotransmitter effect of NO in bovine retractor penis muscle [146]. Hydroxocobalamin has also been shown to reverse the hypotension in an LPS-induced rat model of endotoxic shock and to reduce mortality in LPS-treated mice [145], suggesting a possible role in the therapy of septic shock.

The chemical evidence for the interaction of NO with hydroxocobalamin (Co(III)) is limited though the reaction is theoretically plausible. A recent investigation of the chemical interaction between hydroxocobalamin and NO using both UV/visible and EPR spectroscopy has shown that NO can react with Co(II) oxidizing it to Co(III) and forming nitrite [147]. NO was also able to react with Co(III) to form a Co(III)-NO bond, though this binding was found to be both weak and reversible compared with the Fe-NO bond formed between deoxyhemoglobin (Fe(II)) and NO. Therefore hydroxocobalamin is unlikely to be as potent a scavenger of NO as hemoglobin.

The reactions in a biological environment are further complicated as the Co(III) of hydroxocobalamin can undergo partial autoreduction to Co(II) at neutral pH in oxygenated solution [147,178]. Furthermore, NO may also react in vivo with a superoxide-cobalt(III) complex. These numerous alternatives will therefore complicate any reactions between hydroxocobalamin and NO in the biological environment. A number of differences have been observed in the ability of hydroxocobalamin and hemoglobin to inhibit nitrergic transmission [148]. These observations have led to the suggestion that nitrergic transmission may be mediated by other NO species such as nitrosothiols. However, these differences could be explained by the aqueous chemistry of hydroxocobalamin and the coexistence of Co(II) and Co(III) species [147].

5.3. Metal Complexes

An alternative to the high molecular weight inorganic biomolecules, while still exploiting the properties of NO as a ligand for metals, is to use low molecular weight transition metal complexes capable of reacting with NO. Water soluble small molecules have the added advantage that they are frequently rapidly cleared from the body by urinary excretion, thus limiting their toxicity. Iron is the biological target for NO and a number of iron complexes have been studied both as spin traps for detecting NO using EPR and as potential therapeutic NO scavengers. Diethyldithiocarbamate-Fe(II) was used to demonstrate the production of NO in aqueous solution by cells and tissues. Incubating cells with diethyldithiocarbamate results in the accumulation of intracellular iron to form (DETC)$_2$Fe(II) in situ. This complex then traps NO to give the resultant mononitrosyl (Fig. 7), which has a characteristic three-line EPR spectrum that can be used to demonstrate the presence of NO. The disadvantage with DETC is that it is poorly water-soluble; therefore a number of dithiocarbamate derivatives with improved water solubility have been investigated as possible spin traps for quantifying NO release. These include sarcosine dithiocarbamate, iminodiacetic acid dithiocarbamate, di(hydroxyethyl) dithiocarbamate, and N-methyl-D-glucamine dithiocarbamate (MGD) (Fig. 7). DETCs are also known potent inhibitors of superoxide dismutase (SOD) and the MDG derivative was shown to be a weaker SOD inhibitor than DETC. The iron-MDG complex can also bind NO in both the presence and absence of oxygen. These properties, plus increased water solubility, led to this derivative being selected for in vivo studies [149].

The NO trapping efficiency of this complex was demonstrated using the NO donor PAPA-NONOate (see Sec. 3.1). The iron-MDG complex has subsequently been successfully used to monitor NO production by LPS-activated peritoneal macrophages in vitro over a period of 6 h post activation [150]. Furthermore, the ability of the Fe-MDG to trap

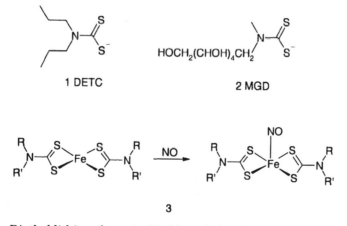

FIG. 7. Diethyldithiocarbamate (1), N-methyl-D-glucamine dithiocarbamate (MGD) (2), and the reaction of the reaction of iron(II) diethyldithiocarbamates with nitric oxide (3).

NO in vivo was demonstrated by intravenous administration of the iron complex simultaneously with the NO donor SNP. The characteristic three-line EPR signal was detected in the blood of the tail vein by immobilizing the tail vein inside the resonator and the time course of the generation and decay of the EPR signal monitored (Fig. 8). This technique therefore provided a method of noninvasively monitoring NO generation in real time in an animal. An additional interesting observation was that, as would be expected, SNP alone led to a drop in MAP. This was partially reversed on administering the iron spin trap [151].

Nitric oxide has been detected in a model of LPS-induced septic shock in mice using this technique [152,153]. The route of administration of the iron spin trap dramatically influenced the trapping of NO. An EPR signal could be detected 6 h after LPS administration when the Fe-MDG was given subcutaneously; however, when the Fe-MDG was given intravenously no signal was observed in the tail vein. The Fe-NO adduct could be detected in liver and kidney when the spin trap was administered subcutaneously or intravenously. These differences may be due to the pharmacokinetic behavior of the spin trap when given by different routes [153]. The NO detected by the Fe-MDG was shown to come from L-arginine, the substrate for NO synthase, by using ^{15}N-arginine [152]. In addition, NMMA reduced the intensity of the EPR signal in the mice with septic shock. The Fe-MDG complex was also able to prevent acetyl choline induced vasorelaxation in rabbit aortic rings [154]. Compounds of this class have been shown to be able to prolong survival in a model of allograft rejection and to have a beneficial effect in a model of ischemic stroke [155]. These compounds are now being developed as nitric oxide scavengers in NO mediated disease.

A range of metal chelators and chelates were screened in a murine model of septic shock induced by a combination of bacteria (*Corynebacterium parvum* and *Escherichia coli*) and LPS [156]. Two iron(III) complexes, iron(III)-diethylenetriaminepentaacetic acid (DTPA), and a complex of the naturally occurring siderophore ferrioxamine B (HDFB) (Fig. 9), were found to be the most effective, giving the highest decrease in mortality in this model. The DTPA complex was more effective than the HDFB complex, though in a comparison of the free ligands the reverse was true with the siderophore being more effective than DTPA. Electrochemical studies employing cyclic voltammetry showed that the two complexes had different mechanisms of action. The Fe(III)(DTPA)

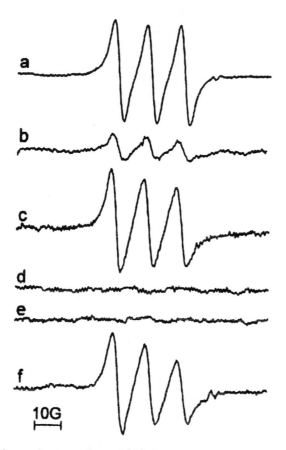

FIG. 8. In vivo spin trapping of NO in mice using an iron-N-methyl-D-glucamine dithiocarbamate complex. Characteristic three-line EPR spectrum of $(MGD)_2Fe(II)NO$ obtained after reaction with NO in Tris buffer, pH 7.4 (a). EPR spectra taken of the circulation through a mouse tail vein (b–e); after injection of $(MDG)_2Fe(II)$ and 5 mg/kg sodium nitroprusside (b), and 20 mg/kg sodium nitroprusside (c); 20 mg/kg sodium nitroprusside alone (d); $(MDG)_2Fe(II)$ alone (e). EPR spectrum of a blood sample taken 20 min after injection of $(MDG)_2Fe(II)$ and 20 mg/kg sodium nitroprusside (f). (Reproduced with permission from [151].)

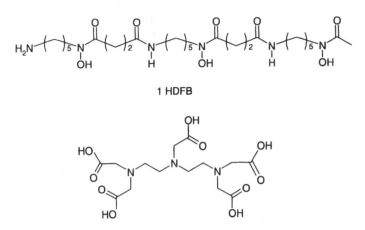

1 HDFB

FIG. 9.   Structure of the ligands ferrioxamine B (HDFB) (1) and diethylenetri-aminepentaacetic acid (DTPA) (2).

complex was reduced to the Fe(II) complex, which then bound NO with a 1:1 stoichiometry. A similar compound, Fe(III)(EDTA), has also been shown capable of reacting with NO to give Fe(II)(EDTA)NO. A postulated mechanism for this reaction is first a partial ring opening of the EDTA ligand, followed by coordination of a labile water molecule, which is then substituted by the incoming NO [157]. In contrast, the Fe(III)HDFB complex is stable and appears to act as an efficient electrocatalyst for the conversion of NO to N_2O. Though both compounds exerted a protective effect in the septic shock model, the Fe(III)(DTPA) complex was more effective than the HDFB complex, even though the latter acted via a catalytic mechanism. This may be due to the different potentials at which the iron center is reduced, with the polyaminocarboxylate complex being reduced at a lower and more biologically relevant potential [156].

The Fe(III)(DTPA) complex was subsequently shown to be protective in two acute models of sepsis, in mice and baboons, in which sepsis was modeled by injection of a lethal dose of live *E. coli* [158]. The course of septic shock in the baboon has been reported to be similar to that in humans. Hypotension developed 2 h after onset of bacterial infusion, which was not reversed by administration of Fe(III)(DTPA). There was, however, a decrease in organ damage and a reduction in nitrite/nitrate levels compared with untreated animals. The mechanism of protection is not clear and there is evidence to suggest that, as well as removing NO, the iron complex may react with ROS such as hydroxyl radical and superoxide. Preliminary pharmacology and toxicology studies suggest that this compound has a suitable safety and pharmacokinetic profile and may be a candidate for clinical trials against septic shock.

The previous examples utilize the properties of physiologically active metals, predominantly iron. In our own laboratory we have exploited the properties of another transition metal, ruthenium [115].

6. RUTHENIUM COMPLEXES

6.1. Drug Design

Several factors have to be taken into consideration when designing new drugs. First the compound must have the desired pharmacological activity, i.e., it must possess an appropriate pharmacophore capable of recognizing and interacting with the molecular target. It must also have the correct physicochemical properties for appropriate biodistribution and pharmacokinetics (see Sec. 3.4). Finally, it must also have low toxicity so that it has a wide therapeutic window, the difference between the concentration of compound required for optimal therapeutic activity and the toxic dose. Pharmacokinetics and toxicology are interrelated and are both dependent on physicochemical properties such as solubility and chemical reactivity. The chemical properties of the transition metals are well suited to drug design [159,160]. They have variable oxidation states, as well as the ability to coordinate ligands in a precise three-dimensional configuration and to undergo ligand exchange reactions. These factors, together with the diversity of opportunities to chemically modify ligands and thus control chemical properties, are well suited for the design of molecules with the capability to interact with biological target molecules. In spite of these opportunities for drug design, the majority of successful metal-based pharmaceuticals have been discovered by chance rather than design. We have exploited the properties of NO as a ligand for metals to design scavenger molecules.

The coordination chemistry of NO is similar to that of carbon monoxide, as both are π-acceptor ligands. The existence of metal nitrosyl complexes has been known for over a century, but they have not been as extensively studied as their carbonyl counterparts. This is because NO is a stronger π acid than CO and therefore forms more stable complexes, which are generally less reactive than the corresponding carbonyl derivative [160].

The chemistry of NO is determined by the π^* electron. This electron can be lost to form the nitrosonium ion, NO^+, to give linear, terminal, M-N-O, complexes. Alternatively, NO can form metal-nitrosyl complexes with a bent, terminal, M-N-O bond, having bond angles of 120–140°. In these cases the ligand is deemed to be NO^-. This, however, is an oversimplification of the situation, as the bonding of NO to a metal center involves a synergic interaction between the NO ligand orbitals and the metal orbitals. This simplified description does, however, provide us with a useful way of discussing the reactivity of NO with metals. This reactivity is the basis for the mechanism of action of both NO as a biological signaling molecule (via its interaction with the heme iron of guanylate cyclase) [25,26,161], and of the inorganic NO scavengers described in the previous section.

Ruthenium readily forms nitrosyl complexes, and there are more known nitrosyl complexes of ruthenium than any other metal [162]. The existence of ruthenium nitrosyl complexes has been known for well over 100 years; $K_2[Ru(NO)Cl_5]$, for example, was isolated in the mid-nineteenth century. The Ru-NO bond is generally very stable and NO as a ligand is kinetically inert and not easily displaced, persisting through a variety of substitution and redox reactions. Ruthenium(III) (d^5) will react with NO to form six-coordinate mononitrosyl complexes with a linear Ru-N-O group. The Ru(III) is reduced to Ru(II) with the ligand being formally described as NO^+ [162]. Ruthenium is therefore an ideal metal center around which to develop scavengers for NO.

The ideal requirements for an NO scavenger include a ligand set that is itself stable but that promotes specific binding of NO, and confers suitable physicochemical properties on the compound, thus giving it the desired pharmacokinetic and toxicological profile. Ruthenium has a rich coordination chemistry providing a variety of complexes with ample opportunity for chemical modification [163]. Generally, ligands around ruthenium(III) are kinetically inert and have the additional benefit that the formation of the Ru-NO bond further stabilizes the trans ligand in the resultant Ru(II) complex. A strategy has been adopted whereby the metal ion is chelated with suitable ligands that confer water solubility, rapid in vivo clearance and low toxicity while providing an available binding site for NO with an overall high affinity and rapid rate of NO binding (see Sec. 3.4). We have screened a number of ruthenium(III) complexes and found that polyaminocarboxylates satisfy most of these requirements as ligands. The ethylenediaminetetraacetic acid (EDTA) complex of ruthenium(III), K[Ru(HEDTA)Cl] (AMD1226), has proven to be a good model compound to investigate the pharmacological properties of Ru(III) polyaminocarboxylates.

6.2. Ruthenium(III) Polyaminocarboxylates and Reaction with Nitric Oxide

The EDTA ligand in the Ru(III) complex K[Ru(HEDTA)Cl] is pentadentate, leaving one free coordination site available for substitution (Fig. 10) [164]. The ability of AMD1226 to bind NO was demonstrated by the introduction of a known volume of NO gas (3 cm^3, 1.3×10^{-4} moles) into the headspace above a stirred, aqueous solution of K[Ru(HEDTA)Cl] (25 cm^3, 1×10^4 moles) in a closed apparatus under argon. Absorption of the NO gas was measured using a manometer, and after complete absorption the reaction mixture was freeze-dried and the product examined by infra red spectroscopy. The IR spectrum of the reaction product had an absorbance peak at 1897 cm^{-1}, characteristic of a linear Ru-NO bond, confirming the formation of a ruthenium(II) mononitrosyl (Fig. 11). An identical result was obtained when AMD1226 was allowed to react with an equimolar concentration of the NO donor compound S-nitroso-N-acetylpenicillamine (SNAP) [165].

The chloro complex K[Ru(HEDTA)Cl] is immediately aquated on dissolution to give the neutral complex Ru(HEDTA)H$_2$O (AMD6245) [166]. The rapid aquation of the chloro complex was confirmed by comparing the UV-visible absorption spectra for the two complexes, which were identical [167]. On addition of an aqueous solution of NO a change was observed in the near-UV region, leading to a loss of absorbance at 290 nm and an increase in absorbance below 270 nm with an isosbestic point at 272 nm (Fig. 12). The absorbance change at 290 nm was utilized to determine the stoichiometry of the reaction. A plot of fractional saturation versus NO concentration indicated formation of a 1:1 ad-

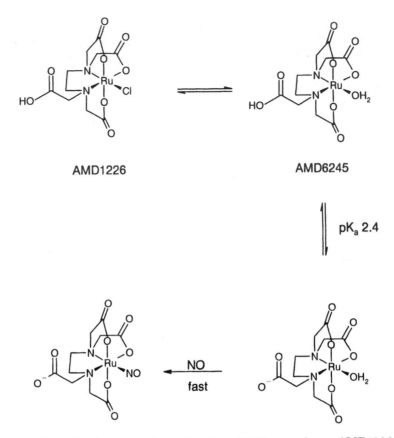

FIG. 10. Reaction between the ruthenium-EDTA complexes AMD1226 and
AMD6245 and NO in aqueous solution.

duct between AMD1226 and NO, and an estimated binding affinity of
more than 10^8 M^{-1}, consistent with the formation of a strong Ru-NO
bond [168].

The reaction kinetics of both the chloro complex (AMD1226) and
the aqua complex (AMD6245) were studied using stopped-flow spectrometry by monitoring the decrease in absorbance at 290 nm on formation of the ruthenium(II) nitrosyl. The binding was very rapid and occurred within the dead time of the instrument at room temperature. A
second-order rate constant of 2×10^7 M^{-1} s^{-1} was obtained at pH of
7.4 by performing the experiments at 7°C (Table 2). Based on these data

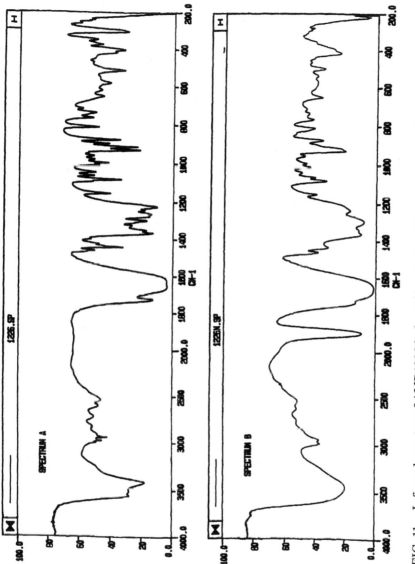

FIG. 11. Infrared spectra of AMD1226 before (A) and after (B) reaction with NO. The product has a sharp absorbance peak at 1897 cm^{-1} characteristic of a linear Ru(II)-NO adduct. Reaction of AMD 1226 with the NO donor compound SNAP gave a product with a similar spectrum. (Reproduced with permission from [165].)

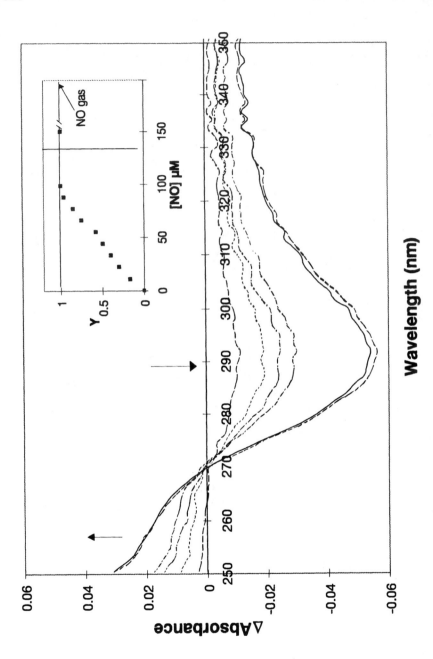

TABLE 2

Calculated Second-Order Rate Constants
from Stopped-Flow Experiments
for the Reaction of Ru(III)EDTA
Complexes with NO[a]

Compound	pH	T (°C)	$10^{-7}\, k$ ($M^{-1}\, s^{-1}$)
AMD1226	8	7.7	3.33 ± 0.17
AMD1226	7.4	7.3	2.24 ± 0.05
AMD1226	6.5	7.4	1.39 ± 0.01
AMD6245	8	7.2	3.29 ± 0.14
AMD6245	7.4	7.2	1.95 ± 0.06
AMD6245	6.5	7.5	2.18 ± 1.06

[a]Errors represent standard deviation, $n = 6$.

the rate constant would be estimated to be of the order of 10^8 $M^{-1} s^{-1}$ at physiological temperature. Similar results were obtained with both AMD6245 and AMD1226, again confirming the equivalence of these two compounds in aqueous solution [167].

The rapid rate of substitution of Ru(HEDTA)H$_2$O has been extensively studied. Studies of the solution chemistry of this complex indicate that there are two pK_a values, one at pH 2.4 due to deprotonation of the pendant carboxylate to give [Ru(EDTA)H$_2$O]$^-$, and one at pH 7.6 due to deprotonation of the water molecule to give [Ru(EDTA)OH]$^{2-}$. The rapid rate of substitution has been shown to be dependent on the presence of the deprotonated pendant carboxylate. Replacement of the EDTA ligand with either N-(hydoxyethyl)ethylenediaminetriacctate (HEDTRA) or N-methylethylenediaminetriacetate (MEDTRA) results

FIG. 12.   Difference spectrum of AMD1226 in phosphate buffer, pH 7.4 titrated against an aqueous solution of NO. The inset is the binding curve of AMD1226 (100 μM) and NO determined from the absorbance change at 290 nm. The fractional saturation, Y, is defined as $(A - A_0)/A_\infty - A_0)$, where A_0 and A_∞ are absorbances in the absence and presence of saturating NO, and A is the absorbance after addition of a subsaturating concentration of NO. The data indicate the formation of a 1:1 Ru-NO complex. (Reproduced with permission from [167].)

in a decrease in the rate Ru(HEDTRA) > Ru(MEDTRA) [169]. The lability of the water in the Ru(III) complex and of substitution reactions, where the relative rates of substitution are Ru(EDTA) $\gg$ the rapid rate of substitution, has been postulated to be due to hydrogen bonding between the pendant carboxylate and the coordinated water molecule. This could result in the creation of an open area and accessible site for associative substitution, and possible partial labilization of the water molecule [166]. The reaction mechanism for the slower substitution of the equivalent ruthenium(II) complex $[Ru(II)(EDTA)H_2O]^{2-}$ is dissociative [164]. This is consistent with the known substitution behavior of Ru(III) (d^5) compared with Ru(II) (d^6). Similar observations have been reported for the dimeric ruthenium(II) polyaminocarboxylate complex $[Ru_2(TTHA)(H_2O)]^{2-}$ (where TTHA is triethylenetetraminehexaacetate), which has been demonstrated to react with NO with a rate constant of 22.7 M^{-1} s^{-1} [170].

6.3. Scavenging NO in In Vitro Biological Systems

The ability of ruthenium-EDTA complexes to scavenge NO in biological systems was first investigated using a murine macrophage cell line, RAW264 [165]. This cell line possesses many of the characteristics of normal macrophages, such as the ability to pinocytose neutral red, phagocytose zymosan, secrete lysozyme, lyse antibody-coated sheep red blood cells, and kill tumor target cells. These cells will also express inducible NOS, and synthesize NO, upon stimulation with IFN-γ and LPS. Nitric oxide production can be easily assessed by measuring nitrite, the oxidation product of NO in an aqueous environment, in the cell culture supernatant using the Griess reaction. The assay involves a diazotization between nitrite and sulfanilamide to form a diazonium salt. This is followed by azo coupling with naphthylethylenediamine dihydrochloride. The absorbance of the azo product can be measured at 540 nm. Both AMD1226 and AMD6245 reduced nitrite accumulation by activated macrophages in culture, comparable with the addition of an NOS inhibitor, L-NMMA (Fig. 13a).

Cell culture studies were extended to examine the effect of the ruthenium complexes on an NO mediated response. Macrophages play a key role in the immune response, and once activated they are cytotoxic and cytostatic toward a variety of pathogens. Nitric oxide has been shown to be one of the mediators of murine macrophage cytotoxicity

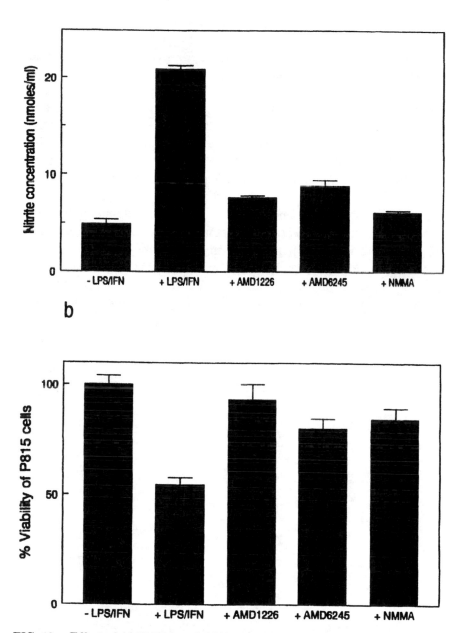

FIG. 13. Effect of AMD1226 and AMD6245 on (a) nitrite production by LPS/
IFN-γ-stimulated RAW264 macrophages and (b) NO mediated cell killing of
the P815 target cells by LPS/IFN-γ-stimulated RAW264 macrophages. (Reproduced with permission from [165].)

toward tumor cells in vitro. The RAW264 macrophages were cocultured with the murine mastocytoma cell line P815. The latter is a non-adherent cell line and can be separated from the adherent RAW264 cells. Using the MTT assay for cell viability it was shown that LPS- and IFN-γ-activated macrophages could kill P815 target cells. The primary effector molecule was shown to be NO; this therefore represents a useful cell culture model of an NO-mediated effect [171]. Both K[Ru-(HEDTA)Cl] and Ru(HEDTA)H$_2$O were able to protect the P815 cells against NO mediated cell killing by activated RAW264 macrophages (Fig. 13b). Similar results were obtained in control experiments with L-NMMA.

The effect of the ruthenium NO scavengers on NO-mediated vaso-dilatation was investigated using an isolated rat tail artery preparation [165]. The arteries were perfused with Krebs solution and test com-pounds either added as a bolus injection immediately upstream of the vessel or introduced into the internal perfusate. Perfusion pressure was monitored by a transducer placed upstream of the artery. This system has been used to investigate the effect of NO donor compounds on the

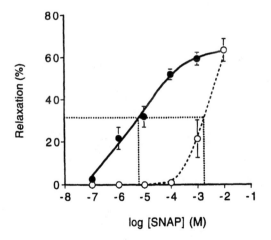

FIG. 14.   Vasodilator response of phenylephrine precontracted isolated rat tail arteries elicited by bolus injections of the NO donor compound SNAP. Blood vessels were perfused either with Krebs solution (●) or with Krebs solution containing 100 μM AMD6245 (○). Addition of AMD6245 strongly attenuated the response to SNAP. (Reproduced with permission from [165].)

vasodilator response to NO [100,172]. Arteries precontracted with phenylephrine were shown to respond to bolus doses of the NO donor compound SNAP in a dose-dependent manner. The vasodilator response to SNAP could be attenuated by addition of either AMD1226 or AMD6245 to the artery internal perfusate (Fig. 14). The ED_{50} for SNAP in control studies was 6.0 μM but was increased to 1.8 mM in the presence of 100 μM AMD6245, a 300-fold increase. A similar effect could be demonstrated in control studies with 5 μM oxyhemoglobin which increased the ED_{50} for SNAP to 200 μM.

Isolated arteries were also taken from LPS-pretreated rats. These arteries were shown to have a decreased responsiveness to the vasoconstrictor phenylephrine compared with those from control animals. The response to the vasoconstrictor could be restored with L-NMMA, which is consistent with reported data. The responsiveness to phenylephrine could also be restored by addition of either AMD1226 or AMD6245 (Fig. 15).

6.4. Pharmacological Activity in In Vivo Disease Models

Having demonstrated the ability for ruthenium(III) polyaminocarboxylates to scavenge NO in in vitro biological systems, the next stage in the process was to look for pharmacological activity in in vivo models of disease. Rodent models of septic shock have been used extensively to study NOS inhibitors and to demonstrate the involvement of NO in the hypotension associated with endotoxic shock. The majority of these models have used surgically operated, anesthetized animals. Because anesthesia causes hemodynamic changes, we have employed a model of endotoxemia using conscious rats [165].

The time course of the development of endotoxemia was monitored in male Wistar rats given a single bolus injection of LPS. Blood pressure of conscious rats was monitored using a tail cuff apparatus before and after LPS treatment. Nitrite and nitrate levels in plasma were measured using the Griess assay, and NOS expression was analyzed by Western blot in lung, heart, liver, and tail arteries. Finally, the hyporesponsiveness of isolated rat tail arteries to vasoconstrictors was monitored over time. The mean arterial pressure of the LPS-treated animals decreased by 28.5% over 20 h and then gradually returned to

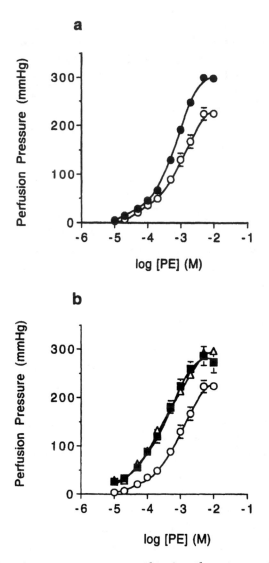

FIG. 15. (a) Log-dose response curves showing the vasoconstrictor effect of bolus injections of phenylephrine on arteries isolated from rats injected with LPS 24 h previously (○), compared to vessels form animals injected with sterile saline only (●). (b) Log-dose response curves to phenylephrine from rats pretreated with LPS 24 h previously (○), and after perfusion of the arteries with either AMD1226 (■) or L-NMMA (△). (Reproduced with permission from [165].)

normal over the next 24–28 h. The hypotension was accompanied by an increase in nitrite and nitrate levels. These levels reached a maximum at 24 h and then returned to pretreatment levels by 48 h. Western immunoblots of tail artery homogenates revealed time-dependent changes in iNOS expression, preceding the increase in NO_x, reaching a maximum at 12 h and then decreasing to below the detection limit by 72 h. The hyporesponsiveness of the arteries to phenylephrine followed the hypotension and NO_x levels reaching a maximum at 24 h and then returning to control levels. These data thus established this system as a valid paradigm for testing the ability of the ruthenium NO scavengers in vivo.

Endotoxic rats were treated with AMD1226 at the nadir of the hypotension, 20 h after LPS injection. In contrast to the control animals, there was a rapid recovery following administration of AMD1226 manifested by an accelerated rise in blood pressure which returned to normal 9 h after drug treatment (Fig. 16).

The results from the rodent model demonstrated the ability of the ruthenium(III)-EDTA complex to reverse NO-mediated hypotension. However, rats are known to produce far more NO after challenge with endotoxin than that seen in human disease. In order to investigate the potential efficacy of the ruthenium NO scavengers in a more clinically relevant model a porcine model of endotoxemia was used. Endotoxemia in the pig model mimics several of the manifestations of septic shock and adult respiratory distress syndrome in humans. These include profound decreases in mean arterial pressure and systemic vascular resistance index, and severe pulmonary hypertension. The pig can also be resuscitated by fluid infusion, restoring cardiac index, another feature of the clinical situation where aggressive fluid resuscitation is part of the treatment protocol [139,173,174]. The porcine model therefore provides a useful analogy with the clinical setting, both for the hypotension seen in septic shock and for the associated acute lung injury.

Male, random-bred Yorkshire swine were anesthetized, surgically instrumented, and given LPS by infusion over a period of 1 h. At this point animals were treated with a single-bolus injection of AMD6245 (5 mg/kg). Hemodynamic parameters, arterial and venous blood gases, and pulmonary compliances were followed for the time course of the study up to 3 h post drug administration. Treatment with AMD6245 restored MAP to control levels but had no effect on pulmonary arterial

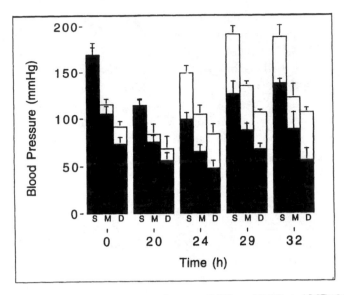

FIG. 16.   Hemodynamic responses of rats to LPS and LPS + AMD1226. Blood pressure measurements are shown for rats immediately before (t = 0) and 20 h after a single bolus intraperitoneal injection of LPS. At t = 20 h one group received saline (□) and the other group 100 mg/kg AMD1226 (■). Blood pressures recovered more rapidly in the AMD1226-treated group. S = systolic arterial pressure, D = diastolic arterial pressure, M = mean arterial pressure. (Reproduced with permission from [165].)

pressure. However, the ruthenium compound did improve both pulmonary compliance and pulmonary shunt as measured by intrapulmonary venous admixture (Table 3). This improvement in lung function is indicative of protection against endotoxin-associated acute lung injury [175].

Though hypotension is an obvious symptom of septic shock, restoration of blood pressure is no guarantee of survival. Nitric oxide synthase inhibitors have been shown to reverse the hypotension associated with septic shock but have demonstrated little or no improvement in survival in animal models. The most deleterious pathological effect of shock is the tissue damage, which ultimately results in mortality. This is probably mediated in part by the interaction of NO with ROS such as superoxide. The multiple organ dysfunction seen in sepsis is possibly associated with a tissue oxygen deficit, despite an observed increase in mixed venous oxygen saturation. This is due to a reduction in oxygen

TABLE 3

Selected Physiologic Parameters Recorded 3.5 h
after Administration of Saline, LPS alone
or LPS with AMD6245 to Anesthesized Pigs

Parameter	Saline control	LPS control	LPS ± AMD6245
MAP	106 ± 10.3	88 ± 26.2	108 ± 18.8*
MPAP	13 ± 2.4*	35 ± 7.8	38 ± 6.2*
PaO_2	244 ± 27*	100 ± 71.7	121 ± 55.5
$\%\Delta C_{dyn}$	3.9 ± 8.8*	−38.4 ± 16.7	−26.7 ± 5.9*
Urine	12 ± 5.4	4 ± 5.3	13 ± 19.5
Q_s/Q_t	1.5 ± 5.8*	35.9 ± 28.8	17.4 ± 19.4*

MAP, mean arterial pressure (mm Hg); MPAP, mean pulmonary
arterial pressure (mm Hg); PaO_2, partial pressure of oxygen in arterial blood (mm Hg); $\%\Delta C_{dyn}$, % change in pulmonary compliance;
urine, urine flow (mL/30 min); Q_s/Q_t, intrapulmonary venous admixture (%), errors represent standard deviation, $*p < 0.05$ vs. LPS
group using Newman-Keuls test.

utilization. These apparently contradictory features have been attributed to the shunting of oxygenated blood away from capillary beds
[176], as is the case in hypoxic pulmonary vasoconstriction. Nitric oxide
may be a key component in this mechanism [141]. Though NOS inhibitors and crosslinked hemoglobin can reverse the hypotension associated with septic shock, they exacerbated the acute lung injury in this
model. Seen in this context the ability of AMD6245 to ameliorate the
LPS-induced lung injury, as evidenced by an improvement in compliance and pulmonary shunt, is potentially significant.

These results show that the ruthenium-EDTA complexes AMD1226
K[Ru(HEDTA)Cl)] and AMD6245 Ru(HEDTA)H_2O are effective scavengers of NO in a variety of biological test systems of increasing pathological complexity. This class of compounds, the ruthenium(III) polyaminocarboxylates, represents an alternative therapeutic strategy to
NOS inhibitors for the treatment of NO-mediated disease.

7. CONCLUSIONS

The awareness of the significance of nitric oxide as a biological mediator in physiological systems as diverse as cardiovascular control,
neurotransmission, and the immune response, and its involvement
in the pathophysiology of numerous diseases, has provided the impetus for pharmacological research into nitric oxide as a therapeutic
target. The need for new methods of delivering NO in a controlled
fashion in cardiovascular and pulmonary diseases has led to the search
for new NO donor molecules and investigation into the use of inhaled
NO gas.

Increased NO production is now known to be a key component of
the autoimmune and inflammatory responses in diseases such as rheumatoid arthritis and diabetes. It is responsible for the hypotension in
septic shock and possibly other forms of circulatory shock such as
hemorrhagic shock. There are two main strategies for therapeutic intervention in diseases associated with overproduction of NO: the development of selective enzyme inhibitors of iNOS and the introduction
of molecules that scavenge the excess NO. While the former has been
the approach of organic medicinal chemistry, the role of NO as a ligand
for metals has provided a new opportunity for inorganic medicinal
chemistry. The reactivity of NO toward the iron porphyrin center of
hemoglobin has led to the development of hemoglobin derivatives as
therapeutic NO scavengers, notably the pyridoxalated hemoglobin
polyoxyethylene conjugate. An alternative approach has been the synthesis of NO reactive inorganic complexes such as iron complexes of
N-methyl-D-glucamine dithiocarbamate or diethylenetriaminepentaacetic acid.

An alternative metal to iron is ruthenium. There are more known
nitrosyl complexes of ruthenium than any other metal, and in general
the Ru-NO bond is kinetically stable. The ruthenium polyaminocarboxylate complexes K[Ru(HEDTA)Cl] (AMD1226) and Ru(HEDTA)H_2O
(AMD6245) can bind NO both tightly and rapidly. These complexes are
also able to scavenge NO in a variety of disease models of increasing
biological complexity. These data indicate a potential role for this class
of compound in the therapy of NO-mediated disease.

ACKNOWLEDGMENTS

The author thanks Drs. B. R. Cameron, N. Davies, F. W. Flitney, and E. Slade for assistance with figures; Dr. M. Mitchell for reading the manuscript; and G. Kreye for assistance in preparation of the manuscript. The author also thanks Drs. L. Molina (Molichem Medicines), J. DeAngelo (Apex Bioscience), C.-S. Lai (Medinox), and M. S. Miller (Louisiana State University) for information and reprints.

ABBREVIATIONS

ADP	adenosine 5′-diphosphate
ARDS	adult respiratory distress syndrome
bFGF	basic fibroblast growth factor
BH_4	tetrahydrobiopterin
cGMP	cyclic guanosine monophosphate
DCLHb	diasprin-crosslinked hemoglobin
DETC	diethyldithiocarbamate
DTPA	diethylenetriaminepentaacetic acid
ED_{50}	concentration of the NO donor SNAP giving a 50% vasorelaxant effect in the isolated rat tail artery preparation
EDRF	endothelium-derived relaxing factor
EDTA	ethylenediamine-N,N,N',N'-tetraacetic acid
EPR	electron paramagnetic resonance
ESR	electron spin resonance
Fe(III)HDFB	iron(III)-ferrioxamine
GTP	guanosine 5′-triphosphate
Hb	hemoglobin
HbNO	nitrosylhemoglobin
HDFB	ferrioxamine B
HEDTRA	N-(hydroxyethyl)ethylenediaminetriacetate
IFN	interferon
IL	interleukin
L-NAA	N^G-amino-L-arginine

L-NAME	N^G-nitro-L-arginine methyl ester
L-NIL	L-N^G-(1-iminoethyl)lysine
L-NIO	N-(iminoethyl)-L-ornithine
L-NMMA	L-N^G-monomethyl-L-arginine
L-NNA	N^G-nitro-L-arginine
LDL	low-density lipoprotein
LPS	lipopolysaccharide
MAP	mean arterial pressure
MEDTRA	N-methylethylenediaminetriacetate
MGD	N-methyl-D-glucamine dithiocarbamate
MTT	3-[4,5-dimethylthiazol-2-yl]-2,5-diphenyltetrazolium bromide
NADPH	nicotinamide adenine dinucleotide phosphate (reduced)
NANC	nonadrenergic, noncholinergic nerves
NMDA	N-methyl-D-aspartate
NOS	nitric oxide synthase (e, endothelial; i, inducible; n, neuronal)
NSAID	nonsteroidal antiinflammatory drug
PAF	platelet-activating factor
PAPA/NO	(Z)-1-[N-(3-aminopropyl)-N-(n-propyl)amino]diazen-1-ium-1,2-diolate, 1-hydroxy-2-oxo-(3-aminopropyl)-3-propyl-1-triazene
PARS	poly-ADP ribosyltransferase
PHP	pyridoxalated hemoglobin polyoxyethylene conjugate
PTI	2-phenyl-4,4,5,5-tetramethylimidazoline-1-oxyl
PTIO	2-phenyl-4,4,5,5-tetramethylimidazoline-1-oxyl-3-oxide
rHb1.1	recombinant human hemoglobin
ROS	reactive oxygen species
SNAP	S-nitroso-N-acetylpenicillamine
SNOHb	S-nitrosohemoglobin
SNP	sodium nitroprusside
SOD	superoxide dismutase
SVRI	systemic vascular resistance index
TNF	tumor necrosis factor
TTHA	triethylenetetraminehexaacetate
VEGF	vascular endothelial growth factor

REFERENCES

1. S. Moncada, R. M. J. Palmer, and E. A. Higgs, *Pharmacol. Rev.*, *43*, 109–142 (1991).

2. C. Nathan, *FASEB J.*, *6*, 3051–3064 (1993).

3. R. F. Furchott and J. V. Zawadzki, *Nature*, *288*, 373–376 (1980).

4. L. J. Ignarro, G. M. Buga, K. S. Wood, R. E. Byrns, and G. Chaudhuri, *Proc. Natl. Acad. Sci. USA*, *84*, 9265–9269 (1987).

5. R. M. J. Palmer, A. G. Ferrige, and S. Moncada, *Nature*, *327*, 524–526 (1987).

6. J. Garthwaite, S. L. Charles, and R. Chess-Williams, *Nature*, *336*, 385–388 (1988).

7. D. S. Bredt and S. H. Snyder, *Proc. Natl. Acad. Sci. USA*, *86*, 9030–9033 (1989).

8. T. M. Dawson, V. L. Dawson, and S. H. Snyder, *Ann. Neurol.*, *32*, 297–311 (1992).

9. J. S. Gillespie, X. Liu, X., and W. Martin, in *Nitric Oxide from L-Arginine: A Bioregulatory System* (S. Moncada and E. A. Higgs, eds.), Elsevier, Amsterdam, 1990, p. 147 ff.

10. M. A. Marletta, P. S. Yoon, R. Iyengar, C. D. Leaf, and J. S. Wishnok, *Biochemistry*, *27*, 8706–8711 (1988).

11. J. B. Hibbs, Jr., Z. Vavrin, and R. R. Taintor, *J. Immunol.*, *138*, 550–565 (1987).

12. J. B. Hibbs, Jr., R. R. Taintor, Z. Vavrin, and E. M. Rachlin, *Biochem. Biophys. Res. Commun.*, *157*, 87–94 (1988).

13. D. J. Stuehr, and C. F. Nathan, *J. Exp. Med.*, *169*, 1543–1555 (1989).

14. M. A. Marletta, *J. Biol. Chem.*, *268*, 12231–12234 (1993).

15. R. G. Knowles and S. Moncada, *Biochem. J.*, *298*, 249–258 (1994).

16. U. Forstermann, I. Gath, P. Schwarz, E. I. Closs, and H. Kleinert, *Biochem. Pharmacol.*, *50*, 1321–1332, (1995).

17. D. S. Bredt and S. H. Snyder, *Proc. Natl. Acad. Sci. USA*, *87*, 682–685 (1990).

18. J. S. Pollack, M. Nakane, L. K. Buttery, A. Martinez, D. Springall, J. M. Polak, and F. Murad, *Am. J. Physiol.*, *265*, C1379–C1387 (1993).

19. J. M. Hevel, K. A. White, and M. A. Marletta, *J. Biol. Chem.*, *266*, 22789–22791 (1991).

20. D. J. Stuehr, H. J. Cho, N. S. Kwon, M. F. Weise, and C. F. Nathan, *Proc. Natl. Acad. Sci. USA*, *88*, 7773–7777 (1991).

21. A. R. Butler and D. L. H. Williams, *Chem. Soc. Rev.*, 233–241 (1993).

22. D. A. Wink, J. F. Darbyshire, R. W. Nims, J. E. Saavedra, and P. C. Ford, *Chem. Res. Toxicol.*, *6*, 23–27 (1993).

23. H. H. Awad and D. M. Stanbury, *Int. J. Chem. Kinet.*, *25*, 375–381 (1993).

24. L. J. Ignarro, J. M. Fukuto, J. M. Griscavage, N. E. Rogers, and R. E. Byrns, *Proc. Natl. Acad. Sci. USA*, *90*, 8103–8107 (1993).

25. Y. Henry, M. Lepoivre, J.-C. Drapier, C. Ducroq, J.-C. Boucher, and A. Guissani, *FASEB J.*, *7*, 1124–1134 (1993).

26. L. J. Ignarro, *Pharmacol. Toxicol.*, *67*, 1–7 (1990).

27. B. Freeman, *Chest*, *105*, 79S–84S (1994).

28. J. S. Beckman, T. W. Beckman, J. Chen, P. A. Marshall, and B. A. Freeman, *Proc. Natl. Acad. Sci. USA*, *87*, 1620–1624 (1990).

29. N. Hogg, V. M. Darley-Usmar, M. T. Wilson, and S. Moncada, *Biochem. J.*, *281*, 419–424 (1992).

30. J. R. Lancaster, Jr. and J. B. Hibbs, Jr., *Proc. Natl. Acad. Sci. USA*, *87*, 1223–1227 (1990).

31. J. B. Hibbs, Jr., in *Iron in Central Nervous System Disorders*. Key Topics in Brain Research (P. Reiderer and M. B. H. Youdim, eds.), Springer-Verlag, Vienna, 1993, pp. 155–171 ff.

32. A. Hausladen and I. Fridovich, *J. Biol. Chem.*, *47*, 29405–29408 (1994).

33. L. Castro, M. Rodriguez, and R. Radi, *J. Biol. Chem.*, *47*, 29409–29415 (1994).

34. M. W. J. Cleeter, J. M. Cooper, V. M. Darley-Usmar, S. Moncada, and A. H. V. Schapira, *FEBS Lett.*, *345*, 50–54 (1994).

35. C. Szabo, B. Zingarelli, M. O'Connor, and A. L. Salzman, *Proc. Natl. Acad. Sci. USA*, *93*, 1753–1758 (1996).

36. A. Calver, J. Collier, and P. Vallance, *Exp. Physiol.*, *78*, 303–326 (1993).

37. J. B. Warren, F. Pons, and A. J. B. Brady, *Cardiovasc. Res.*, *28*, 25–30 (1994).

38. D. D. Rees, R. M. J. Palmer, H. F. Hodson, and S. Moncada, *Br. J. Pharmacol.*, *96*, 418–424 (1989).

39. B. J. R. Whittle, J. Lopez-Belmonte, and D. D. Rees, *Br. J. Pharmacol.*, *98*, 646–652 (1989).

40. P. L. Huang, Z. Huang, H. Mashimo, K. D. Bloch, M. A. Moskowitz, J. A. Bevan, and M.C. Fishman, *Nature*, *377*, 239–242 (1995).

41. L. M. Ruilope, V. Lahera, J. L. Rodicio, and J. C. Romero, *J. Hypertens.*, *12*, 625–631 (1994).

42. P. W. Sanders, *J. Nephrol.*, *5*, 23 30 (1992).

43. U. Forstermann, A. Mugge, U. Alheid, A. Haverich, and J. C. Frolich, *Circ. Res.*, *62*, 185–190 (1988).

44. A. H. Chester, G. S. O'Neil, S. Moncada, S. Tadjkarimi, and M. H. Yacoub, *Lancet*, *336*, 897–900 (1990).

45. L. Sobrevia and G. E. Mann, *Exp. Physiol.*, *82*, 423–452 (1997).

46. S. Moncada and A. Higgs, *N. Engl. J. Med.*, *329*, 2002–2012 (1993).

47. J. F. Kerwin, Jr. and M. Heller, *Med. Res. Rev.*, *14*, 23–74 (1994).

48. R. C. Bone, *Ann. Int. Med.*, *115*, 457–469 (1991).

49. C. Thiemermann, *Gen. Pharmac.*, *29*, 159–166 (1997).

50. R. G. Kilbourn, S. S. Gross, A. Juburan, J. Adams, O. W. Griffith, R. Levi, and R. F. Lodato, *Proc. Natl. Acad. Sci. USA*, *87*, 3629–3632 (1990).

51. C. Thiemermann and J. Vane, *Eur. J. Pharmacol.*, *182*, 591–595 (1990).

52. T. Evans, A. Carpenter, A. Silva, and J. Cohen, *J. Infect. Dis.*, *169*, 343–349 (1994).

53. T. Evans, A. Carpenter, H. Kinderman, and J. Cohen, *Circ. Shock.*, *41*, 77–81 (1993).

54. A. Petros, D. Bennett, and P. Vallance, *Lancet*, *338*, 1557–1558 (1991).

55. J. P. Crow and J. S. Beckman, *Curr. Top. Microbiol. Immunol.*, *196*, 57–73 (1995).

56. T. R. Billiar, R. A. Hoffman, R. D. Curran, J. M. Langrehr, and R. L. Simmons, *J. Lab. Clin. Invest.*, *120*, 192–197 (1992).

57. C. Szabó, *New Horizons*, *3*, 2–32 (1995).

58. M. Stefanovic-Racic, J. Stadler, and C. H. Evans, *Arthritis Rheum.*, *36*, 1036–1044 (1993).

59. X. Wei, I. G. Charles, A. Smith, J. Ure, G. Feng, F. Huang, D. Xu, W. Muller, S. Moncada, and F.Y. Liew, *Nature*, *375*, 408–411 (1995).

60. J. A. Corbett and M. L. McDaniel, *Diabetes*, *41*, 897–903 (1992).

61. J. M. Cunningham and I. C. Green, *Growth Regul.*, *4*, 173–180 (1994).

62. M. L. Lukic, S. Stosic-Grujicic, N. Ostojic, W. L. Chan, and F. Y. Liew, *Biochem. Biophys. Res. Commun.*, *178*, 913–920 (1991).

63. I. I. Singer, D. W. Kawka, S. Scott, J. R. Weidner, R. A. Mumford, T. E. Riehl, and W. F. Stenson, *Gastroenterology*, *111*, 871–885 (1996).

64. P. Garside, A. K. Hutton, A. Severn, F. Y. Liew, and A. McI. Mowat, *Eur. J. Immunol.*, *22*, 2141–2145 (1992).

65. V. Kolb-Bachofen, K. Fehsel, Günther, and T. Ruzicka, *Lancet*, *344*, 139 (1994).

66. Q. Hamid, D. R. Springall, V. Riveros-Moreno, P. Chanez, P. Howarth, A. Redington, J. Bousquet, P. Goddard, S. Holgate, and J. M. Polak, *Lancet*, *342*, 1510–1513 (1993).

67. D. C. Jenkins, I. G. Charles, L. L. Thomsen, D. W. Moss, L. S. Holmes, S. A. Bayliss, P. Rhodes, K. Westmore, P. C. Emson, and S. Moncada, *Proc. Natl. Acad. Sci. USA*, *92*, 4392–4396 (1995).

68. P. Edwards, J. C. Cendan, D. B. Topping, L. L. Moldawer, S. Mackay, E. M. Copeland, III, and D. S. Lind, *J. Surg. Res.*, *63*, 49–52 (1996).

69. L. L. Thomsen, F. G. Lawton, R. G. Knowles, J. E. Beesley, V. Riveros-Moreno, and S. Moncada, *Cancer Res.*, *54*, 1352–1354 (1994).

70. L. L. Thomsen, D. W. Miles, L. Happerfield, L. G. Bobrow, R. G. Knowles, and S. Moncada, *Br. J. Cancer*, *72*, 41–44 (1995).

71. M. Takahashi, K. Fukunda, T., Ohata, T. Sugimura, and K. Wakabayashi, *Cancer Res.*, *57*, 1233–1237 (1997).

72. G. M. Tozer, V. E. Prise, and D. J. Chaplin, *Cancer Res.*, *57*, 948–955 (1997).

73. K. Doi, T. Akaike, H. Horie, Y. Noguchi, S. Fujii, T. Beppu, M. Ogawa, and H. Maeda, *Cancer*, *77*, 1598–1604 (1996).

74. P. J. Wood, I. J. Stratford, G. E. Adams, C. Szabo, C. Thiemermann, and J. R. Vane, *Biochem. Biophys. Res. Commun.*, *192*, 505–510 (1993).

75. M. Ziche, L. Morbidelli, A. Parenti, and F. Ledda, *Adv. Prostaglandin Thromboxane Leukotriene Res.*, *23*, 495–497 (1995).

76. M. Ziche, L. Morbidelli, E. Masini, S. Amerini, H. J. Granger, C. A. Maggi, P. Geppetti, and F. Ledda, *J. Clin. Invest.*, *94*, 2036–2044 (1994).

77. S. J. Leibovich, P. J. Polverini, T. W. Fong, L. A. Harlow, and A. E. Koch, *Proc. Natl. Acad. Sci. USA*, *91*, 4190–4194 (1994).

78. G. D. Kennovin, D. G. Hirst, M. R. L. Stratford, and F. W. Flitney, in *Biology of Nitric Oxide*, Vol. 4 (S. Moncada, M. Feelisch, R. Busse, and A. E. Higgs, eds.), Portland Press, London, 1994, pp. 473–479.

79. M. Ziche, L. Morbidelli, R. Choudhuri, H.-T. Zhang, S. Donnini, H. J. Granger, and R. Bicknell, *J. Clin. Invest.*, *99*, 2625–2634 (1997).

80. G. Montrucchio, E. Lupia, A. De Martino, E. Battaglia, M. Arese, A. Tizzani, F. Bussolino, and G. Camussi, *Am. J. Pathol.*, *151*, 557–563 (1997).

81. S. J. Leibovich, P. L. Polverini, T. W. Fong, L. A. Harlow, and A. E. Koch, *Proc. Natl. Acad. Sci. USA*, *91*, 4190–4194 (1994).

82. E. Pipili-Synetos, E. Sakkoula, and M. E. Maragoudakis, *Br. J. Pharmacol.*, *108*, 855–857 (1993).

83. E. Pipili-Synetos, E. Sakkoula, G. Haralabopoulos, P. Andriopoulou, P. Peristeris, and M. E. Maragoudakis, *Br. J. Pharmacol.*, *111*, 894–902 (1994).

84. H. Maeda, Y. Noguchi, K. Sato, and T. Akaike, *Jpn. J. Cancer Res.*, *85*, 331–334 (1994).

85. J. Wu, T. Akaike, and H. Maeda, *Cancer Res.*, *58*, 159–165 (1998).

86. E. Pipili-Synetos, A. Papageorgiou, E. Sakkoula, G. Sotirpoulou, T. Fotsis, G. Karakiulakis, and M. E. Maragoudakis, *Br. J. Pharmacol.*, *116*, 1829–1834 (1995).

87. K. Xie, S. Huang, Z. Dong, M. Gutman, and I. Fidler, *Cancer Res.*, *55*, 3123–3131 (1995).

88. J. B. Mitchell, D. A. Wink, W. DeGraff, J. Gamson, L. K. Keefer, and M. C. Krishna, *Cancer Res.*, *53*, 5845–5848 (1993).

89. V. N. Verovski, D. L. Van den Berge, G. A. Soete, B. L. Bols, and G. A. Storme, *Br. J. Cancer*, *74*, 1734–1742 (1996).

90. J. A. Cook, M. C. Krishna, R. Pacelli, W. DeGraff, J. Liebmann, J. B. Mitchell, A. Russo, and D. A. Wink, *Br. J. Cancer*, *76*, 325–334 (1997).

91. J. B. Hibbs, Jr., C. Westenfelder, R. Taintor, Z. Vavrin, C. Kablitz, R. L. Baranowski, J. H. Ward, R. L. Menlove, M. P. McMurry, J. P. Kushner, and W. E. Samlowski, *J. Clin. Invest.*, *89*, 867–877 (1992).

92. D. Miles, L. Thomsen, F. Balkwill, P. Thavasu, and S. Moncada, *Eur. J. Clin. Invest.*, *24*, 287–290, (1994).

93. A. Orucevic and P. K. Lala, *Br. J. Cancer*, *72*, 189–197 (1996).

94. L. Brunton, *Lancet*, *2*, 97–98 (1867).

95. W. Murrel, *Lancet*, *1*, 80–81 (1879).

96. T. J. Anderson, I. T. Meredith, P. Ganz, A. P. Selwyn, and A. C. Yeung, *J. Am. Coll. Cardiol.*, *24*, 555–566 (1994).

97. J. Reden, *Blood Vessels*, *27*, 282–294 (1990).

98. J. F. Kerwin, J. R. Lancaster, Jr., and P. L. Feldman, *J. Med. Chem.*, *38*, 4343–4361 (1995).

99. C. M. Maragos, D. Morley, D. A. Wink, T. M. Dunams, J. E. Saavedra, A. Hoffman, A. A. Bove, L. Isaac, J. A. Hrabie, and L. K. Keefer, *J. Med. Chem.*, *34*, 3242–3247 (1991).

100. F. W. Flitney, I. L. Megson, D. E. Flitney, and A. R. Butler, *Br. J. Pharmacol.*, *107*, 842–848 (1992).

101. J. L. Wallace, W. McKnight, P. Del Soldato, A. R. Baydoun, and G. Cirino, *J. Clin. Invest.*, *96*, 2711–2718 (1995).

102. C. G. Frostell, H. Blomquist, G. Hedenstierna, J. Lundberg, and W. M. Zapol, *Anesthesiology*, *78*, 427–435 (1993).

103. M. A. Marletta, *J. Med. Chem.*, *37*, 1899–1907 (1994).

104. G. J. Southan, C. Szabó, and C. Thiemermann, *Br. J. Pharmacol*, *114*, 510–516 (1995).

105. K. Narayanan and O. W. Griffith, *J. Med. Chem.*, *37*, 885–887 (1994).

106. R. B. Silverman, H. Huang, M. A. Marletta, and P. Martasek, *J. Med. Chem.*, *40*, 2813–2817 (1997).

107. W. M. Moore, R. K. Webber, G. M. Jerome, F. S. Tjoeng, T. P. Misko, and M. G. Currie, *J. Med. Chem.*, *37*, 3886–3888 (1994).

108. T. P. Misko, W. M. Moore, T. P. Kasten, G. A. Nickols, J. A. Corbett, R. G. Tilton, M.L. McDaniel, J. R. Williamson, and M. G. Currie, *Eur. J. Pharmacol.*, *233*, 119–125 (1993).

109. G. J. Southan, B. Zingarelli, M. O'Connor, A. L. Salzman, and C. Szabó, *Br. J. Pharmacol.*, *117*, 619–632 (1996).

110. W. M. Moore, R. K. Webber, K. F. Fok, G. M. Jerome, J. R. Connor, P. T. Manning, P. S. Wyatt, T. P. Misko, F. S. Tjoeng, and M. G. Currie, *J. Med. Chem.*, *39*, 669–672 (1996).

111. R. G. Kilbourn, C. Szabó, and D. L. Traber, *Shock*, *7*, 235–246 (1997).

112. C. E. Wright, D. D. Rees, and S. Moncada, *Cardiovasc. Res.*, *26*, 48–57 (1992).

113. F. M. Robertson, P. J. Offner, D. P. Ciceri, W. K. Becker, and B. A. Pruitt, *Arch. Surg.*, *129*, 149–156 (1994).

114. R. G. Kilbourn, J. DeAngelo, and J. Bonaventura, in *Yearbook of Intensive Care and Emergency Medicine 1997* (J.-L. Vincent, ed.), Springer-Verlag, Heidelberg, 1997, p. 230 ff.

115. S. P. Fricker, *Platinum Metals Rev.*, *39*, 150–159 (1995).

116. S. Archer, *FASEB J.*, *7*, 349–360 (1993).

117. J. R. Lancaster, Jr., J. M. Langrehr, H. A. Bergonia, N. Murase, R. L. Simmons, and R. A. Hoffman, *J. Biol. Chem.*, *267*, 10994–10998 (1992).

118. T. Akaike, M. Yoshida, Y. Miyamoto, K. Sato, M. Kohno, K. Sasamoto, K. Miyazaki, S. Ueda, and H. Maeda, *Biochemistry*, *32*, 827–832 (1993).

119. N. Hogg, R. J. Singh, J. Joseph, F. Neese, and B. Kalyanaraman, *Free Rad. Res.*, *22*, 47–56 (1995).

120. H. Maeda, T. Akaike, M. Yoshida, and M. Suga, *J. Leukocyte Biol.*, *56*, 588–592 (1994).

121. M. Yoshida, T. Akaike, Y. Wada, K. Sato, K. Ikeda, S. Ueda, and H. Maeda, *Biochem. Biophys. Res. Commun.*, *202*, 923–930 (1994).

122. H. Maeda, Y. Noguchi, K. Sato, and T. Akaike, *Jpn. J. Cancer Res.*, *85*, 331–334 (1994).

123. J. Wu, T. Akaike, and H. Maeda, *Cancer Res.*, *58*, 159–165 (1998).

124. E. Lilley and A. Gibson, *Br. J. Pharmacol.*, *119*, 432–438 (1996).

125. M. J. Rand and C. G. Li, *Br. J. Pharmacol.*, *116*, 1906–1910 (1995).

126. K. Paisley and W. Martin, *Br. J. Pharmacol.*, *117*, 1633–1638 (1996).

127. M. P. Doyle and J. W. Hoekstra, *J. Inorg. Biochem.*, *14*, 351–358 (1981).

128. Y. Henry, C. Ducrocq, J.-C. Drapier, D. Servent, and A. Guissani, *Eur. Biophys. J.*, *20*, 1–15 (1991).

129. L. Jia, C. Bonaventura, J. Bonaventura, and J. Stamler, *Nature*, *380*, 221–237 (1996).

130. R. G. Kilbourn, G. Joly, B. Cashon, J. DeAngelo, and J. Bonaventura, *Biochem. Biophys. Res. Commun.*, *199*, 155–162 (1994).

131. F. Rioux, G. Drapeau, and F. Marceau, *J. Cardivasc. Pharmacol.*, *25*, 587–594 (1995).

132. J. Loscalzo, *J. Lab. Clin. Invest.*, *129*, 580–583 (1997).

133. S. C. Schultz, B. Grady, F. Cole, I. Hamilton, K. Burhop, and D. S. Malcolm, *J. Lab. Clin. Med.*, *122*, 301–308 (1993).

134. A. C. Sharma and A. Gulati, *J. Lab. Clin. Med.*, *123*, 299–308 (1993).

135. M. T. Heneka, P.-A. Löschmann, and H. Osswald, *J. Clin. Invest.*, *99*, 47–54 (1997).

136. S. R. Fischer, H. G. Bone, and D. L. Traber, in *Yearbook of Intensive Care and Emergency Medicine 1997* (J.-L. Vincent, ed.), Springer-Verlag, Heidelberg, 1997, p. 424 ff.

137. D. Looker, D. Abbot-Brown, P. Cozart, S. Durfee, S. Hoffman, A. J. Matthews, J. Miller-Roehrich, S. Shoemaker, S. Trimble, G. Fermi, N. H. Komiyama, K. Nagai, and G. L. Stetler, *Nature*, *356*, 258–260 (1992).

138. J. E. Chappell, S. R. Shackford, and W. J. McBride, *J. Neurosurg.*, *86*, 131–138 (1997).

139. J. S. Aranow, H. Wang, J. Zhuang, and M. P. Fink, *Crit. Care Med.*, *24*, 807–813 (1996).

140. H. G. Bone, P. J. Schenarts, M. Booke, R. McGuire, D. Harper, L. D. Traber, and D. L. Traber, *Crit. Care Med.*, *25*, 1010–1018 (1997).

141. M. Booke, J. Meyer, W. Lingnau, F. Hinder, L. D. Traber, and D. L. Traber, *New Horizons*, *3*, 123–138 (1995).

142. S. R. Fischer, H. G. Bone, C. Powell, R. McGuire, L. D. Traber, and D. L. Traber, *Crit. Care Med.*, *25*, 1551–1559 (1997).

143. A. R. Butler and C. Glidewell, *Chem. Soc. Rev.*, *16*, 361–380 (1987).

144. A. R. Butler, C. Glidewell, A. S. McIntosh, D. Reed, and I. H. Sadler, *Inorg. Chem.*, *25*, 970–973 (1986).

145. S. S. Greenberg, J. Xie, J. M. Zatarain, D. R. Kapusta, and M. S. Miller, *J. Pharmacol. Exp. Ther.*, *273*, 257–265 (1995).

146. M. A. S. Rajanayagam, C. G. Li, and M. J. Rand, *Br. J. Pharmacol.*, *108*, 3–5 (1993).

147. L. G. Rochelle, S. J. Morana, H. Kruszyna, M. A. Russell, D. E. Wilcox, and R. P. Smith, *J. Pharmacol. Exp. Ther.*, *275*, 48–52 (1995).

148. K. M. Jenkinson, J. J. Reid, and M. J. Rand, *Eur. J. Pharmacol.*, *275*, 145–152 (1995).

149. C.-S. Lai and A. M. Komarov, in *Bioradicals Detected by ESR Spectroscopy* (H. Ohya-Nishiguchi and L. Packer, eds.), Birkhäuser Verlag, Basel, 1995, p. 163 ff.

150. Y. Kotake, T. Tanigawa, M. Tanigawa, I. Ueno, D. R. Allen, and C.-S. Lai, *Biochim. Biophys. Acta*, *1289*, 362–368 (1996).

151. A. Komarov, D. Mattson, M. M. Jones, P. K. Singh, and C.-S. Lai, *Biochem. Biophys. Res. Commun.*, *195*, 1191–1198 (1993).

152. A. Komarov and C.-S. Lai, *Biochim. Biophys. Acta*, *1272*, 29–36 (1995).

153. C.-S. Lai and A. Komarov, *FEBS Lett.*, *345*, 120–124 (1994).

154. G. M. Pieper and C.-S. Lai, *Biochem. Biophys. Res. Commun.*, *219*, 584–590 (1996).

155. M. Cooper, K. Dembny, P. Lindholm, C.-S. Lai, G. Moore, C. Johnson, M. Adams, G. Pieper, and A. Roza, *Surg. Forum*, *48*, 500–503 (1997).

156. W. M. Kazmierski, G. Wolberg, J. G. Wilson, S. R. Smith, D. S. Williams, H. H. Thorp, and L. Molina, *Proc. Natl. Acad. Sci. USA*, *93*, 9138–9141 (1996).

157. V. Zang, M. Kotowski, and R. van Eldik, *Inorg. Chem.*, *27*, 3279–3283 (1988).

158. L. Molina, S. Studenberg, G. Wolberg, W. Kazmierski, J. Wilson, A. Tadepalli, A.C. Chang, S. Kosanke, and L. Hinshaw, *J. Clin. Invest.*, *98*, 192–198 (1996).

159. C. F. J. Barnard, S. P. Fricker, and O. J. Vaughan, in *Insights*

into Speciality Inorganic Chemicals (D. Thompson ed.), Royal Society of Chemistry, Cambridge, 1995, p. 35 ff.

160. G. G. Richter-Ado and P. Legzdins, *Metal Nitrosyls*, Oxford University Press, New York, 1992.

161. A. Tsai, *FEBS Lett.*, *341*, 141–145 (1994).

162. F. Bottomley, *Coord. Chem. Rev.*, *26*, 7–32 (1978).

163. E. A. Seddon and K. R. Seddon, *The Chemistry of Ruthenium*, Elsevier, Amsterdam, 1984.

164. T. Matsubara and C. Creutz, *Inorg. Chem.*, *18*, 1956–1966 (1979).

165. S. P. Fricker, E. Slade, N. A. Powell, O. J. Vaughan, G. R. Henderson, B. A. Murrer, I. L. Megson, S. K. Bisland, and F. W. Flitney, *Br. J. Pharmacol.*, *122*, 1441–1449 (1977).

166. H. C. Bajaj and R. van Eldik, *Inorg. Chem.*, *27*, 4052–4055 (1988).

167. N. A. Davies, M. T. Wilson, E. Slade, S. P. Fricker, B. A. Murrer, N. A. Powell, and G. R. Henderson, *Chem. Commun.*, 47–48 (1997).

168. A. A. Diamantis and J. V. Dubrawski, *Inorg. Chem.*, *20*, 1142–1150 (1981).

169. H. C. Bajaj and R. van Eldik, *Inorg. Chem.*, *29*, 2855–2858 (1990).

170. Y. Chen and R. E. Shepherd, *J. Inorg. Biochem.*, *68*, 183–197 (1997).

171. S. P. Fricker, E. Slade, and N. A. Powell, *Biochem. Soc. Trans.*, *23*, 231S (1995).

172. F. W. Flitney, I. L. Megson, J. L. M. Thomson, G. D. Kennovin, and A. R. Butler, *Br. J. Pharmacol.*, *117*, 1549–1557 (1996).

173. M. P. Fink, B. P. O'Sullivan, M. J. Menconi, P. S. Wollert, H. Wang, M. E. Youssef, and J. H. Fleisch, *Crit. Care Med.*, *21*, 1825–1837 (1993).

174. P. K. Gonzalez, J. Zhuang, S. R. Doctrow, B. Malfroy, P. F. Benson, M. J. Menconi, amd M. P. Fink, *J. Pharmacol. Exp. Ther.*, *275*, 798–806 (1995).

175. A. G. Baggs, S. Fricker, M. Abrams, C. Lee, and M. P. Fink, *Surg. Forum*, *48*, 84–86 (1997).

176. D. M. Rosser, R. P. Stidwill, D. Jacobson, and M. Singer, *J. Appl. Physiol.*, *79*, 1878–1882 (1995).

177. J. R. Lancaster, Jr., J. M. Langrehr, H. A. Bergonia, N. Murase, T. E. Starzl, R. L. Simmons, and R. A. Hoffman, in *The Biology of Nitric Oxide*, Vol. 2 (S. Moncada, M. A. Marletta, J. B. Hibbs, Jr., and E. A. Hibbs, eds.), Portland Press, London, 1992, p. 216 ff.

178. L.-P. Lee and G. N. Schrauzer, *J. Am. Chem. Soc.*, *90*, 5274–5276 (1968).

21

Therapeutics of Nitric Oxide Modulation

Ho-Leung Fung, Brian P. Booth,
and Mohammad Tabrizi-Fard

Department of Pharmaceutics, School of Pharmacy,
Faculty of Health Sciences, University at Buffalo,
State University of New York,
Buffalo, NY 14260-1200, USA

1. INTRODUCTION

As has been pointed out elsewhere in this book, nitric oxide (NO) is an important endogenous regulator of many of the body's physiological functions and pathological states. Therefore, compounds that can affect the availability of this critical molecule should be extremely useful in altering the outcome of many diseases. However, in spite of the intense research activities that have followed the identification of NO as endothelium-derived relaxing factor [1–3], and the prior existence of several known NO donors in cardiovascular therapy, notably nitro-glycerin (NTG), and sodium nitroprusside (SNP), few new NO modulators have advanced to clinical use in the last decade.

Of course, the time delay in drug discovery and development necessary to demonstrate safety and efficacy for regulatory approval requires considerable time. However, the ubiquitous involvement of NO in physiology and pathology, as well as its multiplicity of pharmacological and toxicological actions, have made therapeutic selectivity of NO modulation difficult to achieve in practice. For example, the beneficial effect of nitroglycerin in improving penile erection is diminished by its systemic hypotensive effect [4] and the occurrence of headaches (a side effect presumably due to cerebral vasodilation). On the other hand, the systemic use of nitric oxide synthase (NOS) inhibitors in septic shock has yet to meet with overwhelming clinical success, in part because the existing generations of NOS inhibitors may not be sufficiently specific for NOS II in vivo. To realize the potential benefits of NO regulation in therapy, new NO modulators or delivery modes must be designed to confer the appropriate degrees of selectivity and efficacy of action.

Pharmacological selectivity and efficacy can be constructed, in principle, by appropriate manipulation of the physicochemical, biochemical, and pharmaceutical properties of the drug candidate. The historical example of replacing amyl nitrite, which is highly volatile and therefore has to be encased in glass spheres, with the less volatile NTG allowed the formulation of a convenient pharmaceutical dosage form, the sublingual tablet, which could be easily used by patients. We now know that the sublingual route of administration optimizes the pharmaceutical properties of this drug by allowing rapid drug absorption (necessary to alleviate acute chest pains) and by avoiding the extensive first-pass hepatic metabolism associated with oral administration of this drug. This example illustrates that for an effective NO donor to be used in therapy, other properties besides the ability to release NO should be effectively optimized.

This chapter attempts to examine the therapeutics of NO modulation from a system perspective. Therapeutic efficacy of NO modulators is considered not only from the traditional pharmacological viewpoints of potency and enzyme specificity, but also from those of absorption, metabolism and biodistribution, and pharmacokinetics. These properties jointly contribute to the overall therapeutic efficacy, safety, and selectivity of any drug, and therefore must be taken into consideration if useful NO modulators are to be developed for therapeutic use in the future.

2. DIVERSITY OF MOLECULAR FORMS OF NITRIC OXIDE

The design of NO therapeutic modulators has been further complicated by the recent realization that NO action may not come from a single molecular entity, i.e., the free radical form, NO•. Stamler et al. [5,6] raised our awareness of the different redox forms of NO that exist in equilibrium with each other. The complexity of NO chemistry can be seen in Fig. 1. The distribution of NO species is condition-dependent, and possibly concentration-dependent as well with the latter possibly accounting for the observation that the cellular signaling role of NO is concentration-dependent [7]. Other investigators have also demonstrated that different NO redox forms can bring about diverse pharmacological effects. For example, Lipton and associates [8,9] showed that NO• and NO^+ forms react differently with thiol groups at the receptor level, leading to contrasting neurotoxic and neuroprotective effects. Campbell et al. [10] recently showed that $ONOO^-$ and NO^+ can exert effects opposite to those produced by NO• on the L-type calcium channel in ferret ventricular myocytes.

Diversity of NO action on vascular activity has also been observed, but the potential role played by different NO forms has yet to be explicitly recognized. For example, Sellke et al. [11] observed that coronary vessels of diverse sizes dilate differently toward NTG, NO gas, and S-nitrosocysteine. These findings were interpreted to suggest different availability of sulfhydryl groups in blood vessels of different sizes. We [12], among others, have shown that NTG induces in vitro vascular tolerance, but the blood vessels are not cross-tolerant to RSNO nor NO gas. Again, this result was interpreted to suggest that tolerant vessels may have diminished enzyme activity to metabolize NTG to NO, while

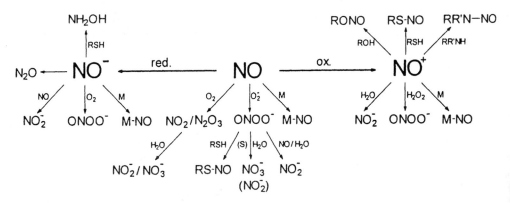

FIG. 1.   Nitrogen oxide species and their reactions. (Taken with permission from [100].)

RSNO and NO do not require metabolic activation. We have also shown [13] that NTG, but not RSNO, induced hemodynamic tolerance in a rat model of congestive heart failure (CHF). We suggested that RSNO can produce balanced vasodilation (arterial and venous), while NTG induces venodilation primarily. We subsequently showed that the use of an arterial dilator, hydralazine, can overcome in vivo hemodynamic tolerance of NTG in CHF [14].

These examples show that NO donors cannot be considered as a monolithic class of therapeutic agents and that their pharmacological actions may differ substantially. To summarize, several important principles must be considered in designing NO donor drugs, i.e.:

NO donors do not produce the same compositions of NO redox forms.

NO redox forms may give rise to substantially different pharmacological responses.

Therefore, the therapeutic action of NO donors could differ significantly, and it may be possible to design different responses (and therefore possibly selectivity of action) among NO donors.

3. INTERACTIONS BETWEEN NITRIC OXIDE AND CALCITONIN GENE–RELATED PEPTIDE (CGRP)

The interaction of NO with the heme site of guanylate cyclase to produce cyclic guanosine 3′,5′-monophosphate (cyclic GMP) is well known [15,16]. Interactions of NO with sulfhydryl sites are also well described [17,18]. Recently, we conducted several studies to explore the interplay between NO and an important neuropeptide, i.e., calcitonin gene-related peptide. This pathway of NO action may represent a novel molecular target for designing selectivity in NO therapy.

3.1. NO Donor-Specific Release of CGRP

We have been interested for some time in the interaction between NTG and CGRP, primarily because NTG has been shown to release CGRP from cerebral arterioles [19] and plasma levels of CGRP were increased during NTG-induced cluster headache attack [20]. If CGRP release is a contributing mechanism to the occurrence of headaches due to NTG administration (a frequent side effect that could be debilitating to some patients), then the mechanisms of this interaction should be clarified.

Recent studies also suggested that CGRP may possess antiplatelet activities. Kitamura et al. [21] used washed platelets as a bioassay to detect CGRP in crude thyroid fractions, based on the knowledge that CGRP stimulates intraplatelet cyclic adenosine 3',5'-monophosphate (cAMP) elevation, which serves as an inhibitory second messenger in platelets [22]. Bull et al. [23] later demonstrated that CGRP can inhibit aggregation of washed platelets in vitro, though the mechanism was not explored. We therefore hypothesized that perivascular release of CGRP in the systemic circulation may be involved in the antiaggregatory effects of NO donors.

We recently reported the results of a study that tested this hypothesis [24]. Briefly, we showed that CGRP is a potent antiaggregatory agent in its own right, with an in vitro IC_{50} of 62.1 nM in whole blood. As anticipated, this antiaggregatory effect is completely blocked by $CGRP_{8-37}$, a specific and competitive antagonist of CGRP. N-Methyl-L-arginine (L-NMMA) also completely blocked this action. Over the concentration range of 50–220 nM, NTG by itself has little in vitro antiaggregatory activity of its own. However, addition of aortic tissue to NTG in whole blood induced nearly complete platelet inhibition, suggesting that vascular tissue interacted with NTG to produce antiaggregatory mediator(s). There is little doubt that one of these mediators is an NO form, which is metabolically activated from NTG. However, addition of $CGRP_{8-37}$ significantly reduced the antiaggregatory effects of NTG in the presence of aortic tissue. The latter finding therefore suggests that NTG releases CGRP from aortic tissue and that CGRP plays a substantial role in the in vivo antiaggregatory activity of NTG. Since both $CGRP_{8-37}$ and L-NMMA also blocked the antiplatelet effects of NTG (in the presence of aorta tissue in whole blood), the data are consistent with a role of CGRP in NTG-induced antiaggregation that is in part mediated via NOS.

Interestingly, when SNP was used as the NO donor in place of NTG, we saw little blockade with $CGRP_{8-37}$ [25]. This finding indicates that CGRP release from NO may be highly dependent on the specific

NO form. We confirmed these observations using another biological preparation and criterion, i.e., vasodilation of the isolated rat thoracic artery. Using conventional vasodilation techniques, we showed that NTG-induced vasodilation was significantly blocked by $CGRP_{8-37}$, which however has no effect on the vasodilatory effect of SNP. In contrast, vascular relaxation by Piloty's acid (which releases NO^- in addition to $NO^\bullet$) is also partially blocked by $CGRP_{8-37}$ (unpublished data).

3.2. Mechanisms of NO-CGRP Interaction

Figure 2 depicts our current understanding of the interaction among NTG, CGRP, and NOS III in blood and the vascular wall. Once administered, NTG quickly penetrates into vascular and humoral cells [26]. In the endothelial and smooth muscle cells, NTG is metabolized to produce NO, which then produces vasodilation. Our studies indicate that the vascular production of NO from NTG also contributes to platelet inhibition [24]. NTG also stimulates the release of CGRP from perivascular nerves, which have been shown to contain this neuropeptide. CGRP then diffuses into the vascular wall and the lumen of the blood vessel. In the rat aorta, CGRP interacts with a specific receptor expressed on the surface of the endothelial cell. The CGRP-receptor interaction activates adenylate cyclase via a G protein, leading to the production of cAMP [27]. NOS III is also activated, though the relationship between cAMP and NOS III has yet to be elucidated. The subsequent production of NO can then contribute to vasodilation [27] and platelet inhibition.

It is also possible that CGRP might act directly on the platelet. The studies of Matsumoto et. al. [28] have demonstrated the specific binding of CGRP to human platelets, leading to elevations of cAMP, which is a potent second messenger in platelets. Since platelets also contain NOS III [29,30], it is also possible that CGRP may stimulate cGMP production in platelets. In addition, previous studies have shown that the interaction of cAMP with cGMP is synergistic [31,32], and this may also contribute to the inhibitory action of CGRP on platelets. Our experiments with $CGRP_{8-37}$ confirmed that the action of CGRP is receptor-mediated, and our studies with L-NMMA suggest that the

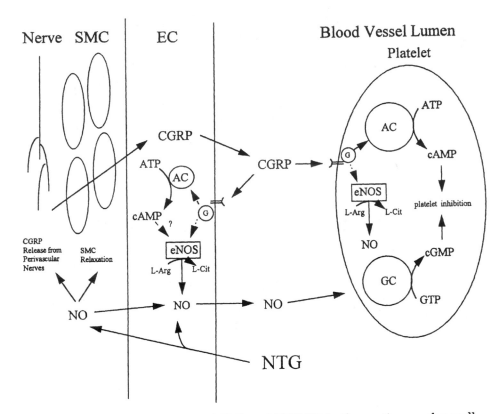

FIG. 2.   Interaction of NTG, CGRP, and NOS III in the aortic vascular wall and whole blood. AC, adenylate cyclase; GC, guanylate cyclase; G, G-protein; eNOS, NOS III; EC, endothelial cells; SMC, smooth muscle cells.

CGRP-NOS III interaction may have important physiological implications.

These studies, therefore, highlight a potential role of CGRP (and possibly other neuropeptides) in mediating the action of NO. However, not all NO donors can elicit this mechanism of action, possibly owing to the specific mixture of NO redox forms that they can each produce. The specific interaction between some redox forms of NO and CGRP can, in principle, be exploited to confer some specificity in the design of NO donors.

4. THERAPEUTICS OF NITRIC OXIDE ENHANCEMENT

Two approaches have been used to enhance NO availability in NO-deficient diseases. The first approach is to augment NO production endogenously by supplying L-arginine, which is metabolically converted by NOS to NO and citrulline. The second approach is to administer NO donors, which release NO either spontaneously or via enzymatic reactions. These latter agents liberate NO independent of NOS activity and therefore mediate their actions regardless of the functional state of NOS.

4.1. Therapeutic Effects of L-Arginine

L-Arginine has been used experimentally to treat a variety of conditions in which NOS dysfunction is believed to mediate a significant role. Based on some of these studies, L-arginine therapeutics have been extended to clinical experiments for selected diseases, such as improving immune function [33], and endothelium-dependent vasodilation in hypercholesteremic patients [34–37]. The use of L-arginine to improve vascular function in hypercholesteremic conditions is the most extensively tested area. Previous studies have demonstrated that hypercholesteremia mediates a significant role in the generation of atherosclerotic plaques, especially of the coronary and carotid arteries, predisposing individuals to ischemic heart diseases and stroke. These studies have also demonstrated that the response to endothelium-dependent vasodilation is impaired in arteries with atherosclerotic lesions, which are also prone to increased platelet thrombosis [37]. More importantly, sections of arteries distal to an atheroma that possess no angiographically demonstrable lesions also have impaired responses to nitrergic stimulation [38]. Therefore, the rationale of therapy has been to improve endothelial NO production by L-arginine supplementation.

There are conflicting reports about the effect of acute intravenous infusions of L-arginine on basal blood pressure. In some cases, supplemental L-arginine induced significant but moderate decreases in systolic as well as diastolic blood pressure in healthy volunteers [39,40]; however, other investigators have reported no changes in mean arterial pressure following L-arginine infusions in volunteers or hypercholesteremic patients [35]. Acute treatment with short infusions of L-arginine, however, uniformly augments endothelium-dependent dilation (EDD), such as that induced by acetylcholine or its analogs, and changes in blood flow to which NOS III activity is sensitive, both in coronary [34,41] and noncoronary vascular beds [36,39,40].

In addition to vasodilation, it has been shown by several studies that treatment with L-arginine has also resulted in reduced ex vivo platelet aggregation [39] and monocyte adhesion [37], two important components that contribute to ischemic heart disease. In hypercholesteremic patients and volunteers, long-term oral supplementation of L-arginine had no significant effect on basal blood pressure but did induce moderate improvements in brachial EDD in hypercholesteremic patients [36, 37].

Collectively, these studies have revealed that although L-arginine may produce beneficial cardiovascular effects, its role is mainly adjunctive. Studies of the oral administration of L-arginine with maximum tolerable doses revealed only a moderate improvement in EDD (4–12%) following treatment that doubled the baseline plasma concentrations of L-arginine [36,37]. These modest responses are not likely to be sufficient for the prophylactic treatment of ischemic heart disease secondary to atherosclerosis. Intravenous infusions of L-arginine improved EDD responses to a greater extent (12–300%) than those produced by the oral route, but large increases in plasma concentrations (at least 10-fold) are required to achieve these effects. These results indicate that physiological NOS III activity is not primarily limited by circulating L-arginine concentrations.

4.2. Therapeutic Effects of NO Donors

NO can, of course, be directly administered through inhalation (in a diluted state) for the treatment of pulmonary hypertension. For example, 5–40 ppm NO has been found to reduce pulmonary arterial pressure by 16% and increase oxygen tension by 44% following long-term administration [42]. However, for the treatment of nonpulmonary diseases, systemic NO donors need to be administered. Currently, four types of NO donors are in clinical use: (1) the organic nitrates, e.g., NTG, isosorbide dinitrate, isosorbide-5-mononitrate, and nicorandil; (2) ferrous nitroso complexes, the only example of which is SNP; (3) the organic nitrites, e.g., isobutyl and isoamyl nitrite; and (4) sydnonimines such as molsidomine. Only the organic nitrates are in extensive use. Although SNP is a potent drug that provides both arterial and venous vascular dilatation, its pharmaceutical and toxicological properties are poor: it is highly photo-labile (therefore has a short shelf-life), not orally bioavailable, and gives rise to toxic cyanide ions upon metabolism. The organic nitrites have largely fallen into disuse because (as discussed earlier) their volatile nature has precluded the formulation of a suitable pharmaceutical preparation. Molsidomine is orally bioavailable and is a potent vasodilator. However, recent awareness that it can produce the destructive peroxynitrite anion may raise questions about its toxicological effects.

Among the organic nitrates, NTG is the most potent, but its oral effectiveness is limited by extensive hepatic metabolism during absorption. The formulation of NTG into transdermal patches popularized clinical use of this drug in the 1980s, but questions about pharmacological tolerance necessitated a relabeling of its dosage regimen to include a 12-hour drug-free period. The use of therapeutic NO donors for the treatment of various diseases is briefly described below.

4.2.1. Stable Angina Pectoris

Traditionally, NTG has been the drug of choice for the treatment of stable angina pectoris. Though its exact mechanism(s) of action remains unclear, NTG exerts at least two distinct effects that may account for its antianginal efficacy. First, NTG acts largely on the peripheral circulation and venous vasculature, which reduces the preload on the heart [43]. This reduction in preload reduces the oxygen requirements of the myocardium [44], which aids the oxygen supply versus oxygen requirement quotient and thus alleviates anginal attacks. In addition to this action, some researchers have shown that NTG induces the vasodilation of large coronary conductance arteries, including atherosclerotically obstructed arteries [11], which leads to the improvement of myocardial oxygenation.

4.2.2. Unstable Angina Pectoris

The etiology of this disease is characterized by the fissure of a coronary atherosclerotic lesion that then elicits the extensive formation of a platelet thrombus, which frequently leads to complete occlusion of the coronary artery and sudden death [45]. In 1970, it was determined that NTG could inhibit platelet function in addition to inducing vasodilation [46]. These findings have since been extended to other organic nitrates and most NO donors in general. Currently, NTG is administered as an intravenous infusion to patients suffering from unstable angina. Although the duration of infusion is known to give rise to hemodynamic tolerance, indirect indicators have suggested that continuous NTG administration can provide sustained benefits in unstable angina. We have demonstrated in an animal model that the platelets continue to respond to NTG infusions despite the appearance of hemodynamic tolerance to NTG [47]. These results have since been confirmed by other investigators [48], and collectively these data suggest that platelets may not be subjected to the same tolerance-inducing mechanisms that are operating in the vasculature.

In vitro studies have demonstrated that the concentrations of NO donors that are required to inhibit platelet functions are 1000-fold higher than the plasma concentrations that can be obtained in vivo. This discrepancy might be explained by recent in vitro experiments demonstrating that both NTG [24] and SNP [25] are dependent on NO production from vascular tissue to inhibit platelet aggregation. These findings suggested that the plasma concentrations in vivo are in fact adequate to inhibit platelet aggregation.

4.2.3. Congestive Heart Failure

Several NO donors have been successfully used to treat acute and chronic congestive heart failure. SNP appeared to provide the best profile by inducing significant reductions in both preload and afterload [49]. However, due to its pharmaceutical limitations and potential toxicity, it is limited to acute clinical situations. NTG administered either as an infusion or transdermally also produces beneficial reductions in preload [43], thereby reducing both myocardial oxygen demands and pulmonary congestion. Again, the utility of NTG is limited by the appearance of hemodynamic tolerance, and the necessity of a nitrate-free period relegates NTG to the role of an adjunct to therapy in CHF. However, the duration of therapy with NTG can be extended and improved by the simultaneous administration of the putative arteriodilator hydralazine, which apparently prevents the onset of tolerance as well as induces beneficial reductions in afterload [14].

4.2.4. Acute Myocardial Infarction

During the mid-1970s, several centers experimented with NTG following acute myocardial infarction [50]. The results of these early studies revealed that infusions of NTG titrated to induce modest reductions in blood pressure without tachycardia significantly reduced the severity of the infarct. In addition to these effects, experimental evidence now suggests that the nitrates may also provide significant long-term benefits by inhibiting the cardiac remodeling that occurs during the progression of CHF [51]. The use of SNP for these purposes is contraindicated because this NO donor can increase intracoronary "steal" and exacerbate acute myocardial infarction by shunting blood flow away from the damaged myocardium [52].

4.2.5. Controlled Hypotension During Surgical Procedures

Both SNP and NTG infusions are frequently used during cardiac and noncardiac surgeries, as well as during hypertensive emergencies to control blood pressure. Both drugs can be easily titrated to produce the desired reduction in blood pressure, while the short half-lives of these agents ensures a rapid offset of action in the event of severe hypotension [53]. As mentioned earlier, SNP must be used with caution, as significant quantities of cyanide anions are released, and several perioperative deaths have been attributed to cyanide toxicity. Protection from cyanide toxicity can be afforded by the coinfusion of thiosulfate, which chelates CN moieties.

NTG is generally well tolerated during these procedures, but therapeutic failures have occurred as a result of drug adhesion to intravenous sets [54] and medical apparatus [55]. Furthermore, as NTG is a prodrug that requires metabolic activation, and clinical procedures and drugs that affect its metabolism can alter its pharmacodynamics.

For example, Booth et al. [56] have shown that the vasodilatory effects of NTG can become attenuated during hypothermic cardiopulmonary bypass, apparently as a result of reduced enzymatic NO production.

4.2.6. Improving the Therapeutic Scope and Efficacy of NO Donors

Historically, the treatment of disease with NO donors has been largely restricted to cardiovascular conditions. However, our understanding of NO pharmacology and NOS biochemistry would suggest that NO can be used in other diseases such as penile erectile dysfunction, restenosis, osteoporosis, and neurodegenerative diseases.

A major problem with the therapeutic use of organic nitrates in stable angina and CHF is the development of hemodynamic tolerance. The mechanism(s) of this phenomenon is complex and is likely to involve multiple factors [57]. At the present time, no clinical approach has been developed to sustain nitrate action around the clock. However, several experimental approaches have shown some promise. One approach involves the use of other NO donors such as an S-nitrosothiol or an organic nitrite. We have used a rat model of CHF and showed that continuous infusion of these NO donors did not lead to hemodynamic tolerance when left ventricular end-diastolic pressure was used as a therapeutic index [13,58]. The mechanisms responsible for the dissimilar tolerance properties of diverse NO donors are not known but may involve the nature and composition of the NO forms produced by these NO donors.

Another approach to minimize nitrate tolerance involves the use of nonclassical dosing regimens. Using the rat model of CHF, we recently demonstrated [59] that a time-variant dosing scheme can effectively overcome the development of hemodynamic tolerance with NTG. This dosing scheme was arrived at through the construction and use of a pharmacokinetic-pharmacodynamic (PK/PD) model of in vivo NTG action based on tolerance development and the occurrence of withdrawal rebound [60,61].

An example of site-specific delivery of NO has recently been provided by Saavedra et al. [62]. These investigators designed an NO prodrug that was activated solely by hepatic metabolism. The release of NO was limited to the hepatocytes, which prevented liver apoptosis without any untoward effects on systemic hemodynamics. These examples illustrate how NO donors with the proper biochemical and biopharmaceutical characteristics can provide effective and selective therapeutic action.

5. THERAPEUTICS OF NITRIC OXIDE INHIBITION

In many pathological conditions such as septic shock, multiple sclerosis, and rheumatoid arthritis, overproduction of NO by the inducible isoform of NOS (NOS II) has been established as the major cause of the disease. Therefore, selective inhibition of NO production by NOS II has been viewed as a potential therapeutic modality for treatment of these diseases. Many inhibitors of NOS isoforms have been examined in both animals and humans, but successful clinical candidates have yet to emerge. Often, the apparent in vitro selectivity observed for some NOS inhibitors has not been directly translated to potential therapeutic benefits in vivo. For some compounds, this poor in vitro/in vivo extrapolation may arise from the presence of suboptimal biopharmaceutical properties, e.g., poor tissue penetration, inappropriate biophase residence time, etc., which limit the selectivity of these inhibitors in a whole biological system.

In this section, we shall focus on an examination of the pharmacokinetic and pharmacodynamic properties of these compounds and define how these properties may affect their in vivo efficacy. For a comprehensive discussion on the biochemistry and pharmacology of NOS inhibitors, readers are referred to recent reviews on this subject [63,64].

5.1. NOS Inhibitors

NOS inhibitors have been employed as an important tool for the mechanistic understanding of the degree of NO involvement in many biological systems. Many analogs of L-arginine, particularly those substituted at the guanidino nitrogen atom, are prototypical NOS inhibitors. Although the mechanism of inhibition of NOS isoforms by L-arginine analogs varies, most of the L-arginine analogs are considered as competitive inhibitors of these isoforms.

5.1.1. Nitro-L-Arginine (L-NA)

Among the L-arginine analogs, L-NA is considered as a potent inhibitor of the constitutive isoforms of NOS (I and III). Although it was recently suggested that L-NA was an irreversible inhibitor of the constitutive NOS isoforms both in vitro and in vivo [65], it is now evident that the onset of in vitro inhibition of brain NOS by L-NA is slow and slowly reversible, with a long dissociation half-time of 10–20 min [66,67]. The slow dissociation of L-NA from the enzyme, which has made the isolation of the L-NA-NOS complex possible [66], is consistent with the apparent irreversibility of in vitro enzyme inhibition during short incubation. Furthermore, the long duration of effect described for L-NA in vivo [65,68] is consistent with the long pharmacokinetic half-life (20 h) observed for this compound in rats [69]. Interaction of L-NA with NOS II appears to be different from that observed with the constitutive isoforms, underlying the differences in the active site for the different isoforms [70]. In all cases, however, inhibition of NOS isoforms by L-NA appears to be reversible by the endogenous substrate L-arginine, consistent with the competitive nature of this inhibition.

5.1.2. Nitro-L-Arginine Methyl Ester (L-NAME)

Due to its greater solubility, L-NAME, the ester prodrug of L-NA, has been employed in many studies. L-NAME quickly breaks down in vivo to its parent compound L-NA. For example, after intravenous administration of L-NAME in anesthetized rabbits, plasma concentrations of the inhibitor decreased rapidly below the detection limit over a 30 min interval. On the other hand, L-NA plasma concentrations were detectable 30 s after administration of L-NAME [71]. Incubation of L-NAME with canine blood resulted in the metabolism of this inhibitor to L-NA, which did not undergo any further metabolism [72]. L-NAME has been reported to elicit in vitro antimuscuranic properties, which may complicate data interpretation when this NOS inhibitor is used [73].

5.1.3. N-Methyl-L-Arginine (L-NMMA)

L-NMMA is one of the most commonly used inhibitors of NOS. In contrast to L-NA, L-NMMA is considered to be an irreversible mechanism-based inactivator of NOS II with a pseudo-first order kinetics of inactivation (0.04–0.07 min^{-1}) and a K_I of 2.6–2.7 μM [70,74]. L-NMMA is also reported to be a reversible inactivator of the NOS I (k_{inact} = 0.02 min^{-1}), whereas NOS III was not reported to be inactivated by the inhibitor [70]. In addition, L-NMMA is an alternate substrate for NOS that can be metabolized by the enzyme to generate NO. However, the rate was reported to be slower than that for L-arginine metabolism [74]. Interestingly, L-NMMA and other methylated arginines such as asymmetrical dimethyl-L-arginines have been detected in human plasma and urine; these compounds have been suggested to contribute to the underlying pathology in some disease conditions [75].

5.1.4. Other NOS Inhibitors

Other amino acid derivatives also have recently been employed as NOS inhibitors. Derivatives of L-citrulline such as thiocitrullines and the alkyl derivatives of thiocitrulline (s-ethyl- or s-methylthiocitrulline) are potent inhibitors of all NOS isoforms although the alkyl derivatives show more selectivity for the constitutive isoforms of NOS [76,77]. Other arginine derivatives such as amino-L-arginine (L-NAA) have been shown to produce a time- and concentration-dependent inactivation of all NOS isoforms [78]. L-NAA administration in animals results in immediate increase in mean arterial pressure that is reversible by administration of L-arginine [79]. However, L-NAA has been found to cause seizures in conscious dogs [80]. Recently, L-iminoethyllysine (L-NIL) was shown to be both a potent and selective inhibitor of mouse NOS II in vitro [81]. Similar to L-NIL, L-iminoethylornithine was reported to be a potent inhibitor of NOS II but, in contrast to L-NIL, it lacked the selectivity for this isoform [81]. Further studies in animals are required to examine the potential therapeutic usefulness of L-NIL in vivo.

Among non-amino acid-based NOS inhibitors, guanidine derivatives such as aminoguanidine have been examined extensively and are shown to be more selective inhibitors of NOS II [82]. Aminoguanidine, a nucleophilic hydrazine compound, is considered to be a mechanism-based inactivator of all NOS isoforms, although in vitro it exhibits specificity for the inactivation of NOS II. In animal studies, aminoguanidine has been shown to be beneficial in the treatment of diabetic nephropathy by inhibiting the formation of advanced glycosylation end-products and preventing vascular complications [83,84]. In addition, aminoguanidine has received some attention for its potential to treat inflammatory conditions. Pretreatment of endotoxemic mice with amino-guanidine resulted in an increase in the mean survival time [85]. How-ever, application of aminoguanidine as a selective NOS inhibitor has been limited since this agent can elicit effects that are not specifically related to NOS inhibition. In particular, aminoguanidine inhibits the activity of copper-containing proteins such as diamine oxidase and catalase [86,87] and interferes with histamine catabolism [88].

5.2. Pharmacokinetic and Pharmacodynamic Issues

Although many NOS inhibitors have been examined as potential drugs in the clinical management of the disabling hypotension observed in states of septic shock that are nonresponsive to the traditional vaso-constrictor agents, their use has been limited because of the presence of unwanted side effects. NOS inhibition in animals and patients with septic shock restores arterial pressure, but with a concomitant delete-rious effect on the cardiac output. In many animal studies, the in vivo dosage regimens chosen for NOS inhibitors did not take the pharmaco-kinetics of these agents into account, making data interpretation dif-ficult.

Recently, we examined the pharmacokinetic properties of L-NA in rats and showed that following intravenous administration of differ-ent bolus doses of the drug, plasma L-NA declined in a biexponential manner [69,89]. The mean residence time, i.e., the time at which 50% of the drug remains in the body, of the inhibitor was about 30 h and the apparent elimination half-life was 20 h. This slow elimination is consis-tent with the prolonged duration of action of L-NA seen in some in vivo systems [65,68]. The pharmacokinetics of L-NA in female rats was similar to that observed in male rats, indicating that L-NA pharmaco-kinetics was not gender-dependent (unpublished observations).

In contrast to L-NA, L-NMMA elimination appears faster, with a half-life of about 70 min and a mean residence time of about 90 min (T. Maurer et al., personal communication). These data are in agree-ment with the shorter in vivo duration of action reported for L-NMMA

[90,91]. The differences in the elimination properties of these two inhibitors arise mainly from metabolism since urinary excretion was not a significant pathway of elimination [69]. Recent reports suggest that the human vasculature and rat tissues possess an enzyme that metabolizes L-NMMA to citrulline but the metabolism of L-NA by these tissues has not been reported [92]. Furthermore, unlike L-NMMA, L-NA does not appear to be a substrate for NOS and therefore not to be metabolized by this enzyme [66,67]. We recently examined the possible metabolism of radioactive L-NA to L-arginine and L-citrulline in liver and kidney homogenates using a thin-layer chromatographic method and observed no significant change in the amounts of L-arginine, L-NA, and L-citrulline over a 4-h incubation interval (unpublished data).

The in vivo selectivity of NOS inhibitors depends on, among other factors, their transport to the relevant tissues and cells. Different amino acid transporters have been shown to mediate the cellular uptake of the NOS inhibitors in vitro [93,94]. L-NMMA appears to be transported by the cationic amino acid (Y^+ system) carrier that is responsible for the transport of L-arginine, whereas L-NA is transported by the neutral amino acid transporter (L system). Pharmacokinetic studies revealed that the apparent volume of distribution of L-NA was large in rats [69]. Indeed, distribution studies conducted in this animal model indicated that, at steady-state plasma concentrations (30 µg/mL), L-NA was extensively distributed into various tissues such as liver and kidneys. Skeletal muscle, which is reported to be rich in the constitutive isoforms of NOS [95], appeared to represent a major portion of the distribution volume at steady state [89].

Our studies further revealed that L-NA could distribute into the central nervous system. The cerebrospinal fluid concentration of L-NA was found to be about 5 µM at steady-state plasma concentrations of 130 µM. Thus, L-NA may exert some pharmacological action in the central nervous system since the K_I values for NOS I as well as NOS III are reported to be in the micromolar ranges [70]. L-NMMA, like L-NA, appears to be well distributed to various tissues, as inferred from the large V_{ss} observed in rats (about 1.0 L/kg), which is much greater than the plasma volume (30–35 mL/kg).

Recent studies in our laboratory indicate that the relationship between L-NA plasma concentration and various hemodynamic effects measured in normal rats appears complex and time-dependent. The temporal aspects observed in these relationships may indeed be due to the slow dissociation of L-NA from constitutive isoforms of NOS and reversible inactivation of these isoforms by the inhibitor. These results indicate that the biodistribution of NOS inhibitors into tissues that are rich in the NOS enzyme may in part determine the scope and intensity of their in vivo action. Unfortunately, information on pharmacokinetic and pharmacodynamic properties as well as the biodistribution characteristics of many NOS inhibitors is not presently available.

5.3.   Selectivity and Other Applications

Since many pathological states may result from unwanted induction of NOS II by proinflammatory cytokines such as tumor necrosis factor, interleukins, and bacterial lipopolysaccharide, much attention has been paid to designing NOS inhibitors specific for NOS II. In general, L-NA is considered to be a more selective inhibitor of the constitutive isoforms of NOS (K_I (μM) = 0.01 NOS I > 0.02 NOS III $\gg$ 8.1 for NOS II) [70]. In addition, L-NA reversibly inactivates the constitutive isoforms of NOS isoforms. Selectivity of L-NA for NOS I along with its ease of access into the brain has made L-NA an important pharmacological probe for understanding the role of NO in the central nervous system and in conditions where inhibition of NOS I may be of therapeutic benefit, e.g., neurological injury in central nervous system due to overproduction of NO. Due to its lack of selectivity for NOS II, application of generally large doses of L-NA and L-NAME in experimental septic shock has been shown to be detrimental, resulting in excessive vasoconstriction, a decrease in cardiac index, and multiple organ toxicities [96].

Similar to L-NA, L-NMMA also appears to be a more selective inhibitor of NOS I (K_I (μM) = 2, NOS I > 0.7 NOS III > 2.6 for NOS II) [70]. However, unlike L-NA, L-NMMA is a mechanism-based inactivator of NOS II [74]. This is indeed an important characteristic that can amplify the apparent selectivity of L-NMMA for NOS II. Although in vivo studies in animals indicated that application of large doses of L-NMMA can be detrimental, application of low doses of the inhibitor in animals and septic patients was shown to elevate blood pressure with no apparent toxicity [96]. The beneficial effects observed with L-NMMA in septic shock at lower doses may in part be due to the selective inactivation of inducible isoform of NOS by L-NMMA. Indeed, the time-dependent inactivation of NOS II may enhance the apparent selectivity of L-NMMA for this isoform.

Other disease states may also benefit from selective NOS inhibition. Several recent studies have demonstrated an involvement of NO overproduction in certain experimental autoimmune diseases, including inflammatory arthritis [97], immunologically induced diabetes [98], and experimental autoimmune encephalomyelitis [99]. It is possible that administration of suitable NOS inhibitors may interrupt the progression of these diseases.

ACKNOWLEDGMENTS

Our research in NO modulation has been supported in part by the National Institutes of Health, grants HL22273 and GM42850.

ABBREVIATIONS

cAMP	cyclic adenosine $3',5'$-monophosphate
cGMP	cyclic guanosine $3',5'$-monophosphate
CGRP	calcitonin gene-related peptide
$CGRP_{8-37}$	calcitonin gene-related peptide fragment 8–37
CHF	congestive heart failure
EDD	endothelium-dependent dilation
IC_{50}	concentration of an inhibitor that induces half of the maximal response
L-NA	nitro-L-arginine
L-NAME	nitro-L-arginine methyl ester
L-NMMA	N-methyl-L-arginine
L-NAA	amino-L-arginine
L-NIL	L-iminoethyllysine
NO	nitric oxide
NOS I	neuronal nitric oxide synthase
NOS II	inducible nitric oxide synthase

NOS III   endothelial nitric oxide synthase
NTG   nitroglycerin
SNP   sodium nitroprusside
V_{ss} volume of distribution of a drug at steady state

REFERENCES

1. R. F. Furchgott and J. V. Zawadzki, *Nature*, *288*, 373 (1980).
2. R. M. J. Palmer, A. G. Ferrige, and S. Moncada, *Nature*, *327*, 524 (1987).
3. L. J. Ignarro, G. M. Buga, K. S. Wood, R. E. Byrne, and G. Chaudhuri, *Proc. Natl. Acad. Sci. USA*, *84*, 9265 (1987).
4. H.-L. Fung and J. A. Bauer, U.S. Patent 5,646,181, July 8, 1997.
5. J. S. Stamler, D. J. Singel, and J. Loscalzo, *Science*, *258*, 1898 (1992).
6. D. R. Arnelle and J. S. Stamler, *Arch. Biochem. Biophys.*, *318*, 279 (1995).
7. T. M. Lincoln, T. L. Cornwell, P. Komalavilas, and N. Boerth, *Meth. Enzymol.*, *269*, 149 (1996).
8. S. A. Lipton, Y. B. Choi, Z. H. Pan, S. Z. Lei, H. S. Chen, N. J. Sucher, J. Loscalzo, D. J. Singel, and J. S. Stamler, *Nature*, *364*, 626 (1993).
9. Z.-H. Pan, M. M. Segal, and S. A. Lipton, *Proc. Natl. Acad. Sci. USA*, *93*, 15423 (1996).
10. D. L. Campbell, J. S. Stamler, and H. C. Strauss, *J. Gen. Physiol.*, *108*, 277 (1996).
11. F. W. Sellke, P. R. Myers, J. N. Bates, and D. G. Harrison, *Am. J. Physiol.*, *258*, H515 (1990).
12. E. A. Kowaluk, R. Poliszczuk, and H. L. Fung, *Eur. J. Pharmacol.*, *144*, 379 (1987).
13. J. A. Bauer and H. L. Fung, *J. Pharmacol. Exp. Ther.*, *256*, 249 (1991).
14. J. A. Bauer and H. L. Fung, *Circulation*, *84*, 35 (1991).
15. L. J. Ignarro, J. N. Degnan, W. H. Baricos, P. J. Kadowitz, and M. S. Wolin, *Biochim. Biophys. Acta*, *718*, 49 (1982).

16. L. J. Ignarro, J. B. Adams, P. M. Horwitz, and K. S. Wood, *J. Biol. Chem.*, *261*, 4997 (1986).

17. L. J. Ignarro, B. K. Barry, D. Y. Gruetter, J. C. Edwards, E. H. Ohlstein, C. A. Gruetter, and W. H. Baricos, *Biochem. Biophys. Res. Commun.*, *94*, 93 (1980).

18. L. J. Ignarro, P. J. Kadowitz, and W. H. Baricos, *Arch. Biochem. Biophys.*, *208*, 75 (1981).

19. E. P. Wei, M. A. Moskowitz, P. Boccalini, and H. A. Kontos, *Circ. Res.*, *70*, 1313 (1992).

20. M. Fanciullacci, M. Alcoolandri, M. Figini, P. Geppetti, and S. Michelacci, *Pain*, *60*, 119 (1995).

21. K. Kitamura, K. Kangawa, M. Kawamoto, Y. Ichiki, H. Matsuo, and T. Eto, *Biochem. Biophys. Res. Commun.*, *185*, 134 (1992).

22. S. M. O. Hourani and N. J. Cusack, *Pharmacol. Rev.*, *43*, 243 (1991).

23. H. A. Bull, J. Hothersall, N. Chowdhury, J. Cohen, and P. M. Dowd, *J. Invest. Dermatol.*, *106*, 655 (1996).

24. B. P. Booth, T. D. Nolan, and H.-L. Fung, *Br. J. Pharmacol.*, *122*, 577 (1997).

25. B. P. Booth and H.-L. Fung, *J. Cardiovasc. Pharmacol.* (in press).

26. H.-L. Fung, S. C. Sutton, and A. Kamiya, *J. Pharmacol. Exp. Ther.*, *228*, 334 (1984).

27. D. W. Gray and I. Marshall, *Br. J. Pharmacol.*, *107*, 691 (1992).

28. Y. Matsumoto, S. Ueda, S. Matsushita, T. Ozawa, and H. Yamaguchi, *Jpn. Circ. J.*, *60*, 797 (1996).

29. M. W. Radomski, R. M. J. Palmer, and S. Moncada, *Br. J. Pharmacol.*, *101*, 325 (1990).

30. J. L. Mehta, L. Y. Chen, C. Kone, P. Mehta, and P. Turner, *J. Lab. Clin. Med.*, *125*, 370 (1995).

31. V. Mollace, D. Salvemini, and J. Vane, *Thromb. Res.*, *64*, 533 (1991).

32. G. Anfossi, P. Massucco, E. Mularoni, F. Cavalot, L. Mattiello, and M. Trovati, *Prostanglandins Leukotrienes and Essential Fatty Acids*, *49*, 839 (1993).

33. J. Brittenden, K. G. M. Park, S. D. Heys, C. Ross, J. Ashby, A. K. Ah-See, and O. Eremin, *Surgery*, *115*, 205 (1994).

34. H. Drexler, A. M. Zeiher, K. Meinzer, and H. Just, *Lancet*, *338*, 1546 (1991).

35. M. A. Creager, S. J. Gallagher, X. J. Girerd, S. M. Coleman, V. J. Dzau, and J. P. Cooke, *J. Clin. Invest.*, *90*, 1248 (1992).

36. P. Clarkson, M. R. Adams, A. J. Powe, A. E. Donald, R. McCredie, J. Robinson, S. N. McCarthy, A. Keech, D. S. Celermajer, and J. E. Deanfield, *J. Clin. Invest.*, *97*, 1989 (1996).

37. M. R. Adams, R. McCredie, W. Jessup, J. Robinson, D. Sullivan, and D. S. Celermajer, *Atherosclerosis*, *129*, 261 (1997).

38. L. Kuo, M. J. Davis, S. Cannon, and W. M. Chilian, *Circ. Res.*, *70*, 465 (1992).

39. S. M. Bode-Boger, R. H. Boger, A. Creutzig, D. Tsikas, F.-M. Gutzki, K. Alexander, and J. C. Frolich, *Clin. Sci.*, *87*, 303 (1994).

40. S. M. Bode-Boger, R. H. Boger, H. Afke, D. Heinzel, D. Tsikas, A. Creutzig, K. Alexander, and J. C. Frolich, *Circulation*, *93*, 85 (1996).

41. Y. Hirooka, K. Egashira, T. Imaizumi, T. Tagawa, H. Kai, M. Sugimachi, and A. Takeshita, *J. Am. Coll. Cardiol.*, *24*, 948 (1994).

42. L. M. Bigatello, W. E. Hurford, R. M. Kacmarek, J. D. J. Roberts, and W. M. Zapol, *Anesthesiology*, *80*, 761 (1994).

43. J. Abrams, *Am. Heart J.*, *110*, 216 (1985).

44. E. Braunwald, *Am. J. Cardiol.*, *27*, 416 (1971).

45. P. K. Shah, *Cardiol. Clin.*, *9*, 11 (1991).

46. P. Synek, K. Rysanek, H. Spankova, and M. Mlejnkova, *Activitas Nervousa*, *12*(Suppl), 77 (1970).

47. B. P. Booth, S. Jacob, J. A. Bauer, and H.-L. Fung, *J. Cardiovasc. Pharmacol.*, *28*, 432 (1996).

48. D. Hebert, J.-X. Xiang, and J. Y.-T. Lam, *Circulation*, *95*, 1308 (1997).

49. R. R. Miller, L. A. Vismara, D. O. Williams, E. A. Amsterdam, and D. T. Mason, *Circ. Res.*, *39*, 514 (1976).

50. P. Reid, J. Flaheerty, and D. Taylor, *Circulation*, *48*(Suppl 4), 207 (1973).

51. B. I. Jugdutt, and M. I. Khan, *Circulation*, *89*, 2297 (1994).

52. M. Chiarello, H. K. Gold, R. C. Leinbach, M. A. Davis, and P. R. Maroko, *Circulation*, *54*, 766 (1976).

53. W. A. Buylaert, L. L. Herregods, and E. P. Mortier, *Clin. Pharmacokinet.*, *17*, 10 (1989).

54. T. D. Sokoloski, C. C. Wu, and A. M. Burkman, *Int. J. Pharmaceut.*, *6*, 63 (1980).

55. B. P. Booth, M. Henderson, B. Milne, F. Cervenko, G. S. Marks, J. F. Brien, and K. Nakatsu, *Anesth. Analg.*, *72*, 493 (1991).

56. B. P. Booth, J. F. Brien, G. S. Marks, B. Milne, F. Cervenko, J. Pym, J. Knight, K. Rogers, T. Salerno, and K. Nakatsu, *Anesth. Analg.*, *78*, 848 (1994).

57. J. Abrams, U. Elkayam, U. Thadani, and H.-L. Fung, *Am. J. Cardiol.*, *81*(1A), 3A (1998).

58. J. A. Bauer, T. Nolan, and H.-L. Fung, *J. Pharmacol. Exp. Ther.*, *280*, 326 (1997).

59. J. A. Bauer, J. P. Balthasar, and H.-L. Fung, *Pharm. Res.*, *14*, 1140 (1997).

60. J. A. Bauer and H. L. Fung, *Pharm. Res.*, *10*, 1341 (1993).

61. J. A. Bauer and H. L. Fung, *Pharm. Res.*, *11*, 816 (1994).

62. J. E. Saavedra, T. R. Billiar, D. L. Williams, Y. M. Kim, S. C. Watkins, and L. K. Keefer, *J. Med. Chem.*, *40*, 1947 (1997).

63. G. J. Southan and C. Szabo, *Biochem. Pharmacol.*, *51*, 383 (1996).

64. J. M. Fukuto and G. Chaudhuri, *Annu. Rev. Pharmacol. Toxicol.*, *35*, 165 (1995).

65. M. A. Dwyer, D. S. Bredt, and S. H. Snyder, *Biochem. Biophys. Res. Commun.*, *176*, 1136 (1991).

66. E. S. Furfine, M. F. Harmon, J. E. Paith, and E. P. Garvey, *Biochemistry*, *32*, 8512 (1993).

67. P. Klatt, K. Schmidt, F. Brunner, and B. Mayer, *J. Biol. Chem.*, *269*, 1674 (1994).

68. P. B. Persson, J. E. Baumann, H. Ehmke, B. Nafz, U. Wittmann, and H. R. Kirchheim, *Am. J. Physiol.*, *262*, H1395 (1992).

69. M. A. Tabrizi-Fard and H. L. Fung, *Br. J. Pharmacol.*, *111*, 394 (1994).

70. D. W. Reif and S. A. McCreedy, *Arch. Biochem. Biophys.*, *320*, 170 (1995).

71. S. Schwarzacher and G. Raberger, *J. Vasc. Res.*, *29*, 290 (1992).

72. K. Krejcy, S. Schwarzacher, and G. Raberger, *Naunyn-Schmiedebergs Arch. Pharmacol.*, *347*, 342 (1993).

73. I. L. Buxton, D. J. Cheek, D. Eckman, D. P. Westfall, K. M. Sanders, and K. D. Keef, *Circ. Res.*, *72*, 387 (1993).

74. N. M. Olken and M. A. Marletta, *Biochemistry*, *32*, 9677 (1993).

75. P. Vallance, A. Leone, A. Calver, J. Collier, and S. Moncada, *Lancet*, *339*, 572 (1992).

76. C. Frey, K. Narayanan, K. McMillan, L. Spack, S. S. Gross, B. S. Masters, and O. W. Griffith, *J. Biol. Chem.*, *269*, 26083 (1994).

77. K. Narayanan and O. W. Griffith, *J. Med. Chem.*, *37*, 885 (1994).

78. D. J. Wolff and A. Lubeskie, *Arch. Biochem. Biophys.*, *325*, 227 (1996).

79. R. G. Kilbourn, S. S. Gross, R. F. Lodato, J. Adams, R. Levi, L. L. Miller, L. B. Lachman, and O. W. Griffith, *J. Natl. Cancer Inst.*, *84*, 1008 (1992).

80. J. P. Cobb, C. Natanson, W. D. Hoffman, R. F. Lodato, S. Banks, C. A. Koev, M. A. Solomon, R. J. Elin, J. M. Hosseini, and R. L. Danner, *J. Exp. Med.*, *176*, 1175 (1992).

81. W. M. Moore, R. K. Webber, G. M. Jerome, F. S. Tjoeng, T. P. Misko, and M. G. Currie, *J. Med. Chem.*, *37*, 3886 (1994).

82. D. J. Wolff, and A. Lubeskie, *Arch. Biochem. Biophys.*, *316*, 290 (1995).

83. M. Brownlee, H. Vlassara, A. Kooney, P. Ulrich, and A. Cerami, *Science*, *232*, 1629 (1986).

84. M. Kihara, J. D. Schmelzer, J. F. Poduslo, G. L. Curran, K. K. Nickander, and P. A. Low, *Proc. Natl. Acad. Sci. USA*, *88*, 6107 (1991).

85. C. C. Wu, S. J. Chen, C. Szabo, C. Thiemermann, and J. R. Vane, *Br. J. Pharmacol.*, *114*, 1666 (1995).

86. H. Tamura, K. Horiike, H. Fukuda, and T. Watanabe, *J. Biochem.*, *105*, 299 (1989).

87. P. Ou and S. P. Wolff, *Biochem. Pharmacol.*, *46*, 1139 (1993).

88. T. Bieganski, J. Kusche, W. Lorenz, R. Hesterberg, C. D. Stahlknecht, and K. D. Feussner, *Biochim. Biophys. Acta*, *756*, 196 (1983).

89. M. A. Tabrizi-Fard and H. L. Fung, *Drug Metab. Disp.*, *24*, 1241 (1996).

90. W. G. Haynes, J. P. Noon, B. R. Walker, and D. J. Webb, *J. Hypertension*, *11*, 1375 (1993).

91. A. Petros, G. Lamb, A. Leone, S. Moncada, D. Bennett, and P. Vallance, *Cardiovasc. Res.*, *28*, 34 (1994).

92. R. J. MacAllister, S. A. Fickling, G. S. Whitley, and P. Vallance, *Br. J. Pharmacol.*, *112*, 43 (1994).

93. K. Schmidt, P. Klatt, and B. Mayer, *Mol. Pharmacol.*, *44*, 615 (1993).

94. K. Schmidt, B. M. List, P. Klatt, and B. Mayer, *J. Neurochem.*, *64*, 1469 (1995).

95. L. Kobzik, M. B. Reid, D. S. Bredt, and J. S. Stamler, *Nature*, *372*, 546 (1994).

96. R. G. Kilbourn, C. Szabo, and D. L. Traber, *Shock*, *7*, 235 (1997).

97. N. McCartney-Francis, J. B. Allen, D. E. Mizel, J. E. Albina, Q. W. Xie, C. F. Nathan, and S. M. Wahl, *J. Exp. Med.*, *178*, 749 (1993).

98. J. A. Corbett, A. Mikhael, J. Shimizu, K. Frederick, T. P. Misko, M. L. McDaniel, O. Kanagawa, and E. R. Unanue, *Proc. Natl. Acad. Sci. USA*, *90*, 8992 (1993).

99. T. Brenner, S. Brocke, F. Szafer, R. A. Sobel, J. F. Parkinson, D. H. Perez, and L. Steinman, *J. Immunol.*, *158*, 2940 (1997).

100. M. Feelisch and J. S. Stamler, *Methods in Nitric Oxide Research*, John Wiley and Sons, Chichester, 1996.

Subject Index

A

AAS, *see* Atomic absorption
spectroscopy
  F-, *see* Flame atomic absorption
    spectroscopy
Absorption bands and spectra (*see
also* Colorimetry, Infrared
  spectroscopy,
  Spectrophotometry, *and* UV
  absorption), 112, 113, 116, 293,
  699
  transient, 214, 217, 241
Aceruloplasminemia, 319, 324
Acetaminophen, 392
  -derived radical, *see* Radicals
Acetate (or acetic acid), 113
  buffer, *see* Buffer
  nickel, 194

[Acetate]
  12-*o*-tetradecanoylphorbol-13-, 194
  trifluoro-, 113
Acetonitrile, 117, 530
Acetophenone
  o-hydroxy-, 490
Acetylacetonate, 638
  cobalt complex, 498–513
Acetylcholine, 569–572, 666, 667,
  681, 684, 687, 690, 732
  esterase, 270
N-Acetylcysteine, 259, 392, 393,
  397, 405, 406
Acetylsalicylic acid, 460
Acidity constants, 491, 553, 557,
  559, 578, 579, 601, 602, 627,
  696, 699
  hydrogen peroxide, 90
  hydroxyl radical, 105

751